ISBN 978-1-5278-7911-9
PIBN 10924120

FB &c Ltd, Dalton House, 60 Windsor Avenue, London, SW19 2RR.
Company number 08720141. Registered in England and Wales.

For support please visit www.forgottenbooks.com

PUBLICATIONS

OF

Cornell University

MEDICAL COLLEGE

STUDIES

FROM THE

Department of Anatomy

VOLUME I

1910

NEW YORK CITY

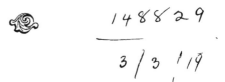

CONTENTS

Being reprints of studies issued in 1910.

11. FURTHER STUDIES ON THE GAMETOGENESIS OF PANDARUS SINUATUS, SAY By **J. F. McClendon.**

Arch. f. Zellforschung. Bd. V, 229-234, 1 text-figure and Pl. XVII.

12. ON THE DYNAMICS OF CELL DIVISION. I. THE ELECTRIC CHARGE ON COLLOIDS IN LIVING CELLS IN THE ROOT TIPS OF PLANTS By **J. F. McClendon.**

Arch. f. Entw'-Mech, XXXI, 80-90, 2 text-figures and Pl. III.

13. ON THE DYNAMICS OF CELL DIVISION. II. CHANGES IN PERMEABILITY OF DEVELOPING EGGS TO ELECTROLYTES By **J. F. McClendon.**

Amer. Jour. of Physiology, Vol. XXVII, 240-275.

14. ON ADAPTATIONS IN STRUCTURE AND HABITS OF SOME MARINE ANIMALS OF TORTUGAS, FLORIDA By **J. F. McClendon.**

Publication of the Carnegie Institution No. 132, 55-62, 1 text-figure and 2 plates.

THE INFLUENCE OF ALCOHOL AND OTHER ANÆSTHETICS ON EMBRYONIC DEVELOPMENT

CHARLES R. STOCKARD

Anatomical Laboratory, Cornell University Medical School, New York City

WITH TWENTY TEXT FIGURES

The adult nervous system is peculiarly sensitive in its responses to the influence of alcohol and other anæsthetics. The writer has found it to be equally true that alcohol and anæsthetics exert a most striking influence over the development of the central nervous system and the organs of special sense. There is considerable variation in the way in which the several anæsthetics act on the developing animal; some of them, such as ether and chloretone, producing effects of a general nature, while alcohol and magnesium are more localized or specific in their action. A similar statement is true for the actions of different anæsthetics on the adult body.

In attempting to explain the occurrence of asymmetrically monophthalmic, cyclopean and blind individuals among fish that had been developed in solutions containing magnesium, the writer advanced the hypothesis that the anæsthetic property of Mg was the causal factor. Many reasons for such a view were put forward in a paper on the artificial production of these monsters (1909). To experimentally test this hypothesis various other anæsthetic agents have been used and all of them to a higher or lower degree inhibit the development of the optic vesicles in fish embryos, and thus give rise to various ophthalmic defects. Alcohol is most decided in its action, causing in some experiments as high as 90 to 98 per cent of abnormal eyes, generally cyclopean, which far surpasses the highest results obtained with Mg.

The effect of alcohol on the general development of the nervous system is more pronounced than that of Mg, and only a few of the alcoholic specimens ever develop sufficiently to hatch and swim about as do the Mg embryos. An explanation for this may be that Mg exerts an influence to inhibit dynamic processes, such as the out-pushing of the optic vesicles, while alcohol acts more especially on the nervous tissues. Mayer (1908) has shown that Mg inhibits muscular contractility without affecting in any way the nervous impulse or nervous rhythm.

The eye defects, it must be remembered, have only been obtained in solutions of one or another anæsthetic; the many other salt and sugar solutions which have been experimented with during four years ('06 and '07) have failed entirely to produce similar results.

The most important outcome of these experiments has been to prove conclusively that many monsters which occur in nature may be artificially produced by changing the environment of the normally developing eggs. The present experiments will demonstrate that this may be done even after development has proceeded for some time. These anomalous structures being the results of external influences and not germinal variations are to some extent within scientific control. A promising field is thus opened in the devising of means to control or regulate the development of the embryo and possibly to obviate certain monstrous conditions at least. Such possibilities were of course beyond our reach if defective germ cells were actually the cause of these monsters.

Mall ('08) has brought forward evidence to show that improper placentation or unfavorable developmental environment is responsible for most human monstrosities, many of which are aborted before reaching term. There is evidently much need of investigation aiming toward the control and regulation of the developmental environment of mammals.

METHOD AND MATERIAL

The eggs of the fish, Fundulus heteroclitus, were used in all of the experiments. The method of treatment varies somewhat for the different solutions employed, so that it is best to describe each separately.

Alcohol solutions were prepared in sea-water on the percentage basis. The strength used being 2, 3, 4, 5, 6, 7, 8, 9, 10, 12, 14, 16, 18 and 20 per cent, 60 cc. of each solution was poured into finger bowls and from 60 to 100 eggs, in the early cleavage stages, four and eight cells, were placed in each bowl. The stronger solutions killed all of the eggs, and those from 3 to 9 per cent gave the best results. In the 3 per cent alcohol solutions at times as many as 90 in every 100 embryos showed abnormal conditions of the eyes, being either eyeless, asymetrically monophthalmic or cyclopean, while in one experiment in a 5 per cent solution of alcohol in sea-water, there were 146 ophthalmic monsters against only 3 individuals with two separate eyes.

Chloretone of 0.1 per cent and 0.066 per cent in sea-water caused abnormalities similar to those produced by other anæsthetics. This substance is more general in its anæsthetic action than either alcohol or Mg, as will be seen in the discussion to follow.

Ether and chloroform also produce rather general effects on the developing embryo, yet a small percentage of cyclopean monsters occur among the embryos which are treated with 60 cc. of sea-water to which 2,2.5, and 3 cc. of ether has been added. In solutions of chloroform of about the same proportions and slightly weaker, a few monsters occurred of the type common to the other anæsthetics. Chloroform is rather toxic in its action on these eggs, large numbers dying in the weaker solutions, while others are so inhibited in their development, that various abnormal conditions follow.

Eggs were not exposed to any of the above anæsthetics for more than twenty-four or thirty-six hours. They were then placed in pure sea-water and continued development showing the abnormal conditions of the eyes and central nervous system that had been induced by their sojourn in the unusual environment.

Similar Mg solutions to those formerly employed were again used. A gram-molecular solution of $MgCl_2$ in distilled water was titrated and kept, to be diluted with sea-water to the proper strength just before the eggs were placed in it. The most favorable results were obtained in solutions of 16, 17, 18, 19, 20, 21 and 22 cc. of molecular $MgCl_2$ made up to 60 cc. by the addition of a sufficient amount of pure sea-water; e.g., 44 cc. of sea-water was added to the 16 cc. of molecular $MgCl_2$ and 43 cc. of sea-water to the 17 cc. of $MgCl_2$. The solutions are, therefore, $\frac{16}{60}$, $\frac{17}{60}$, $\frac{18}{60}$, etc., parts $MgCl_2$ to sea-water. In the $\frac{21}{60}$ solution 66 per cent of the embryos were cyclopean in one of the experiments. Eggs were exposed to the action of Mg shortly after fertilization and at various other times until they reached an early periblast stage, or were fourteen hours old, all with similar results. Although the most favorable time for introducing the eggs into Mg solutions is during the eight-celled stage. The developing embryos were returned to pure sea-water after the third day. The Mg is so slightly toxic that eggs may be kept in it and will continue to develop; the embryos actually hatch and swim about in the solution, being, however, slightly slower in their developmental rate and not so hardy as the specimens which are returned to the sea-water.

The Action of Alcohol on Development

Weak solutions of alcohol exert a most decided effect on the developing fish embryos, causing deformities of the central nervous system, the eyes, and ears in a very large percentage of the specimens.

a. Defects of the eyes

Typical cases of cyclopia showing in the different specimens all gradations, from merely closely approximated eyes, hour-glass eyes with two pupils and two lenses, oval eyes having the two component intimately associated, typical median cyclopean eyes with scarcely an indication of their double nature and extremely small ill-formed cyclopean eyes, were present in the weak alcohol solutions. All of these have been fully described in a former paper ('09) on the artificial production of cyclopea as a result of the action of Mg. The alcohol monsters in some cases also present various degrees of the monophalmicum asymmetricum defect which was common in the Mg experiments. Individuals may have one normal eye and the other eye in different conditions of arrested development from slightly small and defective to entirely absent, see figs. 1 and 2. An important point that was brought out by the alcohol monsters which was not noticed in the magnesium specimens is the fact that some of the embryos have both eyes equally small and defective, figs. 3, 8, 9, 10, and 11. The two eyes are symmetrically defective and the head appears to have small eye-spots instead of normal eyes; compare figs. 3, and 7, 9, 10 and 11 and 12. Finally, as in Mg solutions so also in the alcohol, many eyeless individuals are present .

The eye in some of the alcohol monsters is rather different from that found in the Mg embryos, and may possibly serve to indicate something of the condition of the eye anlagen in the brain. Many embryos possess optic cups with their concave surfaces facing almost directly toward the median sagittal plane of the head. In life the eye presents a heavily pigmented solid convex surface to the side of the head instead of the usual open pupil through which the lens may be seen within the cavity of the optic cup. Fig. 5 and 6 show front and lateral views of such an embryo; the side view indicates the peculiarly solid convex object seen when looking towards the lateral eye. Fig. 13 represents a section through this eye. The choroid coat of the eye-ball is pressed close against the body wall on the sides of the head and the concave retinal surfaces which should face outward are turned directly towards one another; a lens lies between the two eye components which are really separate except along their dorsal borders. Fig. 4 shows a somewhat similar specimen in which the eyes are entirely separate yet they have an arrangement almost identical to that just described. The eyes face the mid-plane of the head and turn their convex choroid coats out against the body wall. The only place at which the inner face of the eyes touches the ectoderm is the ventral body wall and from this a lens has arisen and lies between the two eyes. Fig. 14 illustrates a section through these eyes, in which a most peculiar arrangement exists. Optic fibers probably arising from the ganglionic layer of the retina, (although this can not be positively demonstrated in the sections),

Camera Drawings of Living Fundulus Embryos

Fig. 1. An embryo twenty days old which was treated with 5 per cent alcohol. Only one eye is formed and it faces in a ventro-median direction, instead of towards the lateral wall of the head.

Fig. 2. An asymmetricum nonophthalmicum monster with one almost typical eye while the other eye is small and poorly formed, from a 4 per cent solution of alcohol in sea-water.

Fig. 3. An embryo with both eyes small and closely approximated; the eyes face ventrally. This type is common in all of the anæsthetic solutions.

Fig. 4. A nineteen day embryo from 5 per cent alcohol. The two eyes are not connected yet their convex surfaces are turned out against the lateral walls of the head and the pupils face the median plane. A single lens lies between the two eyes. Fig. 14 shows a section of these eyes.

Fig. 5. A nineteen day embryo from 5 per cent alcohol. This front view shows the two eyes joined dorsally and facing one another with the lens between them.

Fig. 6. shows a lateral view of the same head with the convex choroid surface of the eye-ball close against the side of the head. Fig. 13 illustrates a section through these eyes.

Fig. 7. A normal Fundulus embryo when eight days old drawn to the same scale as the monsters.

Fig. 8. A twenty day embryo from 5 per cent alcohol; the eyes are small and defective as shown in the section fig. 10.

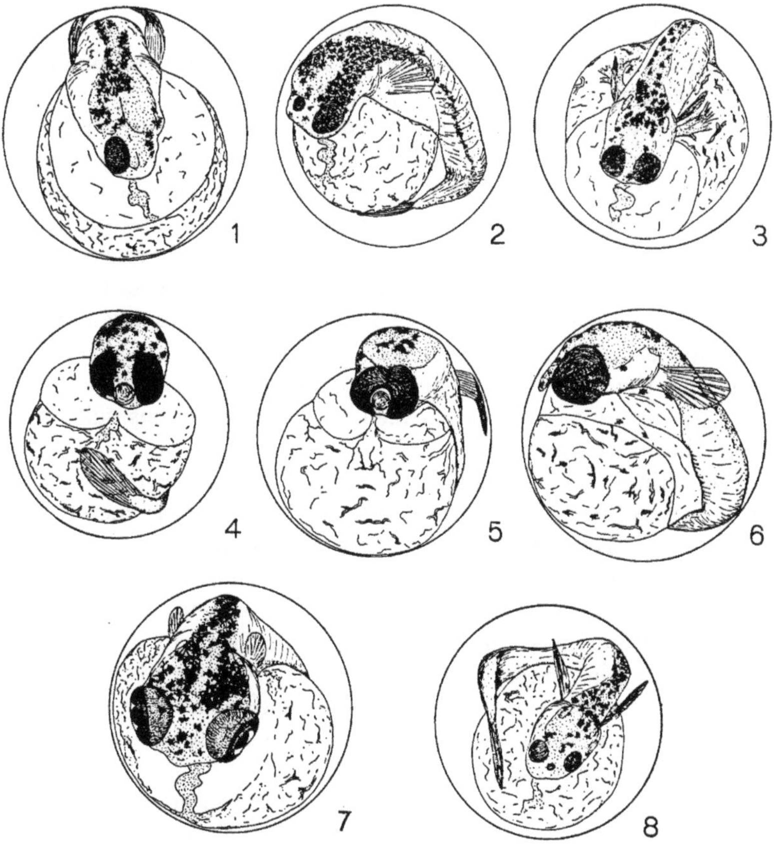
1
2
3
4
5
6
7
8

are collected into optic nerves which pursue extraordinary courses; instead of passing through the outer retinal layers and the choroid coat they take an almost directly opposite course and run across what should be the optic cup cavity (the humor cavity in these specimens is filled with loose cellular tissue) and out through the wide-open "pupil," forming a perfect cross and then passing into the base of the brain to end in the optic lobes.

The position of the lens must not be supposed to determine the actual pupil region of these eyes, as the lens clearly lies between them; see fig. 4 of the living embryo. The wide open pupils of the two eyes face or lead directly into one another. The position might be taken that the entire arrangement represents one large eye; this is not true however since the eyes are entirely separate in all of the sections and the optic cross could scarcely be expected to exist within the base of the eye itself as would be the case if this were one huge eye. The eyes really hang down from the brain as two large retinal disks. Fig. 11 represents a similar case with the eyes rather more flattened out laterally, and the existence of all gradations between the two conditions substantiates the above statement.

The retinal layers of these eyes which are nineteen days old, are poorly differentiated, the inner layer consisting merely of indefinitely arranged cells; a better differentiation is usually attained in normal specimens by the sixth or seventh day. A comparison of Figs. 13 and 14 with Fig. 12 illustrates in a way the more definite orientation of the inner retinal cells in the normal individual when compared with the defective eyes. It will also be noticed that the lenses in the two-eyed specimen are surrounded by clear humor spaces while the lens in the defective eyes lies buried in loosely arranged cellular tissue.

The right retina of the monster faces in the same direction as the left retina of the ordinary individual yet there is no indication of a reversal of the layers which might possibly be imagined in such a case.

The optic stalk was scarcely formed in these eyes, which is not infrequently true in cyclopia. The path of the optic nerve is, therefore, evidently not that usually taken along the

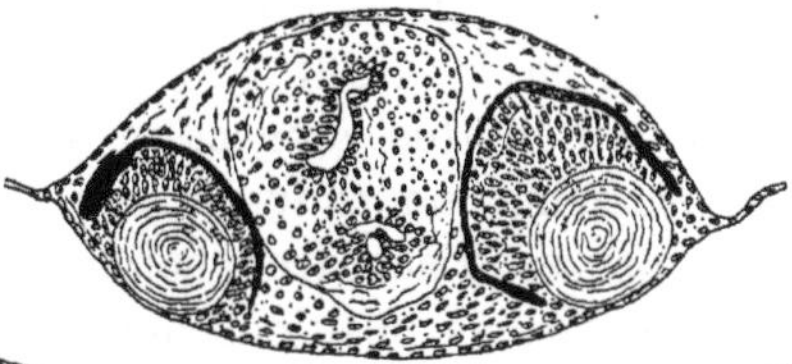

9

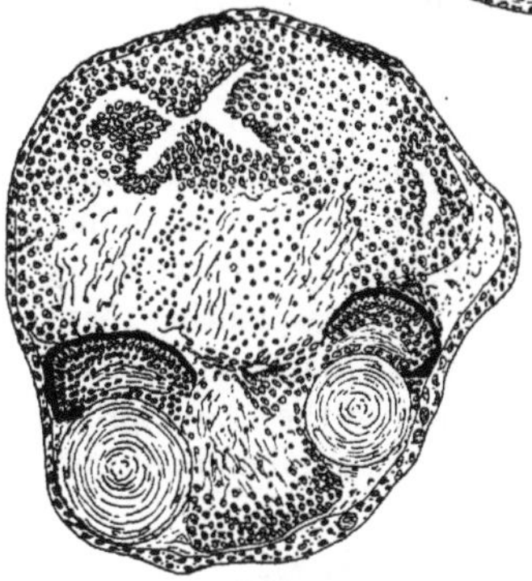

10

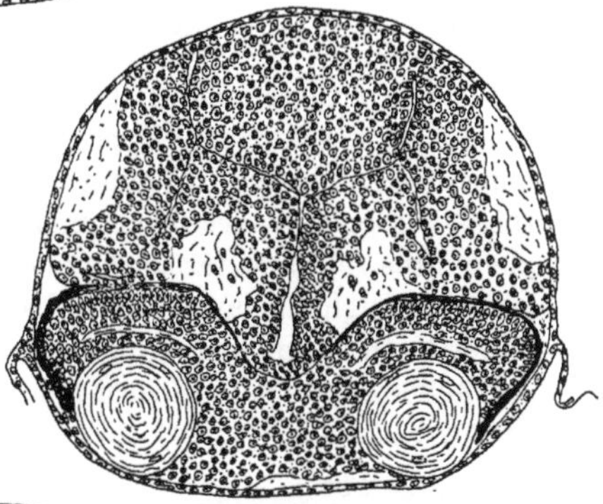

11

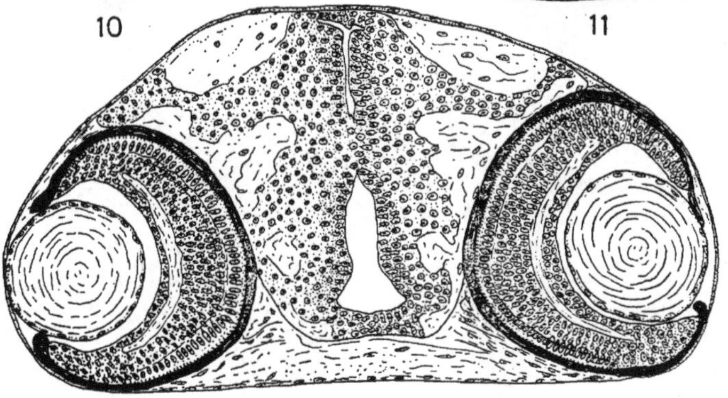

12

FIG. 9. A section through the small defective eyes of an eight day embryo, after treatment with a 4 per cent solution of ether in sea-water. The brain is narrow and poorly developed.

FIG. 10. A section through the head of the embryo shown in fig. 8. Both of the eyes are small and defective with ill-fitting lenses and face in a ventral direction.

FIG. 11. A section through the eyes of a monster commonly found in the alcohol solutions. The eyes are joined beneath the bilateral brain and face ventrally.

FIG. 12. A section through the eyes of a normal thirteen day embryo.

optic stalk, but the optic nerve fibers grow directly into the brain from the region of the eye cavity itself.

It is interesting to find in this connection that Lewis ('70) describes in his experiments on tadpoles a strikingly similar course pursued by the nerves arising from some of the optic cups he had transplanted to various positions along the hind brain region of the embryos. He states that

In a few of these somewhat irregular transplanted eyes the optic nerve takes a very curious course, passing across the cup cavity from the ganglionic layer, through the pupil and then into the mesenchyme, ending there. In both of these experiments a small bundle of optic nerve fibers pierces the retina as far as the pigment layer. In transplanting these eyes the ganglionic layer was probably injured in such a way as to interfere with the normal path of the nerve fibers, and so they have probably followed the path of least resistance through the pupil and out into the mesenchyme.

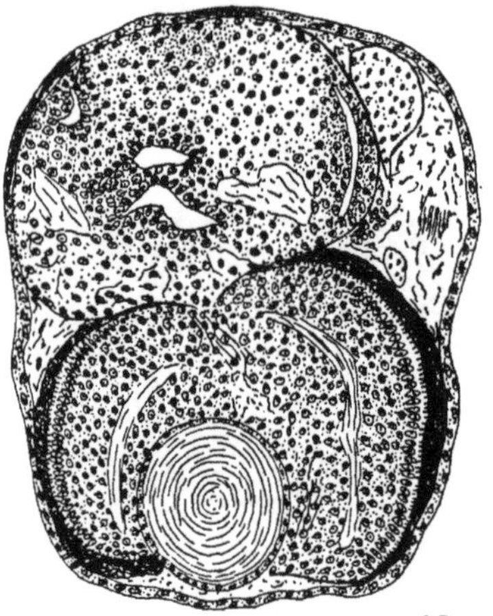

13

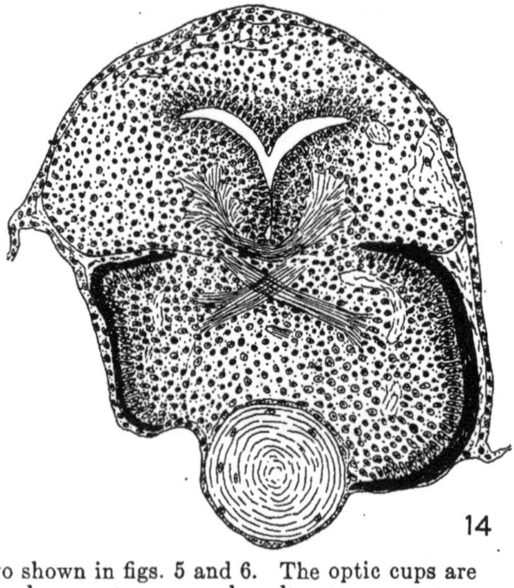

14

Fig. 13. A section of the embryo shown in figs. 5 and 6. The optic cups are joined dorsally and face the median plane of the head; a lens lies between them and is surrounded by loose cellular tissue instead of by the humor.

Fig. 14. Section of the eyes in the embryo shown in fig. 4. The optic cups are not joined yet they face towards one another with their convex choroid surfaces pressed close against the head wall. The optic nerves run across what should be the humor chamber of the eyes and out through the wide pupils to form a perfect cross; the optic fibers then enter the brain floor. A lens lies between the eyes.

Lewis' illustrations are similar to Fig. 14 in so far as the course of the optic nerve is concerned. These experiments would seem to indicate that the direction taken by the optic nerve fibers is not firmly fixed but that they may pursue an almost reverse direction from that generally followed. Many of the eyes in the writer's specimens show various conditions of this kind and he must agree with Lewis in the conclusion that

> It would seem to me impossible to explain these various conditions of the optic nerve on any other basis than that they are outgrowths of nerve cells of the ganglionic layer of the retina.

Direct evidence is thus furnished for the outgrowth theory of the nerve fiber which has been so ably supported in the last few years by Harrison's ('08) experiments.

The present experiments warrant the following explanation for incomplete cyclopean eyes, or double eyes, when compared with the usual condition.

In normal development the eye anlagen push out from the ventro-lateral borders of the brain and turn dorsally as indicated in the diagram, fig. 15 A. The abnormal individuals with two eyes facing the median plane also have them more ventrally situated in relation to the brain, and it may be supposed that when the eyes arose from the brain their formation was directed ventrally instead of dorsally, fig. 15 B. This causes the eyes to hang below the brain and face one another as already shown in figs. 4, 5, 6, 13, and 14, instead of turning dorsally and facing outward as in figs, 11 and 12.

Similar conditions are also found in the development of a single eye. Fig. 1 shows an embryo with an eye on the right side only, yet this eye faces the median plane and is unusually ventral in position; it probably arose as indicated in the diagram fig. 15 D, where as other single-eyed individuals, the commoner type, have an eye looking out from the usual lateral position, fig. 15 C.

From these conditions we may determine whether cyclopia is brought about by a failure of certain central tissues of the brain to develop, thus allowing the eye anlagen to come together as Lewis ('09) has suggested, or whether through a lack of developmental energy necessary for the optic cups to grow dorsally and outward to meet the ectoderm as the writer ('09) has supposed. Considering the case of the single eye it might be held that the failure of certain central tissues of the brain to develop would cause the eye to arise too near the median line, but this lack of central tissue does not explain why the eye faces the median plane instead of the lateral head wall, and it is much less able to account for the absence of the eye on the opposite side of the head. On the other hand, if the conditions are due to a lack of the necessary developmental energy or an anæsthesia produced by the experiment, then it is evident that although one eye does succeed in pushing out from the brain it might not have sufficient developmental energy to grow dorsally and outward to the lateral body wall, but droops, as it were, into a more ventro-median position and faces in toward the median plane. Thus one-half of an incomplete cyclopean eye is formed. The other eye was entirely suppressed, lacking the energy necessary to push itself out from the brain. This inequality in the developmental powers of the two eyes is indicated by their frequent asymmetrical condition.

The two eye components do not always face the median plane and in such cases the eyes merely fail to grow out laterally. They come off ventrally from the brain and either face in a ventral direction or grow so as to face outward.

The experiments fail to give any definite clue as to where the optic anlagen are located in the brain before they become visible, although Lewis' operations on the embryonic shields of older embryos would seem to indicate that at that time the anlagen occupied somewhat lateral positions.

It is clear from the foregoing consideration that alcohol has the power to induce the same typical ophthalmic defects that were formerly described in the embryos from the Mg solutions. The property common to both Mg and alcohol is their anæsthetic effect on animals. The writer concludes that cyclopia, monophthalmicum asymmetricum and entire absence of eyes, all of which are more or less arrested or inhibited condition of development, result from anæsthesia during certain embryonic stages. Of course this may not be the sole cause of such defects; on the contrary the fact that they are produced in this way would indicate

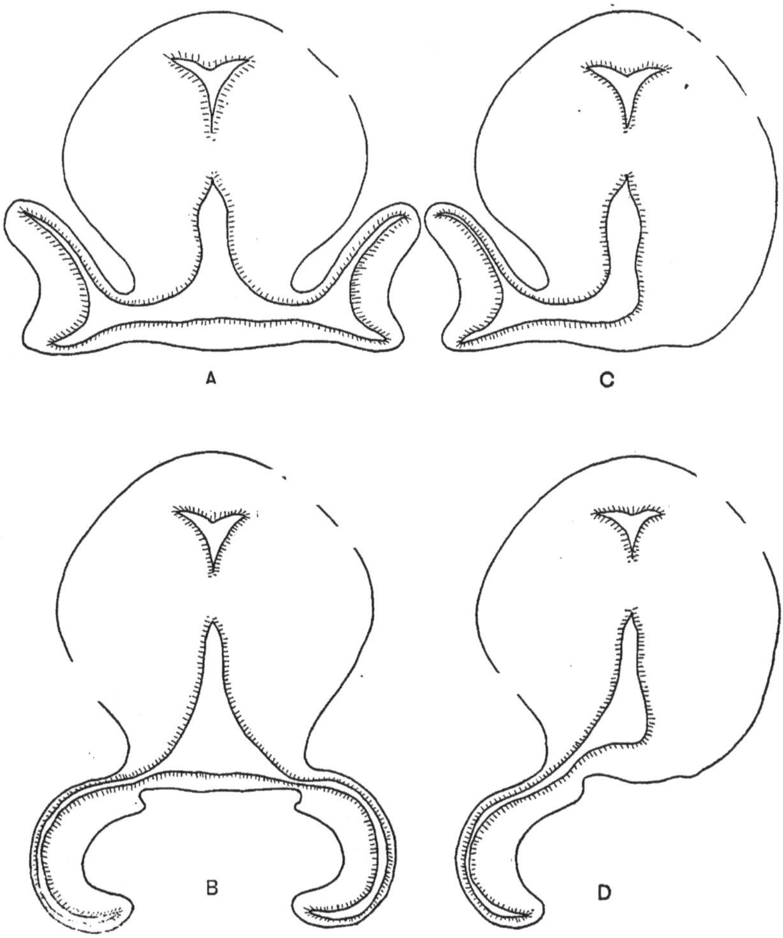

15

FIG. 15. A diagram illustrating several positions taken by the optic cups in development. A, the usual case in which both cups push out from the brain and turn dorsally so as to face the lateral head wall with their convex choroid surfaces towards the brain; see fig. 12. *B*, the optic cups push out from the brain but instead of growing dorsally, they hang ventrally and face in towards the median plane, with their convex choroid surfaces against the outer head wall; see figs. 13 and 14. *C* and *D*, the same conditions are often realized when only one optic vesicle arises from the brain.

that any factor which might come in during early development to lower the developmental energy could possibly induce similar defects. In mammals such monsters probably arise as the result of some weakening or debilitating influence of the environment during early developmental stages, which need only have acted for a short space of time.

b. Defects of the auditory organs

A very pronounced suppression in the development of the auditory apparatus is often noticed in the embryos which have been treated with weak solutions of alcohol. In many individuals only one ear exists. When this condition is found in an embryo with only one eye, two unequally developed eyes or a cyclopean eye with asymmetrical components, it is of interest to find that in all cases observed, the ear is on the same side of the head as the better formed eye. In rare cases both ears are absent, and again it often happens that the ears are apparently normal while the eyes are deformed. Fig. 16, a horizontal section through the head, shows two small abnormal eyes with a lens between them and two perfect ears with cartilaginous capsules, near the hind brain. Fig. 18, which is a section through the ear region of the embryo shown in fig. 4, illustrates two poorly formed ears; on the right side the ear is small and two semicircular canals are represented only by their ampullæ, the epithelial lining of which forms papillæ of cells with long hair-like processes growing from them as is indicated in the drawing. The left ear is almost entirely absent, its median section showing only the small cavity and ampullary papilla seen in the figure. Both ears, however, are surrounded by well formed cartilaginous capsules.

A remarkably abnormal ear is seen in fig. 17. The auditory vesicles have united so as to occupy a dorsal position above the posterior end of the brain. Only two semicircular canals are developed on each side. The cartilaginous capsules in this case seem unable to meet the situation and extend for only a portion of the way around the huge auditory cavity. This union of the lateral auditory vesicles, although formed by an entirely different principle, suggests the large double cyclopean eye.

The final persistence of the ampulla-like cavities seems to be the rule, these structures being present even when all other portions of the internal ear are absent. The ampullæ of the canals are perhaps particularly useful as organs of equilibration in these animals, and their stuborn persistence may be indicative of an ancient origin and suggests the primary function of the ear as an organ of equilibrium.

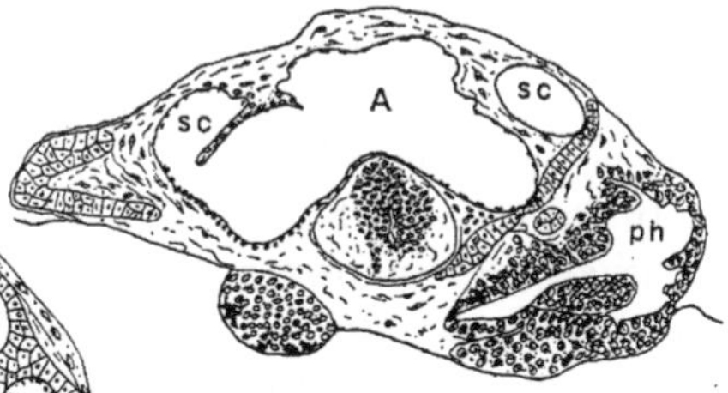

17

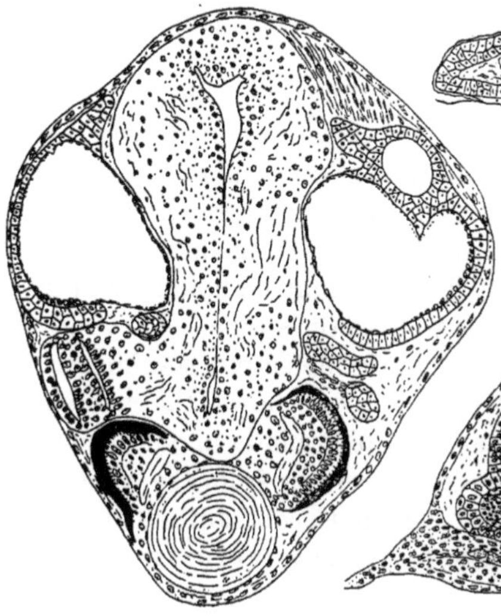

16

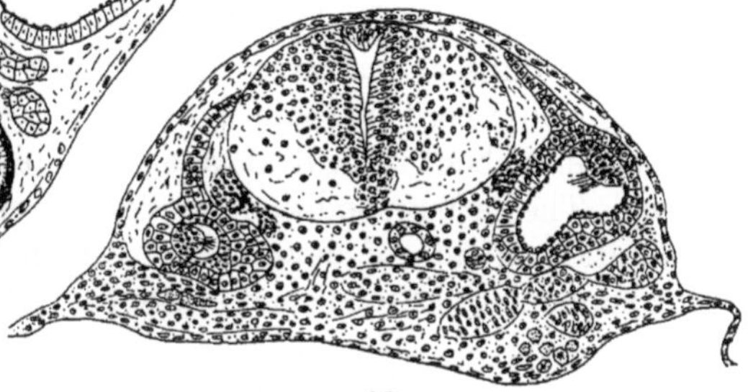

18

FIG. 16. A horizontal section through the head of a thirteen day embryo from a 5 per cent alcohol solution. Two very small and defective eyes have alens between them, while more posteriorly two perfectly formed auditory vesicles are seen with cartilaginous capsules surrounding them.

FIG. 17. A section through the auditory region of an eight day embryo treated with a 4 per cent ether solution. The two auditory vesicles unite dorsally to form a huge cavity above the medulla and spinal cord. This embryo also shows an incomplete cyclopean eye and spina-befida. sc, semicircular canal; a, auditory vesicle,; ph, pharynx.

FIG. 18. A section through the auditory vesicle of the embryo shown in fig. 4 and section fig. 14. The entire auditory vesicle is suppressed except the ampullæ of the semicircular canals. The right ampulla is larger and shows a papilla with hair-like fibers, while the left ampulla is almost completely closed, yet, it too, shows the papilla with projecting hairs. The cartilaginous capsules are small and thus adjusted to the tiny ear parts.

Although all parts of the ear are absent the cartilaginous capsules are present. The shape and size of the capsules, however, seem to be adjusted to that of the auditory vesicle when any part of it exists, as is indicated on the two sides of fig. 18.

c. Defects of the central nervous system

The abnormalities of the brain shown by the specimens treated with alcohol might easily form the subject of an extensive monograph so various and numerous are they. Only a few of them will be briefly mentioned.

In rare cases the brain is almost normal; the fore brain, however, is usually very narrow and gives to the head a characteristically pointed appearance. Dorsal herniæ at times occur in the region of the optic lobes and the hind brain. The histological structure of the brain is often peculiarly abnormal in both the arrangement and the appearance of the cells. The cells may be hyaline and in the region of the central cavity fail to take the stains. They may even be diffusely scattered in peculiarly defective specimens.

The spinal cord in some individuals also shows the hyaline appearance about its central canal, and spina-bifida in not uncommon. The latter condition no doubt results from the general inhibition in rate of development which is constantly true for the specimens in the alcohol solutions. The germ ring is slow in surrounding the yolk and consequently the trunk region of the early embryo is abbreviated. This condition interferes with the median cell proliferation forming the spinal cord so that a split or divided cord results and may extend for various distances in the trunk region. Fig. 19 shows a section through a trunk region with a double cord, *ch*, the notocord, *nch*, is also divided. Fig. 20 is a more posterior section of the same embryo and shows the cord and notocord again single as they are in more anterior region than that shown by fig. 19.

Many of the defects of the central nervous system are of a general nature and almost any substance that inhibits or interferes with the normal developmental rate may cause them. The writer does not intend to convey the idea that these are characteris-

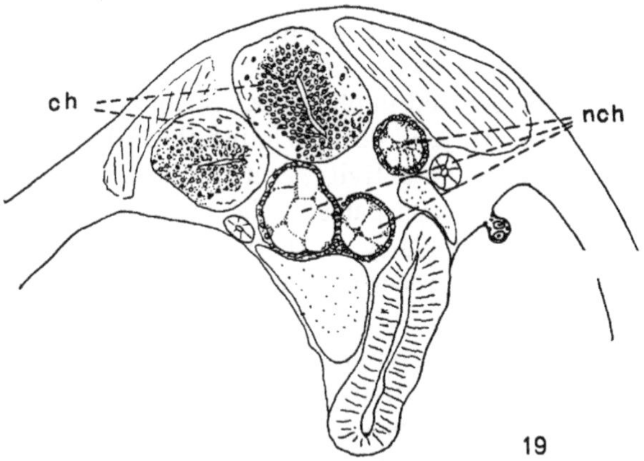

19

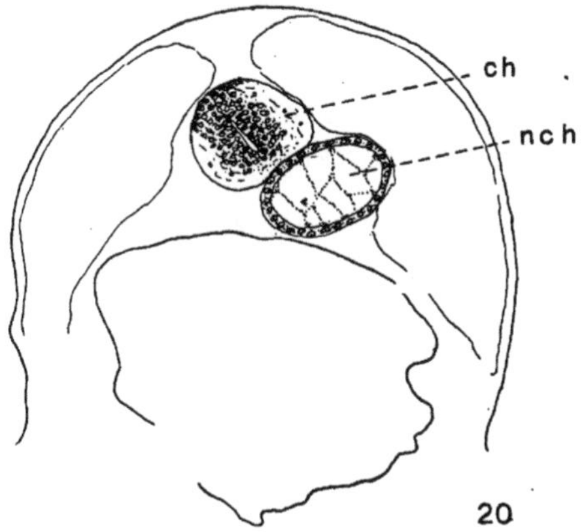

20

Fig. 19. A section through the trunk region of an eight day embryo from a 4 per cent ether solution. The spinal cord, **ch**, is double (spina-bifida) and the notochord, **nch**, is divided into three parts.

Fig. 20. A more posterior section of the same embryo, showing the spinal cord and notocord again single as they are in regions more anterior than fig. 19.

tic anæsthetic effects, but they are strikingly common in the alcohol solutions. On the other hand, similar abnormalities of the nervous system are really infrequent in the weaker Mg solutions.

The slight effects of Mg on the development of the central nervous system are interesting when compared with the marked effects of other anæsthetics on these tissues. In this regard it is important to remember that in physiological experiments on nerve-muscle preparations Mayer ('08) has found that Mg acts directly as an inhibitor of muscular activity, exerting little if any effect on the activity of the nervous parts. The action of Mg in these experiments is not particularly on the nervous system but more largely on the dynamic processes concerned in the outpushing of the eyes.

The occurrence of the several ophthalmic anomalies is common to all of the anæsthetic solutions and similar conditions have not been found in embryos treated with various salts and sugars ('06, '07) which may inhibit general development and induce many other abnormalities.

The Effects of Chloretone, Ether and Chloroform on Development

Chloretone, ether and chloroform when employed in weak solutions influence developing eggs in a way somewhat similar to that described for alcohol. The action of these substances is not so pronounced and may be described more as an inhibition of the general developmental processes. The embryos are usually small and recover very slowly from the inhibiting effects after being returned to sea-water. A few of the individuals in all these solutions exhibit the cyclopean defect in its various degrees just as it has been described in the specimens treated with alcohol and Mg.

The ophthalmic defects, cyclopia, monophthalmicum asymmetricum and entire absence of optic vesicles, are all conditions of arrested or inhibited development and are prevalent among embryos treated with solutions having anæsthetic properties.

The writer is led to conclude, therefore, that his former hypothetical explanation of why cyclopia occurred in embryos treated with Mg solutions was correct. The evidence strongly indicates that the ophthalmic abnormalities produced in these experiments are the result of an anæsthetic action during the early developmental stages.

THE PERIOD OF DEVELOPMENT AT WHICH CYCLOPIA MAY BE INDUCED BY CHEMICAL AGENTS

The Mg experiments were repeated mainly to ascertain at how late a period in development eggs could be subjected to the solutions and subsequently develop into cyclopean monsters. The original experiments seemed to demonstrate the fact that cyclopia was due to external influences acting on the development of the optic vesicles, and not in any sense to a germinal variation. Nevertheless, H. H. Wilder ('08) in the face of these results, advanced a germinal theory to account for the origin of cyclopia and attempted to explain away the obstacle offered in the experiments referred to by claiming that the eggs were subjected to the solutions at so early a stage in development that germinal variations might still have been induced. This is impossible, as the writer has pointed out elsewhere ('09b), since germinal variations may only be induced before embryonic development has begun. After the two-or four-cell stage (the time at which the eggs were subjected to Mg) is reached any thing done to the egg has its effect on the developing embryo to cause this or that abnormal condition. The only germinal variation possible at such a period would be in the primordial germ cells of the developing individual, a variation which would not manifest itself until the next generation of individuals.

The following experiments prove beyond doubt that cyclopia may be produced by the action of environmental influences.

When eggs in the eight-cell stage, or three and one-half hours after fertilization, are subjected to the action of Mg solutions many of the resulting embryos will show the cyclopean defect.

If eggs be placed in Mg solutions five hours after fertilization, when in the sixteen or thirty-two-cell stage, an almost equally large number of cyclopean individuals will occur. The same is true when they are subjected to the solutions after having developed in pure sea-water for seven and one quarter hours and reached the sixty-four or higher cell stages.

Eggs that have developed eleven hours in pure sea-water and are in the early periblast stage with a somewhat flattened blastodermic cap, may be put into Mg solutions and caused to form cyclopean monsters. The percentage of cyclopean embryos arising from eggs treated at this late period is small, yet even after developing for fifteen hours in pure sea-water some eggs may be induced to form cyclopean monsters by treatment with $MgCl_2$ in sea-water solutions.

Eggs that were older than this before being introduced into the solutions failed to respond, all developing into ordinary two-eyed individuals. This is due to the fact that a considerable amount of time is necessary for the Mg to act upon the body substances of the early embryo and prevent normal eye development. The optic vesicles begin to push out from the brain before the thirtieth hour in development. Thus, after the first six or seven hours, the longer the eggs have been allowed to develop naturally the smaller the proportion of cyclopean individuals that may be artificially induced. After fifteen hours no embryo will be so affected, since an insufficient amount of time exists for the Mg to act on the eye anlagen.

The solutions are effective up to a stage in development preceding the formation of the germ ring and embryonic shield, and the action of the Mg on the eye anlagen probably takes place while the embryonic shield and outline of the embryo are forming.

There can be no further doubt that cyclopean monsters are caused by the action of a strange environment on the developing fish embryos. With such evidence at hand it is also highly probable that mammalian cyclops are due to the action of external influences on the embryo and not to an abnormal germinal tendency.

OTHER CASES OF CYCLOPIA IN FISH

The Italian observer, Paolucci ('74), has described a most remarkable cyclopean ray. The monster was almost adult in size, probably two years old, and measured 47cm. across the pectorial fins and 20cm. in length not including the long whip-like tail. Paolucci states that this cyclopean monster was captured in the Adriatic Sea near the shore and had evidently been able to cope with its surroundings and grow into a vigorous ray. So far as the writer knows this case of a cyclopean monster in nature being able to sustain itself and reach the adult stage, is unique.

Paolucci's specimen proves the correctness of the writer's statement ('09) that the cyclopean eye is not necessarily associated with a single instead of a double brain, or with any other serious defect in the brain region. This fact was clearly shown in the brain structure of many of the cyclopean embryos studied, as well as by their apparently normal behavior after hatching from the egg.

The cyclopean Funduli have been kept living for more than one month, which is as long as the experiment was tried. They would doubtless have lived much longer, as they were hardy and able to obtain an abundance of food from the vegetable particles in the sea-water. Paolucci's observation would indicate that the Fundulus monsters might be reared to maturity and possibly interbreed.

Gemmel ('06) has described four cases of cyclopia in newly hatched trout collected from a fish-hatchery in England. The conditions of the eyes and brains in these monsters are exactly similar to those in the artificially produced Fundulus monsters.

The developing trout's egg demands water of such high purity that trouble is often experienced in the hatcheries, and monstrous embryos commonly occur. These may result from weakened developmental forces due to an insufficient oxygen supply or to the accumulation of injurious chemicals about the eggs.

Gemmel ('06b) in describing cases of supernumerary eyes in the trout embryos records that in one case of an aborted twin head the lens alone of all the eye structures was present. Free lenses were also described and figured in other individuals. Free lenses occur very commonly in heads showing various eye abnormalites; a full consideration of these cases is recorded elsewhere.

SUMMARY

1. When the eggs of the fish, Fundulus heteroclitus, are subjected during early stages of development to the action of weak solutions of alcohol the resulting embryos show marked abnormalities in the structure of their central nervous system and organs of special sense.

The eyes in such individuals are either both small with poorly differentiated retinæ, cyclopean, asymmetrically monophthalmic or entirely absent. These ophthalmic defects sometimes occur in as many as 98 per cent of the specimens. Such anomalies are closely similar to those previously induced with Mg, and in both cases are probably due to the anæsthetic property of the substances acting upon the eggs.

Alcohol tends to suppress the development and differentiation of the auditory vesicles. A few specimens are entirely without ears, others have one ear more or less perfectly developed while the opposite ear is scarcely formed at all and still other individuals have both ears extremely defective. In all cases examined the better ear is invariably on the side with the better developed eye if the eyes are also asymmetrically formed. The most persistent portion of the internal ear, or that part which exists when all other parts are wanting, is a cavity with an epithelial lining resembling closely in structure an ampulla of the semicircular canals. This fact may be interpreted to mean that the ampulla is one of the most ancient or fundamental parts of the ear, and it might further be considered indicative of the archaic function of the ear as an organ of equilibrium since this is the chief function of the ampullæ.

The brain is usually narrow and pointed in embryos that have been treated with alcohol. It occasionally has a dorsal hernia and shows regions of poor differentiation. The cell arrangement in the spinal cord are abnormal in many cases and spina-bifida is not infrequent. These conditions of the central nervous system might result from any cause that tends to retard development and are not particularly characteristic of anæsthetic solutions as the eye anomalies are; yet the defects of the central neryous system are commoner in these anæsthetics than in any other solutions with which the embryos have been treated.

2. Chloretone, chloroform and ether induce much the same structural deformities in these embryos as does alcohol. They act, however, as more general anæsthetics, causing a retardation in development. The characteristic eye and ear defects are not nearly so common, though they do occur as a result of treatment with these three substances.

3. The effects of Mg on the developing fish's egg have been previously considered. This substance is even more local in its action than alcohol, the principal defects resulting from its use being various anomalous conditions of the eyes, whereas the nervous system generally may be in many cases structurally normal. The embryos on hatching from the egg are able to swim in the usual manner and live for more than one month in aquaria, which is as long as any effort was made to keep them. The latter fact would seem to indicate that the nervous system also functionates normally.

Magnesium was used to test at how late a period in develop ment the eggs might be introduced into the solutions with the subsequent development of the cyclopean condition. It was found that after normal development had proceeded for two, four, six, eight, ten, eleven, twelve or even fifteen hours, if the eggs were then placed in $MgCl_2$, solutions, many of the resulting embryos showed the cyclopean defect. At fifteen hours the eggs have reached the periblast stage and the blastodem is flattening down upon the yolk. The germ ring arises shortly after this time and begins its downward growth over the yolk mass.

Whenever eggs are allowed to develop beyond the fifteen hour period before being introduced into the solutions of $MgCl_2$ they invariably give rise to normal two-eyed individuals. The occurrence of cyclopia is less frequent when eggs are subjected at later stages than when introduced into the $MgCl_2$ solutions during the four or eight-cell stage. This is doubtless due to the fact that a considerable period of time is necessary for the Mg to act upon the substances of the embryo and influence the origin of the optic vesicles. When an insufficient time intervenes between the period at which the eggs are subjected to the action of the solution and that at which the optic vesicles are given off from the brain the Mg is unable to influence the tissues so as to induce the cyclopean condition.

The production of cyclopia by the action of Mg at such late stages in development proves beyond doubt that this deformity is due to the action of external or environmental conditions on the developing animal. Any explanation of cyclopia based on germinal hypothyses such as that recently advanced by H. H. Wilder must be reconstructed so as to conform to these facts.

Accepted by the Wistar Institute of Anatomy and Biology January 12, 1910. Printed June 21, 1910.

BIBLIOGRAPHY

GEMMEL, J. F. On cyclopia in osseous fishes. Proc. London Zoöl. Society, i, pp. 1906 443–449.

1906b Notes on supernumerary eyes, and local deficiency and reduplication of the notocord in trout embryos. Proc. London Zoöl. Society, i, pp. 449–452.

HARRISON, R. G. Embryonic transplantation and the development of the ner- 1908 vous system. Anat. Record, ii, pp. 385–410.

LEWIS, W. H. Experimental evidence in support of the theory of outgrowth 1907 of the axis cylinder. Am. Jour. Anat. vi, pp. 461–472.

1909 The experimental production of cyclopia in the fish embryo (Fundulus heteroclitus). Anat. Record, iii, pp. 175–181.

MALL, F. P. A study of the causes underlying the origin of human monsters. 1908 Jour. Morph. xix, pp. 1–361.

MAYER, A. G. Rhythmical pulsation in scyphomedusae. Carnegie Institute of 1908 Washington, Pub. No. 102, pp. 113-131.

PAOLUCCI, L. Sopra una Forma Mostruosa della Myliobatis Noctula. Atti della 1874 società Italiana di Sc. Naturali. xvii, pp. 60–63.

STOCKARD, C. R. The development of Fundulus heteroclitus in solutions of lith- 1906 ium chlorid, with appendix on its development in fresh water. Jour. Exp. Zoöl., iii, pp. 99–120.

1907 The influence of external factors, chemical and physical, on the development of Fundulus heteroclitus. Jour. Exp. Zoöl., iv, pp. 165–201.

1909 The development of artificially produced cyclopean fish, "the magnesium embryo." Jour. Exp. Zoöl., vi, pp. 285–338.

1909b The origin of certain types of monsters. Am. Jour. Obstetrics. lix, no. 4.

WILDER, H. H. The morphology of cosmobia. Am. Jour. Anat., viii.pp. 355- 1908 440.

THE INDEPENDENT ORIGIN AND DEVELOPMENT OF THE CRYSTALLINE LENS

CHARLES R. STOCKARD

From the Department of Anatomy, Cornell Medical School, New York City

WITH TWENTY-EIGHT TEXT FIGURES AND TWO PLATES

INTRODUCTION

The crystalline lens normally arises in the embryo from ectoderm overlying the optic vesicle and continues its development in close association with the optic cup. This fact suggests that some correlation exists between the manner of development of the optic cup and the optic lens. In case such a correlation does exist, to what extent is the optic vesicle and cup responsible for the origin and subsequent development of the lens; and, on the other hand, what influence, if any, does the presence of the lens exert over the development of the optic vesicle into the cup and its subsequent development into the eye?

If it is proven that the optic vesicle and cup possess the power to derive a lens from the ectoderm the further question then arises is the ectoderm under any condition able to give rise to a lens without the optic vesicle stimulus? If so, is this independent lens-bud capable of self-differentiation to such an extent as to form a perfectly constructed lens?

Many questions of detail, as, for example, the relationship between the size of the optic cup amd the size of the lens, the ability of partial or defective optic cups to stimulate lens formation, whether the lens may be derived only from certain areas of ectoderm or from any ectoderm, and other problems which we will presently consider, also present themselves.

The study of these questions has come to be known as the lens problem. Experimenters have attacked the questions from several sides during the last ten years and the lens problem has in many ways been beautifully analysed, yet today much additional evidence is necessary to entirely clear the situation.

In a previous paper the writer ('09) briefly described the independent origin and self-differentiation of the optic lens in the fish embryo. Since that time further experiments have yielded much additional material and evidence bearing on this subject. The results of these experiments are considered in the present contribution and show in a most convincing manner that the crystalline lens is capable of originating independently from the ectoderm and of subsequent self-differentiation. Definite proof will also show that although the lens may arise independently, nevertheless the optic vesicle invariably has the power to stimulate the formation of a lens from any overlying ectoderm with which it may come in contact.

The action of the optic vesicle on the ectoderm is a much stronger force for the production of a lens than is the innate tendency of the ectoderm to produce an independent lens. Slightly injured ectoderm may be rendered unable to form a free lens while the same weakened ectoderm will respond to a contact stimulus of the optic vesicle by forming a lens. This point is important, for herein the writer believes, after a study of his own experiments and the results of other workers, lies the explanation of many discrepancies in the operation experiments on the lens. When the operation is so performed that the ectoderm must be folded away in order to extirpate the optic vesicle and is then returned to its place, free lenses have failed to occur, although an optic vesicle may still have been able to derive a lens from this replaced ectoderm. On the other hand, when the early open medullary plate is operated on so as to remove the optic vesicle areas, the ectoderm of the head wall is sometimes left uninjured and from it may arise free lenses. The free lenses of King's experiments arose in embryos which were operated on dorsally to burn out the optic vesicle areas of the partially open medullary tube. The lateral head ectoderm was probably uninjured in some of the specimens. In Spemann's more recent experiments the early open medullary plate was operated upon directly to remove the optic vesicle areas; in such experiments free lenses arose from the uninjured ectoderm. Experiments on other species at such stages and in a similar manner will probably give like results.

In the experiment of removing the optic cup and leaving a partially differentiatied lens, this lens may have degenerated or ceased to differentiate on account of the injury it suffered by the operation, the absence of the optic cup not affecting it. It is unquestionably true that in some amphibians and fishes the lens is capable of perfect self-differentiation.

The optic cup does not exert complete control over either the size or shape of the optic lens. Numerous points of detail are also elucidated by the study of the optic organs in artificially produced fish monsters which are either blind or present various eye defects.

The experimental part of this investigation was conducted during the summer of 1909 in the Marine Bioligical Laboratory at Woods Hole, Mass., while occupying one of the rooms of the Wistar Institute.

2 METHOD AND MATERIAL

In all former experiments on the developing lens, except those which the writer recorded ('09), mechanical methods have been resorted to in destroying the early optic vesicle. This has been accomplished by burning the region with hot needles or by cutting away the tissue. It matters not how cleverly such experiments may be performed they are often open to objection, particularly when the experimenter has to remove or injure the overlying ectoderm in order to reach and extirpate the optic vesicle.

The present experiments have been conducted in an entirely different manner. When developing fish embryos, Fundulus heteroclitus, are treated with certain magnesium salts, alcohol, chloretone or other anæsthetic agents, the development of the optic vesicles is prevented entirely in some cases, while in other specimens only one vesicle forms on either the right or left side, and finally a large majority of the embryos present the cyclopean defect with a more or less double ventro-median eye. We have here, therefore, an exceptioal opportunity to study the relationship between the development of an optic vesicle and a lens. In the first case, does a lens or do lenses ever occur in the eyeless specimens? Does a lens ever appear on the eyeless side in the single-eyed monsters? Finally do lenses ever arise in their usual lateral positions when the embryo has a ventro-median cyclopean eye? All of these propositions are affirmatively answered without an operation to injure in any way the ectoderm or tissues in the primary lens-forming region. It may be thought that the action of the chemical or anæsthetic is as severe as an operation but this is probably not true, as the embryos after being treated for the necessary time with magnesium, on being returned to sea-water develop, hatch and swim actively about, living in aquaria as long as I have tried to keep them, more than one month.

The experiments of the past summer have convinced the writer that his former idea that the eye defects are due to the anæsthetic properties of magnesium is correct; and there is no reason for believing that certain tissues usually between the eyes are entirely absent. These results are given in full in another paper. The solutions employed act as anæsthetics preventing the usual outpushing of the optic vesicles from the brain to a greater or less degree, and give exactly the same condition so far as contact influence of the optic vesicle on the lens is concerned, as though the optic vesicle was cut away, without the disadvantages accompanying the operation.

The experiments with anæsthetics furnish a richness of material, hundreds of specimens being obtained, which one would be unable to duplicate from operations without spending days of tedious labor. The crystalline lenses may be seen with the binocular microscope as spherical refractive bodies in the living specimens. The experimenter in this way is enabled to select various condition for study.

The embryos were best preserved for histological study in picro-acetic though many fixitives gave good results. The lenses stain equally well in eosin or picric acid used as a counter stain after Mann's hæmatein or Delafield's hæmatoxylin.

3 THE LENS IN LIVING EMBRYOS

The living embryos at various stages when examined with a binocular microscope show lenses isolated entirely from the optic vesicle or optic cup. Figure 1 illustrates an embryo eighteen days old that has no trace whatever of optic cups yet two perfectly developed transparent lenses occupy almost normal positions on the sides of the head. Sections of this specimen show the two lenses with well differentiated fibers, fig. 5 and plate II, fig. 3. Mesenchymatous tissue lies between the lenses and the brain. We must conclude that these lenses arose in lateral positions and continued development and differentiation with no direct influence whatever from either optic vesicle or the brain tissue. The possibility of optic cups having arisen and degenerated is entirely out of the question, since, in the first place, this has never been known to occur in any of the hundreds of Fundulus embryos that the writer has studied. In the second place, plate I, fig. 1, shows a lens arising from ectoderm on the eyeless side of a seventy-six hour embryo; it is scarcely conceivable that an optic vesicle arose at about the thirty-fifth or fortieth hour, came in contact with the ectoderm and entirely disappeared, leaving no trace of itself by the seventy-sixth hour. Further, the lens in these cases must continue to develop and differentiate independently, which is contrary to the position held by Spemann ('05), Lewis ('07) and LeCron ('07) for the frog and salamander.

Fig. 2 shows a fish nineteen days old with two well formed lenses of normal size lying in contact with two small irregularly shaped masses of heavy black pigment. In sections, figs. 23 and 24, the lenses are found to be perfectly formed and the two dark spots are shown to be masses of choroid tissue or retinal pigment without a definite retina or other associated eye parts. The lenses may owe their existence to the choroid spots but the latter have failed to influence the size or manner of development of the former. If the origin of these lenses was due to the influence of the choroid areas we have a striking example of how very small an amount of optic tissue may call forth a lens. Many instances will be given to show that extremely small amounts of optic tissue touching the ectoderm will stimulate a lens to form, yet these cases will in no way weaken the evidence that lenses do at other times form entirely independently.

A fish is illustrated by fig. 3 with two small defective eyes deeply buried in the head. The right eye possesses a lens but the left faces ventrally into a mass of mesenchyme and is without a lens. In front of the left eye is shown a lens in an extremely anterior position but completely separated from the eye, and the concave surface of the cup is not directed towards this lens. Fig. 4 proves the case by showing two slightly small eyes each possessing its own lens, while somewhat in front and between the two eyes lies a perfectly isolated and independent lens. It is evident that this supernumerary lens is independently formed and not due to a stimulus from either of the eyes, since each has its own lens.

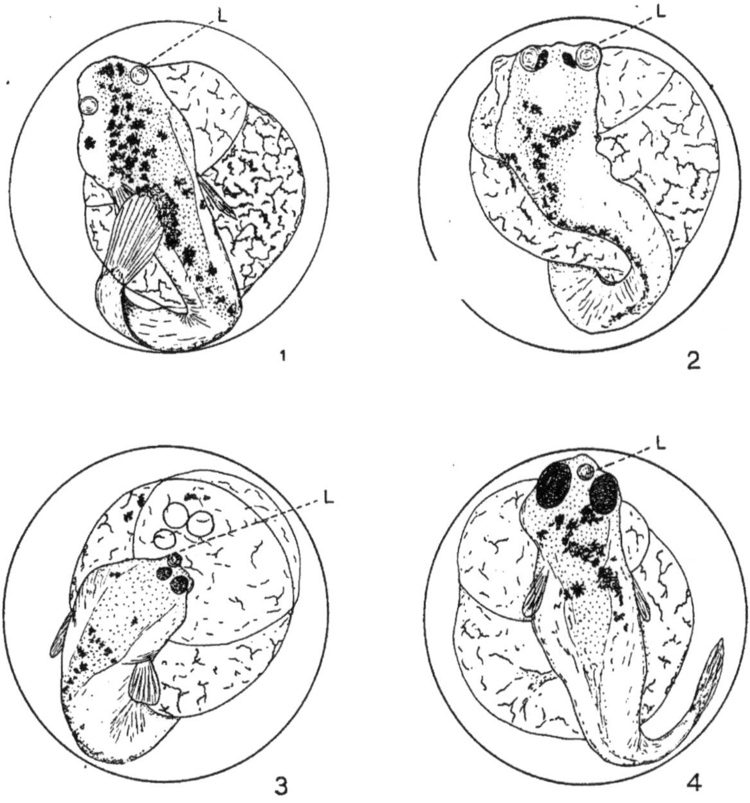

Camera Drawings of Living Fundulus Embryos, Magnified 16 Times

Fig. 1 An embryo eighteen days old. First thirty-six hours after fertilization were spent in 7 per cent alcohol in sea-water. No optic cups formed, yet two perfect crystalline lenses are shown in the sides of the head. See fig. 5 for a section of these lenses.

Fig. 2 An embryo nineteen days old, first thirty-six hours in 9 per cent alcohol. Two perfect lenses near poorly formed eyelike choroid spots. See figs. 23 and 24 for sections of these lenses and eye spots.

Fig. 3 An embryo of same age and lot as fig. 2; a free lens is seen in an extremely anterior position in front of an ill-formed eye. Lens. L.

Fig. 4 Another embryo of the same lot, showing a free lens between two somewhat defective eyes, each of which contains its own lens. Sections of this embryo, figs. 9 and 10, show also a second free lens which was hidden in the living specimen.

4 THE EMBRYOS STUDIED IN SECTION

a Is the Origin of the Lens from the Ectoderm Dependent upon a Contact Stimulus from the Optic Vesicle?

Spemann ('01) and Herbst ('01) first introduced the view that the lens originates from ectoderm only when a contact stimulus from the optic vesicle is present, although Spemann ('07) has since modified his opinion for one species of frog at least. Lewis ('04 and '07) has found this to be true in his experiments on frog tadpoles, and LeCron ('07) on salamanders. King ('05) claims, however, that such is not the case in her experiments and holds the view that the lens arises independently of the contact stimulus by the optic cup. Lewis ('07) has brought objections to the method employed by King and so criticises her results, but the writer believes her method has a real advantage in that she burnt out the optic vesicle areas of the still open medullary tube from the dorsal side and thus may not always have injured the lateral ectoderm of the future lens-forming region.

The writer's experiments on the fish embryos clearly demonstrate that the origin of the lens from the ectoderm may be entirely independent of the contact stimulus of the optic vesicle. He continues to use the expression "contact stimulus of the optic vesicle" since this is what has been deemed necessary for the origin of the lens, although he believes that a lens may arise without any stimulus whatever from the optic vesicle either by contact or from a distance.

In specimens lacking optic vesicles entirely it is difficult to imagine that tissues are present in the brain which possess the power to form substances characteristically formed by optic vesicles and that these substances diffuse until they reach the ectoderm and stimulate it to form a lens. In the case of isolated supernumerary lenses the optic cups possess lenses but still other lenses arise at a distance.

Again referring to fig. 1 of plate I, a section through the eye region of a seventy-six hour embryo, the ectoderm on the eyeless side is forming a lens which is somewhat slower in development than the lens in the eye on the other side, yet this bud shows distinct lens character.

A perfect lens is seen in fig. 2, plate I, to be entirely separated from the brain and no optic cup exists. Fig. 13 shows a lens in a small choroid cup and a second free lens lying near. Fig 7 illustrates a similar case. Figs 8 and 12 show extremely anterior lenses in eyeless individuals and again fig. 5 and fig. 3, plate II, show two beautiful lateral lenses in another eyeless specimen. Fig. 6 shows a section with three well differentiated lenses all free from contact with an optic vesicle; a more posterior section of this embryo, fig. 27 shows a choroid cup deeply buried in brain tissue and without a lens. This cup does not come in contact with either of the three lenses shown in the more anterior sections.

Finally, a most remarkable case of supernumerary lenses is illustrated by figs. 9 and 10 and the outline fig. 11 shows the position of these lenses in the entire head (see also plate II, fig. 4). Two defective eyes each possessing a lens are shown in section, fig. 10, and a third lens lies between the eyes. In a more anterior region, fig. 9 and plate II, fig. 4, is found another section of this third lens, C, and a fourth additional protruding lens lies below it.

These cases might be enumerated and illustrated until they ran into the scores, but sufficient evidence has been given to prove that the crystalline lens in these embryos does *not* depend upon a contact stimulus of the optic vesicle for its origin from the ectoderm but originates independently.

b *Is the Lens-Plate or Lens-Bud Capable of Differentiating into a Lens without Contact with the Optic Cup?*

The above question is convincingly answered in the affirmative by the evidence given in the foregoing discussion. In the older embryos it is clearly shown that supernumerary lenses are as highly differentiated and as perfectly formed in all respects as are the lenses in the eyes. Eyeless individuals, as figs. 1 and 5 and plate II, fig. 3, indicate, may possess perfectly formed transparent lenses which appear in the living specimen as clear refractive bodies.

SECTIONS OF WELL DIFFERENTIATED FREE LENSES

FIG. 5 A section of two free lateral lenses in the nineteen day embryo shown by fig. 1.; no trace of optic cups exists. This embryo has two ears of unequal size.

FIG. 6 Section of the head of nineteen day embryo, first thirty-six hours in 9 per cent alcohol. Three free lenses, A. B. and C are shown, a defective optic cup completely separated from the lenses is deeply buried in the head tissues of a more posterior region, see fig. 27.

FIG. 7 Section of the anterior tip of the head of a nineteen day embryo from 9 per cent alcohol. The upper right lens protrudes from a defective eye shown in more posterior sections, while the lower lens, F., is free, being in no way associated with an eye.

FIG. 8 A somewhat sagittal section of a similarly treated embryo of same age, showing another free lens, p, a pigment spot.

FIG. 9 An anterior and fig. 10 a more posterior section through the eye region of a nineteen day embryo treated with alcohol. The diagram, fig. 11, shows the plane of both sections. Fig. 4 shows the same embryo from life; the eyes in the sections are reversed by the microscope. Two optic cups are present each with a lens, A. and D, while two other perfectly differentiated lenses, B and C, are not connected with an optic part.

FIG. 12 A section through the anterior tip of a pointed-headed eyeless embryo nineteen days old. The lens is well differentiated; the ears in this specimen are scarcely formed.

FIG. 13 A section of a nineteen day embryo showing a small defective choroid cup with a lens, and a second accessory lens is near by.

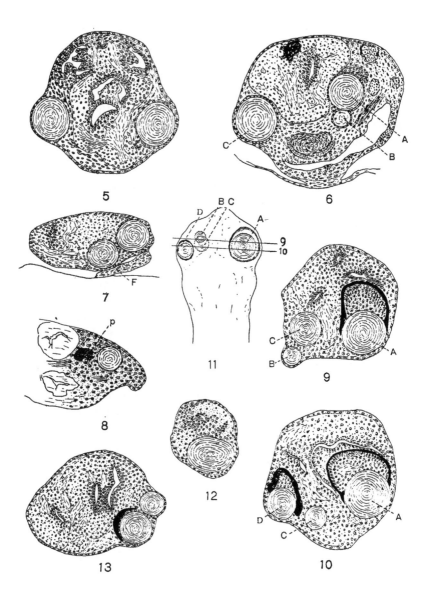

5

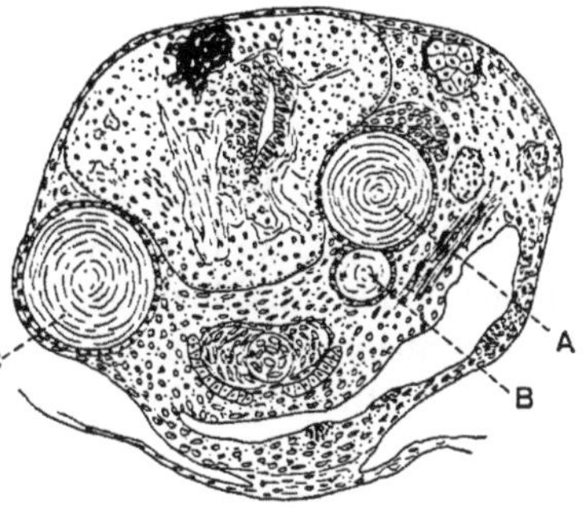

6

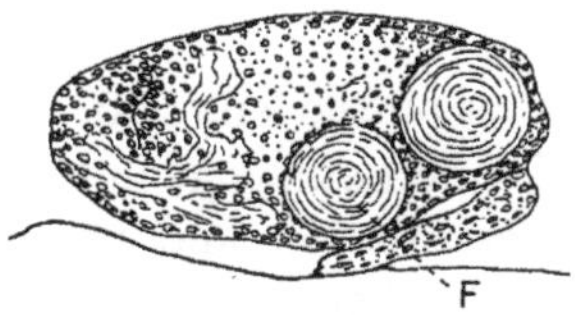

7

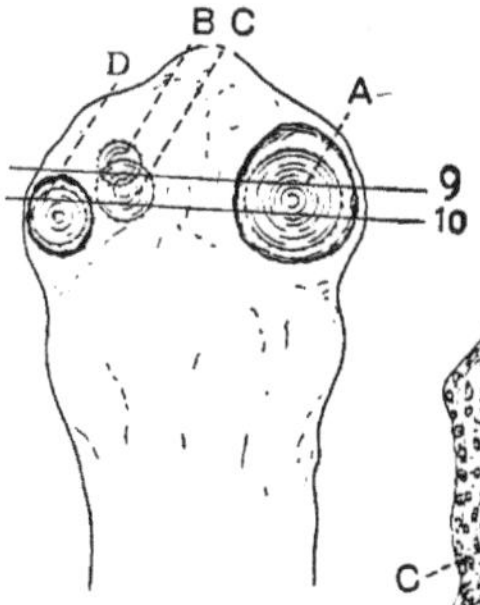

11

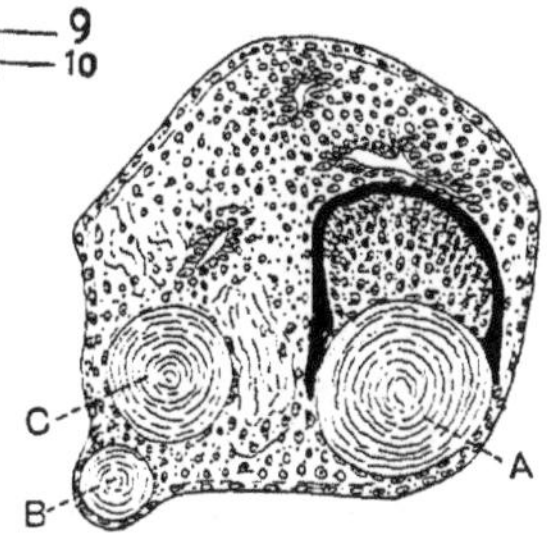

9

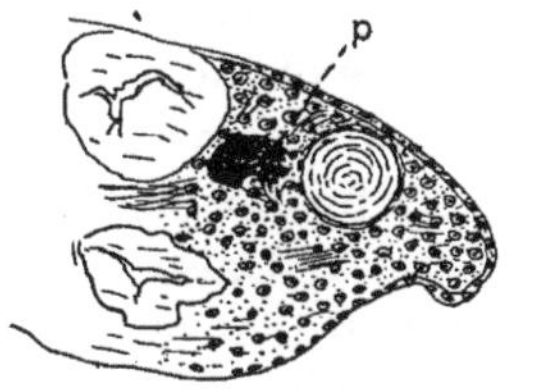

8

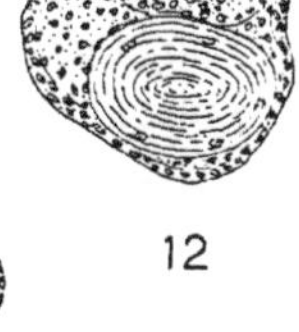

12

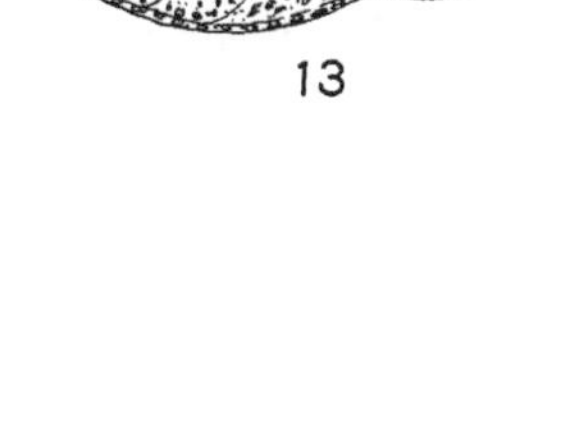

13

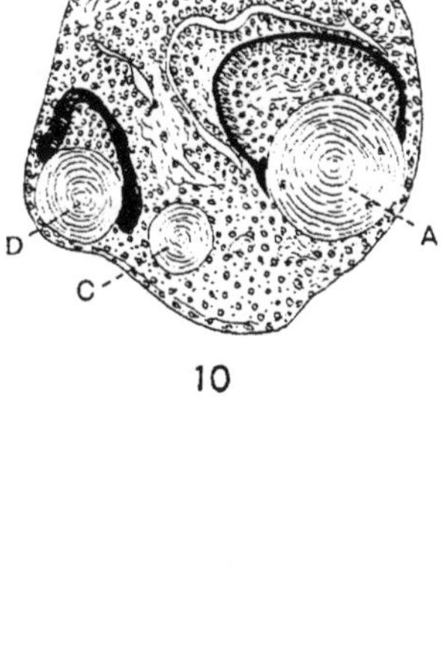

10

The possibility of the action of some substance given off by a distant optic cup is entirely aside, since other experimenters have claimed that when the optic vesicle or cup is in any way separated from the lens the latter organ begins to degenerate and usually disappears.

Figs. 6, 7, 9, 10, 12, 13 and 21 and plate I, fig. 2, plate II, figs. 3 and 4, all go to show that in the fish embryo the lens-plate or lens-bud is capable of self-differentiation, finally forming a perfectly transparent refractive body even though completely isolated from any other eye-like structure.

c Does the Size of the Optic Cup Regulate the Size or Shape of its Associated Lens?

Lewis ('07) has stated in his more recent paper on the lens that, "The lens is neither self originating nor self differentiating, but is dependent for its origin, its size, its differentiation and its growth on the influence of the eye." The writer had also independently been led to think from his first experimental study of cyclopia ('07), which was based on a limited supply of material, that the size of the lens varied directly with the size of the optic cup. He is now able to show that while normlly the lens and optic cup are properly adjusted as to size this is not by any means constantly true of ill-formed eyes. Here the size and also the shape of the lens is often greatly out of accord with that of the optic cup. In normal eyes the optic cup has a definite size and so does the lens. The sizes accord, yet this may be incidental or entirely without correlation, as is suggested by the fact that optic cups of unusual shape and size are not able to regulate the development of the lens so as to adjust it to their strange proportions.

Many of the illustrations of cyclopean eyes given in the writer's recent paper ('09) show misfits between the cups and lenses. Remarkable cases are also shown by figs. 9 and 10 and plate II, figs. 4 and 5, in which the lenses are clearly too large for the associated cups. Fig. 15 shows the two components of an incomplete cyclopean eye with one normally proportioned lens between them. This lens is scarcely large enough to function with the unusually wide double cup. In fig. 20 is seen a somewhat similar double cup with an elongated and slightly constricted lens; fig. 18 also shows one half of a double cyclopean eye with a double lens, the other half eye is in a more posterior section. In figs. 14 and 17, on the other hand, we find illustrated double lenses in one of two closely approximated eyes. Fig. 14 also shows a tiny additional lens still further within the same cup which possesses the large double lens. Fig. 16 shows the extreme anterior tip of a cyclopean eye with two minute lenses protruding from it. This section is only fifty micromillimeters from the anterior end of the head. Numerous other examples of misfitting lenses might easily be given.

These facts force us to conclude that the size of the optic cup does not fully regulate either the size or shape of the associated lens. It is, therefore, evident that the usual harmonious adjustment between the optic cup and the lens may not be so entirely due to a dominating influence of the optic cup on the lens as one might be led to believe from previous contributions to the subject. That some influence or interaction exists, the writer does not deny, and will show in the following parts of this paper the remarkable ability possessed by the optic vesicle to obtain a lens from any part of the ectoderm with which it may come in contact.

d Is the Optic Vesicle, Normal or Defective, always Capable of Stimulating Lens-Formation from the Ectoderm at Some Stage of its Development?

Of all the embryos which the writer has examined not one failed to have a lens in a normal optic cup when the cup came in contact with the ectoderm. If, however, the cup fails to reach the outer body wall, although it may possess well differentiated retinal layers and other parts, it is invariably without a lens, fig. 26. The convex side of the choroid coat or pigment layer of the retina does not cause a lens to arise even though it be closely applied to the ectoderm, as is shown by fig. 26 and many other illustrations in which the choroid touches the body wall.

Defective optic cups when deeply buried and separated from

Ill-Adjustment of Optic Cups and Lenses in Thirteen Day Fish Embryos.

Fig. 14 A section of an embryo treated with 5 per cent alcohol. Both optic cups are defective, the left one contains a spherical lens, while the right cup has a large lens, in shape a constricted oval, and a tiny spherical lens placed further within the same cup. E, ear.

Fig. 15 A poorly formed eye of the incomplete cylopean type. Each component faces in a ventro-median direction and a spherical lens lies between them An ear, E. is shown on the side of the better component, the other ear is absent.

Fig. 16 A section 50 microns from the anterior tip of an embryo. The anterior border of a cyclopean eye is shown with two protruding lenses of very minute size. ec, ectoderm.

Fig. 17 Section through the anterior tips of two small *closely neighboring eyes*, whose median planes are in more posterior sections. The smaller eye has a protruding double lens. This embryo possesses only one ear located on the side with the better eye.

Fig. 18 Part of an incomplete cyclopean eye (other half in more posterior sections) containing a double lens.

Fig. 19 Two *adjacent eyes* facing ventro-medianly with two lenses.

Fig. 20 An ovoid lens in an incomplete cyclopean eye.

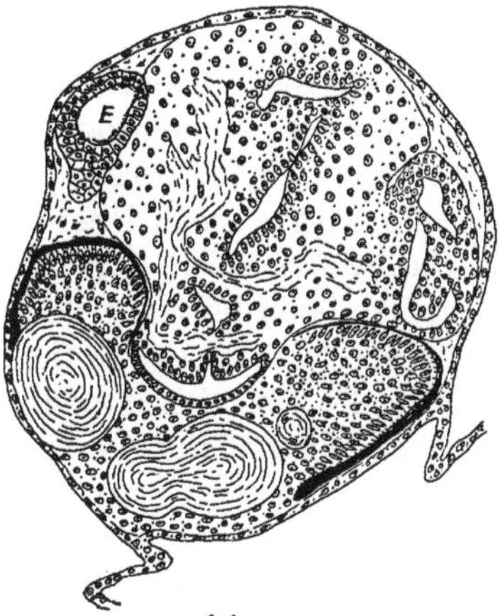

14

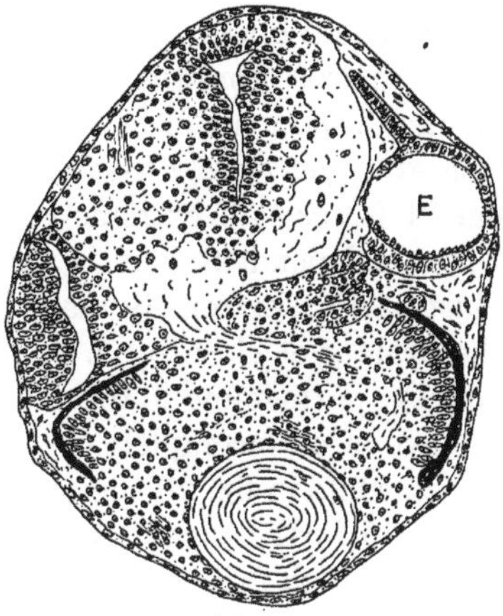

15

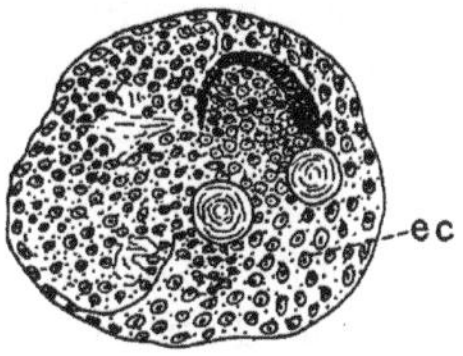

16

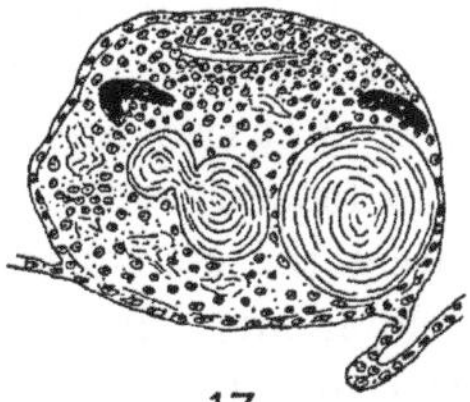

17

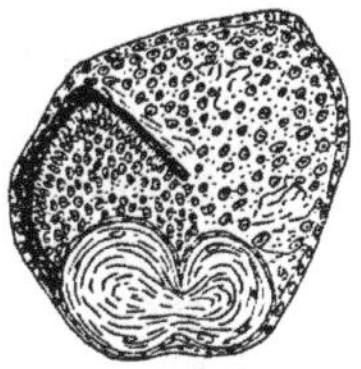

18

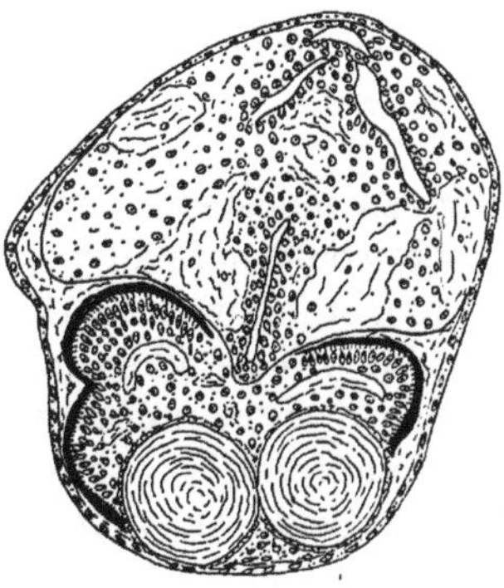

19

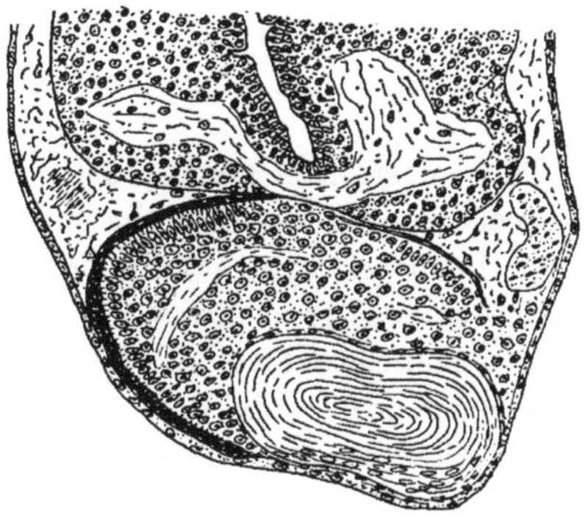

20

the ectoderm also lack a lens, as is illustrated on the left of figs. 27 and 28. On the other hand, it is remarkable how small and ill-formed an optic cup-like structure has the power of stimulating a lens to arise from the ectoderm. Figs. 2, 23 and 24 and plate II, fig. 5, show small choroid cups with no retinal differentiation whatever, yet closely associated with perfectly formed lenses. In fig. 21 is seen an extremely defective cup with a small lens; a larger independent lens is shown on the eyeless side. Fig. 22 illustrates an extremely insignificant eye-like body buried within the brain, yet close by is a small crystalline lens; these are the only eye parts found in this embryo. In fig. 19, a section through the eye of an incomplete cyclops, each component of the eye has a lens, while in fig. 20 the eye components are closer together and in fig. 15 further apart, yet each of these possesses only a single lens, although it is elongate in fig. 20. It is difficult to say why such eyes occasionally possess two lenses. After an examination of a large number of such eyes no general rule is found. It may be due in some way to the manner in which the optic cup periphery meets the ectoderm, whether as a circle, an oval or at times a much constricted oval so that two areas of ectoderm are separately stimulated to form lenses.

This consideration forces the conclusion that an optic cup at some stage in its development, whether normal or defective, invariably possesses the power to stimulate lens-formation from the ectoderm with which it comes in contact.

e *May the Optic Vesicle Cause Lens-Formation from Ectoderm Other than that which Normally Forms a Lens?*

This question is answered by the cyclopean monsters. It is scarcely conceivable that the ectoderm which would normally lie over the lateral eyes has the power to migrate, or follow the optic vesicle so exactly as always to lie just over the vesicle wherever it may chance to develop. Many embryos have eyes in unusually anterior positions and derive their lenses from anterior ectoderm, while others possess ventral eyes with lenses derived from ventral ectoderm. It occurs in a few cases, as the writer previously

21

22

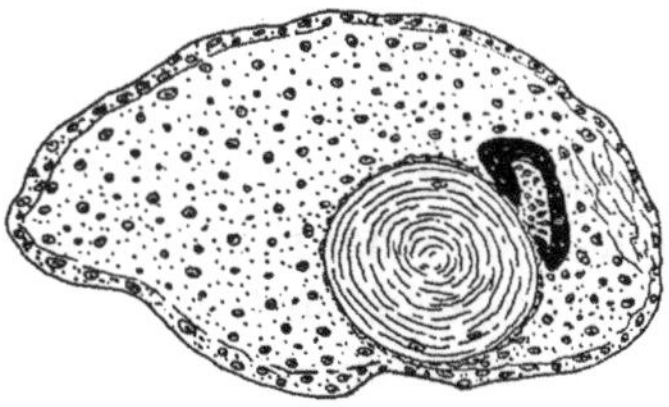

23

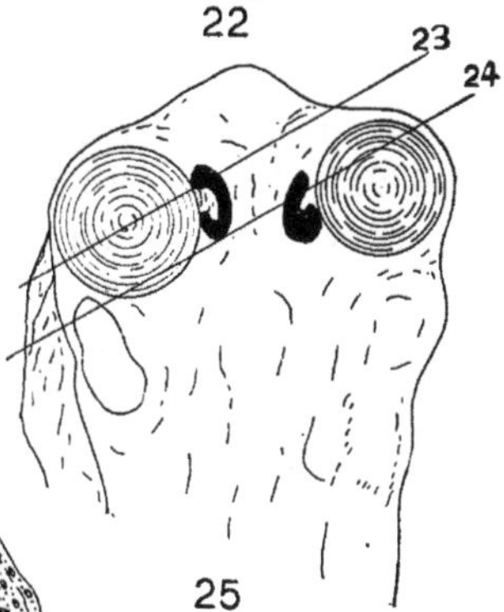

25

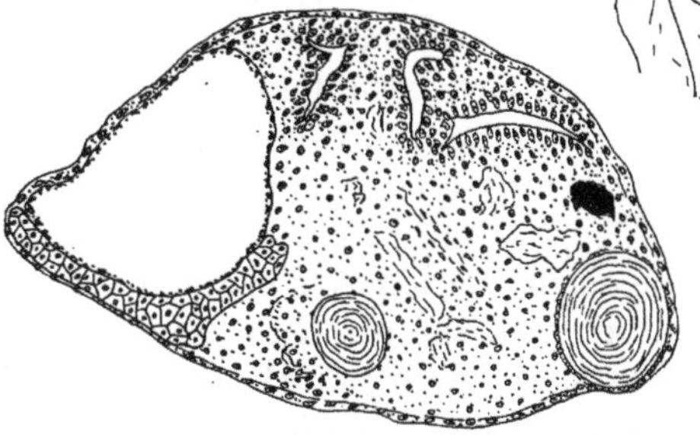

24

EXTREMELY DEFECTIVE OPTIC CUP-LIKE BODIES ASSOCIATED WITH PERFECT LENSES

FIG. 21 An anterior section, as is indicated by its size, of a nineteen-day embryo. A free lens is shown on the left while on the right a very small defective eye contains a small lens.

FIG. 22 A small lens near a vesicle-like structure, E, in the brain which might represent an abortive optic body.

FIGS. 23 and 24 Sections through the embryo shown in fig. 2. In the diagram, fig. 25, is shown approximately the planes of the sections. The large well-formed lenses are associated with defective optic cups which lack any sign of differentiation.

recorded ('09), that free lenses arise in their usual lateral positions while the cyclopean eye possesses its lens of anterior origin.

In the fish embryo the optic vesicle may cause a lens to form from ectoderm far removed from the usual lens-forming area, and rarely in such cases it happens that free lenses may also arise from the usual lens-forming region.

f Does a Deeply Buried Eye have the Power to Regenerate or Form A Lens from its own Tissues?

It has been shown by many experimenters Colucci ('91), Wolff ('95 and '01), Müller ('96) and Fischel ('02), that the salamander's eye regenerates a new lens from the posterior surface of the iris when the old lens is removed, or as Fischel found, if the old lens be merely pushed back out of its usual place in the eye. When the iris was injured in two places during the extirpation of the lens, two lenses arose within the single eye, one growing from each injured area of the iris. It has also been found that a fish's eye would regenerate a lens under certain conditions: when the fish is young and when a sufficiently long time is allowed for the lens to regenerate.

Lewis ('04) finds that the deeply buried eyes in Rana palustris which fail to come in contact with the ectoderm are unable to form lenses from their own tissues. While on the other hand he states that in a second species, Rana sylvatica, the optic cup readily gives rise to a lens from its own tissues if prevented from stimulating the formation of a lens from the ectoderm of the body wall.

The fish embryos which are now being considered, act in a similar manner to Rana palustris and are unable to form lenses from the tissues of their optic cups. Whenever the optic cup is deeply buried and fails to reach the ectoderm it also fails to possess a lens, as is illustrated on the left side of figs. 26, 27 and 28. In this connection it may be mentioned that Morgan was unable to obtain the regeneration of a lens in adult specimens of Fundulus, although as mentioned above, lenses do regenerate in the eyes of another species of fish.

26

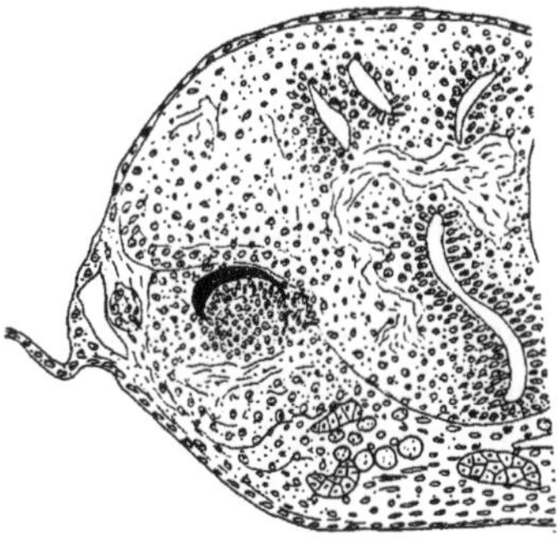

27

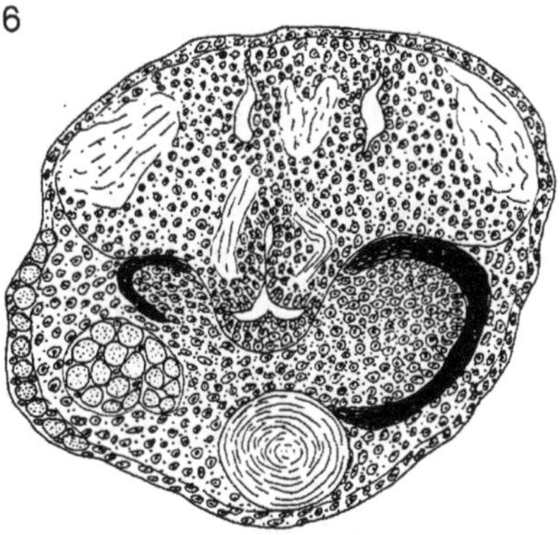

28

DEEPLY BURIED EYES WITHOUT LENSES

FIG. 26. A section of a thirteen day embryo which was treated with alcohol. One eye faces outward and contains a lens, while the other faces in toward the median plane of the head and is without a lens, probably never having come in contact with ectoderm.

FIG. 27 A section of a nineteen day embryo more posterior in the same series than section, fig. 6. The deeply buried defective eye has no lens.

FIG. 28 Section of thirteen day embryo after treatment with alcohol. Section passes in front of the median plane of the right eye which is well differentiated and contains a lens. The other eye is small and scarcely differentiated, buried in the tissue below the brain and contains no lens. A normal ear exists on the side with the large eye and a small defective ear is on the other side in more posterior sections.

5 DISCUSSION OF PREVIOUS OBSERVATIONS AND EXPERIMENTS

Before drawing final conclusions from the cases discussed above it may be well briefly to review the modern work and experiments which have been directed towards a solution of the so-called lens problem.

Rabl ('98) in his study of the structure and development of the lens, found in one case, at least, that a lens-like body was present, although far removed from the optic vesicle. This observation has been criticised by Lewis, who suggested that the ectoderm with the newly forming lens had shifted away from the small optic vesicle. Again, Lewis states that Rabl's case, and also a case to be considered below that was shown by Mencl, prove neither one side nor the other, since the experimental evidence is all directly for the idea that a lens will not arise from the skin without the stimulus of the optic vesicle. At the present time, however, the experimental evidence points in an opposite direction, and the cases of Rabl and Mencl can scarcely be disposed of in so brief a manner. On the contrary Rabl's example must be considered the initial illustration of the origin of an early lens without a stimulation from the optic vesicle.

Following this single case a strong tide turned towards the idea of the dependent origin and development of the optic lens. Herbst's paper ('01) on "Die formative Reize in der thierischen Ontogenese" and Spemann's pioneer experiments ('01) on the development of the lens seemed most convincing evidence in favor of a correlation in development between the optic vesicle and the lens, a correlation in which the latter played a dependent rôle. Herbst claimed the optic lens to be a "Thigmomorphose" originating only by a contact between the optic vesicle and the epidermis. His reason for such a position being that in the case of cyclopean monsters the median optic vesicle always derives a lens from the overlying ectoderm, while no lenses arise in the usual lateral positions. "If the lens is an independent organ why does it not arise in the lateral eye region of cyclopean monsters?" In former experiments the writer ('09) produced just such a case as Herbst thought necessary to show the independent origin of lenses, that is, a cyclopean monster with lenses in the usual lateral positions.

After a study of a great number of such monsters I am now able to reaffirm that free lenses do occasionally arise in the lateral eye regions. Herbst's clever argument based on pathological embryos is, therefore, rendered invalid.

Spemann ('01) was the first to clearly attack the problem experimentally. He injured or destroyed the optic vesicles of Triton embryos by means of hot needles and electric cauterizers. The method was not altogether satisfactory since such an operation often injures much of the surrounding tissue, yet Spemann's results were of the highest value, and stimulated an active interest on the part of many experimenters. His conclusions furnished strong support for Herbst's idea of the dependent origin of the lens. He found that whenever the optic vesicle was so injured that it failed to come in contact with the head ectoderm, the ectoderm failed to form a lens. The lens was, therefore, dependent for its origin on a contact stimulus of the optic vesicle upon the ectoderm. Later Spemann ('05) also concluded that the lens was not self differentiating but that a durable contact stimulus of the optic cup was necessary for it to form lens fibers. Thus the "Herbst-Spemann theory of dependent lens formation," as Mencl has termed it, was developed. We shall see below, since it seems best to consider these papers in a more or less chronological order, that Spemann's own later work has helped materially to overthrow this theory.

Barfurth ('02) also operated with a hot needle to destroy the lens anlage and the optic cup. The embryos were examined after five or six days. One specimen showed on one side a poorly regenerated optic cup that did not come in contact with the ectoderm and yet a lens still connected with the ectoderm was present on this side. Barfurth was inclined to accept the Herbst-Spemann idea of lens formation and so attempted to harmonize his case with the theory as follows. Some sections of the embryo, 9 to 12, showed the optic vesicle lying very near the ectoderm but not in contact with it. Barfurth supposes that at an earlier period in development it may have been in direct contact with ectoderm and at such a time stimulated the lens to arise. This is merely a conjecture and the more recent experiments of King ('05) and Spemann ('07) on the frog, and the writer's on the fish, would warrant equally well an interpretation of independent origin for this lens.

Returning to observations on monsters we may consider the case reported by Mencl ('03) which has called forth so many explanations and criticisms. This report described the existence of an independent lens on each side of the head in an anadidymus embryo of Salmo salar which he had possessed since 1899, his interest in it being aroused by the contributions on the dependent origin of the lens. One of the lenses in this monster was closely applied to the brain wall, and in fact lay in a depression in the side of the brain; the other lens, however, was completely separated from the brain by mesenchyme. Spemann and Lewis have tried to explain this case by assuming the existence in the brain wall of some optic vesicle tissue or substance which when brought in contact with the head ectoderm possessed the power to cause the formation of a lens. There was absolutetly no evidence of optic tissue in this brain wall and further, the explanation could apply only to one of the lenses, though it scarcely explains the origin of either. The second lens, entirely free and with mesenchymatous tissue separating it from the brain wall, was as perfect, though not so large, as the other which lay against the brain. This lens probably arose and developed freely, just as did so many lenses in the fish embryos the writer has studied.

Mencl ('08) has more recently obtained other anadidymus Salmo embryos and confirms his former observations, finding that in such monsters free lenses are present in 25 per cent of the cases.

Gemmill ('06) has also recorded the frequent occurrence of free lenses in the heads of monster trout embryos.

Schaper ('04) in considering a case of a typical lens development comes to the theoretical conclusion that the lens is by nature a primitive sense body similar to the sinnesknospe of Amblystoma, and has secondarily taken on its present function. It must, therefore, arise independently of the optic vesicle, with which it is only recently associated. Such speculation is not entirely out of accord with the present facts of lens formation.

Lewis thinks that the free rudimentary lenses in Schaper's experiments were undoubtedly caused by the shifting of the ectoderm and lens away from the optic vesicle. The writer is unable to see any reason why such a shifting is supposed to have taken place. On the contrary, the facts seem to lend themselves more readily to the interpretation of a free origin of the lens.

The experiments of Lewis ('04) seemed to show convincingly that the lens was dependent for its origin and development upon a contact stimulus of the optic vesicle on the ectoderm. Lewis devised a method far more refined than any previously used in similar experiments. He operated on tadpoles under a binocular microscope with needles and small scissors and was able to cut the ectoderm and fold it forward so as to expose the brain and early optic vesicle, which could now be cut away. The ectoderm was then folded back in place. From these experiments the following are some of the conclusions which were drawn:

Neither a lens nor a trace of a lens will originate from the ectoderm which normally gives rise to one, if the contact of the optic vesicle with the skin is prevented.

There is no predetermined area of the ectoderm which must be stimulated in order that a lens may arise. Various parts of the skin when stimulated by optic cup contact may and do give rise to a lens.

In normal development the lens is dependent for its origin and differentiation on the contact influence or stimulation of the optic vesicle on the ectoderm.

The conclusions are clearly stated and the experimental method employed is most skillful, but not entirely free from objection. The conclusion regarding the entire dependence of the lens on an optic vesicle stimulus will not, the writer believes, be supported by future experiments, and indeed at the present time such an idea has a vast amount of evidence against it, unless the particular species experimented with shows a specific action in lens formation. The later work of Spemann ('07) which contradicts his former conclusion shows that in some species of frogs the lens may arise and differentiate independently of the optic vesicle stimulus. In these experiments it must be noted that Spemann operated with glass needles to remove the optic vesicle regions from the open medullary plate. Such an operation does not injure the lens-forming region of the ectoderm as Lewis' experiment does and it is on this account, the writer believes, that the free lenses arise.

The tendency of the ectoderm to form a lens independently is so delicately adjusted that a very slight injury or disturbance may suppress it, yet the same ectoderm may still have the power to form a lens in response to the stronger stimulus of the optic vesicle. So in experiments where the ectoderm has been cut or injured it loses the power to form free lenses even though the optic vesicle can stimulate a lens to arise from it. When the ectoderm is uninjured, as in some of King's specimens, Spemann's ('07), and the writer's, then free lenses do occur.

We may imagine the lens to represent an ectodermal organ formerly of independent importance. However, it has now become so closely associated with the nervous portions of the eye that it arises whenever such a part meets the ectoderm, yet the lens retains to a feeble degree its impulse to arise independently of other eye parts. When it has once arisen it is perfectly capable of differentiation. Future experiments of removing the optic vesicle without injuring the ectoderm will probably demonstrate further this tendency of the lens to arise independently, just as Mencl's observations and the writer's experiments show for the fish, and many of the experiments mentioned show for amphibians.

King (05) states that she had begun her series of experiments in 1900. She destroyed the optic vesicles from the forebrain region in the closing medullary tube with hot needles. Many of the embryos died as a result of the operation. The conclusions reached by King are entirely opposed to those of Lewis and Spemann's earlier results regarding the dependent origin of the lens. King found that in some of the embryos a lens-like body arose from the ectoderm on the side with an injured or unregenerated optic cup which did not come in contact with either the ectoderm or the lens-like body. Some of the specimens show a very suggestive lens-like structure which is entirely free from any contact with an optic vesicle. The inaccuracy of the operation may account for these lenses, on the ground that in exceptional cases the outer ectoderm of the lens region was too far distant at the early stage of development to have been injured. This is probable when it is noted that the operation did not enter through the lateral ectoderm, but through the partly open medullary tube from above. The present evidence also goes to strengthen King's position on the subject.

Spemann ('05), Lewis ('07) and LeCron ('07) have all claimed that the lens after its origin, was not self differentiating but depended upon a durable contact with the optic cup for its future development. All of the writer's experiments and the observations of Mencl are clearly contradictory to such a view, and Spemann ('07) has more recently found the lens to be self-differentiating in Rana esculenta.

In a former paper the writer ('07) described the development of the lens in the blind Myxinoid, Bdellostoma stouti. At that time he presumed that this animal furnished support for the experimental evidence that the lens was not self-differentiating. In Bdellostoma embryos the optic vesicle is well formed at first and reaches out from the brain to the ectoderm; at this time the ectoderm forms a localized thickening suggesting a lens-plate. The optic vesicle then ceases to maintain its progressive development and loses connection with the ectoderm, finally giving rise to the poorly formed optic cup of the adult. The early lens thickening of the ectoderm degenerates and is entirely absent from older embryos. The writer suggested that this degeneration was due to the loss of contact with the optic vesicle, since such reasoning seemed correct in the light of the experiments up to that time. He has entirely changed his view, however, regarding the matter of the lens in Bdellostoma and now believes that it represents a rudimentary organ which has been lost in the adult and appears only for a short time during embryonic life. As Eigenmann ('01) records of Amblyopsis, the lenses of other blind fishes appear at certain stages in the developing embryo and then degenerate. A case not so far advanced in the degeneration of the eye and lens is that of the Florida burrowing lizzard, Rhineura. In the adults of Rhineura, Eigenmann finds lenses sometimes present and sometimes absent and when present they are very variable. So many examples are known of the traces of rudimentary organs in the embryo which are entirely lacking in the adult that the above cases of the lens are not at all strange and at present it seems that this is the most satisfactory explanation of their behavior.

The more recent experiments of Spemann ('07) so frequently referred to, may be briefly described. Embryos of Rana esculenta were operated upon with glass needles so as to remove the optic vesicle areas from the open medullary fold (his former experiments were made with hot needles, which doubtless injured more of the surrounding tissues).

After allowing these embryos to develop for several days and then studying them in section it was found that in one case on the side of the head lacking an optic vesicle a true lens-bud was still in connection with the ectoderm. In an older embryo a free lens vesicle was found, and finally, in a still older specimen a lens was buried in connective tissue on the eyeless side of the head yet it possessed fully formed lens-fibers. Thus, Spemann, five years after his first paper on the dependent origin of the lens, now concludes that the lens in Rana esculenta is self-originating and self-differentiating. He also accepts Mencl's case of the independent lens in Salmo salar.

We have thus seen in a rather cursory survey how the lens problem has arisen and how experiments have built up first one side of the question and then the other, and many may also feel that the final word is yet to be added. Nevertheless, it is true that at this stage of the investigation it has been clearly demonstrated in several groups of animals that the ectoderm possesses the power independently to originate a lens which subsequently develops into the transparent refractive organ usually found within the mature eye. The manner of origin and differentiation of the lens may easily differ in different animal species, and the statement would not be warranted that all of these facts apply to animals in general.

The lens is self-originating and self-differentiating yet it is more emphatically proven that the optic vesicle or optic cup has the power without exception to cause a lens to form from the ectoderm with which it comes in contact, even though this ectoderm may be far distant from the usual lens-forming area. There is probably no strictly limited region of ectoderm from which the optic vesicle must stimulate lens-formation. The self-originating lenses, however, have invariably occurred in the head region, though not always in their usual lateral positions. It would, therefore, seem that the ectoderm of the head is more predisposed to the formation of lenses than that of the other body regions.

6 SUMMARY AND CONCLUSIONS

1 When the developing eggs of Fundulus heteroclitus are subjected to the action of Mg salts, alcohol, or other anæsthetic agents, the normal outgrowth of the optic vesicles is generally inhibited. Embryos are obtained either entirely without optic cups, with small deeply buried eyes, with only a single eye on one side of the head or, finally, with a median more or less double cyclopean eye. These specimens furnish exceptional material for a study of the relationship between the development of the optic vesicle and the optic lens. The embryos with the nervous eye parts entirely lacking are similar to specimens with the optic vesicles mechanically cut out of the brain. At the same time, the injury to the ectoderm which usually results from the operation is avoided, and this is a most important advantage. When Spemann ('07) operated on Rana esculenta embryos with open medullary plates he was able to remove the optic vesicle areas from the future inner side of the tube without injuring the ectoderm of the lens-forming region and in such cases independent lenses arose on the eyeless side. King ('05) obtained free lenses in a somewhat similar experiment. Lewis' embryos might possibly respond in a like manner if operated upon in this fashion. The optic vesicle may stimulate a lens from slightly injured ectoderm, but the free origin of lenses is a more delicate process and one more easily interfered with.

2 The crystalline lens may originate from ectoderm without any direct stimulus whatever from an optic vesicle or cup. These self-originating lenses arise from regions of ectoderm that are not in contact with either optic vesicle, the brain wall, or any nervous or sensory organ of the individual (see plate I).

3 The lens-plate or lens-bud is capable of perfect self-differentiation. No contact at any time with an optic vesicle or cup, is necessary. These lenses finally become typical transparent refractive bodies exactly similar in histological structure to that in the normal eye (see plate II).

4 The size and shape of the lens is not entirely controlled by the associated optic cup. Lenses may be abnormally small for the size of the cup, or entirely too large, so that they protrude; or, finally, peculiarly shaped oval or centrally constricted lenses may occur in more or less ordinarily shaped optic cups. The lens is by no means always adjusted to the structure of the optic cup as has been claimed by some observers.

5 An optic vesicle or cup is invariably capable at some stage of its development of stimulating the formation of a lens from the ectoderm with which it comes in contact. It is remarkable how extremely small an amount of optic tissue is capable of stimulating lens formation from the ectoderm (plate II, fig. 5).

6 The optic vesicle may stimulate a lens to form from regions of the ectoderm other than that which usually forms a lens. This is shown by the fact that a median cyclopean eye always stimulates a lens to form from the overlying ectoderm. It is scarcely possible that the lateral normal lens-forming ectoderm could follow the cyclopean optic cup to the many strange situations it finally reaches.

7 The ectoderm of the head region is more disposed to the formation of lenses than that of other parts of the body, since the free lenses invariably occur in this region.

8 A deeply buried optic vesicle or cup may fail to come in contact with the ectoderm; in such cases it lacks a lens. The tissues of the embryonic cup itself are unable to form or regenerate a lens. This is not true in all embryos, as Lewis has shown for one species of frog.

9 The optic lens may be looked upon as a once independent organ (possibly sensory or perhaps an organ for focusing light on the brain wall, before the vertebrate eye had arisen) which has become so closely associated with the nervous elements of the eye that it has to some extent lost its tendency to arise independently, although still capable of doing so under certain conditions. The lens now arises much more readily in response to a stimulus from the optic vesicle, a correlated adjustment which insures the almost perfect normal accord between the optic cup and the lens. In experiments on the origin of the lens one must guard against disturbing the ectoderm from which an independent lens might arise. Although an optic vesicle might have the power to stimulate a lens from injured ectoderm, the innate tendency of the ectoderm to form a lens is a more delicate impulse and may be suppressed by a slight injury or disturbance of the ectoderm during the operation.

7 BIBLIOGRAPHY

BARFURTH, D. Versuche über Regeneration des Auges und der Linse beim Hühner-
1902 embryo. *Verh. d. Anat. Gesellsch.* Halle.

COLUCCI, V. S., Rigenerazione parziale dell'occhio nei Tritoni. Mem. Accad. Sc.
1891 Bologna, Series 5, Vol. i.

EIGENMANN, C. H. The History of the Lens in Amblyopsis. *Proc. Indiana*
1901 *Acad.* Sc., 1901.

FISCHEL, A. Weitere Mitteilungen über die Regeneration der Linse. *Arch. f.*
1902 *Entw'Mech.*, Bd. xv.

GEMMILL, J. F. Notes on Supernumerary Eyes, and Local Deficiency and Re-
1906 duplication of the Notocord in Trout Embryos. Proc. Zoöl. Soc. London, Vol. i.

HERBST, C. Formative Reize in der thierischen Ontogenese. Ein Beitrag zum
1901 Verständniss der thierischen Embryonalentwicklung. Leipzig, H. Georgi.

KING, H. D. Experimental Studies on the Eye of the Frog. *Arch. f. Entw' Mech.*,
1905 xix.

LECRON, W. L. Experiments on the Origin and Differentiation of the Lens in
1907 Amblystoma. *Am. Jour. Anat.*, vi.

LEWIS, W. H. Experimental Studies on the Development of the Eye in Am-
1904 phibia. I. On the Origin of the Lens. Rana palustris. *Am. Jour. Anat.*, iii.

1907 Experimental Studies, etc. iii. On the Origin and Differentiation of the Lens. *Am Jour. Anat.*, vi.

Accepted by the Wistar Institute of Anatomy and Biology January 12, 1910. Printed June 21, 1910.

MENCL, E. Ein Fall von beiderseitiger Augenlinsenausbildung während der
1903 Abwesenheit von Augenblasen. *Arch. f. Entw' Mech.*, xvi.
1908 Neue Tatsachen zur Selbstdifferenzierung der Augenlinse. *Arch. f. Entw' Mech.*, xxv.
MÜLLER, E. Regeneration der Augenlinse nach Extirpation be Tritonen. *Archiv*
1896 *f. Mikrosk, Anat.*, xlvii.
RABL, K. Ueber den Bau und die Entwickelung der Linse, I. *Zeitsch. f. Wiss-*
1898 *Zoöl.*, Bd. 63.
SCHAPER. A. Über einige Fälle atypischer Linsenentwickelung unter abnormen
1904 Bedingungen. Ein Beitrag zur Phylogenie und Entwickelung der Linse. *Anat. Anz.*, xxiv.
SPEMANN, H. Ueber Correlationen in der Entwickelung des Auges. *Verhandl.*
1901 *der Anat. Gesellsch.*, 1901.
1905 Ueber Linsenbildung nach experimenteller Entferung der primären Linsenbildungzellen. *Zoöl. Anz.*, xxiii.
1907 Neue Tatsachen zum Linsenproblem. *Zoöl. Anz.*, xxxi.
STOCKARD, C. R. The Artificial Production of a Single Median Cyclopean Eye
1907*a* in the Fish Embyro by Means of Sea-Water Solutions of Magnesium Chlorid. *Arch. f Entw. Mech.*, xxiii.
1907*b* The Embryonic History of the Lens in Bdellostoma stouti in Relation to Recent Experiments. *Am. Jour. Anat.*, vi.
1908 The Question of Cyclopia. *Science*, n. s. xxviii.
1909 The Development of Artificially Produced Cyclopean Fish, "The Magnesium Embryo." *Jour. Expr. Zoöl.*, vi.
WOLFF, G. Regeneration der Urodelenlinse. *Arch. f. Entw.'—Mech.*, I und xii.
1895 and 1901.

Accepted by the Wistar Institute of Anatomy and Biology January 12, 1910. Printed June 21, 1910.

Explanation of Plates I and II

Fig. 1 A photograph of a section of a seventy-six hour Fundulus embryo; the brain is almost bilateral and perfect, with a well formed left optic cup, no indication of the right cup exists, yet the ectoderm on that side has formed a well pronounced lens-bud, L.

Fig. 2 A section through the eye region of a thirty day embryo, there is no optic cup in the specimen. A perfect lens is in the usual lateral position, L, a band of muscle, M, lies between it and the brain.

Fig. 3 A section through the head of a Fundulus embryo eighteen days old. The specimen is shown in text-figure 1, no eyes are present, yet two perfectly differentiated crystalline lenses are seen in the sides of the head. A close examination of the photograph will reveal mesodermal cells between the lens and the brain.

Fig. 4. A section through one defective eye which possesses a lens of disproportionate size, while on the other side of the head two free lenses are seen, the lower small one protruding from the head. A second defective eye with a lens is found in a more posterior region of the head. Text-figure 4 represents this embryo as it appeared in life.

Fig. 5 A section through one of the very defective optic vesicles shown in fig. 2. The photograph illustrates the extreme lack of proportion which may exist between the size of the optic vesicle and its associated lens. It indicates also how small an amount of optic tissue may stimulate lens formation from the ectoderm.

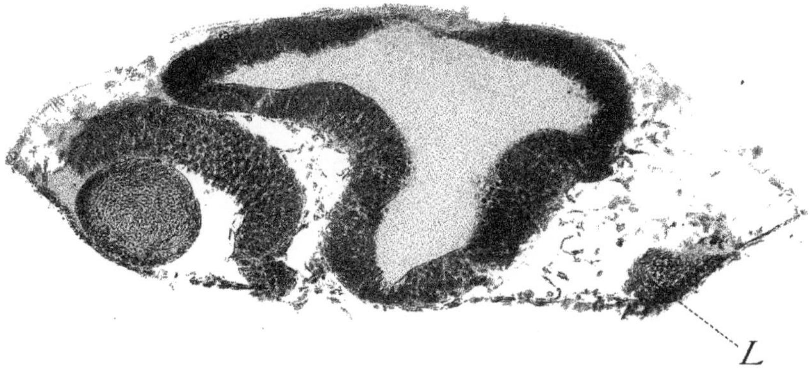

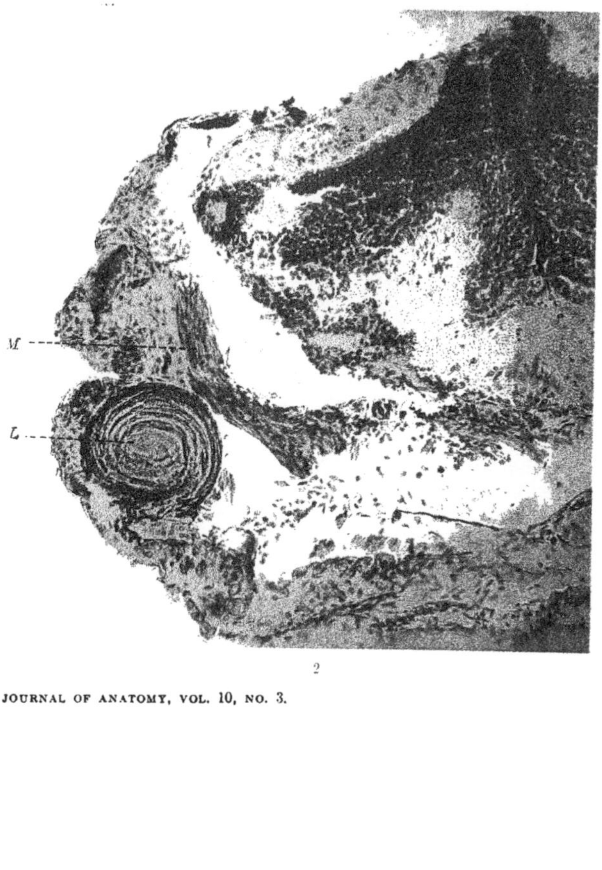

2

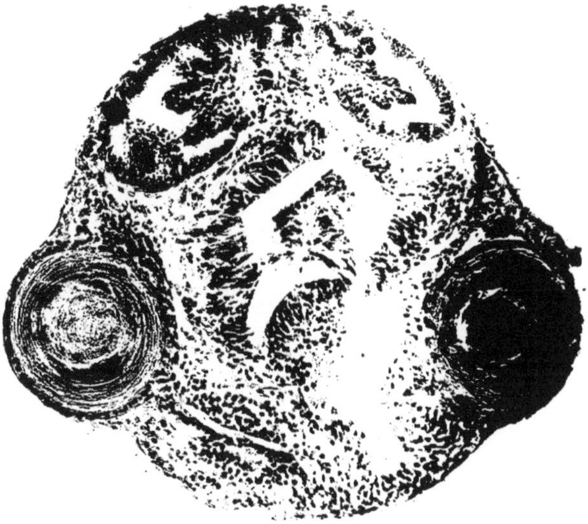

3

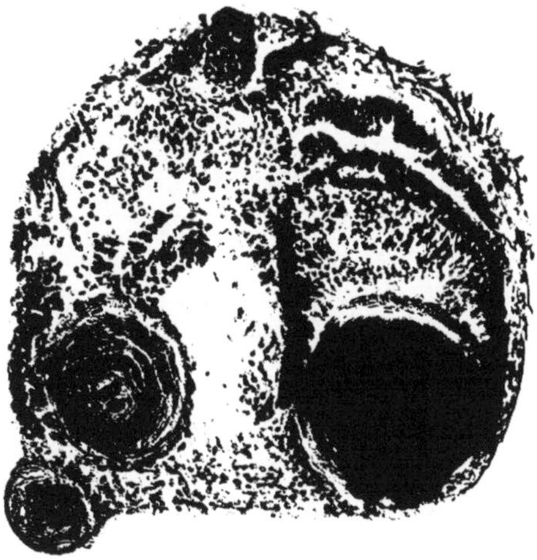

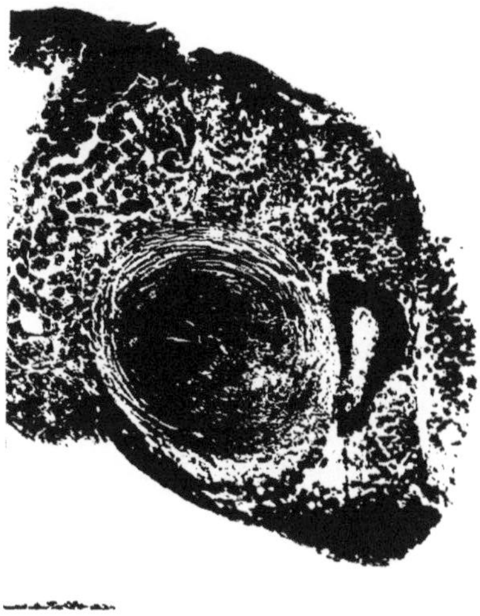

4

Studies of Tissue Growth.

III. The Rates of Regenerative Growth in Different Salt Solutions.
IV. The Influence of Regenerating Tissue on the Animal Body.

By

Charles R. Stockard,
Cornell University Medical School, New York City.

With 4 figures in text.

Eingegangen am 18. Oktober 1909.

III. The Rates of Regenerative Growth in Different Salt Solutions.

The changes in the various physiological actions of the body induced by changing the inorganic salt constituents of the blood suggests that such inorganic salts may also be of importance in determining the rate and manner of growth. Loeb ('05) has shown that regenerative growth does not take place at all in the absence of the K ion. Beebe ('04) found that rapidly growing malignant tumors contained an excess of K while benign tumors showed excesses of Ca. An excess of Ca, however, is usually present in old or degenerating tissues and so may accumulate during inactivity instead of being the cause which produces the inactive state. Loeb ('04) also tried the influence of dilute sea water on regenerating hydroids and found the hypotonic solution to cause a more rapid rate of regeneration than occured in normal sea water. He further tried some of the inorganic salts with indefinite results.

In the first of this series of studies I ('08) recorded the action of the four important metalic ions of sea water, Na, K, Ca, and Mg, on regeneration in the medusa. The experiments were not extensive enough to draw final conclusions yet they accorded with what might have been expected from the related work of Loeb and Beebe mentioned above. The Na ion retarded regeneration, and in some solutions of $CaCl_2$ regeneration did not begin for several days and always proceeded slowly, while in the weaker KCl solutions regeneration occured at a more rapid rate than in the control specimens. The Mg solutions gave indefinite results.

It must be remembered that in such experiments the entire animal is kept immersed in the solution. The effects of the salt may, therefore, be systemic exhilarating or depressing the entire animal body and through such conditions secondarily affecting the regeneration rate. Should salts be injected into the circulation of higher animals the same complexity presents itself, yet this fact in no way lessens the importance of such experiments since all chemical actions and processes in the body secondarily effect other parts than those in which they occur. The experimenter, however, must carefully guard against using a dose of the salt which would be sufficient to perminantly weaken or injure the body since this would necessarily lessen the rate of regeneration as well as other normal processes.

The present series of experiments were arranged in order to carry regenerating animals for long periods of time in strange salt solutions and so determine whether there was any definite effect on the regeneration rates, and if such an effect was sufficiently marked to be of advantage in experimentation. The spotted salamander, *Diemyctylus viridescens,* was used for the experiments since it readily regenerates new legs and tail after the old ones are amputated. This gives two somewhat different structures for observation the legs being complex while the tail is almost uniformly metermeric. The experiments were conducted for the Huntington Fund for Cancer Research in the Pathological Laboratories of Cornell Medical School.

Five groups each consisting of fifteen salamanders were selected and weighed. The groups A, B and D each weighed 39 grams and groups C and E weighed 38 grams thus the groups were practically of the same average size. The individual salamanders averaged about 2.5 grams and were closely alike in size and general body condition. The operations consisted of cutting the left front leg off at the elbow joint and the distal one-third of the tail (Fig. 1 *A*). The tails were measured from the posterior end of the cloacal apature to the tip, the usual length was from 45 to 48 mm. making the removed third measure from 15 to 18 mm.

The first few days following the operation the arm stump is generally held close to the body and not used in swimming or walking upon the bottom of the vessel. After this time the stump is used occasionally at first but later it is brought into continual use.

The vessels containing the salamanders were kept upon the same table and subjected to similar light and temperature conditions. The animals were fed on alternate days with finely chopped beef each individual receiving approximately the same amount.

Fig. 1.

A. B.

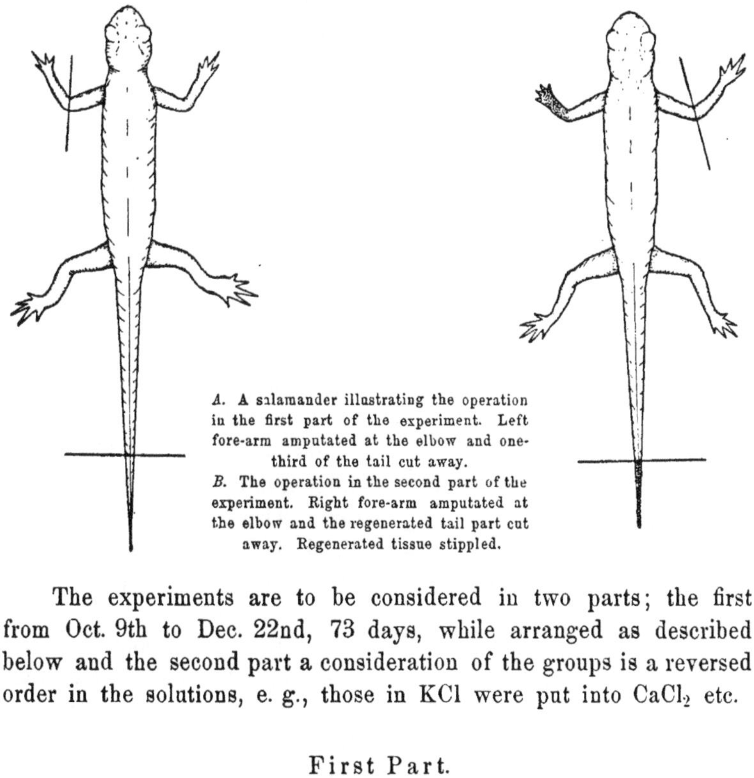

A. A salamander illustrating the operation in the first part of the experiment. Left fore-arm amputated at the elbow and one-third of the tail cut away.
B. The operation in the second part of the experiment. Right fore-arm amputated at the elbow and the regenerated tail part cut away. Regenerated tissue stippled.

The experiments are to be considered in two parts; the first from Oct. 9th to Dec. 22nd, 73 days, while arranged as described below and the second part a consideration of the groups is a reversed order in the solutions, e. g., those in KCl were put into $CaCl_2$ etc.

First Part.

Three days after the operation the animals were placed in the following solutions. Group A, the control, in ordinary tap water. Group B into $MgCl_2$ $^2/_{125}$ m, or 16 cc of a molecular solution of $MgCl_2$ to 984 cc of tap water. Group C in $CaCl_2$ $^2/_{125}$ m, group D in KCl $^2/_{125}$ m and group E into a solution of 8 cc molecular $MgCl_2$ + 8 cc molecular $CaCl_2$ added to 984 cc of tap water. The solutions were changed daily.

Table I.

Average records of regeneration from salamanders kept in salt solutions.

Time in days	Group A (control)			Group B ($MgCl_2$ $^2/_{125}$ m)			Group C ($CaCl_2$ $^2/_{125}$ m)			Group D (KCl $^2/_{125}$ m)			Group E ($MgCl_2$ $^1/_{125}$ m + $CaCl_2$ $^1/_{125}$ m)		
	Body length in mm	Leg-bud in mm	Tail-bud in mm	Body length in mm	Leg-bud in mm	Tail-bud in mm	Body length in mm	Leg-bud in mm	Tail-bud in mm	Body length in mm	Leg-bud in mm	Tail-bud in mm	Body length in mm	Leg-bud in mm	Tail-bud in mm
35	[illegible] 8	3 - 8 D	4.9	43.9	2.2 - 3 D	4.3	44.2	1.9 - 3 D	3.6	45	2.8 - 7 D	4.3	45.	2.5 - 8 D	4.6
52	45.4	42 [illegible]D	6.8	46.4	3.7 - 10 D	6.2	46.	3.4 - 6 D	5.	47	5.1 - 13 D	6.8	47.	4.7 - 14 D	6 9
73	[illegible] 4	5.3 - 13 D	9.2	[illegible]	4.9 - 10 D	8.2	47.4	4.6 - 6 D	6.7	48	5.7 - 13 D	8.3	47.4	5.7 - 14 D	8.8
97	[illegible]	6. - 13 D	10.6	[illegible]	5.3 - 10 D	9.5	47.4	5.1 - 9 D	8.4	47.9	6.7 - 14 D	10.	48.4	6.6 - 14 D	10.1
199	[illegible]	7.4 - 13 D		49.3	6.9 - 14 D		49.8	6.7 - 9 D		49.7	8.2 - 14 D		50.	7.8 - 14 D	

The effects of these salts were tested on other salamanders. It was found that a $^1/_{10}$ m solution of either KCL, $CaCl_2$, or $MgCl_2$ would kill the animals in less than two days. The strengths used were, therefore, about one sixth of the fatal dose but no appreciable injury to the general body health could be detected in such solutions.

Thirty-five days after the operation the newly regenerated tissue and body lengths were measured. The averages for each group are shown in table I. The average body lengths are almost equal, or little different in consideration of the entire length. The control is regenerating the leg and tail faster than any other group at this time. The figures 8 D, 3 D etc. in the leg-bud columns indicate the number of leg-buds which show differentiation or the formation of toes. Those groups in KCl and $MgCl_2$ + $CaCl_2$ are almost up with the control in the lengths of new regenerated buds while the group in $MgCl_2$ is slower and the specimens in $CaCl_2$ are slowest of all. It is recalled that the mixture of $MgCl_2$ and $CaCl_2$ consists of only one half as much of each salt as is used in the unmixed solutions and while both of these salts have retarded regeneration in the strong unmixed solutions the mixture has not lowered the rate.

The next line of the table shows that after fifty-two days the relative body lengths have remained about the same. The new leg-buds in KCl are now much longer than the control, about 25 %. Those in the $MgCl_2 + CaCl_2$ have also overtaken the control, while on the other hand, the specimens in $MgCl_2$ are behind the control and those in the $CaCl_2$ are still further behind. The tail-buds in KCl and $MgCl_2 + CaCl_2$ have also grown faster than the control since the thirty-fifth day. Those salamanders in $MgCl_2$ have kept their original slow rate while the tail-buds in $CaCl_2$ like the arm-buds in this solution are the shortest of all, the tail-buds particularly have fallen still further behind since the thirty-fifth day.

Seventy-three days after the operation the arm-buds are in much the same relative condition, while the tail buds of the control have grown more rapidly than those of the other groups. The $CaCl_2$ group being slowest.

Since the experiment started one individual in each group has been lost through accident, the groups now consist of fourteen individuals each. At this time, Dec. 22nd, the groups were again weighed. The lines below allow a ready comparison with their original weights.

Groups	A	B	C	D	E
	Control	$MgCl_2$	$CaCl_2$	KCl	$MgCl_2 + CaCl_2$
Oct. 9	39 gr.	39 gr.	38 gr.	39 gr.	38 gr.
Dec. 22	41.4	45.2	44.1	46.1	47.3

The original weights were for the fifteen individuals in each group while the lower line for the fourteen thus shows a steady gain in weight during the experiment due no doubt to the regular feeding. It will be noted by referring to table I that the increase in weight and body length of the groups does not correspond, e. g., group A has increased very much less in weight than group E while on the other hand group E has increased less in body length than group A. The increase in weight in these salamanders is due to an enlargement of the body organs rather than to a storing up of fat. The safer criterion in determining actual growth is the increase in body length rather than increase in weight since the later fluctuates so readily, and as Morgan ('06) states a normal rate of regenerative growth is kept up even though the body be in a thin emaciated condition.

The general result of the experiment thus far seems to show that the salamanders are capable of regeneration while living in the salt solutions. The specimens in the KCl regenerate at a rate equal to and often ahead of that shown by the control, while those in $CaCl_2$ and $MgCl_2$ are decidedly inhibited in their rates of regenerating both legs and tails. These facts accord with what might be expected from BEEBE's chemical analysis of tumors referred to above, but we must look further in the experiment for final results.

After the seventy-third day the animals were put into pure water so as to remove all excess of the several salts to which they had been subjected. After being in the water for twenty-four days the groups were again measured and the results are shown in the 97 day line of table I. The measurements stand in much the same relation to one another as that existing while they were in the solutions, this might have been expected since the time is scarcely sufficient for the effects to have been entirely overcome. After the salts had been thoroughly removed the animals were again operated upon and treated as recorded in the

Second Part.

The five groups, now consisting of fourteen salamanders each, were operated upon so as to remove the right front leg at the elbow joint and the entire new tail-bud (Fig. 1*B*). The arm operation gives a new cut for regeneration while all of the tail operations give rise to a second regeneration from the same level as that from which the previous new tail-buds arose.

Jan. 22nd, four days after the operation the animals were put into the salt solutions so that the groups which were in the retarding solutions ($MgCl_2$ and $CaCl_2$) during the first part of the experiment were now put into the solutions which had seemed to accelerate regeneration (KCl) and visa versa.

Group A (Control). now Control.
Group B ($MgCl_2$ $^2/_{125}$ *m*) now KCl $^2/_{125}$ *m*.
Group C ($CaCl_2$ $^2/_{125}$ *m*) now KCl $^2/_{125}$ *m*.
Group D (KCl $^2/_{125}$ *m*) now $CaCl_2$ $^2/_{125}$ *m*.
Group E ($MgCl_2$ $^1/_{125}$ m + $CaCl_2$ $^1/_{125}$ *m*) now $MgCl_2$ $^2/_{125}$ *m*.

A summary of the records of the animals in these solutions is shown in table II.

After thirty-nine days the arm-buds are all short and no differentiation has taken place. It will be noted that these buds are in most cases only about one half the length of the thirty-five day buds in table I. The tail-buds have regenerated at a rate nearer that

Table II.

Average records of regeneration from the same salamanders as in table I. Kept in salt solutions in reversed order.

Time in days	Group A (Control)			Group B (KCl $^{2}/_{125}$ m)			Group C (KCl $^{2}/_{125}$ m)			Group D ($CaCl_2$ $^{2}/_{125}$ m)			Group E ($MgCl_2$ $^{2}/_{125}$ m)		
	Body length in mm	Leg-bud in mm	Tail-bud in mm	Body length in mm	Leg-bud in mm	Tail-bud in mm	Body length in mm	Leg-bud in mm	Tail-bud in mm	Body length in mm	Leg-bud in mm	Tail-bud in mm	Body length in mm	Leg-bud in mm	Tail-bud in mm
39	47.1	1.6 - 0 D	4.	47.7	1.2 - 0 D	3.4	47.9	1.7 - 0 D	3.8	48.8	1.4 - 0 D	4.2	48.9	1.6 - 0 D	4.7
61	47.4	4. - 9 D	6.4	47.5	3.5 - 2 D	5.1	47.7	3.8 - 8 D	6.1	49.2	3.9 - 8 D	6.6	49.	4.4 - 10 D	7.1
78	48.3	5.5 - 10 D	7.2	48.5	4.2 - 2 D	6.1	48.7	4.5 - 11 D	7.5	49.7	4.5 - 9 D	8.	49.4	4.7 - 13 D	8.5
100	49.	6.5 - 10 D	8.7	49.3	4.8 - 4 D	7.3	49.8	5.6 - 13 D	9.3	49.7	6.2 - 9 D	10.	50.	6.3 - 14 D	10.2

shown in the first part of the experiment. The groups that were formally slow in $MgCl_2$ and $CaCl_2$ are still behind the control although now in KCl. After sixty-one days the right arm-buds are about equal to the length the left buds had attained at fifty-two days. In the first part of the experiment the D group in KCl was considerably ahead of all the others, it has now lost this advantage in the $CaCl_2$ solution. Groups B and C which were originally regenerating behind the control in $MgCl_2$ and $CaCl_2$ respectively have failed to improve significantly in the solutions of KCl. Group E which maintained a rate higher than the control when in a mixture of $MgCl_2 + CaCl_2$ is now slightly ahead of the control in $MgCl_2$ alone although after this time it falls behind the control, yet regenerates faster than any other group in the salt solutions.

When in these solutions for seventy-eight days the control is well ahead in length of the arm-buds. Group B in KCl after having formerly been in $MgCl_2$ shows the poorest record and has only two leg-buds that have differentiated toes, while the other four groups have well formed toes on almost all of the new legs. Group C which previously showed the weakest regeneration in $CaCl_2$ is now, while in the KCl solution, to be favorably compared with any other group in the salt solutions. Eleven of the fourteen leg-buds have well differentiated feet and toes.

These results would seem to indicate that the influence a salt exerts depends somewhat upon the former salts to which the animal has been subjected, at any rate, in this case the KCl acts more favorably when used after $CaCl_2$ than when following $MgCl_2$.

The tail-buds after seventy-eight days, like the legs, are slowest in group B but groups C, D and E are ahead of the control.

The last line of the table which shows the condition after one hundred days is not significant since growth in all the groups is very slow and almost complete at this time, the processes of differentiation being primary.

The animals were again weighed on Apr. 28, about seven months after the experiment was started. The weights at the periods indicated are given below.

Groups	A	B	C	D	E
	Control	$MgCl_2$ KCl	$CaCl_2$ KCl	KCl $CaCl_2$	$MgCl_2 + CaCl_2$ $MgCl_2$
Oct. 9	39 gr.	39 gr.	38 gr.	39 gr.	38 gr.
Dec. 22	41.4	45.2	44.1	46.1	47.3
Apr. 28	46.5	44.5	51.1	53.3	50.8

The increase in weight of the animals shows that they were in a healthy condition and not suffering from the treatment. From Dec. 22 to Apr. 28 group B lost slightly in weight and this is of interest when it is remembered how very poorly the individuals of this group regenerated in KCl after having formerly been treated with $MgCl_2$. The evidence indicates that treatment with $MgCl_2$ produces a condition in the animals which renders them unfavorable to treatment with KCl solutions.

A final point of interest to be noted in the reversed second part of the experiment is the condition of the new legs in group C. The left first leg in five of the fourteen specimens had not differentiated in the $CaCl_2$ solution to which they were subjected in the early part of the experiment and actually after seven months two have legs with conical terminations with one toe in one case, while three others have only slight indications of differentiation. The right first legs in these same specimens had been growing only during the second part of the experiment (100 days) yet with a single exception they were perfectly differentiated in the KCl solution. This fact shows in a decided way the different influences exerted by the $CaCl_2$ and KCl solutions on two legs of the same individuals. The bottom line of table I gives the measurements and differentiation of the left legs after 199 days when the experiment ended, these may be compared with the right legs in table II.

Summary.

Considering both parts of the experiment one may conclude that the effects of the salt solutions are not strongly pronounced, yet the following statements seem to be supported.

1) The processes of regenerative growth in *Diemyctylus* are favorably affected by weak doses of KCl while $CaCl_2$ inhibits the rate of growth and differentiation of the regenerating part.

2) Solutions of $MgCl_2$ inhibit both the rate of growth and differentiation of regenerating parts, yet not so strongly as $CaCl_2$. Mixtures of half doses of $CaCl_2$ and $MgCl_2$ do not influence either growth rate or differentiation.

3) If salamanders having had their regenerating powers retarded by $MgCl_2$ be then placed in KCl instead of this salt stimulating growth it further depresses the regeneration of both legs and tails and little if any differentiation takes place.

4) When specimens that had regenerated slowly in $CaCl_2$ were placed in solutions of KCl their rates of regeneration and powers of differentiation were improved. The effect exerted by a salt solution is, therefore, dependent to some extent upon the salts to which the animal has been previously subjected, even though some time may have elapsed since the former treatment was applied.

5) Salamanders that have regenerated at a fair rate in solutions of KCl are less depressed by treatment with $CaCl_2$ than others which have not been previously treated with KCl.

6) All actions of salt solutions on the body of a regenerating animal are probably very complex and the above statements are more to be taken as suggestions than final conclusions; they may serve best to indicate the need of extensive investigation along such a line which may possibly open a way to the control of growth processes.

Zusammenfassung zu Teil III.

Wenn man beide Seiten des Versuches in Betracht zieht, so könnte man zu dem Schlusse kommen, daß die Wirkungen der Salzlösungen nicht stark ausgeprägt sind; immerhin scheinen die im folgenden ausgesprochenen Sätze genügend gestützt zu werden:

1) Bei *Diemyctylus* werden die regenerativen Wachstumsprozesse durch schwache Dosen von KCl in günstigem Sinne beeinflußt, während $CaCl_2$ den Wachstums- und Differenzierungsbetrag in den regenerierenden Teilen vermindert.

2) Lösungen von $MgCl_2$ vermindern sowohl den Wachstums- wie den Differenzierungsbetrag in den regenerierenden Teilen, aber nicht so stark wie $CaCl_2$. Mischungen der halben Dosen von $CaCl_2$ und $MgCl_2$ beeinflussen weder den Wachstumsbetrag noch die Differenzierung.

3) Werden Salamander, deren Regenerationsfähigkeiten durch $MgCl_2$ gehemmt sind, nachträglich in KCl versetzt, so übt dieses Salz anstatt seiner sonstigen wachstumfördernden Wirkung eine noch weiter deprimierende auf die Regeneration sowohl der Beine als der Schwänze aus, und es findet wenig oder gar keine Differenzierung statt.

4) Wenn Exemplare, welche in $CaCl_2$ langsam regenerierten, in Lösungen von KCl gebracht wurden, so wurden ihre Regenerationsgeschwindigkeit und ihre Differenzierungsfähigkeit verbessert. Die von einer Salzlösung ausgeübte Wirkung ist demnach in gewisser Ausdehnung abhängig von der Art der Salze, denen die Tiere vorher unterworfen waren, selbst wenn einige Zeit seit der ersten Behandlung verstrichen sein sollte.

5) Salamander, die mit großer Geschwindigkeit in Lösungen von KCl regeneriert haben, erleiden durch Behandlung mit $CaCl_2$ eine geringere Depression als andre, welche vorher nicht mit KCl behandelt worden sind.

6) Alle Einwirkungen von Salzlösungen auf den Körper eines regenerierenden Tieres sind wahrscheinlich sehr kompliziert und die oben ausgesprochenen Sätze müssen mehr als Vermutungen wie als endgültige Schlüsse aufgefaßt werden; sie können aber gut dazu dienen, auf die Notwendigkeit ausgedehnter Forschungen in einer Richtung hinzuweisen, die möglicherweise einen Weg zur Beherrschung der Wachstumsvorgänge eröffnet.

Übersetzt **Gebhardt.**

IV. The Influence of Regenerating Tissue on the Animal Body.

When the adult animal body begins to regenerate new tissue in order to replace some lost part or when abnormal secondary growths arise the condition of growth equilibrium is disturbed and such a disturbance is followed by changes which effect the usual physiological condition of the body. The question arises whether the changes following or accompanying normal regenerative growth are in any way similar to those affects resulting from malignant or abnormal growths. If one believes that cancerous formations are growths induced by some derangement in the normal growth states and are not of infectious origin then normal secondary growths, in some stages at least, should effect the body in a manner somewhat similar to that resulting from an active tumor growth. The emaciated or cachectic condition of the body resulting from cancerous growths are not always attributable to toxins or products produced in the cancer and taken into the circulation, but at times seem due to the excessive approperation of nutriment by the rapidly growing tumor itself. The malignant tumor continues to grow and so finally kills the body, while on the other hand, the regenerating part although rapidly growing at first gradually decreases in growth rate and begins to differentiate and function thus diverting the energy previously used in the growth processes until finally growth ceases when the body has reëstablished its former condition.

I ('09) showed in the second of these studies that the medusa-disk of *Cassiopea xamachana* decreased rapidly in size while regenerating new oral-arms and that the rate of decrease was faster in those specimens regenerating the greater number of parts. In these experiments a sourse of error might have existed since those specimens with six or eight oral-arms removed have been deprived of more reserve food held in the mouth arms than had the individuals which lost fewer arms. I determined to control this possibility by operating on medusae so as to remove the same number of oral-arms from all and to increase the amount of new regenerating tissue in some individuals by removing also a part of the disk. The specimens were kept under identical conditions and were not fed during the time of the experiment. Thus any difference in their responses is due only to the additional amount of regeneration imposed upon those individuals with the cut disks.

Emmel ('06) has contributed an observation which is most interesting in connection with these experiments. He found that larval lobsters when regenerating new legs molted after longer intervals than normal individuals and increased in size at a rate sometimes 24% slower than the non-regenerating specimens. Most important was his observation that when the removal of legs was not followed by regeneration such specimens grew in size faster than the regenerating individuals and in most instances actually faster than the control. These observations clearly show that the process of regeneration itself and not the injury inflicted is responsible for the retardation of growth in the regenerating lobsters.

The experiments here recorded were performed upon the scyphomedusa, *Cassiopea xamachana*, which is so abundant at the Tortugas Islands, Florida. Healthy individuals of medium size were selected and operated upon as described below.

In the first experiment two groups of twenty individuals of the same average size were used. Group A had five of the eight oral-arms cut from each specimen (Fig. 1). Group B also had five oral-arms cut in a similar manner from each of the twenty individuals and in addition each medusa had a peripheral strip cut from its body-disk which included one third of the circumference and extended in radially beyond the oral zig-zag muscular layer shown in Fig. 2. The specimens were then allowed to regenerate for thirty-four days their disk diameters being carefully measured at intervals so as to determine the difference in body decrease of the two groups.

Fig. 1.

Fig. 2.

Fig. 3.

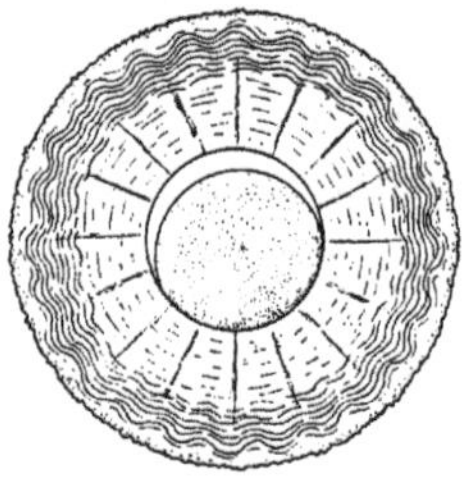

Fig. 1. *Cassiopea xamachana* with five of its oral-arms amputated at their bases.
Fig. 2. Five oral-arms and one-third of the disk periphery removed.
Fig. 3. A medusa with all of its oral-arms and the central stomach mass cut away.

Table I shows the records for group A, the first column giving the original disk diameters, the second column the diameters after twelve days, the third column the lengths of the individual new arm-buds regenerated during the twelve days. Column four gives the diameters after twenty days and column five the lengths of the new arm-buds at this time. Columns six and seven show the same after twenty-eight days and columns eight and nine after thirty-four days when the experiment stopped. A line of averages at the foot of the table shows the general result.

Table II gives the same data for group B and a ready comparison of the tables is facilitated by table III of averages.

The individuals of both groups averaged 81.5 mm in diameter at the beginning of the experiment and after twelve days the specimens of group A were 67.5 mm in diameter while those in group B which were regenerating the disk tissue in addition to the five oral-arms were only 64.3 mm in average diameter. In other words they averaged 3.2 mm smaller than the ones growing only the five arms.

Table I.

Record of regeneration and decrease in body size in *Cassiopea xamachana* when regenerating five oral-arms.

Disk Diam. in mm May 15	Disk Dia. in mm May 27	Length of arm-buds in mm May 27	Disk Dia. in mm June 4	Length of arm-buds in mm June 4	Disk Dia. in mm June 12	Length of arm-buds in mm June 12	Disk Dia. in mm June 18	Length of arm-buds in mm June 18
95	80	3—3—4—4—4	73	4—4—4—5—5	65	5—5—6—6—6	60	6—6—6—6—7
85	75	3—3—4—4—5	67	5—5—5—6—6	61	5—6—6—6—7	57	6—7—7—8—9
82	68	4—4—4.5-5—5	60	4—5—6—6—7	55	5—6—6—6—7	51	6—7—7—7—8
90	75	4—5—5—5—5	64	5—5—5—5—6	58	6—7—7—7—8	54	6—7—7—8—9
90	73	4—4—4—5—5	64	5—5—6—6—6	58	5—6—6—6—7	55	6—7—7—7—8
75	63	4—4—4—4—4	53	4—5—5—5—5	47	5—5—5—6—6	44	5—5—6—6—6
80	68	3—4—4—5—5	60	5—5—5—5—6	53	5—5—5—6—6	48	5—5—5—6—6
75	63	4—4—5—5—5	55	5—5—5—5—6	49	5—5—5—5—6	45	5—5—6—6—6
75	59	4—5—5—5—5	52	5—5—6—6—6	47	4—5—5—6—7	44	5—6—6—6—7
80	72	4—4—4—4—4	64	5—5—6—6—6	57	6—6—6—7—7	54	6—6—7—7—7
75	61	4—5—5—6—6	54	5—6—6—6—6	49	5—5—6—6—6	44	4—5—5—6—6
72	60	5—5—5—5—6	53	4—5—5—5—5	48	5—6—6—6—6	44	6—6—6—7—7
80	64	5—5—5—6—6	56	6—6—7—7—7	51	8—8—8—8—9	47	9—9—9—9-10
80	63	5—5—6—6—6	57	7—7—7—7—8	53	8—8—8—8—9	49	9—9—9—9-10
75	60	5—5—5—6—6	50	5—5—5—5—5	44	7—7—7—7—8	41	7—8—8—8—8
85	71	4—4—5—5—5	64	5—5—5—6—6	58	5—6—6—7—7	54	6—7—7—8—8
75	61	3—5—5—5—6	54	5—5—5—5—6	47	5—6—6—6—6	44	5—6—6—6—6
90	76	5—5—5—5.5-5	66	5—6—6—6—7	61	6—6—7—8—8	57	7—8—8—9—9
70	58	4—5—5—5—6	49	5—5—6—6—7	44	6—6—6—7—7	40	6—6—6—7—7
100	80	4—5—5—5—5	70	5—6—6—6—6	66	6—7—7—8—9	63	8—9—9-10-10
Av. 81.5	67.5	4.7	59.3	5.5	53.5	6.3	49.7	6.9

Table II.

Record of regeneration and decrease in body size in *Cassiopea xamachana* when regenerating five oral-arms and a piece of the body-disk.

Disk Dia. in mm May 15	Disk Dia. in mm May 27	Length of arm-buds in mm May 27	Disk Dia. in mm June 4	Length of arm-buds in mm June 4	Disk Dia. in mm June 12	Length of arm-buds in mm June 12	Disk Dia. in mm June 18	Length of arm-buds in mm June 18[1]
95	78	3 — 3 — 4 — 4 — 4	70	4 — 5 — 5 — 6 — 6	63	7 — 8 — 8 — 8 — 8	59	9 — 9 — 9 — 9 — 9
85	70	3 — 3 — 3 — 3 — 4	62	5 — 5 — 6 — 6 — 6	56	6 — 6 — 6 — 7 — 9	51	8 — 8 — 8 — 9 — 9
82	65	4 — 4 — 4 — 5 — 5	58	5 — 5 — 6 — 6 — 6	52	5 — 6 — 7 — 7 — 8	47	7 — 8 — 8 — 8 — 8
90	69	4 — 4 — 3 — 3 — 3	59	4 — 5 — 5 — 5 — 5	53	4 — 5 — 5 — 5 — 5	48	4 — 4 — 4 — 5 — 5
90	73	3 — 3 — 3 — 4 — 5	65	3 — 3 — 3 — 4 — 4	59	4 — 4 — 4 — 5 — 6	57	4 — 4 — 4 — 5 — 5
75	59	3 — 3 — 4 — 4 — 5	54	6 — 6 — 6 — 6 — 6	50	5 — 5 — 5 — 6 — 6	46	6 — 6 — 6 — 7 — 7
80	60	2 — 4 — 4 — 4 — 5	55	4 — 5 — 6 — 6 — 6	49	5 — 5 — 6 — 7 — 7	45	6 — 6 — 7 — 7 — 7
75	60	4 — 4 — 5 — 5 — 6	54	5 — 6 — 6 — 6 — 6	48	5 — 5 — 6 — 6 — 6	44	6 — 7 — 7 — 7 — 8
75	55	3 — 3 — 4 — 4 — 4	50	5 — 5 — 6 — 6 — 6	45	5 — 6 — 6 — 6 — 7	41	5 — 5 — 6 — 6 — 6
80	63	4 — 4 — 4 — 5 — 6	57	5 — 5 — 5 — 6 — 6	54	5 — 6 — 7 — 7 — 7	49	7 — 7 — 7 — 7 — 7
75	60	3 — 3 — 4 — 4 — 5	56	6 — 6 — 6 — 6 — 6	51	6 — 7 — 7 — 7 — 8	47	7 — 7 — 7 — 7 — 8
72	55	5 — 5 — 5 — 5 — 5	50	6 — 6 — 6 — 6 — 7	46	6 — 7 — 7 — 7 — 7	42	6 — 8 — 8 — 8 — 8
80	66	4 — 4 — 5 — 5 — 5	60	6 — 6 — 7 — 8 — 8	54	6 — 6 — 7 — 7 — 8	49	6 — 7 — 7 — 8 — 8
80	57	5 — 5 — 5 — 5 — 5	50	6 — 6 — 6 — 6 — 7	45	6 — 6 — 6 — 7 — 7	41	5 — 5 — 6 — 6 — 7
78	57	4 — 5 — 5 — 5 — 6	51	5 — 6 — 6 — 6 — 6	45	7 — 7 — 7 — 7 — 7	42	7 — 7 — 7 — 7 — 7
87	70	2 — 3 — 4 — 4 — 5	62	3 — 4 — 5 — 5 — 6	56	3 — 5 — 6 — 6 — 7	52	5 — 7 — 7 — 8 — 8
72	59	3 — 5 — 5 — 5 — 6	52	5 — 6 — 6 — 7 — 7	48	7 — 8 — 8 — 9 — 9	45	6 — 9 — 9 — 9 — 9
90	70	3 — 3 — 4 — 4 — 5	62	3 — 3 — 3 — 4 — 4	58	2 — 3 — 4 — 5 — 5	53	2 — 4 — 5 — 6 — 7
70	57	4 — 5 — 5 — 5 — 5	6	5 — 5 — 6 — 6 — 6	44	7 — 7 — 8 — 8 — 8	42	7 — 7 — 8 — 8 — 8
00	84	[2] 0 - 3 — 3 — 4 — 4	76	[2] 0 - 4 — 4 — 5 — 6	71	[2] 0 - 5 — 6 — 6 — 7	67	[2] 0 - 8 — 8 — 9 — 9
Av. 81.5	64.3	4.1	57.6	5.4	52.3	6.25	48.3	6.9

[1] Linear measurements do not indicate entire growth at this time, since the arms are branching.
[2] Not included in the average.

After this time, however, group B did not decrease as rapidly since the disk injury had almost completely regenerated; thus after twenty days the A group was only 1.7 mm larger than B, after twenty-eight days only 1.2 mm larger and after thirty-four days there was still only 1.4 mm difference in average size.

Table III.

Comparison of averages when medusae are injured to different extents
A summary of tables I and II.

Group	Disk diameters in mm at stated intervals					Length of arm-buds in mm			
	Original Dia.	After 12 days	After 20 days	After 28 days	After 34 days	After 12 days	After 20 days	After 28 days	After 34 days
A	81.5	67.5	59.3	53.5	49.7	4.7	5.5	6.3	6.9
B	81.5	64.3	57.6	52.3	48.3	4.1	5.4	6.25	6.9

The experiment clearly shows that while the B group was regenerating the cut disk part in addition to the five oral-arms the individuals of B were decreasing in body size as a result of this additional regeneration more rapidly than the specimens in A which were regenerating only the five oral-arms. The regenerating tissue through an excessive capacity for the absorption of nutriment draws upon the old body tissues and causes them to. decrease in size in a manner very much as it may be supposed the rapidly growing tumor imposes upon the substances of the surrounding body. It is certainly clear that in both cases the growing tissue causes the old body to become weak and emaciated while the growth itself continues in a vigorous manner.

The above experiment is of further value in regard to the influence of the degree of injury on the rate of regenerative growth. I have previously shown ('09) that the rate of arm regeneration in this medusa is independant of the degree of injury as is also the case in the brittle-star, *Ophiocoma riisei*; while *Ophiocoma echinata* regenerates each arm the slower the greater the number of removed arms. These results are contrary to the idea advanced by Zeleny ('05) that the greater amount of injury would be followed by a faster regeneration rate.

The two groups of individuals A and B are each regenerating five oral-arms but the group B is the more extensively injured since a portion of the disk was also cut away. If the additional injury or regeneration imposed upon the B group exercises any influence over the rate of regeneration it should be seen by comparing the rates of growth of the arm-buds in B with those of group A. The right side of table III gives a ready comparison of the two groups. Those specimens with the disk uncut or the less injured ones regenerated slightly more rapidly during the first twelve days but after this time the rates were practically equal. These facts show in a most convincing manner that more extensive injury to the medusa fails to give an increase in the subsequent regeneration rates.

A second experiment differed somewhat in manner of operation from the above yet the results are in perfect accord. Twenty-eight healthy medusae were arranged in two groups of fourteen individuals each and operated upon as follows. The specimens of group I had all of their oral-arms and the central stomach mass entirely removed, leaving only the medusa disk (Fig. 3). Such a preparation lives and pulsates in a normal manner and regenerates new tissue to cover over the central stomach space, then begins to bud new oral-arms from this tissue until finally the medusa regains its normal organs and parts. The central space is first covered by a thin veil of tissue which tears repeatedly and reforms until it begins to thicken and then the new arm-buds first appear. The regenerative growth is, therefore, very vigorous from such specimens during the early part of the experiment and later becomes much less. Group II was operated upon in the same manner as the specimens of group B in the above experiment, five oral-arms and a part of the medusa disk were cut away, Fig. 2.

Table IV.

Decrease in size of medusae regenerating different amounts of tissue.

Group	Disk diameters in mm at stated intervals				Length of arm-buds in mm		
	Original Dia.	After 14 days	After 22 days	After 28 days	After 14 days	After 22 days	After 28 days
I	88.6	62	55.3	51			
II	88.	69	63.	59	4.1	5.2	6.

Table IV facilitates a ready comparison of the averages from these specimens. The original diameters of the groups averaged 88.6 mm and 88 mm, after fourteen days group I was only 62 mm in diameter while group II was 69 mm, or 7 mm larger. After twenty-two days they were 55.3 mm and 63 mm and after twenty-eight days

51 mm and 59 mm. It will be noted however that group I ceased to decrease rapidly after the first fourteen days when its rapid regeneration also ceased and from this time on it decreased almost as slowly as group II, since in the last six days of the experiment it lost only 4.3 mm while group II lost 4 mm.

The rate of growth for the arm-buds in group II is practically the same as from the specimens in the previous experiment.

Groups I and II again show that when the animal regenerates a certain amount of tissue in a given time such an individual suffers a loss in body size which is greater than the loss from other specimens regenerating a less amount of tissue. Of course the animals must be subjected to the same food conditions, in these experiments they were all unfed. Regenerating tissue, therefore, consumes the old body substance and has an effect that would finally so weaken the body as to cause death should the regeneration continue for a sufficient time. A method which could eliminate the factors that cause growth to cease when an organ attains a certain size would allow the organ to grow at the expense of the other body parts until death would follow in a manner closely similar to that by which a malignant tumor growth finally kills the body containing it. The absence of certain growth inhibiting substances in the body may be responsible for indefinite cancer growths, and experiments that in any way lead to a determination of the controling factors in normal primary or secondary growths are of importance in this regard.

Summary.

The medusa, *Cassiopea xamachana*, when unfed decreases in body size. This decrease is greater in regenerating individuals, and the larger the amount of tissue an individual is regenerating the more rapidly does it decrease in size. The new regenerating tissue grows at a vigorous rate on account of its excessive capacity for the approperation of nutriment from the old body tissues, and it is this fact which causes the body to decrease in size and become weak and emaciated. A close similarity to such an action is seen in the case of certain malignant growths.

Zusammenfassung zu Teil IV.

Die Qualle *Cassiopea xamachana* nimmt im Hungerzustand an Körpergröße ab. Diese Abnahme ist bedeutender bei regenerierenden Individuen, und mit der Größe des zu regenerierenden Gewebsbetrages wächst die Geschwindigkeit der Größenabnahme des Individuums. Das neue regenerierende Gewebe wächst mit großer Energie nach Maßgabe seiner außerordentlichen Fähigkeit zur Nahrungseinverleibung auf Kosten der alten Körpergewebe, und dieser Umstand ist es, welcher die Größenabnahme, das Schwächerwerden und Abmagern des Körpers veranlaßt. Eine weitgehende Ähnlichkeit mit einer derartigen Wirkung beobachtet man in den Fällen von gewissen malignen Tumoren.

Übersetzt **Gebhardt.**

Literature cited.

BEEBE, S. P., '04, The Chemistry of Malignant Growths. II. The Inorganic Constituents of Tumors. Am. Journ. Physiol. XII. pp. 167—172.

EMMEL, V., '06, The Relation of Regeneration to the Molting Process in the Lobster. 36th Annual Report, Inland Fisheries, R. I.

LOEB, J., '04, Über den Einfluß der Hydroxyl- und Wasserstoffionen auf die Regeneration und das Wachstum der Turbellarien. Arch. f. d. ges. Physiol. 101. H. 7/8. S. 340—348.

—— '05, Studies in General Physiology. Univ. Chicago Press.

MORGAN, T. H., '06, The Physiology of Regeneration. Journ. Exp. Zool. III. pp. 459—500.

STOCKARD, C. R., '08, Studies of Tissue Growth. I. An Experimental Study of the Rate of Regeneration in Cassiopea xamachana. Carnegie Institution. Pub. 103, and Science. N. S. XXVII. p. 448.

—— '09, Studies of Tissue Growth. II. Functional Activity, Form Regulation, Level of the Cut, and Degree of Injury as Factors in Determining the Rate of Regeneration. The Reaction of Regenerating Tissue in the Old Body. Journ. Exp. Zool. VI. pp. 433—469.

ZELENY, C., '05, The Relation of the Degree of Injury to the Rate of Regeneration. Journ. Exp. Zool. II. pp. 347—369.

Sonderdruck
aus Archiv für vergleichende Ophthalmologie Bd. I, H. 4, Seite 473—480.

The experimental Production of various Eye Abnormalities and an Analysis of the Development of the primary Parts of the Eye.

By

Charles R. Stockard,
Cornell University Medical School, New York City, U. S. A.

With two figures in the text.

While studying the influence of various substances on development the writer found that it was possible to produce at will a number of ophthalmic defects by the use of Mg, alcohol, cholreton, ether and other anæsthetics. The action of these substances seems to weaken or distroy the dynamic processes necessary for the optic vesicles to push out from

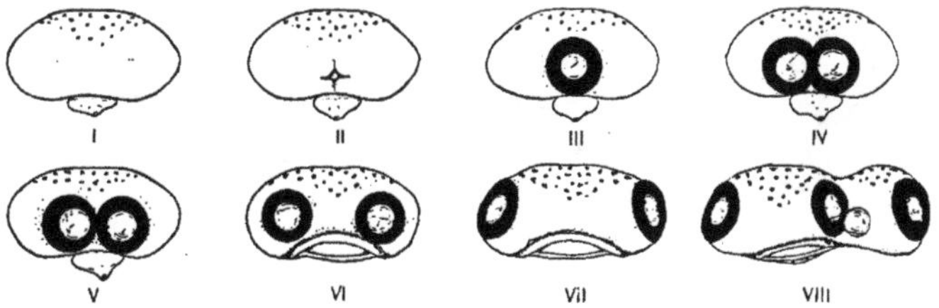

Fig. 1. Anterior views of fishes heads showing different eye conditions produced by treatment with Mg solutions.

the brain or to subsequently grow out to their lateral positions at the sides of the head. In consequence of this, various degrees of the cyclopean condition often occur among the fish embryos with which I have experimented.

The cyclopean fish embryos are in all respects exactly comparable to the human cyclops. One eye exists in the middle of the face an the nasal pits are often represented by a single or double pit in front of the eye. The eye conditions, as illustrated in the diagram Fig. 1. shwo all steps in a series beginning with two eyes unusually close together.

VI, two approximated eyes, V, a double eye with two lenses, two pupils, etc., IV, a laterally broad eye with a double retinal arrangement, and a single lens and pupil and a typically single eye showing no indications

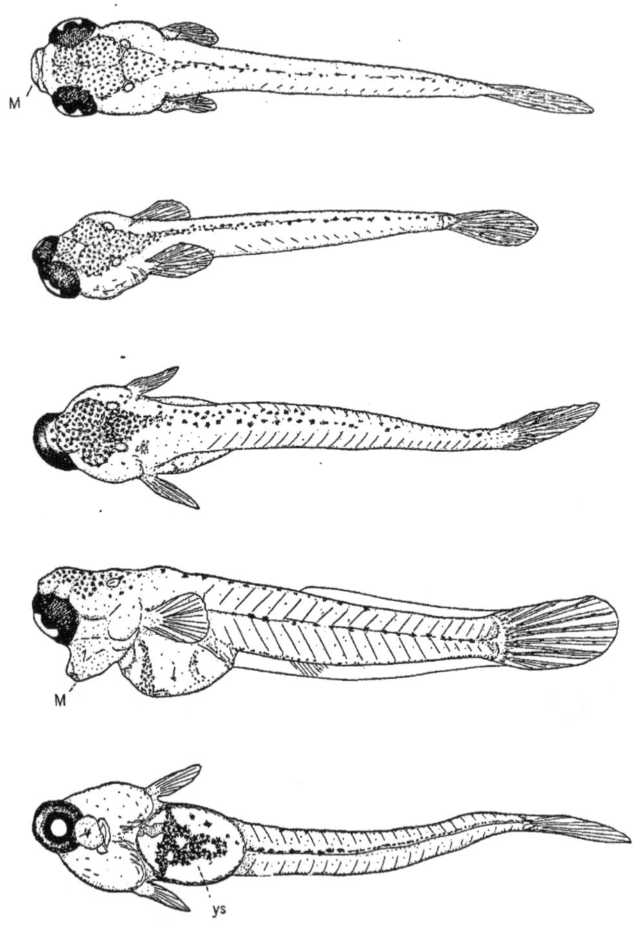

Fig. 2 Young fish, a normal individual above and cyclopean monsters below.

of its double nature, III. The later condition may be termed typical or perfect cyclopia, from this we pass to extreme cyclopean eyes which are unusually small, II, sometimes deeply buried in the head, others with small optic cups and illfitting lenses which protrude beyond the eye, and, finally, all retinal or optic cup portions of the eye may be absent, with independent lenses present, or both optic cup and lens may fail to form and eyeless creatures result, I. Many illustrations of all these stages have been found and studies among the hundreds of cyclopean fish which have been produced by these methods.

Some of the embryos present perfectly normal bilateral brains and show no abnormality other than the cyclopean eye and characteristic probocis-like mouth. The cyclopean eye occupies an antero-ventral position, Fig. 2, and many fish with such an eye hatch from the egg and swim about for a month or more in a perfectly normal fashion, the cyclopean eye functioning as an effecient organ of sight.

The development of the cyclopean eye in human monsters has been difficult to interpret on account of the scarcity of material and want of early stages of the defect. Such abnormalities are not readily explained from later stages. In the summer of 1906, when these monstrous fish were first produced, I secured only later stages and on finding all degrees of union between the eyes concluded that the cyclopean condition resulted from a more or less intimate fusion of the two eye components after they had arisen from the brain. This position has been held by other workers both before and since my study. A more careful investigation, however, of the earliest stages of cyclopea in the living eggs and in sections shows that the final condition of the eye is foreshadowed in the first appearance of the optic anlage from the brain. The early eye is either perfectly single or duble from the start, and the union of the two components does not become more intimate during development, even though the eye may develop partially within the brain itself.

In addition to the above series showing the various degrees of cyclopia, another series of ophthalmic defects were induced by the same chemical substances. These individuals have one perfectly normal eye in its usual position on either the right or left side of the head while the eye of the other side may be either smaller than usual, or very small and defective in structure, or deeply buried in the head tissues as a choroid vesicle entirely failing to form a typical eye, or finally, the optic vesicle of one side may not arise from the brain and thus the individual has a perfect eye on one side and no indication of an eye on the other. I have called such monsters as these „Monstrum Monophthalmicum Asymmetricum" as distinguished from the cylopean monster with its one symmetrically placed eye.

This type of monster seems rarer in nature than the well known cyclops, although they do occur. I have known a child with one perfect eye and the other small and defective, and Professor T. H. Morgan has shown me a pigeon that presented a similar condition.

The asymmetrical monophthalmic monsters are possibly due to the differing degrees of resistance to the anæsthetics possessed by the anlagen of the two optic vesicles. The doses used are of course very delicately adjusted. One eye begins to push out from the brain slightly before the other (in studying normal embryos it is often found that one organ of a bilateral pair forms and develops ahead of the other for a time) the weaker or slower eye is affected by the solutions while the stronger is not. The different degrees of abnormality shown by one eye may be an index to the more or less wide difference in developmental energy possessed by the two eyes of the same individual.

It is of interest to record here that in some cases the development of the auditory vesicle or ear is affected by the alcoholic solutions. At times one ear is injured and the other not, in such individuals when the eyes also exhibit an asymmetrical condition it invariably happens that the defective eye is on the side with the defective ear. The rates or strengths of development of the sides of the embryo are different, in other words development is not perfectly bilateral, one side may go ahead of the other for a time then the other catches up and may actually get ahead and so on.

Still other cases occurred in which both eyes were small and defective although they pushed out from the brain and reached the sides of the head. The two eyes of these individuals show the various degrees of imperfection found in the small eye of the asymmetrical monophthalmic monsters.

Alcohol was most effective of all the anæsthetics employed in producing these conditions. Proper strength solutions of alcohol may cause as many as 98 per cent of the eggs to develop into young fish exhibiting the various ophthalmic defects mentioned above. Magnesium salts in solution were next in efficiency, in one experiment 66 per cent of the embryos were cyclopean or asymmetrically monophthalmic. The fish were in better general condition in the Mg solutions than in other substances. Mg effects the development of the eyes without causing, in many cases, defects or weaknesses in other parts of the central nervous system. Alcohol on the other hand disturbs the development of the central nervous system as a whole so that embryos are rarely strong after treatment with it.

These experiments prove that many of the eye malformations met with in nature are probably due to some abnormal condition in the developmental environment having acted upon the early embryo. Anæsthesia tends to weaken or lower the dynamic processes of development and it is probable that other causes which would interfere with normal nutrition might cause similar effects. This would apply especially to the mammalian egg where the yolk has been lost and the embryo depends upon a perfect placentation for its proper nourishment from the mother. I suggest, therefore, that cyclopea and other ophthalmic defects in mammals are due to poor placentation or a diseased condition of the mother which subjects the embryo to an abnormal environment during development. In man such defects are probably often due to an alcoholic mother. There is no evidence to indicate that these defects are the result of a peculiar or abnormal germ cell, and against such a view the experimenter has the power to cause at will perfectly normal eggs to develop into cyclopean monsters by the use of alcohol and other anæsthetic agents.

In a recent paper in the American Jour. of Anatomy, X, p. 369, I have discussed the anatomical conditions of cyclopia and shall not mention them here.

Development of the Primary Parts of the Eye.

The embryos discussed above furnish excellent material for a study of the relationship between the development of the optic vesicle and the crystalline lens; the two primary parts of the eye which arise in the embryo from different sources, the vesicle from the brain wall and the lens from the head ectoderm. It has been claimed by several experimenters that the optic-vesicle was entirely independent of the lens in its development, while on the other hand, the lens was entirely dependent upon the optic vesicle for its origin from the ectoderm as well as for its later differentiation into the clear refractive lens of the eye.

A study of embryos having no optic vesicles, others with a vesicle on only one side and finally those with a median cyclopean eye show the following facts regarding the relationship between the development of the optic vesicle and optic lens.

The crystalline lens may originate from ectoderm without any direct stimulus whatever from an optic vesicle or cup. These self-originating lenses arise from regions of ectoderm that are not in contact with either optic vesicle, the brain wall, or any nervous or sensory organ of the individual.

The lens-bud is capable of perfect self-differentiation. No contact at any time with an optic vosicle or cup is necessary. These lenses finally become typical transparent refractive bodies exactly similar in histological structure to a lens in the normal eye.

The size and shape of the lens is not entirely controlled by the associated optic cup. Lenses may be abnormally small for the size of the cup, or entirely too large, so that they protrude; or, finally, peculiarly shaped oval or centrally constricted lenses may occur in more or less ordinarily shaped optic cups. The lens is by no means always adjusted to the structure of the optic cup as has been claimed by some observers.

An optic vesicle or cup is invariably capable at some stage of its development of stimulating the formation of a lens from the ectoderm with which it comes in contact. It is remarkable how extremely small an amount of optic tissue is capable of stimulating lens formation from the ectoderm.

The optic vesicle may stimulate a lens to form from regions of the ectoderm other than that wich usually forms a lens. This is shown by the fact that a median cyclopean eye always stimulates a lens to form from the overlying ectoderm. It is scarcely possible that the lateral normal lens-forming ectoderm could follow the cyclopean optic cup to the many strange situations it finally reaches.

The ectoderm of the head region is more disposed to the formation of lenses than that of other parts of the body, since free lenses invariably occur in this region.

A deeply buried optic vesicle or cup may fail to come in contact with the ectoderm; in such cases it lacks a lens. The tissues of the embryonic cup in the fish are unable to form or regenerate a lens. This is not true for all embryos as has been shown for one species of frog.

The optic lens may be looked upon as a once independent organ (possibly sensory or perhaps an organ for focusing light on the brain wall, before the vertebrate eye had arisen) which has become so closely associated with the nervous elements of the eye that it has to some extent lost its tendancy to arise independently, although still capable of doing so under certain conditions. The lens now arises much more readily in response to a stimulus from the optic vesicle, a correlated adjustment which insures the almost perfect normal accord between the optic cup and the lens.

Finally it may be stated in brief, that the optic vesicle or cup always has the power to stimulate a lens to arise from any ectoderm with which it may come in contact during certain stages of its development. Secondly, the ectoderm of the head region also has the power under proper conditions to form an independent lens which will differentiate perfectly without the stimulus or presence of an optic vesicle or cup.

Kurze deutsche Inhaltsangabe zu vorstehender Arbeit:

Die experimentelle Erzeugung verschiedenartiger Abnormitäten des Auges nebst einer Erörterung über die Entwicklung der Hauptteile des Auges.

Stockard studierte den Einfluß des Magnesiums, Alkohols, Chloräthyls, Äthers und anderer Anästhetika auf die Entwicklung des Fischauges. Er erhielt so u. a. verschiedene Grade cyklopischer Mißbildungen, die in allen Hinsichten der menschlichen Cyklopie vergleichbar waren. Auf der einen Seite der Reihe stehen die Fälle, wo zwei getrennte Augen, die nur näher als normal beieinander liegen, vorhanden sind, den Übergang bildet ein median gelegenes, äußerlich einfaches Auge mit doppelter Retina, der höchste Grad wird durch einen medianen hochgradigen Mikrophthalmus oder durch völligen Mangel der Augenanlage dargestellt. Die Wirkung der erwähnten Chemikalien ist also eine hemmende. Es zeigte sich übrigens bei frühen Stadien, daß die Cyklopie keinen Verschmelzungsprozeß vorher isolierter Anlagen darstellt, sondern daß solche Augen von vornherein einfach oder doppelt angelegt sind.

Eine andere seltenere Art von Mißbildungen ist durch Vorhandensein eines normal gelagerten, gut entwickelten und eines symmetrisch gelegenen unvollkommen entwickelten Auges charakterisiert (Monstrum monophthalmicum asymmetricum). Es ist anzunehmen, daß in diesen Fällen die eine Augenanlage weniger widerstandsfähig ist oder etwas später auswächst als die andere und daß die Giftwirkung gerade hinreicht, diese letztere zu schädigen.

Zuweilen trifft auch mit einseitiger Mißbildung des Auges eine ebensolche des Ohrapparates zusammen, dann finden sich beide Störungen auf der gleichen Körperseite, was obige Auffassung von der verschiedenen Qualität der Kopfhälften bestätigt.

Andere Fälle zeigen zwei richtig gelegene, in verschiedenem Grade mißgebildete Augen.

Starker Alkohol rief über 98 % Mißbildungen hervor: Magnesium in einem Falle 66 % bei besserem Allgemeinzustande des Tieres, während beim Alkohol die Schädigung des Zentralnervensystems größer ist, so daß die Tiere selten kräftig sind.

Diese Versuche zeigen, daß viele Augenmißbildungen von ungeeigneter Ernährung des Embryos herrühren können (z. B. Alkoholismus der Mutter), während kein Beweis für eine abnorme Beschaffenheit der Keimzelle vorhanden ist.

Entwicklung der Hauptteile des Auges.

Besonderes Interesse bietet das gegenseitige Verhalten von Augenbecher (bzw. -blase) und Linsenanlage.

Im Gegensatze zu bisherigen Annahmen zeigte sich, daß das Kopfektoderm ganz unabhängig von dem Vorhandensein einer Augenblase imstande ist, eine völlig entwickelte Linse zu liefern, auch bezüglich der Dimensionen besteht keine feste Abhängigkeit der Linse vom Augenbecher.

Jede Augenanlage hat die Fähigkeit, das Ektoderm zur Linsenbildung anzuregen, auch diejenigen Teile derselben, die gewöhnlich keine Linse produzieren. Wenn die Augenanlage tief im Kopfe liegt und das Ektoderm nicht erreicht, so fehlt die Linsenanlage.

Die Linse kann als ein Organ angesehen werden, das einmal selbständig war und diese Selbständigkeit nur ausnahmsweise einmal wiedererlangt, während es jetzt gewöhnlich in Abhängigkeit von den nervösen Teilen des Auges ist. *G. Freytag* (*München*).

Reprinted from THE AMERICAN JOURNAL OF ANATOMY, VOL. 11, NO. 2
January, 1911

THE ANATOMY OF THE THYROID GLAND OF ELASMOBRANCHS, WITH REMARKS UPON THE HYPOBRANCHIAL CIRCULATION IN THESE FISHES

JEREMIAH S. FERGUSON

Assistant Professor of Histology, Cornell University Medical College

TWENTY FIGURES

INTRODUCTION

Even a casual reference to the literature of the thyroid gland is sufficient to indicate that the organ has been more carefully studied in most all other classes of animals than in the Elasmobranchs.

The organ may be said to make its first appearance in the Ascidians, Amphioxus, and Cyclostomes as a depressed groove, trough, or series of recesses in the ventral floor of the pharynx, usually known as the endostyle, or the hypobranchial or hypopharyngeal groove, which, as first shown by W. Müller ('71) who studied Myxine glutinosa and Petromyzon, is to be considered the homologue of the median thyroid of the vertebrates. In the Cyclostomes the structure, relations and development of the primitive thyroid have been more recently studied by Guiard ('96), Cole ('05), Schaffer ('06), and Stockard ('06). The structure of the organ is very simple and only partially resembles the thyroid of higher vertebrates. Possibly the thick tenacious secretion formed by the endostyle, upon the presence of which the function of the organ very largely depends, may well be taken to bear a relation to the colloid material which is so characteristic of the mammalian gland. Inasmuch as the retention of an albuminous secretion within the glandular lumina of the animal body, a condition frequently observed by the pathologists and normally present in other glands as well as the thyroid, *e.g.*, mammary gland, kidney, hypophysis cerebri and parathyroid gland, leads to the accumulation of a colloid material bearing a more or less striking resemblance to the colloid material in the follicles of the thyroid gland, the deduction from the phylogenetic standpoint, that the retention within the follicles of the thyroid of a once free mucous secretion would account for the colloid character of the follicular content, would not seem inappropriate. The character of cells which pour forth the free secretion of the endostyle or hypobranchial thyroid of the Ascidians, Amphioxus and Cyclostomes is not so very different from the colloid secreting cells of the thyroid follicles of mammals.

In the Teleostei, Wagner ('53) has studied the form and location of the thyroid gland and directed attention to the similarity of its structure to that of mammals. Simon ('44) and Baber ('81) have given extended descriptions of the thyroid gland in several species of bony fishes; Maurer ('86) described the structure and studied fully the development of the thyroid gland of the carp and trout. In the more recent literature the structure of the thyroid in fishes seems not to have received the attention which it apparently deserves.

Extended descriptions of the thyroid gland of reptiles are found in the articles by Simon ('44) and Baber ('81). De Meuron ('86) studied the organ in Lacerta. Van Bemmelën ('87) described the gland in Hatteria and Lacerta as being transversely placed over the trachea near the heart, and as forming a small, thin unpaired organ. Guiard ('96) has also discussed the structure of the organ in reptiles.

In the Amphibia the work of Maurer ('88) and the excellent descriptions by Wiedersheim ('04) apparently leave little to be desired, though the organ has been much studied in this class of animals.

In Aves the thyroid gland has been extensively studied by Simon ('44), Peremischko ('67) who considered the histology as well as the gross anatomy of the organ, Baber ('81) and De Meuron ('86). The literature of the structure and development of the thyroid in the chick is extensive.

In most of the Mammalia the anatomy of the thyroid gland is well known and its literature has acquired voluminous proportions. Its review does not fall within the scope of the present paper.

REVIEW OF THE LITERATURE

A careful study of the available literature has revealed, with the exception of the work of Guiard, only casual references to the anatomy of the thyroid gland in the Elasmobranchs. Simon ('44) studied the Selachian (Squalus) and the skate (Raia). He describes the thyroid gland as "a single organ, situated in the median line, in connection with the anterior surface of the cartilages which bind together the branchial arches of opposite sides of the body," and he states that it may lie in contact with the "lingual bone," or may be more or less distant from the mouth, but that it is "always at the spot where the great trunk of the branchial aorta distributes its terminal branches. It lies at the angle of this bifurcation . . . ; it is covered by the sterno-hyoid or sterno-maxillary muscle, and also by the myo-hyoid and genio-hyoid, when these are present." His description I find to hold good for Raia, but it does not entirely correspond to the position of the thyroid gland of Squalus, Mustelus, or Carcharias. As to its vascularization, Simon states that the gland receives its blood supply by means of a recurrent branch given off by the first branchial vein, while yet within the gill, and that "it never receives the smallest share of supply from the branchial artery with which it is in contact." The last portion of this statement is precisely correct for all the species which I have examined, though apparently at variance with the observations of some authors: the first portion, as to the origin of the thyroid artery, would appear to be not very accurately expressed, for it never arises from the afferent branchial vessel which leaves the dorsal end of the branchial arch, but, on the contrary, arises from the ventral end of the nutrient or efferent loop. The relation of the thyroid gland to the bifurcation of the ventral aorta is so intimate as to readily suggest the error of other observers who have presumed that the organ received some blood directly from the first pair of branchial arteries. Moreover, the thyroid artery passing from its origin lies directly under, and in contact with the first pair of afferent branchial arteries so that until these latter vessels have been carefully dissected out of their sheath it is impossible to determine with certainty that they have no connection with the thyroid vessels. In the skate the gland lies directly upon the aortic bifurcation and the pulsating blood-vessels are readily seen through the transparent organ as soon as it is exposed.

Müller ('71) speaks of the thyroid gland of Raia clavata as a flattened brownish-red body, lying at the point of division of the branchial artery. It possesses a connective tissue capsule with trabecula which divide the organ into a small number of lobes, within which the connective tissue penetrates between the lobules. The follicles possess a thin membrana propria and a cylindrical epithelium; they contain a homogeneous, gelatinous yellowish mass. The epithelium possesses a shiny cuticular border and appears to send processes into the lumen. The description given by Müller holds good for Raia, the genus which he studied, but it does not correspond to the condition of the thyroid gland of Mustelus, Squalus or Carcharias, the difference being chiefly due to the fact that in the Batoidei the connective tissue forming the thyroid trabecula and interfollicular septa is apparently much more abundant than in the Selachii.

Balfour ('78) discusses very briefly the early development of the thyroid gland of Elasmobranchs prior to the appearance of a lumen within its follicles. He does not consider the anatomy of the organ in the adult.

Baber ('81) says that "in the skate the gland is single (with the exception of a few detached vesicles) and forms a yellow, flattened, lobulated body, occupying the median line at the bifurcation of the branchial artery. Anteriorly it sometimes presents a narrow process of gland-tissue running forward, and behind it is limited by the bifurcation of the branchial artery." The contents of its vesicles consisted of coarsely granular masses or globules of various sizes which "correspond with the 'colloid substance' of authors." He surmises the non-existence of lymphatic vessels and says that in both the skate and the conger-eel "an extensive system of vessels lined with epithelium becomes injected by the method of puncture"; he considers that these are blood-vessels. The narrow process of glandular tissue which Baber says is occasionally present in Raia is more frequently seen in the Selachii; it is constantly present in all the examples of Carcharias which I have dissected. It extends forward until it comes into contact with the anterior margin of the basi-hyal cartilage (lingual bone) which presents a depression, frequently amounting to a complete foramen, for the reception of the anterior extremity of the glandular process with the connective tissue by which it is heavily invested. This process is obviously analogous to the pyramidal lobe of the mammalian thyroid, and as it extends all the way to the pharyngeal submucosa in many instances it may well be considered as indicative of a phylogenetic connection of the gland with the cavity of the pharynx, a condition which is also indicated by the ontogeny of the organ in all the orders, and which appears to be permanent in the Ascidians, Amphioxus, and the Cyclostomes.

Baber's suspicion of the non-existence of lymphatics within the thyroid gland appeared to the writer to be such a remarkable observation and so out of harmony with the known anatomy and physiology of the organ in the higher animal orders as to require further study. This study was pursued by means of a considerable series of careful dissections with many injection experiments and did not appear to confirm Baber's opinion. Baber's observation that the blood-vessels in these animals could be injected by the method of puncture is quite accurate, but it does not by any means disprove the existence of lymphatics. His results were apparently dependent upon the fact that the smallest veins and capillaries are of very considerable caliber and readily admit of injection, while the lymphatics are very minute and are entered only with difficulty and not frequently when the needle is thrust into the substance of the gland.

Balfour ('81) referring to the development of the organ in Scyllium and Torpedo says that at first it is solid and attached to the esophagus. "Eventually its connection with the throat becomes lost, and the lobules develop a lumen."

Dohrn ('84) in his plate XI, fig. 5, indicates by outline the thyroid gland of Ammocetes, but does not illustrate or describe the thyroid gland of the Selachii, though in his text he includes an extended description of the thymus of the latter animals. His outline of the thyroid of Ammocetes does not conform to that the the gland in the Selachii.

De Meuron ('86) says that in Scyllium the thyroid is elongated, in Galeus and Acanthias, much flattened, in Raia pyramidal or rounded. It lies just above the terminal bifurcation of the branchial artery. In Myelobates it lies behind the os hyoideus, beneath the sterno-mandibularis muscle, and is triangular in shape, short, flattened, transversely elongated, and has a length of 2 cm. In Acanthias the thyroid gland presents an irregular contour, certain groups of follicles being even completely detached, and placed around the principal group. The observations of De Meuron would appear to be accurate as far as they go but are possibly founded upon the examination of too few individuals. Thus in Squalus acanthias I found the thyroid frequently broken as described by De Meuron for Acanthias but other individuals presented a thyroid which was perfect, not the least broken up or irregular in contour, and in the closely related Mustelus canis irregularity of contour is certainly the exception, not the rule. In Scyllium he says the thyroid is elongated and I find that superficial examination of the related species Carcharias, would indicate a similar condition, but if the semi-opaque white mass of connective tissue, in which the thyroid gland of Carcharias is heavily clothed and so closely invested that it seems to form paart of the gland, be dissected out and held, stretched in its normal form, between the bright sun and the eye of the observer there is readily seen within the reddish-white connective tissue mass the outline of the yellowish-orange thyroid gland, which instead of having the elongated form of the outward mass is flattened, transversely elongated, and of the same peculiar triangular or shield-like shape which is characteristic of the organ in the dogfish and closely simulated by that of Raia. It seems possible that the elongated gland observed by De Meuron in Scyllium might be susceptible to a similar analysis.

Guiard ('96) studied six species of the Selachii and five of the Batiodei. In Scyllium he found the thyroid gland of pyriform shape, the anterior extremity being prôlonged forward as far as the anterior margin of the lingual cartilage ("copule"), where it passes between the two lateral halves of the coraco-hyoid muscle. This description, as given by Guiard, corresponds with the position and form of the gland which I find in Carcharias and which, as regards the anterior prolongation, appears to be analogous to the pyramidal process of mammals. But Guiard's fig. 1, in the absence of specific contradiction in his text, might be taken to indicate that the thyroid gland had been found beneath the coraco-hyoid muscle; this is not the case in any of the species which I have examined and I presume it is not the case in Scyllium catulus, from which species his figure was drawn. In each species I have found the gland lying, without exception on the ventral surface of the coraco-hyoideus, between it and the coraco-mandibularis, except that at the anterior portion the gland lies between the coraco-hyoid muscles of the two sides, the divergence of the two muscles exposing the ventral surface of the cartilage at this point. In the Batoidei the coraco-hyoidei are so widely separated that the whole thyroid gland may come to lie directly upon the basi-hyal cartilage, the aortic bifurcation and the coraco-branchialis muscles, which are successively exposed from before backwards by the separation of the coraco-hyoids, but in this case the fascia which covers the ventral surface of the coraco-hyoids dips beneath the dorsal surface of the thyroid gland.

In Acanthias vulgaris and Mustelus lœvis Guiard as did De Meuron, notes the tendency of the thyroid to present detached vesicles, its contour being very irregular. In Galeus canis the thyroid gland lies rather farther forward and is partially covered by a fold of the buccal mucosa. In Carcharias glaucus the coraco-mandibularis muscle is relatively very broad and completely hides the thyroid gland; the gland is described as reniform and voluminous. With the exceptions recorded the position of the thyroid in the various species of Selachii corresponds fairly well with that described for Scyllium.

Of the Batoidei, in Raia alba the coraco-hyoidei are small and widely separated, and between these muscles the first pair of branchial arteries emerge. The thyroid gland is described as lying between the bifurcation of the aorta and the hyoid arch; it is a very large globular organ and its deeper surface is slightly prolonged as far as the arterial bifurcation. In Raia oxyrhynchus the thyroid gland in transversely elongated. In Raia pastinaca the coraco-hyoidei approach one another and the gland is longitudinally elongated; in this particular it corresponds to the Selachian type. In this last species it is a large pyriform organ with its broad end in relation with the hyoid cartilage, and its point extending nearly to the bifurcation of the branchial artery; at its point the gland presents a prolongation "which descends between the branchial sacs to a depth of about 0.5 cm." I desire to call attention to the fact that in Carcharias a posterior prolongation apparently also exists and is constantly present, but so far as I am able to observe it consists solely of connective tissue and contains no glandular substance; it can not, therefore, be in any way analogous to the anterior prolongation of the processus pyramidalis. I think it is to be connected with the fascia of the thyroid sinus, which will be discussed later on, rather than with the gland itself.

Guiard sums up his work on the morphology of the thyroid gland in the rays by saying that the organ lies beneath the coraco-mandibularis, between the coraco-hyoids, is always globous, of more of less pyramidal form, and with a prolongation backward to the bifurcation of the "branchial artery." This corresponds very well with the condition which I find in Raia Erinacea except that in this species, at least, the gland constantly overlies the bifurcation of the ventral aorta (branchial artery), and that it is always somewhat flattened, its ventro-dorsal axis being shortened. The organ is relatively thicker than in the Selachians because of the presence of an increased amount of connective tissue between its vesicles.

On page 26 Guiard says that the thyroid gland "of fish" is always unpaired; it is quite obvious that this remark should apply only to the Elasmobranchs, the only order of fishes which Guiard appears to have studied.

Bridge ('04) passes the thyroid gland with the brief statement that "in adult Elasmobranchs the thyroid is represented by a moderately large compact organ, situated near the anterior end of the ventral aorta." Although he describes the gland as one of the "blood glands" in connection with the vascular system, he does not mention, nor indicate in any way, the source of its blood supply. The statement of its intimate relation with the aortic bifurcation might well lead one to erroneously suspect a supply from this source. In quite another place (page 332) he speaks of "a remarkable system of arteries for the supply of nutrient blood to the gills and heart," which takes origin from the ventral ends of the loops about the gill slits, the commissural vessels forming by their union the "median longitudinal hypobranchial artery which lies beneath the ventral aorta." He fails to mention the ultimate ramifications of this system of vessels or its relation to the thyroid gland, falling into the same error in this particular as T. J. Parker, from whose plates Bridge takes his figures, and upon whose description he appears to have largely based his text.

The literature upon the blood supply of the Elasmobranch thyroid begins with Hyrtl ('58) who first described the hypobranchial arterial system in the Batoidei, if we except the very incomplete description by Monro (1787). Hyrtl described the thyroid artery as the "Ramus thyreoideus seu submentalis" which takes origin from the "vein" of the second gill sac, and which gives off muscular branches to that part of the oral mucosa which lies between the inferior maxilla and the tongue bone as well as the Glandula thyreoidea." Hyrtl did not at this time describe a median hypobranchial artery, this vessel being represented in his description by two anastomosing vessels on either side of the median line which arise from the second and third arches and which pass backward to supply the anterior coronary vessels. Hyrtl very clearly pointed out that the posterior coronary vessels arise from the subclavian artery in the Batoidea and in 1872 he showed that these vessels (posterior coronaries) were absent in the Selachii, a point emphasized at considerable length some years later by G. H. Parker and Davis ('99). In 1872 Hyrtl extended his description of the hypobranchial arterial system to the Selachii. He found the thyroid gland to be supplied by the "Arteria thyreo-maxillaris seu submentalis" which supplied the floor of the mouth and thyroid gland as in the Batoidei but which took origin from the "veins" of the first gill sac, rather than from the second as he had previously described for the Batoidei. He also described the "Arteria cardio-cardiaca," called later the "commissural" and "longitudinal commissural" (T. J. Parker) and the commissural and "lateral hypobranchial" (G. H. Parker and Davis), which fused in the median line to form a median vessel (median hypobranchial) and from which the coronary vessels were derived. At this time Hyrtl emphasized the absence of the posterior coronary branches of the subclavian in the sharks and called attention to an anastomosis from the subclavian forward to the median vessel from which the coronary arteries arose. This anastomotic vessel has since been called by T. J. Parker the "hypobranchial artery."

Turner ('74) injected the conus arteriosus and studied the course of the afferent and efferent branchial vessels; the course of these vessels is now well known. He neither mentioned nor excluded any relation to the thyroid gland, apparently not recognizing this organ, nor did he work out the ultimate connections of any of the smaller cervical vessels.

T. J. Parker ('80) described the venous system of Raia nasuta and called attention to "the extraordinary number of transverse anastomoses it [the venous system of the skate] presents, the results being to produce numerous 'venous circles,' comparable to the circle of Willis in the arteries of the mammalian brain, and the circulus cephalicus in the arterial system of bony fishes. There is also a direct passage from the sinus venosus and back again, in four different ways, namely: (1) by the hepatic sinus; (2) by the anterior part of the cardinal vein and the cardinal sinus; (3) by the whole length of the cardinal veins and their posterior anastomosis;

(4) by the lateral veins and the prolongation of the cardinal sinus into which they open."

I would like to direct attention to the presence of similar venous anastomoses in the Selachii as well as in the Batoidei. These anastomoses in the cervical region are frequent and voluminous. A complete circular anastomosis surrounds the mouth close to the maxillary and mandibular cartilages. Though not so readily seen in the Selachii, it follows the same course as in the skate in which fish it is visible through the skin and oral mucosa; it ends in a maxillary sinus at either angle of the mouth, which is connected with the orbital sinus and with the jugular vein. The hyoid sinuses are similarly connected across the median line near the ventral surface, two anastomotic vessels, the anterior the larger, connecting the opposite sides. This anastomosis bears a most important relation to the thyroid gland. The anterior vessel is so large as frequently to almost envelop the gland as in a capsule, the vessel is subdivided by fibrous partitions, or consists, rather, of a mass or series of vessels within a common sheath, and from its relation to the thyroid gland, in its more or less dilated condition it is more truly a sinus than a vein; it is conveniently designated the *thyroid sinus*. It fills and empties with each movement of the mouth and gills as water is forced through the branchial clefts, thus functioning with the aid of extrinsic muscles after the manner of a venous heart. When the fish is examined out of the water the violent movement of the gills so distends the sinus as often to wholly obscure the thyroid gland by the volume of its contained blood. It is almost impossible to reach the gland by dissection from the ventral surface without cutting the sinus or some of its numerous tributaries. The thyroid sinus receives the veins from the thyroid gland, most of these vessels leaving the dorsal surface or posterior margin of the organ.

T. J. Parker ('86) offers a description of the larger blood-vessels of Mustelus antarcticus, which is, however, deficient as regards the ultimate distribution of the smaller arteries. Exceptions may also be taken to his statement of the distribution of the arteries which constitute the rather remarkable hypobranchial arterial system, which as already mentioned, bears an important relation to the thyroid gland. Parker's description of the venous system is quite accurate. The hyoid sinus is shown to empty into the jugular vein with a valve at the orifice. The anastomosis between the hyoid sinuses is considered, but its relation to the thyroid gland is not discussed; since no mention of the thyroid gland is made it would appear that this important organ was either overlooked or ignored. Unless one is specially looking for the gland, in the effort to separate the muscles without injury to the venous channels the organ is easily broken up, and once disintegrated its particles are readily lost amongst the mass of muscular tissue.

The tributaries of the hyoid sinuses are stated by Parker to include the submental, posterior facial, internal jugular, and the nutrient veins from the first hemibranch.

Parker's description of the hypobranchial system follows that of the ventral and dorsal aorta and begins with the subclavian artery, which, he says, gives off the branchial and hypobranchial arteries. Apparently he omits to mention the large lateral or epigastric artery, whose course parallels that of the lateral vein, though the beginning of the vessel is indicated but not named in some of his figures. The hypobranchial artery described and figured as a continuation of the subclavian, after giving off the antero-lateral artery—which I find to be distributed to the pericardium and adjacent muscles—unites with its fellow of the opposite side, passes forward 2 cm. in front of the conus arteriosus, and forms a plexus from which are given off the coronary arteries posteriorly, and anteriorly the median hypobranchial artery. The plexus communicates laterally by two commissural arteries on either side with the longitudinal commissural vessels uniting the ventral ends of the efferent branchial loops. One gathers from the description that the course of the circulation is from the subclavian artery through the hypobranchial to the efferent branchial loops, a direction which may be thus tabulated:

- branchial
- hypobranchial
    - coronary (paired)
    - median hypobranchial (azygos hypobranchial)
    - commissural (two pairs)
- antero-lateral (paired)

Mention is not made of the lateral nor of all the coronary arteries. Parker's observations were made on Mustelus antarcticus. I have dissected three species of the Selachii and one of the Batoidei, I have not only been unable to confirm the course of the circulation as indicated but I find that beyond the so-called hypobranchial artery the course of the circulation is in the opposite direction, viz., from the efferent branchial loops to the coronary vessels and systemic capillaries, and the hypobranchial artery serves as a relatively unimportant anastomosis which, in these species is not even constantly present. In addition to the pair of coronary arteries distributed to the ventricle I have in my specimens observed a dorsal artery which ramifies largely in the wall of the auricle. The mandibular artery as described and figured by Parker, is the one from which in my preparations the thyroid artery is sometimes, though not constantly, derived. His description leaves one somewhat in doubt as to the origin of this vessel, but he has figured it correctly as coming from the first efferent branchial loop. His coraco-mandibular artery, derived from the mandibular, is a vessel which apparently corresponds with that which distributes its main branches, in my preparations, within the thyroid gland and only incidentally gives small branches to the coraco-mandibular and coraco-hyoid muscles; I have therefore called this vessel the *thyroid artery*.

G. H. Parker and Davis ('99) in an article on "the blood-vessels of the heart in Carcharias, Raia, and Amia" repeated the work of Hyrtl ('58 and '72) so far as it immediately concerned the origin of the coronary vessels, but being concerned only with the cardiac vessels they made no mention of the thyroid artery or other derivatives of the first hemibranch, nor of the gastric and pharyngeal branches which arise in close relation to the anterior coronaries. They described "the irregular longitudinal artery by which the ventral ends of some or all of the efferent branchial arteries of a given side are brought into communication," hitherto referred to as "longitudinal commissural" vessels (T. J. Parker) or as part of the Arteria cardio-cardiaca (Hyrtl), and called the vessels the "lateral hypobranchial artery," reserving for the name commissural "those arteries which leave the lateral hypobranchials on their median sides and, after more or less tortuous courses, unite with one another in the median plane" to produce by their union the median hypobranchial artery. The ventral continuation of the subclavian artery they call the "coracoid artery." Concerning the anastomosis of this vessel with the median hypobranchial formed by the hypobranchial artery of T. J. Parker they speak as follows: "Moreover neither of these vessels [median and lateral hypobranchial] can be properly considered a dependency of the subclavian, for the branch which leaves that artery, and which T. J. Parker regarded as their root, may be connected with them, as Hyrtl ('58, p. 17, Taf. 2) has shown, by only a relatively small vessel. The union, then, is not in the nature of a continuous trunk, but an anastomosis, and the vessel posterior to this union must be considered in the light of an independent artery. This we have called the coracoid artery."

MATERIAL AND METHODS

For the purposes of the present study I have dissected 32 specimens of Mustelus canis, 10 of Carcharias litoralis, 3 of Squalus acanthias, and 14 of Raia erinacea. In addition to these I have had access to a number of sections from various Elasmobranchs prepared by my late assistant, Dr. Guy D. Lombard. The most of these animals were dissected through the courtesy of The Wistar Institute of Anatomy at the Marine Biological Laboratory at Wood's Hole, Massachusetts. My thanks are due these institutions for the opportunity afforded.

The form and position of the thyroid gland was carefully observed in each instance and its vascular connections determined both by dissection and by various methods of injection. The injections were made chiefly with a hypodermic needle of very fine caliber, though finely drawn-out glass tubes were used with some success. For pressure an aspirating syringe was used for routine work and served very well; air pressure was also used at times. For tracing the lymphatics, injections were frequently made into the substance of the thyroid gland and into the connective tissue about the thyroid blood sinus and the other cervical blood vessels. For tracing the blood-vessels injections were made into both remote and near-by vessels, the points selected including the hyoid and thyroid sinuses, the thyroid artery, the efferent branchial loops and commissural arteries, the median hypobranchial artery, the coronary arteries, the ventral aorta, conus arteriosus, cardiac ventricle and auricle, the caudal artery and vein, the mesenteric artery and the dorsal aorta. Injections from these various points were made not only because of expediency in a given species but for the special purpose of determining the direction of flow and the relation of the vessels to the thyroid circulation; hence, injections were made from both sides of the branchial circulation, in the direction of the flow in the veins and the arteries while other injections were made in a direction opposite to the usual course of the circulation on the arterial side, though this was, of course, impossible in the veins because of the presence of valves.

Many of the thyroid glands were cleared and mounted *in toto*. This was best accomplished with those from Mustelus, in which species the gland is very thin. Some of the others were cut free-hand into thick sections. These preparations gave very good pictures of the lymphatics and blood-vessels except in the case of the very thick glands. Still other thyroid glands were sectioned for histological study.

THE ANATOMICAL RELATIONS OF THE THYROID GLAND[1]

The thyroid gland is more or less closely related to most of the structures of the ventral cervical region, a region included between the mandible in front, the coracoid arch or shoulder girdle behind, and the branchial clefts on either side. This region forms the ven-

[1] The fact that the thyroid gland may be readily overlooked in the Selachii is amply demonstrated by the frequency with which this region has been studied and the almost entire absence of any adequate description of the gland. A brief description of the methods of dissection which may be relied upon to locate and expose the gland is offered in the hope that it may materially aid future investigation of this organ.

The thyroid gland of Elasmobranchs can be readily reached from either the oral or the cutaneous surface. By the cutaneous route two methods are especially serviceable, the one by longitudinal, the other by transverse incision.

tral wall of the pharynx and contains the whole course of the ventral aorta and its immediate branches. In Raia the heart is contained within this area at its posterior margin, lying in the median line just in front of the cartilaginous arch, but in Mustelus, Squalus and Carcharias the heart has been pushed backward and lies just beneath the coracoid arch.

The skin of the ventro-cervical region is thin but very tough; laterally it is folded upon itself at the branchial clefts, on the inner surface of which it becomes continuous with the mucosa covering the gills. The fibers of the mylohyoid and geniohyoid muscles are attached to the derma over the greater part of the ventral cervical area. These muscles form a thin sheet, thickest and most prominent in Carcharias, thinnest and frequently almost wanting

The first method is the more applicable in Raia, where the skin is loosely attached and the coraco-mandibular muscle is small, thin and easily lifted. With some variations I have followed the method outlined by Lombard ('09). A longitudinal incision is made in the median line through the skin and membranous constrictor pharyngis. The skin and adherent muscle are dissected away and retracted laterally, exposing the coraco-mandibularis. (Fig. 1 B.) A probe is passed beneath the muscle which, after being well freed, is divided midway between the mandible and the coracoid arch. The anterior flap is grasped and reflected forward, at the same time dissecting away from its dorsal surface the deep cervical fascia in which the thyroid gland is embedded. The gland is easily recognized by its deep yellowish orange color and its peculiar rounded or triangular form. In those exceptional instances when the thyroid gland is displaced forward in Raia the anterior division of the coraco-mandibularis will have to be reflected forward all the way to its mandibular insertion before the gland is fully exposed; ordinarily the organ will be found directly over the aortic bifurcation about midway between the mandible and the point at which the muscle was bisected.

The above method is less easily applied to the Selachii for the reason that the skin and the constrictor pharyngis are much more firmly adherent to the underlying structures than in Raia; moreover, in reflecting forward the anterior division of the coraco-mandibularis one is almost certain to injure the thyroid sinus, deluging the part with blood, before the gland can be exposed. In Carcharias one encounters the added disadvantage that the thyroid gland is quite firmly united to the coraco-hyoideus and the surface of the basi-hyal cartilage, and the gland is buried in a mass of connective tissue by which it usually is entirely obscured even after the coraco-mandibularis has been completely reflected away from its surface. The second method is, therefore, the more applicable in Mustelus and Squalus and is very much more certain in Carcharias. One blade of a blunt scissors is inserted into the first branchial cleft on the right and then on the left side and the clefts lengthened to their extreme ventral limits, the ends in Raia, and subject in all species to great individual variation in volume. The muscles take origin posteriorly from the coracoid arch, anteriorly from the hyoid arch and mandible, and laterally from the outer surfaces of the branchial arches. Acting from these "fixed points" upon the more movable, but tough and inelastic skin, these muscles form a very powerful constrictor of the pharynx and collectively are very properly termed the "constrictor pharyngis" (fig. 1, A). In addition to this constriction the muscular contraction at the same time tends to draw open the branchial clefts, thus permitting the more ready passage of water during the rhythmic pharyngeal contraction or respiratory movement. The muscular fibers of the constrictor pharyngis are intimately adherent to the derma.

of the incisions exposing the margins of the coraco-hyoideus muscle. One blade of the scissors is then pushed beneath the skin where it readily passes between the coraco-hyoideus and coraco-mandibularis muscles (fig. 1); the incision is continued across the median line from side to side. This divides the coraco-mandibularis; its anterior portion is grasped with the forceps, lifted, and a longitudinal incision through the skin and fascia carried forward along either margin of the muscle. In the dogfish the divided muscle with the attached skin is easily raised and the loosely attached deep cervical fascia dissected away from its dorsal surface, exposing the thyroid gland. In Carcharias it is better to dissect the deep cervical fascia away from the ventral surface of the coraco-hyoideus muscle, rather than from the coraco-mandibularis; the thyroid gland is then raised with the latter muscle and dissected out from the mass of connective tissue which envelopes it. Finally, the gland must be dissected away from its anterior attachment to the margin of the basi-hyal cartilage, or, in Carcharias, to a median depression, in the ventral surface of this cartilage, which corresponds to the foramen caecum linguae of mammals; occasionally this depression is a true foramen, in which case the thyroid process becomes obviously analogous to the lobulus pyramidalis of the mammalian thyroid gland. This lobule is represented in Mustelus by a short triangular projection, not constantly present, which overlies a shallow median groove in the anterior margin of the cartilage. In Raia a similar condition is much less frequently present.

The thyroid gland is readily accessible from the oral cavity. A needle passed through the oral mucosa just in front of the basi-hyal cartilage—"lingual bone" —enters the substance of the thyroid gland if directed backward in Mustelus and Squalus, well backward and close to the cartilaginous surface in Carcharias, or backward and slightly ventralward in Raia. A transverse incision through the oral mucosa, parallel to and just in front of the basi-hyal cartilage, exposes the anterior margin of the thyroid gland and it may then be readily dissected out from Mustelus or Raia, though with greater difficulty from Carcharias or Squalus.

Removal of the integument with the adherent constrictor muscles exposes the coraco-mandibularis (fig. 1, B), a slender paired muscle, its two sides intimately fused in the median line, which takes origin by a tendinous fascia from the ventral surface and anterior margin of the coracoid arch. The paired muscle passes forward to its insertion, ending in short, rounded and slightly divergent tendons which are attached to the posterior margin of the inferior mandible. The muscle is inclosed within the folds of a superficial cervical fascia, which forms its aponeurosis and extends laterally to the surface of the branchial arches, but on either side of the muscle, the aponeurosis fuses with the deep cervical fascia with which it is in more or less close contact.

On lifting the coraco-mandibularis with its superficial cervical fascia the coraco-hyoid muscle (fig. 1, C) is exposed; it is similar in shape and appearance to the coraco-mandibular, but is much broader, its lateral margin projecting from beneath the coraco-mandibularis, and in Mustelus, Squalus and Carcharias extending laterally almost to the ventral ends of the branchial clefts, or even overlapping them somewhat. In Raia the gills are more widely separated, leaving a broad portion of the floor of the pharynx exposed at the side of the coraco-hyoideus in the anterior portion of the cervical region. Posteriorly the several divisions of the coraco-branchialis muscle cross this exposed portion of the pharyngeal floor to be inserted into the branchial arches.

The ventral surface of the coraco-hyoideus is smooth, its dorsal surface separates into several muscular processes, the musculus coraco-branchialis (*M. c. br.*, fig. 5), to be inserted into the membranous floor of the pharynx by a tendinous fascia overspreading and firmly adherent to the fibrous pharyngeal submucosa and the surfaces of the cartilaginous branchial arches. The divergent portion of the coraco-hyoidei, on either side of the median line, are similarly inserted into the movable basi-hyal cartilage (lingual bone), so that the combined coraco-hyoid and coraco-branchial muscles, arising from the anterior border of the coracoid arch, form a very powerful dilator of the mouth and pharynx.

As the coraco-hyoideus does not extend forward beyond the hyoid arch it exposes the membranous floor of the oro-pharynx

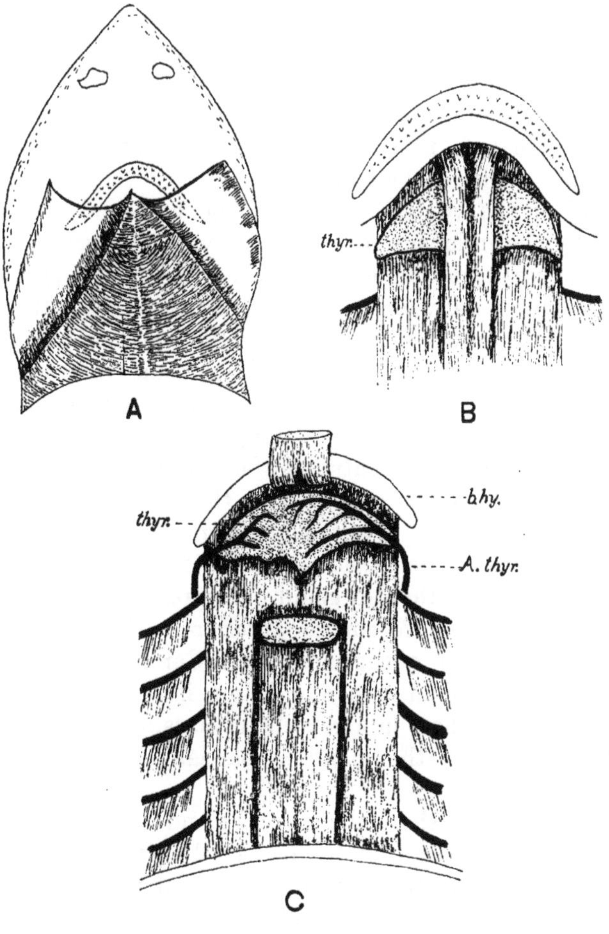

Fig. 1. Dissection of the thyroid gland of Mustelus canis. In *A* the skin has been reflected to show the superficial "constrictor pharyngis" muscle. In *B* the "constrictor pharyngis" has been removed and the coraco-mandibularis exposed. In *C* the coraco-mandibularis has been divided, exposing the thyroid gland lying upon the coraco-hyoideus. (See page 209 for explanation of abbreviations used in all figures.)

between this point and the inferior mandible. The cartilages of the hyoid arch except only the dorsal surface of the basi-hyal, are loosely adherent to this membranous floor so that when the mouth has been closed and the pharynx contracted the tongue-like basi-hyal cartilage is pushed forward, producing a deep fold in the oral mucosa. A needle thrust through this fold in the median line, passing dorsal to the cartilage, penetrates directly into the thyroid gland.

Not all of the ventral surface of the basi-hyal cartilage is covered by the insertion of the coraco-hyoideus muscle; the portion of the cartilage thus exposed varies in different species and to some extent in individuals. In Carcharias and Raia the muscle is inserted into only a small portion of the cartilaginous surface, while in Mustelus all but a narrow anterior margin is covered by the muscle fibers. The thyroid gland typically lies upon this exposed cartilaginous surface, extending backward for a greater or less distance upon the ventral surface of the coraco-hyoideus (fig. 1, C). In Mustelus and Squalus the ventral surface of the basi-hyal cartilage has a raised arciform anterior margin with a very slight medial depression, in Raia it is nearly flat, and in all these species the thyroid gland overspreads the cartilage like a thin membrane whose convex anterior border nearly corresponds with the outline of the cartilage; in Carcharias the cartilage presents a deep median groove or furrow into which the thyroid gland sinks, lying there in a gelatinous mass of connective tissue so voluminous that the gland is partially, sometimes wholly, obscured.

In the dogfish and shark the thyroid gland is rarely pushed forward beyond the margin of the basi-hyal cartilage; in Raia the organ may extend farther forward so that it rests in part upon the membranous floor of the oro-pharynx. In one of the skates I found the gland carried so far forward that it lay wholly in front of the cartilage. In Mustelus and Carcharias individual variations are much less frequent than in Raia, but in Raia in the majority of individuals the gland lies directly upon the bifurcation of the ventral aorta (fig. 7).

THE THYROID VESSELS

The anterior margin of the thyroid gland is firmly attached by a dense fibrous fold of fascia to the antero-ventral margin of the basi-hyal cartilage. Its ventral surface is in contact with the coraco-mandibular muscle, to which it is firmly united by a fascia. Its dorsal surface rests upon the basihyal cartilage and the insertion of the coraco-hyoid muscle as described in the preceding section. The lateral angles and posterior margin of the thyroid gland are continuous with a fold of the deep cervical fascia which is placed between the coraco-mandibular and coraco-hyoid muscles, loosely attached to the opposed surfaces of these muscles, but more firmly fixed to their lateral margins, so as to form an aponeurosis for each. This fold of the fascia is much thickened anteriorly where it approaches the margin of the thyroid gland: at this point it incloses a transverse anastomosing vein which connects laterally with the hyoid sinuses.

At the posterior margin of the gland the fascia splits, a thin layer passing dorsally between the gland and the coraco-hyoid muscle, a thicker portion extending forward over the ventral surface of the organ, between it and the coraco-mandibular muscle. This ventral division is of special importance; it contains the large thyroid sinus consisting of an intricate net of veins and lymphatics, which connects laterally with the hyoid sinus and in the median line spreads over the ventral surface of the thyroid gland so that when distended with blood it entirely obscures the organ.

The thyroid sinus is surrounded with connective tissue containing a network of lymphatic vessels. Ink injected in the living animal into the space between the lateral margin of the thyroid sinus and the ventral end of the first branchial cleft will after a few minutes be found filling many of the lymphatic vessels of the thyroid sinus as well as many other perivascular lymphatics in relation with most of the cervical veins and the arteries of the hypobranchial system (fig. 8). The lymphatics of the thyroid plexus are so mumerous and anastomose so freely that when filled with ink they form a sac-like investment entirely obscuring the sinus and the thyroid gland (fig. 2). An attempt to inject with a fine needle directly into the thyroid sinus may force the fluid into either the venous or the lymphatic plexus, according as the one or the other system of vessels happens to be entered. I am convinced, as a result, of my injection experiments, that the lymphatics open freely into the veins of this part, after the manner of the "vasa lymphatica" of Favaro (*vide infra*). These observations are of interest in connection with Baber's inability to find lymphatics in the thyroid gland as indicating the relation between the vascular systems. I find, however, that it is not possible for fluids injected into the thyroid sinus or its plexus of lymphatics to pass in any quantity into the vessels within the thyroid gland; this is presumably because of the presence of valves.

The alternate expansion and contraction of the mouth and pharynx, forcing the stream of water through the branchial clefts, alternately fills and empties the thyroid sinus, so that by means of these respiratory movements the sinus acts somewhat after the manner of a venous heart. In this connection it is interesting to consider the observation of Favaro ('06), as quoted by Sabin ('09), that the relation of the veins and lymphatics in fishes is much more primitive than in mammals and that both lymph-hearts and vein-hearts may be present in these animals. The emptying and filling of the veins can readily be seen in the Selachii or Raia on removing the skin, or even through the integument in the living skate, the colored blood showing readily through the vascular walls. The relation of the lymphatics to the blood-sinus is so intimate that they must also be emptied and filled in the same way, though since they contain a colorless fluid they can not be so readily observed. I have, however, demonstrated that a colored fluid injected into these lymphatic vessels will overspread the thyroid region in ten to fifteen minutes and will almost entirely disappear within the next fifteen minutes; the lymphatic circulation must therefore proceed with considerable rapidity.

The thyroid artery approaches the organ from either side; in Mustelus and Squalus it enters at the extreme lateral angle of the triangular gland (fig. 19). In Raia, where the organ is of a more rounded form it enters near the middle of the lateral border. The branches of the artery ramify upon the surface of the organ send-

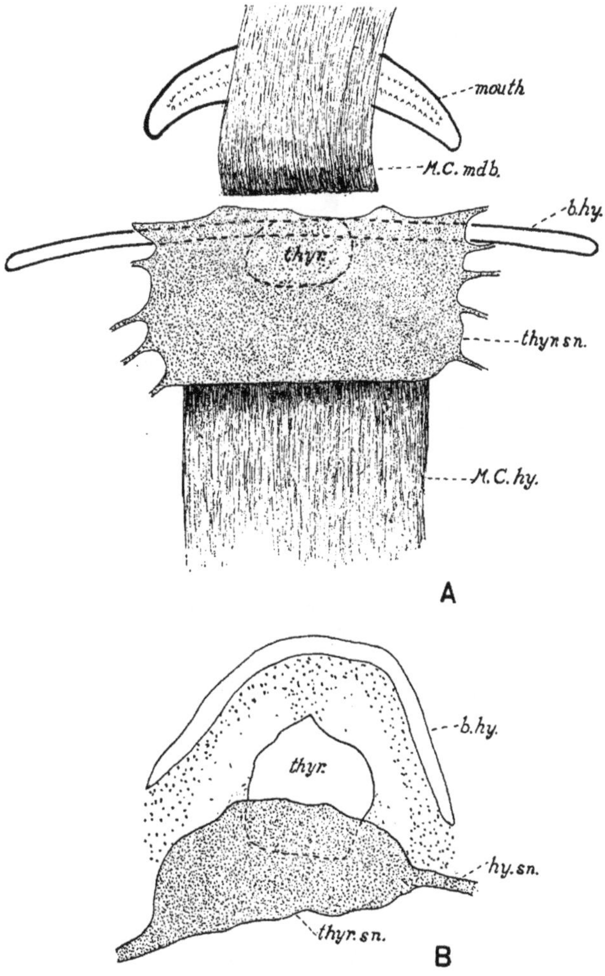

Fig. 2. Showing the form and relations of the thyroid sinus when injected with ink. *A*, in Raia. *B*, in Carcharias.

ing finer twigs into the interior. In this particular they offer an interesting analogy to the condition found by Major ('09) in man. Major says (page 484), "in the human the branching of the large arteries takes place mostly upon the surface of the gland, and having by their branching obtained their approximate distribution, the smaller branches are sent in." In the Elasmo-

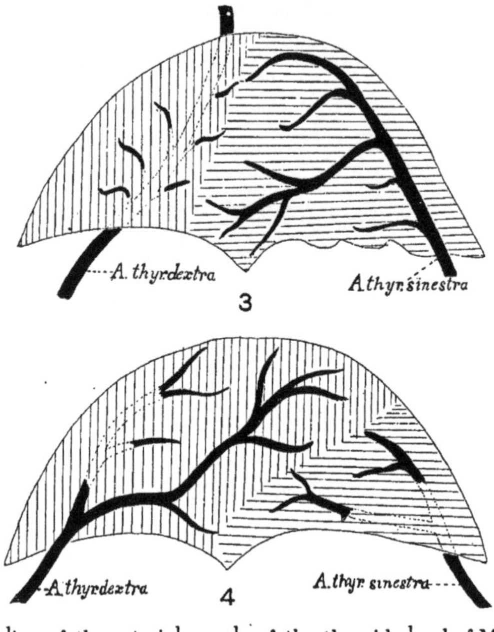

Fig. 3. Outline of the arterial supply of the thyroid gland of Mustelus canis as usually found. The vertical shading indicates the area supplied by the right thyroid artery, the horizontal shading that supplied by the left.

Fig. 4. Showing the less usual distribution of the thyroid arteries; the shading as in fig. 3.

branchs the condition might well be described in the same words; this is the more remarkable inasmuch as Major states that this is not the condition in other mammals, *e.g.*, the dog and cat. The Elasmobranchs, therefore, seem to harmonize with the human rather than the lower mammalian condition as regards the distribution of the main branches of the thyroid arteries.

The left thyroid artery usually supplies a greater portion of the gland than the right (fig. 3), though the relative area is subject to extreme variation and in occasional instances the ratio may be reversed (fig. 4). The area of distribution in the great majority of individuals is approximately as indicated in fig. 3.

THE HYPOBRANCHIAL CIRCULATION AND THE ORIGIN OF THE THYROID VESSELS

In attempting to trace the circulation of the thyroid gland by means of injection experiments I was at once struck with the difficulty of reaching the gland by means of injections into the gill arteries, the ventral aorta or the heart. It was obvious that the thyroid artery has no direct connection with the ventral aorta or the afferent branchial vessels, a fact which seems to have been first observed by Simon ('44) who stated, without further explanation or any outline of his reasons therefor, that the thyroid gland in Raia "never receives the smallest share of supply from the branchial artery with which it is in contact." This fact seems to have been seldom recognized and never sufficiently emphasized by later writers.

The thyroid artery arises either from the mandibular artery, or by a common trunk with this artery, from an arterial sinus at the ventral extremity of the first branchial cleft (figs. 5 and 6); this sinus forms the ventral portion or connecting vessel of the efferent vascular loop contained in the hyoidean hemibranch and first holobranch. From the dorsal extremity of this same loop the first efferent branchial artery passes to the dorsal aorta. It is therefore necessary for fluid injected into the ventral aorta or its immediate branches to pass through the gill capillaries before it can enter the thyroid arteries, and few injection fluids readily pass through capillary vessels. On the other hand, fluid injected into the thyroid artery or, as I later found, into any portion of the hypobranchial system passes readily into the vessels of the thyroid gland; such injections I repeatedly made, into the sinus at the ventral end of the first efferent branchial loop, into the thyroid artery, the median hypobranchial artery, and even into the coronary artery taking care to prevent the escape of the fluid through the coronary vessels into the sinus venosus.

The hypobranchial system of vessels is so important for the thyroid gland as to deserve more than passing mention. T. J. Parker ('86) has described this system in connection with his much quoted work on the circulation in Mustelus antarcticus, but he makes no mention of its relation to the thyroid gland, in fact, he mentions neither the gland nor the thyroid artery; the gland was apparently not observed.

The main trunk of the hypobranchial system, in the species which I have examined is the median hypobranchial artery; it is formed by the commissural arteries coming from the ventral ends of the loops formed by the efferent branchial vessels which receive blood from the gill capillaries. These loops surround each branchial cleft, and within the gill they lie parallel to the afferent branchial arteries; they are just antero-internal to the afferent vessels. The ventral efferent vessels are very much smaller than either the dorsal efferent or the afferent (fig. 5). Opposite the ventral end of the second and third branchial clefts (sometimes only the second or the third) each loop gives off a commissural branch which passes inward and somewhat backward to the ventral surface of the ventral aorta where these vessels, with considerable variations, unite with their fellows of the opposite side to form a median hypobranchial artery which is frequently double so as to form a sort of elongated arterial circle. Sometimes the vessels fail to unite in front so that instead of a median hypobranchial there is a right and left hypobranchial artery, one on either side of the ventral aorta (fig. 5). Frequently the vessels so unite as to form an annular anastomosis which encircles the aorta, but portions of the ring may be absent. Some of these variations are indicated in figs. 5 and 6. The varied arrangements of these vessels are all indications of a more or less complete fusion of the commissural vessels to form a median hypobranchial artery. This vessel terminates posteriorly in a small sinus-like dilatation, single or double as the case may be. From this sinus the coronary vessels arise either as a median vessel which promptly divides, or as two or three independent vessels. From this same sinus a small paired

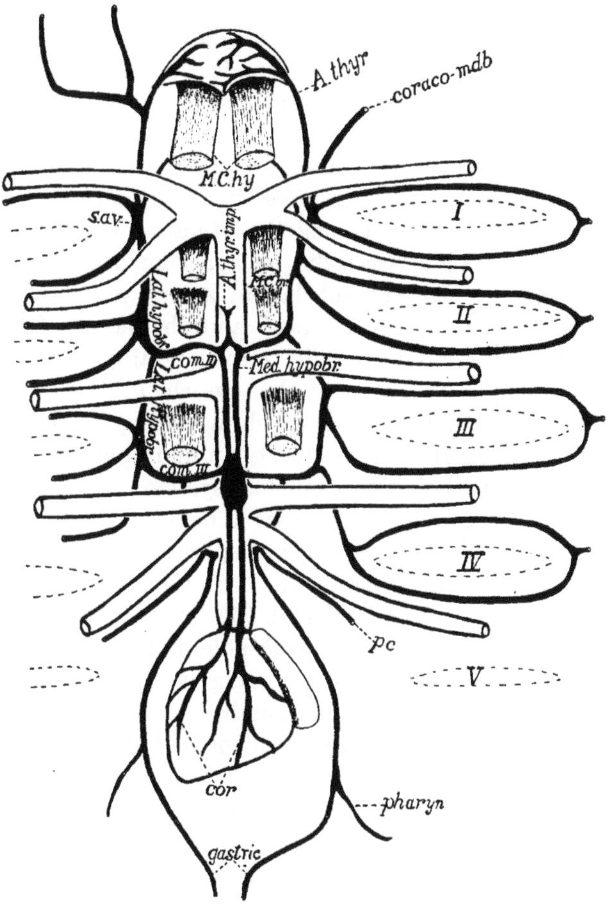

Fig. 5. Diagram of the hypobranchial arterial circulation in Mustelus canis; the insertions of the ventral divisions of the coraco-branchialis muscle are also shown. Ventral view. Anastomosis with the subclavian artery was wanting in this specimen.

artery passes backward on either side of the median line beneath the dorsal portion of the pericardium at the lateral margin of the cartilaginous floor of the pharynx formed by the basi-branchial cartilage; after anastomosing with its fellow of the opposite side beneath the apex of the cardiac ventricle it distributes its terminal branches to the wall of the esophagus and stomach near the cardia (figs. 5 and 6).

From the loop at the ventral end of the fourth branchial arch a very small anastomotic branch (less frequently arising as in fig. 6, *hypobr'*) passes backward along the lateral wall of the pericardium and penetrating between the precaval sinus and the coracoid arch anastomoses with the subclavian artery just prior to its division into the brachial (axillary) and the lateral (or hypogastric, a large artery lying parallel to the lateral vein. This anastomotic branch is undoubtedly that which T. J. Parker ('86) describes as the hypobranchial, which, according to his description receives blood from the subclavian and supplies the coronary arteries and whole hypobranchial system. Such is not the case, however, in any of the species I have studied and Hyrtl ('72) in his careful study of various species of the Selachii did not so find it, nor did Parker and Davis ('99). The hypobranchial is a very small artery, so small that its connection with the median hypobranchial is scarcely traceable, and in some individuals is entirely wanting (fig. 5), there being in these cases a small branch from the subclavian and a similar vessel from the median hypobranchial which follow the usual course but never unite, the subclavian branch distributing its blood to the muscles while the anterior division supplies the lateral pericardial wall and the adjacent muscles in front of the coracoid arch. Certainly where the vessel is wanting the flow of blood can not be in the direction indicated by T. J. Parker. Parker and Davis ('99), as already quoted, found the hypobranchial artery insignificant, though they did not record its absence.

If the median hypobranchial artery of Mustelus be injected the major portion of the fluid passes into the coronary arteries and thence through the coronary veins to the sinus venosus and auricle, while at the same time very little passes through the hypobranchial artery; the fluid pours into the sinus venosus and auricle very freely before it has even reached the subclavian by way of the hypobranchial artery. This is the case whether the fish has been previously bled or not. This experiment would certainly show

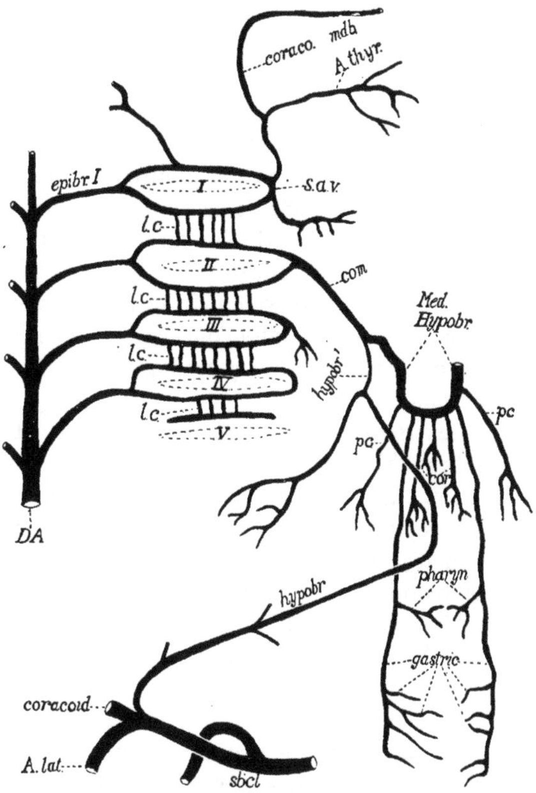

Fig. 6. Diagram of the hypobranchial arterial circulation in Squalus acanthias. Lateral view.

that the hypobranchial is of too small a caliber to supply the blood necessary to fill the coronary arteries, and if it can not supply this much it certainly is still less competent to supply blood for the whole hypobranchial system, which T. J. Parker's description would seem to indicate was the case. The whole system may be readily injected from any one of the gill-loops with which it is connected. The true direction of flow is therefore from the efferent gill-loops through the commissural arteries to the median hypobranchial and from it to the muscular, pericardial, gastric, esophageal, and coronary branches, with only a relatively insignificant and inconstant anastomotic supply from the subclavian artery.

Anastomoses in both the arterial and venous systems, forming "circles" about the body wall and the viscera are of very frequent occurrence in this class of fishes as was pointed out for the venous system by T. J. Parker in 1880 (*vide supra*). The subclavian artery forms such a circle beneath the coracoid arch and several similar "circles" are formed by anastomosis between the two sides in the hypobranchial system as well as in other parts of the body with which we are not now specially concerned. The arrangement in the arterial system is therefore very similar to that which Parker found in the venous.

At the ventral extremity of each efferent gill-loop, at the point where the hypobranchial commissural arteries arise, is a small sinus-like dilatation (figs. 5 and 6, *s. a. v.*) which obviously serves as a reservoir where the blood coming from the two sides of the loop, which are in adjacent branchial arches, will intermingle, and from this sinus blood is distributed through the commissural arteries ("lateral hypobranchial" of Parker and Davis) anteriorly, posteriorly, or to the median hypobranchial in such proportion as the caliber of the several vessels and the course of the circulation dictate. The arterial sinus at the ventral end of the first gill-loop (first ventral sinus) is usually a trifle larger than the others. The thyroid artery arises from the anterior end of this sinus or from the adjacent portion of its anterior limb in the hyoidean hemibranch. It arises either as a separate and independent vessel or as a conjoined trunk with the mandibular artery; more frequently, in the specimens I have dissected, it was independent. The artery passes directly forward and inward to the extreme lateral border or angle of the thyroid gland. It continues its path along the surface of the thyroid gland (figs. 3, 4 and 19) near its anterior border, distributing its main branches to the substance of the gland and small collateral branches to the floor of the pharynx in front of the hyoid arch and to the anterior third of the coraco-hyoideus and coraco-mandibularis muscles. A small median unpaired vessel arising from the left thyroid artery (less frequently from the right), penetrates the thyroid gland, divides, and enters the coraco-hyoideus to supply the antero-median portion of this muscle. This is very probably homologous with the anterior portion of the arteria thyroidea impar, derived from the median hypobranchial as described by Hyrtl ('72).

The left thyroid artery is usually larger, longer, and more extensive as to its area of distribution than the right. Lombard ('09) dissected a number of specimens of Mustelus and Raia and found that the left thyroid artery more frequently entered the dorsal surface, and the right the ventral surface, of the thyroid gland. I have found a somewhat similar condition, though I very frequently find both vessels coursing upon the ventral surface and sending their branches dorsally into the substance of the gland. Occasionally the right thyroid artery enters the dorsal surface of the gland and the left the ventral (fig. 3 and 19). The position and distribution of the thyroid arteries is, however, subject to considerable variation and, as I have already pointed out, the right may even supply a greater portion of the gland than the left thyroid artery (figs. 3 and 4).

The thyroid artery as described very probably in part corresponds to the vessel which was recognized by T. J. Parker ('86) as the coraco-mandibular artery. This latter is an obviously inaccurate designation, for the coraco-mandibular branches are insignificant as compared with the other ramifications of the artery. Hyrtl ('72) recognized and more accurately described the thyroid artery as arising from the "veins of the first gill-arch" in conjunction with the submental artery; his description appears to be accurate with the exception that the two arteries in my dissections appear more frequently to arise independently.

The observations which I have recorded concerning the origin

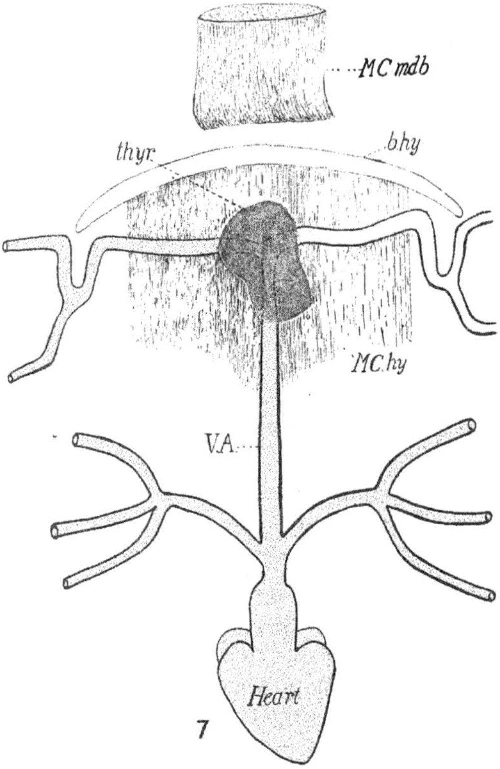

7

8

Fig. 7. The position and relations of the thyroid gland in Raia.

Fig. 8. Injected lymphatics in the tunica adventitia of the right commissural artery at its junction with the median hypobranchial. The specimen is from Carcharias, the fish having been injected with ink at a point just ventral to the right hyoidean hemibranch. The injection followed the lymphatics and was traced as far as the median hypobranchial artery in one direction and into the thyroid gland in the other.

and course of the thyroid artery and the hypobranchial system were carefully worked out with specimens of Mustelus and Squalus and verified in all their important particulars in Carcharias and Raia.

VEINS AND LYMPHATICS OF THE THYROID REGION

The numerous small veins of the thyroid gland discharge into the "thyroid sinus," which connects together the hyoid sinuses of the two sides. The conformation of the hyoid sinuses and their tributaries and connections have been well described by T. J. Parker ('86). In addition to the transverse anastomosis formed by the thyroid sinus, the hyoid sinus receives a submental vein from the region of the mandible and numerous small muscular branches from the neighboring muscles. This sinus and its connecting vessels can be most readily observed in Raia. The submental vein is seen to begin as a double transverse anastomosis; the larger, anterior, tributary lies close behind the cartilage of the inferior mandible; the smaller, posterior vessel arches across the floor of the mouth just in front of the hyoid arch and the anterior border of the thyroid gland. At the angle of the jaw these vessels unite in a small sinus which also receives a transverse anastomosis from in front of the maxilla, so that the mouth is thus encircled by an annular venous sinus. The thyroid sinus similarly forms a double transverse anastomosis, rather more deeply placed, behind the hyoid arch at the posterior border of the gland. These vessels convey the blood from the ventral cervical region to the hyoid sinus.

The thyroid veins open into the thyroid sinus as several small branches, the largest of which are a median vein, leaving the organ near the middle of its dorsal surface, and two anterior veins which leave the same surface near the anterior margin of the organ, but a little to either side of the median line. Other smaller veins leave the lateral margins of the organ passing either to the thyroid or the hyoid sinus. The thyroid veins must contain valves, for although the vessels can be readily traced with the dissecting microscope and even with the naked eye, it is with difficulty that fluid injected into the thyroid or hyoid sinus can be forced back into the venous channels of the thyroid gland; an extreme pressure will accomplish this result to a limited extent only. I have been able to find some traces of valves in microscopical sections.

The hyoid sinus passes around the base of the hyoidean hemibranch to connect with the jugular vein, through which a portion of its blood is transmitted to the precaval sinus beneath the coracoid arch, and thence to the sinus venosus and auricle. The flow through the hyoid sinus in this direction is quite intermittent, and, as already indicated, it is chiefly dependent upon the muscular force of the pharynx as it alternately relaxes and contracts to force water through the gill-openings.

Blood is also transmitted from the hyoid sinus to the heart by the more ventral and direct path through the inferior jugular (anterior cardinal) vein. This vessel maintains a more constant flow, receiving blood from the ventral cervical region and the branchial arches, along the ventral ends of which it courses to terminate in the precaval sinus.

In Raia the thyroid sinus is thin, and its investment of connective tissue containing the lymphatic plexus is less pronounced than in the other species studied, so that, except when the sinus is fully distended with blood, the gland in Raia is not much obscured. In Mustelus the sinus is larger and the fascia about it is more voluminous so that the gland is usually more or less obscured, though there is much individual variation: the same is true of Squalus. In Carcharias the vascular walls in the sinus are so thick, and the connective tissue about it so abundant that in most of the animals examined the outline of the thyroid gland contained within this mass could only be discerned on holding up the stretched membranous mass between the eye and the bright sun so that the intense transmitted light showed the yellowish orange gland contained within the connective tissue mass.

The thyroid lymphatic plexus forms an extensive group ot vascular channels surrounding the gland and the vessels of the blood sinus. It is contained in a fold of the deep cervical fascia which stretches across from side to side between the ventral ends of the first branchial clefts: it is broad in the mid-portion, but tapers from the postero-lateral angle of the thyroid gland outward to the tissue surrounding the hyoid sinus. The vessels form perivascular lymphatics about the venous sinuses. Ink or a colored fluid injected into the connective tissue about the hyoid and thyroid sinuses readily fills the anastomosing vessels forming a sheetlike mass of peculiar form (fig. 2, *thyr. sn.*). Ink thus injected can also be traced into the perivascular lymphatics of the hypobranchial arterial vessels (fig. 8) as far backward as the walls of the coronary arteries; it can likewise be found in small perivascular lymphatics in the walls of the thyroid arteries and to some extent in the broad venous spaces between the vesicles of the thyroid gland, indicating that the lymphatic vessels to some extent may open into the veins of the thyroid. The vessels of the lymphatic plexus in the cervical fascia are apparently connected with the blood-vessels of the thyroid sinus, for excessive pharyngeal contraction in the living fish forces blood into areas which otherwise appear to be occupied only by lymphatic vessels. The blood-vessels may without doubt be classed as "venae lymphaticae" and the lymphatics as vasa lymphatica" after the terminology of Favaro ('06), who says that the same vessel may in fishes carry either blood or lymph at the same or different times so that these vessels may in this sense be either vasa or venae lymphaticae. Fluid injected into the lymphatics spreads so rapidly over so great an area that it seems almost impossible to trace a connection with the blood sinus by means of injections; the fluid enters the blood-vessels so readily that one is unable to exclude the possibility of an intra-venous injection.

The statement by Baber ('81) that he was able to demonstrate no lymphatics in the thyroid gland of Elasmobranchs led me to pay special attention to the study of these vessels by injection methods. As I have already pointed out, Baber states that "in both the skate and the Conger-eel an extensive system of vessels lined with epithelium becomes injected by the method of puncture." He then injected the blood vessels of a Conger-eel with Berlin blue through the "efferent branchial vein" and "dorsal aorta and thereupon states that "in the Conger-eel at least, there is no evidence of any system of lymphatic vessels," emphasizing his statement by the use of italics. I can confirm that portion of Baber's statement which says that an extensive system of vessels within the thyroid gland can be readily injected by the method of puncture, but I would maintain that neither that procedure, nor the injection of the dorsal aorta or efferent branchial vessels with Berlin blue, would demonstrate the absence of lymphatics; that they may still be present, I have demonstrated both in microscopical sections and by injection (figs. 9, 10 and 11).

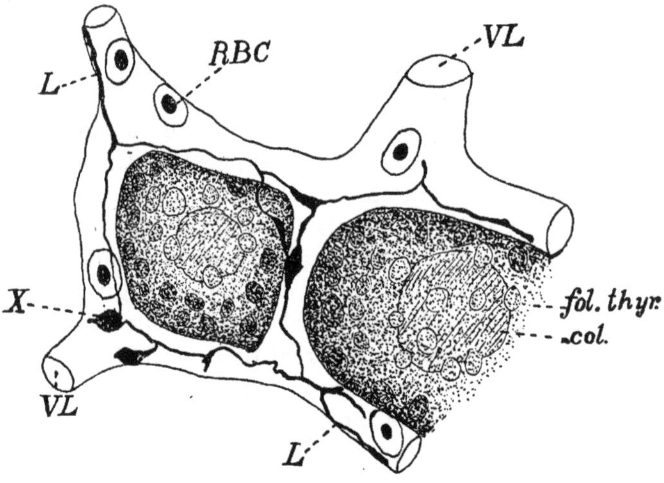

Fig. 9. Lymphatics, "vasa lymphatica," and veins, "venae lymphaticae," of the thyroid gland. The "vasa lymphatica" have been injected with ink and the thyroid gland cleared and mounted *in toto;* the lumen of the follicle and the follicular epithelium are only indistinctly seen. At X in the specimen the two sets of vessels anastomose.

Injection by puncture does not always fill the extensive system of vessels observed by Baber. If one takes care to use only a very gentle pressure, this system, which completely surrounds each vesicle, is only filled near the point of injection, while at the margins of the injected area the fluid spreads through more minute vessels which lie in closer contact with the vesicular epithelium (fig. 9, L). I believe these last are true vasa lymphatica in the sense of Favaro and I find the contents of the vesicles apparently secreted into them as in the mammalian thyroid (fig. 10). The larger vascular channels then are venae lymphaticae, readily injected by puncture if the pressure is excessive, the veins being easily entered because of their large caliber and extremely thin walls; they transmit only lymph when the intravenous blood-pressure is low within the gland, but fill with blood when from any cause the pressure is raised. I have invariably found some blood cells in the venae lymphaticae; I have never found them filled with blood in all the three score animals I have examined except in one case in which as a result of injury the thyroid gland was greatly congested. In this case they were filled to distension. In microscopical sections I have been able to trace the connection of the vasa with the venae lymphaticae (fig. 10). I have been unable to demonstrate positively the presence of any valves at the orifices of these vessels, but the extreme obliquity of the anastomosis considered in conjunction with the very thin vascular walls might well serve a valvular function when the blood pressure is low, though with increased pressure and venous distention some blood would be forced back into the vasa lymphaticae and even into the vesicles. The frequent occurrence of red blood corpuscles within the vesicles of all animals is well known and in the Elasmobranchs it is thus accounted for. The intimate relation between the venous and lymphatic systems pointed out by Sabin ('09) would possibly suggest that an homologous vascular relation may account for the presence of red blood corpuscles within the vesicles of the mammalian thyroid gland.

THE HISTOLOGY OF THE ELASMOBRANCH THYROID GLAND

The thyroid gland in Elasmobranchs consists of a mass of vesicular follicles (figs. 10, 11 and 13 to 18) which very closely resemble those of the mammalian gland. The vesicles are lined by epithelium of a low columnar type, contain more or less colloid material, and are loosely bound together by a connective tissue framework which is very richly supplied with blood-vessels.

The shape of the gland in Mustelus canis (fig 1, C) is sufficiently

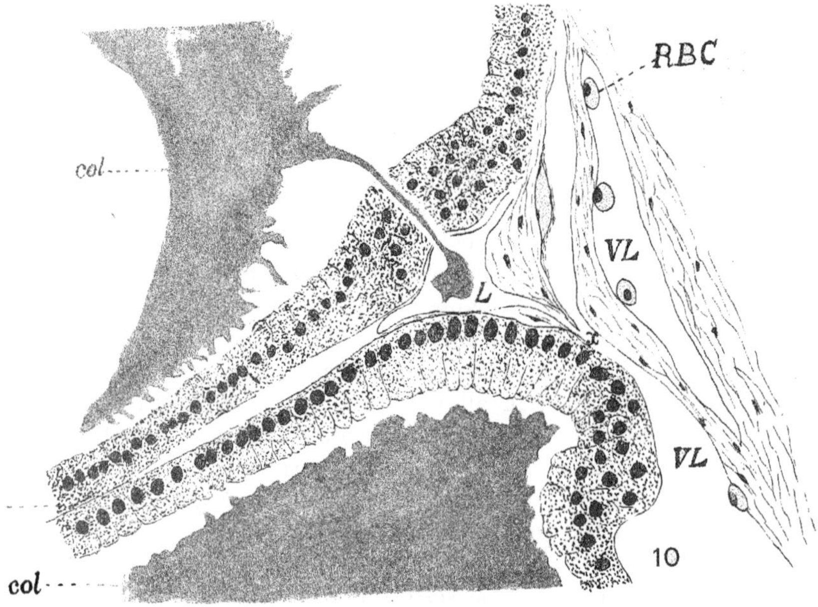

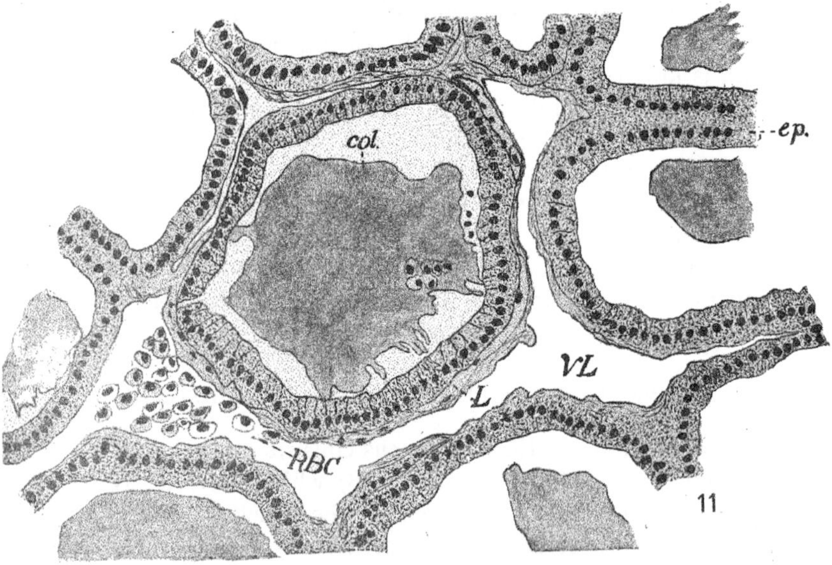

Fig. 10. Section of the thyroid gland of Mustelus canis showing the anastomosis at the point X of the vasa and venae lymphaticae.

Fig. 11. Typical section of the thyroid gland of Mustelus canis; the intimate relation of the epithelium of the thyroid follicles to the lymphatics and blood vessels is accurately shown.

peculiar to deserve passing mention. It may be described as consisting of two triangles whose bases are fused in the median line, the apices directed outward, the anterior borders convex and conforming to the anterior margin of the basi-hyal cartilage, the posterior borders concave and free, except for their attachment to the deep cervical fascia. In the median line the conjoined bases are prolonged backward to form a short median projection; anteriorly a shallow notch separates the two lateral triangular halves. The gland is approximately bilaterally symmetrical (figs. 1, 3 and 4).

The thin, almost membranous character of the gland in Mustelus canis offers an excellent opportunity for the recognition of a lobar or lobular structure if such exists, for the whole gland is frequently no more than four or five follicles in thickness and may be stained, cleared and mounted *in toto*, giving very excellent microscopical pictures of the entire organ. I have not been able to find any indication of definite lobes or lobules. Portions of the thyroid substance are here and there wanting, as observed by Lombard ('09), and these deficient areas occur more frequently in the posterior than in the anterior half of the gland. In one of the thirty-two fishes of this species the deficiencies were so great that the gland was only represented by a few specks which were positively identified as portions of the thyroid only after microscopical examination. A similar case was found in Squalus, and one gland from Carcharias consisted of three small pieces.

Occasionally the posterior border presents a notched deficiency in or near the median line. There may be one, two or three such notches, either symmetrically or asymmetrically disposed. Deficiencies of the thyroid tissue also occur within the gland and may, or may not, be connected with the notches in the posterior border. These deficiencies are all of inconstant occurrence, irregular location, and could scarcely be taken to indicate any suggestion of definite lobes. They seem rather to be due to the extreme thinness of the gland and in many of the thicker specimens they are in no way indicated. When present they are occupied by connective tissue continuous with the glandular capsule. Frequently they transmit the larger thyroid vessels.

Bits of thyroid tissue of inconstant form or location are occasionally separated from the body of the gland by narrow partitions of connective tissue; they are most frequently found near the border of the gland or adjoining an area in which the thyroid substance is deficient. Since they possess no constant relation to the vascular supply, the detached masses can not correspond in any sense to true anatomical lobules. The arteries branch irregularly, for the most part after a somewhat dicotymous fashion (fig. 12), the arterial twigs passing off at acute angles. Partially injected specimens in which the injection fluid has passed through the arteries but has not penetrated in quantity into the veins

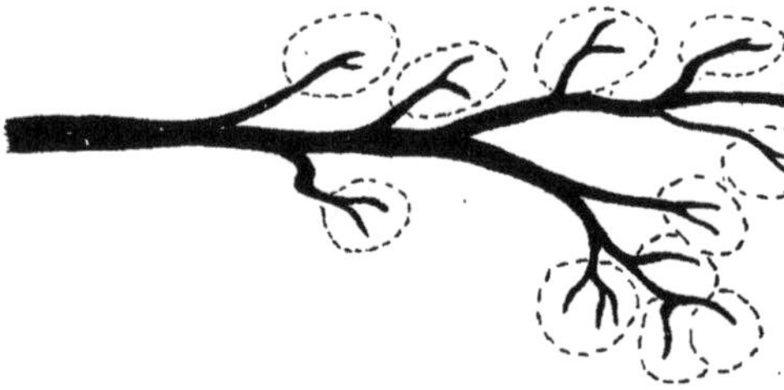

Fig. 12. Terminal divisions of the thyroid artery. The area occupied by injected capillaries, "venae lymphaticae," directly connected with each terminal arteriole is roughly indicated by the dotted lines.

show areas of injected capillaries surrounding the terminal arterioles (fig. 12), but the extent of these injected areas and their relation to the artery seems to be dependent rather on the pressure of the injection than on any constant or characteristic relation to the vascular system. I can not recognize any probable vascular or anatomical unit which might in any sense serve as an anatomical lobule or structural unit, as described for various other glands by Born, Mall and others.

In Raia the occurrence of partially detached groups of thyroid follicles is more frequent than in the other species, but the number of such groups present in a gland varies from two or three to a score or more. The groups are outlined by connective tissue in which broad venous spaces to a certain extent encircle the quasi-lobule. The veins thus lie at the periphery while the artery on reaching the group promptly breaks up into a plexus of broad capillary spaces—venae lymphaticae—which surround the follicles within the quasi-lobule. The number of follicles in the group varies from four or five to several score.

In Carcharias the condition is similar to that in Mustelus, there being no indication of lobular groups except about the occasional irregular deficiencies in the thyroid mass. Except for the anatomical disintegration of the gland in one fish there was similarly no indication of lobulation in Squalus, but as none of my specimens from this species were prepared as total mounts I can not speak with the same certainty as in the other species.

The form of the thyroid follicles is subject to considerable variation, but, in general, they may be said to be of ovoid shape, and, as pointed out for the mammalian thyroid by Streiff ('97), they present frequent diverticula. The Elasmobranch thyroid differs from those described by Streiff in that they show very little tendency to branch and no indication of a tubular character when the whole follicles are examined in total mounts of the gland (fig. 13). In cut sections diverticula are of frequent occurrence and are apparently the result of pronounced infoldings of the follicular wall rather than of any protuberance, or of any tendency of the follicle to branch. Fig. 13 shows characteristic follicles from all four species; the figures are of whole follicles and differ from the cut sections in that only the largest infoldings of their wall are visible. As already indicated, diverticula are more apparent in sections than in the preparations (total mounts) from which the drawings have been made. The particular follicles drawn from Carcharias present rather greater infoldings than those from the other species. I have not, however, observed that this is characteristic of Carcharias. In the figure the magnification is the same for the several Selachian species but less by one half for Raia; the follicles of Raia are, therefore, relatively about twice as large as shown.

The size of the follicles is subject to considerable variation as regards the individual follicles, the different thyroids, and the

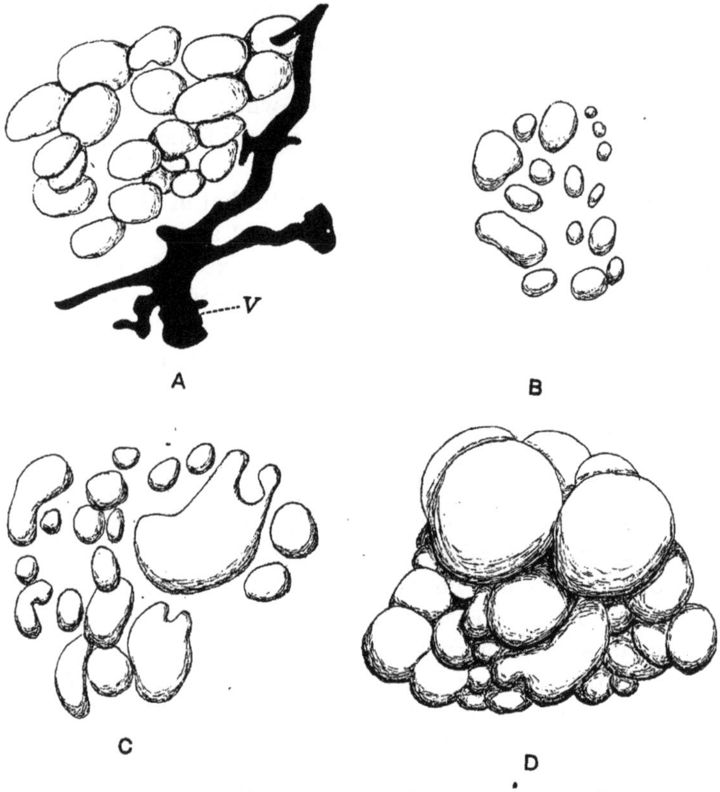

Fig. 13. Outline of the follicles of thyroid glands as seen in total mounts. *A*, from Mustelus canis, × 152. *B*, from Squalus acanthias, × 152. *C*, from Carcharias litoralis, × 152. *D*, from Raia erinacea, × 80.

various species. I have tabulated the results of some of the measurements.

SPECIES	MAXIMUM DIAMETER OF FOLLICLE	MINIMUM DIAMETER OF FOLLICLE	AVERAGE DIAMETER OF FOLLICLE	AVERAGE LENGTH OF FISH	RATIO FOLLICLE TO FISH	AVERAGE DIAMETER THYROID GLAND
Mustelus.........	.160mm.	.017mm.	.067mm.	67.8cm.	101.7	13.6mm.
Squalus..........	.079	.013	.047	53.7	115.	no data
Carcharias......	.273	.023	.100	117.2	117.2	13.7
Raia.............	.340	.053	.167	46.2	27.7	6.5

The very large relative size of the follicles of Raia is at once apparent. They are approximately four times as large, relatively to the length of the fish, as in the case of any other species. When compared with the diameter of the gland the ratio is again increased, but this difference is in part compensated for by the increased thickness of the gland in Raia as compared with the other species. The thyroid gland of Raia is 1.5 times as thick as that of Carcharias, and 2 to 2.5 times the thickness of the gland in Mustelus.

The column of ratios in the above table would indicate that in the Selachians the size of the follicle is in approximate proportion to the size of the fish, but that in Raia the relative size of the follicle is many times as great; the actual number of follicles in the thyroid gland of Raia is only a small fraction of those in the gland of any of the other species. It is readily susceptible of mathematical proof that the combined circumference of large follicles contained in a given area is less than the combined circumference of smaller follicles in the same area; hence the gland of Raia with its larger follicles will contain proportionately less epithelium than the glands of the other species. I estimate that the difference is just about sufficient to render the volume of secretory epithelium in the gland of Raia relative to the size of the fish equal to the volume of secretory epithelium in each of the other species.

But it is equally susceptible of mathematical proof that the cubical contents of the combined follicles is greater in the gland having the larger follicles; hence there is in Raia a greater volume of intrafollicular space than in the other species. It is scarcely susceptible of proof but entirely reasonable to suppose that the epithelium of different individuals of the same or different species so closely allied and the Batoidei and the Selachii is approximately equally active as regards its secretory function. There is ample evidence that the fluid secreted by the thyroid epithelium into the cavity of the follicle finds its way through the wall of the follicle to the neighboring vascular spaces so that the direction of flow must, in part at least, be from the epithelium into the follicle and thence through the follicular wall to the vessels. In view of these facts the rate of this secretory flow in Raia, with its relatively large intrafollicular space must be slower than in the other species with their relatively small intrafollicular cavities, or, to express it differently, there is relative stagnation in intrafollicular secretory flow in the case of the thyroid follicles of Raia. It is well known that an albuminous secretion which is rendered relatively stagnant within the epithelial cavities of the body tends to produce colloid masses whose microchemical reactions more or less closely resemble those of the colloid material of the thyroid gland; this occurs, *e.g.*, in the ducts and tubules of the resting mammary gland and in dilated cystic tubules in the kidney. We would therefore expect that in the thyroid follicles of Raia with their relatively stagnant secretory flow we should find an increased amount of colloid material. This I find to be the case, the proportionate volume of colloid present in the follicles of Raia being decidedly greater than in the other species. Similarly I find it the rule that the larger follicles contain relatively more colloid than the small follicles in the same gland. The volume of colloid contained in the thyroid follicles, therefore, can not be regarded as an index of the activity of the secretory epithelium; it would rather appear as a sort of by-product whose volume was dependent upon the rate of flow in the fluid from which it was formed. This view harmonizes the appearance of colloid material in the thyroid gland with the occurrence of similar material in the other glandular portions of the body, and with those theories of thyroid secretion which regard the colloid as a by-product rather than as the secretion. Moreover the great variations in the amount of colloid in the thyroid follicles are then explicable upon the basis of variations in the rate of secretory flow which, in turn, is dependent upon the physiological factors of blood and nerve supply as well as upon the anatomical factors.

It is also interesting to observe that the volume of secretory epithelium in the several species examined remains in each case approximately proportionate to the size of the thyroid gland and to the size of the fish. The relation of the epithelium to the follicular content and to the blood vessels and lymphatics seems to me to indicate most clearly that the secretion is poured out from the epithelial cells so as to find its way, on the one hand directly into the vessels, and on the other hand into the follicular cavity, whence it eventually passes through the follicular wall to reach the vessel. It is in the course of the latter flow that the colloid appears and its volume is dependent upon the rate of flow.

The question arises as to whether or not the colloid may serve for the storage of secreted materials, somewhat after the manner in which the hepatic glycogen may be considered as stored carbohydrate to be delivered as the needs of the economy necessitates; if so the colloid material should show further evidences of change, at least, under certain conditions. That the colloid does undergo changes is evidenced by the appearance within its otherwise homogeneous mass of such structural alterations as vacuolation, basophile degeneration, and disintegration into granules of greater or less size, changes which are frequently observed (figs. 14, 15 and 18). As to the physiological nature of these changes in colloid, and their possible connection with a storage function I can offer no conclusive proof, but it seems to me quite possible that such a relation exists.

The thyroid follicles are lined by a simple columnar type of epithelium (fig. 11) whose cells show considerable variation in height. In the same gland the epithelium lining certain vesicles measured as much as .010 mm., others only .006 mm. The epithelium of occasional vesicles was even lower, but was possibly open to the criticism of mechanical distortion since the colloid was often crowded against one side instead of lying in the middle of the follicular lumen, even though the tissues had been prepared with the greatest care. Being anxious to avoid any possible distortion of the tissue, I removed nearly all of the glands studied without allowing them to be touched by either instrument or fingers, the knife or scissors being passed through the muscle beneath, and the gland, supported on a thin layer of muscle, dropped bodily into the killing fluid.

As a rule those follicles which were well filled with colloid possessed low epithelium, in those with taller epithelium the reverse was the case. In making this comparison the surface area of the sections of colloid mass was compared with that of the containing follicle. The average height of the epithelium of a number of follicles in Mustelus which were well filled with colloid was .077 mm., while in a similar number of follicles which were either devoid of colloid, or nearly so, the height of the epithelium averaged .087 mm.

Each epithelial cell possesses a fairly distinct cell-wall. Many cells appear to have a well marked cuticular border which appears to be more highly refractive than the endoplasm, but wherever the colloid lies in contact with the surface of the cell the cuticular border is obscured. I have also noticed that it is less pronounced in the thinner sections so that I am inclined to regard it as an optical diffraction line rather than a true cuticular membrane. The conformation of the free ends of the epithelial cells tends to confirm this opinion. These cells project slightly into the lumen of the follicle by means of a somewhat convex free border so that the height of a cell is greater in its axis than at its margin. Thus the lower portions of a cell will, in the thicker sections (.010 mm. or more), show through the taller central or axial portions and so account, at least, for a portion of the cuticular appearance.

The exoplasmic membrane is specially distinct in the epithelium of the Elasmobranch thyroid gland. Baber ('81) called attention to this fact, and described it as an intercellular network enveloping the cells and connecting the lumen of the follicle with the surrounding tissue spaces. If the lining epithelium of the follicle be cut parallel to the surface the resulting sections will show the membrane as a distinct mosaic within whose meshes the cells are apparently contained. It appears to me that this mosaic, which is distinct from the intercellular colloid observed by Langendorf (see page 200), is rather to be regarded as a cell membrane than as an intercellular substance, for there are many portions where in thin sections a narrow intercellular space is distinctly apparent and is bounded on either side by the exoplasmic membrane of adjacent cells. Occasionally the cells are separated by wider intervals through which the follicular lumen is placed in direct communication with the surrounding tissue spaces.

There appears to be no distinct basement membrane upon which the follicular epithelium may rest. At intervals the cells are invested with a very small amount of loose connective tissue (fig. 11), but in large part the epithelium rests directly upon the walls of the venous channels and lymphatic vessels (fig. 14). Thus the relation of the epithelium to the vascular lumen is a very intimate one.

The cytoplasm of the "chief" cells is relatively clear, but contains a coarsely granular eosinophile reticulum. Some cells appear much more granular than others. In such cells as are filled with colloid, "colloid cells," the granular reticulum is entirely obscured (fig. 14, A).

The nuclei of the chief cells are spheroidal, vesicular, and are placed near the base of the cell. From the apices of many of these cells threads of secretion extend to the central colloid mass. The apices of many of the chief cells appear ragged, frayed, and often shrunken, so that the height of the cell is decreased. Such cells present an appearance suggestive of an advanced stage of secretion. Other cells contain granules at the distal ends which are arranged in vertical rows, giving this portion of the cell a somewhat rodded appearance; such cells are usually well filled with granules. Occasionally a similarly ragged and rodded appearance is seen at the base of the cell and it suggests that secretion may also be discharged at that point. Such a possibility is rendered more probable by the absence of basement membrane and the intimate relation to the lymphatics and blood vessels, these cells often resting directly upon the vascular endothelium. Laterally the epithelial cells frequently are separated from one another, leaving considerable spaces or channels through which secretion may find its way from the follicular lumen to the neighboring vessels; such channels are often occupied in part by colloid and in a few cases I have traced the colloid in a continuous line from the intrafollicular mass to the interior of the vasa and venae lymphaticae (fig. 10).

The above observations suggest that secretion may either be discharged from the chief cells into the lumen of the follicle and thence find its way through the follicular wall to the blood and lymphatic vessels, or that it may be discharged from the cells directly into the vessels; this is in harmony with the conditions indicated in the thyroid gland of mammals.

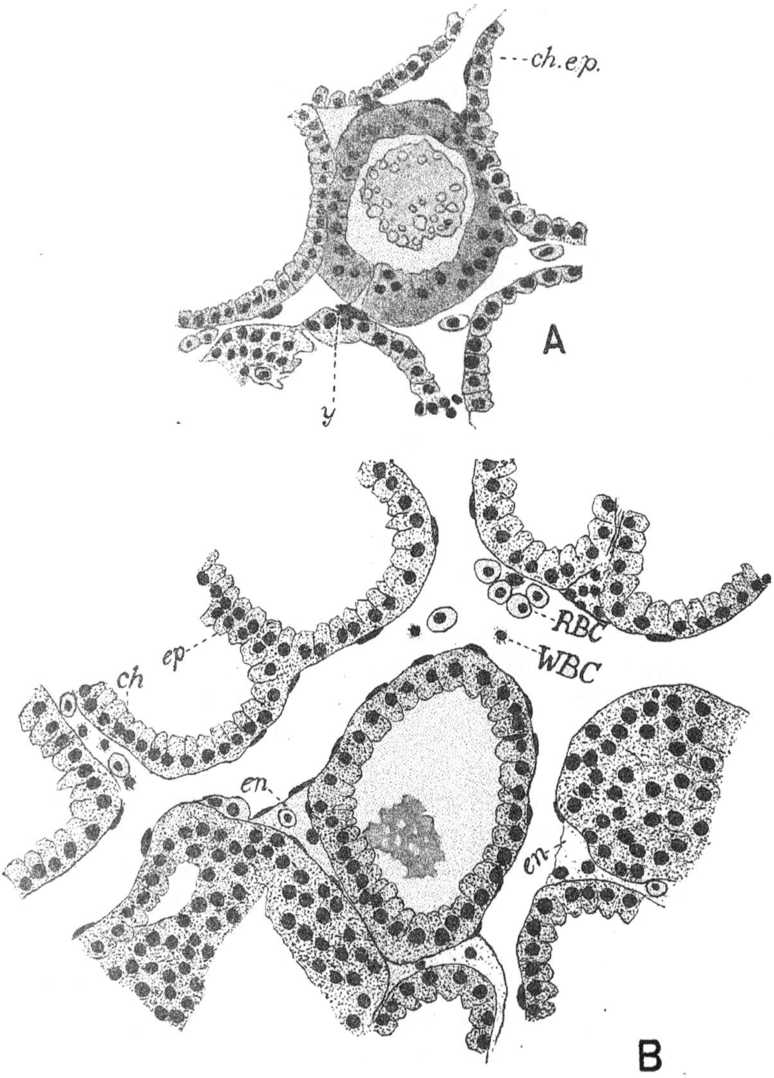

Fig. 14. *A*, section of the thyroid gland of Mustelus showing a follicle completely surrounded by "colloid cells." At *y* a carmine granule, derived from the injection mass, lies in the vena lymphatica. *B*, section of the thyroid gland of Mustelus showing follicles lined entirely by "chief cells."

The colloid cells described by Langendorf ('89) are remarkably distinct in most of the sections of Elasmobranch thyroids and constitute one of the most characteristic features of the thyroid gland of these fishes. The colloid cells are distinctly acidophile and are easily recognized in specimens stained with hematoxylin and eosin if the eosin is used in dilute solution and allowed to act for one-half hour or more. They present a glistening, highly refractive colloid appearance, which is in marked contrast to the granular chief cells. The colloid cells occasionally occur singly, but are more frequently disposed in groups along one side of the follicle. One such group (fig. 14, A), more extensive than the others, was seen to include fully three-fourths of all the epithelium in its follicle. The groups are often in contact with the central colloid mass, and the colloid within the cell may then appear continuous with that within the follicle. Occasionally a group of such epithelium appears to have been completely engulfed by the colloid mass, the epithelial nuclei then appearing well within the colloid. Such appearances might suggest mechanical distortion, but as the surrounding follicles show no evidences of injury, and, as already stated, the tissues were very carefully prepared, I am more inclined to agree with Bozzi ('95) that these appearances are the result of vital phenomena.

The nuclei of the colloid cells are small and deeply stained, so deeply, in fact, that in the usual preparations they frequently show neither nuclear wall nor karyosomes. Unlike the nuclei of the chief cells they are usually situated near the inner extremity of the cell rather than at its base. The greater the cell is distended with colloid the farther its nucleus is pushed toward the cell's apex; in the most distended cells there was frequently some distortion and even fragmentation of the nucleus. A further continuation of this process would account for at least a portion of the extruded and disintegrating nuclei found within the intrafollicular colloid masses (figs. 11 and 18).

The intrafollicular colloid closely resembles that of the mammalian thyroid gland. It is strongly acidophile and is usually homogeneous or very finely granular in appearance. Frequently a minor portion of the mass, *e.g.*, one side, is finely granular while the major portion is clearly homogeneous. The well-known filaments pass at frequent intervals from the colloid mass to the epithelial surface. Some of these filaments can be traced to the free surface of the epithelial cell while others quite clearly enter the intercellular spaces, where, in tangential sections of the follicle, they form intercellular masses simulating the net-work described by Baber ('81) and interpreted by Langendorf ('89) as the ramification of colloid cells.

Occasionally the colloid mass appears to have been disintegrated into small spherules .007 to .008 mm. in diameter (fig. 15). The size of these spheres is suggestive of the red blood cells of mammals, but the red cells of Elasmobranchs are ovoid and larger. Of the spherules some are distinctly acidophile but many are slightly basophile, none, or very few, are strongly basophile. All the spherules are homogeneous, and I have observed in the more basophile no tendency to chromatolysis such as one might expect to find if the spherules of this type were thought to represent degenerating nuclei of the red cells, nor have I been able to trace stages of transition from the nucleus to the basophile spherule. Since all the blood cells of the species studied are nucleated one could not well infer that the acidophile spherules could represent any stage in the disintegration of red blood cells, for none of these spherules contain even traces of chromatin. On the other hand, both red and white blood cells can occasionally be found within the colloid quite independently of the spherules I have described; in this particular the Elasmobranch thyroid is in accord with the well known structure in other vertebrate orders. The appearance, location, disposition and reactions of the spherules indicate their origin from the solid colloid masses, from which they would appear to be formed by disintegration with progressively increasing basic reaction. That the reverse process occurs, viz., that the spherules may represent intermediate stages in the formation of the colloid masses, is contraindicated by the fact that only very few follicles contain spherules, nor does there appear to be any indication of a tendency of the spherule to fuse. On the other hand, a tendency to further disintegration is quite apparent, and the possibility is suggested that the colloid in this way may be trans-

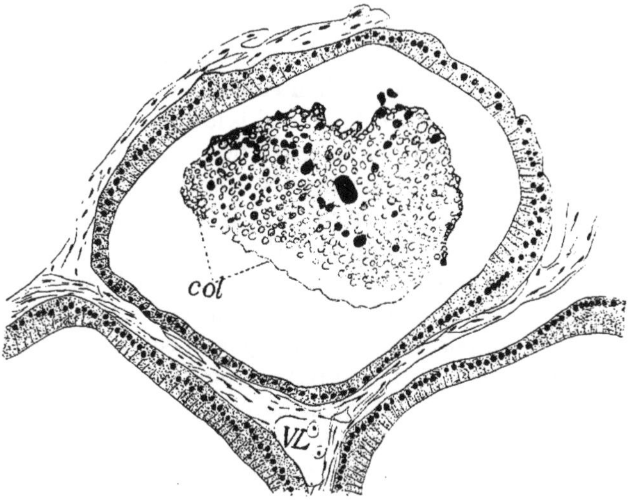

Fig. 15. Section of the thyroid gland of Raia showing the colloid within a follicle disintegrated into spheroidal masses of varying size and depth of stain.

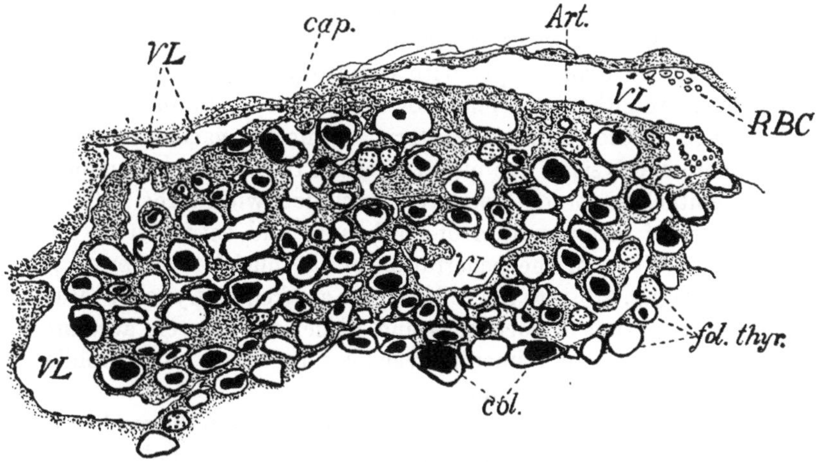

Fig. 16. Section through the ventral margin of the thyroid gland of Mustelus with sections of the very broad venae lymphaticae of the thyroid sinus simulating an endothelial capsule about the gland.

formed to such a state that it may be secreted through the wall of the follicle, in which case the intrafollicular colloid would presumably assume the nature of stored secretion which is first poured out from the cells as a fluid, is then condensed through retention within the follicle, to form colloid, and is later disintegrated, passing out of the follicle with the secretory flow. The chemical analyses of the thyroid gland, showing the common relation of iodin to the colloid and to the active principle of thyroid secretion would harmonize with such an hypothesis.

Vacuolation of the colloid mass (fig. 14) is of frequent occurrence; it may result either from the inclusion within the intrafollicular colloid of portions of the peripheral cup-like impressions which Langendorf has surmised result from the secretion pouring out from the surface of the epithelial cells, or it may be further evidence of disintegration of the colloid with formation within its substance of fluid droplets, rather than of solid spherules. The vacuoles are filled with a clear fluid and occasionally contain basophile granules. A colloid mass may contain many small vacuoles mostly at or near the periphery of the mass and containing few, if any, basophile granules, or it may contain one or more vacuoles of relatively large size which occupy the interior of the mass and may be more or less completely filled with the granules. These granules stain deeply with hematoxylin and similar dyes, and they are either amorphous or somewhat crystalline in form. The origin of the chromatic material within the vacuoles may be from chromatolysis of the nuclei of either the disintegrating follicular epithelium or of blood cells included within the colloid. Evidences of disintegration of epithelium and extrusion of the nuclei as well as of the penetration of the nuclei together with blood cells (fig. 20) into the colloid mass are frequently seen. But the disintegration of such cells and nuclei can scarcely account for the much more numerous vacuoles in which no basophile chromatic substance is found.

The follicles of the thyroid are supported within the meshes of a connective tissue stroma in which the blood-vessels lie. The volume of connective tissue is never great, much less than in the mammalian gland. There is more connective tissue in the gland of Raia than in the other species examined; in Mustelus and Squalus there is so little that one wonders at the relative compactness of the organ. In these Selachii the epithelium rests directly upon the walls of the blood-vessels, and they in turn consist of little else than endothelium and form broad sinuses rather than capillaries or venules.

The gland is inclosed by a very thin connective tissue capsule and its tissue is thus always sharply defined from the surrounding structures. In Mustelus and Squalus, and to some extent in the other species, the broad vascular channels of the thyroid sinus are in direct contact with the capsule, so that in sections the ventral surface of the gland often appears clothed with an endothelial coat derived from these vessels (figs. 16 and 17); a similar disposition of the vascular endothelium of collapsed blood-vessels is also occasionally seen on the margins and dorsal surface of the gland.

The blood vessels have been in each case carefully studied by dissection, injection, sections, and transparent total mounts of the gland. Both blood vessels and lymphatics were demonstrated beyond doubt, though lymphatics have not hitherto been observed in these fishes and their existence was denied by Baber ('81). Fig. 9 shows the lymphatics filled with injection mass, lying between the blood channels and the follicular epithelium; they appear as perivascular lymphatics in the wall of the venae lymphaticae. Similar vessels, perivascular lymphatics, are found in the walls of the arteries and veins of the thyroid, the thyroid sinus, and the arteries of the hypo-branchial system (fig. 8).

The course of the larger blood-vessels was readily followed in injected specimens in which the whole gland was examined under the microscope. The arteries course upon the surface of the gland, the major portion of them being always on the ventral surface. Fig. 19 shows the distribution of the arteries in the thyroid of Mustelus, and figs. 3 and 4 indicate the relative area of the gland supplied by the arteries of the right and left side, the left thyroid artery, as in figs. 3 and 19 usually supplying the greater part of the organ, though occasionally the major part, as in fig. 4, is supplied from the right side. Twigs from the superficial branches here and there penetrate the gland, break into arterioles, and

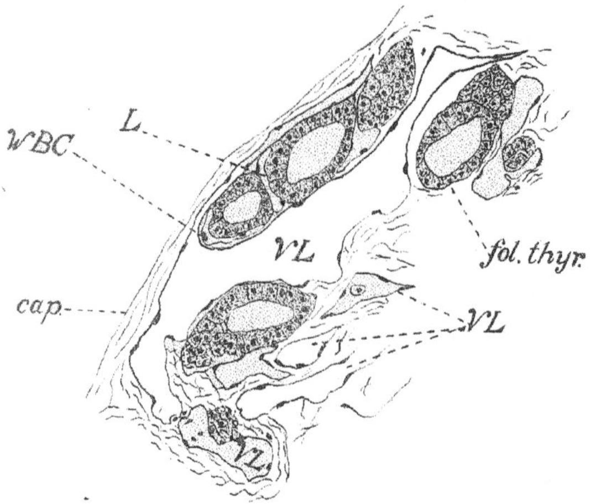

Fig. 17. Section through the ventral margin of the thyroid gland of Squalus showing peripheral venae and vasa lymphatica.

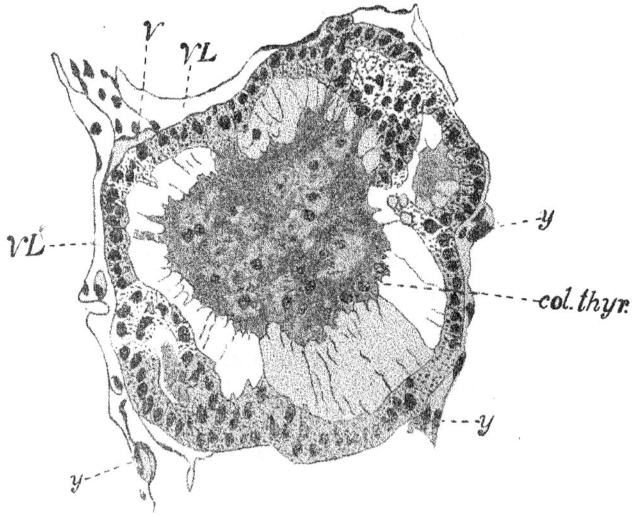

Fig. 18. Section of the thyroid gland of Carcharias showing the disintegrating nuclei of leucocytes or red blood cells within the intrafollicular colloid.

promptly empty into groups of broad interfollicular, blood capillaries, the venae lymphaticae (figs. 9, 11, 12), which envelop the follicles on all sides. From the venae lymphaticae the veins pass out of the gland at its posterior and lateral borders and dorsal surface to enter the thyroid sinus; a few veins from the lateral border of the gland pass directly to the hyoid sinus.

The course of the lymphatics was much less easily determined than that of the blood vessels. "Stick injections," as ordinarily made, spread so rapidly through the loose connective tissue of the gland and so easily entered and filled the venae lymphaticae that they entirely obscured the smaller vasa lymphatica. The venae lymphaticae thus injected form a dense almost opaque mass, showing that the thyroid may well occupy the place in these fishes assigned to it by Tscheuwsky ('03) as the most vascular of mammalian glands. After several futile attempts to inject the lymphatics in the ordinary way the method was so altered as to inject only minute areas under a very low pressure. In this way it was found that at the margins of the injected area the fluid which had entered the vessels traveled farther than that in the connective tissue spaces, and in many cases the vasa lymphatica were filled beyond the limits of the injected venae lymphaticae, so that in the outermost zone of the injected area the true lymphatics could be readily studied, the venae lymphaticae in this zone being either empty or only partially filled.

The vasa lymphatica are, for the most part, perivascular channels (fig. 9), but they are also in direct contact with the epithelial walls of the follicles (fig. 10). The vasa lymphatica could not be followed for any great distance through the injected zone, for, on the one hand, they entered the area of opaque injection mass, and in the other direction they ended abruptly, often with a small knob-like dilatation. By means of serial sections I was able to determine in uninjected specimens that the vasa lymphatica opened at the points of terminal dilatation directly into the venae lymphaticae (fig. 10, *X*). Having demonstrated the connection between the two sets of vessels in uninjected specimens, showing the true relation of vasa and venae lymphaticae, many points in the injected specimens could be readily found at which it seemed quite certain that the injection mass was

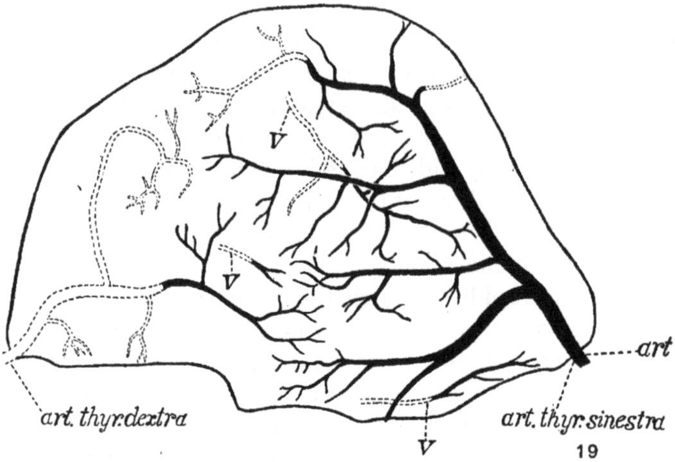

19

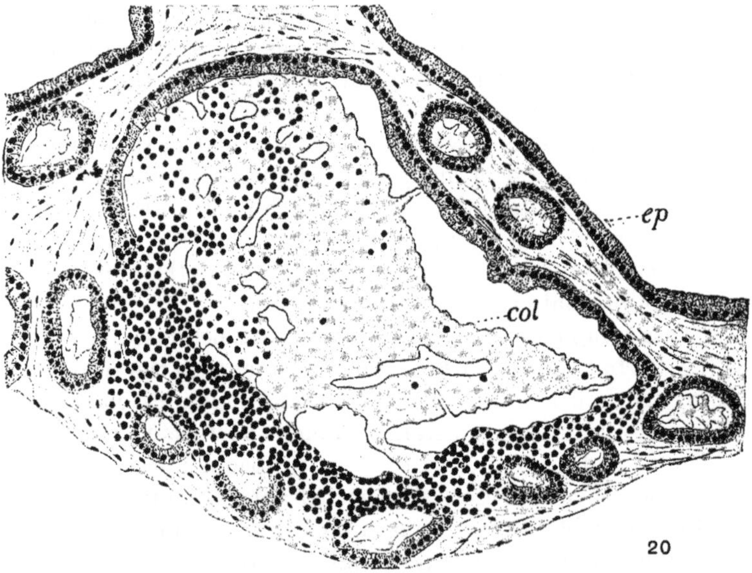

20

Fig. 19. Drawn from a total mount of the thyroid gland of Mustelus canis, showing the course and distribution of the thyroid arteries and the origin of some of the veins of exit. The vessels on or near the ventral surface are indicated by the solid black lines, those on or near the dorsal surface of the gland by dotted lines.

Fig. 20. A section from the thyroid gland of Carcharias, showing invasion of epithelium and colloid by leucocytes.

passing from the vasa lymphatica directly into the venae. The relatively intimate relation between the veins and lymphatics in fishes is well known; Wiedersheim ('07) and Favaro ('06) have recently emphasized the fact so far as the tail vessels of fishes were concerned. This intimate relationship seems to be quite as obvious in the thyroid vessels and in those of the region occupied by the thyroid sinus (*vide supra*).

That the vasa lymphatica are true lymphatics and not blood-vessels is shown by the fact that in many cases they are of altogether too small caliber to transmit the large red blood-cells of the Elasmobranch fishes. Moreover, the quasi valvular nature of their anastomosis with the veins, as already described, renders highly improbable the regurgitation of blood-cells into the vasa lymphatica even in the larger vessels.

In mammalian thyroid glands, especially in dogs, one now and then observes instances where the colloid has accumulated beyond the bounds of the follicles, giving to sections of the organ the appearance of a tissue completely infiltrated by the waxy colloid substance. No such appearance was found in the Elasmobranch thyroids which were studied, unless the follicles lined chiefly by colloid cells could be so interpreted.

SUMMARY

1. The Elasmobranch thyroid gland closely simulates the human, both in the form and structure of its follicles and the distribution of its blood-vessels.
2. The gland rests upon the basi-hyal cartilage whose anterior margin forms an excellent guide to its location.
3. The pyramidal lobe of mammals is often represented in Elasmobranchs by a process passing forward and reaching the floor of the pharynx through a notch in the anterior margin of the basi-hyal cartilage; this notch is sometimes converted into a foramen.
4. Baber's opinion that lymphatics are not present in the thyroid of Elasmobranch fishes was founded on insufficient evidence and is incorrect.
5. Lymphatics are present in considerable numbers both in and

about the thyroid gland and can be demonstrated by injection and in sections; they are true "vasa lymphatica."

6. The blood-vessels within the thyroid gland terminate in a network of "venae lymphaticae" which invest the follicles, receive the vasa lymphatica, and transmit either or both blood and lymph, under varying conditions of blood-pressure.

7. The thyroid artery arises from the ventral end of the efferent hypobranchial arterial loop contained in the hyoidean hemibranch and the adjacent half of the first holobranch by an independent origin or by a common stem with the mandibular or submental artery.

8. The thyroid veins in these fishes for the most part enter the "thyroid sinus," a mass of veins and lymphatic vessels which pour their blood into the hyoid sinuses.

9. The rhythmic respiratory movements of the pharyngeal wall cause the thyroid sinus to act somewhat after the manner of a "vein-heart" or "lymph-heart."

10. The hypobranchial arterial system is formed as described by Hyrtl in 1858 and 1872, and the direction of the flow of its blood is from the gill vessels toward the coronary and other terminal arteries, as indicated by Hyrtl and again by Parker and Davis in 1899, and not from the subclavian artery toward the coronaries, as described by T. J. Parker in 1886. The hypobranchial artery of T. J. Parker, forming an anastomosis between the subclavian and median hypobranchial arteries, is of insignificant importance and is frequently wanting.

11. The relative volume and distribution of the "colloid" in the glands of different species indicates that this substance is a retention product, formed from the albuminous secretion of the follicular epithelium.

12. The further changes occurring in the "colloid" indicate the possibility of its usefulness as a sort of stored-up secretion.

13. The follicular epithelium contains both "chief" and "colloid" cells, the latter being even more numerous and characteristic than in the mammalian thyroid.

14. The parenchymal epithelium is in intimate relation with the vasa and venae lymphaticae; it rests directly upon or in close proximity to the endothelial wall of these vessels.

EXPLANATION OF FIGURES

Abbreviations

I—V, first to fifth branchial clefts
A. lat., lateral artery
A. thry., thyroid artery
A. thyr. imp., arteria thyreoidea impar
Art., artery
b. hy., basi-hyal cartilage
cap., fibrous capsule of the thyroid gland
ch. ep., chief cells of the follicular epithelium
col., colloid
com., commissural artery
cor., coronary arteries
coraco-mdb., coraco-mandibular artery
coracoid, coracoid artery
D A., dorsal aorta
en., vascular endothelium
ep., epithelium of the thyroid follicles
epibr., epibranchial artery
fol. thyr., follicles of the thyroid gland
gastric, gastric arteries
hy. sn., hyoid sinus
hypobr., hypobranchial artery
hypobr.', its anastomosis with the commissural artery.
L., vasa lymphatica
l. c., lateral commissural arteries
Lat. hypobr., lateral hypobranchial artery
M. c. br., coraco-branchial muscle
M. c. hy., coraco-hyoideus muscle
M. c. mdb., coraco-mandibularis muscle
Med. hypobr., median hypobranchial artery
p. c., pericardial arteries
pharyn., pharyngeal arteries
R B C, red blood cells.
s. a. v., ventral arterial sinus of the first hypobranchial loop.
sbcl., subclavian artery
thyr., thyroid gland
thyr. sn., thyroid sinus
V. vein
V A, ventral aorta
V L, venae lymphaticae
W B C, white blood cells
x., point of anastomosis of vasa and venae lymphaticae
y., injected carmine granules in the venae lymphaticae

BIBLIOGRAPHY

BABER, E. C. 1881 Phil. Trans., Roy. Soc. London, 172, 577.

BALFOUR, F. M. 1881 Treatise on Comp. Anat., London 2, 626.

BRIDGE, T. W. 1904 Cambridge Nat. Hist., London, 7.

COLE, F. J. 1905 Anat. Anz., 27, 323.

DE MEURON, P. 1886 Rec. zool. suisse, 3, 517.

DOHRN, A. 1884 Mitth. a. d. zool. Station z. Neapel, 5, 102.

ECKER U. WIEDERSHEIM. 1904 Anat. des Frosches, Braunschweig, 205.

FAVARO, 1906 Quoted by Eisler, Schwalbe's Jahresb., 12, 323.

GUIARD, J. 1896 Thèse Paris.

HYRTL, J. 1858 Denks. d. k. Akad. Wissen., Wien, Math. Nat. Cl., 15, 1.

1872 Denks. d. k. Akad. Wissen., Wien, Math. Nat. Cl., 32, 263.

LAGENDORF, O. 1889 Arch. f. Physiol., Suppl. Bd., 219.

LOMBARD, G. D. 1909 Biol. Bull., 18, 39.

MAJOR, R. H. 1909 Am. J. Anat., 9, 475.

MAURER, F. 1886 Morph. Jahrb., 11, 129.

1888 Morph. Jahrb., 13, 296.

MÜLLER, W. 1871 Jena. Zeitschr., 6, 428.

PARKER, G. H. AND DAVIS, F. K. 1899 Proc. Bost. Soc. Nat. Hist., 29, 163.

PARKER, T. J. 1880 Trans. and Proc. New Zealand Institute, 13, 413.

1886 Phil. Trans., Roy. Soc. London, 177, 685.

PEREMISCHKO 1867 Zeitschr. f. wis. Zool., 17, 279.

SABIN, F. R. 1909 Am. J. Anat., 9, 43.

SCHAFFER 1906 Anat. Anz., 28, 65.

SIMON 1844 Phil. Trans., Roy. Soc. London, pt. 1, 295.

STOCKARD, C. R. 1906 Anat. Anz., 29, 91.

TSCHEUWSKY 1903 Arch. f. d. ges. Physiol., 97, 210.

TURNER 1874 J. Anat. and Physiol., 8, 285.

VAN BEMMELEN, J. F. 1887 Zool. Anz., 10, 88.

WAGNER, R. 1853 Handwörterbuch d. Physiol., Braunschweig, 4, 111.

WIEDERSHEIM, R. 1907 Comp. Anat. of the Vertebrates, p. 432.

Reprinted from JOURNAL OF MORPHOLOGY, VOL. 21, No. 4, Supplement, February, 1911

THE THYREOID GLAND OF THE TELEOSTS

J. F. GUDERNATSCH

From the Department of Embryology and Experimental Morphology, Cornell University Medical School, New York City

TWENTY-ONE TEXT FIGURES AND FIVE PLATES

During the summer of 1909 at the suggestion of Dr. C. R. Stockard[1] I undertook the study of the distribution of thyreoid tissue within the gill region of Teleosts, especially of the trout. It seemed important to clear up certain doubtful facts in this connection, since this organ of the trout is liable to disease and at present is attracting considerable attention in cancer research. The examination of but a small number of species brought such an abundance of interesting material to light that I determined to carry out a comparative study of the anatomy and histology of the thyreoid gland in a large number of Teleosts, and to summarize our entire knowledge of the organ in this group of vertebrates. At the same time I attempted, by comparing the present results with the facts known of the thyreoid in other classes, to define more clearly the features of this organ in the entire vertebrate group.

It gives me pleasure to express my best thanks to The Wistar Institute of Anatomy and to Prof. F. R. Lillie for the use of a room in the Marine Biological Laboratory at Woods Hole, Mass. Some of the species were obtained from the New York Aquarium, for which I wish to thank the Director, Mr. C. H. Townsend.

Twenty families of Teleosts including twenty-nine species were investigated, the detailed description of which I give in the special

[1] My thanks are due to Professor Stockard for many suggestions during the work and for carefully revising this paper.

part of this paper. The results demonstrate the possibility of unusually wide variation in the thyreoid gland and establish a continuous series of transitions from one form to the other.

The general part deals with the anatomy, histology and embryology of the organ, and is based on the facts gained from a systematic comparison of all the species.

GENERAL PART

ANATOMY OF THE GLAND

The literature relating to the structure and the location of the thyreoid gland in Teleosts and fishes in general is comparatively meagre. Long after the presence of this interesting organ had been known in the higher vertebrates it was also found in fishes. Simon was the first to demonstrate the existence of a thyreoid gland in this lowest class of vertebrates, but as he supported his statements only by macroscopic examination, it is not certain whether everything he regarded as thyreoid really belongs to this organ.

The peculiar anatomy of the gland renders it at times very difficult, if not entirely impossible, to diagnose tissues in the gill region as thyreoid without a microscopic examination. I have several times, especially when particles of the thyreoid were diffusely scattered, regarded small masses of tissue as thyreoid which under the microscope did not prove to be such. Simon's reports on the different locations of the gland are, therefore, not to be relied upon and it is for the above reason especially that his discoveries in the Gadidae, Cyprinus, Anableps, Esox and Exocoetus are to be very strongly doubted. In these species he locates the thyreoid gland far *dorsally* in the region of the soft palate. Maurer has since demonstrated the true position of the thyreoid in the carp at least, and although Simon's statements have not been contradicted for the other species, they seem to be, from the phylogenetic point of view, entirely untenable. In some other species, however, Merlangus, Anguilla etc., Simon appears to have found the organ in his dissections, as he places it near the bifurcation of the first gill arteries. It seems strange, on the other hand, that he was unable to locate the thyreoid gland in some other fish, among them those in which it is easily visible to the naked eye. Thus he failed to see it in Perca, Mugil, Trigla, Scomber, Tinca, Salmo, (Salmo fario), Clupea, Pleuronectes, Hippoglossus, Rhombus, Solea, Cyclopterus, Gymnotus and Balistes. In all these species, as far as they were at my disposal, I could demonstrate the thyreoid gland as a well developed organ, and feel certain that it is also present in the others. There seems to be little, if any doubt that the thyreoid gland is an organ belonging without exception to all fishes and vertebrates in general.

Of the later investigations describing the thyreoid gland in Teleosts, Baber's work (1881) should be mentioned, although he studied the gland of only one species, the conger eel. A later paper by Maurer (1886) deals principally with the embryology of this organ in the trout, but also contains some remarks on its comparative anatomy.

The thyreoid gland of the vertebrates, as is well known, is closely connected in its development with the median ventral wall of the pharynx and the gill arches. These relationships are only slightly different whether the gills persist throughout life as in the fishes, or are found only in larval stages as in the Amphibia, or finally transform in later embryonic life as is the case in the Amniota. The location of the thyreoid in the animal body is therefore dependent upon the gill region. While in the higher vertebrates the topographical arrangements become more or less changed or rendered indistinct during the progress of development we find them in the fishes typically marked.

The region of distribution of the thyreoid gland in the fish extends below the floor of the pharynx, in the body of the tongue, between the gill arches and back posteriorly behind the origin of the third and fourth branchial arteries from the ventral aorta. Maurer limits the region of distribution anteriorly by the bifurcation of the aorta and posteriorly by the last branchial aortic arch, but these lines are certainly too limited, especially in the anterior direction. In some species the main body lies in front of the aortic bifurcation. The posterior limit, however, only exceptionally occurs behind the last branchial arch. The single parts of the basibranchiale define this 'thyreoid region' dorsally, while ventrally the paired musculus sternohyoideus is spread out beneath the organ. This region is at the same time that of the ventral aorta and its branches to each of the gills.

Thus the thyreoid gland is located along the trunk through which the blood for the entire body is pumped from the heart into the respiratory organs. The narrow cleft between the bony parts of the floor of the pharynx dorsally and the muscles ventrally is completely filled with thyreoid tissue except for the space occupied by the large arterial trunks. This region, as we see from the extensive literature on the visceral skeleton and musculature in fish, in the manifold development of its bony (cartilaginous) and muscular parts shows a decided tendency to vary. It is only natural that this tendency should be found in the thyreoid gland also; since it has to accommodate itself to the configuration of the tissues just mentioned, a pronounced adaptability must be of the greatest benefit to it. The property of variation is possessed by the thyreoid gland of the Teleosts to a most striking degree and within the same species rather remarkable differences are found. Twelve weak-fish (Cynoscion) for instance, all differed in the extension and position of their respective thyreoids. Similar conditions were observed in other species. This variability within the species may indicate that in the thyreoid gland we have a very unstable organ, which perhaps in vertebrate phylogeny has not yet acquired its final condition. We know that in the higher classes of vertebrates there is the same variability among individuals regarding the development of their thyreoid glands. In mammalia the individual variation is very great. The lobes may have different forms, and give to the organ a paired appearance, or there may be a more or less well developed isthmus between them. Interesting comparisons have been made especially in man.[2] In the phylogenetically younger epithelial bodies the variation is still larger. All of these facts indicate that the gill slits and their

[2] Marshall ('95) examined the thyreoid glands in sixty children in which he found all possible variations.

derivates are still easily modifiable and do not yet represent a permanent condition.

The thyreoid gland of Teleosts is not a single compact organ, as we find it in the higher vertebrates, where the small parts of the gland, the follicles, are united into one complex and enclosed by a common capsule of connective tissue. Only in such a case would the term 'gland' be justified, since here numerous anatomical elements possessing the same physiological function are closely connected. The Teleosts, however, possess numerous elements, whose totality from a physiological standpoint one must regard as a thyreoid gland, while anatomically we are unable sharply to define the organ in question in this group of vertebrates. In most cases we can speak of thyreoid follicles only, or groups of follicles, in pointing out the distribution of the organ. Thus in plates I and II, not the thyreoid glands of the respective species, but the regions of distribution of the thyreoid follicles themselves are figured. On plate III, however, which indicates the variation in the location of the gland within the species, real parts of the thyreoid are indicated, so far as they were macroscopically visible.

In the more closely defined region we may find thyreoid tissue in all parts. It is usually of a brownish yellow color. The follicles are generally most densely located in the neighborhood of the ventral aorta or of the branchial arteries. Those places, particularly, are favored where the branchial arches arise from the aortic stem. Follicles are most abundant at the origin of the second gill arteries, that of the first being next, and finally the roots of the last branches have fewest thyreoid follicles about them. A more or less dense accumulation is always found along the stem of the ventral aorta, which may be completely surrounded by thyreoid tissue. In other cases dense accumulations of follicles are located either dorsally or ventrally to the aorta.

The glandular tissue is usually distributed so that it is more densely accumulated in the central region, while towards the periphery a more and more pronounced dissolution and scattering of the follicles takes place. These conditions cannot, however, be strictly generalized, as we find cases in which, even in the most central portions, the follicles are not more closely arranged than in the peripheral. In some instances, as in Cynoscion, rather well developed central portions are found which can be recognized by the naked eye. But even in these cases numerous follicles lie well separated from the main body. Thus we have all transitional stages from a perfect dispersion of the follicles to a rather compact union of them. It is possible that further investigations may show cases in which the organ is still more compact than in those thus far examined, and so present a structure similar to that in higher vertebrates. Judging from my observations, however, this does not seem probable, and at present I am inclined to regard the conditions found in Sarda (*vide* Special Part) as the limit of compactness in the series.

The cephalad and caudad extensions of the thyreoid gland vary very much. In general, it might be said that a spreading out toward the tip of the tongue takes place in all cases, while towards the heart the distribution is not so uniform. Far cephalad of the first aortic bifurcation we find single follicles scattered below the hyoid bones. The caudal limit of the thyreoid gland usually lies between the second and third aortic branches, or at the third. Rarely does it go beyond this point, and if so, with a few exceptions, only scattered follicles are found in the posterior region. Thus we find an accumulation of the glandular elements around the anterior part of the ventral aorta, with follicles scattered towards the head and the heart.

The organ decreases in mass in an anterior-posterior direction. In one instance, Siphostoma, just the reverse is the case.

The embryonic center from which the thyreoid starts to grow lies between the first and second gill branches, this place in the adult animal is near the aortic bifurcation, and in many cases we find the main part of the organ in this region; while in other cases, Cynoscion and Tautogolabrus, it is exclusively there. From the region of the aortic bifurcation the thyreoid elements travel so as to occupy the different positions which are described in the special part of this paper. The migration of thyreoid follicles occurs more particularly in the direction of the heart, although a cephalad migration is decidedly pronounced. The development of the hyoid bones obstructs the anterior spreading of the follicles to some extent. The follicles apparently tend to go around the basihyale and along its sides towards the tip of the tongue. In the dogfish and shark where the hyoid region offers a comparatively free space we find it occupied by the compact thyreoid, which is pushed slightly forward of the aortic bifurcation (Ferguson). In these animals also the thyreoid is originally placed in the bifurcation of the truncus; later, according to W. Müller, the anlage moves forward and becomes encapsuled. In Raja, on the other hand, it remains in the bifurcation.

The dorso-ventral, in combination with the lateral, extension of the thyreoid follicles seems to be more dependent upon the configuration of the pharyngeal floor than does the cephalo-caudad extension. In fishes, in which the isthmus region is deep and narrow, as in Brevoortia, we find, as might be expected, the dorso-ventral distribution of follicles far surpassing the lateral. While in other species, for instance, Tautoga, in which the floor of the pharynx is very broad, the lateral extension is the important one. In general it may be said that the lateral outweighs the dorso-ventral distribution of follicles.

Dorsally the follicles are usually found between the ventral aorta and the copulae of the gill arches. In cases where those skeletal parts come close to the vessel the follicles are forced away laterally, and sometimes intrude into the spaces between the copulae and hypobranchialia. When the parts of the basibranchiale lie well separated the follicles extend up between the copulae and come to lie close to the mucous membrane of the pharyngeal floor.

The main mass of the organ lies almost exclusively above, or dorsal, to the ventral aorta. This is opposed to Maurer's statement of the case. Below the aorta is usually found the smaller part of the gland and the follicles are also more loosely scattered. The development of the thyreoid gland above the aorta should be expected since there is usually much more open space between the aorta and the gill arches than is found below the aorta where the muscles lie close to the vessels. The thyreoid elements endeavor to intrude below the aorta as much as possible, and when this vessel, in the region of the third branchial arteries, sinks deeper into the musculus sternohyoideus, the follicles follow in the course and are thus distributed far ventrally within the muscle. The *number* of follicles along the sides of the aorta is always less than either above or below it. This by no means contradicts the statement made above that the lateral *extension* of follicles outweighs the dorso-ventral one. Only in the genus Fundulus, especially in majalis, less so in heteroclitus, where a transverse interbranchial muscle pushes the vessels away from the skeletal parts, is the dorsal extension small, or sometimes lacking entirely. Here the follicles extend directly away from the aorta towards the bases of the gills.

Along the aortic stem between the bases of the gill arteries the lateral extension is somewhat limited and reaches its height along the branchial arteries. The vessels seem to serve as bases along which the follicles migrate. The free space about the gill arteries becomes narrower and narrower as the gill arches are approached and therefore the number and size of the follicles decrease towards these points until there is no more room for extension. When, however, there exists an especially open passage along the vessels the follicles may even extend into the gill arches, to a considerable distance beyond the point of their origin. Such cases are common in trout (text fig. 7*D*).

The peculiar distribution of the thyreoid elements forces us to regard the organ in bony fishes as unpaired, a view also supported by embryology. This statement should be especially emphasized, as in Wiedersheim's Comparative Anatomy, 1907, the author still speaks of the thyreoid gland in the Teleosts as a paired organ, although Maurer in 1886 (p. 134) criticizes this statement in an earlier edition. On another page, (p. 140) Maurer himself claims that the main bulk of the organ at a certain stage is not paired, while later single portions of the gland lying in the median line, as well as on both sides of the truncus arteriosus, take a paired arrangement. This is certainly incorrect, since the follicle groups on the sides of the aorta are not only unpaired but are also not bilaterally arranged.

The relationship between the thyreoid gland and the stem of the ventral aorta is purely anatomical and without any physiological importance. The thyreoid does not receive its blood supply from this group of vessels, since they carry only venous blood, and the arterial blood which nourishes the gland comes from a special thyreoid artery. In Petromyzon, however, Cori claims that the arteria thyreoidea arises from the truncus arteriosus; this is probably an error, as he also finds the ventral carotid connected with the truncus. The thyreoid artery, as Silvester demonstrated in Lopholatilus and twenty other species of Teleosts by his perfected method of injection, arises as a dorsal branch from the united right and left fourth commissural arteries. The latter vessels originate from the second efferent branchial arteries and unite in the median line below the thyreoid and the aorta as the hypobranchial artery. Shortly after the union of the fourth commissural arteries the thyreoid artery branches off from the dorsal side and immediately enters the gland in its posterior region. Whether the widely scattered follicles all receive their capillaries from this one vessel cannot at present be stated, though it would seem very doubtful especially in the case of the more anteriorly isolated follicles.

In Selachians, where the thyreoid is pushed far forward, the arteria thyreo-spiracularis (Dohrn) originates in the first aortic arch from the arteriae efferentes of the hyoid gill. In Teleosts, also, the first aortic arch breaks up into a capillary network. Dohrn, therefore, speaks of an arteria thyreo-spiracularis. Perhaps it is from this vessel that the most cephalad parts of the thyreoid gland receive their blood supply. The artery pointed out by Silvester seems, however, to supply the bulk of the organ, and the term arteria thyreoidea as applied to it is apparently justified.

It is of interest to recall Simon's statement, which was also supported by others, that the thyreoid gland is placed in the blood system so as to regulate the supply of blood to the brain. This, in a way, was a foreshadowing of our present views that the physiological action of the thyreoid gland exerts an important influence on the central nervous system.

The venous blood from the thyreoid gland passes into the thyreoid vein, a vessel, which also collects the veins from the musculature below the aorta and carries the blood directly into the sinus venosus.

Little is known about the relation of the thyreoid gland to the lymph system. This is largely due to the fact that in the fishes the lymph vessels are in a much closer connection with the venous system than in the higher vertebrates. It is almost impossible to distinguish between veins and lymphatics by the injection method.

In many species large cavities lie around the aorta, two dorsal ones being constant in trout. These are extraordinarily large and lined with endothelium and although they often contain blood corpuscles there is little doubt that they are lymph sinuses. The corpuscles probably come in from the venous system, or possibly by traumatic haemorrhages. A further fact in favor of their being lymph sinuses is that no descriptions of large veins in this region has come from the numerous injections of the circulatory system of Teleosts. In some species there is only one large 'lymph sinus' which surrounds the aorta dorsally and laterally.

DEVELOPMENT OF THE THYREOID

Only one contribution deals with the embryology of the thyreoid in Teleosts, this is by Maurer ('86) who traces its development in the brcok trout. The gland arises in much the same manner as it does in the other classes of vertebrates.

The thyreoid develops very early in the Teleosts, after the first gill slit has broken through,[3] as an unpaired evagination of the stratified epithelium on the ventral side of the pharynx between the first and second gill pockets. It is thus placed in the curve of the S-like tubular heart before the gill arteries have developed, with the exception of that to the hyoid arch. The vesicular thyreoid anlage very soon separates itself from the pharynx and enlarges by budding. The organ lies close to the tubular heart, but only remains for a short time near the place of its origin. With the development

[3] In Hertwig's Handbuch d. Entwicklungslehre II, 1, 1906, Maurer states, however, that the anlage of the thyreoid appears in all Gnathostoma before the breaking through of the first gill slits.

and shifting of the heart and aorta as well as by its own growth the thyreoid gland comes to lie far from its original position. The absence of a capsule of connective tissue similar to that in higher vertebrates admits the loosening and separation of the thyreoid follicles in the bony-fish.[4]

The Teleosts show a condition of the thyreoid gland somewhat similar to that in Myxine glutinosa, as W. Müller, Cole, Schaeffer and Maurer state. The follicles in Myxine, partly isolated, partly in groups, are found between the pharynx and the truncus arteriosus throughout the gill region. In the Teleosts, however, the gland also extends below the truncus. In the skate Baber observes "a single body and a few detached vesicles"; in the Amphibians separated particles have also been described. Yet in both of these groups the thyreoid possesses a capsule, which sends septa into the inner portions, as W. Müller has shown for Acanthias and Raja. Maurer finds a delicate connective tissue capsule in the Urodela. These observations on Selachii and Amphibia, however, are exceptional and the small detached particles can only be looked upon as 'aberrant thyreoids' the main thyreoid in all cases being a sharply defined body. Maurer observed in the Urodela that a breaking up of the thyreoid into smaller parts occasionally occurred. These accessory thyreoid glands were parts of the former isthmus which, after the anlage had divided, persisted and remained in their original position, while the true halves moved in a postero-ventral direction. In the Ophidia the organ is compact and encapsuled; but in the Saurii, according to W. Müller, the interstitial tissue increases so much through the accumulation of fat, that the glandular tissue proper is broken up into irregular groups which are sometimes completely disconnected. We have here a dissolution within the capsule suggesting that the connective tissue capsule is the only factor in other vertebrates which prevents the thyreoid elements from becoming scattered about as they are in Myxinoids and Teleosts.

[4] The elements of the Teleost pancreas are similarly scattered in the mesenterium.

During development of the Teleosts some follicles cling to the wall of the aorta and are in later life found along it. Usually the follicles become arranged into several distinct groups, forming different centers of growth, as is shown by Cynoscion in plate III. With the branching off of the gill arteries from the aorta thyreoid material is carried out laterally towards the gills and spreads in this region. This accounts for the larger lateral extension of thyreoid follicles along the gill vessels rather than in intermediate regions. The larger vessels form a substratum upon which the follicles migrate as do also the smaller vessels and especially the lymphatic vessels. The larger vessels are means for the antero-posterior dispersion while the smaller ones allow the migration of follicles from the denser central thyreoid portions towards the periphery. Even the most peripheral follicles are usually found near blood capillaries although they do not necessarily come in close contact with them. The way in which these isolated follicles function is not clear. They certainly seem normal and contain colloid.

The growth of the connective tissue and fat in which the follicles are imbedded favors their dispersion from the central portions; thus a combination of influences are at work to widen the thyreoid region as much as possible. W. Müller regards the immense development of the 'interstitial' tissue as alone responsible for the dissolution of the thyreoid into isolated groups. He no doubt refers to fat and connective tissue, as we shall see below that the term 'interstitial' is not properly used in this case. Although the growth of these tissues may be an important factor I do not regard it as primary, since in the first place, even in young individuals, the follicles are found isolated, and secondly, the breaking apart of a formerly compact organ by excessive growth of connective tissue would certainly not account for the carrying of the follicles into the muscles and gills.

The follicles actually seem to overcome the obstruction offered by other tissues in their course and may even penetrate into them. In trout and Micropogon the thyreoid follicles are at times imbedded in the muscle tissue, into which they creep between the connective tissue lamellae or along the blood vessels. Real activity on the part of the follicles is most unlikely, and the probability is that they are simply passively pushed or pulled as circumstances may have it. The forces in development unite to make it possible for the thyreoid gland to spread, and so form a greater amount of functional tissue than could be contained in a compact organ situated in the narrow space between the basihyale and the ventral musculature.

Little is known of the manner in which the thyreoid gland grows and forms new follicles, and contradictory statements are also found in the literature regarding the primary anlage of the organ. It is scarcely conceivable that a vesicular anlage should exist in all fishes except Ceratodus in which Greil observed a solid one. Before the solid outpushing in Ceratodus separates from the pharyngeal wall it is said to become vesicular, a process exactly the contrary to the usual one.

Amphibians are believed to have a solid bud-like thyreoid anlage. Maurer states that two days after its evagination the thyreoid is solid in the Anura, and W. Müller observed a solid anlage in Rana temporaria and platyrrhinus, in which the first lumen appeared in 25 mm. larvae, after the gland had divided into two halves. In the Urodela Maurer records a solid epithelial bud, Livini finds the same in Salamandrina perspicillata and Muthmann in Triton alpestris. Platt claims that Maurer's description does not apply in all the Urodela, as is shown by the condition in Necturus.

The reptiles, birds and mammals are said by the majority of observers to show a vesicular thyreoid anlage, which changes into a compact organ from which follicles later originate. Kölliker, however, observed in the rabbit a thickening in the ventral wall of the pharynx, from which a wart-like solid process was cut off. Born also records the same for the pig. (Both authors quoted from Streckeisen).

This point is of importance in phylogenetic interpretations since our present views regarding the ancestry of the thyreoid gland are mainly based on a similar evagination, that for the endostyle, found in the Tunicates and Amphioxus.

Maurer finds in the trout a primarily globular vesicle stretching in an antero-posterior direction, and on the 41st day of development lying ventral to the stem of the aorta. If these statements apply to all Teleosts, the thyreoid must first originate dorsal to the aorta and early migrate ventrally and later return to a position dorsal to the vessel, since it usually occurs there in the adult. The condition in the Teleosts is similar to that in the Myxinoids, where Stockard describes the origin of the thyreoid as a median down-pushing from the ventral floor of the pharynx throughout the entire gill area, and consisting, in newly hatched Bdellostoma, of diffusely scattered alveoli below the pharynx and above the median branchial artery (ventral aorta).

In the trout, where development is rather slow, Maurer observes that 35 days after fertilization, when the embryo is about 6 mm. long the first thyreoid vesicle begins to pinch away from the pharynx. While originally the evagination, visible on the 28th day, possesses a stratified epithelium, it has on the 35th day a single layer of cuboidal cells. Three weeks later the whole stem of the aorta is surrounded by follicles. I find in rainbow-trout, one month old, or only 30 days older than those mentioned by Maurer, that the majority of follicles, and the larger ones, lie above the aorta.

Maurer also observes that in the brook-trout shortly after the first follicles have appeared the organ grows so rapidly that for a considerable period it surrounds the aorta as a compact mass. "In very late embryos, the growth of the thyreoid does not keep pace with that of the artery; thus the gland breaks away from the aorta and separates into a number of irregular clusters of different sizes lying either laterally partly paired or dorsal or ventral to the aorta, always, however, in its immediate neighborhood." He records the main mass of the gland in trout of even 25 cm. as being compact and situated ventrally between the second and third branchial arteries, and it is only in animals of 30–40 cm. that the thyreoid breaks up into the clusters of follicles characteristic of the adult. Maurer describes the same conditions in a number of other species, of which only the eel was at my disposal. The age of the eel I examined was unknown, though according to Maurer it must have been rather old, yet it measured only 30 cm. long.

Maurer's observations do not accord with the conditions I find in rainbow-trout four weeks old, nor in 25–30 cm. brook trout. As before stated, there is no paired arrangement of the thyreoid clusters, and the follicles are also in many cases distantly removed from the stem of the aorta. Other differences may be either due to specific or individual variations. Maurer's statements would indicate that the thyreoid gland tends to preserve its original unity, being finally broken up by force. My observation, however, seems to show the contrary, at least in Salmo irideus and fontinalis. In individuals one year old the follicles are more densely packed than in those one month old, although the intervening spaces have grown larger. The follicles also have become more numerous. This seems to warrant the supposition, that the thyreoid elements are disassociated at an early stage and subsequently multiply.

The multiplication of the follicles is described by Maurer as being very simple. While the epithelial cells are increasing in number after the forty-first day (in trout) solid buds appear on the primary vesicle, which very soon form central cavities and then pinch away. We do not know whether a similar process is maintained in later life, follicles coming from follicles, or whether new follicles are derived only from primary epithelial cells multiplying and forming a lumen. The latter supposition would more readily explain the scattering of thyreoid elements, germ elements I might say, to distant regions. L. Müller believes the new follicles to originate from old ones by buds from the epithelium which are subsequently pinched off. Baber contributes an interesting observation in the conger eel where in the wall of large follicles small ones sometimes lie imbedded so deeply that the epithelium between them is flattened out. Baber thinks that at times the wall breaks through and the two lumina are united. In other cases, however, the small imbedded follicles grow out and become independent.

The epithelial tubes found in the thyreoid of higher vertebrates as transitory growth stages are absent in the Teleosts. In the

Urodela a solid cylinder of epithelial cells from which the follicles pinch away, exists for from two to four weeks. These cylinders are observed in sheep and pig embryos up to the 20 cm. embryos and in man up to the 24 cm. embryo. The follicles in fish are formed comparatively earlier, and perhaps the gland functions earlier.

Thus a rapid multiplication of follicles occurs in the Teleosts without the formation of cell cylinders. This is an exception to W. Müller's claim that the thyreoid gland in all vertebrates passes through three stages: (1) a severing of the anlage from the pharynx; (2) formation of a network of tubes of glandular epithelium; and (3) the formation of follicles from these tubes. In the Teleosts and also in the Myxinoids, as Stockard has shown for Bdellostoma, the second stage seems to be suppressed or absent.

The first appearance of colloid in the thyreoid gland is generally thought to occur early in lower vertebrates but very late in the higher ones, towards the end of fetal life or often not until extra-uterine life.

Maurer reports colloid in the trout thyreoid on the forty-first day of embryonic development. How far this early appearance of colloid is connected with the function of the organ is unknown. From a comparative physiological standpoint it would seem that in the lower forms the thyreoid might function much earlier than in the Placentalia, where in intra-uterine life the gland of the mother might supply the needs of the developing embryo.[5]

HISTOLOGY OF THE GLAND

The histological structure of the thyreoid gland in Teleosts has been little studied. The meagre observations made by Baber in 1881, describing some features of the thyreoid in the conger eel were the first reported. Maurer later ('86) mentions a few points regarding the histology of the thyreoid in the trout and carp.

The microscopic appearance of the gland varies as much as does its anatomical structure. In sections from some specimens the

[5] In young mammalian embryos Peremeschko found no colloid, in older embryos it occasionally existed, while in young animals colloid was present in the majority of follicles and in old ones in all of them.

follicles are closely arranged and so densely packed that apparently only lymph spaces exist between them, in others we find the follicles more loosely connected and suspended in the connective tissue; while again in other specimens they lie so far apart that they can scarcely be thought of as belonging to one organ. The histological appearances also differ very much within the individual, depending upon the region from which the section is taken.

When the arrangement is such that the thyreoid may be dissected out and then sectioned, the follicles are found to be rather densely packed (pl. V, fig. 17). By this method, however, we are unable to get a correct idea of the extension of the thyreoid and the arrangement of its follicles, since it is only possible to remove the somewhat denser masses around the stem of the aorta, usually near the base of the second aortic arch, and all the particles in front and behind this region still remain. Properly to study the general distribution of the thyreoid follicles serial sections through the entire gill region are absolutely essential.

The spaces between the muscles, branchial arteries and gill arches are filled by wide-meshed connective and fatty tissue. In these tissues the follicles are suspended. The connective tissue is, therefore, not so directly a part of the thyreoid organ in these fish as it is in the encapsuled organ of mammals. The primary object of this tissue is to form a connection between the muscles and bones without regard to whether there may be thyreoid tissue in the region or not. True interstitial tissue, as such, is not found in this diffusely scattered thyreoid organ. Of course, the tissue in which the follicles lie imbedded performs the same function as does the capsule in higher vertebrates: in both cases it serves to support the follicles. In glands, where many follicles are accumulated in one mass, as in Cynoscion, or in the central portions of some others, for instance the trout, the supporting tissue may be regarded as part of those masses, but not as part of the entire thyreoid gland; here also the formation of connective tissue is the primary process, and the suspending of the follicles only a secondary one.

The supporting tissue is simple except in two species, Salvelinus and Sarda (pl. V, fig. 21) where smooth muscle fibres are freely suspended in the connective tissue. These muscle fibres are found especially below the aorta, where they approach the follicles and at times surround them. This is accomplished by the fibrillae of a bundle loosening up a little, then enclosing a row of follicles and finally uniting again.[6]

Regarding the number, size and form of the follicles, all variations exist which have been demonstrated by comparative investigations in the other classes of vertebrates. The size of the follicles is, in general, in reverse proportion to their number. The size, however, is not of great importance, since the chief factor in the activity of the gland is the epithelial surface; this will be the larger, the greater the number of small follicles contained in a given region. Biometric calculations would be interesting in this direction as experiments have shown that the functional value of the thyreoid gland varies with the individual. Glands are found in which the size of the follicles is uniform; in such cases the follicles are usually large. As a rule, however, the follicles are of various sizes as would be expected in view of the process of formation of new follicles. In many cases I have observed that a few (three or four) follicles are unusually large.

The follicles lying in the central parts are generally larger than those towards the periphery. This seems quite natural in view of the mode of extension of the gland. In only one case, Sarda, do the central portions consist of nests of numerous small follicles while larger follicles lie peripherally (text fig. 12). This condition resembles somewhat that in birds (Baber), and mammals (Anderson, Forsyth) and if it be due to the fact that in Sarda the gland is almost as compact (of course without a capsule) as in the higher vertebrates (with capsule) then we must suppose that in such a case new follicles are formed in the centre and are pressed out towards the periphery, while in the breaking up of the gland minute parts are continually carried towards the periphery and there form new follicles. In the first case the peripheral follicles would be the oldest and in the second the youngest ones.

[6] Streiff finds muscles between the glandular tissue in the thyreoid of the cat, Zielinska in a young dog, Wölffler in a child (cit. from L. Müller), L. Müller in an adult woman. The muscle must have migrated into the gland during the first half of the embryonic period, before the capsule was formed.

The form of the follicles is also variable, most typical perhaps are the globular or elliptical and tubular types. The smaller follicles are nearly always circular in section (pl. V, figs. 10–12), especially when they are free. The shape of the more closely packed follicles is influenced by pressure, and may be flat, indented, or irregular in outline. When the follicles lie next to the cartilages or muscles they are usually oblong-oval, with the longer side towards the tissue. Single follicles lying in the supporting tissue, if large, are rarely prefectly circular, but have irregular outlines due to pressure from the fibres of the substratum. The shape of these follicles indicates the existence of actual pulling forces in the supporting tissue.

Not only small irregularities are found in the surface of the follicles, but also deep invaginations of the epithelium as well as long evaginations. The follicle may consist of a central body with sprouts or branches of cylindrical and globular shapes (pl. V, figs. 15, 16). How far these irregularities in form are connected with the cutting off of smaller follicles from larger ones could not be determined. Anderson doubts the multiplication of follicles by such a process.

It is now generally accepted that no communication from follicle to follicle exists; the follicles are closed on all sides and perfectly separated from each other. Sometimes, however, as many as five follicles are observed in a section, apparently perfectly separated, but on tracing through the series of sections they all unite into one follicle (pl. V, fig. 16). This is due to evaginations from the follicular wall somewhat like the fingers of a glove, which when cut across, give the appearance of several independent follicles, while in reality there is only one lumen. In Anguilla chrysopa, however, there really seemed to be a communicating duct between two follicles; the lumen of the tube was much narrower than that of the follicles and the epithelial cells of it were much higher (pl. V, fig. 15). This closely resembles a 'Schaltstück' as seen in other glands. This was not due to a waist-like constriction of the epithelium, but to a far reaching evagination from one follicle with a globular swelling on the free end representing a second follicle. There was no colloidal substance in the 'intercalary' duct.

Branched follicles are particularly abundant in some species. In Muraenoides all follicles seem to branch. Baber states that in young animals the follicles are much more ramified than in older ones; he, therefore, regards this branching as the method of follicle multiplication. Anderson, on the other hand, holds the 'melting' of the epithelium (a process about which I shall speak later) at the point where two follicles meet responsible for the communication between several lumina; this of course is an opposite process from that of budding. Anderson, therefore, believes that in old animals there are more irregular follicles than in the young. It seems to me that the ramification of the follicles does not depend so largely upon age, but rather on the species.

The follicular epithelium varies but little with the species, perhaps the number of cells may differ in follicles of the same capacity. The epithelium is of the form usually found in the thyreoid glands of higher vertebrates. All transitions exist from a pavement epithelium of very low broad cells, through cuboidal cells as high as broad, to very high and narrow cylindrical cells (pl. V, fig. 10).

The form of the epithelium is probably connected with the age of the specimen, as it undoubtedly flattens with increasing age. (In very old human subjects only perfectly flat cells have been found.) Age, however, can scarcely be the only factor, as in some species different forms of epithelium appear at the same time. This may be due to the different ages of the follicles, though it cannot be regarded as an absolute rule that the older follicles have a lower epithelium than the younger. Hürthle definitely states that these two factors are independent of one another. Langendorff points out that the follicles increase in size, not by a flattening out of the epithelium, but by multiplication of cells. I should say that both processes may be simultaneously involved since we often find large follicles with high epithelium, yet karyokinesis is rarely observed in the epithelial cells. The latter fact led Stockard to suppose that amitotic cell division might occur in the growing thyreoid tissue of Bdellostoma.

The different types of epithelium might be accounted for in still another way by supposing the follicles to be in different stages of activity. Here again we meet with difficulties since the same follicle sometimes shows high cylindrical epithelium on one side and a flattened epithelium on the other (pl. V, fig. 11). Hürthle considers the flattening and stretching of the cells to be the final stage in the process of colloid formation. He finds this type of cells not sporadically, but always in the larger groups of follicles. The low epithelial cells are still alive and according to Hürthle may again transform into high ones. Biondi claims that when the follicle has reached a certain size, the epithelium partially flattens and vanishes, thus establishing a communication between the follicular lumen and the lymph spaces and allowing the colloidal material to be poured into the lymph system. The emptied follicle is said to collapse and from its cell mass a new follicle originates. Anderson, also, thinks that by a 'melting' of the epithelium a connection is formed between the lymph space and follicle, but the individual follicle is not destroyed.

In the conger eel Baber finds oval cells between the cylindrical ones and attributes to them the formation of new follicles, an idea which I think is incorrect. The two classes of cells could not be found in the common eel. Baber also finds in the conger eel 'club-shaped' cells between the epithelial cells. They are much narrower than other cells and possess elongated nuclei. Their free ends project above the general surface and are expanded 'fan-like;' the bases may also show a similar condition. Baber regards them as branched cells, often existing in pairs, and forming stomata which play an important part in absorption and secretion. If this be true such cells should exist in all glands. I was unable to find these and think perhaps they may have been consequences of his alcohol preservation.

The form of the nucleus changes with the form of the epithelial cell. It is usually circular or somewhat oval in cross section. When the cell is either cylindrical or flattened, the nucleus becomes more and more elliptical in shape, with its long axis parallel to that of the cell. Thus in the first case the long axis of the nucleus is vertical to the free surface of the cell and in the second parallel with it. When the nuclei are oblong in spite of the cells being broad and cubical (for instance in some follicles of Muraneoides) they always present the long side to the free cell surface. Frequently narrow cells with oblong nuclei are seen between the cuboidal cells.

In trout degenerating epithelial cells of small size were observed with compact nuclei, deeply staining or pyknotic.

The nucleus usually lies at the base of the cell (pl. V, fig. 11) but may sometimes, especially in an epithelium with many cells, move a little towards the lumen (pl. V, fig. 10). Nuclei may lie at different altitudes, in an alternating fashion. One or two nucleoli are visible.

The shapes of the nuclei usually give no indication of the state of activity of the cell as Anderson has claimed. Even pyknotic nuclei usually have regular outlines. An exception to this is seen in the trout where often, in some varieties almost exclusively, epithelial cells show nuclei of very irregular shapes, as indicated in pl. V, fig. 10. The nuclei are elongated with more or less bent corners—horse-shoe shape. These were generally found in lower cells; they may have been degenerating, since they did not stain as deeply as the normal ones in other parts of the gland, when such were present. It seems, however, scarcely conceivable that the epithelium of the entire gland should degenerate, unless from some pathological condition. (These animals were all reared in the N. Y. Aquarium.)

The cytoplasm of the cell appears granular and sometimes stains slightly darker in the basal region. There is no cuticle lining the lumen, but the refraction of light in this region has misled some authors. The base of the cells is usually rather smooth, though in cases where vessels come into close contact with the follicle the straight basal line becomes somewhat interrupted through the influence of the surrounding structures. In Brevoortia the epithelial cells are nearly all drawn out as if they possess projections. Those of one follicle approach very closely those of others and it seems almost as if a connection between the follicles were established (pl. IV, fig. 1, 2). Other somewhat broader cells possess pedicel-like bases which are sometimes branched, giving the impression that the cells are sending out pseudopodia. The processes disappear in the interfollicular tissue in close contact with the blood and lymph capillaries. The process cells are limited in number and lie close together. Their cell body is swollen with foamy cytoplasm containing several deeply stained highly refractive granules. Perhaps these cells are in a state of degeneration, probably colloidization, although their plasma does not show any acidophilia (pl. IV, fig. 5, E). Peremeschko observes somewhat comparable features in the thyreoid glands of birds and mammals, especially in that of the rabbit. Some of the epithelial cells possess at their basal end from one to ten small projections, and thus resemble the tassel cells Pflüger has described in salivary glands, except that in the thyreoids the processes are shorter. In some cases Peremeschko found such cells in fresh material and could isolate these follicles, which appear to be surrounded by a fringe. Pflüger regarded the cell processes as nervous, but Peremeschko correctly believes them to come from the cytoplasm of the epithelial cells.

The function of the follicles can be much more easily studied in other groups of vertebrates than the teleosts. In dissecting out the follicles, as far as they are macroscopically visible, and fixing them, it is almost impossible to avoid destroying the finer structures. Hemorrhages are almost unavoidable in cutting open the gill region. On the other hand, in fixing the entire floor of the pharynx the fixation fluid does not penetrate sufficiently fast to preserve the finest details, and the general structures are unfavorably influenced by the decalcification process. Microchemically, therefore, little can be done and I limit myself to what could be determined from studies of general structures.

Hürthle's colloid cells were seldom seen in the thousands of follicles observed. Whether they are not generally formed, or whether they appear and are emptied in so short a time as to be rarely preserved I am unable to say. In Clupea (pl. V, fig. 20, Coz.) they were limited to four or five neighboring follicles and in these all of the epithelial cells were so swollen that in some cases they met in the center of the follicle, obliterating the lumen. The nuclei were compact and deeply staining and occurred directly under the free surface of the cell. The cytoplasm was homogeneous, highly eosinophile, and sharply distinguished from that of other epithelial cells, and thus the colloid forming zone was well defined. This agrees with Hürthle's account, which states that the colloid cells always appear at the same time in a large portion of the wall of a follicle or in several neighboring follicles. The size of the follicles has nothing to do with their appearance. In one case, Siphostoma, the epithelium of all follicles consisted of cuboidal swollen cells, the nuclei of which were near the cell center or towards the lumen and the cytoplasm was highly acidophile, (pl. IV, fig. 8).

The normal contents of the follicles is the colloid. It is found in all thyreoids and usually all the follicles contain it; only in a few cases were the majority of them empty.

In spite of the various ideas expressed in the literature regarding the surface irregularities of the colloid there is little doubt that they are caused by shrinkage in fixation. In the majority of follicles the surface of the colloid was perfectly smooth, in some a little retracted from the epithelium, but in others completely filling the lumen. In some follicles the colloid showed surface indentations. These differences can scarcely be connected with the age of the organ, as they were observed in different stages. In all the young trout, however, the colloid filled the follicles completely. One possibility is that the content of the follicles does not always possess the same chemical composition, and is influenced by the same fixation fluid in different ways.

The view has been held that the true secretion of the cells is hyaline and that it appears in the form of small droplets which are set free on the surface of the cells. This process is thought by some to be responsible for the irregular surface of the colloid. Two kinds of surface irregularities must be distinguished, first, the large ones which do not correspond with depressions of the epithelial cells. These are without doubt due to the fixation. The connecting threads of colloid between the central portion and the epithelium seem to run between the cells and to take hold there. Tangential sections through the follicle wall, cutting the epithelium just under the free surface, show that the cells do not always lie closely placed in their upper portions and a network of colloidal threads is shown between them. In higher vertebrates these large surface irregularities in the colloid seem more common than in fish. The second, smaller irregularities might resemble secreted droplets. They give to the surface of the colloid, especially from a top view, the appearance of being beset with oil drops. In some places there are merely slight depressions in the free margin, some distance apart, while in others the whole surface is corrugated, but these irregularities do not appear in all of the follicles. Whether they are really physiological products of the cells is not determined. The irregularities may be more easily explained on the theory that where the free ends of the cells do not come in close contact, the colloid which fills the follicles is pressed into the intercellular spaces and surrounds the top of the cells like a cap. In shrinkage from fixation the caps would be pulled from the cells, leaving on the surface of the colloid the impressions. Anderson regards these 'droplets' as well as the numerous vacuoles, which he finds within the colloid even of living glands, as "cavities lined with a hyaline membrane and containing the 'chromophobe' secretion, a part of the secretory activity of the gland." Langendorff and others more correctly regard them as artefacts, having no physiological significance. Vacuoles within the colloidal substance are seldom seen, (pl. V, fig. 13, V).

The colloidal material seems to become denser with age, as far as this can be determined by its staining capacity. In young trout it is rather pink so that it can scarcely be distinguished from the blood serum in the vessels. Both structures show the same microscopic appearance. In older trout, however, the colloid stains very deeply with acid dyes. These observations agree with those of Schmid on dogs of different ages. Anderson, Boéchat, Peremeschko and others also state that the number of follicles containing a slightly staining finely granular colloid diminishes with age, being small in old individuals. I failed to find some follicles distinguished by a greater affinity for the stain than others, as was claimed by Hürthle, but did find that sometimes within the same follicle the colloid stained differently in different places.

The structure of the colloid varies with the species and is the same through all the follicles. A perfectly homogeneous colloid exists in cases, in others it is granular, and finally in some fish it is of a lumpy consistency. The consistency of the colloid also varies with age, in old animals being rather cloudy in appearance and evidently very brittle after fixation. Occasionally the outer portion of the colloidal mass stains a trifle lighter which is the only indication of a concentric structure. This alone, however does not argue for the view that the colloid is a by-product of the active thyreoid, which collects and remains in the follicle. Langendorff first presented such an idea which of course called forth great opposition.

Blood corpuscles are occasionally found in the Teleost thyreoid and sometimes completely fill the lumen of the follicles or may be scattered or bunched together. Blood is also occasionally found in the human follicles. Baber was no doubt mistaken when he spoke of a real flow of blood into the follicles, as such does not occur. How the corpuscles enter the follicles is not known, though it is probable that somewhere, by pressure or tension, the delicate wall of a capillary lying next to the epithelium is ruptured and the corpuscles find their way into the follicular lumen through an injured wall. Hürthle believes the 'melting' of the epithelium responsible, when it occurs at a place where capillaries lie deeply imbedded. It is evident that whenever blood corpuscles do enter the lumen they are destroyed, and they may be seen in all stages of disintegration until finally pyknotic shadows of nuclei alone exist with no indications of their cell bodies. The scattered corpuscles lie within the colloid, which must therefore, be rather liquid. The content of the follicles has a haemolytic property without being itself of haematogenic (Baber) origin.

Cells from the follicular epithelium also form a part of the follicle content. These are pushed off either singly or several together into the lumen and there destroyed; they also lie within the colloid. Two kinds of cells are distinguished, those with a small body and dense cytoplasm, resembling somewhat epithelial cells of the ordinary type and those with a swollen body, and clear cytoplasm (pl. V, fig. 13, Coz), which may resemble Hürthle's colloid cells. The two types are probably different stages in the same process. At times the cell body is broken up into pieces before being transformed into colloid. The nucleus is always destroyed last.

A part of the colloid is therefore formed by degenerating epithelial cells which are either destroyed in their primary position or after being pushed into the lumen. Anderson believes that this is invariably the fate of the cells after several periods of secretion. Hürthle, also, noticed this 'melting' of the epithelium and was able to trace the complete disintegration of the cytoplasm, though the fate of the nuclei remained doubtful. They, too, are unquestionably destroyed within the colloid, and as a matter of fact I could observe cell nuclei, such as those of the red blood corpuscles, in all stages of disintegration, (pl. V, fig. 13, N). L. Müller regards the formation of colloid material from disintegrated cells as of slight importance. Hürthle remarks that in follicles of mammals with a flattened epithelium, which he considers the final secreting stage, cell remnants or defects in the wall can rarely be found. This is equally true in Teleosts.

I have never seen the signs of degeneration described by Maurer in old carp. He records a swelling of the epithelial cells which breaks down the follicles, permitting lymphatic elements to enter and form lymph nodules, similar to processes in Anura. Perhaps my specimens were not old enough to show these phenomena.

Pigment was not observed within the follicles, though outside of them brown pigment is often found in the supporting tissue. This is probably of haematogenic origin. Baber found brown pigment granules within the colloid in the thyreoid gland of the conger eel. I also fail to find crystals in the follicles or around them as has been reported by some investigators. They are undoubtedly postmortem products.

In the conger eel Baber observed a reticulum between the epithelial cells, in which they were partially imbedded. He states that this reticulum is formed by coagulated intercellular substance and has nodal thickenings. At the thickened places the 'club shaped' cells described above are located and may be clearly distinguished from the ordinary epithelial cells. I did not find such a reticulum, and it is possible that the filling of the intercellular spaces with colloid substance as before mentioned, may have been what Baber observed. He states that the reticulum (intercellular substance) stained with hematoxylin, which makes it very different from the highly eosinophile network observed by me. Baber's technique, however, seems to have failed to produce the proper differentiation, since he actually succeeds in staining parts of the colloid with nuclear dyes.

The disputed membrana propria was not observed. W. Müller, Kölliker and others claim to have seen it everywhere while Schmid and others definitely deny its existence. The connective tissue approaches the follicles and surrounds them but this loose connective tissue sheath, which is by no means always present, could scarcely be called a propria.

The blood supply of the thyreoid gland is abundant and varies somewhat with the species. Baber is the only observer who has studied the conditions in the Teleost thyreoid by aid of the injection method, and unfortunately he used only one specimen of the conger eel.

The capillaries often approach the follicles so closely as to seem imbedded between the epithelial cells. This is best shown when both follicles and vessels are cut in cross section, (pl. IV, fig. 6). Hürthle describes a similar condition in the thyreoid glands of young dogs and pictures them in plate II, fig. 6. The epithelial cells often partly surround the capillaries by means of processes, thus forming deep impressions. Baber speaks of small intercellular projections from the capillaries which seem to serve in retarding the circulation of the blood.

There is usually a network of capillaries around each follicle, four or five often being seen in cross section just outside the epithelium, (pl. IV, fig. 5, *a*). In longitudinal section, the capillaries at times surround almost the entire periphery of a follicle. Such specimens illustrate how closely epithelium and endothelium are neighbors without a separating basement membrane, (pl. IV, fig. 5, E, Ca).

Where the follicles are densely packed, numerous spaces and channels run between them. The smallest of these seem to have no endothelial wall, so that the lymph flows directly against the epithelium of the follicles. In other cases the lymph vessel is indicated by two parallel endothelial lines running between the follicles. This does not agree with L. Müller's view that the blood capillaries are in close contact with the epithelium while the lymph vessels are separated from the follicles by blood vessels or connective tissue. The follicles are sometimes, as described in the anatomical part, situated directly on the big lymph spaces around the ventral aorta as the text figures 4 to 7 show. (See also pl. IV, figs. 2, 3.)

In the conger eel and skate Baber was unable to detect the lymph vessels. Since he injected the venous system which is connected with the lymph vessels he thus regarded the lymph capillaries as veins. Ferguson has been more successful in distinguishing between these two sets of vessels in the dogfish.

In some instances, less often, however, than it occurs in higher vertebrates, a substance was found in the lymph vessles, which had apparently the same structure as the contents of the follicles. The lymph spaces were filled with this substance in one instance and showed many smaller channels running together into the larger ones (pl. V, fig. 18, *L*). According to Anderson the colloid in the lymph vessels undergoes a change, becoming diluted and finely granular and is difficult to distinguish from blood serum.

The way in which the colloidal material leaves the follicle is not made clear by my study. Attention may be called to the varying views of different authors, especially those of Biondi and Anderson, given in their description of the follicular epithelium. It must be mentioned also that Hürthle believes in temporary intercellular channels which form between the cells for the passing of the colloid. I saw in a very few cases a colloidal pseudopodium, as it were, push through the epithelium.

I am also unable to state from the thyreoid gland in the Teleosts whether the veins contain colloidal substances and carry them into the circulation.

RESUMÉ

The anatomy of the thyreoid gland of the Teleosts is decidedly different from that of most other vertebrates. It is not an anatomical unit. The term 'thyreoid gland,' therefore, is scarcely appropriate. Physiologically isopotent units are distributed over a wide area. Physical influences must be made responsible for this distribution, which is due to mechanical conditions of pressure and pull.

If the thyreoid gland of the Teleosts really have its prototype in the endostyle of the Tunicates, its phylogeny is somewhat as follows. We have at first a uniform organ with a given function, later a change of structure and function takes place, and the organ loses its unity (Myxinoids and Teleosts). In higher forms the new function is maintained but the organ retains its original uniformity and integrity.

The development of the organ from its anlage to the mature state seems to be simpler in Teleosts than in higher vertebrates.

The histology of the glandular elements of the thyreoid in the Teleosts is but little simpler than in higher vertebrates. It shows many parallels to the different features observed by numerous authors in other thyreoid glands.

The function of the thyreoid, concluding from its microscopical appearance, must be closely the same in Teleosts as it is in other vertebrates.

SPECIAL PART

The species examined were:

ORDER	FAMILY	SPECIES
Apodes	Anguillidae	Anguilla chrysypa.
Isospondyli	Clupeidae	Clupea harengus.
		Brevoortia tyrannus.
	Salmonidae	Oncorhynchus kisutch.
		Salmo mykiss.
		Salmo irideus.
		Cristivomer namaycush.
		Salvelinus fontinalis.
	Argentinidae	Osmerus mordax.
Hemibranchii	Gasterosteidae	Apeltes quadracus.
Lophobranchii	Syngnathidae	Siphostoma fuscum
Haplomi	Poecilidae	Fundulus heteroclitus.
		Fundulus majalis.
		Fundulus diaphanus.
Acanthopteri	Atherindae	Menidia notata.
	Mugilidae	Mugil cephalus.
	Scombridae	Sarda sarda.
	Pomatomidae	Pomatomus saltatrix
	Serranidae	Morone americana.
	Sparidae	Stenotomus chrysops.
	Sciaenidae	Cynoscion regalis.
		Micropogon undulatus.
	Labridae	Tautogolabrus adspersus.
		Tautoga onitis.
	Tetraodontidae	Spheroides maculatus.
	Triglidae	Prionotus carolinus.
	Batrachoididae	Opsanus tau.
	Blenniidae	Muraenoides gunellus.
	Pleuronectidae	Pseudopleuronectes americanus.

ANGUILLA CHRYSYPA RAFIN

The thyreoid gland in young eels, 30 cm. long, has a transverse and not a dorso-ventral extension as one might expect in a species with a narrow floor of the pharynx. It begins far forward in the arterial bifurcation lying close under the basihyale and extends back to the second gill arteries (plate I, fig. 11.) Close behind the anterior end of the gland the transverse distribution of follicles becomes rather wide, (fig. 1, *A*), extending over the

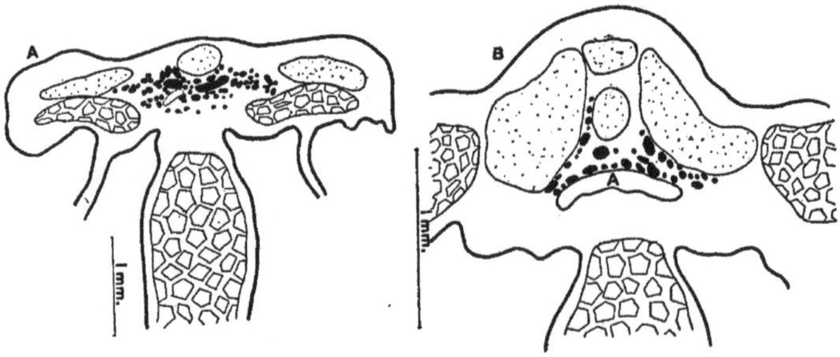

Fig. 1. Sections through the thyreoid gland of Anguilla. *A*, anterior to the aortic bifurcation; *B*, between the first and second branchial arteries. Thyreoid follicles in all figures shown in solid black. Transverse muscles lined. Long-muscles in polygons. Skeletal parts stippled. Arteries heavy line. Veins light line. Lymph sinus broken line. *A*, ventral aorta. A_I, A_{II}, A_{III}, branchial arteries.

entire space between the first gill arches, about 2.5 mm. The layer of follicles is very thin so that the dorso-ventral extension is slight. Near the union of the two first gill arteries the follicles are somewhat more dispersed, and reach out dorsally along the sides of the basibranchiale. Some follicles actually lie dorsal to the skeletal parts. The thyreoid is in contact with the first gill arteries for a short distance, and here it reaches its maximal extension. Further back it is limited to the neighborhood of the ventral aorta.

Behind the aortic bifurcation the follicles lie closely above and to the sides of the aortic stem and extend along it to the second gill arteries. A string of follicles lies separated between the first and second arterial branches. Baber states that in the conger eel the gland is in the first bifurcation and forms a reddish flattened body. This would correspond to the region of maximal dispersion of thyreoid follicles in the species here mentioned.

The follicles exhibit a variety of shapes, elliptical ones being in the majority. They are rather small, 100μ representing the average diameter of the circular follicles. A few very large follicles are present; these 'giant' follicles as they might be called, are of elliptical shape measuring 600μ in the long and 200μ in the short axis. Baber observed in the conger eel follicles of very large size.

Some follicles send out branches which widen near their end to form secondary cavities, (pl. V, fig. 16). In the series from which one section is figured, (pl. V, fig. 15), may be found a large follicle sending out a branch, and further along two follicles (*F. f.*) connected by a tube (*D*) of high cylindrical epithelium. The tube represents the branch of the former section and in another section both follicles are entirely separated. Going further in the series the small follicle increases in size while the large one sends out a second branch. Thus around a larger follicle as a center may be grouped several smaller ones connecting with the original follicle by 'ducts' as it were. These ducts might be compared with the intercalary portions of other glands. Baber likewise observed branching follicles in the conger eel. Baber claims that new follicles arise from groups of cells somewhat rounded in form and situated in the epithelial wall of the larger ones. I was unable to observe such processes. Lymphatics are present in the thyreoid gland of the eel, although Baber denies their existence.

Baber records the follicular epithelial cells as highly columnar in form. I find cuboidal epithelial cells measuring from 10 to 15μ high.

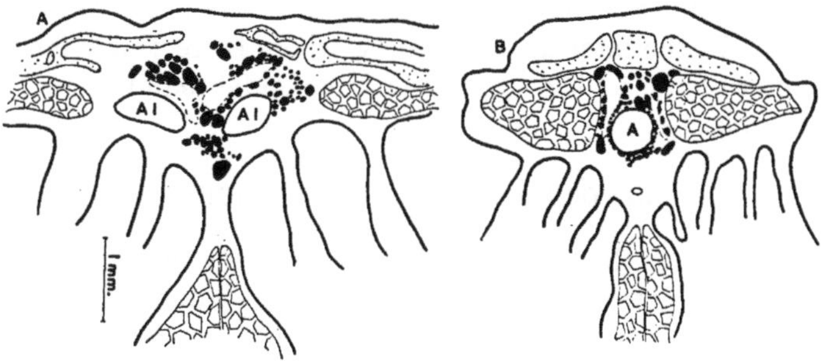

Fig. 2. Sections through the thyreoid gland of Clupea. *A*, in the aortic bifurcation; *B*, between the first and second branchial arteries.

CLUPEA HARENGUS L.

In the herring (specimens 30 cm. long) the thyreoid gland is well developed (pl. I, fig. 2). The triangular region formed between the floor of the pharynx, bases of the first gill arches and a ventrally lying cartilage is entirely filled with follicles. The distance between the floor of the pharynx and the ventral musculature is considerable, while the cartilages of the basibranchiale are only slightly developed; there is thus sufficient space for a dorso-ventral distribution of the thyreoid follicles (fig. 2, A, B). In certain places the first gill arteries are completely surrounded by follicles; this is also true of the ventral aorta behind the anterior bifurcation (fig. 2, *B*). Back of the second branchial arteries the extension of the gland diminishes, and only small follicles make a complete ring around the aorta, from which rays of follicles go out towards the cartilages and muscles.

The average size of the follicles is about 200μ in diameter; very large ones are not seen. The follicular epithelium is in general rather high and varies between narrow cylindrical cells to broad cubical ones. The cells are not very densely arranged. In some regions are found a few neighboring follicles with high epithelial cells which almost obliterate the central follicular space (pl. V, fig. 20, *Coz*). Other follicles have lower cells and all stages exist between these and the normal ones. This suggests a zone of Hürthle's colloid-forming cells. The cytoplasm is highly eosinophile and the nuclei are located near the inner surface of the cells. In the intermediate stages, where there is a lumen in the center of the follicle we find in it colloidal material and red blood corpuscles.

BREVOORTIA TYRANNUS LATROBE

In this species (length 40 cm.) there are very interesting conditions in the extension of the thyreoid gland, due to the enormous elongation of the gill region. The distance from the heart to the anterior aortic bifurcation measures about 5 cm. and with this stretching of the ventral aorta the thyreoid becomes extended over a long region. The front end of the gland lies well beyond

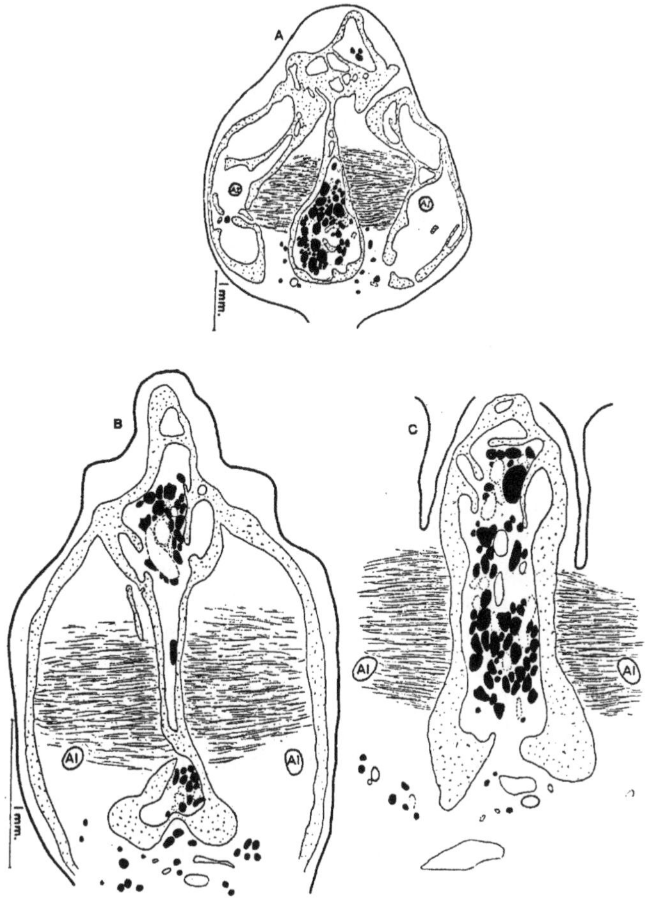

Fig. 3. Sections through the thyreoid gland of Brevoortia, all anterior to the aortic bifurcation.

the aortic bifurcation and the posterior end is at the second branchial arch, (pl. I, fig. 3).

The floor of the pharynx is very narrow, thus there is no chance for a lateral extension and the thyreoid follicles become dispersed only in a dorso-ventral direction. This extension is sometimes 4 mm. high (fig. 3).

The parts of the basibranchiale and hypobranchiale are not compactly developed, the floor of the pharynx being supported by a scaffold of osseous lamellae, between which a wide-meshed fatty tissue appears. There is also an osseous tube open at both ends (representing perhaps a sub-copula) and enclosing the most anterior portion of the ventral aorta (fig. 3, *A*). In this tube the thyreoid gland extends from the branchial vessels towards the tip of the tongue. In transverse section the gland appears to be lying within a bony ring which completely separates it from the parts outside. At certain places, however, there are openings in this capsule through which the follicles escape into the outside tissue. Within the capsule are found osseous lamellae dividing it into several compartments, and thus three or more bunches of thyreoid tissue may be seen separated by bone.

At the anterior end the follicles lie outside the osseous capsule and are scattered far apart. They are always located in the neighborhood of either blood or lymph vessels and probably follow the vessels as paths of dispersion. The follicles are not always, however, in direct contact with blood vessels.

The osseous capsule lies above the ventral aorta, and we find thyreoid tissue only above the vessel. The first gill arteries for a short distance are completely surrounded by very small follicles. From the aortic bifurcation the follicles extend far forward into the capsule although there are no large vessels, thus there seems to be a tendency towards a forward migration. This is really the only available space into which the thyreoid can expand, unless it enter the ventral musculature.

The histology of the gland is somewhat different from that in other fish. The follicular epithelial cells are drawn out into long processes which come into contact with those arising from the cells of near-by follicles (pl. IV, fig. 1). This suggests that the cells of one follicle might communicate through these processes with those of the adjacent follicles. The only explanation for this phenomenon is as follows: originally the follicles lie close together, with their epithelial cells touching, and when the space between the skeletal parts becomes wider the meshes of the fatty tissue, in which the follicles are suspended, are pulled somewhat apart, carrying the follicles with them. The cells, which were in contact with others or with blood and lymph vessels may have held fast to them, becoming drawn out into long processes. They thus form a network between the follicles. These bridges often surround the capillaries.

There are only a few follicles which have a regular epithelium with a smooth outline. Outside the bony ring, described above, the follicles have the usual epithelium with a smooth surface.

In places the epithelium was found to be disintegrating. The association of the cells seemed rather loose, their surfaces were also drawn out into long processes like pseudopodia which sometimes divided into two and disappeared in the interfollicular tissue (pl. IV, fig. 5, *E*). These cells did not show any distinction between nucleus and cytoplasm, and their contents was of a foamy nature and showed two or three compact deeply staining granules. They were probably cells which having completed their secretion period were disintegrating.

SALMO IRIDEUS GIBBONS

Specimens 4 cm. long, one month old. In the young rainbow-trout the thyreoid gland begins in the aortic bifurcation and extends almost to the third gill arteries, (pl. II, fig. 22). There is little space for a lateral extension, as the cartilages of the basi- and hypohyalia form a rather narrow arch, and limit the gland to the space immediately above and below the aorta. At the aortic bifurcation the copula comes close to the vessel, so that the follicles are pressed away from the median line, and lie close to the sides of the cartilage. Later the skeletal parts move back, the space between them becoming somewhat clearer. Half-way between the first and second gill branches the thyreoid gland also extends below the aorta, and a large number of the follicles lie near the second arterial branches. These ventral follicles are smaller than the dorsal ones (fig. 4, *B*). Towards the posterior limit the follicles become smaller and fewer and are again limited to the region above the aorta. Only two or three follicles are seen in a cross section and at the third gill arteries they have entirely disappeared.

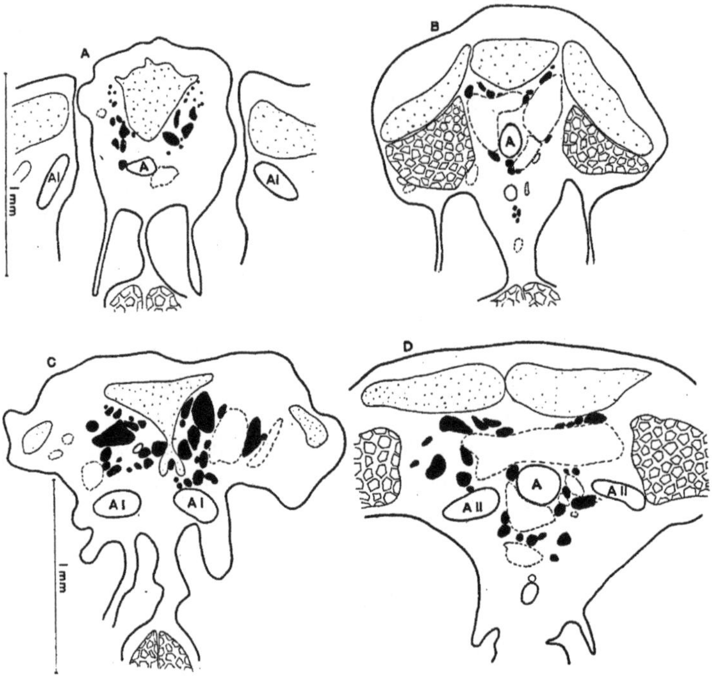

Fig. 4. Sections through the thyroid gland of Salmo irideus. *A* and *B*, from a specimen one month old. *A*, just posterior to the aortic bifurcation; *B*, near the second branchial arteries. *C* and *D*, from a specimen one year old; *C*, in the aortic bifurcation; *D*, at the second branchial arteries.

The follicles are usually circular but there are also irregular forms, due to pressure from the surrounding tissues. The diameters of the circular follicles vary between 10 and 80μ, the larger ones being rare. The follicular epithelium is everywhere low, the cells measuring about 3μ. The nuclei are all placed with their broad side towards the lumen, which might be called the flat cell position to distinguish it from the cylinder cell position in which the nuclei point towards the lumen.

Almost all the follicles are filled with clear homogeneus colloid, which has only here and there retracted a little from the wall. Blood capillaries belonging to the follicles, as observed in other species, are not visible. Around the aorta there are rather large veins or lymph vessels with extremely thin walls, close to which the follicles lie (fig. 4, *B*). There is no tissue (basement membrane) between the epi- and endothelium, the first being almost as thin as the latter. The nuclei of these epithelial cells are spindle-shaped and lie far apart.

Specimen one year old. In this fish the distribution of the thyreoid is about the same as in the younger one, (pl. II, fig. 23). The anterior end is pushed further forward in the aortic bifurcation, and the posterior end still lies close to the third gill branches. The mass of thyreoid tissue is much enlarged. The follicles are much larger in the bifurcation, and in a section there are more than three times as many as in a one month old individual. They are packed more densely and completely fill the space between the cartilages and arteries. The process of the copula mentioned above, which comes down to the level of the aorta, here divides the thyreoid into a right and left half. While in the younger trout the lateral extension of the follicles was less than the dorso-ventral, at this age the floor of the pharynx has become broader through a widening out of the gill arches, and the lateral distribution is more than twice as extensive as the dorso-ventral, although the follicles still go high up along the cartilages (fig. 4, *C*). The follicles also extend some distance along the first branchial arteries. Here the entire thyreoid lies dorsal to the blood vessel and is grouped around two or more large lymph spaces (fig. 4, *C*). Immediately behind the aortic bifurcation the lateral and then the dorso-ventral dispersion of the follicles decrease, so that they lie more densely packed and are fewer in number.

The hypobranchialia approach closer and closer to the copula as we pass backward and force the thyreoid to a more ventral position. Finally the aorta lies almost on the cartilages and the thyreoid shows only one or two follicles in the section. This restriction of the thyreoid zone (pl. II, fig. 23) between the first and second branchial arteries is typical for all salmonids. It may also occur in some other species but is never so pronounced as in the trout.

Near the second branchial arteries the skeletal arch becomes flattened again, the copula does not reach so far down, and first the dorso-ventral and later the lateral distribution of the follicles again increases. Comparatively few follicles now appear below the aorta (fig. 4, *D*). Behind the second branchial arteries the follicles decrease in number and size, and completely disappear before the third branchial arteries are reached.

The follicles are circular, oval or irregular in cross section. The diameters of the circular ones vary between 40 and 200μ, the larger ones are more numerous, especially in the anterior region. Branched follicles occur, sometimes as many as five follicles leading into a larger one.

Here also the follicular epithelium is low, almost flat, and the follicles are completely filled with homogeneous colloidal substance. Sometimes, however, the colloid contains particles, probably destroyed blood corpuscles or epithelial cells. The blood supply is rich, many capillaries lying close to the follicles. There seems to be a comparatively better circulation here than in the younger stages.

SALMO MYKISS WALBAUM

Specimen 11 cm. In the black spotted trout the thyreoid gland shows a great antero-posterior extension. The posterior limit is about that shown by Maurer in a 20 cm. trout, species not named, apparently a brook trout. However, the main part of the gland is situated above the aorta, not below it as Maurer claimed. The anterior limit of the gland lies well in front of the aortic bifurcation and the posterior end behind the third branchial arteries (pl. II, fig. 21). The dorso-ventral distribution is also more pronounced than in most of the other species, especially as to the number of follicles below the aorta. The main mass of the organ lies in the aortic bifurcation (fig. 5, *A*). The copula reaches far down and divides it into two halves. Along this cartilage the follicles extend dorsally close up to the floor of the pharynx. Laterally also the extension of the follicles goes as far as possible.

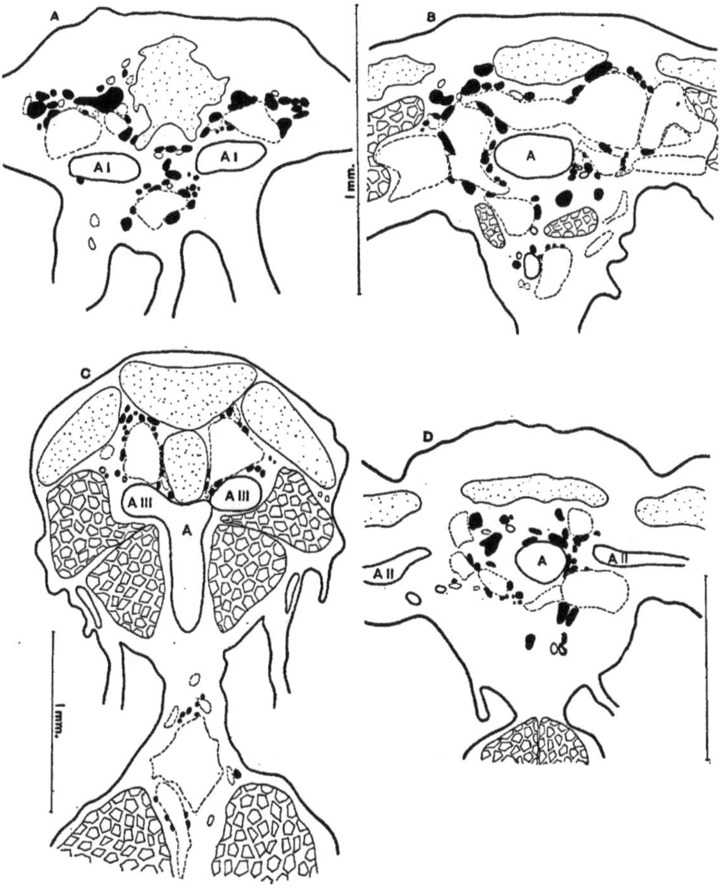

Fig. 5. *A* to *C*. Sections through the thyreoid gland of Salmo mykiss. *A*, in the aortic bifurcation; *B*, closely anterior to the second; *C*, at the third branchial arteries. *D*. Section through the thyreoid gland of Cristivomer at the second branchial arteries.

The follicles do not lie directly on the first branchial arteries, but are grouped around large veins or lymph sinuses, the dorsal follicles being much larger than the ventral ones. Behind the first arterial branches the dispersion of follicles is very much reduced. They are forced away from the dorsal region by the development of the basi- and hypobranchialia and occur only laterally and ventrally of the aorta. Further back even the lateral follicles disappear and only a few small ventral ones are grouped around a small vessel. Space again becomes available towards the second branchial arteries, since the skeletal parts retract more and more, and the follicles reappear in their former locations. The lateral extension however is not as great as in the region of the first arteries, since longitudinal muscle bundles prevent it (fig. 5, *B*). Dorsally the follicles again reach up to the pharyngeal floor. Close to the second branchial arteries the dorsal extension again diminishes and almost disappears when the second gill arteries are reached. Here the follicles lie far below the aorta, as they are forced away from the vessel by a longitudinal muscle. From this place backward a few follicles again appear above the aorta; they are small and scarce, five or six in a section, and widely scattered. Behind the third arteries the ventral aorta lies buried far beneath a muscle, between which and the skeletal parts a portion of the thyreoid lies. Another portion lies below the aorta between the third and fourth aortic branches, and here once more the amount of thyreoid tissue is slightly increased. A small mass of follicles disconnected from the main mass appears behind this place, lying below the aorta.

Cross sections through the follicles are usually circular, some are irregular. Their size decreases from the anterior towards the posterior end of the thyreoid region. The diameters vary between 10 and 60μ in a single section. The epithelium is rather flat, though some follicles have a cubical epithelium 3 to 5μ high. The nuclei of the flat cells show a peculiar feature; in all other cases they are either round or oval, but here with a few exceptions they are bent, taking forms ranging from wide arches to perfect horse-shoe shapes and are from 8 to 10μ long. In all probability they are degenerating, since they do not stain as deeply with nuclear dyes as do the round nuclei. Many of the follicles do not contain colloid.

The blood supply of the thyreoid zone is rich but there are no capillaries to the follicles proper, although there are smaller blood vessels in the region. Large veins and lymph vessels lie around the aorta and the follicles lie close to their walls (fig. 5, *B*).

CRISTIVOMER NAMAYCUSH WALBAUM

Length of specimen, 12 cm. The outlines of the thyreoid region in the great-lake trout are about the same as in the former species, but the ventral and posterior extensions are more limited. The anterior end lies in front of the aortic bifurcation, the posterior end at the third branchial arteries, (pl. II, fig. 24). The conditions from the aortic bifurcation to the second branches are the same as described in the species above but at the second arteries the accumulation of thyreoid material is rather large. Here also are found the largest follicles. The lateral extension is wider than at the first branches. The aorta is surrounded by follicles (fig. 5, D) but they do not lie very close to its wall. Posteriorly the extension decreases, three to four follicles being seen in a section above and below the aorta. The ventral follicles soon disappear and at the third aortic branches the dorsal ones also run out.

The follicles are a little larger than those of the black trout. Irregular and circular cross sections of the follicles are seen, the latter 20 to 100μ in diameter. The epithelial cells are generally cubical, about 6μ high. The nuclei are circular, 3μ in diameter, oval or somewhat irregular. The bent nuclei described in the black trout are present, but not so numerous. Some follicles show only regular nuclei, others only irregular, so that one might imagine these forms associated with different physiological stages. Almost all the follicles contain colloid.

There are many capillaries in the fatty tissue in close contact with the follicles. The follicles are not located on large veins and only a few lie close to the lymph sinuses.

SALVELINUS FONTINALIS L.

Length 4 cm., age 1 month. In this young brook trout the thyreoid gland has not developed very far, certainly not so far as Maurer describes for this stage. The follicles are scarce, the most anterior lying in the aortic bifurcation. Between the first and second branchial arteries there are a few follicles in each section, situated above the aorta; near the second a few appear below the aorta (fig. 6, *A*).

Length 25 to 30 cm. In the brook trout a condition of remarkably wide distribution of thyreoid material is seen. The region of the thyreoid in this species is comparatively larger than in any other fish. The anterior end of the gland is far in front of the aortic bifurcation and small follicles extend to the floor of the pharynx (fig. 6, *B*).

The first branchial arches are completely surrounded by thyreoid follicles. In the aortic bifurcation the follicles are very numerous, densely packed and occupy a rather large field. They reach up to the dorsal edge of the copula and laterally to the gill bases. On both sides of the aorta they are scattered between the fibres of longitudinal muscles (fig. 6, *C*). The follicles force their way through the muscle tissue along blood vessels and connective tissue fibres. Below the aorta their arrangement is less dense. Close behind the aortic bifurcation the amount of thyreoid tissue is reduced in the typical way, the copula extending down to the aorta. By this arrangement three, more or less separated, thyreoid masses are formed, two dorsally to the right and left of the copula and one below the aorta. The ventral part decreases, then the dorsal masses, the arrangement of the follicles becoming looser. Although the dorsal space becomes more open the follicles still decrease in size and are scattered far apart, indicating that this is a zone between two accumulations of thyreoid tissue, those around the first and second aortic branches. Two centers of growth may easily be determined.

Just before reaching the second branchial arteries the lateral extension becomes very great (fig. 6, *D*). The follicles migrate into the first gill arches along the branchial arteries and occur at the base of and extend into the second gill arches. This wide distribution of thyreoid elements is certainly the most remarkable feature of the organ in the Teleosts. Follicles not only lie at the base of the gills, but are distributed along the laminae at the base of the villi.

At the second branchial arteries the thyreoid gland, as mentioned above, once more shows an extensive development. Above

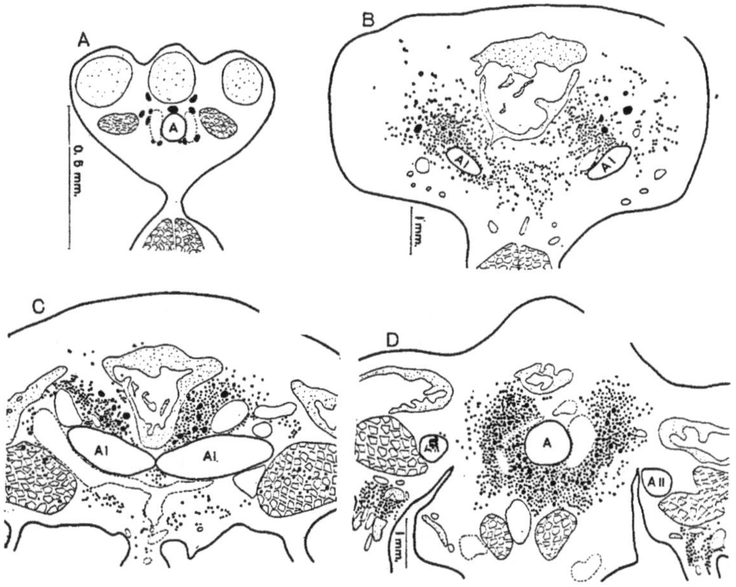

Fig. 6. *A*. Section through the thyreoid gland of a one month old specimen of Salvelinus fontinalis, between the first and second branchial arteries.

B to *D*. Sections through the thyreoid gland of an adult specimen. *B* and *C*, in the aortic bifurcation; *D*, near the second branchial arteries.

the aorta a dense arrangement of follicles is seen and below the number increases. Laterally the follicles extend along the arteries. Behind this place the aortic stem is entirely surrounded by thyreoid tissue which completely fills the space between bones and muscles. Small follicles are imbedded in the adventitia of the aorta. There is a dissolution of the dense arrangement in the peripheral zones, especially ventrally. Close behind the second branchial arteries the mass of thyreoid tissue decreases very suddenly and only a thin ring of follicles surrounds the aorta. Towards the third branchial arteries the aorta becomes buried between the ventral muscles, and the ventral follicles disappear sooner than the dorsal ones which continue and surround the third branchial arches for a short distance. Behind the third branchial arteries a small accumulation of thyreoid tissue once more appears.

A second series shows conditions similar to those above described. The follicles in the anterior region are less densely arranged. The basibranchiale comes very close to the aorta and separates to some extent two portions of thyreoid tissue along the aortic stem. The follicular mass is little reduced behind the aortic bifurcation but a little in front of the second branchial arteries the typical restriction is found. At this place the first ventral follicles appear. When the second aortic branches are reached the lateral extension of follicles becomes very wide. The mass of thyreoid tissue is here much increased, and the ventral portion is well developed but not so far as in the trout described above. The thyreoid stops close behind the second branchial arteries.

In a third series the separation of follicles in the anterior portions is still greater than described in either the first or second. The dorsal limit reaches to the upper edge of the basihyale, where there is an accumulation of follicles on both sides. The first branchial arteries are for a long distance completely surrounded by follicles, but the number of follicles decreases visibly towards their union; thus in this case there is an accumulation of follicles in front of the first aortic bifurcation. The ventral follicles appear at the first branchial arteries and disappear before reaching the second. It seems that here the entire thyreoid mass is pushed much farther towards the head than in the other trout described. Between the first and second branchial arteries the conditions are similar to those in the other specimens, the distribution of the follicles being restricted. There is no pronounced increase of thyreoid tissue or lateral distribution at the second branchial arteries and the posterior limit of thyreoid follicles is in front of the third branchial arteries.

High epithelial cells were predominant in the follicles of all the thyreoids. The cubical cells measure 9 to 10μ broad and 12μ high, and the narrow cylindrical cells are 2 to 3μ broad and 20μ high. The nuclei are usually large and round, except in the very high cells where they are compressed. In a few places not all the nuclei of a follicle show the same structure or the same reaction towards the stain and thus may be in different physiological stages. In addition to normal, large nuclei with distinct nucleoli and granular structure we find compact deeply staining nuclei which sometimes contain a vesicle. There are also small pyknotic nuclei in small (degenerating) cells. Often such compact nuclei with a halo of colloid are found within the lumen and it seems then that the epithelial cells have emptied their entire content. These masses can be easily distinguished in the colloid even after their outlines become indistinct as they have a different refractive index. Maurer describes somewhat similar structures in trout and carp. In other cases several neighboring cells with much swollen bodies have been pushed off from the epithelium and may be seen in the colloidal substance (pl. V, fig. 13).

The general form of the follicles is globular, though the surrounding fat and muscle tissue influences the outlines to some degree (pl. V, figs. 10-12).

Smooth muscle fibres are found in the entire thyreoid region; in one case (the first specimen) only ventral to the aorta. They run in all directions in the interfollicular tissue. The follicles are often arranged along them or are surrounded by them. Where the follicles lie in clusters of five or ten or more, smooth muscle fibres are found running between them. The muscle fibres with the follicles, their capillaries and the connective tissue fibres form a somewhat compact structure.

The blood supply to the secreting epithelium is extremely rich, several capillaries going to each follicle (pl. V, figs. 10, 12 *Ca*).

The thyreoid gland in two other species was dissected out as far as it was visible macroscopically. In this way of course one does not get the scattered follicles but only the main masses. Figs. 28 and 29 of plate III from these two dissections as well as figs. 25 to 27 of plate II, which are from specimens cut in serial sections, show that the distribution of the thyreoid in the trout is very variable.

ONCORHYNCHUS KISUTCH WALBAUM

Specimen 6 months old, 7 cm. long. The thyreoid gland in the silver salmon extends further back than in most of the trouts, reaching beyond the fourth branchial arteries (pl. V, fig. 20). Another feature in the arrangement is that the follicles lie rather close together, surrounding the stem of the ventral aorta through-

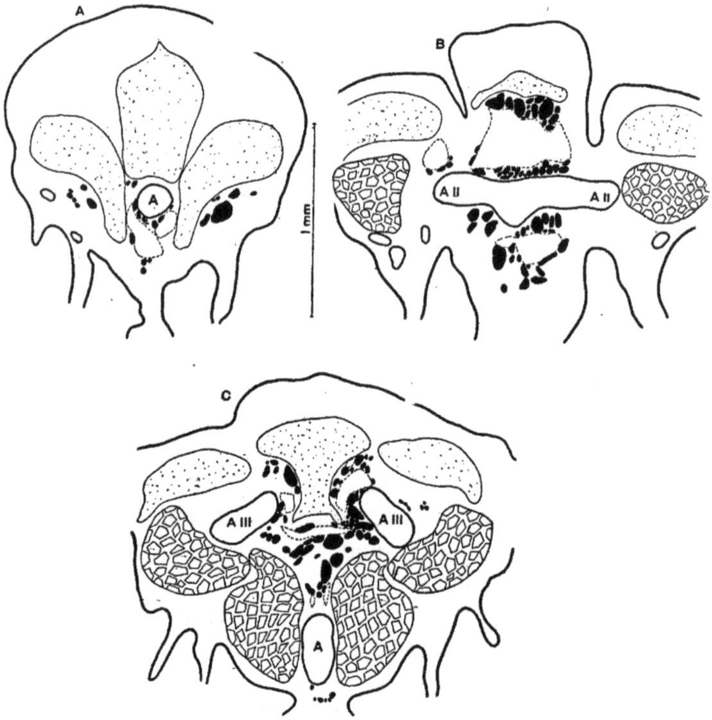

Fig. 7. Sections through the thyreoid gland of Oncorhynchus. *A*, just posterior to the aortic bifurcation; *B*, at the second branchial arteries; *C*, posterior to the third branchial arteries.

out almost its entire length. The amount of thyreoid tissue is small at the aortic bifurcation and between the first and second branchial branches (fig. 7, *A*). At the second gill artery the thyreoid tissue is most abundant.

In front of the aortic bifurcation the basi- and hypobranchialia reach down and here only a few follicles are found on both sides of the hypohyalia. Back of the place where the cartilages have retracted the longitudinal muscle bundles prevent the lateral expansion of the thyreoid (pl. V, fig. 14). At the second branchial arteries however, the mass of thyreoid tissue is very much increased, again surrounding the aorta (fig. 7, *B*). The ventral extension is pronounced. As a rule in the trouts no follicles lie directly against the aortic wall but here there is a complete ring of them around it. Above this ring lies a large lymph sinus and between it and the skeletal parts thyreoid tissue is again found. Towards the third aortic branches the cartilages again compress the aorta, and here the follicles lie around the aorta and along the outlines of the cartilages. Further back the aorta sinks down between the muscles, the ventral follicles disappear and the dorsal ones do not follow the vessel, but increase in number and group themselves around a subcopula between the third branchial branches. This dorsal rather compact group of follicles extends back behind the fourth branchial arteries. Below the aorta a small cluster of four or five follicles appears as is seen in other species of trout (fig. 7, *C*). At the level of the fourth aortic branches the copula extends so far ventrally that the dorsal follicles are pressed between the muscles and again come down into contact with the aorta. The posterior end of the thyreoid is in this region of the fourth arch.

The form of the follicles is usually elliptical, though circular cross sections are also found, ranging from 15 to 100μ in diameter. The larger ones are more abundant.

The follicular epithelium is very flat (pl. V, fig. 14), and in most of the cells are again seen the irregular nuclei described in some of the above species of trout (pl. IV, fig. 9, *N*). The majority of follicles are in close relation with large lymph sinuses, epi- and endothelium being in contact (fig. 7, *B*, *C*). There are no capillaries to the follicles proper.

OSMERUS MORDAX MITCHILL

A smelt 20 cm. long. This smelt presents the thyreoid conditions described below. A few follicles appear far in front of the aortic bifurcation (fig. 8, *A*), and further back more are arranged around the copula. At the bifurcation every section shows twelve to fifteen follicles between the stem of the aorta and the

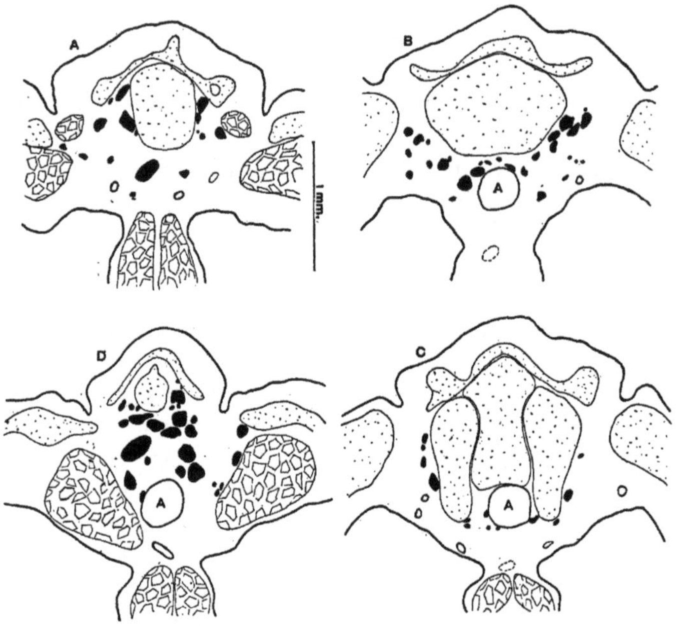

Fig. 8. Sections through the thyreoid gland of Osmerus. *A*, anterior to the aortic bifurcation; *B*, just posterior to it; *C*, between the first and second; and *D*, close to the second branchial arteries.

copula. The follicles vary in size from 40 to 200μ. Behind this the main mass of thyreoid tissue lies above the aorta, a few follicles lie to either side, and ventral to the aorta they are very scarce. Further back the basibranchiale comes nearer and nearer the aorta, finally reaching it, so that the follicles are forced out laterally (fig. 8, *C*). In the region of the hypobranchialia are seen only a few follicles far to the sides of the aorta and skeletal parts. Behind this the two hypobranchialia have retracted a little from the copula and very small follicles appear in the crevices between them. As the copula retracts from the aortic stem, more follicles appear on the dorsal side of the aorta. At the second branchial arteries the follicles cease (pl. I, fig. 4). The follicular epithelial cells are low cuboidal with the longer axis parallel to the base. The colloid appears homogeneous.

SIPHOSTOMA FUSCUM STORER

Specimen 30 cm. long. In the pipe-fish the thyreoid gland consists of entirely isolated follicles, lying above and to the sides of the aorta (fig. 9). The external form of the fish influences, of course, the form of its inner organs. The thyreoid gland has not found room for dorsal, ventral or lateral expansion and therefore extends far backwards as a rather narrow streak. The anterior end lies at the aortic bifurcation and the posterior end close to the bulbus arteriosus (pl. I, fig. 5). Thus we have a condition in which the organ reaches further towards the tail than usual and where the thyreoid region tapers towards the head end, while as a rule the reverse is true. The number of follicles is not very large, five or six to the section behind the aortic bifurcation. The number decreases towards the second branchial arteries and still more so towards the third, where a transverse muscle occupies the space between the bones and the aorta. At this place there are only one or two follicles in a section, yet there is a continuous chain of them. Near the third branchial arteries the aorta goes down ventrally, the transverse muscle has decreased, and thus the thyreoid finds more space for development. There are six or eight follicles in a section and they lie between the third gill branches which run dorso-laterally. Behind this place the dispersion of follicles increases (pl. IV, fig. 7), although the aorta lies far ventrally, a fact showing that the thyreoid follicles do not necessarily use the aortic stem as a migration path. On each side of the median line a muscle runs in an antero-posterior direction upon and under which the thyreoid follicles lie. The ventral group consists only of small follicles which have traveled downwards along a vein running between the two halves of the muscle. Further back the muscle bundles separate and here the greatest mass of thyreoid tissue is found. The space between the pharynx and bulbus is well filled by follicles which lie in a chaos of capillaries (pl. IV, fig. 7, *Ca*, *F*).

Histologically this gland is as different from that in other species as it is anatomically. The gland, when fixed, may have been

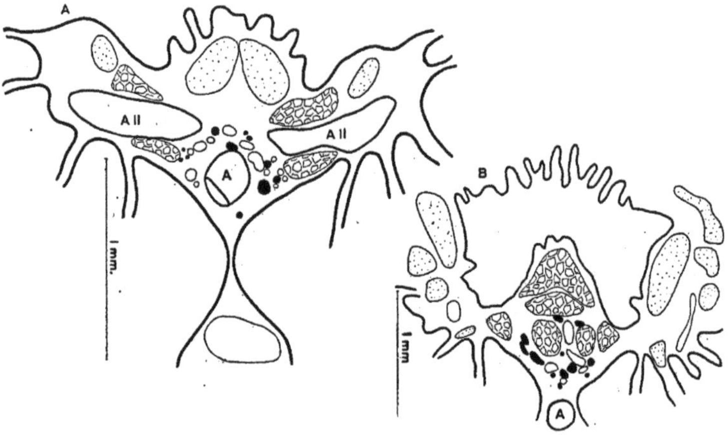

Fig. 9. Sections through the thyreoid gland of Siphostoma. *A*, at the second; and *B*, posterior to the branchial arteries.

in a peculiar state of function since there are no reasons to assume that the histological structures observed are permanent. Colloid was not found in any of the follicles, at least not as a uniformly compact mass. Certain follicles contained highly acidophile lumps about the size of epithelial cells. The epithelium, however, seemed in a state of colloidalization. The cells were high, cuboidal and swollen, with bulged out bases and surfaces (pl. IV, fig. 8). The nuclei were centrally located or towards the lumen. Thus they seemed to be typical colloid forming cells. The nuclei are in some cases round and massive, usually however they are very irregular. In some it seemed as though amitosis was taking place.

The formation of colloid ordinarily occurs in only a part of the thyreoid at a time. Here, however, the entire gland seemed to be in a similar physiological state.

FUNDULUS HETEROCLITUS L.

Specimens 10 cm. in length. The follicles in the region of the aortic bifurcation are grouped around a vein, most of them lying to the sides of it and under a transverse muscle. The elliptical shape of the vein in sections indicates the pressure between this muscle and the m. sternohyoideus which forces the follicles out from the median line. The follicles become more numerous towards the aortic bifurcation and they extend part way out along the first branchial arteries, and more on their ventral than dorsal side. Between the first and second gill branches follicles are found under and above the transverse muscles around which they have traveled. The ventral aorta in this region is completely surrounded by thyreoid tissue, more being found on the sides than either dorsally or ventrally (fig. 10, *B*). At the second gill branches the follicles again spread out laterally. Behind this place only a few scattered follicles are found (pl. I, fig. 6).

The size of the follicles varies extremely. The smallest are found at the anterior end and the largest in the middle of the thyreoid region. They are either circular in cross section, oval or with irregular evaginations.

The epithelial cells are usually cubical, but in very small follicles sometimes columnar, while in large empty follicles the cells are flat. Narrower cells with spindle shaped nuclei are seen in places.

The colloid is granular, and in some regions is seen to leave the follicle. Whether this is due to artificial pressure cannot be stated. Occasionally two neighboring epithelial cells will flatten out somewhat as if they were about to form a passage between them.

The blood supply to the thyreoid region is rich. The follicles are almost completely surrounded by a net of capillaries. These vessels are so pressed against the follicle that they form grooves in it (pl. IV, fig. 6). The projections of the epithelium between

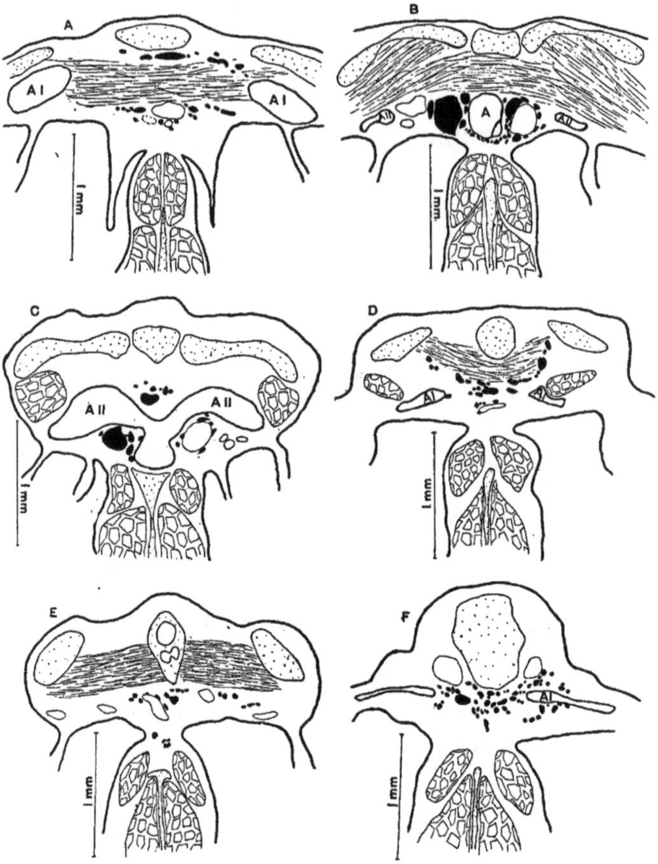

Fig. 10. Sections through the thyreoid gland of Fundulus. *A* to *C*, F. heteroclitus. *A*, in the aortic bifurcation; *B*, anterior to; *C*, at the second branchial arteries. *D*, F. diaphanus, in the aortic bifurcation. *E* and *F*, F. majalis, anterior to the aortic bifurcation.

the capillaries show narrower and longer cells, and some of these cells entirely lose their communication with the follicular lumen. The connective tissue sometimes forms an almost complete sheath around the follicles and their capillaries. Red blood corpuscles in all stages of disintegration are found in many follicles (see General Part). How these corpuscles get into the lumen could not be determined. Erythrocytes are often seen partially imbedded in the follicular epithelium as if they would force their way in between two cells. In other places corpuscles are found pressed against an epithelial cell which has so flattened out that only a thin layer of cytoplasm separates the corpuscle from the lumen.

FUNDULUS DIAPHANUS LE SUEUR

A specimen 9 cm. long. The main mass of the thyreoid is located a little nearer the tip of the tongue than in F. heteroclitus (pl. I, fig. 7). The posterior end lies at the second branchial arteries where the follicles become scarce and scattered. A further difference from heteroclitus is that the main mass of follicles always lies above the aortic stem, only a few small ones lying below. The lateral extension is here also unimportant.

The floor of the pharynx is narrow and the connection between it and the ventral musculature is only a narrow streak. In heteroclitus the lateral pharyngeal axis is the longer one, therefore, the lateral thyreoid extension prevails; while in diaphanus the dorso-ventral axis is longer, and here the extension of the thyreoid is mainly in this direction. Ventrally, however, it is prevented by the narrow isthmus, and follicles are mainly found above the aorta (fig. 10, *D*). In this way the distribution of the follicles may be figured out mechanically in almost every case.

The follicles are of all sizes, though not so large as in heteroclitus. There are more elliptical or irregular ones and these have a longer axis. The cuboidal cells of the follicular epithelium are not as high as in heteroclitus and cylindrical ones are not found. The colloid is homogeneous and the blood supply is not rich.

FUNDULUS MAJALIS WALBAUM

Length of specimen 9 cm. The follicles spread out laterally much further than in the other two species (fig. 10, *E*, *F*). They extend for a distance along the first aortic branches. Between the first and second branches there is only a narrow streak of thyreoid tissue, but the main mass of the organ lies at the second gill branches and here the greatest lateral extension occurs under a transverse muscle. The vertical extension is small and there are no follicles below the aorta. Behind the second gill branches is found the posterior limit of the gland (pl. I, fig. 8).

The follicles are still smaller than in diaphanus and more uniform in size. The circular type predominates and they are more numerous than in the other species. The colloid is homogeneous and the follicular epithelium similar to that in diaphanus.

MENIDIA NOTATA MITCHILL

Length of specimen 10 cm. The thyreoid mass is rather small (pl. I, fig. 9). The follicles are extremely small, 20–25μ, and are scattered along the stem of the aorta between the first and second branchial arteries and out along the second arteries. The lateral extension is greater than the antero-posterior.

MUGIL CEPHALUS L.

A mullet 15 cm. long. Small follicles are found in the anterior end of the thyreoid region and are grouped around a vein (fig. 11, *A*). At the aortic bifurcation the organ is better developed, but is hardly in contact with the gill vessels (fig. 11, *B*). The thyreoid lies above the aorta, and at the second branchial arteries it comes into contact with the vessel. Here the gland is well developed with numerous large follicles. The follicles disappear towards the third aortic branches (pl. I, fig. 13).

The size of the follicles varies between 30 and 140 μ. In section they are slightly oval. In the follicular wall are found transitions from flat to high epithelium. The height of the cells varies within the same follicle, showing that it is independent of follicle size. The height of the cells rather depends upon outside pressure, *e.g.*, a follicle pressed into oval shape by cartilage shows low epithelium on the longer sides and higher cells on the short sides.

SARDA SARDA BLOCH

Length of specimen 50 cm. The thyreoid gland of the Spanish mackerel shows the most remarkable conditions of all fish thyreoids.

The mass of the organ is enormously large and the dorso-ventral and cephalo-caudad extensions are unusual. The relation of the thyreoid gland to other tissue is singular, and could be compared only with that in Brevoortia. There exists such an intermingling of thyreoid, bone, cartilage, smooth and striated muscle fibres, fat and connective tissue that it is impossible sharply to define

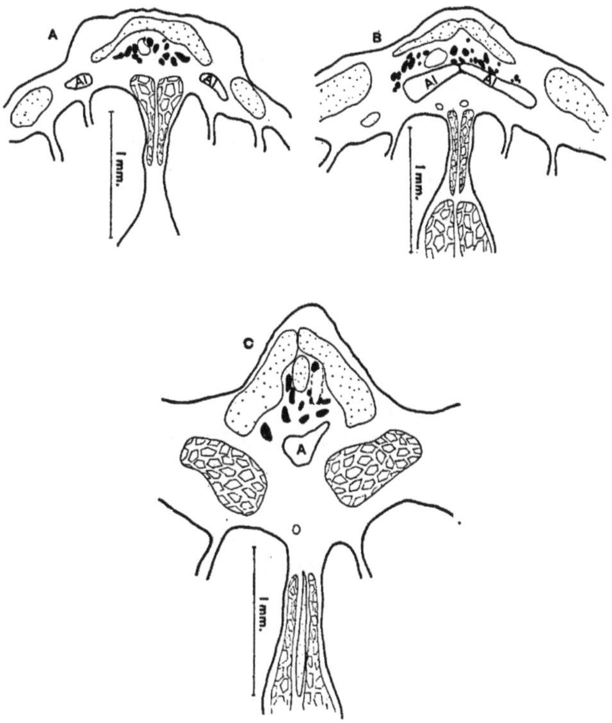

Fig. 11. Sections through the thyreoid gland of Mugil. *A*, anterior to; *B*, at the aortic bifurcation; *C*, near the second branchial arteries.

the organ. Yet on the other hand, there can hardly be found a group of follicles detached from the main thyreoid body.

The isthmus is long, as in Brevoortia, and hence the thyreoid region is much elongated (pl. I, fig. 10), measuring 4 cm. in length. It is not, however, as narrow as in the menhaden, having a wide lateral extension. The anterior end is pushed far forward, 2.5 cm. in front of the aortic bifurcation, so that it also comes to lie in front of the hyoid arch. The entire development of the organ takes place more cephalad than usual and the main mass lies in front of the aortic bifurcation (sharks!), deeply buried in the body of the tongue, as a consequence of the ventral extension of the copulo-hyoid (fig. 12, *A*). It occupies a more ventral position than any other fish thyreoid. The follicles are located around a large vein and are rather closely arranged. As the basi- and hypohyalia retract the follicles creep into the clefts between them and thus the thyreoid mass assumes the shape of a horse-shoe, the two arms of which point dorsally (fig. 12, *B*, *C*). The smooth muscle fibres of this region are completely invaded by follicles (pl. IV, fig. 21), as are also the bones of the gill arch, especially the copula, in regions where they lose their compactness and break up into lamellae. The thyreoid takes the form of three masses converging ventrally, and as we pass back it expands more and more on the sides, 6 to 7 mm., while the median branch becomes smaller. About one cm. in front of the aortic bifurcation the most extensive region of the gland is reached. In cross section the mass is rhomboidal, the diagonals being about 7 and 4 mm. The lateral extension decreases while the ventro-median mass increases, from which two branches tend dorsally along the edge of the copula. Thus again the sections show a horse-shoe shape, with a broad middle piece and narrow dorsally converging arms, in which the follicles are oval with their longer axis parallel to the line of extension. On reaching the first branchial arteries, which run in this species towards a ventro-lateral zone and do not come into contact with the follicles (fig. 12, *C*) we pass to their union where a few follicles surround them (fig. 12, *D*). The central portion of the gland becomes smaller and lies separate in the aortic bifurcation while

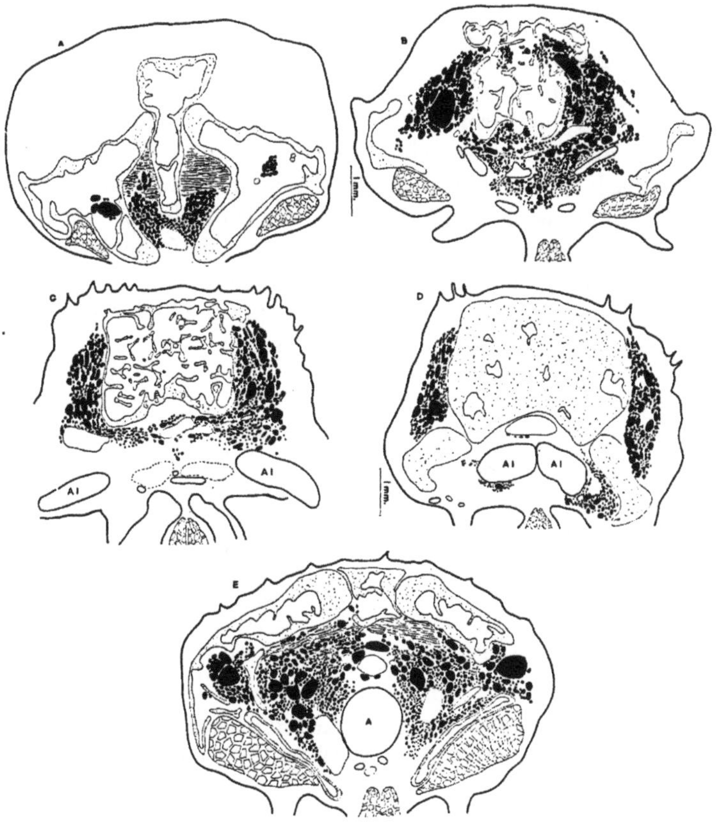

Fig. 12. Sections through the thyreoid gland of Sarda. *A*, near the anterior end of the gland; *B*, in the region of greatest extension anterior to the aortic bifurcation; *C* and *D*, close to the aortic bifurcation; *E*, near the second branchial arteries, the region of greatest extension.

the lateral parts increase posteriorly. Thus in the sections there are three portions of thyreoid which become more and more separated by the enlargement of the copula (fig. 12, *D*). Behind the aortic bifurcation some follicles appear below the aorta; the middle mass again enlarges and the three parts unite. One branch again extends into the copula and soon becomes smaller, while the lateral portions increase. The posterior end of the gland is found behind the second gill branches.

The follicles are usually circular or oval in cross sections, though many are polygonal from pressure. Their size varies between 30 and 350μ medium sizes being most abundant. Giant follicles reach 800μ long by 400μ in short diameter. Many follicles are without colloid, while in others the colloid is much more shrunken than usual. The colloid is homogeneous. The follicular epithelium is of high cylindrical cells or cubical ones. The cytoplasm is stained more darkly in the basal portions; in the higher parts it is sometimes reddish. The blood supply is rich.

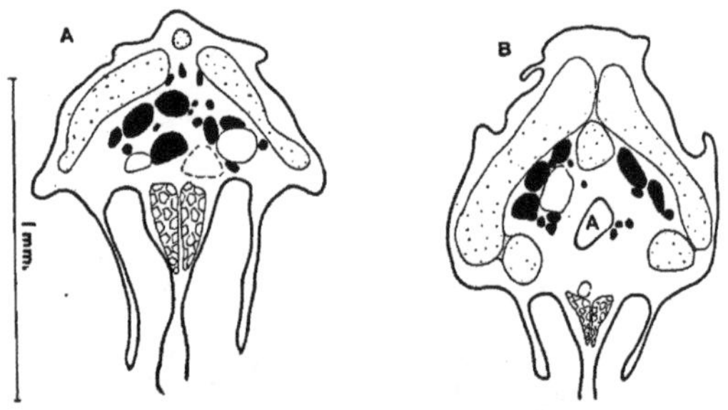

Fig. 13. Sections through the thyreoid gland of Pomatomus. *A*, anterior to the aortic bifurcation; *B*, between the second and third branchial arteries.

POMATOMUS SALTATRIX L.

Young bluefish 30 cm. long. In this species the dispersion of the thyreoid follicles is prevented in both a lateral and dorso-ventral direction, since the arch formed by the basibranchiale and hypobranchialia is very narrow (fig. 13). The gland is thus a long narrow streak (pl. I, fig. 11). At the aortic bifurcation there are only a few follicles, some of which lie close to the first gill arteries, just in front of their point of union. The thyreoid mass reaches its maximum extension above the ventral aorta and between the first and second gill arteries, especially towards the second. But even here there are only ten or fifteen follicles in a cross section. Some follicles lie close to the base of the second gill arteries and from this point the gland extends, with from six to ten follicles in a cross section, to a little behind the third branchial arteries where it ends. The follicles are generally dorsal to the ventral aorta (fig. 13, *B*), only a few being below it.

The form of the follicles is irregular, but approaches the globular type. Their size ranges from 15–100μ in diameter though some are far above this size (giant follicles). The minute histology shows no peculiarities. The epithelium is usually cubical, the cells being 6μ high.

MORONE AMERICANA GMELIN

Specimen 35 cm. in length. The thyreoid gland of the white perch is characterized by the enormous size of nearly all the follicles as well as by their unusually loose arrangement. Cephalad of the aortic bifurcation there is little room for dispersion since the copula reaches far down and the skeletal arch is rather narrow. Behind the bifurcation (fig. 14, *B*) this arch becomes wider and from here to the second gill arteries the main mass of the thyreoid is situated (pl. I, fig. 12). From the second branchial arch towards the third two narrow lines of follicles run along the sides of the aorta. The entire length of the thyreoid region measures 3.5 cm. The majority of the follicles lie above the aorta except in the anterior region.

The size of the follicles varies from 120 to 600μ in diameter, the very large ones are most abundant especially in the more anterior region. In cross sections the follicles are almost all circular. The epithelial cells are low, 3 to 4μ high. In these follicles there are no indentations in the colloid, it either fills out the lumen completely or is retracted from the epithelium and has a smooth edge. (The differences in the colloid of different species may be of some physiological significance.)

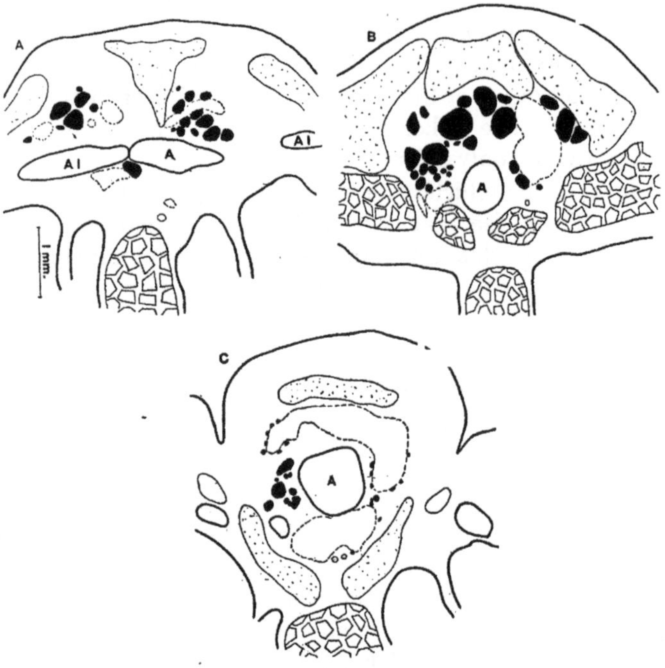

Fig. 14. Section through the thyreoid gland of Morone. *A*, at the aortic bifurcation; *B*, between first and second; *C*, close to the second branchial arteries.

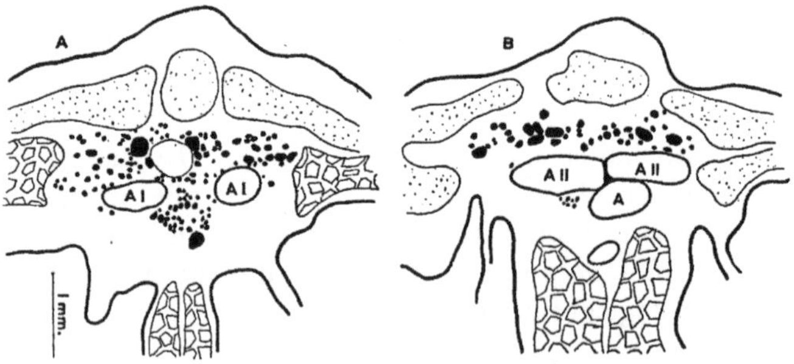

Fig. 15. Sections through the thyreoid gland of Stenotomus. *A*, in the aortic bifurcation; *B*, at the second branchial arteries.

STENOTOMUS CHRYSOPS L.

Length of specimen 25 cm. The scup presents the thyreoid gland as a rather continuous organ, only one group of follicles lying below the aorta is isolated from the main mass. The largest expansion of the gland is in (fig. 15, *A*) and immediately behind the aortic bifurcation; here it measures 3 mm. in width, and dorso-ventrally over 1 mm. This expansion is followed by a restriction, the follicles always lying above the aorta. At the second branchial arches another increase in the thyreoid tissue occurs, and here a few follicles appear below the aorta (pl. I, fig. 14).

The size of the generally circular follicles varies from 20 to 300μ in diameter, a few reaching 400μ.

CYNOSCION REGALIS BLOCH

Specimens of 60 cm. in average length. Twelve specimens of the squeteague were examined and they serve to show a series of variations in the thyreoid gland within the species. The region of the gland extends from in front of the aortic bifurcation to the third branchial arteries. The majority of follicles always lie either dorsal or lateral to the aortic stem and in only two cases were any follicles found below the aorta. In one case the aortic stem between the first and second branches was surrounded. The region of the second aortic branches is commonly filled by the gland. The tendency to extend from this place anteriorly is more often expressed than in the opposite direction. The lateral extension is greater along the branchial arteries than in intermediate regions (pl. III, figs. 31–41).

In some of the specimens there were two (pl. III, figs. 33, 36, 37, 40) or even three and four (pl. III, fig. 34) well developed isolated portions of the gland lying on different branches of the gill vessels. Macroscopically they appear to be separated, but on tracing the entire region in serial sections it is found that follicles spread out and connect the several masses, although the follicles are small and scattered so thinly that they were not seen with the naked eye.

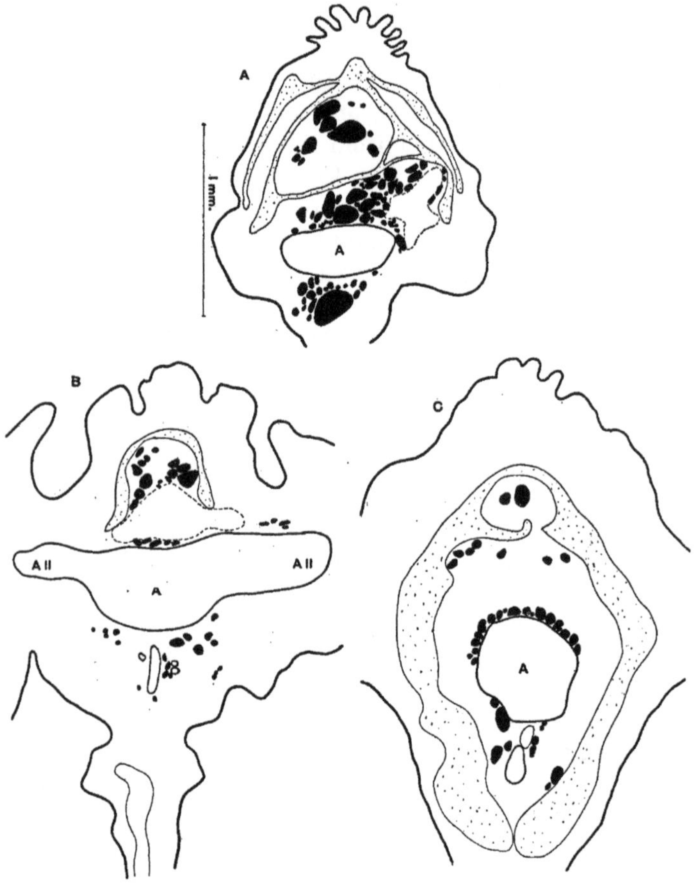

Fig. 16. Sections through the thyreoid gland of Micropogon. *A*, between the first and second; *B*, at the second; *C*, near the third branchial arteries.

The mass of thyreoid tissue, roughly judging, differed in the specimens, although they were of about the same size, yet the fish may have been of different ages.

The shape of the follicles varies from very irregular to circular. Their size also varies extremely. Those lying nearest the vessels are huge and irregular, while the small peripheral ones approach a globular shape. This indicates that the shape of the follicles is a result of the pressure directions. The follicular arrangement is rather compact in the central portions. (pl. V, fig. 17.)

The epithelial cells vary from low cuboidal to high cylindrical shapes. The smaller follicles seem to have a little higher epithelium, though it is rather uniform in the same individual and varies more among the several specimens. It may seem therefore that the entire gland is in the same physiological stage.

MICROPOGON UNDULATUS L.

A croaker, 30 cm. long. The thyreoid extends from the first to the third branchial arteries (pl. II, fig. 15). The dispersion of follicles is largely dorso-ventral, since laterally they are hindered by the narrowness of the isthmus (fig. 16, *A*, *C*). For this reason also a considerable part of the gland lies below the aorta, yet not so large a portion as above, though the dorsal follicles are less densely arranged.

There are only a few follicles in front of the aortic bifurcation, yet at the bifurcation and behind it lies the main mass of the gland. The follicles completely fill the spaces between bones and vessels (fig. 16, *A*). Towards the second gill branches the copula extends further and further down and forces the follicles into a somewhat lateral position. The ventral mass is larger in this region. At the second arterial branches there is no special increase in mass, the number of ventral follicles having decreased (fig. 16, *B*), the dorsal ones increasing and soon extending to the epithelium of the pharyngeal floor. The follicles lie rather loosely arranged, but have not noticeably increased in size. A small line of follicles above the aorta extends from here towards the third gill branches, others are scattered irregularly around the aorta. The aorta has sunk into the ventral muscle and carries the posterior follicles with it.

The thyreoid gland of Micropogon is characterized by rather small follicles of almost uniform size, though in some regions large ones appear. The diameters range from 10 to 300μ, but those of 30 to 50μ are most abundant.

The epithelium is rather low, even in the smallest follicles. Branched follicles are numerous. The blood supply is rich, many capillaries being present around the follicles. There are several larger veins running through the thyreoid region.

TAUTOGOLABRUS ADSPERSUS WALBAUM

Length of specimen 25 cm. In the cunner the thyreoid gland occupies a unique position, almost resembling that in the sharks. It is pushed far forward in the aortic bifurcation, and touches both the first branchial arteries laterally (fig. 17, *B*), but does not extend far enough back to come into contact with the ventral aorta (pl. II, fig. 16). The main mass is, as it were, imbedded in a bony capsule. The follicles are grouped around a median vein (fig. 17, *A*). Dorsal to the aortic bifurcation the copula and a transverse muscle are well developed, so that the thyreoid is forced forward. The follicles are not numerous, and are all more or less irregular. Their diameters measure from 15 to 200μ. The follicular epithelium is cuboidal.

TAUTOGA ONITIS L.

Specimen 35 cm. long. In the closely related tautog the thyreoid gland also occupies a rather cephalad position (pl. II, fig. 17). It extends back from within the aortic bifurcation almost to the second branchial arteries. It lies chiefly above and to the sides of the aorta. The anterior, main part, is imbedded in an osseous capsule which is square in cross section, and is formed by three branchial bones above and a ventral supporting bone (fig. 18, *A*). At the aortic bifurcation the capsule becomes incomplete and the follicles are widely dispersed over 6 mm. (fig. 18, *B*). The follicles follow the dorsal side of the first arterial branches out to the base of the gills. Behind the first branchial arteries the lateral extension decreases and the follicular dispersion is in a dorso-ventral direction. The follicles are loosely arranged, and yet globular ones are rare, most of them being polygonal in outline. The average size is 150μ in diameter, but there are a few giant folli-

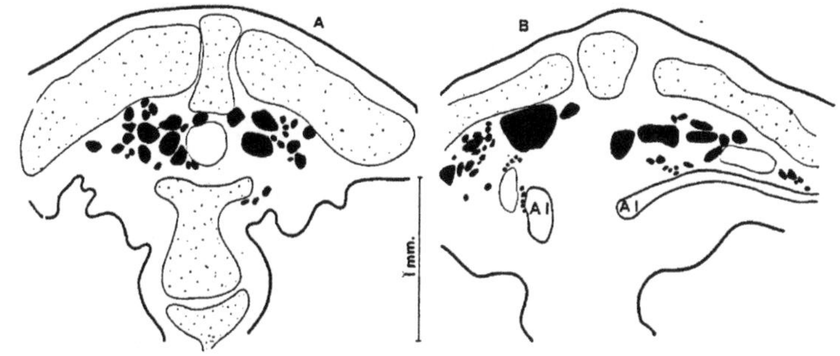

Fig. 17. Sections through the thyreoid gland of Tautogolabrus. *A* and *B*, anterior to the aortic bifurcation.

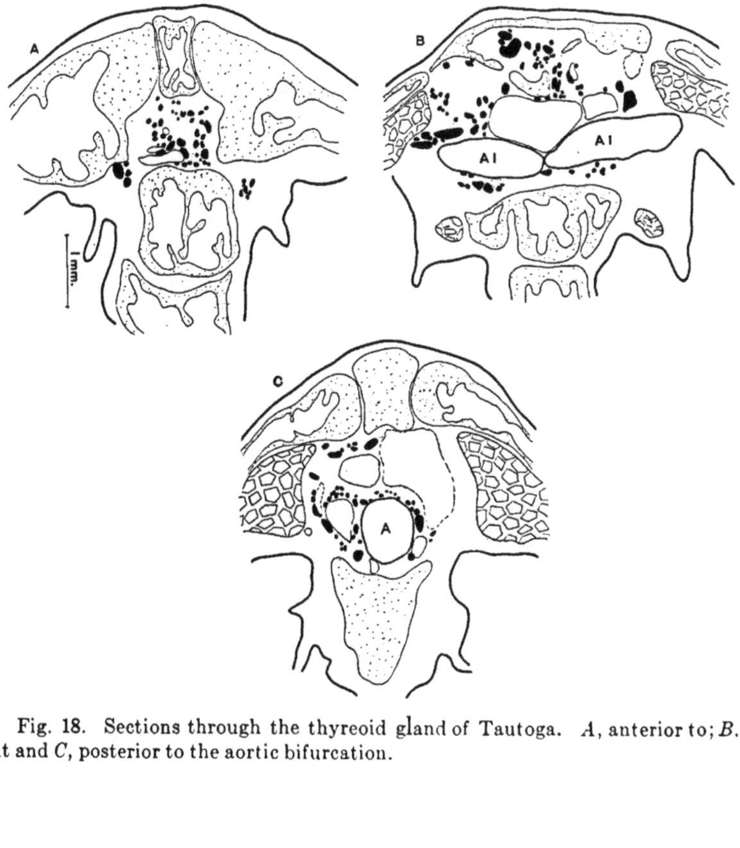

Fig. 18. Sections through the thyreoid gland of Tautoga. *A*, anterior to; *B*, at and *C*, posterior to the aortic bifurcation.

cles, 700μ long by 400μ broad. Branched follicles are numerous. The epithelial cells are cuboidal in shape.

PRIONOTUS CAROLINUS BLOCH

In a sea-robin 30 cm. long the thyreoid gland seemed to show a pathological appearance. The invasion of the surrounding tissues by thyreoid follicles was extraordinary, but may be abnormal. For this reason it can only be stated that the gland in this species occupies a posterior position, close to the origin of the truncus arteriosus.

OPSANUS TAU L.

A toadfish 30 cm. long the gill region in this species is extremely shortened, and therefore the thyreoid region begins rather far forwards. Anteriorly the largest follicles lie on both sides of a process of the copula which extends ventrally (fig. 19, *A*). Towards the aortic bifurcation the size of the follicles decreases and the two lateral portions unite in the median line, at the same time the lateral extension (fig. 19, *B*) of the follicles increases remarkably (pl. II, fig. 18). Some follicles appear below the aortic stem. Between the first and second branchial arches the number of follicles decreases above the aorta, while ventrally they disappear entirely. Along the second branchial arteries the follicles again reach laterally and also again appear ventrally. Behind this point the aorta sinks more and more and the space around it becomes freer. Yet there is no special increase of thyroid tissue in this region, there being only loosely scattered small follicles. A few follicles accompany the aorta in its course into the space between the musculus sternohyoideus. The caudal end of the thyreoid lies behind the third gill branches.

The arrangement of the follicles is loose, and they are usually circular in cross sections. Some are flattened between the bony and muscular surrounding tissues. Their size varies extremely. The largest ones, 600μ in diameter, lie in the anterior end, which is the reverse of the general rule for other species. In other regions the follicular diameters vary from 50–400μ, the median size follicles being the most abundant.

The follicular epithelium is always cuboidal. Colloid is present in almost all the follicles, and is very brittle and homogeneous. In the larger follicles the colloid stains much lighter than in the

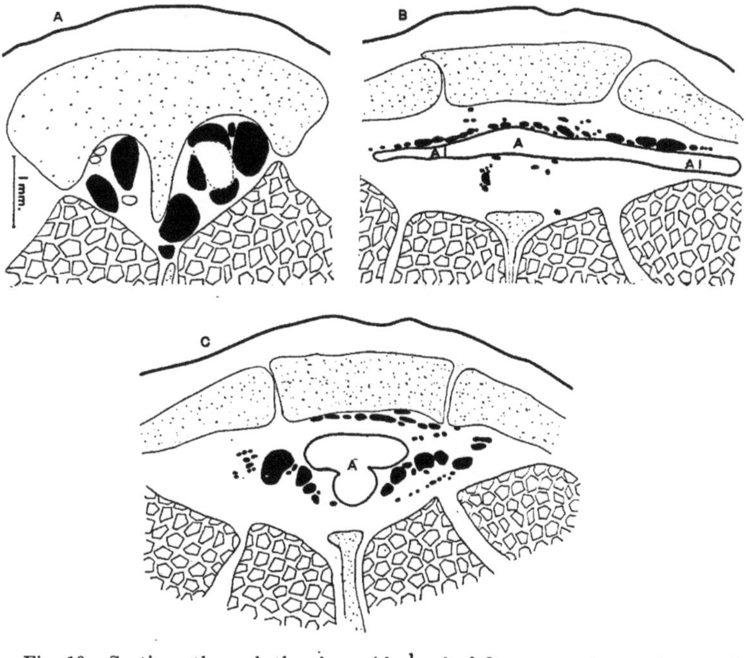

Fig. 19. Sections through the thyreoid gland of Opsanus. *A*, anterior to; *B*, at the aortic bifurcation; *C*, at the second branchial arteries.

smaller. Lymphocytes are numerous within the follicles. The blood supply of the thyreoid region is very poor.

MURAENOIDES GUNELLUS L.

Length of specimen 40 cm. The thyreoid gland in the butterfish reaches a considerable size (pl. II, fig. 19). The anterior end lies in front of the aortic bifurcation and consists only of small

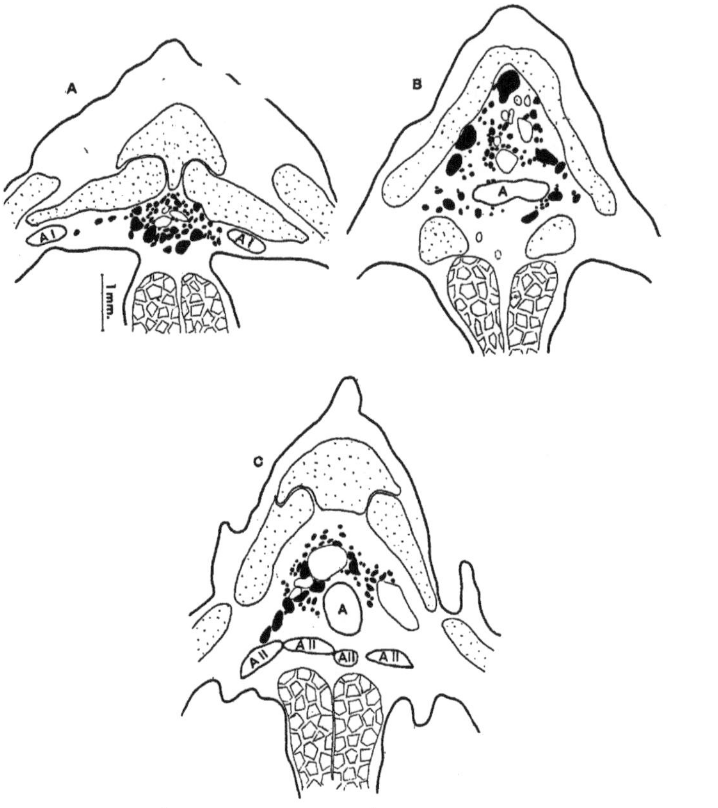

Fig. 20. Sections through the thyreoid gland of Muraenoides. *A*, anterior to the aortic bifurcation; *B*, between the first and second; *C*, close to the second branchial arteries.

scattered follicles. In the middle portion of the gland the follicles are numerous and closely arranged. At the aortic bifurcation they lie around a large vein and completely fill out the space between the gill arch and muscles. The lateral extension of the follicles is small as compared with the dorso-ventral, since the isthmus is narrow (fig. 20, *B*). A cross section through the thyreoid area measures about 1 mm. Taking 1 mm. as the average width we would have 10 cubic mm. of thyreoid tissue in this species.

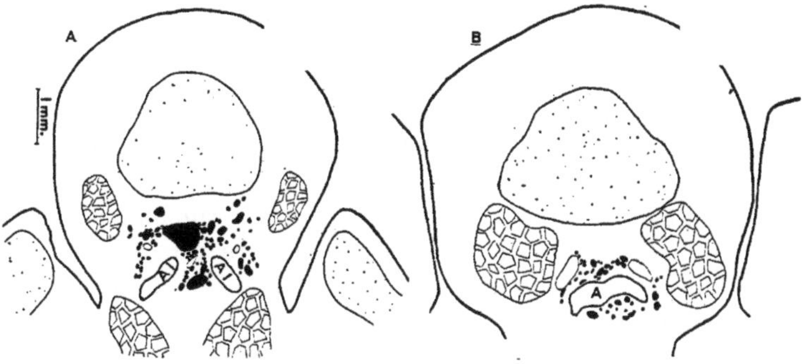

Fig. 21. Sections through the thyreoid gland of Pleuronectes. *A*, in the aortic bifurcation; *B*, just posterior to it.

The first branchial arteries are partly surrounded by follicles. Behind the aortic bifurcation there is an open space for lateral extension, but not for ventral, since the aorta rests on the musculature. The caudal end of the gland lies a little behind the second branchial arteries, and consists again of small scattered follicles (fig. 20, *C*).

The follicles are of globular or long ovoid shape, some are very irregular. The circular cross sections vary from 20 to 500μ in diameter. The very large ones lie at the second arterial branches. Branched follicles with connecting channels between them are numerous, so that almost all follicles may be traced in sections as evaginations of others.

The follicular epithelial cells vary from highly cylindrical, narrow shapes to broad cuboidal. Flattened epithelium is rare. The cells are extremely numerous and densely arranged. The nuclei are located near the base of the cells, even in the higher ones, and the cytoplasm stains darker about the nucleus. Sometimes it appears as if there were a cuticle on top of the cells, as many authors have described. This, however, is nothing else than a refractive appearance of the cell margin from which the cytoplasm has slightly retracted. The blood supply is extremely rich.

(A parasitic worm was found in this thyreoid and had caused a considerable hemorrhage.)

PSEUDOPLEURONECTES AMERICANUS WALBAUM

Length of specimen 45 cm. The position of the thyreoid gland in the flounder varies (pl. III, figs. 30, 31). In one case it formed a rather compact nodule between the first and second branchial arteries, while in another the main mass was found in the aortic bifurcation between and surrounding the first branchial branches (fig. 20, *A*). Behind the aortic bifurcation there were only smaller follicles dorsal and lateral to the aorta. The broad base of the deep reaching copula permits only a lateral extension; thus the thyreoid presents itself as a transverse streak. Small detached follicles lie close to the base of the gills.

The size of the follicles varies from 15 to 1000μ in diameter, those of about 200μ being in the majority. There are also a few 'giant' follicles. The epithelial cells of the follicles are closely arranged and rather high. The nuclei are oval. The blood supply is rich and lymphatic vessels are well developed.

BIBLIOGRAPHY

ANDERSON, O. A. 1894 Zur Kenntnis der Morphologie der Schilddrüse. Arch. f. Anat. u. Phys., Anat. Abt., 177.

BABER, S. C. 1876 Contributions to the minute anatomy of the thyroid gland of the dog. Phil. Trans. R. Soc., London, 166, Part 2, 557.

1881 Researches on the minute structure of the thyroid gland. Phil. Trans. R. Soc., London, 172, 577.

BOÉCHAT, P. A. 1873 Recherches sur la structure normale des corps thyroïde. Paris.

BORCEA, J. 1907 Observations sur la musculature branchiostégale des Teleostéens. Ann. Sc. Univ. Jassy, 4, 203.

COLE, F. J. 1905 Notes on Myxine. Anat. Anz., 27, 324.

CORI, C. J. 1906 Das Blutgefässytem des jungen Ammocoetes. Arb. Zool. Inst. Wien, 16, 217.

DOHRN, A. 1885 Studien zur Urgeschichte der Wirbeltiere. Mitt. Zool. Stat. Neapel.

ERDHEIM, J. 1903 Zur normalen und pathologischen Histologie der Glandula thyreoidea, Parathyreoidea und Hypophysis. Ziegler's Beitr. z. path. Anat. u. allg. Path., 33, 158.

FERGUSON, J. S. 1911 The anatomy of the thyreoid gland of Elasmobranchs with remarks upon the hypobranchial circulation in these fishes. Am. Jour. Anat., Vol. 11, No. 2.

FORSYTH, D. 1908 The comparative anatomy, gross and minute, of the thyroid and parathyroid glands in mammals and birds. J. Anat. and Phys., 42, 141, 302.

GALEOTTI, G. 1897 Beitrag zur Kenntnis der Secretionserscheinungen in den Epithelzellen der Schilddrüse. Arch. f. mikr. Anat., 48, 305.

GREIL. 1906 Ueber die Entstehung der Kiemendarmderivate von Ceratodus F. Verh. Anat. Ges., 20. Vers., 115.

GUDERNATSCH, J. F. 1909 The structure, distribution and variation of the thyreoid gland in fish. Am. Ass. Cancer Research, Nov. 27, 1909. (J. Am. Med. Ass., 54, 227.)

HÜRTHLE, K. 1894 Beiträge zur Kenntnis der Secretionsvorgänge in der Schilddrüse. Arch. f. d. ges. Physiol. 65, 1.

JORDAN AND EVERMANN. 1906–'00 The fishes of North and Middle America.

KÖLLIKER, A. 1861 Entwicklungsgeschichte d. Menschen u. d. höheren Tiere, Leipzig.

LANGENDORF, O. 1889 Beiträge zur Kenntnis der Schilddrüse. Arch. f. d. ges. Physiol., Suppl., 219.

LIVINI. 1902 Organi del sistema timo-tiroideo nella Salamandrina perspicillata. Arch. It. Anat. Embr., Firenze, 1, 1.

MARCUS, H. 1908 Beiträge zur Kenntnis der Gymnophionen. I. Ueber das Schlundspaltengebiet. Arch. f. mikr. Anat., 71, 695.

MARSHALL, C. F. 1895 Variation in the form of the thyroid gland in man. Jour. Anat. and Phys., 29, 234.

MAURER, FR. 1886 Schilddrüse und Thymus der Teleostier. Morph. Jahrb., 11, 129.

1888 Schilddrüse, Thymus und Kiemenreste bei Amphibien. Morph. Jahrb., 13, 296.

MÜLLER, L. T. 1896 Beiträge zur Histologie der normalen und der erkrankten Schilddrüse. Ziegler's Beiträge z. path. Anat. u. allg. Path., 19, 127.

MÜLLER, W. 1871 Ueber die erste Anlage der Schilddrüse und deren Lagebeziehung zur ersten Anlage des Herzens bei Amphibien, insbesonders bei Triton alpestris. Anat. Hefte, 26, 1.

MUTHMANN, E. 1904 Ueber die erste Anlage der Schilddrüse. Anat. Hefte, 26, 1.

PEREMESCHKO. 1867 Ein Beitrag zum Bau der Schilddrüse. Zeitschr. f. wiss. Zool., 17, 279.

PLATT, J. 1896. The development of the thyroid gland and of the suprapericardial bodies in Necturus. Anat. Anz., 11.

SCHAFFER, J. 1906 Berichtigung, die Schilddrüse von Myxine betreffend. Anat. Anz., 28, 65.

SCHMID, E. 1896 Der Secretionsvorgang in der Schilddrüse. Arch. f. mikr. Anat., 47, 181.

SILVESTER, C. F. 1905 The blood-vascular system of the tile-fish, Lopholatilus chamaeleonticeps. Bull. Bur. Fish. Washington, 24, 87.

SIMON, J. 1844 On the comparative anatomy of the thyroid gland. Phil. Trans. R. Soc., London, 134, 295.

STOCKARD, CH. R. 1906 The development of the thyroid gland in Bdellostoma stouti. Anat. Anz., 29, 91.

STRECKEISEN, A. 1886 Beiträge zur Morphologie der Schildrüse. Virchow's Arch. f. path. Anat., 103, 131, 215.

PLATES

EXPLANATION OF PLATES I TO III

Plates I and II show the regions of distribution of the thyreoid elements in different species. I, aortic bifurcation or first branchial arteries; II, III and IV, the second, third and fourth branchial arteries.

1 Anguilla chrysypa
2 Clupea harengus
3 Brevoortia tyrannus
4 Osmerus mordax
5 Siphostoma fuscum
6 Fundulus heteroclitus
7 Fundulus diaphanus
8 Fundulus majalis
9 Menidia notata
10 Sarda sarda
11 Pomatomus saltatrix
12 Morone americana
13 Mugil cephalus
14 Stenotomus chrysops
15 Micropogon undulatus
16 Tautogolabrus adspersus
17 Tautoga onitis
18 Opsanus tau
19 Muraenoides gunellus
20 Oncorhynchus kisutch
21 Salmo mykiss
22 Salmo irideus, age 1 month
23 Salmo irideus, age 1 year
24 Cristivomer namaycush
25–27 Salvelinus fontinalis

Plate III, Diagrams of actual portions of the thyreoid gland visible to the naked eye.

28 and 29 Salvelinus fontinalis

30 and 31 Pseudopleuronectes americanus

32–41. Ten specimens of Cynoscion regalis, demonstrating the great variability in extent and position of the organ within the species.

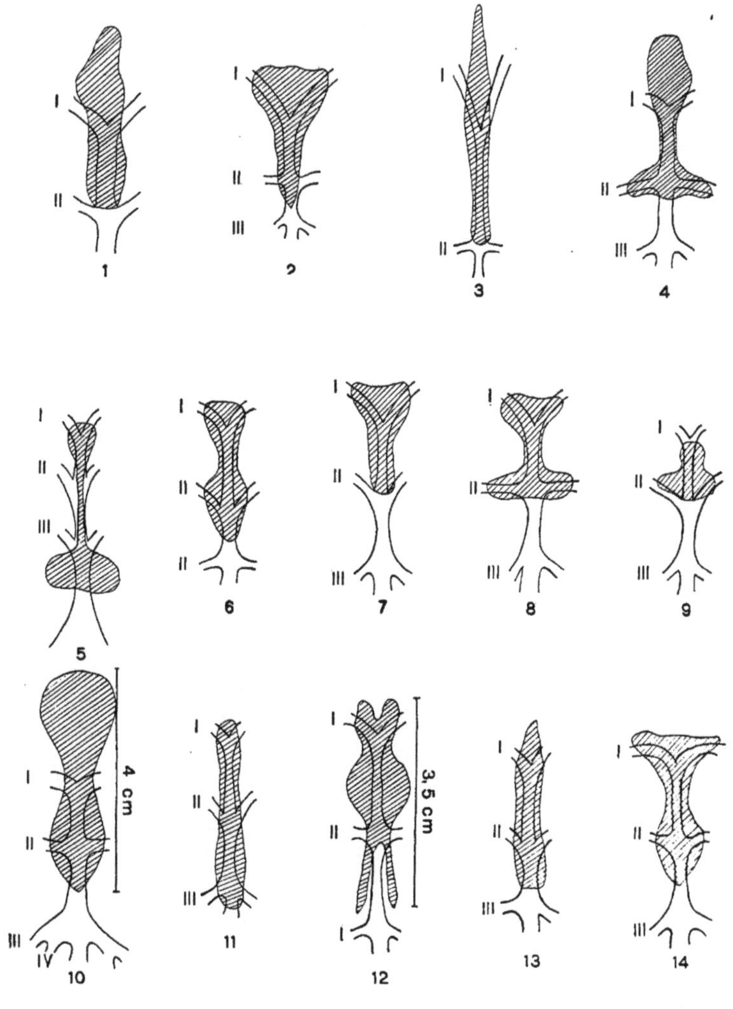
1
2
3
4
5
6
7
8
9
10
11
12
13
14
4 cm
3,5 cm
I
II
III
IV

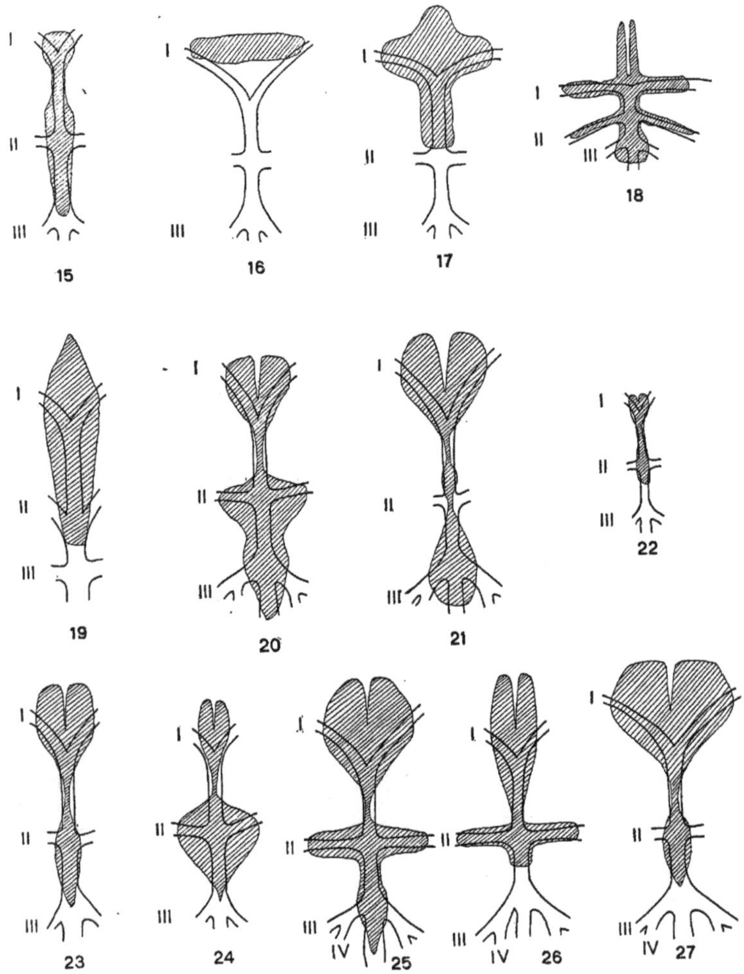
15
16
17
18
19
20
21
22
23
24
25
26
27

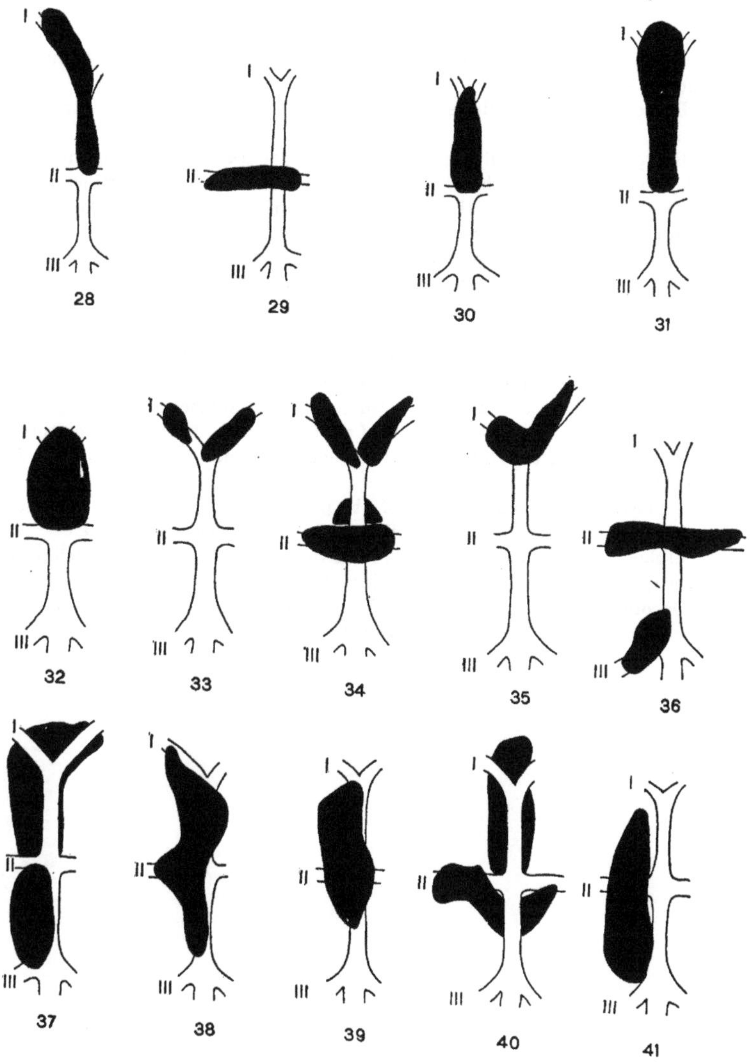
28
29
30
31
32
33
34
35
36
37
38
39
40
41

EXPLANATION OF PLATE IV

Plates IV and V, Photographs of the histological features of the thyreoid gland in different species.

1 Brevoörtia. Two follicles with their epithelial cells drawn out into spinous processes. Dia. 1:165.

2 Brevoortia. Two follicles, *F* (the right one containing colloid), and a large lymph vessel, *L*, between them. The content of the vessel shows similar droplets to those sometimes seen on the surface of colloid, and believed by Anderson to contain the 'chromophobe' secretion. Dia. 1:350.

3 Brevoortia. An isolated follicle, *F*, in the most anterior portion of the thyreoid, with neighboring lymph vessels, *L*. Dia. 1:160.

4 Brevoortia. General view of the thyreoid in the osseous capsule. Dia. 1:60.

5 Brevoortia. Degenerating epithelial cells and their basal processes. *F*, Follicular lumen, *E*, Epithelium. Dia. 1:700.

6 Fundulus heteroclitus. Two pictures showing numerous small capillaries, *Ca*, deeply buried in the epithelium, *E*, of the follicles, *F*. In the lower picture the epithelium has retracted slightly from the endothelium. *A*, ventral aorta. Dia. in a, 1:134; in b, 1:345.

7 Siphostoma. A general view of the posterior end of the thyreoid gland. *Ph*, pharynx; *A*, ventral aorta; *F*, follicles; *Ca*, capillaries. Dia. 1:34.

8 Siphostoma. Single follicles, *F*, with their colloid forming epithelial cells, but containing no colloid. *Ca*, capillaries. Dia. 1:345.

9 Oncorhynchus. Irregular nuclei, *N*, of the epithelial cells. *E*, epithelium of a follicle viewed from the top. Dia. 1:650.

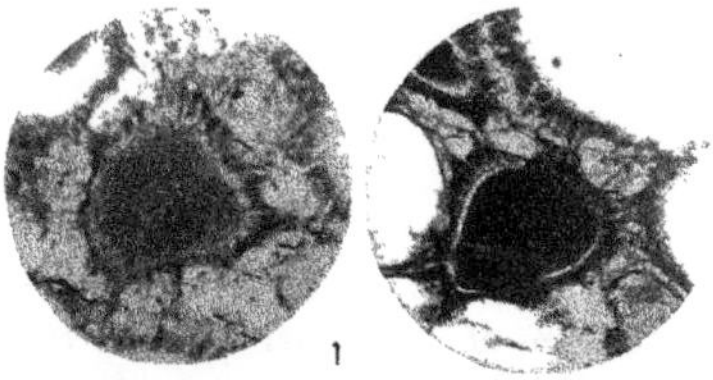

1

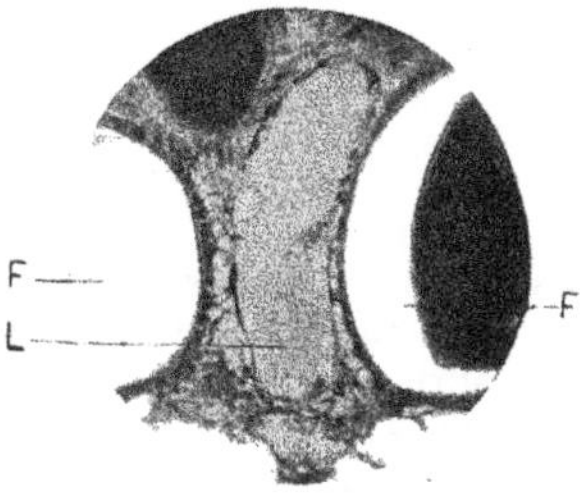

2

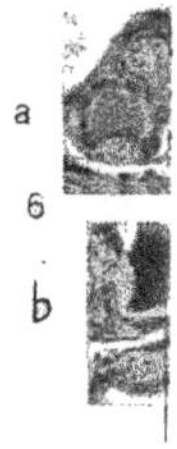

6

L

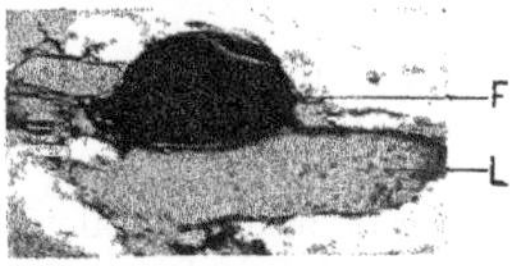

3

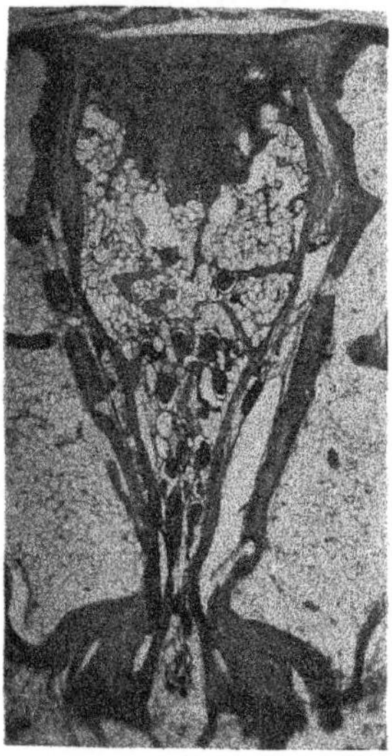

4

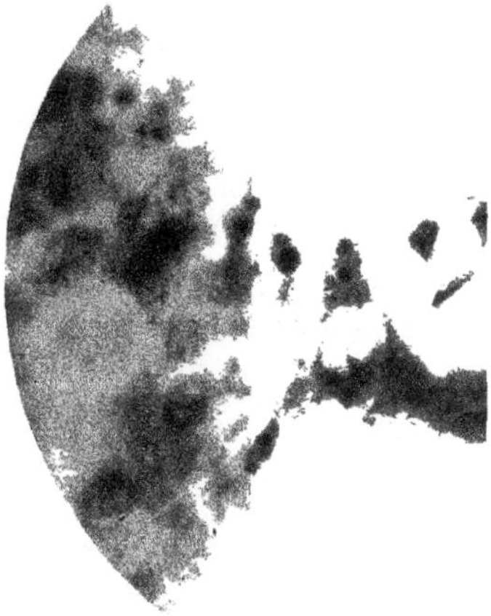

5

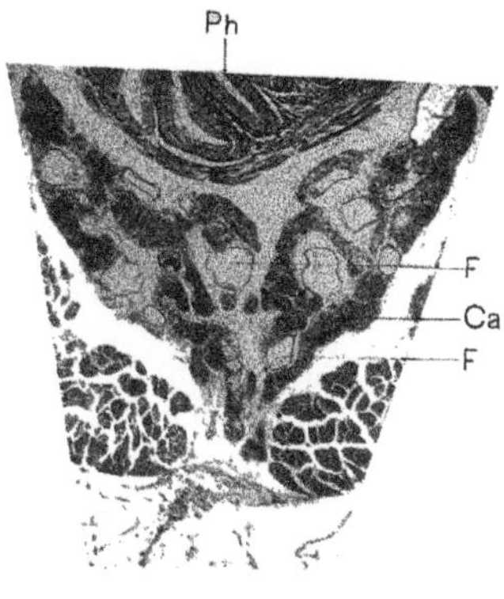

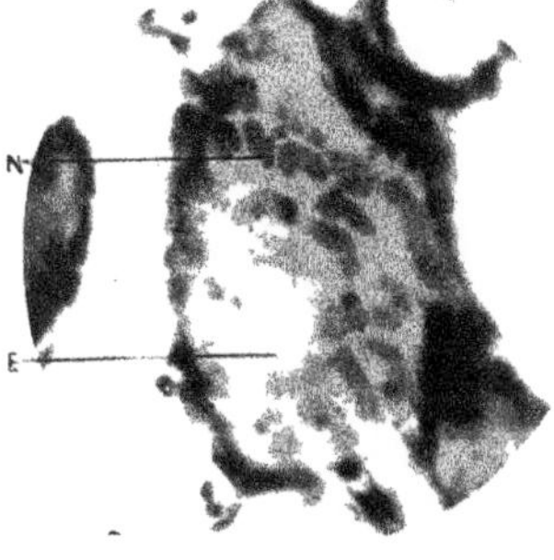

EXPLANATION OF PLATE V

10 Salvelinus fontinalis, spec. no. III. Capillary network, *Ca*, around the follicles. Note the highly columnar epithelium of the upper and the cuboidal epithelium of the two lower follicles. Dia. 1: 134.

11 Salvelinus fontinalis, spec. no. II, showing different heights of the epithelial cells in the same follicle. Dia. 1: 375.

12 Salvelinus fontinalis, spec. no. II. Distribution of the thyreoid elements and their capillaries, *Ca*, in the connective and fatty tissue network. Dia. 1: 64.

13 Salvelinus fontinalis, spec. no. III. Much swollen colloid forming cells, *Coz*, which have been cast off from the follicular wall into the colloid, *Co*. *E*, epithelium; *N*, nucleus; *V*, vesicles in the colloidal substance. Dia. 1: 650.

14 Oncorhynchus. A general view of the thyreoid gland between the first and second branchial arteries. *Ph*, epithelial floor of the pharynx; *F*, follicles surrounding the copula; *A*, ventral Aorta. Dia. 1: 60.

15 Anguilla. A duct, *D*, connecting a small, *f*, and a large thyreoid follicle, *F*. Dia. 1: 165.

16 Anguilla. This photograph shows a complex of follicles which, on tracing through the series, are found to connect with the follicle, *F*, on the right side of the illustration. Dia. 1: 170.

17 Cynoscion. A general view of the densely arranged follicles of this species. *Co*, colloid; *E*, epithelium. Dia. 1: 165.

18 Cynoscion. This picture shows the ramifying lymph spaces, *L*, completely filled with the same substance as the follicles, *F*. Dia. 1: 170.

19 Tautoga. The epithelial wall, *E*, of the follicle, *F*, is cut somewhat tangentially so that the network of anastomosing capillaries, *Va*, enclosing the follicles can be seen. Dia. 1: 145.

20 Clupea. Much swollen colloid forming epithelial cells, *Coz*, in a colloid zone. Dia. 1: 170.

21 Sarda. Showing smooth muscle bundles, *M*, invaded by thyreoid follicles. Dia. 1: 60.

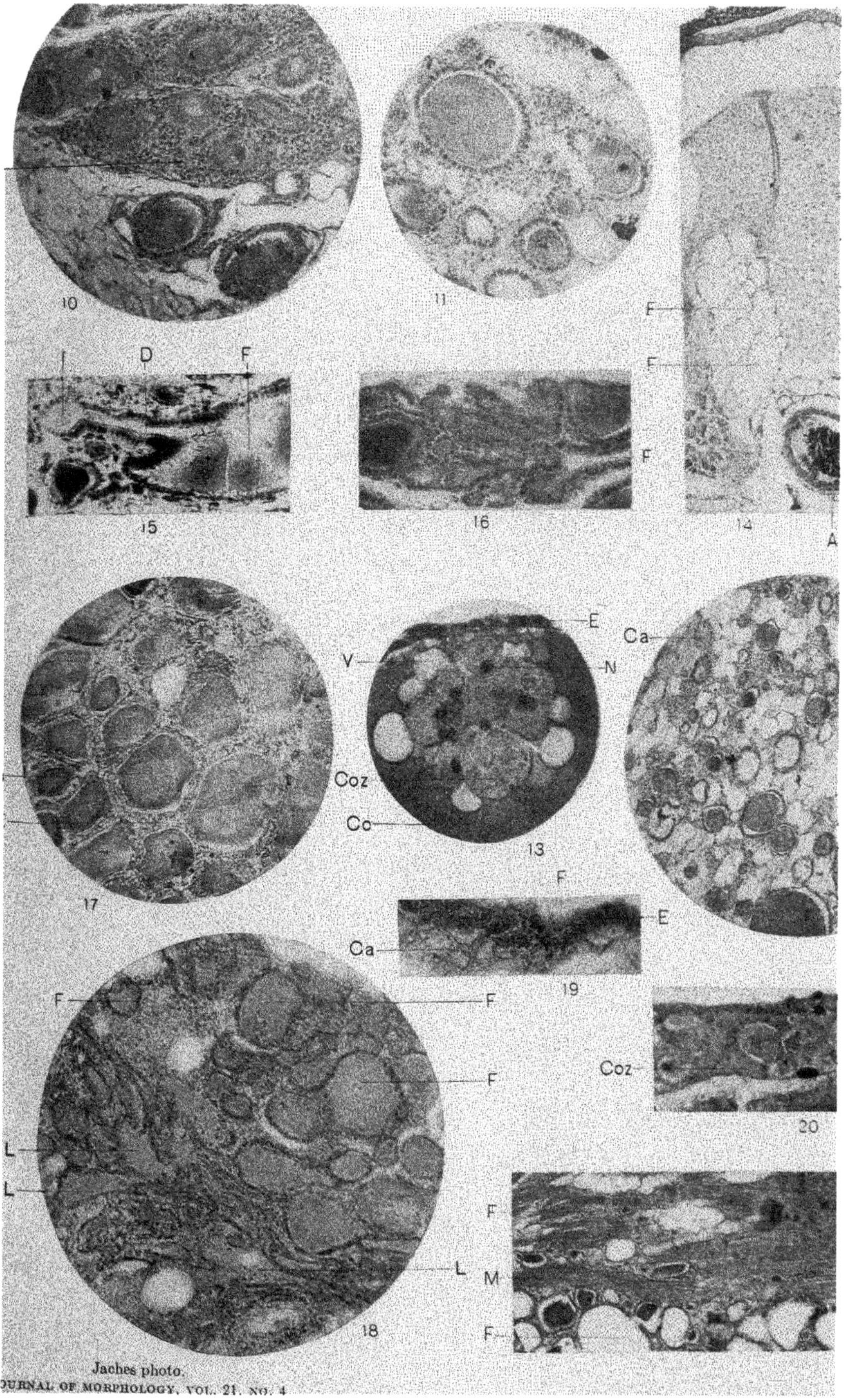

Jaches photo.

[Reprinted from THE AMERICAN NATURALIST, Vol. XLIV., July, 1910.]

ON THE EFFECT OF EXTERNAL CONDITIONS ON THE REPRODUCTION OF DAPHNIA[1]

DR. J. F. McCLENDON

CORNELL MEDICAL COLLEGE

SINCE in the great majority of organisms only the germ cells are capable of reproducing the entire individual, the question as to what differences exist between the germ and body cells, and how they arise, is of general interest. Therefore any change of conditions which affects the germ cells or their relation to the body cells deserves special study. In *Daphnia* external conditions not only affect the relation of the germ cells to the body cells, but they affect the egg cells in such a manner as to determine whether they do or do not need fertilization. The purpose of the present paper is not merely to add the results of my experiments to those of other investigators, but to tentatively arrange the available data under a general working hypothesis in the hope that some more direct method of investigating the relation of the germ and body cells be devised.

Last spring (March 10) I began experiments on the effects of environment on *Daphnia pulex,* De Geer, without knowing that Woltereck was working on the same line. The material came from a small pool investigated by Dr. W. C. Curtis and containing a single strain of this and no other species of *Daphnia*. For the first few weeks some ice remained on the pool and the temperature did not much exceed 4° C.; after this it rose steadily to about 20° by the end of May. Specimens from the pool were examined at intervals as a control on the experiments.[2]

[1] From the Zoological Laboratory of the University of Missouri and the Histological Laboratory of Cornell University Medical College, New York City.

[2] When the daphnids were crowded in dishes of the same pool water they soon began to die, owing to the accumulation of their excretions. When

EFFECT OF ENVIRONMENT ON DIFFERENTIAL GROWTH

By differential growth I mean the unequal growth of different parts, viz., the germ and body cells. Only parthenogenetic females were used, and each was kept separately in the same quantity of water. All measurements were made at sexual maturity, *i. e.*, when the first eggs appeared in the brood pouch. Warren[3] found that under uniform conditions there was a slight variability, but Woltereck showed that these fluctuating variations were very small, though he did find mutations as rare occurrences.

Nutrition.—I had in the laboratory a pure culture of a unicellular green alga which the daphnids ate readily. This alga did not remain entirely suspended in the water, but as the daphnids fed on the bottom as well as while swimming, and stirred up the algæ, it can not be said that most of the food was out of their reach.

Those with a superabundance of food were larger at sexual maturity and had a shorter spine than those with insufficient food, and conversely. The smaller size and longer spine of the starved daphnids are characteristic of immature stages.

Temperature.—Three sets of experiments were

transferred suddenly to artesian tap water many died, though with a gradual change all lived. The composition of the tap water was as follows: Ca, .148 $\times 10^{-8}$ molecular; Mg, .1 $\times 10^{-8}$ molecular; CO_3, .045 $\times 10^{-8}$ molecular; SO_4, .146 $\times 10^{-8}$ molecular; Cl, .055 $\times 10^{-8}$ molecular. Besides these were very small quantities of silica, clay, iron, ammonia and nitrates, and traces of lithium and potassium. Beside the carbonates the water when drawn from the tap was super-saturated with carbon dioxide. In order to find the cause of death from change of water I added various amounts of molecular solutions of certain salts. The toxicity of cations increased as follows: $NA < 1/2\,Ca < K$, and of anions: $Cl < 1/2\,SO_4 < HCO_3 < 1/2\,HPO_4$. But as K became toxic only on 1/100 and HPO_4 on 1/500 molecular concentration there must be other toxic substances than salts in the water. The carbon dioxide in the water killed crayfish and was probably the most toxic constituent to the daphnids.

[3] "An Observation on Inheritance in Parthenogenesis," *Proc. Roy. Soc. London*, 1899, LXV, and "On the Reactions of *Daphnia magna* to certain changes in Environment," *Quart. Jour. Micr. Sc.*, 1900.

started: at 4–10°, 19–20° and 29–31° C. The results were as follows:

	Average Body Length.	Ratio of Average Spine Length to Body Length.	Time from Hatching to Sexual Maturity.
4–10°	22	0.24	35 days
19-21°	20	0..5	14 "
29–31°	17.7	0.27	6 "

1 unit = $\frac{2}{27}$ millimeters.

All were given a surplus of food daily. It may be observed that a higher temperature has the same effect as insufficient food.

Salts.—Since salts have such a marked effect on the development of marine and some fresh-water animals, I placed daphnids in the strongest solutions of various salts that they would live in (without acclimatization). The effects in two months (four generations) were unnoticeable.

Light.—Cultures were kept in the dark and in diffuse and direct sunlight, but no effect was observed.

A number of observers have recorded season-polymorphism in daphnids. Wesenberg-Lund[4] pointed out that when the specific gravity and consequent buoyancy of the water decreased—by heat in summer—the body of the daphnids became smaller or were provided with outgrowths, so as to offer a greater resistance to sinking. Wolfgang Ostwald[5] produced, all at the same time, the forms that occurred in nature at different seasons, by varying the temperature. He emphasized the fact that rise in temperature lowered the internal viscosity of the water. He found that in the warm cultures the daphnids often became productive at an undeveloped stage and, as is true generally, reproduction retarded body growth.

[4] "Ueber das Abhängigkeitsverhältnis zwischen den Bau der Plankton-organismen und den specifischen Gewicht des Süsswassers," *Biol. Centlb.*, 1900, XX, pp. 606–619, 644–656, and "Studier over de Danske söers Plankton," Copenhagen, 1904.

[5] "Experimentelle Untersuchungen über den Saisonpolymorphismus bei Daphniden," *Archiv f. Entwicklungsmech. d. Organismen*, 1904, XVII, p. 415.

Woltereck[6] maintains that Ostwald's results were due to the fact that at a higher temperature the daphnids need more food. Woltereck caused decrease in body length both by starving and by increasing the temperature, but the latter was not effective with an optimum supply of food. He found that more food than the optimum produced effects similar to starving. Raising the internal viscosity of the water by adding quince gum produced no effect. He showed that though feeding influenced the differential growth, there was a cyclical tendency for this to vary, viz., season-polymorphism. However, the effects of prolonged abundant feeding were inherited to some degree.

One might interpret these results in different ways. It is known that the temperature coefficients for various chemical reactions are slightly different. Possibly the mean of the temperature coefficients for the processes in the development of the reproductive organs is higher than the same for the body wall, and at a higher temperature the germ cells would develop faster. However, under adverse conditions the "affinity" of the reproductive organs for nutriment is greater than that of the rest of the body, so with deficient food the body wall is retarded more than the germ cells in development. The higher temperature may be considered an adverse condition since the mortality is greater. In this way starving has the same effect as a higher temperature.

Langerhans[7] found that accumulation of excretions caused shortening of the spine in daphnids. I do not know what relation this bears to the above results.

[6] "Ueber naturliche und kunstliche Varietätenbildung bei Daphniden," *Verh. Deutsch. Zool. Gesell.*, 1908, p. 234; and "Weitere experimentelle Untersuchungen über Artveranderung, speziell über das Wesen quantitativer Artunterschiede bei Daphniden," *ibid.*, 1909, p. 110.

[7] "Ueber experimentelle Untersuchungen zu Fragen der Fortpflanzung, Variation und Vererbung bei Daphniden," *Verh. Deutsch. Zool. Gesell.*, 1909, p. 281.

EFFECT OF ENVIRONMENT ON THE LIFE CYCLE

In most species of daphnids, generations of parthenogenetic females alternate with generations of males and females which produce eggs that must be fertilized, and either frozen, dried or kept a long time before they will develop (resting or "winter" eggs). In different species the number of successive parthenogenetic generations varies. In some all are, and in some none are, parthenogenetic.

I found that heat hastened the appearance of sexual forms, as did starving or the accumulation of excretory products. All of these factors might be combined in the drying up of a pond, as heat would aid in drying, and drying would concentrate the daphnids and their excretions, and concentration of the daphnids would cause them to eat up the algæ faster than they could multiply. However, by keeping the culture cold, fresh or well-fed, or all combined, I could delay but not prevent the appearance of sexual forms.

Kurz[8] said the drying up of the water caused the appearance of sexual forms, and Schmankewitz[9] suggested that it was the increase in salts. Weismann[10] tested both of these hypotheses and concluded that they were wrong. He also tried the effect of food and temperature, with varying results. He concluded that the life cycle was fixed for each species and variety. Issakowitz[11] concluded that cold favored the appearance of sexual forms and warmth favored the parthenogenetic. Also, hunger favored the appearance of sexual forms and abundant food the parthenogenetic. It may be that cold retarded multiplication of the food plant or the

[8] "Dodekas neuer Cladoceren nebst einer kurzen übersicht der Cladocerenfauna Böhmens," *Sitz. Ber. math. naturw. Wien*, 1875.

[9] "Zur Kenntnis des Einflusses der ausseren Lebensbedingungen auf die Organisation der Tiere," *Zeit. wiss. Zool.*, 1877, XXIX.

[10] "Beiträge zur Naturgeschichte der Daphnoiden, VII," *Zeits. wiss. Zool.*, XXXIII, p. 111.

[11] "Geschlechtbestimmende Ursachen bei den Daphniden," *Biol. Centralb.*, 1909, XXV, pp. 529–536.

movements of the daphnids so that they did not keep the algæ stirred up in the water sufficiently to get at them. The parthenogenetic egg arises from four cells, but a large number of cells enter into the composition of the fertilizable egg. If the latter egg is not fertilized it is absorbed and, as Issakowitsch noted, furnishes food for the development of parthenogenetic eggs.

Woltereck[12] found that starving hastened the appearance of sexual reproduction, but a concentration of food above the optimum produced results similar to starving. He found, as Weismann maintained, a cyclical tendency toward the alternation of sexual and parthenogenetic generations which, contrary to Weismann, was temporarily influenced by nutrition, and the effects of constant nutrition over a long period was inherited to some extent.

Langerhans[13] found that the accumulation of excretions caused the decrease in numbers of parthenogenetic females in the autumn and thinks that the appearance of sexual forms is due to the same cause.

The life cycle of a daphnid is, therefore, an hereditary tendency, but can be influenced by nutrition and probably by temperature and the accumulation of excretions. Nutrition is the most important factor, and former experiments on the effect of temperature and the drying up of the water were complicated by secondary effects on concentration of food and excretory products.

Discussion of Results.—Two views might be held as to the origin of the differences between the germ and body cells: the differences might be the result of difference in position in the embryo, or of unequal mitoses. In the parasitic copepods I found the primary germ cell to arise by an unequal mitosis at the fifth cleavage of the egg. The germ cell when first formed is one thirty-second of the total number of cells, but owing to the more

[12] *Loc. cit.*

[13] *Loc cit*

rapid division of the body cells this ratio decreases. In fact the chief difference between the yolk-free body cells and the (yolk-free) germ cells is the slow rate of division of the latter. Finally, the *eggs* will not divide at all unless specially stimulated, by fertilization. The question now arises: what causes the cell to divide. Sacks found that plant cells divide when they have reached a certain size. This rule has been extended to animals, and the final size of the cell found to be determined by the ratio of nucleus to cytoplasm. This rule may apply to the germ cells, since it appears that after the egg cell, primary oocyte, reaches a certain size any additional food absorbed does not cause growth of the protoplasm, but is precipitated as yolk.

If the egg is properly stimulated, rapid growth of protoplasm and cell divisions follow. From the study of artificial parthenogenesis it appears probable that stimuli which lead to development of the egg increases the permeability of its plasma membrane. If this be true we may say that the germ cells are distinguished by the fact that their plasma membranes are poorly permeable and retard those reactions between the cell contents and environment which lead to growth and cell division. In other words, the optimum intensity of stimulation toward growth and division is higher for the germ cells than for the body cells.

This difference is probably due to a difference in the colloids of the cell, which in animals could be explained as the result of an unequal mitosis. This explanation may be modified so as to apply to plants. Klebs has shown that those conditions which are adverse to vegetative growth of plants (too strong stimuli?) call forth flowers. Perhaps there are slight differences in the sensitiveness of plant cells to stimuli, and as the stimuli increase, those initially least sensitive cells acquire further immunity to the stimulus, whereas those initially more sensitive cells are overstimulated and weakened. Thus the difference between germ and body cells is gradually acquired. The Malpighian layer of the skin may be stimulated to proliferate more rapidly, but if the stimuli are too strong the growth will be retarded instead of increased. On gradually increasing the stimulus immunity to it may be acquired.

To apply this hypothesis to the daphnids: conditions which are adverse to the growth of the body cells, such as extremes of temperature (viz., high temperatures) or of concentration of excretory products, or disorded nutrition, either fail to retard the development of the germ cells or stimulate their development, so that in either case the daphnid becomes sexually mature at a less developed stage. Under the less extreme conditions the eggs develop on receiving the slight stimulus incident to their transfer to the brood pouch, but under the more extreme conditions those eggs which develop at all must be stimulated by fertilization before they develop. These two types of eggs may perhaps develop from two kinds of cells, or the sexual egg may arise from the same kind of cell producing the parthenogenetic egg by acquiring an immunity to slight stimuli. Whereas more than one cell goes to make up a single egg; only one nucleus is retained, and it may be said that one cell is the egg cell and the remainder furnish its food.

THE DEVELOPMENT OF ISOLATED BLASTOMERES OF THE FROG'S EGG

J. F. McCLENDON

From the Zoölogical Laboratory of the University of Missouri and the Histological Laboratory of Cornell University Medical College, New York City

WITH TWO FIGURES

After the numerous proofs furnished by various investigators, there is no question that differentiation begins very early in development, but the determination of the exact stage at which various differentiations begin, has been hampered by technical difficulties. The importance of overcoming these difficulties lies in the fact that it is only by an exact study of the early development that we can ever hope to know the mechanics of differentiation. The later stages are so complex that it is doubtful that they could be analysed even though the early stages were understood.

The frog's egg, owing to its large size, has been a favorable object of study. O. Hertwig,[1] after pricking one blastomere, found that a complete embryo was formed.

Roux,[2] in similar experiments obtained a half embryo from the uninjured blastomere, and explains Hertwig's observation by the fact that some of these half embryos became complete larvæ by a process of regulation which he called post-generation. This might occur in the following different ways: first, the half embryo might fold together on the side corresponding to the median plane and be transformed into a whole embryo of half size; second, the injured blastomere might recover from the operation and develop into the missing half; and third, the injured blastomere might be

[1]Arch. f. Mikroscopische Anat., 1893, XLII, p. 662.

[2]Ges. Abh. zur Entwicklungsmechanik der organismen, Leipzig, 1895.

reorganized under the influence of the half embryo into the missing half. The occurrence of post-generation, and especially the third type, has been denied by a number of investigators who repeated this experiment. Laquer[3] has reinvestigated the question.

This production of a half embryo from one blastomere is not due to the inability of the egg to produce more than one embryo, since Schultze[4] produced double forms by inverting the egg in the two-cell stage, and Morgan[5] produced a complete embryo from one blastomere by pricking the other blastomere and then inverting the egg. In order to determine whether this inversion of the egg and consequent stirring up of its contents is necessary for the production of a complete embryo from one blastomere, I attempted to find a frog's egg that would permit the complete removal of one blastomere without death of the egg. No one has hitherto been able to do this, although Roux[6] observed a partial but very abnormal development of extra-ovates of the frog's egg.

Last spring at Columbia, Missouri, I located the breeding places of Rana pipiens and Chorophilus triseriatus. The unsegmented eggs of both species could be collected in the ponds and small streams, or in the frog cages in the laboratory, the eggs of the latter species being more easily obtained.

I tried various methods of removing one blastomere in the two-cell stage. It was possible to suck out one blastomere with a capillary pipette connected with a rubber tube, one end of which was held in the mouth, but it was very difficult to pierce the jelly with the pipette without killing the egg, and the suction often removed both blastomeres. A better way of making a hole in the jelly was the time honored method of piercing it with a hot needle. A very coarse needle was used and heated to a temperature that would coagulate a portion of the blastomere into which it was thrust. On withdrawing the needle, the coagulated parts of the jelly and blastomere were pulled out, thus leaving a large hole

[3]Arch. f. Entwicklungs. 1909, XXVIII, p. 327.
[4]Arch. f. Entwicklungsmech., I. p. 298.
[5]Anat. Anz., 1895, X. p. 523.
[6]Jahresbericht d. Schles. Ges. f. vaterl. Cultur, Juni, 1887.

through which the remaining part of the blastomere could be extracted. The eggs of Chorophilus were more easily handled, since the jelly was softer and the egg itself did not collapse readily. This egg is not so small as to make operations difficult with the low power of the binocular microscope.

When the first cleavage furrow had extended almost around the egg, the greater part of the jelly was carefully removed with filter paper and the puncture made. The egg was then allowed to rest until the first cleavage plane had completely divided it. At this time the remains of the punctured blastomere were removed with the capillary pipette. In case the last remnants could not be removed, the egg was allowed to rest again until the remnants of the punctured blastomere turned pale, which was an indication that they were more brittle and could now be removed with a fine needle.

When the egg was punctured, the perivitelline space was obliterated by the escape of the contained fluid and the jelly pressed close to the egg, an opening was also made through which bacteria might enter; therefore, the eggs laid in the laboratory under a more sterile condition, gave better results than those laid in the ponds. Care was taken to keep the egg normally orientated in reference to the direction of gravity both during and after the operation.

The operated eggs were observed with the binocular microscope and the early cleavage noted. The gastrulation was watched through the sides and bottom of the glass dishes in which the eggs were kept. After the embryos became ciliated, some of them turned in a normal manner so that they could be observed in all aspects, from above.

A large number of eggs of both Rana pipiens and Chorophilus triseriatus were operated on. All of the former died, but of the latter a considerable number gastrulated and several reached the tadpole stage. Two were four days, three were five days and one was eight days old when preserved for sectioning.

The control eggs hatched in less than eight days, but the operated eggs were retarded in development and the resulting tadpoles were too weak to break through the jelly, although the twitching of their tails showed that the muscles were functional.

The development of the operated eggs was very similar to that of the normal, and quite different from that of the uninjured blastomeres in Roux's experiment. The remaining blastomere rounded up gradually after its partner had been removed, and the blastula, gastrula and later stages were *wholes* of half size. As the mortality was great, it is to be expected that defects would be observed in many of the specimens, but these defective specimens could not be interpreted as half embryos. The only defect that I could observe in the oldest specimen was the small size of the suckers,

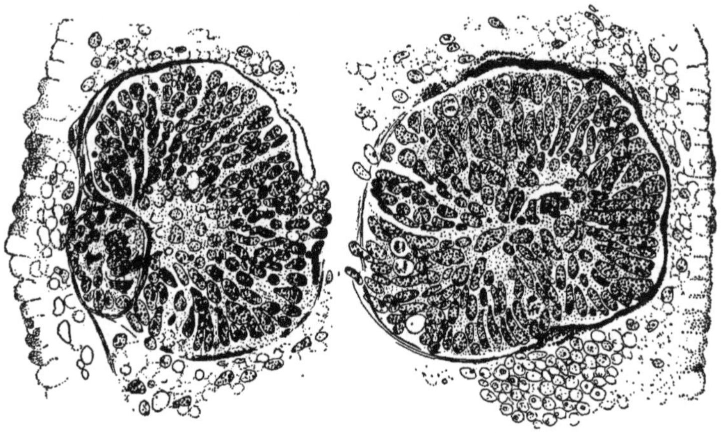

Fig. 1 Fig. 2

Fig. 1. Chorophilus triseriatus: a section of the left eye of a tadpole developed from one of the first two blastomeres.

Fig. 2. A section of the right eye of the same tadpole as in fig. 1.

before the pericardium began to swell so that the heart beats could no longer be observed.

Sections confirmed the observations on the living material and showed that the embryos were complete and not half-forms.

Figs. 1 and 2 represent the right and left eyes seen in transverse sections of the eight day larva. The right eye lacks a lens, and the left eye is not quite normal, yet these defects make no approach to those of a half-larva since they are not very extensive and are not all on the same side. There is some displaced pigment in the center of the lens, and the cells of the optic cup have not preserved their normal arrangement. An examination of the ectoderm showed more or less disintegration, and the defects observed may be explained by the fact that the specimen was dying when it was taken out of the water to be preserved.

These differences in the two eyes of the eighty-day embryo from one blastomere, are the most striking differences between the right and left sides. The ear vesicles are almost identical on the two sides, and the same is true of the other paired organs. Furthermore, no general defect of the anterior, posterior, dorsal or ventral regions of the body could be found. The study of the five other embryos developed from operated eggs, and cut in serial sections also revealed complete forms.

Roux described right and left "hemiembryones laterales" and also "hemiembryones anteriores" developed from one blastomere in the experiments mentioned above.

Since the anterior end of the neural folds lies on one side of the egg between the equator and the animal pole, i. e., about the region in which the gray crescent appeared, and the dorsal side of the embryo is formed chiefly on this side, it is probable that in case the first cleavage plane were transverse to the plane of symmetry, it would divide the egg into the halves that would form the dorsal and ventral parts of the embryo respectively. Therefore the "hemiembryones anteriores" might better be described as "dorsales" as indicated by Przibram. Roux obtained no "posteriores" ("ventrales" of Przibram) and since a number of the eggs died, it is probable that those blastomeres which correspond to the ventral part of the embryo, and therefore contain no part of the gray crescent, are not so well adapted to independent development. In my experiments the position of the grey crescent as marking the future dorso-anterior region of the embryo was not recorded, but it seems probable from the results of Roux's experiment, that those isolated blastomeres containing no part of the grey crescent did not develop.

As noted above, Roux found that half-embryos sometimes became complete embryos by post-generation.

None of the three types of post-generation occurred in my experiments. The isolated blastomeres rounded up in a manner that might be compared to the first type of post-generation, but this regulation occurred before an embryo was formed, and might be distinguished as "immediate."

It appears then that the egg at the time of the first cleavage is differentiated in the direction of the primary axis, and furthermore in the direction of a plane passing through this axis and bisecting the grey crescent, so that the egg may be divided in this plane into two totipotent halves. The halves of the egg divided in any other plane may or may not be totipotent. The formation of a half embryo instead of a whole embryo from one blastomere of the two-cell stage is due to the presence and mechanical interference of the other blastomere.

In this connection may be mentioned the results of Endres, Herlitzka and Spemann on the development of the first two blastomeres of urodeles when separated by constriction with a fine hair. If the constriction was slight, the hair merely marked the plane of the first cleavage, if it was deeper, double monsters occurred, and if it was complete, two perfect embryos resulted.

In from 66 per cent to 75 per cent of the eggs, the first cleavage plane became a frontal plane of the embryo, and in the remainder, it became the median plane. In the latter case complete constriction gave rise to two complete embryos, but in the former case, complete constriction resulted in one complete embryo developed from the prospectively dorsal half of the egg, whereas the prospectively ventral half did not develop beyond gastrulation. Thus it appears that the eggs of the urodeles possess the same organization as the eggs of the anura, with the difference that the first cleavage plane is usually frontal in the former, and median in the latter in relation to the embryonic axes.

Accepted by The Wistar Institute of Anatomy and Biology, January 19, 1910. Printed June 21, 1910.

On the Effect of Centrifugal Force on the Frog's Egg.

By

J. F. McClendon

(From the Histological Laboratory of Cornell University Medical College, New York City, U.S.A.)

With 9 figures in the text.

The centrifuge has become a popular piece of apparatus for the analysis of the structure and developmental processes in the animal egg. The results thus obtained are very interesting, but the interpretation of these results have been divergent in some cases. Therefore it may be of interest to consider certain facts that have a bearing on the disputed points.

On the Cytoplasmic Structure.

In a recent paper, Gurwitsch (09) maintains that centrifugal force destroys the alveolar structure of the frog's egg by separating the contents of the alveoles, enchylemma, as a layer at the animal pole, from the substance forming the walls of the alveoles, hyaloplasm, which then forms a layer just below the enchylemma. Gurwitsch had reported the observation before the Anatomische Gesellschaft at Jena in 1904, and similar results were published by Konopacka (08), with this difference that Gurwitsch, following the alveolar theory of Butschli, called the substance that rises to the top of the egg enchylemma, whereas Konopacka, adhering to the reticular theory of Leydig used the term hyaloplasm to denote the same substance. In order to throw light on the question whether the alveolar structure is destroyed by centrifugal force I have made a study of the cytoplasm of the normal egg.

A small portion of a thin smear of the living ovarian egg of *Rana pipiens* is shown in fig. 1a. The large ovals represent yolk-platelets, the black dots pigment, and the small circles represent spheres of a substance that can hardly be differentiated in the living egg from the substance of the yolk-platelets. Fig. 1b shows these bodies in an egg fixed in osmic acid. The small circles are here striated to denote that they represent bodies that take the characteristic osmic impregnation for fat. There is some difference in the size of these fat droplets in the living cytoplasm and the osmicated section, but none of them in either preparation are as large as the yolk-platelets. If a smear of the living egg be fixed in osmic acid and immediately examined, some of the fat droplets are seen in the act of fusing to form larger drops, fig. 1c. By treating a smear

Fig. 1.

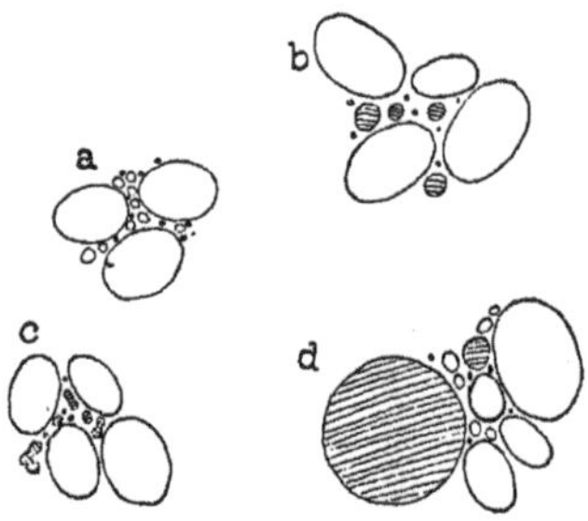

Cytoplasma of ripe ovarian egg of *Rana pipiens*; a. living egg burst under cover-glass; b. section fixed in osmic; c. fresh smear fixed in formaldehyde vapor and stained in alcoholic solution of Sudan III. The large ovals are yolk-platelets; the black dots pigment and the striated bodies gave the reaction for fat; the small circles in a. represent fat, this is also true of the small circles in d. which represent fat droplets which are too small to show a perceptible intensity of stain.

Fig. 2.

Juncture of the fatty layer (above) and protoplasmic layer (below) in the cytoplasm of the egg of *Chorophilus triseriatus* centrifuged in the four-cell stage, centrifugal force = 2771 × gravity for seven minutes.

of the living egg, fixed for a few seconds in formaldehyde vapor, with an alcoholic solution of Sudan III, the fat droplets are stained, but they vary much more in size than in any of the above preparations, some of them being larger than the yolk-platelets, fig. 1d. I attribute this variation in size of the fat droplets in different preparations to the fusion of the original droplets to form larger drops, a process which is due to the rubbing of the smears or to the addition of substances which reduce the surface tension of the droplets. This last point is illustrated by the fact that if the osmicated smear be treated with the alcoholic solution of Sudan III, the fat droplets may be seen to fuse and form larger drops.

By chemical analysis I found (09, 1 and 3) the pigment granules to be composed of a melanin containig 0,6% of sulphur and 9% of nitrogen;

the yolk-platelets to be composed of 6% of lecithin and 94% of batrachiolin, a nucleo-albumin containing 1,2% of phosphorus, 1,3% of sulphur and 15% of nitrogen; and the fat droplets to be composed of a liquid fat, a solid fat melting at 58 degrees, and a yellow lipochrome.

In the living frog's egg, I have not been able to observe any structure to the cytoplasm proper, that is, to the substance that fills the spaces between the yolk-platelets, fat droplets, and pigment granules. GURWITSCH (09) supposed an alveolar structure to exist, but which was obscured by the yolk-platelets. I examined younger ovarian eggs, containing no yolk, fat or pigment, alive under a 2 mm oil immersion lens and could observe no structure in the cytoplasm except clumps of granules in certain regions, the yolk-nuclei. In a former paper (09, 1) in speaking

Fig. 3.

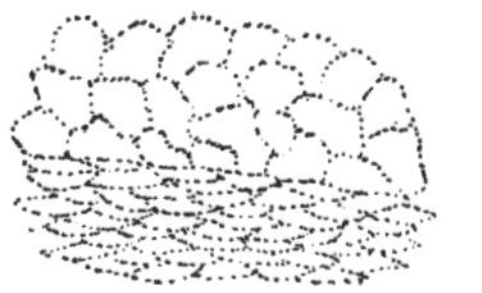

The same structure as in fig. 2, in the ovarian egg of the toad, centrifugal force = 2771 × gravity for two hours. The greater viscosity of the ovarian egg necessitated a longer duration of the force to obtain the same stratification.

Fig. 4.

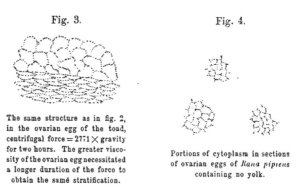

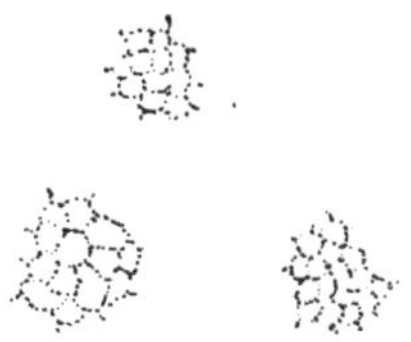

Portions of cytoplasm in sections of ovarian eggs of *Rana pipiens* containing no yolk.

of the alveolar structure of the normal frog's egg, I referred to the spaces filled with yolk-platelets and fat droplets and containing in addition enchylemma, as I supposed. It is hardly probable that spaces entirely occupied by yolk or fat can be rightly called alveoles in BUTSCHLI's sense.

If the frog's egg be centrifuged, I observed (09,1) that the fat droplets rise to the top of the egg and fuse to form larger drops, whereas the cytoplasm proper forms a layer just beneath the layer of fat drops. A small portion of a section showing a part of each of these two layers is represented in fig. 2. Each layer has an alveolar or reticular appearence, the meshes in the fatty layer being larger than those in the layer below, which GURWITSCH calls artefacts. The meshes in the fatty layer are formed of the coagulated cytoplasm that separated the fat drops until the latter were dissolved out with xylol. The meshes in the layer below are probably artefacts as GURWITSCH maintains, but they are about as large as the meshes seen in the cytoplasm of fixed ovarian eggs containing no yolk, fig. 4, and if all these are artefacts, the assertion that an alveolar structure in BUTSCHLI's sense exists in the cytoplasm of the frog's egg is not based on optical evidence (cf. RHUMBLER). The fact remains that the spaces which GURWITSCH supposed to be filled with enchylemma in the centrifuged frog's egg are in reality filled with fat droplets, and the process which he describes as a destruction of the alveolar structure is in reality an aggregation of the fat droplets. Whether the process which GURWITSCH describes as a restitution of the alveolar structure, which occurs in eggs after they have been removed from the centrifuge, is a redistribution of the fat droplets, I cannot positively affirm, as I have not determined that the fat droplets are ever uniformly redistributed, neither do KONOPACKA's figs. 12 and 14, Pl. XXVI indicate such a process, and although on p. 772 she states that the vacuolated layer seen in centrifuged eggs disappears soon after the eggs are removed from the centrifuge, she adds that the stratified arrangement of substances persists during the entire development of the eggs.

On the Cause of the Abnormal Development of Centrifuged Eggs.

GURWITSCH assumes that the abnormal development of centrifuged frog's eggs is due to the failure of the nuclei to fit the portions of the cytoplasm which surround them. This he attributes to the meniscus formation, which did not occur in my experiments. The question arises whether such an assumption is necessary to explain the facts. I concluded (09, 1) that the physical and chemical differences between the layers of the egg were sufficient to account for the abnormal development, and found that if the centrifugal force was sufficiently prolonged (2771 × gravity for 20 minutes) not even the beginning of segmentation ever occured. GURWITSCH maintains that similar explanations are invalidated by the fact that sometimes a portion of one layer will segment while the remainder of the layer is unsegmented, that is to say that the injury to the egg is independent of the stratification. It should be remembered that frog's eggs are easily injured by handling, and that centrifugal force not only stratifies the egg, but causes currents in the egg and presses the eggs one against the other The currents in the egg are indicated by the streaming of the black pigment observed by MORGAN and myself. PFLUGER observed that if the frog's egg be inverted, the small cells appear in the white hemisphere, and BORN demonstrated this to be accompanied by an inversion of the egg contents. It has been proved that the polarity of the frog's egg is determined by the arrangement of egg substance, either deutoplasm or protoplasm proper. The nearly normal development of certain centrifuged eggs has been used as an argument for the position that the deutoplasm moved by the centrifuge has little or no effect on polarity. If this be true the polarity is determined by the protoplasm. If the polarity be determined by the cytoplasm alone, currents might destroy the relation of the polar axes of various regions to one another, just as would be the case if a magnet were melted and its substance stirred.

Fig. 5.

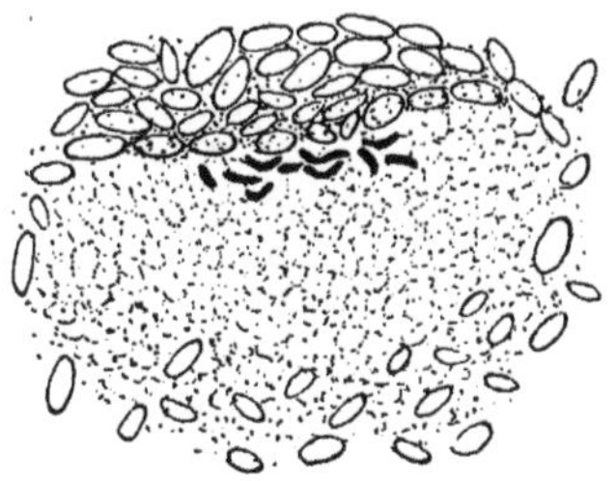

Section through the equator of a mitotic figure in an egg of *Rana pipiens* that was centrifuged in the two cell stage, centrifugal force = 2771 × gravity for one minute. The yolk platelets and pigment granules, in being precipitated, were stopped by the spindle and piled up on it.

Fig 6.

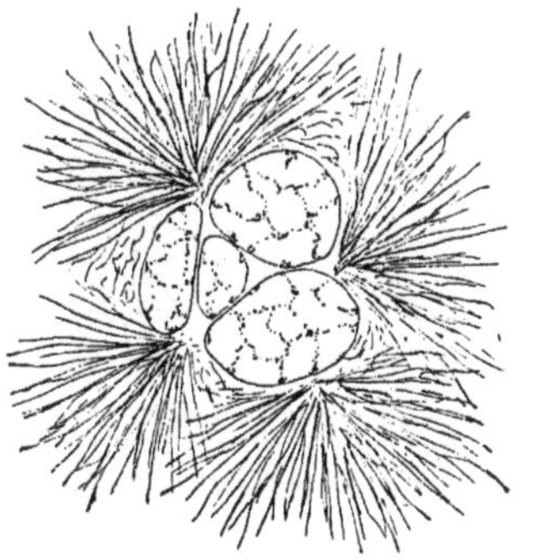

Group of chromosomal vesicles surrounded by four asters in a horizontal section of the egg of *Chorophilus triseriatus* that was placed in the centrifuge in the four cell stage, centrifugal force = 565 × gravity for 45 minutes.

Fig. 7.

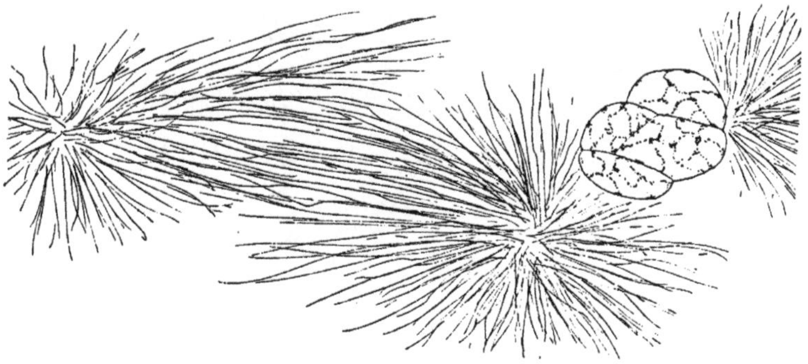

From a section of the same egg shown in fig. 6, showing a nucleus, one of whose asters has apparently divided and one of the products of the division has moved away.

Another cause for the abnormal development of centrifuged frog's eggs might be abnormalities in the mitotic figures. I have previously described (09, 1) the flattening of the mitotic figures in the direction of the centrifugal force. In some cases the yolk and pigment is piled up upon the spindle, as shown in fig. 5, which represents a cross section of the spindle containing the equatorial plate. The longer the eggs remain in the centri-

Fig. 8.

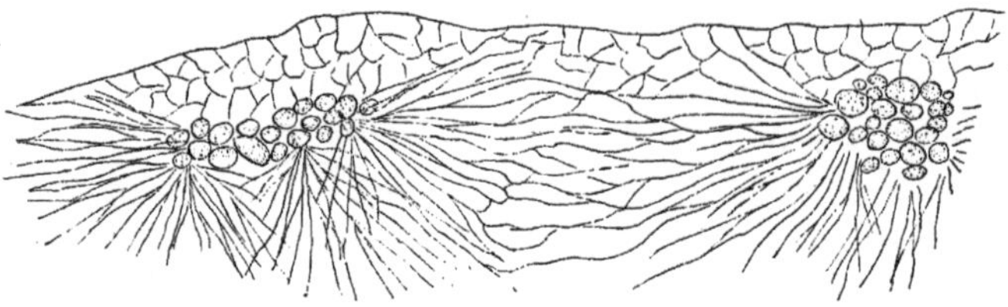

Two groups of chromosomal vesicles, abnormal or incipient nuclei, indicating division of nuclei without cytoplasmic division, in an oblique section very near the animal pole of an egg of *Rana pipiens* that was placed in the centrifuge in the eight cell stage, centrifugal force = 565 × gravity for 4½ hours.

fuge the greater the abnormalities in the mitotic figures. Supernumerary asters appear, that are probably the products of the division of the asters of the mitotic figures. Fig. 6 represents the chromosomal vesicles sur-

Fig. 9.

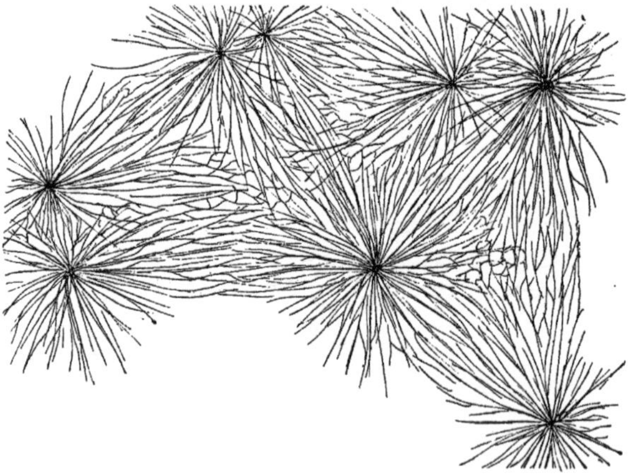

From a section lower down, in the egg shown in fig. 8, representing numerous asters in division and far removed from any chromatin.

rounded by four asters in an egg centrifuged 45 minutes. Fig. 7, from the same egg, represents a nucleus with two asters, but nearby is a third aster which is apparently a product of the division of one of the asters connected with the nucleus. Fig. 8 shows two masses of chromosomal vesicles surrounded by numerous astral radiations in an egg centrifuged one and a half hours. Fig. 9, from a region of the same egg further removed from the animal pole, represents numerous asters apparently in the act of division and not connected with chromosomes or nuclei. During the action of the centrifuge the nuclei remain very near the animal pole, though mitotic figures and supernumerary asters may be found a little further down in the egg. If the centrifugal force is sufficiently strong, division of the nuclei without division of the cytoplasm occurs, which may or may not be followed by division of the cytoplasm after the eggs are removed from the centrifuge. Just what effect these abnormalities in the process of mitosis have on the development of the embryo, I cannot determine, but it is probable that they are partly responsible for the abnormalities in development.

In conclusion to this section, it seems probable that the abnormal development of centrifuged frog's eggs may be due to one of a number of causes or to a combination of causes. Currents in the egg may be partly responsible for its abnormal development, which possibility is rendered more probable by the fact that the egg of *Arbacia*, in which centrifugal force does not produce currents, is little effected in its development thereby, however I have given other reasons for this difference (09, 2).

On the Effect of Centrifugal Force on the Egg at Different Stages of its Development.

It has been recorded by several observers that the degree of the effect of centrifugal force on the frog's egg at different periods of its development is different. Konopacka studied this subject systematically, and found that the ratio of abnormal to normal embryos resulting, increased markedly when the force was applied after the entrance of the spermatozoan, and that there was a second period of increase of this ratio when the force was applied just before the completion of the third cleavage, after which the ratio decreased again. Without knowing of Konopacka's results, in the spring of 1908, I performed similar experiments, which might be of interest since the centrifugal force was greater and acted for a shorter time (1 minute) and the results were denoted by the character of the most perfect embryos. The eggs of *Rana pipiens* laid by one female were divided into lots, and at the end of each ten minute period one lot was centrifuged with a force equal to 2771 × gravity for one minute. Part of each lot was then preserved for sectioning and the remainder placed in a dish of tap water to await depelopment. The highest developmental stages reached by the eggs of the various lots were classified as normal tadpoles (except in coloration), abnormal tadpoles, partial embryos and irregular cleavage stages. In the list below, the numbers refer to the succession of ten minute periods:

1 normal tadpoles
2 » »
3 » »
4 » »
5 abnormal tadpoles
6 » »
7 » »
8 » »
9 » »
10 » »
11 irregular cleavage (♀ pro-nucleus seen in sections)
12 » »
13 normal tadpoles
Commencement of first cleavage furrow
14 normal tadpoles
15 irregular cleavage
16 normal tadpoles
17 partial embryos
18 normal tadpoles
19 » »
Commencement of second cleavage furrow
20 normal tadpoles
21 partial embryos
22 » »
23 » »
Commencement of third cleavage furrow
24 abnormal embryos
25 irregular cleavage.

In a second set of experiments, lots corresponding to lots 20—25 in the above list all gave partial embryos as the best results. Although the results are not quite uniform, it appears that the early stages after the eggs are laid, as indicated by the first ten lots, are less affected by centrifugal force than later stages. I attribute this to the fact that the egg is more viscid in early stages and consequently the stratification of substances and production of currents is less than in later stages. Thus I found that the time required to stratify the ripe ovarian egg by means of a given centrifugal force to be much greater than that required during early cleavage. The same is true for the ovarian egg of the toad, fig. 3. With closely consecutive stages it is much more difficult to make quantitative comparisons.

BIATASZEWICZ found that the frog's egg increased in volume progressively after it was placed in water (except while the perivitelline space was forming) and that this occured whether it were fertilized or not. In this connection might be mentioned the fact recorded by BACKMAN and RUNDSTRÖM that the osmotic pressure of the frog's egg drops from that of frog's serum to that of pond water from the time it leaves the ovary to the first cleavage. This decrease in osmotic pressure is probably due to absorption of water, although BACKMAN and RUNDSTRÖM interpret it differently. This absorption of water, and the mixing of the contents of the germinal vesicle with the cytoplasm, accounts for the decrease in viscosity of the egg after it leaves the ovary.

Literature.

BACKMAN and RUNDSTRÖM, 1909. Physikalisch-chemische Faktoren bei der Embryonalentwicklung. Biochem. Zeitschr. XXII, S. 390

BIATASZEWICZ, 1908. Beiträge zur Kenntnis der Wachstumsvorgänge bei Amphibienembryonen. Bull. de l'Acad. des Sc. Cracovie, Math.-Nat., Okt.

GURWITSCH, 1909. Über Prämissen und anstoßgebende Faktoren der Furchung und Zellvermehrung. Arch. Zellforschung, II, S. 495.

KONOPACKA, 1908. Die Gestaltungsvorgänge der in verschiedenen Entwicklungsstadien zentrifugierten Froschkeime. Bull. de l'Acad. des Sc. de Cracovie, Math.-Nat. Juli, S. 689.

MCCLENDON, 1909. Cytological and Chemical Studies of Centrifuged Frog's Eggs. Arch. f. Entwicklungsmech. XXVII, S. 247.

—— Chemical Studies on the Effects of Centrifugal Force on the Eggs of the Sea Urchin. Am. Journ. Physiol., XXIII, p. 460.

—— On the Nucleo-albumin in the Yolk-platelets of the Frog's Egg, with a Note on the Black Pigment. Ibid., XXV, p. 195.

MORGAN, 1906. The Influence of a Strong Centrifugal Force on the Frog's Egg. Arch. f. Entwicklungsmech. XXII.

RHUMBLER, 1900. Physikalische Analyse von Lebenserscheinungen der Zelle, III. Ibid. IX, S. 63.

Further studies on the Gametogenesis of Pandarus sinuatus, Say.

By

J. F. Mc Clendon.

(From the Zoological Laboratory of the University of Missouri and the Histological Laboratory of Cornell University Medical College, New York City.)

With 1 figure in the text and plate XVII.

In two former papers[1]) I considered certain phases of this subject, but owing to the increasing interest in the part played by the chromosomes in heredity a more detailed study seemed advisable. The present paper considers especially the individuality of the chromosomes during the gametogenesis.

Material and Methods.

During the summers of 1904—6 and 1909 I collected parasitic copepods at Woods Hole. The most favorable species for the study of the gametogenesis seemed to be *Orthagoriscicola muricata* KRÖYER[2]), which occurs on the gills of the moon fish, Selene vomer, but more abundantly on the skin of the sun fish, Mola mola. The only specimens of this species which I was able to obtain were fixed in toto, so that a more abundant species, *Pandarus sinuatus*, which occurs on the skin of the smooth dog fish and the sand shark, had to be used. The integument of the copepods was partly removed or the sexual organs dissected out to insure rapid penetration of the fixing fluid, FLEMMING's. Thin sections were made and stained in iron haematoxylin.

---

1) Biological Bulletin, 1906, XII, p. 37 and 1907, XIII, p. 114.
2) See Proc. National. Museum, XXXIII, p. 473, for description and figure.

The primary germ cell is separated at the fifth cleavage of the egg at the ventral edge of the blastophore. It and its descendants may be distinguished by their position in the embryo and slow rate of division. When the germ cells have increased to four, they migrate to the definitive position of the gonads and continue to multiply, but the sex cannot for some time be distinguished.

Spermatogenesis.

In the former study of the spermatogenesis I was chiefly interested in tracing the development of those „spermatids" which do not become spermatozoa, but are transformed into food reservoirs by the accumulation of plastic products within the nuclei. In the present account the history of the chromatin will be chiefly considered.

The spermatogonia are polyedral cells containing large nuclei the chromatin of which exists in the form of a reticulum associated with one or more plasmosomes. During mitosis sixteen chromosomes may be counted, all of which are apparently equal in size.

The primary spermatocytes are smaller than the spermatogonia but have at first a somewhat similar arrangement of chromatin, fig. 1. If plasmosomes are present they are too small to be distinguished from the chromatin reticulum. The chromatin reticulum soon begins to condense into slender filamentous chromosomes giving the typical leptotane stage, text fig. It is to be expected that some of the filaments might be in contact when first formed, and such seems to be the case, but I have never been able to persuade myself that they ever constitute a continuous spireme such as has often been seen in nuclei. Fig. 2 represents an optical section showing the sixteen filaments. Erroneous impressions may easily result from a study of optical sections, and I do not present this as evidence of the number of filaments, which I have determined by a careful study of whole nuclei, but such a figure gives a better idea of the appearance of the filaments.

The filaments bunch up in the synezesis stage and at the same time become straighter, figs. 3—6. For this reason the chromosomes of this species are more favorable for study at this particular stage than those of the free swimming copepods[1]). During the synezesis stage parallel synapsis takes place, i. e. the chromosomes become associated in pairs with the members of each pair parallel, figs. 3—5. The synezesis zone occupies a considerable part of the testis. I have sectioned and studied

[1]) Cf. LERAT: La Cellule, 1905, XXII, p. 161.

numbers of these testes, yet in no case has it been possible to count the number of chromosomes in the early part of synezesis. The paired arrangement of them, however, is clearly seen. Near the end of this stage all of the filaments can sometimes be distinguished, and are eight in number, thicker than in the leptotane stage, and double at least in some parts, fig. 6.

The eight double filaments now become dispersed toward the periphery of the nucleus, giving the diplotane stage, figs. 7 and 8. The double filaments often interlace so that they cannot be counted, although where counting is possible eight are found. The double chromosomes now shorten, fig. 9, and I once thought that a transverse constriction was present, yet this might possibly be an accidental notch in the ragged outline of the double chromosome, and further study has not cleared up this point.

Leptotane Synezesis-Synapsis Diplotane Shortening Ist. Mat. Spindle

At a little later stage, however, a distinct second division is present, transforming the double chromosome into a tetrad, fig. 10. The shape of the tetrad is such that its longitudinal axis cannot be determined, and therefore it is impossible to decide whether the second division (constriction) is longitudinal or transverse.

The two maturation mitoses then rapidly follow, distributing the parts of each tetrad into the four spermatids, figs. 11—13. Owing to the form of the tetrads it would be difficult to decide whether the reduction in number of the chromosomes occured in the first maturation mitosis (pre-reduction) or in the second (post-reduction), but before going into this question we may briefly consider whether pre- and post-reduction have different effects in heredity.

Obviously in ordinary sexual reproduction if reduction occurs at all, it is immaterial whether it takes place during the first or second maturation mitosis. Weismann found that there is no second maturation division in certain parthenogenetic eggs, and concluded that reduction is thus omitted. Recent investigations show, however, that if synapsis occur, reduction takes place, whether the egg be parthenogenetic or not. Shall this fact prove universally true WEISMANN's hypothesis is erroneous and the question of pre- or post-reduction can have no significance in heredity. For this reason we will not discuss this question in relation to the spermatogenesis of *Pandarus*. The significant fact is that reduction has taken place, i. e. that each spermatid contains numerically one half of the same individual chromosomes that were present in the primary spermatocyte. This process is shown diagrammatically in the accompanying text figure.

Oogenesis.

The oogenesis is similar to the spermatogenesis through the synezesis stage. The primary oocytes are larger than the primary spermatocytes when first formed, and this difference increases as growth proceeds. The leptotane stage is shown in fig. 14, and two cells in the synezesis stage are seen in fig. 15. Immediately after synezesis a plasmosome is found, but whether it existed during synezesis could not be determined owing to the massed condition of the chromosomes.

Eight filaments radiate from the plasmosome and in favorable aspects are seen to be double, thus showing that the diplotane stage has been reached, fig. 16. These eight double filaments arose by a longitudinal synapsis in the synezesis stage, from the sixteen single filaments of the leptotane stage.

The growth period now begins, fig. 16; the cell outlines first disappear but reappear some time after the oviduct is reached. During the syncytial stage the cytoplasm becomes filled with basophile granules. MOROFF[1]) found, in the eggs of free swimming copepods, basophile granules extruded from the nucleus into the cytoplasm, but I have not noted a similar origin of the granules mentioned above. The oocytes have now become much flattened and suggest in their arrangement a pile of coins. The yolk is deposited in the form of proteid bodies ond oil droplets. The plasmosome becomes vacuolated and is probably in process of solution. Fig. 17 shows a nucleus and fig. 18 a section of an entire oocyte on a less magnified scale, at this stage. The black bodies in the cytoplasm are proteid and the light spots are fat.

The yolk accumulates and crowds the nucleus. The nuclear sap becomes basophile and the nuclear wall disappears leaving the eight double filamentous chromosomes radiating from the large plasmosome and sur-

[1]) Arch. f. Zellforschung, II, S. 432.

rounded by a cloud of chromatin granules, in the centre of a yolkfree area, fig. 19. The chromosomes have already begun to shorten. The plasmosome disappears, and the chromosomes continue to shorten until their breadth equals their length, when a second, though only partial, division transforms them into tetrads. The first division can be distinguished as it is the most complete, and the tetrads become arranged on the first maturation spindle so that the resulting mitosis separates whole chromosomes, and thus pre-reduction takes place, fig. 20. This reduction division is easily followed in the related species, *Orthagoriscicola muricata*, fig. 22. The second maturation mitosis is equational and separates the halves of the resulting diads, fig. 21. A diagram of the synapsis and reduction is shown in the accompanying text figure, and illustrates in general both spermatogenesis and oogenesis. The plasmosome and growth of the nucleus in the oogenesis are not represented.

Vs. the Critique of Hagedoorn.

Studies relating to the individuality of the chromosomes have been brought into special prominence owing to the debated question of the relation of the visible parts of the germ cells to the factors discovered by analyzing breeding experiments. I will make no attempt at a general consideration of this subject, but wish to call attention to an argument brought forward in a recent paper on animal breeding. HAGEDOORN[1]) says: „I do not see why we should attach any more value to the inheritance of a chromosome than to that of any other organ or character.“ I should like to emphasize the fact that the chromosomes and the other parts of the mitotic figure are the only „organs“, to use HAGEDOORN's term, that are seen to be transmitted bodily to the daughter cells in cell division.

In order to depreciate the significance of the chromosomes in heredity, HAGEDOORN states further that „GODLEWSKI — succeeded in fertilizing enucleated eggs of the sea-urchin with sperm of a crinoid, with the result that the ensuing larva — developed into a normal sea-urchin gastrula“. Now compare this with his further statements that „All inheritance is Mendelian inheritance“ and „It is impossible for a character to be in a recessive condition“ (p. 33). How can we harmonize these statements? If all inheritance is Mendelian and characters of immature stages are on a par with adult characters, how does he account for the

[1]) Arch. f. Entwicklungsmech., 1909, XXVIII, S. 28.

fact that the sea-urchin ♀ crinoid ♂ hybrid showed only maternal characters? Certainly the crinoid is not distinguished from the sea-urchin merely by the absence of characters. It is very doubtful whether all inheritance is Mendelian, and Godlewski's experiment does not disprove the hypothesis that the chromosomes bear some of the Mendelian factors.

It has not been demonstrated that the characters of the sea-urchin or crinoid larvae obey Mendelian laws, and the purely maternal character of the hybrid might be explained on the ground that the strange spermatozoon did not immediately affect the organization of the egg. Tennant produced hybrid echinoderm larvae showing only paternal characters by decreasing the alkalinity of the sea water.

Explanation of Plate.

Fig. 18 was drawn with the camera lucida with Zeiss apochr. homogen. ob. 2 mm, oc. 4; figs. 6—8 and 14 with oc. 12; the remainder with oc. 18 and reduced $^1/_3$. Figs. 1 to 21 are from *Pandarus sinuatus* Say; fig. 22 is from *Orthagoriscicola muricata* Kröyer.

Fig. 1. Spermatogonium.
Fig. 2. Primary spermatocyte, leptotane stage.
Figs. 3—6. Primary spermatocyte, synezesis stage.
Figs. 7—8. Primary spermatocyte, diplotane stage.
Figs. 9—10. Primary spermatocyte, formation of tetrads.
Figs. 11—12. Primary spermatocyte, first maturation mitosis.
Fig. 13. Secondary spermatocyte, second maturation mitosis.
Fig. 14. Primary oocyte, leptotane stage.
Fig. 15. Primary oocyte, synezesis stage.
Fig. 16. Primary oocyte, diplotane stage.
Figs. 17—18. Primary oocyte, growth period.
Fig. 19. Primary oocyte, first maturation mitosis, prophase.
Fig. 20. Primary oocyte, first maturation mitosis.
Fig. 21. Secondary oocyte, second maturation mitosis.
Fig. 22. Primary oocyte, first maturation mitosis, side view of equatorial plate.

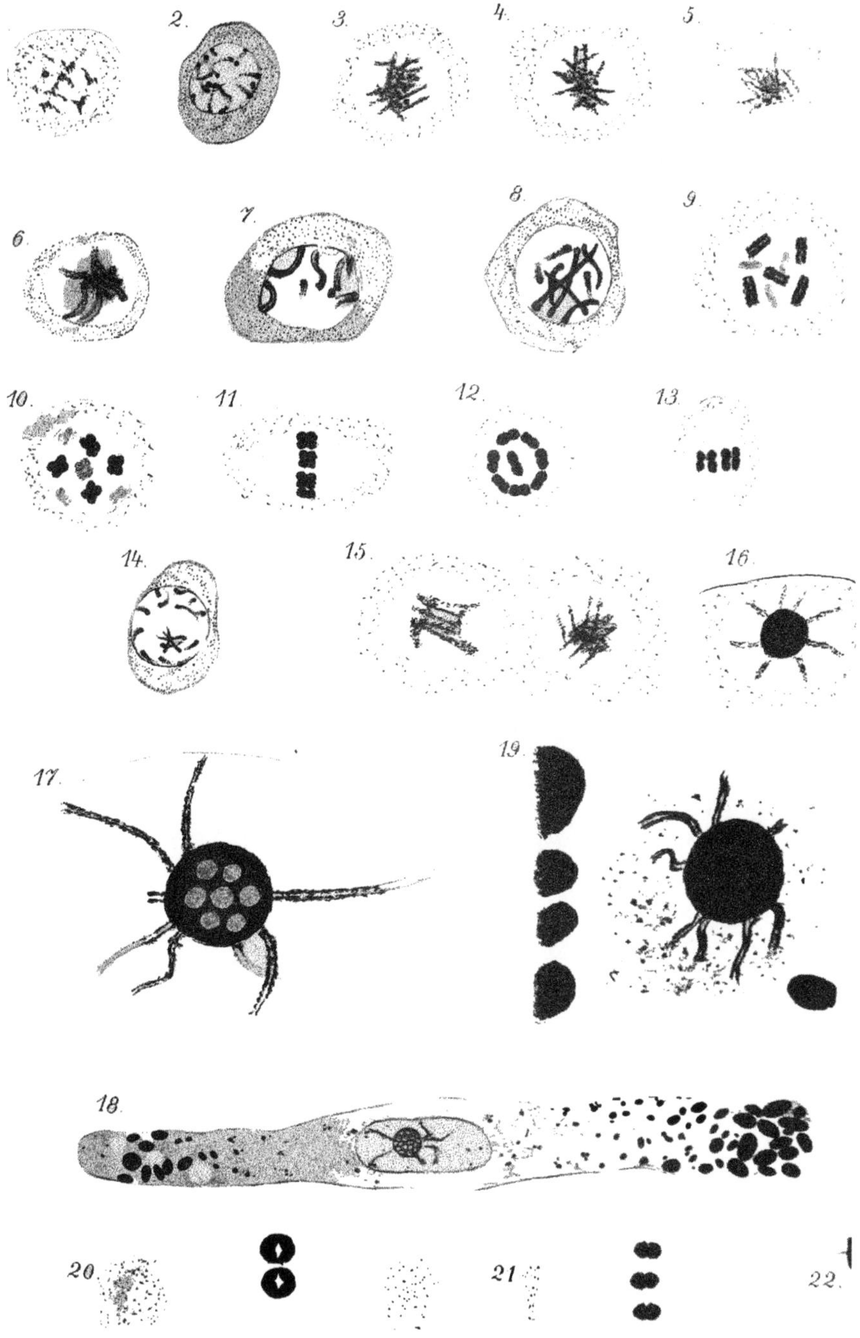
2.
3.
4.
5.
6.
7.
8.
9.
10.
11.
12.
13.
14.
15.
16.
17.
19.
18.
20.
21.
22.

On the Dynamics of Cell Division.

I. The Electric Charge on Colloids in Living Cells in the Root Tips of Plants.

By

J. F. McClendon.

(From the Histological Laboratory of Cornell University Medical College, New York City, U. S. A.)

With 2 figures in text and plate III.

Eingegangen am 26. April 1910.

The electric charges carried by ions have been extensively investigated and have yielded such valuable data that a similar study of colloids seems advisable. Here however we meet with difficulties, for many colloids are precipitated in the absence of electrolytes and the electrolytes may determine the sign of the charge on the colloids. Thus egg albumin is positive in acid and negative in alkaline solution.

Many complex colloids are positive, whereas pepsin is negative[1]). Nuclei usually go toward the anode and cytoplasm to the cathode[2]).

Last summer I began experiments on the cataphoresis of colloids in the living cell in order to test certain theories as to the cause of movement of granules and chromosomes accompanying cell division. These experiments are still in progress. I found that the nucleus as a whole did not bear the negative charge, which was localized in the chromatin, whether the latter was in the form of granules, spireme or chromosomes. The chromatin when carried toward the anode by as strong a current as was tried (in the case of *Diemyctilus*

---

1) Journ. Exp. Med. IX. p. 86 and 254. Biochem. Zeitschr. XXIV. S. 53.

2) R. Lillie, Am. Journ. Physiol. VIII. p. 273. Henri, Compt. rend. Soc. de Biol. LVI. p. 867.

not exceeding .02 amperes to the square mm. for 30 min.) was retained by the nuclear wall when the latter was present. The mitotic figure was sometimes moved as a whole toward the anode, and the chromosomes were never pulled out of it or carried through the cell wall.

Fig. 1.

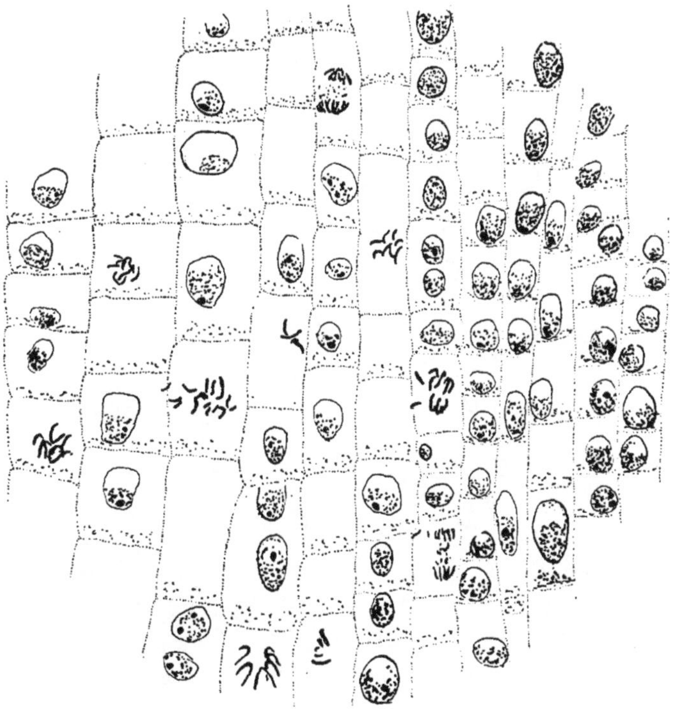

A portion of a longitudinal section of a hyacinth root through which an electric current of .0005 ampere was passed for 30 minutes, stained with safranin. In the figure the anode was below and cathode above. The basophile substances were carried by the current toward the anode. In sectioning the microtome knife passed in the direction of the figure from right to left, and some of the chromosomes were torn out of the spindles by the microtome knife, and carried toward the left.

I found that the nucleoli, pigment granules and yolk platelets of frog's egg also traveled toward the anode.

In the meantime the paper by Pentimalli on the cataphoresis of the chromosomes in the root tips of the hyacinth appeared[1]. He states that the chromatin is more and more affected by the current

[1] Influenza della corrente elettrica sulla dinamica del processo cariocinetico. Arch. f. Entw.-Mech. XXVIII. S. 210.

as the process of mitosis progresses, 'and that with a current of about .00005 amperes lasting 30 minutes some of the chromosomes are torn away from the spindle and carried through the cell wall

Fig. 2.

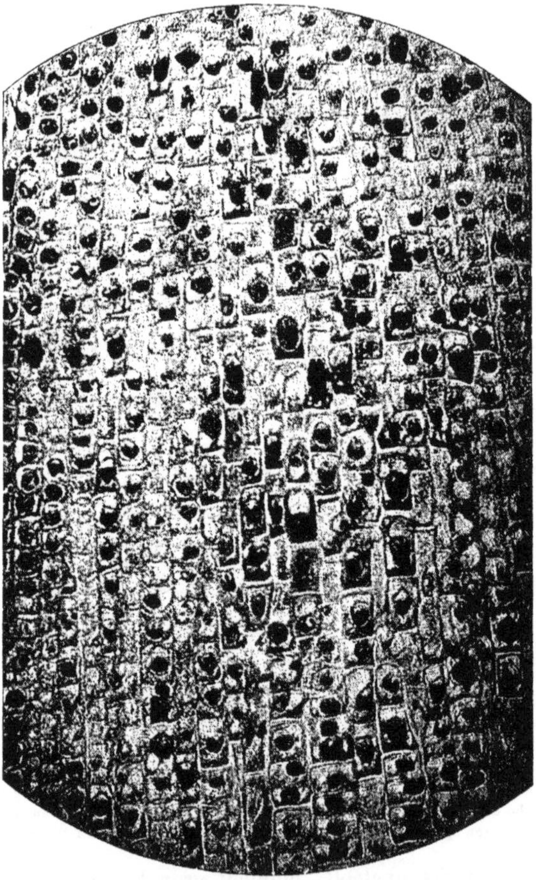

Photograph of another portion of the preparation shown in text fig. 1 and with the same orientation. The mitotic figures are not distinct, but the accumulation of basophile substance in the anodal ends of the cells and nuclei may be observed.

toward the anode. Such results are interesting indeed, and if verified would tend to lend support to dynamic theories of the nature of the mitotic figure. But it might be considered unusual to find chromosomes being carried through cellulose cell walls by such weak currents. The experiments appeared to be of sufficient importance to warrant repetition. After an extensive series of experiments I failed to convince myself that PENTIMALLI's interpretation of his results was correct. In the controls many chromosomes were displaced by the microtome knife, even when the sharpest and smoothest blade was used. The fact that chromosomes were not carried by the electric current through the cell walls of animal cells, in which displacement of chromosomes by the microtome knife seldom occurs, suggested that the chromosomes which PENTIMALLI supposed were moved by the current were in reality moved by the knife in cutting the sections. This suggestion was supported by the fact that with an extensive graded series of currents including those used by PENTIMALLI, many of the mitotic figures showed no displacement of chromosomes, whereas with those currents which moved the chromatin in the resting nuclei, all of the nuclei in the root tip showed this phenomenon, text figs. 1 and 2. PENTIMALLI allowed me to examine one of his preparations, for which my sincere thanks are due him. It did not convince me that my interpretation of my results was erroneous. Since these experiments may so easily be repeated on onion as well as on hyacinth roots and should be included in every course in Cytology, the reader may have a chance to interpret the results for himself.

PENTIMALLI's conclusion that the chromatin bore a negative charge was confirmed by my experiments, which did not however support his conclusion that the cataphoresis increases as the process of mitosis advances. The chromosomes seem to be firmly attached to the spindle so that in order to move them by the current the whole spindle must be moved, a process which, as will be shown later on, requires more current than to move the chromatin granules in the resting nucleus. The charge on the chromatin may increase as the process of mitosis advances as PENTIMALLI supposes, but the resistance offered by the spindle prevents proving such a supposition by this means.

Material and Methods.

Hyacinth and onion bulbs were set in damp sawdust or in tap water until the roots were of sufficient length to be handled. In the onion, more mitosis are found at about 1 p. m.[1]) than at any

[1]) KELLICOTT, Bull. Torrey Bot. Club. XXXI. p. 529.

other hour of the day, but this difference is so slight that it was hardly worth while to delay experiments until that hour. In roots grown in tap water sometimes very few mitoses were to be found, but a very large number of mitoses is a disadvantage and the convenience of this method made it useful.

Various forms of electrodes were tried. One form that may commend itself to those who wish to repeat the experiments without much trouble can be easily set up, provided a direct current of about 110 volts is available, as follows:

One wire of a lamp cord is cut and the cotton insulation removed from 2 inches or more of the cut ends. The rubber insulation is removed from an inch or more of the cut ends, exposing the copper wires, which are then introduced into glass tubes of such size that they may be tightly plugged by the rubber insulation. These tubes, while being adjusted, are completely immersed in copper sulphate solution, and on removal from the solution their free ends are plugged with absorbent cotton soaked in tap water, and fitted to rubber tubes filled with tap water or salt solution. The free ends of the rubber tubes are then plugged with absorbent cotton soaked in the same fluid with which they are filled, and the electrodes are ready to be applied to the tissue, provided no air bubbles have gotten into the tubes.

As the current is determined by the sum of resistances of the lamp, of the fluids in the glass and rubber tubes and of the tissue, it may be varied by varying any of these resistances. The resistance of the copper sulphate solution and of the tissue may be made negligable. To change the current it is most convenient to change the resistance of the lamp or of the fluid in the rubber tubes. This is done by substituting a lamp of a different candle power or changing the length or the salt content of the rubber tubes.

Pentimalli used physiological salt solution to fill the rubber tubes. The roots are not normally exposed to sodium chloride of such high concentration, and it would seem safer to fill the rubber tubes with tap water, to which may be added a very little salt if necessary.

If the rubber tubes are of sufficient length there will be no danger of the copper reaching the tissue, otherwise some proteid should be added to the fluid in the rubber tubes to precipitate the copper, or the more complicated electrodes described by Pentimalli substituted.

The current was measured with a galvanometer[1]) which was sensitive enough for all except the greatest densities of current used, but was not calibrated with very great accuracy. The cross section of the current in the tissue was made approximately constant by using root tips of about the same size and running the current longitudinally, except in a few experiments in which the current was run crosswise.

After the passage of the current, the root tip was fixed in BOUIN's fluid for 30 minutes, sectioned serially and stained in safranin, sometimes followed by methyl blue (wasserblau). Usually the latter was dropped on one end of the slide so that some of the sections had the safranin only while others had the double stain. Care was taken to record the direction of the current and also to have it pass in the plane of the section.

Experiments.

As PENTIMALLI used the hyacinth and as the cells in its roots are about twice the diameter of those in the onion only the results on the hyacinth will be described. The results on the onion present no material difference.

The amount of cataphoresis would depend on the density and duration of the current and could be changed by varying either, but as a change in duration of the experiment might involve a change in the number of phases during which a single cell was subjected to the current, it was thought best to vary the current density and have the time factor constant. In most of the experiments the current was passed for 30 minutes and varied from .00001 to .01 amperes. PENTIMALLI used about .00002 to .00005 amperes for 20—45 minutes.

I first thought it necessary to determine whether the current killed the cell. To do this currents were passed through roots still attached to bulbs placed in a moist chamber, the roots being marked by colored threads. In case the cells of the root tip were killed, the root ceased growing after one or two days and the tip rotted, but when the proximal part of the root was killed, the whole root rotted. It was found necessary to section roots some time after the passage of the current to determine more exactly the extent of the necrosis. Whether these lethal effects of strong currents were due to the migration of ions or of colloids, I have no means of determining. The general result was that if the current was much greater

[1]) For the use of which I am indebted to Dr. MAX MORSE.

than the minimal amount necessary to produce visible cataphoresis of the colloids, some of the cells disintegrated within a few days. I think it safe to assume however that all of the cells maintained many of the properties of living matter during the passage of the current except perhaps with the maximum current used. PENTIMALLI states that in using currrents of .004 to .005 amperes, the resistance of the tissue fluctuated and finally increased so that the current ceased altogether, indicating rupture of cell walls and disintegration and death of the cell. Such changes may indicate death of the cell but I know of no reason for supposing this to be the case, unless the final high resistance be taken to indicate coagulation of the proteids. The resistance of many tissues decreases during stimulation and shows a permanent decrease after death. Root tips sectioned after the passage of the strongest currents which I used, showed considerable plasmolysis of the cells but no rupture of membranes.

In each of the first seven experiments a control root was cut in serial sections. A resting cell is shown in fig. 1, plate III. It will be noted that both cytoplasm and nuclear contents are almost uniformly distributed. The plasmosome shown is excentric, but as in most cells there were two plasmosomes present, one of which does not appear in the optical section. The chromatin is granular and no definite reticulum is observed.

A special search was made for abnormalities, as these might aid in the interpretation of the experiments. Sometimes the plasmosome was dragged out of the nucleus by the microtome knife. More often however the mitotic figures were affected, the microtome knife displacing chromosomes or portions of the spireme, figs. 6—11. In some cases the spindles were diagonal, fig. 9, an unusual but probably not an abnormal occurrence. In a few cases the axis of the spindle was curved, fig. 7.

With the minimum current used no effect is observed. As the current is increased to about .00005 or .0001 amperes, the basophile substances, shown by darker shading in the plate, are carried toward the anode, fig. 2. This effect is more evident in the nucleus, which contains large amounts of basophile substance, the chromatin, but is also distinct in the cytoplasm. As the current is increased to .00015 amperes or more, the basophile substances are still further concentrated in the anode ends of cells and nuclei, and only acidophile substances remain in the cathode ends. Whether the acidophile substances are moved by the current cannot well be determined, as their distribution in the controls is obscured by the intermingling of basophile substances. With a current of about .01 amperes the nuclear membrane is pushed out towards the anode by the pressure of the chromatin, fig. 4, but apparently the cathodal end of the nuclear wall was moved very little if at all. Whether the nucleus changed in volume during the process, was not determined.

It might be objected that this alteration of the appearance of the stained section is not due to the movement of basophile or acidophile substances, but is accounted for by the alteration of the affinities of the different parts for stains, due to the accumulation of acid at the anode and alkali at the cathode, forming acid and alkali albuminates. Such an accumulation of acid and alkali would necessarily take place if the surface layers of cells and nuclei offered more resistance to the passage of ions than the interior, and there is much evidence to indicate that such is the case for cells in general. But the tissue was fixed in a strongly acid fluid, and we should then expect all of the protoplasm to be basophile. Furthermore, in the prophase of mitoses the spireme is carried toward the anode, fig. 5, demonstrating undoubtedly the cataphoresis of colloids, which are in this case perhaps in the »gel« phase. We may assume then that acid appears at the anode and alkali at the cathodes of both cell and nucleus, but this accumulation does not account for the redistribution of basophile and acidophile substances, which is due to the migration of colloids.

As the process of mitosis advances the chromatin is less and less affected by the current. Whereas .0015 amperes are sufficient to move the spireme, fig. 5, .003 amperes are required to move the spindle, fig. 16, which is often hardly affected by 0.1 amperes, fig. 12. However in the cell shown in fig. 12 the movement of the spindle may have been stopped by the cell wall. The cytoplasm is very little affected by the current during mitosis. This may be due to the resistance offered by the mitotic figure to the movement of colloids.

When the current passes during the later phases of mitosis the cell division is displaced toward the anode, fig. 17. This is a demonstration that cataphoresis takes place during the life of the cell, for cell division is a vital process, and the formation of a new cell wall in this case evidently took place after the displacement of the mitotic figure by the current.

The study of the effects of the current on the mitotic figure is complicated by the displacement of the chromosomes by the microtome knife in cutting the sections. In controls a considerable amount of displacement occurs, and in regions with relatively few mitoses it is easily observed that the displacement always occurs in the direction of the movement of the knife. Where mitoses are numerous this is not so evident, owing to the fact that chromosomes of one mitosis may be carried to the vicinity of another mitosis. After passage of the current the same appearances were observed. By picking out individual mitoses in certain slides it is very easy to find numerous instances where the chromosomes are displaced toward the anode, even through cellulose cell walls. But by careful study of many whole series of sections very different observations may be made, especially if the direction of the microtome knife was purposely chosen so as to be at right angles to the direction of the current, as it was in the large majority of my experiments. In text figure 1, it may be noted that the direction of the current as marked by the polarization of cells and nuclei, is toward the top of the page, whereas the displacement of chromosomes is toward the left. On the other hand, after the passage of a current of any density within the limits above mentioned, many spindles showing no displacement of chromosomes were found, figs. 12—15.

Conclusions.

PENTIMALLI's assertion that the effects of the passage of an electric current through the cell demonstrates that the chromatin bears a negative electric charge is confirmed by the experiments described above. If the chromosomes were attracted to the asters or poles of the spindle because the poles are positive, the chromosomes being negative, we would expect to find the yolk platelets and pigment granules of the frog's egg also attracted, as they bear negative charges. The pigment granules in the frog's egg are considered by some observers to be attracted toward the aster, though a considerable area around the centrosome is always free from them. However I think that everyone will agree that the yolk platelets are repelled by the aster. The fact that the negative chromosomes are attracted, and the negative yolk platelets are repelled by the aster shows that the mitotic figure is not simply a bipolar electric field. Furthermore if the poles of the spindle bear positive charges and the chromosomes negative, we should expect the chromosomes to move toward the anode and the poles of the spindle to move toward the cathode on the passage of a current, which was not found to be the case. We should also expect the spindle in the meta- or anaphase to break in two at the equator when the cell contents are subjected to violent mechanical disturbance since there is no attraction between the two halves of each divided chromosome, but on the contrary the mitotic figure resists mechanical distortion to a remarkable degree.

The forces concerned in mitosis may be electric, but in ultimate analysis all forces may finally be reduced to electric components. To suppose that the spindle consists of proteids in the »gel« phase and that it »grows«, does not explain the phenomenen. We know that osmotic forces enter largely into the growth of plants, but osmotic membranes in the mitotic figure would be difficult if not impossible to demonstrate. The reduction of mitosis to a readjustment of surface tension in the walls of demonstrable or hypothetical alveoli, as has been done by RHUMBLER, following the work of BÜTSCHLI, suggests itself as an alternative, and is worthy of further investigation.

The appearance of cells through which electric currents have been passed is similar to that of those which have been centrifuged. In many cases investigated, the substances of greater specific gravity bear the negative charge, and thus the effects of centrifugal force and the electric current appear to be the same, but this is probably a mere coincidence. Whereas the bodies moved by the centrifuge are largely stored materials, the electric current has been shown to move ions, enzymes and various thermo-labile bodies, which may or may not be in solution. In other words, we have in the electric current a little used method of investigating life processes which might possibly be applied as centrifugal force has been. In much of the work with the electric current the end results have been recorded but the intermediate steps not observed. The separation of basophile from acidophile protoplasm by the electric current points out a method of observing the immediate effects of the electric current on tissues. It would be interesting to trace the relation between these immediate effects and the abnormal mitoses which have been observed in tissue subjected some time previously to the electric current.

Summary.

In the cells of root tips of the onion and hyacinth the basophile substances migrate toward the anode on the passage of an electric current, except in case of the mitotic figure, which migrates as a whole toward the anode.

As the process of mitosis advances the effect of the current on the chromatin decreases, contrary to the conclusion of PENTIMALLI, who supposed that it increases.

Chromosomes or other bodies are never carried through the nuclear or cell walls by currents of the density used (.00001—.01 amperes).

Zusammenfassung.

In den Wurzelspitzen der Zwiebel und der Hyazinthe wandern die basophilen Substanzen beim Durchgang eines elektrischen Stromes nach der Anode hin, außer wenn eine mitotische Figur vorhanden ist, welche im ganzen nach der Anode hinwandert.

Mit dem Vorschreiten des mitotischen Prozesses nimmt die Einwirkung des Stromes auf das Chromatin ab, entgegen der Folgerung von PENTIMALLI, welcher annahm, daß sie wächst.

Chromosomen oder andre Körper werden durch Ströme von der verwendeten Stärke (0,00001—0,01 Ampère) niemals durch die Kern- oder Zellwände hindurchgetrieben.

(Übersetzt von **W. Gebhardt.**)

Explanation of Plate III.

All the figures were drawn with a camera lucida, with B. & L. homogen. imm. obj. 2 mm., oc. 1. All are from longitudinal sections of root tips of the hyacinth. The figures are so arranged that the microtome knife passed from right to left in cutting the sections, and in case a current was passed, the anode was below and cathode above, and the duration of the current was 30 minutes. The basophile substances are distinguished by darker shading.

Fig. 1. Control.
Fig. 2. Subjected to a current of .0001 amperes.
Fig. 3. - - - - - .00015 -
Fig. 4. - - - - - .01 -
Fig. 5. - - - - - .0015 -
Fig. 6. Control.
Fig. 7. -
Fig. 8.
Fig. 9.
Fig. 10.
Fig. 11. -
Fig. 12. Subjected to a current of .01 amperes.
Fig. 13. - - - - - .0015 -
Fig. 14. - - - - - .0002 -
Fig. 15. - - - - - .0006 -
Fig. 16. - - - - - .003 -
Fig. 17. - - - - - .01 -

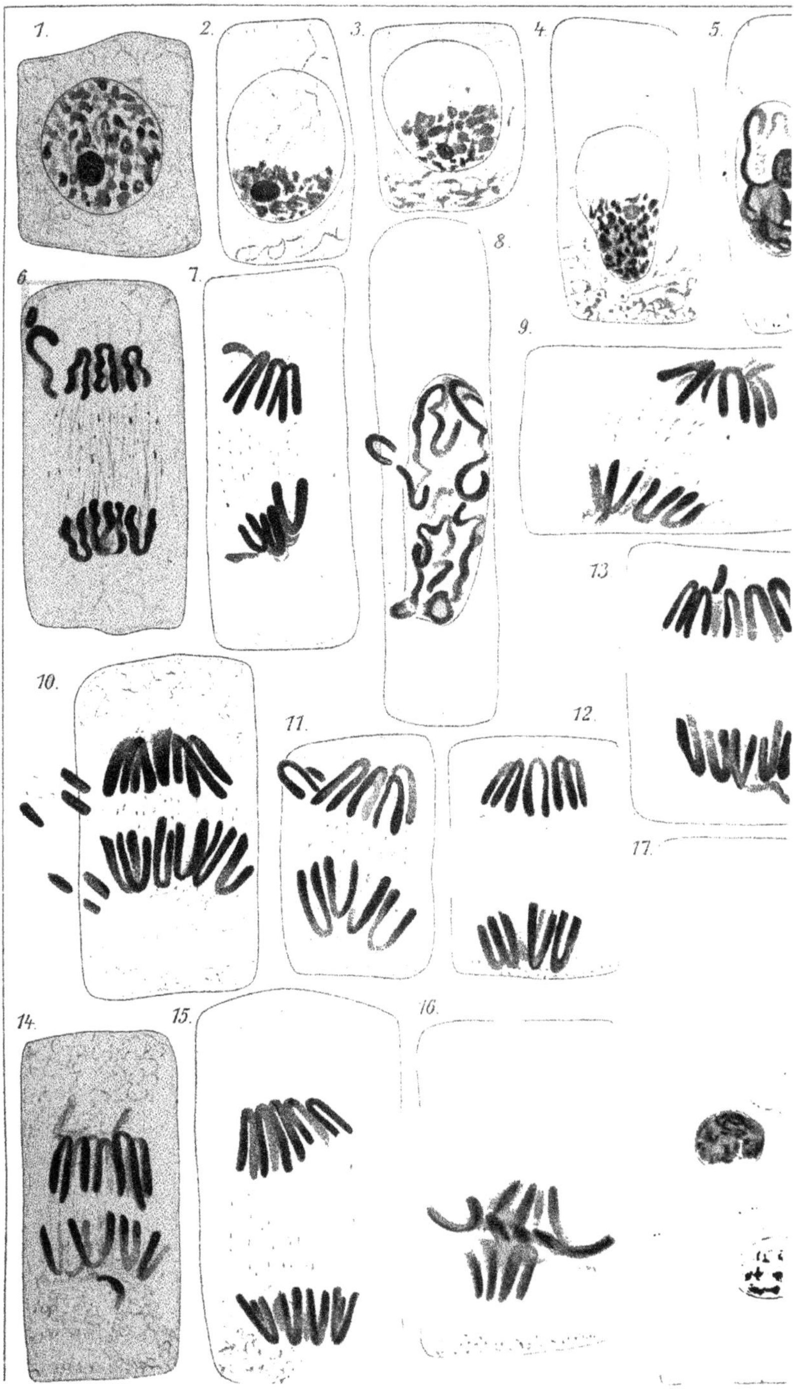
1.
2.
3.
4.
5.
6.
7.
8.
9.
10.
11.
12.
13.
14.
15.
16.
17.

Reprinted from the American Journal of Physiology.
Vol. XXVII. — December 1, 1910. — No. II.

ON THE DYNAMICS OF CELL DIVISION. — II. CHANGES IN PERMEABILITY OF DEVELOPING EGGS TO ELECTROLYTES.

By J. F. McCLENDON.

[*From the Histological Laboratory of Cornell University Medical College, New York City, and the Laboratories of the Carnegie Institution at Tortugas, Fla., and the U. S. Bureau of Fisheries at Woods Hole, Mass.*]

THE process of cell division may be divided into two distinct phenomena, the division of the nucleus and of the cytoplasm. Although these processes are closely interrelated, they can occur separately. Karyokineses may occur without subsequent cytoplasmic division, and cytoplasmic division may occur without the presence of a nucleus or chromatin in any form.[1] The division of the cytoplasm in plant cells is accomplished by the formation of a division wall, but in most animal cells by simple constriction.

The constriction of the cell seems to be a special case of these protoplasmic movements that were shown by Quincke to resemble movements accompanying surface tension changes.[2] If the constriction of the cytoplasm were due to surface tension changes, we should expect a band of greater surface tension to include the cleavage furrow. In this case there would be a flowing of the superficial cytoplasm from the poles of the cell toward the cleavage furrow, and of the deeper protoplasm in the opposite direction.

By a study of fixed material Nussbaum [3] showed movements of the pigment granules in cells of frog's embryos from the interior to the surface and along the surface to the position of the future cleavage furrow. On constriction of the cell the granules were massed in the form of a plate in the cleavage plane. These movements indicate the surface tension changes described above.

[1] McClendon: Archiv für Entwicklungsmechanik, 1908, xxvi, p. 662.
[2] Buetschli: Archiv für Entwicklungsmechanik, 1900, x, p. 52.
[3] Nussbaum: Anatomische Anzeiger, 1893, viii, p. 666.

Conklin [4] found evidence for such movements in the changes in position of certain structures in Crepidula eggs.

The first observation of this process in living cells was made by Erlanger,[5] who saw movements of superficial granules toward the cleavage furrow, and of internal granules toward the poles, of Nematode eggs.

Gardiner [6] observed, in living eggs of Polychœrus caudatus, colored granules move to the surface and then along it to the position in which the cleavage furrow appeared immediately afterward. Fischel's observations are considered below.

The fact that cells usually round up before cleavage, if not previously spherical, indicates a general increase in surface tension, and it is only necessary to assume a greater increase to be localized along the cleavage furrow to account for the constriction.

Robertson [7] floated an olive oil drop on water and laid across it a thread moistened with soap (or soap-forming) solution. After the thread reached the edges of the drop the latter was torn in two. Since soap decreases the surface tension between oil and water, he concluded that cytoplasmic division is due to a decrease in surface tension along the cleavage furrow. As this view has been accepted by Lillie [8] and Loeb,[9] it seems worth whlie to point out Robertson's error.[10] In Robertson's experiment three different surface tension films occur, between air (*A*) and water (*W*), air and oil (*O*), and water and oil (Fig. 1), and an equilibrium is established when the water-air surface tension equals the horizontal components of the air-oil plus the oil-water surface tensions. When the moistened thread is laid across the oil drop, two more films are added, *i. e.*, air-soap solution (*S*) and oil-soap solution (Fig. 2). At opposite edges of the drop where the thread touches the water, the soap would decrease the water-air surface tension, and the undiminished pull on the remainder of the edge of the drop would pull it in two (Fig. 3).

[4] Conklin: Biological lectures at Woods Hole, 1908, p. 69.

[5] Erlanger: Biologische Centralblatt, 1897, xvii, p. 152.

[6] Gardiner: Journal of morphology, 1897, xi, p. 55.

[7] Robertson: Archiv für Entwicklungsmechanik, 1909, xxvii, p. 29.

[8] Lillie: Biological bulletin, 1909, xvii, p. 203, footnote.

[9] Loeb: Chemische Entwicklungserregung des tierischen Eies, p. 5.

[10] This was first pointed out by me before the American Society of Zoölogists, see Science, xxxi, p. 467. It was discussed by A. B. Macallum, Science, 1910, xxxii, pp. 498–500.

I have repeated Robertson's experiment and also modified it by entirely submerging the oil drop. Enough alcohol was added to the water to make the oil sink below the surface, and the soap solution introduced through a capillary pipette, or a piece of solid soap held near the oil drop, or a thread covered with solid soap was wrapped around the oil drop. Very little movement of the oil occurred, but

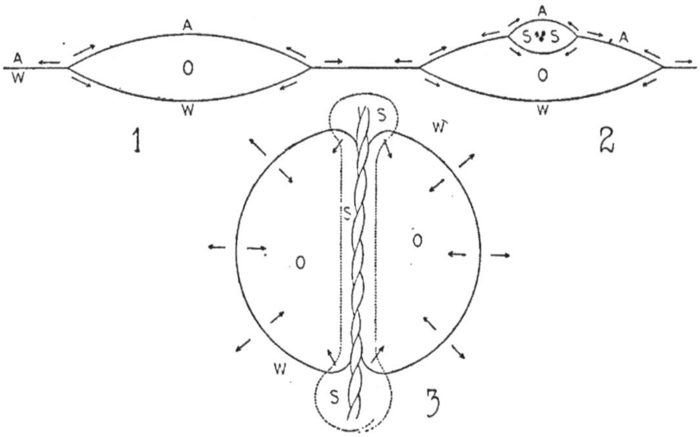

FIGURES 1 to 3. — *A*, air; *O*, oil; *W*, water; *S*, soap solution. The arrows show the direction of the pull of surface tension. The dotted line in Fig. 3 bounds the soap solution. Further explanation in text.

that which did occur was always a bulging toward the soap, and never a constriction or receding from the soap. Similar experiments were also tried on the under side of oil drops floating on water, and unless the soap reached the water-air film, the oil advanced toward the soap and no constriction occurred. We may conclude, then, that the cleavage furrow is a region of *increased* surface tension, as shown by Bütschli and others, and not of decreased surface tension, as Robertson, Lillie, and Loeb maintain.

One may ask why such movements as seen by Erlanger, Gardiner, and others have not been observed in all dividing cells that have been studied alive. The answer to this may be sought in the structure or consistence of protoplasm. If the cytoplasm present an alveolar structure, the spreading of the surface in regions of reduced surface tension would be almost entirely confined to the individual alveoles, and the general effect would be a slow stretching of the surface in areas of less surface tension and a slow contraction of the surface in areas of greater surface tension, as shown by Bütschli's microscopic oil foams. The constriction of the cleavage furrow would then take place as though a rubber band around the cell contracted.

In the constriction of the egg of the sea urchin, Arbacia punctulata, usually just such a number of chromatophores are carried into the cleavage furrow that when the two daughter cells are formed the pigment is evenly distributed over all parts of their surfaces. Under certain abnormal conditions in which the cleavage is more violent the pigment is massed in the furrow. Fischel observed a massing of pigment along the lines of the future cleavage furrows in the eggs of Arbacia pustulosa. The reasons we get so little movement of pigment in the egg of Arbacia punctulata probably are the alveolar structure and the presence during cleavage of the "hyaline plasma layer," or "Verbindungsschicht," to which the surface movements are chiefly confined. The hyaline plasma layer is said to be formed by a recession of granules toward the interior of the egg, leaving the superficial layer free from granules and almost invisible. It is formed before the first cleavage and becomes heaped up in the cleavage furrow.[11] Whenever such an outer layer occurs, as in eggs of sea urchins and ctenophores, it becomes heaped up in the cleavage furrow; this indicates increased surface tension in this region (or decreased surface tension at the poles).

In the cutting off of the micromeres in the Arbacia egg the pigment entirely disappears from the micromere pole, indicating spreading movements due to the surface tension being less here than in the region of the future cleavage furrow. Similar movements of granules have been observed in the cutting off of polar bodies in various eggs, and it may be concluded that for the separation of a very small cell from a large mass of protoplasm a very great difference in surface tension between the pole of the small cell and the cleavage furrow is required.

Changes in surface tension may be the result of the presence of certain substances in one of the two fluids in contact, changes in temperature, or a difference in electric potential across the boundary. In numerous instances electric changes have been found to accompany vital movements. Hyde[12] detected electric changes accompanying

[11] Goldschmidt and Popoff: Biologische Centralblatt, 1908, xxviii, p. 210.

[12] Hyde: This journal, 1904, xii, p. 241.

cleavage, and the question whether the constriction of the cytoplasm is due to electric changes seems worth investigation.

If an electric current were passed through a solution containing a living cell, and if the cell surface offered more resistance to the passage of ions than either the medium or the cell interior, a difference of potential would be produced between the inner and outer sides of the cell surface, and would be proportional to the angle that the surface made with the current lines cutting it, *i. e.*, it would be greatest at the point where the surface was at right angles to the current lines and equal to zero at the point where the surface was parallel to the current lines. The surface tension would be reduced at the poles (the points nearest the electrodes), and the equator would lie in a region of relatively greater surface tension. This would result in the protrusion of the polar regions and constriction of the equator, thus producing the form change of the first stage of cleavage.

When a current of a certain density was passed through an unfertilized Arbacia egg, the surface nearest the anode showed spreading movements and bulged out. We might conclude from this that this egg was less permeable to anions than to cations. The confined anions caused a difference of potential between the two sides of the surface nearest the anode, thus decreasing the surface tension, and spreading of the surface and bulging of the egg followed. At the surface nearest the cathode, the anions that could not enter could pass around the egg, and therefore the difference of potential was not so great as at the opposite pole. If the egg were as poorly permeable to cations, we should expect a reduction of surface tension at the pole nearest the cathode.

It seemed to me that an analysis of artificial parthenogenesis might throw light on the question of cell division, and in the summer of 1909 I began an attempt in this direction.

Artificial Parthenogenesis.

It is well known that the eggs of different individuals of the same species vary in response to stimuli. Investigators usually suppose that they have normal material when a large per cent of the eggs develop in a control to which sperm is added. In order to test this method of controlling experiments and be sure no unknown factor vitiated the results, I made a series of experiments by fertilizing eggs of the same mother with sperm of different fathers, and eggs of different mothers with sperm of the same father, with the following results, giving the percentage of eggs developing:

Fathers.	Mothers.			
	A	B	C	D
1	100	98	93	98
2	100	97	94	96
3	100	98	92	97
4	100	96	90	100

It may be seen by inspecting the table that development depends more on the eggs than on the sperm, so the practice of keeping a fertilized control seems to be a good one. Fertilized and unfertilized controls were kept to all of my experiments, and the experiment thrown out if fertilization did not occur.

The question which concerns us first is, in what ways have cells been caused to divide, and maturation (in most cases), as well as segmentation, is cell division. We may summarize the methods used as follows:

Hypotonic solutions (distilled water, Schücking).

Nearly isotonic solutions made by adding to sea water or to distilled water the following substances:

Acids (Delage, Fischer, Herbst, Lefevre, Loeb, Lyon, Neilson, Schücking, Tennent).

Alkalis (Delage, Loeb, Schücking).

Neutral salts (Delage, Lillie).

Hypertonic solutions:

Acids (Delage, Loeb).

Alkalis (Delage, Loeb).

Neutral salts (Bataillon, Bullot, Delage, Fischer, Hunter, Kostanecke, Loeb, Lyon, Mead, Scott, Treadwell, Wilson).

Non-electrolytes (Delage, Loeb).

Mechanical shock (Delage, Fischer, Mathews, Scott).

Thermal changes (Bataillon, Delage, Greeley, Lillie, Loeb, Schücking).

Electric changes (Delage, Schücking).

KCN or lack of oxygen (Loeb, Lyon, Mathews).

Fat solvents (Loeb, Mathews).

Alkaloids and glucosides (Hertwig, Loeb, Schücking, Wassilieff).

Blood sera (Bataillon, Loeb).

Soap and bile salts were found effective by Loeb when followed by hypertonic solutions.

As all of these agents were not tried and found effective on eggs of the same species, their differences might be thought to correspond to differences in the eggs, and therefore I thought it worth while to try a large number of them on the egg of Arbacia punctulata at Woods Hole, Mass. Segmentation was produced by the following methods: the time indicated is the optimum duration found after a series of experiments, the tables being omitted:

Hypotonic solutions (70 c.c. sea water + 30 c.c. tap water or distilled water, one to one and a half hours).

Nearly isotonic solutions:

Acids (50 c.c. sea water and 3 c.c. 1/10 normal acetic, fifteen to sixty seconds. Or sea water charged with CO_2 in a "sparklet fountain," five to ten minutes).

Alkalis (1.2 c.c. 1/10 normal NH_4OH or NaOH + 50 c.c. sea water, twenty to sixty minutes).

Salts (5/8 normal NaCl, one-half to two hours, this is very slightly hypertonic).

Hypertonic solutions (100 c.c. sea water + 15 c.c. 2½ normal NaCl, one hour. Or sea water boiled down to .76 of its volume, one hour.)

The eggs were also made to segment by placing them in sea water brought from Boca Grande Key, twelve miles west of Key West, Florida, in steamed out, glass-stoppered and paraffine-sealed bottles. The eggs were allowed to lie twenty minutes in Woods Hole sea water, then placed for two and a half hours or four hours in Boca Grande sea water and returned to Woods Hole sea water, or allowed to remain indefinitely in Boca Grande sea water. At the end of nine hours segmentation had occurred in all three lots, about 10 per cent were segmented in that left four hours in Boca Grande sea water.

Woods Hole sea water has a $\Delta = 1.818$, specific gravity $= 1.024$ (Garrey). Boca Grande sea water has a $\Delta = 2.05$, specific gravity $= 1.0248$. The alkalescence of the Boca Grande sea water is greater than that of Woods Hole. The hypertonicity and greater alkalinity may have both aided in producing the segmentation. This experiment producing such poor results (10 per cent), is given so much space merely because the effects were produced by natural sea water. However, this does not prove that Arbacia grown at Boca Grande would be naturally parthenogenetic.

Mechanical shock (shaking in vial, by hand, five minutes. Or pouring from one dish to another every ten minutes for three hours. One effect of agitation is the removal of the jelly-like covering from the eggs, after which, perhaps, the mere contact of the egg with another surface will start development. Mathews supposes the effect of shaking is rupture of the nuclear wall, at least in the starfish egg).

Thermal changes (keeping at 32° C., four minutes; keeping at 1° C., one to eleven hours; or at 10° C., one to twenty hours. By the end of twenty hours some were already segmented).

Electric changes (several entire ovaries were placed in longitudinal series in a glass tube, and an alternating current from a small induction coil was passed through it for two hours. To avoid error from polarization at the electrodes, only the eggs from the central ovary were observed).

KCN or lack of oxygen (a stream of hydrogen eight to twenty-two hours, with or without previous boiling of the sea water. Or 1/500 normal KCN, seventeen to thirty-two hours. Or 1/1000 normal KCN thirty-two hours).

Fat solvents (½ saturated solution of ether in sea water, ten minutes).

Certain combinations such as carbonic or fatty acid followed by hypertonic sea water, or tannic acid + an excess of NH_4OH,[13] or tannic

[13] In making his "ammonium tannate" solution, Delage considered tannin a hexivalent acid, but I can find no confirmation of this view in chemical handbooks. On adding this solution to sea water, a slight precipitate, probably calcium or magnesium tannate, forms, and the fact that Delage obtained as good results by adding the "ammonium tannate" to a sugar solution does not prove that the precipitation of some salt in the sea water solution was not in this case a factor in the production of parthenogenesis.

or acetic acid followed by NH_4OH or NaOH, seemed to produce better results than the single treatments.

The following results were obtained on other species:

First, at Woods Hole, eggs of Cumingea and Mytilus were caused to maturate by treatment with hyperalkaline sea water.

Second, at Tortugas, Fla., where I found the $\Delta = 2.03$, specific gravity = 1.0246, and alkalescence greater than at Woods Hole.

A few of the immature eggs of Ophiocoma rizii maturated when left twelve hours in 50 c.c. sea water + one drop of dilute ammonia, whereas none maturated in the control.

Eggs of Toxopneustes (Lytechinus) variegatus and Tripneustes (Hipponöe) esculentus were made to segment by placing a test tube of sea water containing them for one minute in water at 38°–44° C., then pouring the eggs into a dish of sea water at the normal temperature.

After treatment with sea water carbonated in a "sparklet syphon," followed by hypertonic sea water, development went farther in both species than after any of the single treatments tried. By increasing the duration of the treatment the rate of development was increased (approached or equalled that of fertilized eggs), but the percentage of resulting larvæ decreased, indicating injury to the eggs.

The optimum for Toxopneustes was: carbonated sea water one and one-half to five minutes, followed by 100 c.c. sea water + 16 c.c. 2½ normal NaCl, thirty to forty-five minutes. If eggs had remained a long time in sea water before the beginning of the experiment, a shorter stay in carbonated sea water was required than if they had been just taken from the ovaries. In this connection it is to be remarked that carbonated sea water hastens the solution of the jelly-like coverings, which takes place more slowly in natural sea water.

The optimum for Tripneustes was carbonated sea water ten minutes, followed by hypertonic sea water one hour. If we look for something in common in all of these methods of artificial parthenogenesis, we meet with many difficulties. I concluded that all of the methods of artificial parthenogenesis could not directly initiate any one single chemical reaction in the egg, but must have their first common effect in some physical or physico-chemical change.

The osmotic methods have this in common, that in all there is an *increase* in osmotic pressure. If the eggs are placed in sufficiently concentrated sea water or other hypertonic solution, some may segment while remaining in the hypertonic solution, but if placed in distilled water (Schücking) or diluted sea water (McClendon) they segment only after removal to natural sea water, which means an increase in osmotic pressure of the medium. I do not, however, conclude from this that in the latter instance it is the return to sea water rather than the sojourn in the hypertonic solution that starts the development of the egg.

Traube [14] showed that "fertilization membrane-forming" substances are effective in greater dilution the more they lower the surface tension of water. If the egg surface contain lipoids, such substances will be adsorbed or absorbed by the lipoids in the ratio that they lower the surface tension of water. But in what way can an absorption or adsorption by the lipoids of the cell, of a host of different substances, cause the development of the egg, and how can we explain those methods in which no lipoid soluble substance is used?

Loeb had shown the similarity between methods of artificial membrane formation and hæmolysis, and many eggs segment after artificial membrane formation. But in my opinion Loeb has not given a satisfactory explanation of the mechanism of hæmolysis. It seems to me that the "membrane theory" of hæmolysis, so admirably presented by Stewart,[15] is a satisfactory explanation.

Lillie [16] advanced the view that the essential element in artificial parthenogenesis is the increase in permeability of the plasma membrane to CO_2, allowing the chief end product of oxidation to escape and the rate of oxidation to increase, the more rapid oxidation causing development. But Lillie has never published any determinations of changes in permeability of the egg to anything except pigment, and there is no certain proof that the escape of pigment is due to increased permeability of the plasma membrane, as the pigment must first be liberated from the chromatophores, in which it is held physically or chemically.

I know of no method of determining changes in permeability of the egg to CO_2, but have thought of five methods for detecting changes in

[14] TRAUBE: Biochemische Zeitschrift, 1909, xvi, p. 182.

[15] STEWART: Journal of pharmacology and experimental therapeutics, 1909, i, p. 49.

[16] LILLIE: Biological bulletin, 1909, xvii, p. 188.

the permeability of the egg to electrolytes in general: 1. Electric conductivity of masses of eggs; 2. Electric conductivity of individual eggs as determined by destructive effects of the electric current on the cell; 3. Plasmolysis; 4. Chemical analysis of masses of eggs; 5. Microchemical analysis of single eggs. These will be considered in the order given.

Brown[17] concluded that the membrane of the Fundulus egg is practically impermeable to salts and water during the first eight hours, and becomes most permeable after eighteen to twenty hours. Apparently he refers to the thick membrane which is pushed out after oviposition, but Sollmann[18] observed that the "yolk" swells in distilled water or one-fourth molecular cane sugar, obliterating the perivitelline space, and thus indicating that the plasma membrane is less permeable to salts than is the thick egg membrane or chorion (vitelline membrane of Sollmann).

Biataszewitz[19] found that the absorption of water by the unfertilized frog's egg increased five times for every rise of 10° C., and concluded from this that heat increased the permeability of the plasma membrane to water.

The Electric Conductivity of Masses of Eggs.

This work was done at the Tortugas Laboratory of the Carnegie Institution. The experiments were made on board the yacht "Physalia" anchored off Boca Grande Key, twelve miles west of Key West, Fla. I am indebted to Dr. W. R. Warren of Key West for the use of a centrifuge, as the one provided was left behind. My thanks are also due Dr. Alfred G. Mayer for the unusual facilities at my disposal.

The determinations were made by Kohlrausch's method. A resistance box of 15,000 ohms and a metre sliding resistance were used. A number of difficulties arose in eliminating possible sources of error, and these will be considered in order:

First. The procuring of sufficient quantities of suitable eggs to

[17] Brown: This journal, 1905, xiv, p. 354.

[18] Sollmann: This journal, 1904, xii, p. 112.

[19] Biataszewitz: Bulletin de l'Académie des Sciences de Cracovie, Math. Nat., October, 1908.

make accurate determinations. This was met by going to Boca Grande Key, where the sea urchins, Toxopneustes variegatus and the Tripneustes esculentus, could be picked up by the ton in shallow water. As the ripe eggs of the former species were more abundant, they were used exclusively.

Second. The handling of the eggs with sufficient rapidity to insure their being in normal condition. Washing the eggs repeatedly by allowing them to settle in sea water requires much time, though it will be shown later that for the first washing this is an advantage.[20] But the time required for them to settle into a compact mass for the conductivity determination must be shortened as much as possible, as the continued crowding of the eggs might produce abnormal effects. No suitable conductivity vessel on the market could be placed directly in the centrifuge.

I made a conductivity vessel at the Tortugas Laboratory. It consisted of two glass tubes, the inner one fitting nicely into the outer one. Owing to a very slight curvature of the tubes, a rotation of one in the other would clamp them tightly together. The inner tube was 131 mm. long, 10 mm. inside diameter, and sealed at the lower end. Two small glass tubes were placed longitudinally within the outer tube and sealed into its upper end. One of these projected 25 mm. below the other. Platinum electrodes 6 × 9 mm. were sealed in the lower ends of the two smaller tubes. The electrodes were "platinized" with platinic chloride solution containing a little lead acetate.

The advantages of this conductivity vessel were the comparatively large surface area of the electrodes, 108 sq. mm. each; the distance between them, 25 mm.; the small volume of eggs, less than 3 c.c., required to cover the electrodes, and the short time required for its contents to reach the temperature of the thermostat. The electrodes were plane and vertical, and hence could be pressed down into a mass of eggs with the least possible disturbance of them. The inner tube, containing the eggs, could be placed directly in the centrifuge, and thus the eggs could be washed with sea water or other solutions without removal from the conductivity vessel. By inserting the lower end of the outer tube into a short, closely fitting test tube, the electrode could be protected from drying while the inner tube was in the centrifuge.

[20] The eggs of Toxopneustes are of but very little greater specific gravity than sea water and hence settle much slower than those of Tripneustes or Arbacia.

Third. The temperature of the thermostat, if constant, would on some days be very far from that of the sea surface, which varied greatly (from about 25°–30° C.), and the sudden change of the eggs from one to the other might affect them in some undesirable way, as they could be caused to segment by a change in temperature; also the time required for the eggs to reach this temperature would be greater. This was obviated by making the temperature of the thermostat about one degree higher than that of the sea before each set of determinations and then keeping it constant, within one tenth of a degree. The thermostat held 20 litres of water, was closed at the top and well insulated at the sides, stirred with a paddle attached to a rod going through a small hole in the cover, and regulated by hand with a minute flame beneath.

Fourth. The spaces between the eggs might vary. This was obviated by centrifuging the eggs in the conductivity vessel and marking their upper limit accurately in indelible ink with the finest drawing pen, and before each reading centrifuging them again to the same line.

The egg as it leaves the ovary is surrounded by an invisible gelatinous covering, which I have called the "jelly," since it shows similarity to the jelly-like coat of the frog's egg, a mucin which yields galactosamin (Schulz and Ditthorn). Loeb applies the name chorion to it. This jelly seems to be a mucin which in sea water slowly dissolves. The solution of the jelly is aided by weak or strong alkalis and very weak acids, though it is coagulated by tannin and basic dyes, and seems to contract (coagulate?) on addition of strong mineral acids. Delage says it is lifted from the egg before the formation of the fertilization membrane. It is possible that this appearance was due to contraction caused by the dye used to make it visible.

If eggs bearing this jelly are put in the conductivity vessel and centrifuged, and later mixed with sea water and centrifuged again, much less force is required to precipitate them to the same level, and they will slowly settle below this level by gravity while the eggs are being brought to the temperature of the thermostat. This is due to the washing away of some of the jelly. To prevent this occurrence, the jelly had to be entirely washed off of them before they were first placed in the conductivity vessel, a process accomplished by stirring in large quantities of sea water and repeated centrifuging. The jelly could be removed completely from Toxopneustes eggs and with more difficulty from Tripneustes eggs. Care was taken that no treatment was used that would cytolyze even a few eggs, as cytolysis causes swelling or disintegration which would affect the volume of the eggs and therefore the spaces between them when they were precipitated to a certain level.

The pushing out of fertilization membranes might affect the shape of the spaces between the eggs and thus change the free paths of the ions in the sea water filling those spaces. This was obviated by washing the eggs so long in sea water that no fertilization membranes were pushed out when they were fertilized or treated with the solutions used. This was tested by microscopic examination of control eggs and of the eggs taken from the conductivity vessel after each experiment. Such eggs develop.

It has been objected that a membrane was formed and not pushed out. I found two methods of detecting membranes that lie so close to the egg as not to be distinguishable with the microscope. If the egg be plasmolyzed with a molecular solution of cane sugar, such membranes are often if not always lifted from the egg. Harvey says the fertilization membrane is relatively impermeable to sugar, and one would suppose that sugar would push the membrane closer to the egg. From many of my experiments it is evident that sugar will go through the fertilization membrane, and Harvey has not stated the degree of impermeability to sugar that he observed. The second and more certain method is as follows: If an electric current of sufficient density be passed through, the membrane will be lifted from the cathode end of the egg.

It may be objected that the sugar solution and electric current caused the membranes to be formed, and that they were lifted from the egg in the usual manner. I can only answer to this that the same sugar solution and electric current were tried on normal unfertilized eggs and no membranes appeared.

Under the same conditions as those of the later conductivity experiments, eggs did not form membranes that could be detected by either of the above methods, and similar eggs developed. Harvey [21] says that after an egg stands in sea water twenty-eight hours and is then fertilized, it becomes surrounded by a thick adhering membrane

[21] HARVEY: Journal of experimental zoölogy, 1910, viii, p. 365.

which, on cleavage of the egg, surrounds each blastomere. He calls this a fertilization membrane, and maintains that fertilization membranes are formed on all developing sea urchin eggs. Evidently in the above case and in that of eggs placed in hypertonic or calcium-free sea water he mistook the so-called "hyaline plasma layer" or "Verbindungsschicht" for the fertilization membrane.[22] Harvey states that he saw membranes on Hipponoe eggs on slightly "high focus." A membrane on the surface of a sphere could only be determined positively with so high a power that the optical section was extremely thin in comparison to the diameter of the sphere, and passed exactly through the point of contact, with the sphere, of a tangent drawn from the eye to the sphere. Hence there is only one focus, and "high" or "low" focus means out of focus. One need only try this "high focus" on an air bubble or oil drop to see what appearances of "membranes" are thus obtained.

But it is impossible for me to see how a change in a surface film, of immeasurable thickness, of the egg, thus forming a "membrane" in close contact with the egg, can cause such a change in the conductivity of the inter-egg spaces as to account for the great differences in the conductivity of unfertilized and fertilized eggs which I obtained. I account for the change in conductivity by a change in the surface film of the egg, allowing ions to pass through more easily. Probably such a change would be accompanied by a visible change if microscopic technique were sufficiently developed to detect it, but I would not call this the formation of a new membrane; it is a change in the "*plasma* membrane," a condensed surface tension film or haptogen membrane.

The conductivity of the spermatic fluid and of the acidulated sea water were slightly less than that of natural sea water, so that if they replaced natural sea water between the eggs, they would cause a slight decrease in the conductivity reading, but they were thoroughly washed out with natural sea water before the readings were taken.

Fifth. The electric current passed through the eggs might alter their conductivity. As no measurement of conductivity could be taken before the current began to pass through, I cannot determine this point. But the first reading and later readings taken at short intervals were always the same provided the eggs had been centrifuged down

[22] See GOLDSCHMIDT and POPOFF, Biologische Centralblatt, 1908, xxviii, p. 210.

so compactly that they did not settle further by gravity, and the content of the conductivity vessel was at the same temperature as the thermostat. By using a special induction coil with a rheostat, an alternating current of such high frequency and such low amperage was obtained that when passed through my finger from electrodes wet with sea water, I could not feel it, yet this was the current used in the experiments, and variations in it could easily be detected with the telephone used in the experiments.

Sixth. The increased elimination of carbon dioxide by fertilized eggs might cause an increase in conductivity of the sea water between the eggs. To test this, the conductivity of a sample of sea water was determined. It was then charged with CO_2 in a "sparklet syphon," and no increase in conductivity could be detected with the electrodes used in the experiments with eggs. Perhaps with electrodes specially adapted to good conductors like sea water, a change could be detected, but it would evidently be extremely small and incapable of accounting for the large differences observed in the experiments with eggs.

The Conductivity Determinations.

The conductivity of one sample of sea water at 30° C. was found to be .061, while that of unfertilized eggs washed in the same water and precipitated by gravity was .04655 at the same temperature. The conductivity of the same water at 32° C. was .0624 and of the same eggs at 32° C., .0469. At 26° the conductivity of another sample of sea water was .05535, of spermatic fluid direct from testes, .0439, and of unfertilized eggs precipitated with the centrifuge, .002404. In this case the conductivity of the eggs was about one twentieth that of the sea water, although they still contained some of the latter between them. The conductivity of the eggs would have fallen still lower if all of the sea water had been pressed out from between them.

First lot of experiments. — In all of these preliminary experiments the thermostat was kept at 32° C., which was at times very near, but at other times very much above, that of the sea. The eggs were precipitated by gravity. The supernatant sea water in the conductivity vessel was pipetted off, and the vessel shaken, then readings taken, until successive identical results showed that the eggs were of the temperature of the thermostat, after which less than a drop of sperma was added, the vessel shaken again, and a second series of readings taken. The observed conductivities are given in the following table:

1.	Unfertilized eggs	.05980
	Fertilized eggs	.16950
2.	Unfertilized eggs	.04480
	Fertilized eggs	.04835
3.	Unfertilized eggs	.04690
	Momentarily heated	.05600 [23]

The above figures show a great increase in conductivity on fertilization, or momentary heating to a point that will cause segmentation.

Second lot of experiments. — In this and all subsequent lots of experiments the thermostat was brought before each set of readings to about sea temperature, the eggs were precipitated in the conductivity vessel to such a degree that they were not further precipitated by gravity.

1.	Unfertilized eggs at 27.5° C.	.01524
	Fertilized eggs " "	.01627
2.	Unfertilized eggs " 28° C.	.01076
	Fertilized eggs " "	.01266

Third lot of experiments. — As it was feared that the admixture of but a fraction of a drop of spermatic fluid might raise the conductivity independent of fertilization, and as by this method only a small per cent of the eggs were fertilized, in the third and fourth lots of experiments the eggs were centrifuged in the conductivity vessel and their upper level marked accurately, and the conductivity determined, then they were mixed with sea water containing sperm or acetic acid in the conductivity vessel, and washed by repeated precipitations and precipitated down to the same level, before determining the conductivity of the developing eggs. By this method almost 100 per cent of the eggs could be caused to begin development.

1.	Unfertilized at 26° C.	.002404
	Fertilized " "	.004320

[23] The vessel was set a few moments in water at 45°, then returned to the thermostat; this was shown by control to cause segmentation.

2. Unfertilized at 25.75° C.	.004445
Fertilized " "	.006340
3. Unfertilized at 25° C.	.006523
Fertilized " "	.009544
4. Unfertilized at 26.25° C.	.004900
Fertilized " "	.006390
5. Unfertilized at 28° C.	.008230
Fertilized " "	.009220
6. Unfertilized at 28.17° C.	.006876
Fertilized " "	.007298
7. Unfertilized at 26° C.	.005620
After .006 normal acetic acid in sea water 1½ min. 26° C.	.006000

Fourth lot of experiments, in which all precautions were taken. — In these experiments the eggs were washed so long in sea water that no fertilization membranes could be caused to push out.

1. Unfertilized at 29.5° C.	.01182
Fertilized " "	.01537
2. Unfertilized at 30° C.	.01153
After .006 normal acetic acid in sea water 1½ min. 30° C.	.0127
3. Unfertilized at 30° C.	.00877
After acid sea water 1½ min 30° C.	.00965
4. Unfertilized at 29.5° C.	.005135
After acid sea water 1½ min. 29.5° C.	.00839

From the above experiments I have concluded that there is an increase in electric conductivity of the sea urchin's egg at the beginning of development. The question now arises whether the resistance to the movement of ions through the mass of eggs is the impermeability of the plasma membrane or the presence of fat globules or proteid granules within the egg, or the combination of the egg electrolytes with colloids, forming poorly dissociated or poorly diffusible compounds. I found by centrifuging them that there was as great a volume of fat globules and proteid granules in the sea urchin's egg immediately after, as there was immediately before, fertilization.

On the Internal Conductivity of the Cell.

The majority of my experiments on this subject were made at Cornell Medical College.

Höber [24] has devised a method by which the electric conductivity of the cell interior may be measured without breaking the cell wall. The determinations cannot be made with great accuracy, but the results on blood corpuscles clearly demonstrate that the conductivity of the interior is many times greater than that of the corpuscle as a whole, indicating that the greatest resistance to the current lies in the plasma membrane.

Stewart [25] made certain determinations which I take to indicate that hen's egg yolk is a very much poorer conductor than is a solution of its salts made up to the same volume.

The yolk of the hen's egg before it leaves the ovary forms the bulk of a single cell. The yolk of an egg is often considered as "dead" material; but in conductivity experiments we cannot separate "living" and "dead" portions of the cell. There is a small amount of white yolk, but the major volume is yellow yolk.

The yellow yolk under the microscope presents a fluid matrix containing large globules of another fluid of almost the same specific gravity, viscosity, and refracting index as the matrix, the boundary between the two fluids being the seat of little surface tension. Both matrix and globules contain numerous fine granules. On the addition of alcohol, under the microscope, a substance (or substances) in both fluids disintegrates, setting free lipoids which appear as droplets, which are blackened by osmic acid and colored by Sudan III. If one of the large globules is watched closely as the alcohol is applied, fine, lipoid droplets appear and grow and fuse to form larger drops, and some of them may then migrate toward the periphery and fuse to form a lipoid envelope surrounding the globule. The lipoid droplets appearing in the matrix grow and fuse to form larger ones.

The yellow yolk darkens after the addition of osmic acid, the globules becoming darker than the matrix, but if alcohol and then osmic be applied, the lipoid droplets thus formed, quickly become an in-

[24] Höber: Archiv für das gesammte Physiologie, 1910, cxxxiii, p. 237.

[25] Stewart: Journal of experimental medicine, 1902, vi, p. 257.

tense black. Only the lipoid droplets are colored by Sudan III in 80 per cent alcohol.

The white yolk resembles the yellow yolk that has been treated with alcohol in that it contains lipoid droplets. But as there is very little white yolk in the hen's egg before incubation, there are very few lipoid droplets to impede the electric current. The lipoids in the yellow yolk are probably bound up with proteids, forming combinations which are disintegrated by alcohol, as indicated by the above observations.

In order to determine to what extent the granules impede the electric current, I precipitated them with the centrifuge. The large globules cannot be separated from the matrix by this means. With small quantities I was able to obtain the fluid entirely free from granules. It forms 11/17 of the total volume, dissolves in dilute alkalis, and with slight milkiness in dilute acids, and when shaken with water the insoluble portion forms an emulsion or a coagulum resembling yeast plants.

With large enough quantities to fill the smallest suitable conductivity vessel at hand, the precipitation was so slow that I feared decomposition might commence before a granule-free fluid was obtained, so I contented myself with the comparison of a portion containing a very small per cent of granules with a portion containing a very large per cent of granules.

At 25° C. the conductivity of the granule-poor layer was .00302 and that of the granule-rich layer .00278, showing that the granules impede the current to a great extent.

Since dilution with water breaks up many ion-colloid compounds, I used this method to determine whether the electrolytes in the yolk were bound up with colloids. The conductivity determinations at 25° C. are given in the table below:

Portion.	(Undiluted) Vol. = 1.	(+ 1 vol. H_2O) Vol. = 2.	(+ 3 vols. H_2O) Vol. = 4.	(+ 7 vols. H_2O) Vol. = 8.
Granule-poor	.00302	.00268	.00162	.00096
Granule-rich	.00278	.00278	.00200	.00125

The above table shows that whereas the granule-poor layer decreases in conductivity on dilution with distilled water, at first slowly and later slightly more rapidly (which may be partially accounted for by the more rapid increase in ionization of inorganic salts at the beginning than at the end of the series) the conductivity of the granule-rich layer is not reduced at all by a dilution with one volume of H_2O. It may be said that this is due to the separation of the granules, thus widening the conducting paths, and I have demonstrated that such might occur in an emulsion of oil in soap solution, as shown by the following table of conductivities at 25° C.:

Material.	(Undiluted) Vol. = 1.	(+ 1 vol. H_2O) Vol. = 2.	(+ 3 vols. H_2O) Vol. = 4.
Soap solution	.002490	.001460	.000903
Emulsion of oil, containing 17 per cent soap solution	.000434	.000434	.000335

But how are we to explain the fact that on dilution with one or more volumes of water the conductivity of the granule-rich portion of yolk is greater than that of the granule-poor layer, although the former contains less of the fluid portion of the yolk? Evidently (since there are no inorganic crystals in the yolk) some of the electrolytes must have been bound up in the granules (either by adsorption or chemical combination) and liberated on dilution. Since the fluid portion of the yolk does not entirely dissolve in water, the undissolved portion may impede the current, but this would occur in the granule-rich as well as in the granule-poor portion.

I doubt that Höber's method of measuring the internal electric conductivity of cells is sensitive enough to determine whether the increase in conductivity of the egg is due to liberation of electrolytes in the interior or to increased permeability of the plasma membrane of the egg, but it shows, by exclusion, in case of the cells on which it was used, that by far the greatest resistance to the current lies in the plasma membrane.

Swelling (first stage of cytolysis) of sea urchin eggs causes a decrease in the conductivity of the mass of eggs, as shown by the following determinations of the conductivity:

1.	Unfertilized eggs of Toxopneustes at 27.25° C.	.01354
	After addition of nicotine at 27.25° C.	.01318
2.	Unfertilized eggs at 32° C.	.04850
	After momentary elevation to about 50° C. at 32°	.04780
3.	Unfertilized eggs at 27.5° C.	.01645
	After momentary elevation to about 50° C. at 27.5°	.01626
4.	Unfertilized eggs of Tripneustes at 32° C.	.04730
	After shaking with fraction of a drop of chloroform at 32° C.	.02286

Microscopic examination showed that the addition of nicotine or chloroform, or momentary elevation of temperature in the above experiments, caused the eggs to swell. This could only take place by the absorption of one or more constituents of the sea water between the eggs. If the salts of the sea water did not go into the eggs, the abstraction of H_2O from the sea water would increase the concentration of salts in the sea water remaining between the eggs, and might cause a liberation of electrolytes (by dilution) within the eggs, in the latter case causing increased conductivity of the egg interior without a corresponding decrease in conductivity of the inter-egg spaces. If the membrane became freely permeable to salts, the swelling of the eggs might increase, but should not diminish, the conductivity of the mass. The fact that the conductivity decreased can only be explained by assuming that the salts of the sea water entered the eggs and were adsorbed to or combined with colloids, or that the membrane was very poorly permeable to salts (though perhaps more permeable than the normal egg) and the narrowing of the inter-egg spaces caused the decrease in conductivity. In fact, I think this can be taken as an indication that the egg is a poor conductor not so much because of the low concentration of free electrolytes within it, but chiefly because the electrolytes cannot easily pass the plasma membrane.

As no dilution of the contents (swelling) of the sea urchin's egg occurs at the beginning of development, it is improbable that a liberation of the electrolytes within it, sufficient to account for the increased conductivity, occurs. The only alternative is that the increase in conductivity is due to an increase in permeability of the plasma membrane to electrolytes.

The Electric Conductivity of Individual Eggs.

It is well known that cells may be killed or injured by the passage of electric currents through the media containing them. The current might affect them by raising the temperature, by the passage of ions into or out of the cells, by the accumulation of ions of one sign that are stopped by parts of the cell.

In which of these ways does the current affect the cell most destructively? The heating effect may be practically eliminated. If electrolytes are transported into or out of the cell by the current, they could also diffuse in the absence of a current, but the accumulation of ions of one sign would not occur by *free* diffusion. Therefore I have regarded the accumulation of ions impeded by the cell structures as explanation of the destructive effects, and the destructive effects as an indicator of the resistance to the passage of ions.

The experiments were made at the United States Bureau of Fisheries at Woods Hole, Mass. The 110-volt direct current from the light circuit was used. Cylindrical non-polarizable electrodes of copper in one-half molecular copper sulphate were plugged at their free ends with absorbent cotton and connected to rubber tubes of 4 mm. internal diameter and about one foot each in length, filled with sea water. The free ends of the rubber tubes were plugged with absorbent cotton, which was allowed to protrude sufficiently to conduct the current to the sea water containing the eggs under a cover glass on a slide on the microscope stage. The current was reduced by passage through a 16-candle power light and further regulated by turning the screw of a pinch-cock which was clamped on one of the rubber tubes. The copper sulphate diffusing into the sea water in the rubber tubes reacted with the calcium carbonate, copper hydrate and calcium sulphate being precipitated and carbonic acid being liberated. The copper sulphate solution and sea water were renewed before each experiment. Usually a piece of ash-free filter paper was cut the size of the cover glass and a hole cut in its middle. This filter paper ring was placed on the slide, and sea water containing eggs placed in the hole in the ring, so that when the cover glass was placed on the preparation, the eggs were contained in a cell which was freely permeable to ions at the sides. At other times the eggs were mixed with sea water containing enough cotton fibres to support the cover glass. Eggs of Arbacia punctulata were used.

When the current is passed through the egg, the latter is affected at that surface nearest the anode, as observed by Brown, who placed the eggs in a molecular solution of urea. Changes in surface tension are indicated by bulging or amœboid movements. The pigment suddenly leaves each of the chromatophores in turn and diffuses into the cytoplasm in this region of the egg, which is turned a red or orange hue (it is a deeper red if the chromatophores have been stained with neutral red), showing that the reaction is not alkaline.[26] The anodal

[26] The pigment extracted from the eggs is red or orange in acid according to dilution; it is violet or green and precipitates in alkali. If the eggs, or especially

end of the cell absorbs water and swells, often a blister is formed and masses of granular cytoplasm pass into the blister fluid and dissolve. Gradually these changes extend from the anode end to the cathode end of the egg, the egg swells enormously and may burst.

Very probably this disintegration commencing at the anode end of the egg is due to the accumulation of anions which cannot pass the plasma membrane. If the plasma membrane is poorly permeable to anions in one direction, it is probably so in the other, and it may be asked why they do not accumulate outside the cell at its cathode end. The anions which are unable to enter the egg at its cathode end are free to move around the egg and hence do not accumulate to form as great a concentration as at the anode end.

Since no destruction of the egg of Arbacia, beginning at the cathode end, was observed, we may conclude that the plasma membrane is more permeable to cations than to anions. This is not true of all eggs, as I observed the eggs of Hydractinea begin to disintegrate at the cathode as soon as at the anode end. However, it is true of a number of living cells.[27]

If fertilized and unfertilized Arbacia eggs are placed in an isotonic sugar solution containing little sea water, through which a current of gradually increasing density is passed, the unfertilized eggs begin to disintegrate, at their anode ends, sooner than the fertilized eggs do. We may interpret this as indicating that the fertilized eggs are more permeable to anions, which therefore accumulate in them to a less extent, or the fertilized eggs are more permeable to electrolytes, which therefore have passed out into the sugar solution to a greater extent, and therefore the current passes through them less, than in case of the unfertilized eggs.

I did not obtain the same results on eggs in sea water, but the uncertainty of the material toward the end of the season prevented the determination of the mode of action of the sugar solution. Possibly the heating effect of the current in sea water increased the permeability of the unfertilized eggs. Sea water is so much better a conductor than the eggs that only a small per cent of the current passes through the latter, and in order to produce visible effects on the eggs an enor-

the perivisceral fluid cells containing much pigment, are killed, the nuclei and some other parts absorb the pigment and turn brownish purple.

[27] See VERWORN's Physiology.

mous current must be passed through the sea water. It is known that sugar solutions produce abnormal conditions in eggs, but these experiments were made quickly after placing the eggs in the sugar solutions. The nucleus does not begin to disintegrate as soon as the cytoplasm; this is in harmony with McCallum's view that the nucleus contains no free salts. The nucleus as a whole or the contained nucleoproteids migrate toward the anode.

Plasmolysis with Non-Electrolytes.

Osterhout has obtained shrinkage of marine cells in distilled water, and thinks the action of sugar similar; *i. e.*, first the membrane is made permeable and then the salts diffuse out and the cell contracts by some non-osmotic force. But in the only animal cell in which he has obtained this result there is first a swelling, with formation of blisters, and later shrinkage, with the nucleus becoming homogeneous and distinct, which, I think, denotes death and perhaps coagulation. Since the Arbacia egg in an isotonic sugar solution does not swell first and then shrink, I think this objection may not apply to my experiments.

The following tables show that fertilized (Arbacia) eggs shrink more rapidly than unfertilized eggs in molecular sugar solutions, which are calculated to have only slightly greater osmotic pressure than the sea water at Woods Hole, where the experiments were made. It appears that the plasma membranes of the fertilized eggs are more permeable, allowing the salts to diffuse out of the eggs more rapidly, thus lowering the internal osmotic pressure to a greater extent than is the case with unfertilized eggs. Sollmann [28] observed Arbacia eggs contract in hypertonic, and swell in hypotonic, salt solutions.

In normal sea water fertilized are not smaller than unfertilized eggs.[29] Before the first cleavage the hyaline plasma layer forms, thus taking material away from the more opaque portion of the egg, and it might be supposed that the failure to include this layer in taking the measurements caused the appearance of shrinkage, but such would be the case also in the control in normal sea water, and furthermore the measurements were taken before the hyaline layer was formed or had reached visible thickness.

[28] Sollmann: This journal, 1904, xii, p. 111.

[29] McClendon: Science, 1910, xxxii, p. 318.

Fertilized and unfertilized eggs in a molecular solution of dextrose were placed under the same cover glass, which was supported to prevent compression of the eggs, and sealed to prevent evaporation. The fertilized were distinguished from the unfertilized eggs by the presence of the fertilization membrane. The eggs were observed in the order in which they appeared in the field as the slide was moved so as to observe the whole area under the cover glass once and once only. The diameter of each egg in turn was drawn with the camera lucida, and the drawings were measured later with a rule. In case an egg was irregular, approximately its mean diameter was drawn.

The results of two series of measurements are recorded on page 266. In the first column of figures the diameter of the egg in the unit used for all the measurements is represented. In the second and third columns of figures the frequencies of the occurrence of fertilized and unfertilized eggs of the diameters given in the same horizontal line are represented. The fourth and fifth columns of figures represent a second series of measurements in the same manner.

The table shows that there is considerable variation in the size of the eggs, but that the mean (and also the mode if the curve were plotted) of the diameters of the fertilized eggs is less than the mean of the unfertilized eggs. I did not succeed in making measurements fast enough to determine the rate of plasmolysis.

Chemical Analysis of Masses of Cells.

Fertilized and unfertilized eggs may be placed in solutions differing from sea water, and the passage of substances into or out of them detected by analysis of masses of the eggs. There are three sources of error: 1. The presence of the jelly-like coverings on the eggs; 2. The fluid in spaces between the eggs; and, 3. The large surface for adsorption.

I tried some preliminary experiments on yeast cells at a time when suitable eggs could not be obtained. I found yeast and dextrose placed in .3 molecular $MgCl_2$ eliminated CO_2 more rapidly than in .5 molecular NaCl or .325 molecular $CaCl_2$, all of which are calculated to have approximately the same osmotic pressure. Also the CO_2 elimination was more rapid in the magnesium solution than in a solution of the same concentration of magnesium chloride with either of the

Diameter.	Frequency.			
	Fertilized.	Unfertilized.	Fertilized.	Unfertilized.
110	2	..	..	..
111	0	..	..	..
112	2	..	..	..
113	1	..	1	..
114	2	..	1	..
115	2	..	1	..
116	3	1	3	..
117	5	1	3	..
118	3	3	4	1
119	7	2	4	0
120	10	3	3	2
121	9	6	4	0
122	10	5	5	2
123	7	6	6	3
124	6	6	9	3
125	4	7	6	2
126	4	5	6	2
127	1	8	3	3
128	3	7	4	3
129	0	6	1	2
130	2	7	0	1
131	0	3	0	2
132	0	3	1	5
133	1	3	..	7
134	..	2	..	4
135	..	1	..	3
136	..	1	..	3
137	..	1	..	4
138	..	0	..	2
139	..	0	..	2
140	..	0	..	2
141	..	0	..	1
142	..	0	..	..
143	..	0	..	..
144	..	1	..	..
Mean diameter	121	126	122	131

other salts in addition, or in a solution containing NaCl and $CaCl_2$ in the same concentration as in the respective pure solutions, or in a solution containing all three salts, or in tap or distilled water.

The magnesium must have entered the cell or altered the permeability of the plasma membrane to CO_2, sugar, alcohol, the enzyme, or some other substance. In order to determine whether the magnesium entered the cells, I took two blocks of compressed yeast of the same volumes and weights and mixed one with H_2O and the other with a molecular solution of $MgCl_2$ for five hours, then washed each by rapid precipitation in renewed H_2O several times with the centrifuge. The two lots were ashed and weighed with the results: control, ash = .0466 gm.; ash from Mg culture = .048 gm. Evidently the magnesium did not enter the yeast to any great extent and probably acted on the surface, increasing the permeability to some other substance.

Lyon and Shackell have analyzed fertilized and unfertilized eggs placed in salt solutions, and obtained some results indicating that the salts enter and leave the fertilized more easily than the unfertilized eggs. They found an exception in the case of iodine. Iodine (in potassium iodide solution) is absorbed by the unfertilized more quickly than by the fertilized eggs.[30]

I had intended to work along this line, but was forced to postpone it until another season.

Microchemical Analysis of Individual Eggs.

Lyon and Shackell[30] and Harvey have concluded that certain dyes enter fertilized more easily than unfertilized eggs. Loeb supposes that the dye is chemically combined in the fertilized egg and merely in solution in the unfertilized egg. Unfortunately these dyes belong to the class of substances which Overton found to most easily penetrate plant cells, so that a demonstration that they more easily enter the fertilized than the unfertilized egg does not necessarily indicate that the same is true for electrolytes in general.

Harvey[31] found that eggs became more permeable to NaOH after being fertilized or treated with cytolytic agents.

[30] Lyon and Shackell: Science, 1910, xxxii, p. 250.

[31] Harvey: Science, 1910, xxxii, p. 565.

The Migration of the Chromatophores.

The chromatophores of the egg of Arbacia contain a red substance which I found to have an absorption spectrum similar to McMunn's echinochrome, at least in certain solvents. I have crystallized two derivatives of the Arbacia pigment and perhaps the pigment itself, and a chemical study of it is being attempted.

These chromatophores or pigment plastids show similarities to the chloroplasts of some green plants. Similar plastids occur in the perivisceral fluid cells of Arbacia, where they are so closely packed together in the cytoplasm as to be separately distinguishable only on careful observation. In some of the cells the plastids contain pigment and in others they are colorless.

McMunn, finding that the spectrum of echinochrome in certain solvents was changed by strong reducing agents, concluded that it was respiratory in function. Griffiths [32] briefly states that on boiling with mineral acids echinochrome is transformed into hæmochromogen, hæmatoporphyrin, and sulphuric acid, indicating a relation to hæmoglobin.

I separated the cells from about 50 c.c. of the perivisceral fluid of Arbacia and mixed them with sea water to form 50 c.c. This suspension of cells, and 50 c.c. of sea water as a control, were exhausted under an air pump for six hours, during the last half hour at practically water vapor tension. While in the vacuum, the cells must have exerted a reducing action on the pigment if it can be reduced. Each was then shaken with air for thirty minutes in closed apparatus. The suspension of cells had absorbed 1.25 c.c. of air and the control only 0.8 c.c., at atmospheric pressure. The volume of oxygen used in oxidations in the cells during the shaking was probably partly replaced by CO_2 given off by them, but the difference of about half a cubic centimetre does not demonstrate conclusively that the pigment combined with oxygen. Under somewhat similar conditions dogfish blood absorbed many times as much air as the perivesceral fluid of Arbacia.

The migration of the chromatophores in the egg is evidently *not always* in the direction of greater oxygen concentration, but whether it is *ever* a chemotropism toward oxygen I was unable to determine.

[32] Griffiths: Comptes rendus, 1892, cxiii, p. 419.

In 1908 I observed movements of the chromatophores in the eggs of Arbacia punctulata. As Roux had caused a whitening of the cathodal pole of the frog's egg by passing an electric current through it, I tried in 1909 and again in 1910 to move the chromatophores of the Arbacia egg with the electric current. I observed that in the unfertilized egg the chromatophores are distributed throughout the cytoplasm, but after the egg is fertilized or stimulated artificially the chromatophores migrate to the surface.[33]

Harvey[34] says that the pigment comes to the surface within ten minutes after fertilization, but I found that this process sometimes required half an hour, by which time the cleavage spindle had formed. At each cleavage chromatophores sink into the cleavage furrows of the blastomeres. Just before the micromeres are formed the chromatophores move along the surface of the blastomeres, away from the micromere pole of the egg, so that after the resulting cleavage the micromeres are practically free from pigment. Under abnormal conditions there is a great massing of pigment in the cleavage furrow or other regions of the surface or in the interior of the egg. The sinking of pigment into the cleavage furrows and its retreat from the micromere pole are probably due to surface tension changes as discussed above, and perhaps the abnormal massing of pigment at one portion of the surface is due to a local increase in surface tension.

"Membrane-forming" and parthenogenetic agents, even in concentrations too low to produce membranes or segmentation, cause the pigment to come to the surface. If a few normal unfertilized eggs are kept in a relatively large amount of sea water protected from evaporation, and oxygen is very abundant, it appears that there is more pigment at the surface after twelve or more hours than at the beginning of the experiment, but disintegration commences before all the pigment has reached the surface. In an oxygen vacuum this did not seem to occur. The pigment may all come to the surface in a stream of washed hydrogen, but this may be caused by some impurity.

Fischel,[35] observing similar movements of pigment in the egg of Arbacia pustulosa, concluded that the pigment was repelled by the asters according to the forces described by Rhumbler as moving

[33] McClendon: Science, 1909, xxx, p. 454.

[34] Harvey: Journal of experimental zoölogy, 1910, viii, p. 355.

[35] Fischel: Archiv für Entwicklungsmechanik, 1906, xxii, pp. 526–541.

granules toward or away from asters in the cytoplasm.[36] Bütschli and Rhumbler have shown how the contraction of an area in a foam structure causes aster-like radiations around it, and Rhumbler has shown that such radiations to a limited extent may occur around a rigid sphere inserted into a foam or alveolar structure. Rhumbler assumes that the concentration of the alveolar wall substance would increase its surface tension, and that this increase toward the centre of the aster would reduce the thickness of those alveolar walls perpendicular to the astral rays, both of which assumptions have no facts of which I am aware to support them,. On them rests Rhumbler's explanation of the movement of granules away from asters.

However, if those bodies which seem to be repelled by asters (chromatophores of Arbacia eggs, yolk platelets of frog's eggs) lie within or are larger than the largest alveoles, as I have observed to be the case, aster formation might explain their repulsion. Rhumbler's theoretical asters were made of a central body and of alveoles of a uniform size. If the alveoles were of different sizes, the largest ones would seek the periphery of the aster.

I sectioned eggs that had been so treated artificially that all of the pigment came to the surface but no segmentation occurred, and found no asters, though perhaps asters had formed and disappeared.

After the passage of an electric current of a certain density and duration through unfertilized eggs, some of them have their pigment more abundant toward their cathodal surfaces. If the current exceed a certain density, one by one the chromatophores toward the anodal surface of the egg lose their pigment suddenly. When the current was slowly and carefully increased just to the density required to change the distribution of the pigment, no loss of pigment by the chromatophores toward the anode could be observed, but it is mechanically impossible to watch every chromatophore in the anodal region of one egg. I found it possible to observe a single chromatophore for a long time, and attempted, by noting its distance from the anodal surface of the egg, to record its movements. Each time this observation was attempted the chromatophore appeared to move, but its movement was not constant in direction, and a considerable migration in any one direction was not observed, except rarely in case the chro-

[36] RHUMBLER: Archiv für Entwicklungsmechanik, 1896, iii, p. 527; 1899, ix, pp. 32 and 63.

matophore was very near the surface. In this exceptional case the chromatophore moved along the surface, toward the cathode, which movement was probably due to surface tension changes. In the egg just taken from the ovary the chromatophores are slightly more numerous near the surface than in the interior, and when the current is passed, this difference is increased. The passage of the current causes the anodal surface of the egg to spread (the increased difference of potential between the two sides of the surface reducing the surface tension), sometimes carrying the more superficial chromatophores along the surface toward the cathode. This is not a cataphoresis of the chromatophores, since they do not go in the direction of the current, but is due to surface tension changes, and is therefore a secondary effect of the current.

Fearing that the high viscosity of the cytoplasm might interfere with the movement of the chromatophores by electric convection, I centrifuged both fertilized and unfertilized eggs until the pigment was massed at one pole of each, and passed the current through solutions containing them. No orientation of the eggs to the potential gradient occurred. I then tried to move the perivisceral fluid cells, which are practically masses of chromatophores, by means of the electric current, but my apparatus did not exclude all sources of error, and this experiment was reserved for another season. The pigment may be caused to leave the chromatophores in these cells by the electric current or by chemicals, to which agents these cells are much less sensitive than are the eggs.

We have, then, no evidence that the chromatophores are electrically charged.

Harvey[37] attempted to explain my observation that the chromatophores come to the surface at the beginning of the development of the egg, by assuming that there is a positive charge over the surface of the egg until the commencement of development, when the surface becoming permeable to anions causes a potential gradient between the surface and centre of the egg. He further assumed that the chromatophores are charged negatively and migrate in the potential gradient.

His evidence for the existence of the positive charge over the surface of the unfertilized egg of Arbacia punctulata is the fact that it is not

[37] HARVEY: Science, 1909, xxx, p. 694.

always spherical when it leaves the ovary. His evidence for the loss of the charge is the fact that this egg rounds up more rapidly when it is fertilized than when it is left in sea water without sperm. His evidence for the negative charge on the chromatophores is the fact that they come to the surface after development commences.

My observations indicate that the plasma membrane of the unfertilized egg is less permeable to anions than to cations, which would cause the appearance of the positive charge over the surface provided some electrolyte whose undissociated molecules could not easily pass the membrane was more concentrated in the egg than in the sea water, or was produced with sufficient rapidity within the egg. Carbon dioxide might be this substance. However, my observations seem to indicate that the permeability of the egg is increased suddenly (in less than five minutes) on fertilization, in which case the positive charge over the surface would be lost suddenly; and if the ions within the egg were free to move, the potential gradient would be of momentary duration, whereas the chromatophores require from ten to thirty minutes to come to the surface. Before Harvey made this hypothesis I had attempted, as described above, to move the chromatophores by inducing a potential gradient, in order to determine whether they were electrically charged. Harvey has yet to prove that they are charged, and furthermore that they are negative, and that the potential gradient is of sufficient intensity and duration to move them to the surface. I do not wish to be considered an opponent of his hypothesis, but am merely searching for facts. Garbowski observed chromatophores move toward the centrosomes.

On the Contents of the "Perivitelline" Space.

The assumption has been made by several observers that there exists a colloid between the fertilization membrane and the egg. Here the question arises, what is meant by the surface of the egg? The "hyaline plasma layer," or "Verbindungsschicht," which forms before the first cleavage, is considered by some as part of the egg and by others as a "membrane" outside of the egg. In this section I will not include the hyaline layer in speaking of the egg, as under these experimental conditions the surface of the hyaline layer (if such had formed) could not usually be distinguished, *i. e.*, the presence of this layer could not be ascertained.

When an electric current is passed through the egg of Arbacia punctulata having a "pushed-out" fertilization membrane, the latter is bulged out toward the cathode, and the egg moved in the opposite direction and pressed against the anodal portion of this membrane. When the current ceases, the egg returns to the centre of the "perivitelline space." I first thought that this was caused by anodal electric convection of the egg, due to confined anions, but sometimes the fertilization membrane bursts at its anodal pole and the egg passes out, and should on this hypothesis continue its migration toward the anode. But as soon as the egg is free from the fertilization membrane it stops its migration, even though floating in a fluid of equal specific gravity. Perhaps an invisible colloid having a positive charge fills the perivitelline space, and its migration toward the cathode pushes the egg in the opposite direction.

Loeb postulated a colloid in the perivitelline space as exerting an osmotic pressure which pushed out the fertilization membrane. This may be true, but the membrane must harden in the expanded condition, for if it is burst by passage of the electric current or other means it does not collapse, but remains spherical unless distorted by violence.

When the electric current causes bulging of the fertilization membrane, the perivitelline space exhibits fine striations radially to the egg or parallel to the current lines. Schücking, Goldschmidt and Popoff, Herbst, and others have described striations or fibres in the perivitelline space, or around the fertilized egg, including the spaces between the early blastomeres, usually under abnormal conditions. These striations are probably due to tension of the colloid filling the perivitelline space (including the hyaline plasma layer or "Verbindungsschicht").

The Action of Parthenogenetic Agents on the Plasma Membrane.

Salts, acids, alkalis, shaking and thermal or electric changes might alter the aggregation state of the colloids of the plasma membrane. Fat solvents, alkaloids, glucosides, blood sera, soap and bile salts might alter the aggregation state of the colloids, especially lipoids of the plasma membrane.

Lillie [38] found that pure solutions of sodium salts were effective as parthenogenetic agents in the following series arranged according to the anions: $Cl<Br<NO_3<CNS<I$. This order of anions is reversed in the precipitation of lecithin, and a somewhat similar reversed order of anions occurs in the salting-out of proteids.[39] Hence it appears probable that these salts act by virtue of their dissolving power on the colloids of the plasma membrane.

Alkalis may act not only on the membrane, but by slow diffusion into the egg favor oxidations, as an alkaline reaction favors oxidation in general, and Loeb has shown that an alkaline reaction of the medium is necessary for the normal oxidations in the sea urchin's egg.

KCN or an oxygen vacuum may act by suppressing oxidation until enough of the confined CO_2 can escape to raise the alkalinity within the egg to such a point that when oxygen is readmitted oxidation may proceed with sufficient rapidity to allow the development of the egg.

It seems probable that the undissociated molecules of carbonic acid or CO_2 can diffuse out of the egg at all times. How then could the resistance of the plasma membrane to one or both or its ions so reduce oxidation within the egg as to prevent its development? Perhaps the per cent of CO_2 within the unfertilized egg is sufficient to lower the alkalinity to such a degree that oxidation cannot proceed with sufficient rapidity to allow development.

Loeb [40] finds that an oxygen vacuum or KCN reduces the toxicity of certain poisons to unfertilized and to a greater extent to fertilized eggs (poisons that affect fertilized in less concentration than unfertilized eggs). He concludes that this cannot be explained on the permeability hypothesis (as the mere absence of oxygen would probably not affect the permeability?). It may be that these poisons are toxic because they increase the permeability of both fertilized and unfertilized eggs, but since the fertilized eggs are more permeable before the action of the poison, the additional increase in permeability is fatal because oxidation is abnormally increased. Hence KCN or an oxygen vacuum would be antitoxic.

[38] Lillie: This journal, 1910, xxvi, p. 106.
[39] Hoeber: Zeitschrift für Allgemeine Physiologie, 1910, x, B. p. 178.
[40] Loeb: Biochemische Zeitschrift, 1910, xxvi, p. 288.

Loeb[41] finds that after membrane formation a saccharose solution of much lower osmotic pressure will cause the egg to develop than a pure NaCl solution, which, he says, proves that the membrane is permeable to salts or ions, for the explanation requires that the egg salts diffuse into the sugar in the former case or the NaCl diffuses into the egg in the latter. This shows a certain degree of permeability of the egg to electrolytes after "membrane formation," but proves nothing as regards the normal unfertilized egg. It is not necessary to postulate absolute impermeability, even of the unfertilized egg, to electrolytes, in order to account for development by increased permeability, and I have shown that an increase in permeability follows the action of membrane-forming substances. Furthermore it has not been demonstrated to my satisfaction that the action of the hypertonic solution after membrane formation is purely osmotic.

Lyon has shown that the CO_2 and catalase elimination by the sea urchin's egg increases after fertilization, and Warburg, Mathews, and Loeb and Wasteneys have shown that the oxygen absorption increases. These changes might be due to an increase in permeability.

It is not supposed that an increase in permeability will cause any cell to divide; growth is prerequisite to division. However, permeability might influence growth. Growth is supposed to cause division only when it affects the volume of the cytoplasm more than that of the nucleus. The ratio of the cytoplasm to the nucleus in the egg may be considered sufficient for a number of successive divisions, or the "true" cytoplasm may grow at the expense of the yolk after each division.

[41] LOEB: University of California publications, Physiology, 1908, iii, No. 11, p. 81.

V.

ON ADAPTATIONS IN STRUCTURE AND HABITS OF SOME MARINE ANIMALS OF TORTUGAS, FLORIDA.

By J. F. McCLENDON,
Instructor in Histology, Cornell University Medical College, New York City.

2 plates, 1 text figure.

ON ADAPTATIONS IN STRUCTURE AND HABITS OF SOME MARINE ANIMALS OF TORTUGAS, FLORIDA.

By J. F. McClendon.

In June, 1908, at the laboratory of the Carnegie Institution of Washington, at Tortugas, Florida, I began the study of the habits of some reef animals with a view to some comparative studies of behavior. The results were written up in the Zoological Laboratory of the University of Missouri.

It was found that many of these animals were thigmotatic and remained in glass tubes rather than in the open. They also learned to find the tubes when removed from them. Such was the case with five species of the Alpheidæ, one of the Pontoniidæ, *Typton tortugæ* Rathbun, and *Gonodactylus œrstedii*. All the anemones were thigmotactic on their bases. These same animals were heliotropic. The Crustaceans were negatively heliotropic and the anemones kept their bases from the light, while *Cradactis variabilis* Hargitt hid all but the tips of the fronds and tentacles from the light. In removing its base from the light, *Stoichactis helianthus*, which lives on coral heads, makes snail-like movements similar to *Metridium*,[1] while *Cradactis*, which lives in holes in decayed coral heads, crawls on its tentacles.

ON ADAPTATIONS OF SYNALPHEUS BROOKSI AND TYPTON TORTUGÆ.

In lagoons between the reefs is found the loggerhead sponge, *Hircinia acuta*, which grows to 3 feet or more in diameter, but is of no commercial value. The passages in this sponge are thickly populated by *Synalpheus brooksi* Coutière. These Alpheids are thigmotactic and negatively heliotropic and seldom come outside the sponge, which they do only at night and then rarely leave its surface. The only other animals seen in the interior of the sponge were a small species of Amphipod and a Pontoniid. The Alpheids were several hundred times as numerous as the Amphipods or Pontoniids. Near or at the surface crabs and worms were sometimes found.

Both Alpheid and the Pontoniid, *Typton tortugæ*, have the fourth and fifth pairs of thoracic appendages pincer-like (plate 1, figs. 1 and 3). In the Alpheid the fourth and in the Pontoniid the fifth pair are asymmetrically hypertrophied. In the Alpheid the asymmetry is very great, and the large chela can be snapped with such vigor as to produce a loud, clicking sound. When this claw is removed its mate grows to replace

[1] McClendon, 1906, On the Locomotion of a Sea Anemone, Biol. Bull. 10.

it and the asymmetry is reversed, as first shown by Przibram. It is not known on which side the large claw develops first. I interpret Herrick's records as demonstrating that the large claw develops first on the left side in *Synalpheus minus* (*Alpheus saulcyi*).[1] It was found in my specimens about as frequently on the right as left side in both large and small individuals. Of 50 taken at random, 22 had the large claw on the left and 28 on the right. In another species Przibram found 40 individuals with the large claw on the left and 47 on the right.

The Pontoniid *Typton tortugæ*, as was stated above, has the pincer-like appendages of the fifth thoracic segment well developed. One of these claws is much larger than the other, but the asymmetry is not as great as in the Alpheids. Both of these claws are snapped with a sharp, clicking sound. When the large claw is removed the small one grows to take its place, as in the Alpheids.

The two animals do not perhaps resemble one another as much in general coloration as in general form, though the color varies so much in both animals that these differences are not at first noticeable. The color darkens with age. The Alpheid varies from the color shown in plate 1, fig. 1, to a light brown. Specimens with a claw like fig. 2 may be a dull cream or light brown in general color. The nerve cord and some other organs may be surrounded by red pigment cells. Yellowish, brownish, or reddish glands in thorax or abdomen may show through.

The Pontoniid *Typton tortugæ* varies from the color shown in fig. 3 to an almost colorless condition, or to a light red or a pale bluish. The large claw of the pale specimens is often paler than the small claw in fig. 3. After the large claw has been removed the small one grows to take its place, but for some time retains more or less its general form and color. Often yellow, brown, or green glands show through in the thorax and abdomen.

As these animals pass their entire adult existence in the dark or dim light, it is improbable that their color is of much significance in their struggle for existence; hence it would not be fixed by natural selection. The fact that their eyes are not degenerate might indicate that they sometimes come near the mouths of the passages in the sponge. Perhaps they are forced out when the sponge becomes overcrowded, but I doubt that many of the larger ones would find another sponge before they were eaten by fish. Neither form was found in any other habitat, though Herrick records the Alpheid from reef rocks as well as loggerhead sponges in the Bahamas.

The Alpheid has large eggs, few in number, attached to the swimmerets of the female. The metamorphosis is abbreviated, and in some cases omitted. The young remain attached for a time to the mother, but perhaps always leave the sponge and live a short pelagic life before finding another sponge. The female Pontoniid deposits numerous small eggs on the swimmerets. These hatch into small larvæ which lead a comparatively long pelagic life before acquiring the form and habits of the adult.

1

I studied the habits of these animals as well as I could in dim light in the cavities of pieces cut from the sponge. They appeared to behave normally, whereas in glass tubes in brighter light they remained motionless. Both animals explore the cavities of the sponge with cautious movements unless disturbed, in which case they snap their claws. The Alpheid advances, using its large claw as an antenna and protector. Its antennæ can be extended about as far forward as the large claw. When meeting an Alpheid or a Pontoniid it may try to squeeze past or it may snap its claw. When placed in glass dishes Alpheids cut one another to pieces, but this is seldom if ever done in the narrow passages of the sponge.

The Pontoniid *Typton tortugæ* advances, using both claws as antennæ, the antennæ being very much shorter than the smaller claw. It spreads the claws apart and waves them about, thus exploring the cavity in front of it. On meeting another it behaves as the Alpheid, except that it may snap either or both claws. Both animals try to squeeze through small openings. The chelæ of *Typton* sometimes show what appear to be claw marks. As their claws are more slender, hence more easily grasped and less powerful than those of the Alpheids, it is to be expected that they would show claw marks first in case both species snapped with equal frequency.

Both animals appeared to eat from the walls of the cavities in the sponge, but I did not determine whether they ate the sponge itself or a sediment deposited on it. I did not determine whether they ate one another in the sponge, but they were so numerous that it seems strange that they received sufficient oxygen. The Alpheids are sometimes infested with a parasitic isopod, *Bopyrus*, in the gill cavity.

I do not intend to discuss here the origin of the form or habits of these animals, but it seems to me that we have here a convergence both in form and habit. It is probable that similarity in form and habits made both animals better suited to living in the same habitat, *i.e.*, the sponge, and that accidentally finding the sponge they remained there. However, this does not explain why the young at the end of pelagic life always (or at least usually) select the loggerhead sponge. There are numerous Alpheids living in holes in the reef rocks, and certainly they are more closely related in form and general habits to *Synalpheus brooksi* than is *Typton*.

This *Synalpheus* and the *Typton* select the sponge not because it has holes in it in which they can hide, but on account of some more specific quality, such as taste (smell), color, or outward form. Or when some individuals of these species have established themselves in a sponge the others may be attracted to it by a social instinct (which may not be disproved by the fact that they destroy one another when placed under unnatural conditions). The isolation of these animals in the loggerhead sponge is an example of what Gulick calls habitudinal segregation and may have been a factor in the evolution of the species.

Since the Alpheids occur in far greater numbers than *Typton* we might suppose the former to be much better adapted to living in the eggs, it has a much longer pelagic life than the Alpheid and is much more likely to be eaten or swept out to sea by the tides, where it can not find a sponge when the proper time comes.

The smallest loggerhead sponges I found would not live in a large aquarium with running sea-water more than 2 days before the water began to get foul within the passages in the sponge and the Alpheids and *Typton* began to die. Field observations were very limited. These and other difficulties restricted the investigation to its present limits.

ON ADAPTATIONS OF THE REEF ANEMONE, CRADACTIS VARIABILIS.

Cradactis variabilis Hargitt is an anemone about an inch or two in length when expanded, living in holes in old coral heads or reef rocks. Besides the tentacles, which are few in number and arranged as in *Sagartia*, long outgrowths called fronds extend from the region bearing the tentacles (plate 1, figs. 4, 5; plate 2, figs. 8, 9, 10). The animals may be a moss-green or brown in general color, but the tentacles are always paler and often colorless and transparent at their tips. The fronds may or may not be branched, and may end simply or in pale knobs, as in plate 1, fig. 4, or in curious "eyes," as in fig. 5.

These anemones are usually found in cavities in old coral heads that communicate with the exterior by a number of passages about half an inch or more in diameter. The anemones are attached near enough to these passages to extend the tips of the fronds to the exterior (plate 2, fig. 7). This extension is caused by heliotropism of the fronds. One mistakes them at first for sea-weed, although they do not resemble any particular kind of sea-weed that I have found growing on the reefs. The tentacles are extended about as far as and sometimes a little farther than the fronds, but the fronds tend to conceal the tentacles. At night the fronds are contracted and the tentacles remain extended; therefore it is probable that the fronds are not necessary as breathing organs.

If a bit of crab meat is held near the passage through which the *Cradactis* is extended no response is obtained. But if one of the fronds is touched with the meat the tentacles are extended toward it, while the frond touched may contract slightly. In order to observe the food-taking more minutely, some of the anemones were taken from the rock and allowed to attach themselves to the bottom of an opaque dish filled with sea-water. When a bit of crab meat is placed on the end of a tentacle it adheres and the tentacle and one or more adjacent ones are bent down and the food placed on the mouth and pressed there. Immediately many or all of the tentacles are pressed on the food, hiding it from view until it is swallowed. The fronds may contract more or less during the process. *Cradactis* sometimes swallows filter paper placed firmly on the mid-region of a tentacle or on the disk, but not when placed on the end of a tentacle. This may be a question of degree or extent of stimulation. It disgorges the paper within 10 minutes. It rejects bits of shell, etc., placed on the disk or tentacles.

India ink placed in the water near the anemone showed ciliary the disk running towards the mouth. A secretion sticks the particles together. These currents are useful on both fronds and tentacles in the rejection of particles, and on the tentacles in the placing of food in the mouth, the food being carried to the tip of the tentacle before it is placed in the mouth.

When disturbed by light falling on the base, it sometimes moves with snail-like motion (like *Metridium*) a short distance, but the tentacles catch hold of the substratum on all sides. The tentacles and column sometimes perform writhing movements. More often the animal bends over to one side and catches hold of the substratum with the tentacles, with or without previously elongating the column, the fronds contracting slowly all the while. It then loosens the base, walks on its tentacles to a new place (plate 2, figs. 11, 12), bends over and attaches the base, and lets go its hold with the tentacles. This method of locomotion is much more rapid than that of *Metridium*, but could not be used if the *Cradactis* did not live in holes, as it might otherwise be washed away by the currents that constantly sweep over the reefs.

The resemblance of the fronds to sea-weed leads one to suppose that they act as lures or in hiding the *Cradactis* from its prey (anemones being unpalatable are usually not in need of protection). The fact that the fronds are heliotropic and contracted completely at night is in harmony with this view. I did not cut them off to see whether the anemone would live and reproduce as well without them. The cavities containing the *Cradactis* are inhabited by other animals, especially a small black crab, and one might suppose that the fronds protected the tentacles of the anemone from the legs of the crabs that crawled over it. The crabs are active at night in the least light in which they can be seen (their black color making them hard to see in the holes in the rock). In case they are normally active at night the fronds would serve as a protection from the crabs only half of the time. The anemones sometimes grasp the crabs and hold them until they wrench themselves loose, which they invariably do in a short time. Perhaps the anemone gets part of its food as particles dropped from the crabs' mouths.

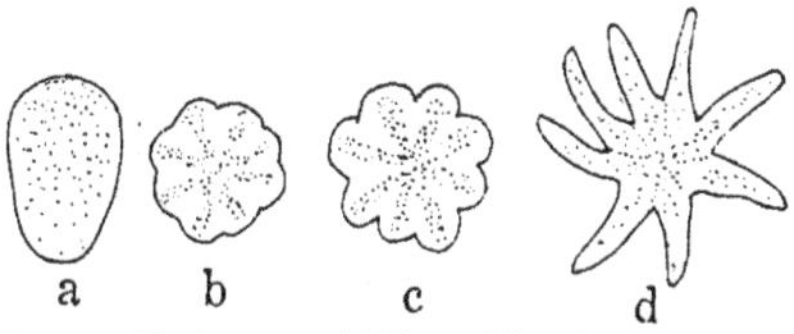

FIG. 1.—*Cradactis variabilis*. *a*, Planula just escaped from cœlentric cavity of mother, oral (pigmented) side uppermost. *b*, The same, second day, seen from oral side; pigment arranged radially. *c*, The same, third day. *d*, The same, fourth day; tentacles elongating and septa becoming distinct. The mouth should be elongated in the plane of symmetry.

Cradactis develops to the planula stage in the cœlenteron of the mother. On being released, the planula swims around for a few hours (text fig. 1, *a*) and attaches itself (*b*) by the smaller end. It gradually develops a mouth and tentacles (*b*-*d*). When first liberated, the planula has 8 mesenteries, and 8 tentacles develop soon after. Individuals were seen with 8, 10, 12, 14, 16, 18, 20, 22, 24, 26, 28 and more tentacles. From this one might conclude that the tentacles (and mesenteries) appear in pairs, but they were often observed to appear in sets of four, symmetrical in relation to the oral plane. The first pair of fronds appear in the 20-tentacle stage as outgrowths of the body-wall just beneath the tentacles and with their axis perpendicular to the oral plane. The second pair of fronds appear in the 28-tentacle stage or later.

SUMMARY.

(1) Convergence in structure and habitat is the cause of commensalism between an Alpheid and a Pontoniid living in the loggerhead sponge.

(2) Abbreviation of its pelagic life accounts for the numerical supersedence of the Alpheid.

(3) The weed-like outgrowths or fronds of a reef anemone, *Cradactis*, probably hide it from its prey.

(4) *Cradactis* is kept just within the mouths of cavities in reef rocks by the combined action of negative heliotropism of its base and positive heliotropism of the fronds. The fronds are entirely contracted in the absence of light.

(5) The fronds possess the sense of taste but do not carry food to the mouth.

(6) *Cradactis* moves from place to place by walking on its tentacles, a phenomenon sometimes seen in *Hydra*.

DESCRIPTION OF PLATES.

(Figures 4 and 5 were redrawn by Mr. Kline, fig. 6, from life, by K. Morita, otherwise the drawings and photographs are the author's.)

Plate 1.

1. *Synalpheus brooksi* Coutière.
2. Chela of same species to show different coloration.
3. *Typton tortugæ* Rathbun commensal with the above.
4. *Cradactis variabilis* Hargitt.
5. The same, showing another variety in color and shape of fronds.
6. *Cradactis variabilis* Hargitt, × 2.

Plate 2.

7. A portion of an old coral head showing the fronds (f) of *Cradactis* protruding from the cavities.

8–10. *Cradactis variabilis*, showing varieties in shape of fronds.

11, 12. *Cradactis variabilis*, walking on its tentacles, with detached base toward the observer.

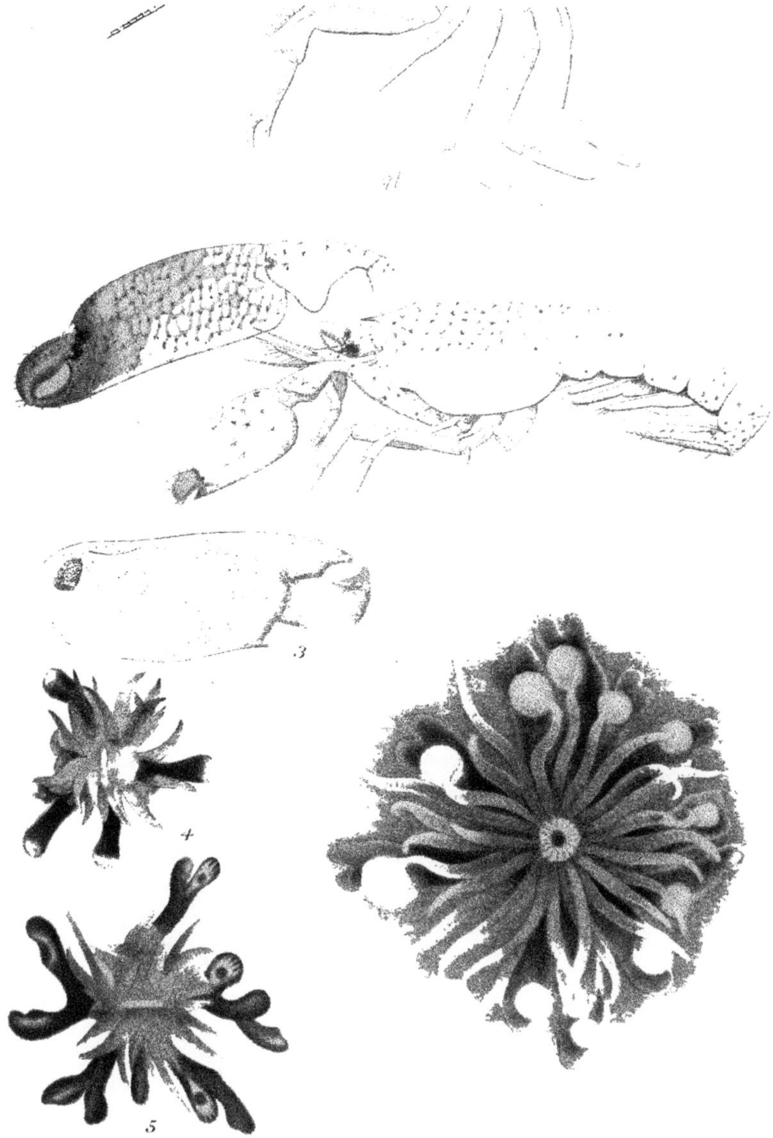

1. Synalpheus brooksi Coutière.
2. Chela of same species to show different colorations.
3. The Pontoniid commensal with the aboVe.
4. Cradactis variabilis Hargitt.
5. Cradactis Variabilis Hargitt.
6. The same, showing a third variation

7

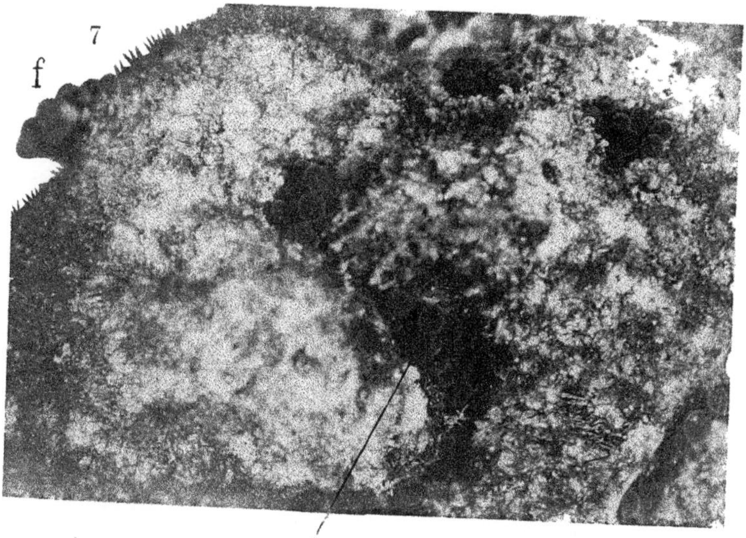

8

11

9

10

12

Fig. 7. A portion of an old coral head

PUBLICATIONS

OF

Cornell University

MEDICAL COLLEGE

STUDIES

FROM THE

Department of Anatomy

VOLUME II

1911-12

NEW YORK CITY

CONTENTS

Being reprints of studies issued in 1911 and 1912.

12. EIN FALL VON HERMAPHRODITISMUS VERUS HOMINIS.
By **J. F. Gudernatsch.**
Verh. des VIII, Internat. Zool. Kongress zu Gratz, 570-574.

13. THE OSMOTIC AND SURFACE TENSION PHENOMENA OF LIVING ELEMENTS AND THEIR PHYSIOLOGICAL SIGNIFICANCE. By **J. F. McClendon.**
Biol. Bull., Vol. XXII, 113-162.

14. AN ATTEMPT TOWARDS THE PHYSICAL CHEMISTRY OF THE PRODUCTION OF ONE-EYED MONSTROSITIES.
By **J. F. McClendon.**
Am. Jour. Physiol. Vol., XXIX., 289-297.

15. THE INCREASED PERMEABILITY OF STRIATED MUSCLE TO IONS DURING CONTRACTION. By **J. F. McClendon.**
Am. Jour. Physiol., Vol. XXIX, 302-305.

16. DYNAMICS OF CELL DIVISION—III. ARTIFICIAL PARTHENOGENESIS IN VERTEBRATES. By **J. F. McClendon.**
Am. Jour., of Physiol., Vol. XXIX, 298-301.

17. HOW DO ISOTONIC SODIUM CHLORIDE SOLUTIONS AND OTHER PARTHENOGENIC AGENTS INCREASE OXIDATION IN THE SEA URCHIN'S EGG?
By **J. F. McClendon.**
Jour. of Biol. Chem., Vol. X, 459-473.

18. A NOTE ON THE DYNAMICS OF CELL DIVISION. A REPLY TO ROBERTSON. By **J. F. McClendon.**
Arch. f. Entw.-Mech., Vol. XXXIII, 263-266.

19. EIN VERSUCH, AMÖBOIDE BEWEGUNG ALS FOLGEERSCHEINUNG DES WECHSELNDEN ELEKTRISCHEN POLARISATIONSZUSTANDES DER PLASMAHAUT ZU ERKLÄREN. By **J. F. McClendon.**
Arch. f. die. ges. Physiologie, Bd. 140, 271-280.

The Fate of Ovarian Tissues when planted on different Organs.

By

Charles R. Stockard,

Cornell Medical School, New York City, U. S. A.

With 2 figures in text and Plates XI—XIII.

Eingegangen am 20. Januar 1911.

It is a well known fact as has been shown in the case of the ovary and thyreoid that an entire organ may be transplanted from one individual to another without discontinuing to grow or function. GUTHRIE and CASTLE have found that fowls and guinea pigs with transplanted ovaries may ovulate and reproduce in an entirely normal manner. MEISENHEIMER has conducted a most elaborate and beautiful series of experiments to show that although the sex glands may easily be transplanted from one to the other sex in caterpillars yet the moths into which these caterpillars metamorphose show the typical secondary sexual characters of their original sex, not being affected by secretions from the transplanted bodies. All of these experiments indicate that there is no antagonistic action towards the organs of a different individual, even of the opposite sex, when planted into the body of another individual of the same species of animal.

Yet we seem to face a different proposition when considering the transplantation of portions of organs or tissues from one animal to another. In most cases such pieces of organs or tissues live and may actually grow for a time but invariably they cease to grow and finally disappear entirely. In this way normal tissues differ from malignant growths which continue to grow sometimes even more actively after transplantation.

It is also true, and I think a point of importance, that the ability of a tissue transplant to live and grow depends largely upon the kind of tissue on which it is planted. The indiscriminate injections of tissue emulsions and tissue pulps of both adult and embryonic tissue as sometimes used in the experimental study of cancer, are most unreliable and rarely give results owing to the hit and miss method employed. All tissue transplants must be carefully made and a circulation established by grafting in minute blood vessels, before deductions are to be drawn from the reactions which follow. It is evident that if the transplant is not properly nourished during the first hours or days it will begin to undergo degenerative changes which will in all cases effect its future behavior.

With these points in view the questions arise: First, do certain transplanted tissues survive equally well when planted on any organ, or do they survive longer and better on certain organs than on others (provided of course that the attachment and circulation is equally good in all cases)? Secondly, if they do survive better on certain organs what relationship exists between the tissues and these organs and what is the cause of their better survival?

LEÒ LOEB and ADDISON showed a few years ago that when guinea pig tissue was planted into other species of animals and into other guinea pigs, that the tissue always grew better in guinea pigs than in any other animal; and it was further indicated that the tissue survived for a shorter time the more distantly related was the species into which it was transplanted. Thus, as might have been expected there is a specific reaction on the part of the body of an animal to the transplanted tissues of other species of animals.

The present experiments, although of a preliminary nature, bear upon the question of a resistance or antagonism between the tissues of one organ and those of another in either the same individual or other individuals of the same species. And also the further question, of antagonism between different tissues, or different cells, in similar organs.

The animals used in these experiments were guinea pigs and the common salamander *Diemyctylus viridescens*. The tissue chosen for transplantation was that of the ovary since it is composed of two so entirely different classes of cells, the stroma tissue and the germ cells or ova, and further as an organ is so interestingly related to the testis of the male. Pieces of the ovary of guinea pigs were transplanted into the testis and into the body wall and liver tusses of the male. They lived better when planted into the testicle and here the artificially established circulation seemed more efficient than in other organs. Nevertheless, all of the experiments with ten guinea pigs were unsatisfactory since in no cases did the transplanted tissue live sufficiently long or well to allow valid comparisons. Ribbert, Lubarsch, Levin, Loeb and others have all had similar experiences in transplanting the tissues of mammals. The tissues may grow for a short while but soon stop and ultimately all disappear. Inflamatory conditions are also commonly produced in the guinea pigs if not operated upon with great care.

With the salamanders the experiments were much more satisfactory, the ovarian tissue is easily transplanted and grows and lives for several months, in many of the cases, and undergoes changes so slowly and uniformly as to permit careful study and comparisons.

Thirty individuals were employed in the experiments and portions of ovary in various degrees of maturity were planted on the liver, lungs, kidney, stomach and body wall of the same individual and on the testis, stomach, kidney, body wall, lungs and liver of male salamanders. The ovarian tissue grew equally well on similar organs of the male as upon those of the female, showing that there is no marked individual reaction against the tissues of other specimens, even though of the opposite sex. The most favorable of all transplants, as will be considered below, was that of ovary tissue on the testis.

The transplanted tissue in all cases was carefully attached to the new organ with the finest silk or gut fibre suture and neighboring blood vessels were dissected out and embedded within and around the tissue, the entire operation being performed under the binocular microscope. The blood vessels in almost every case readily sent out branches and supplied the new tissue.

Ovarian tissue on the testis.

The tissue from the ovary, as stated above, was most persistant and successful in its growth when planted upon the testis. A portion of the ovary containing ova or germinal epithelium and stroma tissue was planted in a pocket or slit cut in the testis, a branch of the spermatic artery was carefully dissected loose and placed around the ovarian mass which was fastened to the testicle by a silk or gut-fibre suture. Ten such operations were entirely successful. The ovarian piece was well nourished and the ova continued to grow, in most cases each ovum having a plexus of capillaries about it as is shown in the camera drawings (Figs. 1 and 2).

Fig. 1.

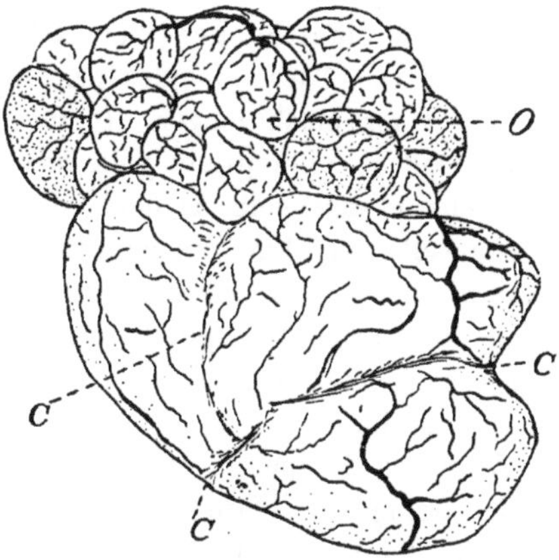

A camera drawing showing part of an ovary planted on the testis of a salamander 41 days after the operation. *O* Ova, all well supplied by capillaries indicated in black lines; *C* scars resulting from the cuts made during the operation.

After forty-one days the entire plant is in an apparently normal condition (Figs. 1 and 2). There are no indications of degeneration in either the egg cells or stroma (Pl. XI, Fig. 1). At no time does there seem to be a tendency for the cells of the testicular stroma to migrate in, or replace the ovarian stroma cells. The separation between the testicular and ovarian tissue remains distinct. As will be mentioned below, this is in contrast to the reaction of other tissues to the ovarian transplant (see Figs. 2 and 4, Plates XI and XII).

Fig. 2.

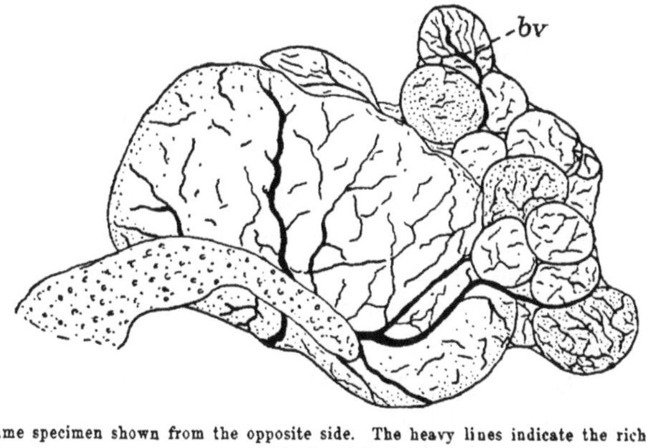

The same specimen shown from the opposite side. The heavy lines indicate the rich supply of blood vessels, *bv*.

The egg cells seem to be unable to continue their growth as

they are highly laden with yolk and need certain conditions for their maturation and further multiplication or development. The yolk granules then begin to loosen apart, and later globules are formed by the fusion of groups of granules, these globules become scattered throughout the ovarian tissue as is seen by comparing transplants of different ages (Plate XII, Figs. 3 and 4, and Plate XIII, Figs. 6 and 7). These globules persist for as long as seven months and might for much longer. It thus seems that the stroma cells do not tend to appropriate yolk bodies as food. This inability of these cells to use yolk as food, or to absorb it, is necessary that the egg cells may have an opportunity to accumulate or form yolk in the presence of the stroma tissue. Yolk granules disappear much earlier in ovarian transplants on other organs since the cells of these organs migrate into the ovarian piece and dissolve or absorb the yolk grains.

After the dissipation of the yolk, as for example, in a seven months' transplant (Plate XIII, Figs. 6 and 7) the ova cells can no longer be distinguished, the entire piece seems of the stroma cell type with yolk globules scattered through the tissue. In rare cases, however, masses of yolk globules are localized and probably represent the persistant remains of an egg cell (Plate XII, Fig. 3). The pigment granules of the egg are lost and cannot be identified in the old plants.

The ovarian stroma persists and is well preserved in the transplants of seven months' duration. They would doubtless have lasted much longer if specimens could have been kept after this time.

It appears then, that there is little antagonistic response between two such organs as the ovary and testis and that the tissues of these organs do not tend to replace or destroy one another, thus they may live and grow side by side. The ova persist perfectly for one or two months and then undergo retrogressive changes probably on account of having passed the period at which they should have matured or undergone some important modification. The stroma of the ovary, however, is not attacked by the testicular stroma and may survive in an apparently perfect condition for longer than seven months, which is as far as the observations extended. Ovarian transplants on all other organs had entirely disappeared long before, neither the ova or stroma being able to exist in contact with the tissue cells of such organs.

Ovarian tissue on the liver.

Ovarian tissue when planted on the liver with a blood supply established persists longer than on any other organ except as discussed above on the testis. The ovarian tissue persists equally well on the liver of the same individual or on that of a male of the same species of salamander. In no case out of fifteen successful transplants on the liver did the blood vessels branch and supply the ovarian piece so efficiently as did the branches of the spermatic artery. This was due to the fact that no artery of the liver could be so nicely placed about the piece as could the spermatic of the testis. The circulation of the liver-plant, however, was very good and easily sufficient to have supplied the piece as was shown by the length of time it persisted. The difference in circulation does not account for the earlier fate of the ovarian piece on the liver, but this difference was due to the different way in which the liver tissue or cells themselves reacted to the strange tissue. While there was no tendency on the part of the testicular stroma to encroach upon the ovarian tissue and replace its cells by wandering testicular cells, the liver very soon reacted in such a manner.

The ova and stroma cells lived well on the liver for several weeks though they did not grow very much. After this time, however, the cells of the liver had encroached and migrated into the ovarian tissue apparently replacing and destroying the tissue before it. The ovarian piece decreases in size and the ova gradually loose their yolk and finally themselves disappear. Fig. 2, Plate XI and Fig. 4, Plate XII show an ovarian piece planted upon the liver; it will be noticed that the liver tissue extends into the ovarian part and many liver cells have migrated far into the transplant, the typical pigment spots of the liver are seen in the ovarian graft in Fig. 4, Plate XII. These figures are photographs of a 42 day transplant and may be compared with Fig. 1 Plate XI which is a transplant of about equal duration on the testis.

Transplanted on the liver the ova show all indications of degeneration and breaking down while on the testis they are in a normal healthy condition. The large blood vessel, *bv*, at the base of the transplant on the liver as well as smaller vessels shown in the sections of the transplanted tissue would indicate that the piece was sufficiently nourished and did not degenerate on account of a poor blood supply.

The liver cells, then, show a kind of antagonistic action against the ovarian tissue which is not shown by the cells of the testis. This we may speak of as the antagonism between two different organs or the antagonism between tissues of different organs. A more pronounced antagonism, mentioned before as shown by LEO LOEB for transplanted tissues and well known from many haemolysis experiments, is that which exists between the tissues or parts of animals of different species, specific antagonism.

Ovarian tissue on lung, kidney, stomach and body wall.

Pieces of ovary planted upon the lungs, kidney, stomach or body wall of the same or of another individual often live for a short time before being absorbed but usually disappear within a week or ten days after the experiment, only three out of more than fifty such transplants lived as long as 45 days. These three were almost completely replaced and would have soon disappeared.

A fair circulation may be established for the transplant on any of these organs, yet the ovarian tissues seem unable to maintain themselves in such an environment and both ova and stroma begin to degenerate and are readily replaced or absorbed by the cells and tissues of the supporting organ.

It is difficult, from the observations at hand, to state whether or not certain of these organs are more antagonistic to ovarian tissue than others. The wall of the stomach is an unfavorable place to make a transplant, but the body wall would seem favorable since a good circulation is easily obtained, yet the tissue readily breaks down and disappears in either place. Transplants on or within the delicate lung tissue often break away but do not thrive even when successfully made, and pieces of ovary on the kidneys readily disappear.

Summary and Conclusions.

From these experiments on thirty salamanders it would seem that the behavior or fate of transplanted tissues depends largely upon the nature of the organ upon which the tissue is transplanted. Ovarian tissue grows and lives incomparably better when transplanted upon the testis than upon any other of the body organs experimented upon.

The next most favorable ground for this tissue was upon the liver, although here the liver cells soon begin to encroach upon the ovarian mass replacing and absorbing its cells. On other organs the ovarian tissue undergoes degeneration and absorption within a very limited time.

Ovarian tissue planted upon the testis persisted with the stroma in good condition for more than seven months. On the liver ovarian tissue was found still to contain ova and stroma after more than 45 days. While on the body wall, lungs, kidney and wall of the stomach the tissue disappears within about two weeks, only three indications, in more than fifty such transplants, were found after 45 days and these had almost disappeared.

It is very important, therefore, to realize that there is a marked difference between the reactions of certain tissues to others, and all transplants should be made in the most careful manner between organs of as near as possible a similar type. The introduction of mixtures or emulsions of adult or embryonic tissue into different parts of the body is a very unreliable method and one likely to give most contradictory results, depending upon the proportion of certain tissues present which successfully plant, and upon the organ of the host into which the tissues are placed or happen to reach.

Just as there is a specific reaction between the tissues of animals of different species which tends to prevent the growth of foreign tissue when planted in their bodies, there seems to be, from this preliminary series of experiments, a reaction between the tissues of different organs of the same or different individuals which causes transplanted tissue to exist to better advantage on one, usually a more closely similar organ, than on another.

Zusammenfassung und Schlußfolgerungen.

Nach den vorstehenden Versuchen an 30 Salamandern hängt das Verhalten oder das Schicksal transplantierter Gewebe anscheinend weitgehend von der Natur des Organs ab, auf welches sie transplantiert wurden. Ovarialgewebe wächst und lebt bei Transplantation auf den Hoden unvergleichlich viel besser, als auf irgend einem andern der in den Versuchen gewählten Körperorgane.

Den nächstgünstigsten Grund für dieses Gewebe stellte die Leber dar, obgleich hier schon die Leberzellen auf die Ovarialmasse überzugreifen und deren Zellen zu ersetzen und zu absorbieren anfangen. Auf andern Organen unterliegt das Ovarialgewebe bereits innerhalb einer sehr beschränkten Zeit der Entartung und Absorption.

Auf den Hoden überpflanztes Ovarialgewebe erhielt sich mit dem Stroma mehr als 7 Monate lang in guter Verfassung. Auf der Leber fand sich Eierstockgewebe noch nach mehr als 45 Tagen im Besitz von Eiern und Stroma.

Während auf der Körperwand, den Lungen, den Nieren und der Bauchwand das Gewebe ungefähr im Laufe zweier Wochen zu verschwinden pflegt, fanden sich im ganzen zweimal Anzeichen desselben, unter mehr als 50 derartigen Transplantationen, nach 45 Tagen, und auch diese bereits im Verschwinden begriffen.

Es ist daher sehr wichtig, festzustellen, daß hier ausgeprägte Unterschiede in den Reaktionen gewisser Gewebe auf andre bestehen, und alle Transplantationen sollten daher so sorgfältig als möglich zwischen Organen von möglichst nahestehendem Typus vorgenommen werden. Die Einführung erwachsener oder embryonaler Gewebe in Gestalt von Mixturen und Emulsionen in verschiedene Teile des Körpers ist eine sehr unzuverlässige Methode, geeignet, zu ganz widersprechenden Ergebnissen zu führen und abhängig von dem Vorhandensein eines entsprechenden Betrages erfolgreich überpflanzbarer Gewebe, endlich von dem Organe des Wirts, in welches die Einpflanzung stattfindet, oder welches sie zufällig erreichen.

Gerade so, wie es eine spezifische Reaktion zwischen den Geweben verschiedenartiger Tiere gibt, welche das Wachstum eingepflanzten fremdartigen Gewebes zu verhindern strebt, so scheint nach dieser vorläufigen Versuchsreihe auch zwischen den Geweben verschiedenartiger Organe desselben oder verschiedener Individuen eine Reaktion zu bestehen, welche einem transplantierten Gewebe vorteilhaftere Existenzbedingungen auf einem, gewöhnlich einem näher verwandten, Organ verschafft, als auf einem andern.

(Übersetzt von **W. Gebhardt,** d. 6. II. 1911.)

Literature cited.

CASTLE, W. E., On the Nature of Mendelian Factors. Read before Am. Soc. of Naturalists. 1909.

GUTHRIE, C. C., Further Results of Transplantation of Ovaries in Chickens. Journ. Exp. Zool. V. 1908.

LEVIN, I., Cell Proliferation under Pathological Conditions with especial Reference to the Etiology of Tumors. Journ. Med. Research. VI. 1901.

LOEB, L., und ADDISON, W. H. F., Beiträge zur Analyse des Gewebewachstums. II. Transplantation der Haut des Meerschweinchens in Tiere verschiedener Species. Archiv f. Entw.-Mech. XXVII. 1909.

MEISENHEIMER, J., Experimentelle Studien zur Soma- und Geschlechtsdifferenzierung. Erster Beitrag. Jena, Fischer, 1909.

ROUS, P., Comparison of Conditions that Regulate the Growth of Transplanted Tumor and Transplanted Embryo. Journ. Am. Med. Assoc. LIV. 1910.

Explanation of Plates.

All figures are microphotographs of sections through transplants of ovarian tissue of various ages upon the testis and liver of the male salamander, *Diemyctylus viridescens.*

Plate XI.

Fig. 1. A 41 days old ovarian graft upon the epididymal end of the testis. The ova and ovarian stroma are still in normal condition. *Ov* ovary, *Ep* Epididymis.

Fig. 2. An ovarian mass of about the same age planted on the liver. The two sections are equally magnified and it is noted that the ovarian parts planted on the liver are much reduced the ova being very small and almost all of the yolk of the egg cells has been lost. *Ov* ovary, *Li* liver.

Plate XII.

Fig. 3. A highly magnified section of ovarian tissue after seven months existance on the testis. An ovum, *O*, is shown in which the yolk granules have massed into globules.

Fig. 4. A higher magnification of the portion of ovary shown in Fig. 2 plate XI which has been planted upon the liver of a male for 42 days. The liver cells and liver pigment are shown to be encroaching upon the ovarian mass. *O* ova, *bv* large blood vessel at the base of the transplanted tissue. Compare the ova, *O*, with that of the seven month transplant in Fig. 3.

Plate XIII.

Fig. 5. A portion of ovary, *ov*, that has been living for seven months upon the testis, *Ep*. All of the ova have disappeared but the stroma is still in good condition.

Fig. 6. A higher magnification of the section of ovary in Fig. 5. Scattered throughout the stroma are seen yolk globules, *y*, which represent the remains of the once large egg cells.

Fig. 7. A highly magnified section from another specimen of ovary after seven months upon a testis. Here also the yolk globules, *y*, are scattered through the stroma.

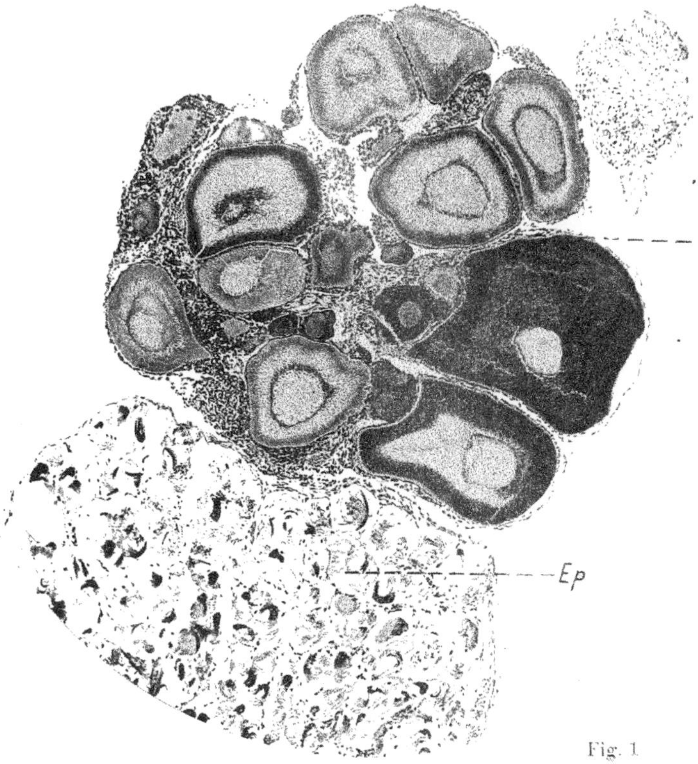

Fig. 1

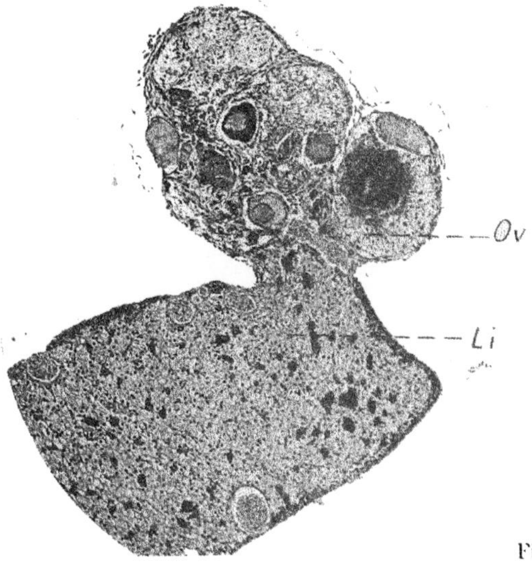

Fig. 2

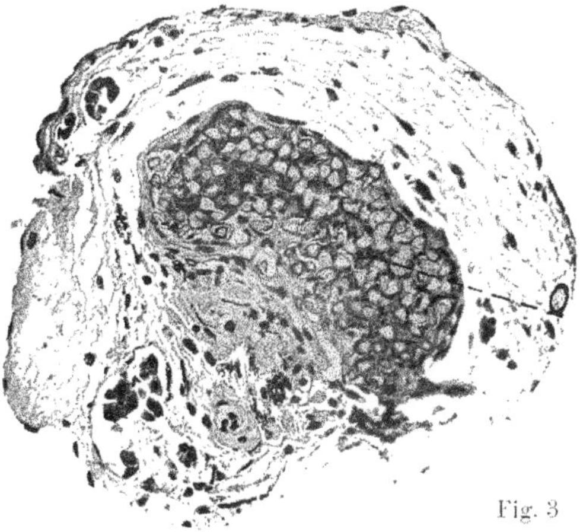

Fig. 3

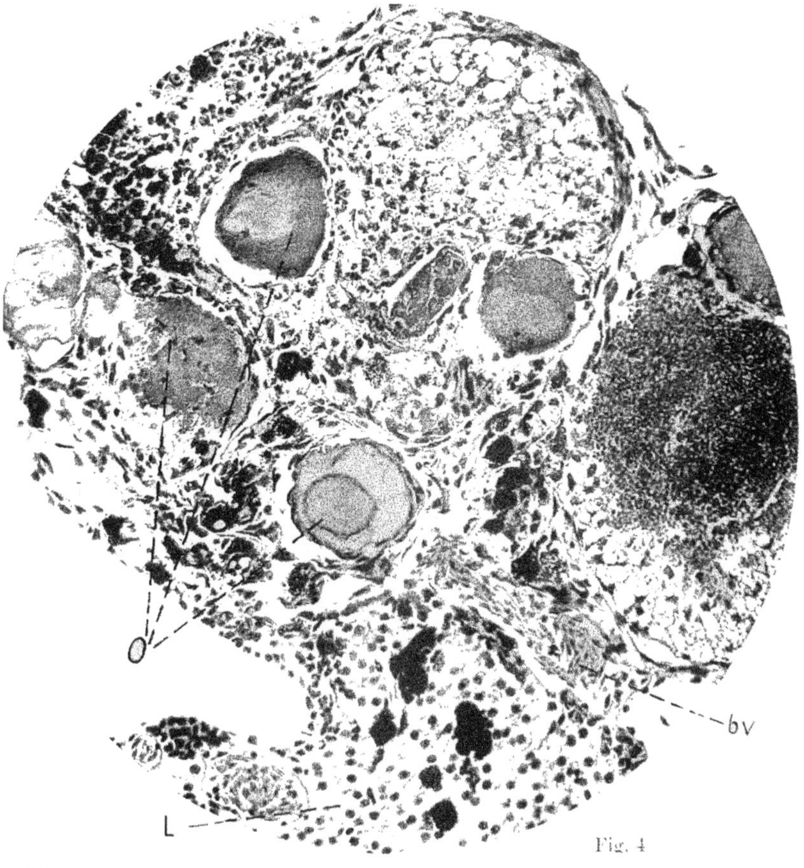

Fig. 4

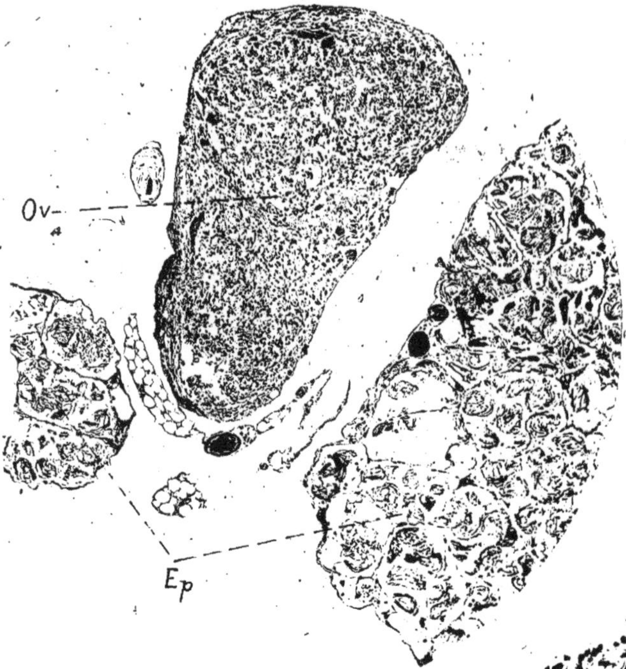

Fig. 5

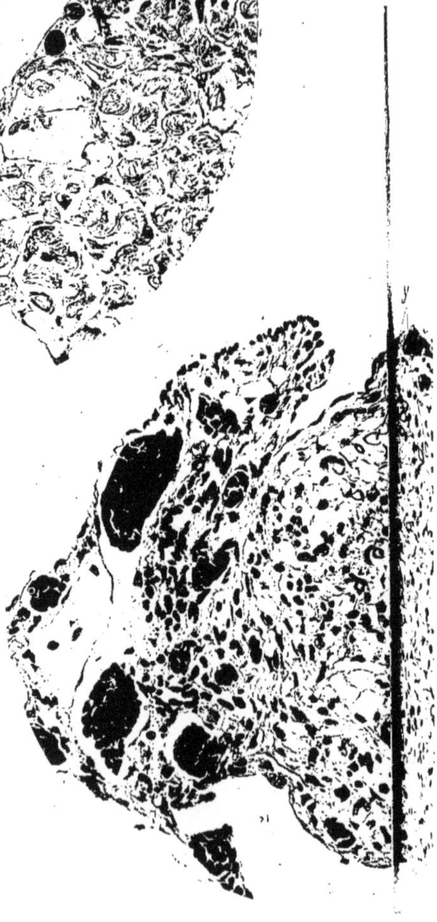

ig

Verlag von Wilhel

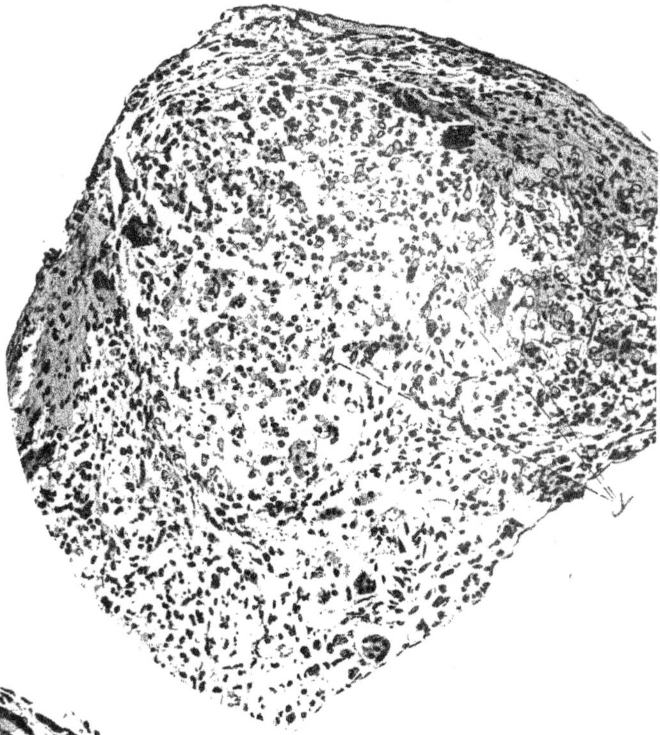

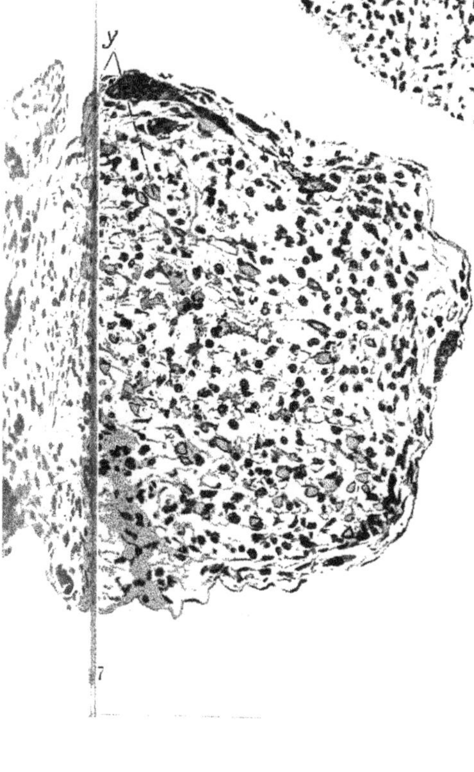

Fig. 6

Reprinted from THE ANATOMICAL RECORD, VOL. 5, NO. 5
MAY, 1911

THE RETICULUM OF LYMPHATIC GLANDS

JEREMIAH S. FERGUSON

Cornell University Medical College, New York City

TEN FIGURES

The silver impregnation method of Bielschowsky[1] in its newest form has with some modifications been applied to the study of tissues other than those of the nervous system for the staining of whose neurofibrils it was originally devised. Maresh[2] and Studnicka[3] seem to have been the first to appreciate its wider applicability. Levi[4] slightly modified the method for use with connective tissues, and his publication stimulated various Italian observers to a study of the collaginous and reticular tissues in a considerable variety of organs. Ciaccio, Cesa-Bianchi, Alagna, Balabio, Favaro, and others have put forward such publications, and the method has been shown to serve as a fairly distinctive stain for the peculiar fibres of reticular tissue. This method applied to nature tissues blackens only the reticular tissue and shows beautifully its distribution.

The inequality in the distribution of the lymphocytes in lymphatic glands with formation of denser areas is well known. The areas of great density are: 1, the cortex as compared with the medulla; 2, the follicles; 3, the cords. The increased density of the cortex is not solely due to the presence of follicles within it, for the intervening areas are of greater density than much of the medulla. In the follicles the periphery is of greater density than the central portion. This is due to the presence of the

[1]Bielschowsky and Pollack, Neurol. Centralbl., 1904, Bd. 23, 387.
[2]Centralbl, f. allg. Path., 1905, Bd. 16, 641.
[3]Studnicka, F. K, Zeitschr. f. wis. Mik., 1906, Bd. 23, 414.
[4]Levi, G., Monitore Zool. Ital, 1907, Anno. 18, 290.

germinal centers of Flemming. In the medulla the leucocytes in the cords are densely packed as compared with those in the intervening sinuses. While the facts regarding the crowding of leucocytes have been generally recognized, the somewhat similar distribution of the reticulum has not been so frequently observed. Silver blackened sections of lymphatic glands show that the reticulum, like the lymphocytes, is crowded at certain points. The denser areas are found in the cortex, the medullary cords, at the periphery of the follicles, and about the blood-vessels and trabecula. The condensation of reticulum beneath the capsule on the borders of the peripheral sinus, and at the boundary line between the cords and trabecula on the one hand and the cortical and medullary sinuses on the other, is so pronounced as to form a distinct membrane. Of two comparable areas it often happens that the denser the lymphoid cells the denser the reticulum. Thus, the reticulum of the cortex is denser than that of the medulla, that at the periphery of a follicle is denser than in its center, that of the medullary cords is coarser and often denser than that of the sinuses: of course so general a rule is not without exceptions, notably in that the cortical tissue between the follicles is frequently provided with a very dense and firm reticulum.

Ciaccio[5] has already called attention to the increased density of the reticulum in the periphery of the follicle as compared with the broad meshes and scant fibres of the germinal centers. This perifollicular plexus of Ciaccio is very characteristic. It consists of course fibres arranged to form peculiar lozenged-shape meshes. Ciaccio states that in the active follicles of the spleen and lymphatic glands the intrafollicular network is rare, while in the inactive it is nearly as dense as the perifollicular plexus. One must however call attention to the possible error resulting from the examination of sections whose planes pass through only the peripheral portions of certain follicles (fig. 2). In such cases the apparently active follicles show a dense network throughout a considerable portion of their diameter but this net is obviously

[5]Anat. Anz., 1907, Bd. 31, 594.

All figures, unless otherwise stated, were drawn with the aid of the Edinger projection apparatus and the magnification accurately measured. All sections from which figures were made had been stained by the method of Bielschowsky.

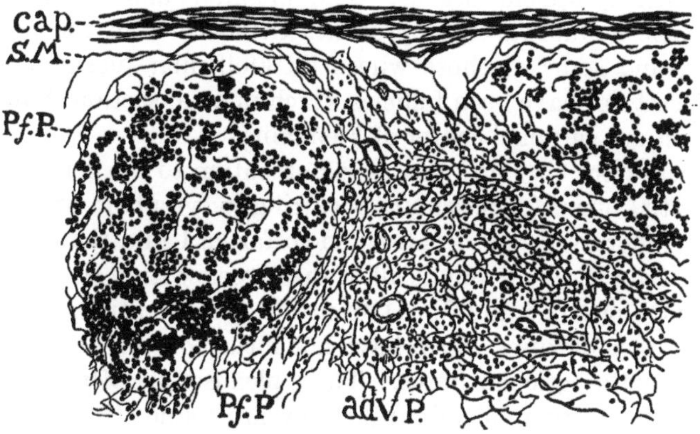

Fig. 1 Follicles at the periphery of a lymphatic gland of a dog. *Avd. P.*, adventitial plexus; *Pf.P.*, perifollicular plexus; *S. M.*, subcapsular membranes. × 200, reduced ¼ in reproduction.

Fig. 2 Two adjacent follicles in the cortex of a lymphatic gland of man. The section passes only through the surface of the follicles and apparently shows a dense intrafollicular plexus which is, however, in reality a surface view of the perifollicular plexus. The coarseness and density of the plexus is apparent on comparison with the intrafollicular plexus shown in fig. 1. × 256, reduced ¼ in reproduction.

that of the peripheral plexus, the germinal center lying outside the plane of the section. This is clearly shown in many of my sections both in the lymphatic glands and in the intestine, spleen and other lymphoid organs. The intrafollicular plexus of a lymphatic nodule rarely approaches the density of the perifollicular, and some of those whose intrafollicular plexus showed the closest net have been follicles which in axial section showed a very distinct germinal center of Flemming.

Ciaccio (loc. cit.) has also called attention to the perivascular adventitial plexus of reticulum occurring in lymphoid organs. I can corroborate his observations in so far as I find that such an adventitial plexus surrounds all of the vessels within the lymphatic gland (figs. 1, 6 and 7). I desire to call attention, however, to two other plexuses which have not hitherto been described as formed by the reticular fibres. These plexuses are often so dense as to form reticular membranes; they are found, the one, pericordal, limiting the surface of the medullary cords and so forming a wall for the lymphatic sinuses of the medulla, and the other a double membrane, the subcapsular, at the surface of the gland beneath the capsule where it forms limiting walls for the peripheral lymphatic sinuses.

A closer examination of the peripheral sinuses shows that throughout a considerable portion of their extent they lie between two distinct membranes of pure reticulum against which the endothelium is applied. The outer subcapsular membrane (*O.S.M.*, fig. 3) is in contact with the capsule and its fibres to some extent intertwine with the innermost collaginous fibres of the capsule. Here and there, especially where it passes from the surface of the capsule to the trabecula, the outer subcapsular membrane may be partially detached from the collaginous tissue upon which it elsewhere lies. At each trabeculum its fibres are continued into the interior as well as upon the surface of the fibrous septum, so that the trabecula contain throughout a considerable proportion of reticulum, whereas in the capsule this tissue is confined to the innermost layer of the membrane.

The inner of the two subcapsular reticular membranes is formed by a condensation of the reticular network of the cortex of the

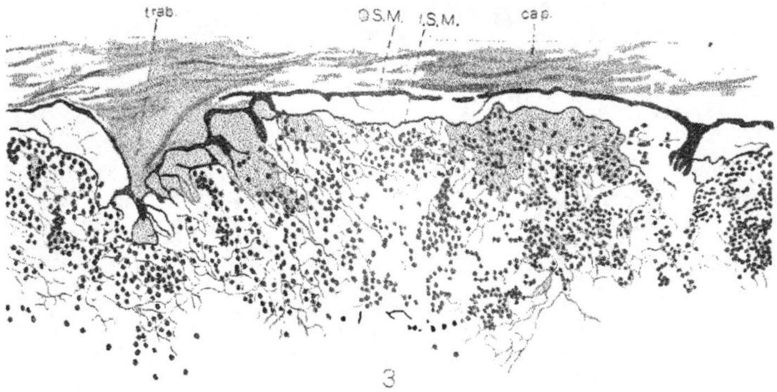

Fig. 3 The peripheral sinus and its reticular subcapsular membranes from a lymphatic gland of a dog. *Cap.*, capsule of the gland; *I.S.M.*, inner subcapsular membrane; *O.S.M.*, outer subcapsular membrane; trab., trabeculum. × 229, reduced $\frac{2}{5}$ in reproduction.

organ, and since it represents a condensation only, it is as one must expect, much more evident at some points than at others. Here and there it forms an almost continuous line along the cortical surface of the organ (*I.S.M.*, fig. 3) while elsewhere one may find areas where the condensation of reticulum is so slight that in transections a membrane can scarcely be discerned.

In the peripheral sinuses, between the two reticular membranes, the reticulum forms a loose network of delicate fibres and very broad meshes. The fibres of this net become directly continuous with those of the reticular membranes.

In the medulla a reticular membrane occurs on the surface of each of the cords, where it forms a limiting wall for the medullary lymph sinuses. Certain lympho-glandulae in which the cords are numerous and characteristic show continuous lines of reticular membranes on the surface of each cord (fig. 4). Here and there the inner subcapsular membrane may be reflected inward along the trabecula to become continuous with the pericordal membranes (fig. 5B) so that the peripheral sinus passes directly into the medullary. But more frequently the inner subcapsular membrane is lost in the cortical reticulum in the vicinity of the trabecula, the fibres of the outer wall of the sinus pass into the

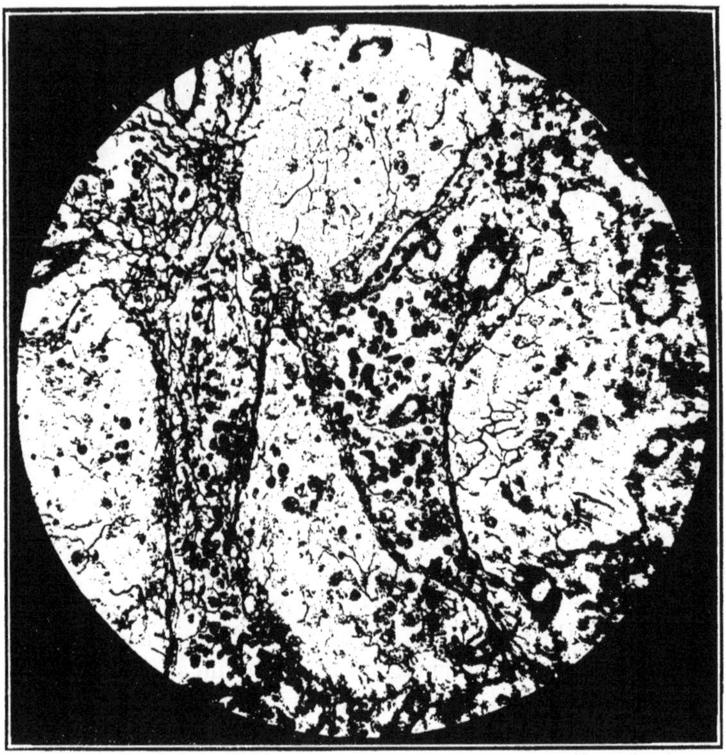

Fig. 4 Photomicrograph of two medullary cords of a dog's lymphatic gland; each is bounded by a distinct pericordal membrane. The small blood-vessels within the cord show well the adventitial plexus. × 250, reduced $\frac{1}{5}$ in reproduction.

trabecula and within the gland become frayed out, as it were, the fibres being continuous with those of the cortical reticulum: in this case the peripheral sinus directs its flow of lymph into the dense cortex, through which lymph must percolate in order to reach the medullary lymph sinuses.

A third series of reticular membranes is formed about the walls of the blood vessels by the adventitial plexuses of Ciaccio, and similar adventitial or peritrabecular plexuses (pericordonal plexus of Balabio) about the nonvascular as well as the vascular portions of the trabeculae. In the vascular trabecula of the lymphatic glands this plexus forms no part of the wall of artery or vein but lies in the substance and upon the surface of the fibrous trabeculum, the fibres of the plexus being continuous with those

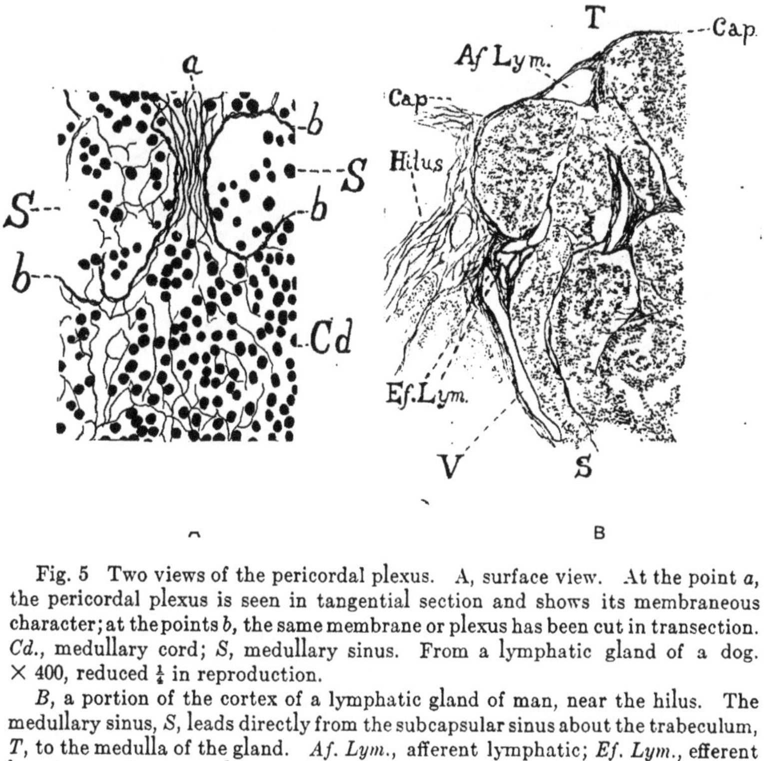

Fig. 5 Two views of the pericordal plexus. A, surface view. At the point *a*, the pericordal plexus is seen in tangential section and shows its membraneous character; at the points *b*, the same membrane or plexus has been cut in transection. *Cd.*, medullary cord; *S*, medullary sinus. From a lymphatic gland of a dog. × 400, reduced ¼ in reproduction.

B, a portion of the cortex of a lymphatic gland of man, near the hilus. The medullary sinus, *S*, leads directly from the subcapsular sinus about the trabeculum, *T*, to the medulla of the gland. *Af. Lym.*, afferent lymphatic; *Ef. Lym.*, efferent lymphatics; *Cap.* capsule of the gland; *V*, vein, making its exit through the connective tissue of the hilus. The sinuses are bounded by the pericordal plexus. × 60, reduced ¼ in reproduction.

of the trabeculum. This is well shown in fig. 6, in which the collaginous adventitia of the artery and of the lymphatic vessels is seen well within the blackened trabecular reticulum.

Fig. 6 The adventitial plexus about the arteries in a vascular trabeculum in the medulla of a lymphatic gland of man. The plexus lies just outside of the tunica adventitia. × 220, reduced ¼ in reproduction.

Balabio[6] states (p. 137) that the adventitial plexus of the blood vessels has not the importance in the lymphatic glands that it has in the spleen. I can verify this statement so far as it concerns the arteries, the periarterial plexus in the spleen being much heavier than in the lymphoglandulae; but the same is scarcely true of the veins (fig. 7). The small venules which Calvert[7] has described at the periphery of the follicles (figs. 1 and 8)stand out prominently in the lymphatic glands because of their coarse, close-meshed, adventitial plexus. With the exception of those vessels lying within the vascular trabecula I find the thin-walled veins of the spleen very deficient in those adventitial fibres which blacken with silver.

Ciaccio (loc. cit.) dismisses the lymphatic capillaries with the statement that they have a structure similar to that of the blood capillaries. On p. 600 he says: "I capillari linfatici presentano

[6]Anat. Anz., 1908, Bd. 33, 135.

[7]Calvert, W. J., Anat. Anz., 1897, Bd. 13, 174; and Johns Hop. Hosp. Bull., 1901, vol. 12, p. 177.

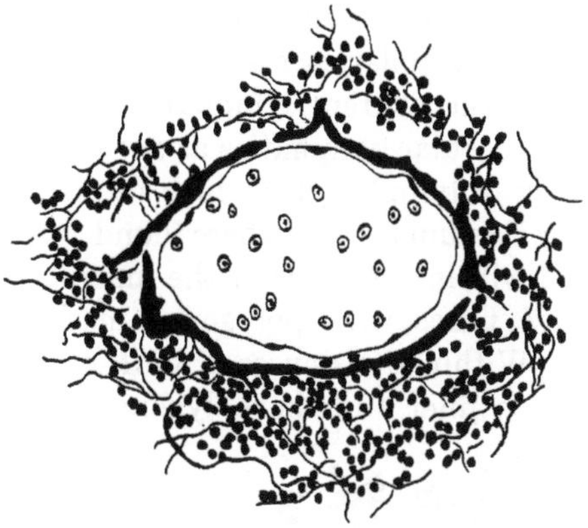

Fig. 7 Periadventitial plexus about the wall of a medullary vein in a lymphatic gland of man. × 255.

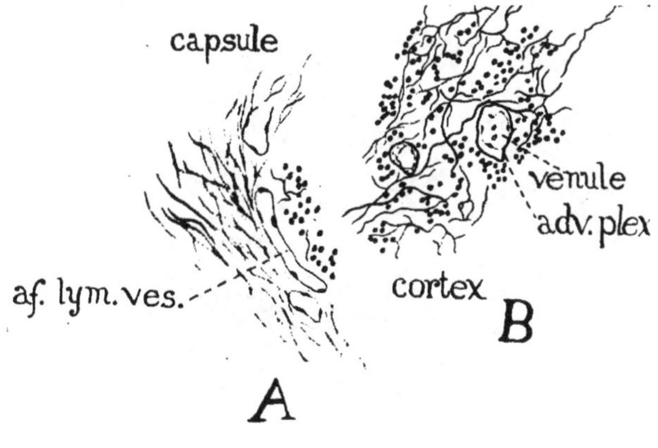

Fig. 8 Lymphatics and venules from the capsule and cortex, respectively, of a lymphatic gland of a dog. *A*, afferent lymphatic vessels in the capsule; *B*, venules in the adjacent cortex of the same section; *adv. plex.*, adventitial plexus; *af. lym. ves.*, afferent lymphatic vessel. × 225.

quasi la stessa struttura dei capillari sanguigni, differendo da questi soltanto per la forma, larghezza e disposizione." From my observations of the lymphatic vessels I am led to believe, however, that the volume of reticulum in the adventitial plexus is considerably less in the lymphatics than in veins of corresponding size. This is specially well shown if one compares vessels somewhat larger than capillaries, *e.g.* the venules at the periphery of the follicles and the afferent lymphatics in the capsule, (fig. 8), or such intratrabecular lymphatics as are shown in fig. 5B.

Within the mass of lymphocytes which forms the parenchyma of the organ the reticulum is everywhere found. At the periphery in the cortex and within the cords of the medulla the network is formed by course fibres with polygonal meshes of relatively small diameter Within the sinuses of both cortex and medulla the fibres are much finer and the polygonal meshes broader so

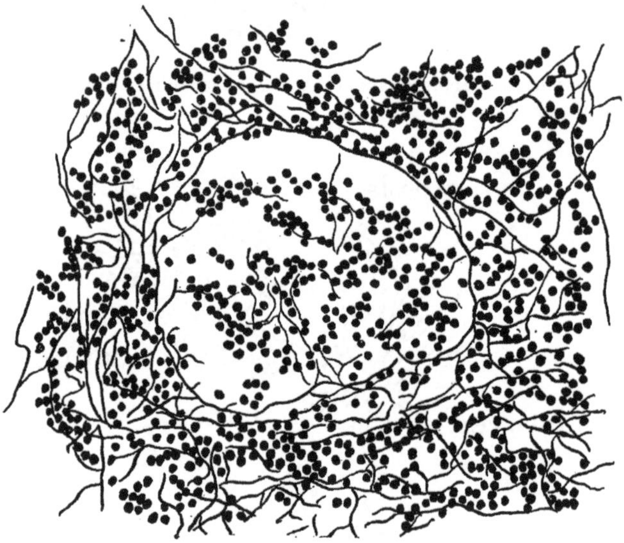

Fig. 9 A medullary sinus in transection, showing its very delicate network of reticular tissue. From a lymphatic gland of man. × 255.

that in the sinuses the total volume of reticulum must be only a small fraction of that in the denser portions of the gland.

The broadest mesh is that in the germinal centers of follicles, the finest fibres those of the sinuses. The shape of the mesh is typically lozenge-shaped in the perifollicular plexus of Ciaccio, and polygonal elsewhere, though here and there the polygonal spaces are considerably elongated in the directions of stress or of lymph

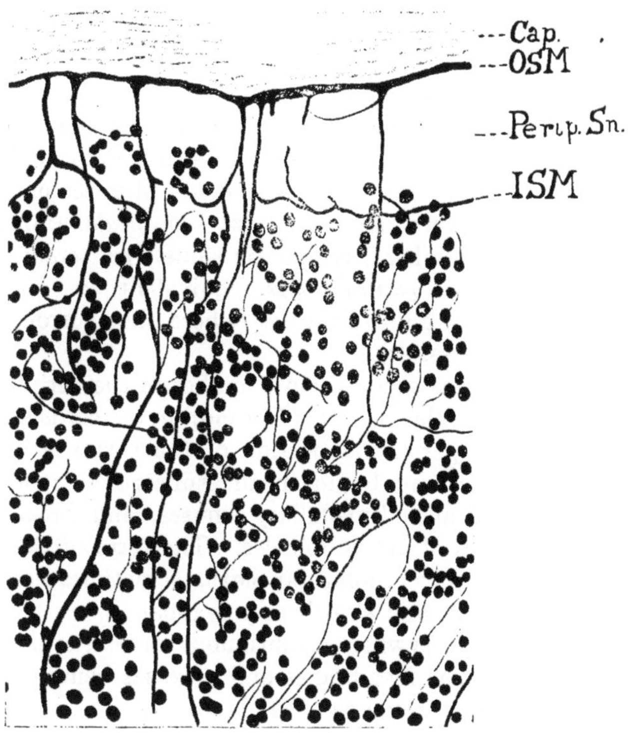

Fig. 10 The reticulum at the periphery of a lymphatic gland of man; the meshes are elongated in a direction perpendicular to the capsule. *Cap.*, capsule of the gland; *ISM*, inner subcapsular membrane; *OSM*, outer subcapsular membrane; *Perip. Sn.*, peripheral sinus. × 625, reduced ⅓ in reproduction. Camera lucida.

flow; this is most noticeable at the periphery of the cortex where beneath the subcapsular membranes the meshes are elongated in a direction perpendicular to the surface (fig. 10). This arrangement is confined to the intervals between the cortical follicles for wherever the surface of a follicle comes into contact with the inner subcapsular membrane of the peripheral sinus the meshes of the perifollicular plexus appear to be flattened against the reticular membrane and the network of reticulum thus rendered more dense than elsewhere.

SUMMARY

The lymphatic glands are everywhere pervaded by a reticulum which blackens with silver impregnation by the method of Bielschowsky. This reticulum is condensed to form more or less complete reticular membranes bounding both cortical and medullary sinuses. Hence one finds these dense plexuses or reticular membranes in several locations, viz., *subcapsular* (inner and outer wall of the peripheral lymphatic sinus), *pericordal* about the medullary cords, *peritrabecular* (equivalent to the 'pericordonal' of Balabio) and *adventitial* (where they surround the small naked blood-vessels).

The reticulum forms a net of close meshes in the periphery of the follicles (perifollicular plexus of Ciaccio), in the cords a cordal plexus, and in the peripheral portion of the cortex a cortical plexus. The meshes are broad in the subcortical zone, still broader in the cortical and medullary sinuses, and broadest in the germinal centers of active follicles. The form of the mesh is lozenge-shaped in the perifollicular plexus, sometimes elongated in the cortical plexus, and elsewhere polygonal.

The reticulum is everywhere a continuous mass of stroma, the fibres of both reticular membranes and plexuses being continuous with each other throughout the gland.

Reprinted from THE AMERICAN JOURNAL OF ANATOMY, VOL. 12, NO. 3
November, 1911

THE APPLICATION OF THE SILVER IMPREGNATION METHOD OF BIELSCHOWSKY TO RETICULAR AND OTHER CONNECTIVE TISSUES

1. THE MATURE TISSUES

J. S. FERGUSON

From Cornell University Medical College, New York City

THIRTEEN FIGURES

The minute structure of true reticular tissue (reticulum, Mall) has long been a matter of more or less controversy. Kölliker and his followers regarded it as composed of branching, anastomosing cells, while Bizzerzo, recognizing and laying more stress upon its fibrous character considered it composed of bundles of fine fibers to which the fixed connective tissue cells are closely applied. As Mall ('96) has pointed out it makes but little difference whether the fibers are within or without the cells provided we understand what is the precise relation.

A few fundamental facts of structure may be taken for granted. That reticulum contains both fibers and fixed connective tissue cells is obvious. That it is more or less infiltrated by leucocytes is well known. That its anastomosing elements are frequently continuous with bundles of white fibers may be readily observed in any section of lymphoid tissue. The points of divergence appear when one attempts to determine the relation: (1) of fixed connective tissue cells to reticular fibers ('Gitterfasern'), (2) of reticular fibers to elastic fibers, (3) of reticular fibers to the white fibers of collaginous tissue, (4) of 'fixed' to 'wandering' cells. The present paper is concerned with the attempt to throw some light upon the basic problems related to the first three of these questions.

The early studies of Kölliker, Ranvier and others, were largely conducted upon teased and unstained reticulum or upon lymphoid tissue from which the lymphocytes had been washed out by various methods. They were followed by the employment of the more recent dye reactions by which the recognition of the fibrillar character of the tissue is rendered somewhat more apparent. It was not until the more exact methods of chemistry and microchemistry were applied by Siegfried, Mall and others that a fairly clear perception of the exact structural relation of the reticular tissue began to be apparent. The introduction of new methods often renders plain certain hitherto obscure facts. This is especially true as a result of the silver impregnation methods of Bielschowsky ('04) when applied to various connective tissues. Even the finer fibers, which are more or less obscure after preparation with other methods, stand out clearly in these preparations and one is thus enabled to draw sharper distinctions than is otherwise possible. It is with the application of this method, and its modification by Maresh ('05), that my observations were largely made; the results have been confirmed by comparison with consecutive serial sections stained by well known methods, chiefly depending on haematoxylin and eosin, Mallory's fibroglia stain, and the combination of haematoxylin with the Weigert and Van Giesen stains described in my Textbook of Histology ('05). The tissues studied have been lymphatic glands, spleen, tonsil, thymus, the lymphoid tissues of the digestive and respiratory tracts, the skin and various other tissues being used for comparison. The material was obtained from man, pig, dog, cat, rabbit, ox, sheep, calf and fish. It was fixed by the various methods in common use and was both mature and embryonal.

The method of impregnation which I have followed has been a variation of the rapid modification described by Maresh ('05). With individual exceptions I have gotten uniformly good results after all the methods of fixation used. The method was applied as follows:

1. Sections cut in paraffin were fixed on the slide and placed 12 to 24 hours in a 2 per cent solution of silver nitrate.

2. Transfer for 15 to 30 minutes to freshly prepared alkaline silver solution (20 cc. of 2 per cent silver nitrate to which are added

three drops of 40 per cent caustic soda and the precipitate redissolved by adding ammonia drop by drop while stirring).

3. Rinse quickly in distilled water and place in 20 per cent formalin for three minutes or till the sections are black.

4. Wash in distilled water and place for ten minutes in an acid gold-bath (10 cc. distilled water to which are added 2 to 3 drops of 1 per cent gold chlorid and 2 to 3 drops of glacial acetic acid).

5. Immerse in 5 per cent hyposulfite of soda ½-1 minute to remove all unreduced silver.

6. Wash in distilled water, dehydrate, clear in xylol and mount in balsam.

Bielschowsky advised leaving tissues in the 20 per cent formalin for 12 to 24 hours but as Maresh has shown this seems to be unnecessarily long, at least when sections are used. To still further shorten the process Woglom ('09, '10) has advised, for the purpose of preventing shrinkage of the tissue, that the initial immersion in 2 per cent silver nitrate solution need not exceed 5 minutes. I have, however, found this time entirely too short in many cases and its use led to much confusion in the interpretation of my early results. In sections insufficiently impregnated the contrast between collaginous and reticular fibers was not sharp, either the reticulum taking a brown instead of a proper black color in lightly toned preparations or if the toning was intensified many of the collaginous fibers became a greyish black instead of the proper golden brown. Moreover, I did not find troublesome shrinkage of tissue in well fixed preparations. I would therefore advise a strict adherence to the 12–24 hour period of immersion recommended by Bielschowsky and Pollock ('04), Levi ('06) and others, rather than the shorter period advocated by Woglom and, with reservations, by Maresh ('05).

In the lymphoid tissues the impregnation brings out most distinctly the reticular fibers; they take on a deep opaque black and stand out prominently against the golden brown of the collaginous fibers. The method has already been applied to connective tissues and the assumption of a more or less specific staining property for reticulum in the liver, lymphatic glands, tonsil, spleen, ovary, and pleura has been casually recorded by Maresh ('05), Ciaccio ('07), Alagna ('08), Balabio ('08), Cesa-Bianchi

('08), and Favaro ('09). Realizing the uncertainty which attends the use of various silver methods one readily appreciates the necessity for careful study of the effects of the method upon the various tissue elements. Of the primary tissues epithelium and other cells are scarcely if at all colored or have a faint brownish tint; red blood cells darken readily and are either opaque black or a deep brown according to the depth of the impregnation and the duration of the toning bath; blood serum and intercellular cement substance blacken, the latter appearing very granular; all nuclei are an intense black, the silver reacting specially to the chromatic portions, viz., nuclear wall, chromatin net and karyosomes; the axis cylinders of nerves are somewhat blackened, though with the method employed the neuraxes are not nearly so opaque as the fibers of reticular tissue. Muscle fibers blacken irregularly depending on the depth of impregnation and they show something of their fibrillar structure; the cross striations and, in smooth muscle, the myofibrillae and intercellular bridges appear beautifully shown in certain instances but it is possible with care to have the muscle almost colorless and the reticular fibers an opaque black. The silver apparently adsorbs somewhat to the surface of muscle cells and elastic fibers and thus frequently fills the interstices between fibers, forming an apparent interfibrillar network in smooth muscle, in epithelium and in dense elastic tissue, e.g., ligamentum nuchae. It is possibly this which accounts for the apparent blackening of intercellular substance. Both because of their reactions and because of their characteristic differences of structure one has little difficulty in differentiating these tissues after impregnation and distinguishing them from the various types of connective tissue fibers.

Before we can regard the method of Bielschowsky, applied to tissues outside of the nervous system, as a specific stain for reticulum, it is necessary to examine more carefully than has been done, into the reaction of connective tissue fibers to the silver impregnation and the differentiation, in sections prepared by this method, of the collaginous, elastic, fibroglia and reticular fibers. Of the fibrous tissues cartilage, bone, and dentine may be set aside because of their characteristic structure, obvious at a glance, though one sometimes encounters difficulties in the transition from the fibrous perichondrium to the cartilage matrix.

The distribution of the fibers blackened by silver is so extensive that one is tempted to question the selective action of the method. They are encountered in all the lymphoid organs, in the mucosa of the digestive tract, in and about the walls of the lymphatics and blood-vessels, in the framework of all the secreting glands, e.g., liver, salivary glands, pancreas, mucous glands of the respiratory and digestive tracts, in the kidneys, ovary, uterine wall, testis, prostate, sweat glands, in the corium of the skin, in the tunicae propriae of the respiratory and digestive apparatus, about the glands of the gastro-intestinal mucosa, and to a limited extent among the fibers of areolar and collaginous tissue wherever it is found. Many of the basement membranes consist largely of these argentiferous fibers.

If one is careful not to overtone the specimens the results in mature tissues are fairly constant; in such preparations (with a few reservations) the method appears quite definitely selective for the blackened fibers of reticular tissue ("reticulum." Mall), the elastic and fibroglia fibers remain colorless and the collaginous fibers assume a brownish tint.

This result was arrived at by a careful comparison of the effect of this and other stains upon the several tissues in locations where each is known to occur. The conclusions are based on the following observations.

A. ELASTIC FIBERS

Sections of the ligamentum nuchae after impregnation show the elastic fibers absolutely colorless and outlined by an intense black fibrous mass which occupies nearly the entire non-vascular area between the elastic fibers, and which at the borders of the elastic bundles shades into the golden brown of the collaginous fibers forming the coarse bands of the framework (fig. 1). In the controls the elastic fibers show the characteristic staining reaction with hematein and eosin and with Van Giesen's stain, and the intervening tissue is colored red by eosin and by acid fuchsin.

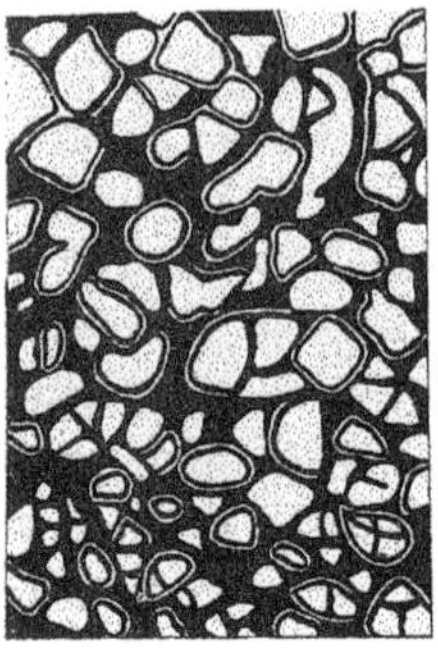

1

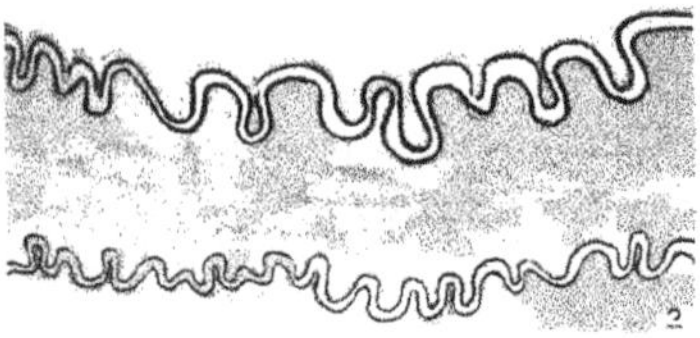

2

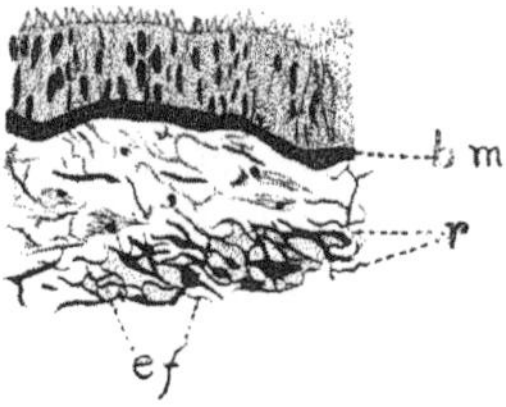

3

Fig. 1 Transection of the ligamentum nuchae of an ox. Bielschowsky stain. The elastic fibers are colorless, the intervening tissue opaque. Camera lucida; occ. 1, obj. $\frac{1}{6}$.

Fig. 2 A pulmonary artery from the human lung showing the colorless internal and external elastic membranes invested by blackened reticulum. Bielschowsky stain. Camera lucida; occ. 1, obj. $\frac{1}{6}$.

Fig. 3 From a primary bronchus of man. Basement membrane (*bm*) is a dense opaque black, its fibers so closely packed as to be indistinguishable. The elastic fibers (*ef*) are colorless and are invested by a black reticulum (*r*). Bielschowsky stain. Drawn with the Edinger projection apparatus, × 255.

In the fenestrated coats of the arteries one sees in the larger arteries of the lungs, stomach, lymphatic glands and many other organs vessels showing in the controls the typical elastic membrane in the tunica intima completely encircling the vessels in transections. The impregnated specimens, when the same vessel is examined in adjacent slides in the series, show the internal elastic membrane colorless, the elastic coat in the larger vessels being invested on either surface with a close net of black reticular fibers (fig. 2). This investment of the elastic tissue by reticular fibers is readily observed and is most remarkable. Similar, though less numerous fibers are seen investing the elastic tissue in the intermuscular spaces of the tunica media, and in the tunica adventitia.

In the smaller arteries and in the small and medium veins the coat of Henle is so thin, and often incomplete, that it is more difficult to determine that the elastic fibers are colorless as distinguished from the blackened reticulum but in view of the constant and obvious condition in the larger vessels one is warranted in assuming that the elastic fibers in the smaller vessels, as in the larger, are colorless and that it is the reticulum, when present, which blackens. The intimate clothing of elastic fibers by reticulum, readily observed in the larger vessels, accounts for the occasional appearance of blackened fibers in the position of Henle's coat in vessels so small as to possess only an incomplete internal elastic membrane.

In the basement membranes of the bronchii one finds only argentiferous reticular fibers. In the larger bronchi the basement membrane is specially distinct and consists of a dense, closely packed mesh of blackened reticular fibers (fig. 3), forming a complete membranous investment continuous with the reticular fibers of the tunica propria and supporting the epithelium. With the Weigert-elastic picro-fuchsin stain the argentiferous fibers take a red color.

In the tunica propria of the trachea and bronchi are large bundles of longitudinal elastic fibers. These fibers remain colorless in the Bielschowsky sections even when the stain has been made so intense as to darken to a considerable extent the collaginous fibers and the muscles. One finds each elastic fiber invested by a distinct coat of blackened reticular fibers forming an intricate net. If one selects a known and readily recognized point for study, consecutive sections stained by different methods show the broad lines of elastic fibers, which in the Bielschowsky sections are colorless, to be flanked on every surface by a blackened reticulum, but clothed in the Weigert-elastic picro-fuchsin section by fuchsin stained fibers. With haematoxylin and eosin the whole breadth of the basement membrane and both elastic and argentiferous fibers in the tunica propria take the characteristic eosin tint, and reticular and elastic fibrils are almost indistinguishable.

Thus wherever the recognition of unquestionable elastic fibers can be made with certainty they are found uncolored by the silver, while giving characteristic reactions to other stains. It is only in those locations where identification of elastic fibers is questionable that one is inclined to suggest their identity with the blackened fibrils, but even then one sees indications of noticeable difference. If a bit of areolar tissue is carefully examined in the wall of the digestive tract, the skin, the peritoneum or elsewhere, and sections of the tissue are also stained by the selective elastic tissue stains, Weigert's, or Unna's, one observes on comparison a difference in the number and arrangement of fibers selected by the compared methods. The orcein and Weigert sections will compare very closely. The Bielschowsky sections of the same series frequently show many more fibers of the blackened type; moreover the blackened or reticular fibers are usually more wavy and of irregular distribution, often having a typical spiral appearance as compared with the relatively straight elastic fibers. This is well shown in sections of collapsed or undistended lung in which the elastic fibers of the alveolar walls and bronchioles are straight while the reticular fibers, from the extreme contraction of the organ, are thrown into a remarkably intricate network of wavy and twisted fibrils, equally as distinct and more abundant than the elastic.

These findings are in confirmation of the views already expressed by Woglom ('10) and others and appear to prove conclusively the non-identity of elastic fibers with those fibers (reticulum) which blacken in these preparations. This view is in accord with that expressed by Mall ('02), who as a result of his comparison of the tissues by chemical methods likewise demonstrated the non-identity of elastic fibers and reticulum, but his studies of mesenchymal tissues showing embryonic stages of the connective tissue, resulted in pictures delineating the first appearance of elastic fibrils which simulate those which I have obtained in similar tissues by the method of impregnation (fig. 4). I shall consider this phase of the subject in a later paper. At this time it is sufficient to say that I consider the fibers referred to to be collaginous in type. One must therefore finally emphasize the fact that in well recognized portions of mature tissues elastic fibers in the silvered preparations remain entirely colorless while reticular fibers

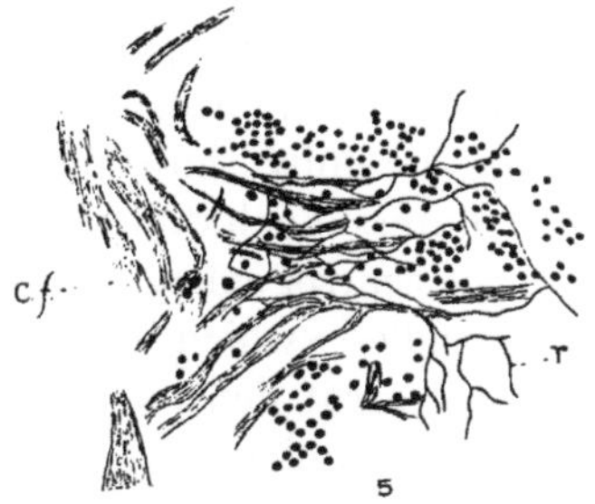

5

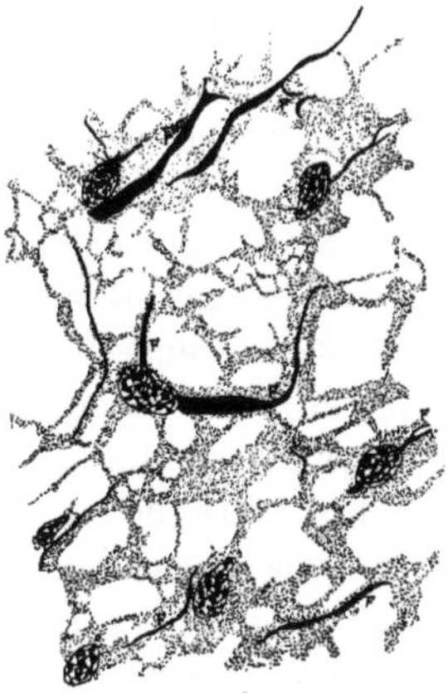

4

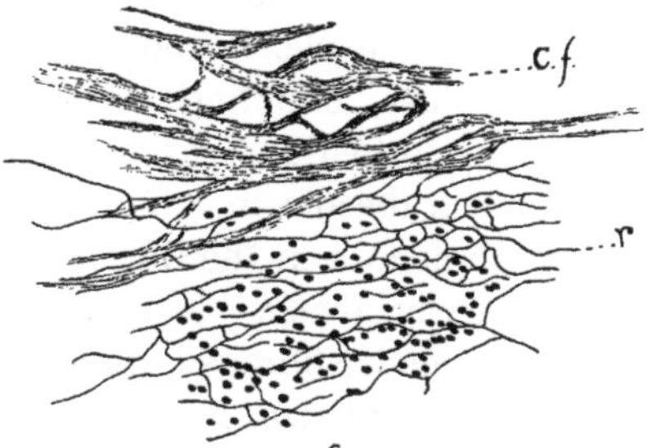

6

Fig. 4 From the subectodermal mesenchyme (corium) of a fetal pig of about 80 mm. neck-breech length, showing blackened fibrils (F) bearing intimate relations to the mesenchymal cells. Bielschowsky stain. Camera lucida; occ. 1, obj. $\frac{1}{12}$ hom. im.

Fig. 5 Reticular and collaginous tissue at the periphery of a lymphatic gland. Observe the sharp outlines of the fine black reticular fibrils (*r*) which intermingle with the bundles of collaginous fibers (*cf*) in a trabeculum. Bielschowsky stain. Camera lucida; occ. 1, obj. $\frac{1}{6}$.

Fig. 6 The figure exhibits the relation between the collaginous (*cf*) and reticular (*r*) fibers in the region adjoining a nodule of a patch of Peyer in the small intestine of man. Note the interlacing of fibers at the border of the lymphoid tissue. Bielschowsky stain. Camera lucida; occ. 1, obj. $\frac{1}{6}$.

blacken; one is justified in assuming that this reaction to the Bielschowsky method, with reasonable care, is constant.

B. COLLAGINOUS FIBERS

That the typical reticular fibers of lymphoid tissue and the typical collaginous fibers of dense fibrous tissue take on different colors after the silver impregnation has been generally recognized, at least by the Italian writers, Levi ('06), Ciaccio

('07), Cesa-Bianchi ('08), Balabio ('08), Alagno ('08). The reticular fibers assume a dense opaque black while the collaginous fibers take on a golden brown when well differentiated. Yet if one examines carefully those points at which the two tissues blend one encounters much difficulty in determining whether the black color of the coarser collaginous bundles is due to the opacity of the brown bundles—which are of considerable size and thickness and often of great density—or to the presence within the coarse collaginous bundles of finer, blackened, reticular fibers. In some locations the latter relation is apparent. For example, in the perifollicular plexus about the lymphatic follicles, described by Ciaccio ('07), one finds the characteristic lozenge shaped meshes of the 'reticulum' extending into the adjacent collaginous tissue of the trabecula in lymphoid organs or of the tunica propria and submucosa in the digestive tract, but there the reticular fibers are nearly always clear and sharp among the collaginous fibers of the smaller fibrous bundles (figs. 5, 6, 7, and 13). As the bundles increase in size, however, the difficulty of distinguishing the exact outlines of the two types of fibers increases.

Another difficulty in the way of exact and positive differentiation is the variable result of silver impregnations. With varying degrees of impregnation, reduction, and toning the collaginous fibers may lose their typical golden brown and acquire an increasingly opaque condition. This is specially prone to occur if the sections are overtoned in the gold chlorid bath. One halts, therefore, between the idea of similarity if not positive identity of collaginous fibers and "reticulum" and the opinion of Mall ('01) which regards reticulum as an independent tissue, distinctly differentiated from the collaginous by its somewhat different chemical reactions, a view not fully accepted by Studnicka ('03) nor yet generally adopted by German authors (Fürbringer, '09). Yet if one uses care with the silver process one can obtain from nearly all tissues quite distinctive preparations. Thus in the lung the fibrous tissue of the pleura, as shown by Favaro ('09), as well as that of the "interlobular septa" appears to be formed by golden brown fibers arranged in bundles having the characteristic wavy course together with but few intermingled black reticular fibers, whereas the reticulum in the walls of the alveoli and smaller bronchi, though often composed of coarse typically spiral fibrils, forms an interlacing mass of discreet fibers, or fiber bundles, among which a limited proportion of finer bundles of brownish collaginous fibers may be recognized. In the vascular trabecula of the spleen (fig. 7) the collaginous fibers of the blood-vessels acquire a typical brown while the close network of reticular fibers take on an intense black and have a characteristic, either somewhat regularly spiral, or a reticular course, very different from the irregularly wavy collaginous fibers.

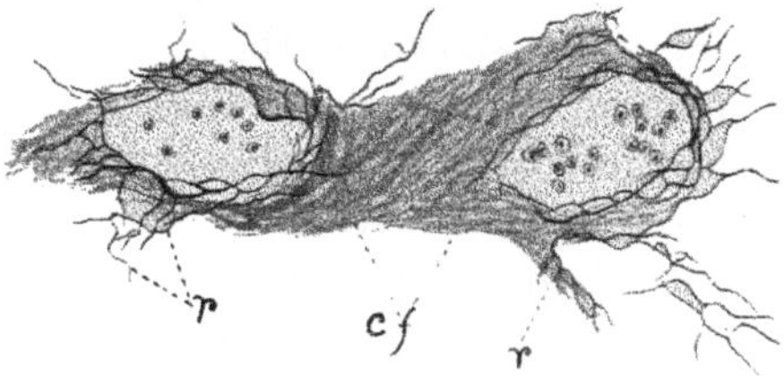

Fig. 7 A vascular trabeculum of a child's spleen. The blackened fibers of reticulum (*r*) show clearly in contrast to the collaginous fibers (*cf*) which in the section are a golden brown. The reticulum surrounds the vessels and is continuous with that of the splenic pulp. Bielschowsky stain Drawn with Edinger projection apparatus, × 255.

In the trabecula of the lymphatic glands the distribution is not so apparent, the collaginous and the reticular fibers pursuing somewhat similar courses, though the latter are apt to be more distinctly spirillar. From careful examination I am led to believe that the relation simulates, in reverse, that already described (see fig. 2) for the elastic fibers, in that it would appear with considerable certainty in many places that the black reticular fibers are invested or enveloped by a sleeve or coat of collaginous fibrils, so that the latter fibrils consequently assume a spiral course corresponding closely to that of the reticular fibrils. Indications of a similar investment of the reticular fibrils can be found wherever reticulum occurs, but it is not always possible to distinguish with certainty between the collaginous fibers and the protoplasm of mesenchymal or fixed connective tissue cells.

Again in the fibrous perichondrium of hyaline cartilage, as Studnička ('06) has pointed out, there is a considerable layer of blackened fibers marking the border of the cartilage and in the younger types extending into its matrix; the outer layers of the perichondrium are clearly, however, collaginous tissue, and in my preparations present the characteristic golden brown color, sharply distinguished from the intense black of the argentiferous fibers. The matrix of the cartilage in the same sections retains a brownish tint except in the younger specimens and at the margins of the cartilaginous plates in the more mature cases. The blackening of the innermost fibers of the perichondrium which mark the "growing surface" of the cartilage may be explained by the increased affinity for silver shown by the fibers of young connective tissue as compared with the mature, a relation which I am not ready to discuss further at this time.

The remarkable differences in reaction to the impregnation in many of the mature tissues, especially such as contain typical reticulum, would tend to refute the German idea of the identity of collaginous and reticular tissue and to confirm the opinion of Mall that reticular tissue or "reticulum" is a distinct entity, though this latter contention cannot yet be established from the standpoint of the method here used until it is viewed in the light of the histogenesis of the connective tissues, for there the sharp lines of demarcation diminish even to the vanishing point.

Such characteristic differences between reticulum and collaginous fibers as may be observed at almost any point in thin sections of the lymphoid tissues impregnated by silver leave little to be desired in the way of morphological differentiation of these two types of fibers. Such areas are well and accurately shown in fig. 5, from the lymphatic gland of man, and fig. 6 from the margin of a Peyer's patch in the human intestine. One feels, therefore, that the separate and distinct character of collaginous and reticular fibrils in the mature tissues as shown by silver impregnations, fortified as it is by the chemical differences demonstrated by Siegfried and Mall, the one, collaginous, yielding gelatin, the other yielding a "reticulin" presenting different chemical reactions, forms at least a satisfactory working basis for the further study of the distribution of these fibers as shown by the Bielschowsky method, a work already begun by Studnička, Ciaccio, Balabio, Alagna, Favaro, Maresh, Cesa-Bianchi and others.

C. RETICULUM

The careful observation of Bielschowsky preparations also yields valuable data as to the finer structure of reticular tissue and the relation of its fibrils to the "fixed" connective tissue cells.

The coarser fibers of 'reticulum' may be readily seen and, where such fibers come into relation with the "knots" of the reticular net, one can observe these fibrils breaking up into a plexus within, or about, the cells as pointed out by Balabio ('08). Somewhat

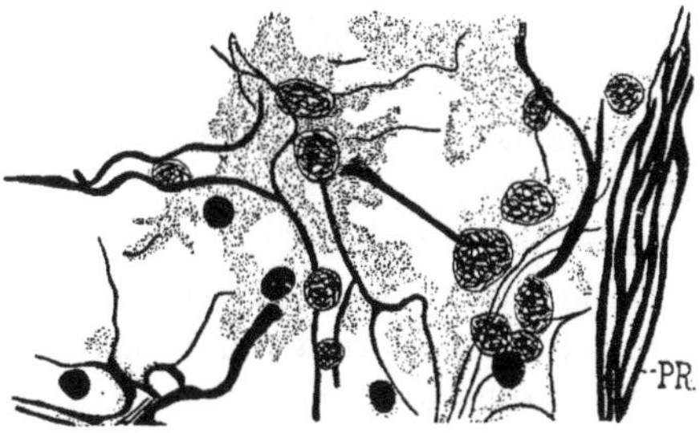

Fig. 8 From a lymphatic gland of man showing the relation of the blackened fibers of reticulum to the branching protoplasm of the fixed connective tissue cells. The small black nuclei are those of lymphocytes. *PR*, perifollicular reticulum. Bielschowsky stain, after stained with acid fuchsin. Camera lucida; occ. 1, obj. $\frac{1}{12}$ hom. im.

of this arrangement is indicated in fig. 8, though in other places the fibers appear to enter the cell and end either abruptly or, more frequently, pass through the cell in close proximity if not in contact with its nucleus. The appearance of abrupt ending might if only occasionally observed, be due to the passage of fibers out of the plane of the section, but it occurs far too often so that this certainly is not always the case. The finer fibrils, as well as many of the coarser ones, appear as single fibrils though because of the complete opacity of the impregnation one cannot say that this is actually the case. Certainly the larger fibers distinctly show indications of fibrillation.

This leads to the question of the relation of reticular fibrils to the "fixed" connective tissue cells. Are the fibers contained within the cells or are they only in surface contact? It seems to me that Mall has given us the key to the situation with his theory of exoplasmic deposit of the fibrils with constant recession of the endoplasm during development. If one regards reticulum as an immature or least differentiated type of connective tissue it is plain to see that the fibers must readily lie now within the cell or endoplasm, and now without the cell, where they are left "high and dry," as it were, by the complete recession of the endoplasm which leaves in mature collaginous tissue only the nucleated cellular remnants. Since certain fibers, or portions of fibers, would thus lie without the anastomosing syncytial mass of endoplasm while certain others would lie quite as plainly within it we have here a possible harmonization of the otherwise conflicting theories of Kölliker and Bizzozero. The facts of the case as I observe them in silver impregnated sections of embryonal as well as mature tissues appear to coincide with this hypothesis.

Ciaccio ('07) attacked this problem casually in connection with his study of the distribution of reticular tissue in the lymphoid follicles of lymphatic glands and observed a relation of contiguity of fibers and cells, the two being independent. Thus, he says, "le fibrille alla loro volta si diramano in tutti i sense e si montrano independenti dalle cellule."

Balabio ('08), cognizant of the work of Ciaccio, approaches the problem circumspectly, and describes the fibers as "superimposed" upon the cells forming a characteristic close and delicate pericellular plexus. He observed that the cellular prolongations "intertwine among" the fibrils but he was not able to determine "with certainty" whether they were superimposed or whether they "anastomosed in the form of a sort of continuous cellular net." He "limits himself," as he says "to emphasize the fact without pronouncing upon the existence or non-existence of true cellular anastomoses." He is inclined to confirm the theory of Bizzozero for he in one place says "Si puo confirmare conscicurezza quanto gia Bizzozero ed altri affermarono che si tratta di rapporti di sola contiguita."

If one assumes that to disprove the theory of Bizzozero one must find the fibrils at all times outside the anastomosing cells, never within, then proof is not forthcoming. On the other hand proof is also lacking if to demonstrate the theory of Kölliker fibers must always be found within the cellular syncytium. But viewing the tissue in the light of its histogenesis, one need not, as pointed out above, be thus limited in either case, for the fibers of reticulum may, according to this interpretation, come to lie now within, now without the syncytial endoplasm. The examination of such appearances as those shown in any of the figs. 8 to 11, which are accurately drawn with high magnification, or yet more

Fig. 9 Reticulum and cells as seen in a thin section through the pulp of the human spleen. No collaginous fibers have been included Fibers and nuclei are black, the cytoplasm is granular. The fibers are surrounded by a halo of cytoplasm especially distinct wherever their cut ends are directed toward the eye of the observer. *Ret*, reticlular fibers; *fc*. cytoplasm: *L*. lymphocyte. Bielschowsky stain. Camera lucida; occ. 1, obj $\frac{1}{12}$ hom im.

truly if one studies the actual preparations, must convince one at a glance that in mature lymphoid tissue the fibrils of the reticulum are not entirely contained within the fixed cells. The burden of proof lies on the other side.

The accurate and sharp delineation possible under high magnification between the opaque black fibrils and the light brown protoplasm of the cells presents appearances in thin sections which seem to me to show unmistakably that some portions of the fibrils are certainly contained within the cytoplasm of fixed connective tissue cells. I do not find any such condition in relation to the lymphocytes which are so numerous in the same vicinity. In fig. 10 the fibrils *a* are, in the case of the lower cell at least, certainly outside the cell at one point, viz., where it ends by passing out of the plane of the section. But at the point *b* each fiber makes a distinct loop which can be followed by change of focus. The granular cytoplasm forms a continuous mass but in the mid-portion of the loop it can be distinctly seen at a level *above* that of the fiber; while at the ends of the loop, in fact at all the solid black portions of the fiber the cytoplasm is distinctly *below* the fiber. It would appear obvious that each fiber has penetrated the cell and must, therefore, during its passage have been found within the cytoplasm. Fig. 10 was drawn from a section of the spleen, but in fig. 11, which is from a lymphatic gland and in which *a* again marks the portion of the fiber above, and *b* that below the cytoplasm, the same condition holds. Such places are extremely abundant, and in thin sections of all the lymphoid tissues examined they can be found with ease, often several in a single field. Again in transections of the coarser fibers, or in oblique sections, the fibers are often seen surrounded on all sides by a light brown halo of cytoplasm. Such appearances are indicated by fig. 9, though it is difficult to depict them accurately even with the aid of the camera lucida because of the extreme fineness of the fibers and the very thin cytoplasmic coat (represented by the stipling) by which they are surrounded. The cut ends of most of the transversely and obliquely cut fibers show the halo of cytoplasm in the actual sections.

The above observations appear to show convincingly that, at least, at times the fibrils lie distinctly within the cells. That they may be so found, as also without the cells, is in harmony with Mall's suggestion as to the ontogenetic relationship of "reticulum" and other connective tissues since he supposes this tissue to represent a less mature type than the collaginous, one in which the primitive relations of endoplasm and exoplasm still persist to a considerable extent. That this is the case is, perhaps, indicated by the fact that in developing mesenchymal tissue one finds fibrils, bearing similar relations to the endoplasm and exoplasm of the connective tissue syncytium, and which like reticulum blacken with the silver impregnation (fig. 4).

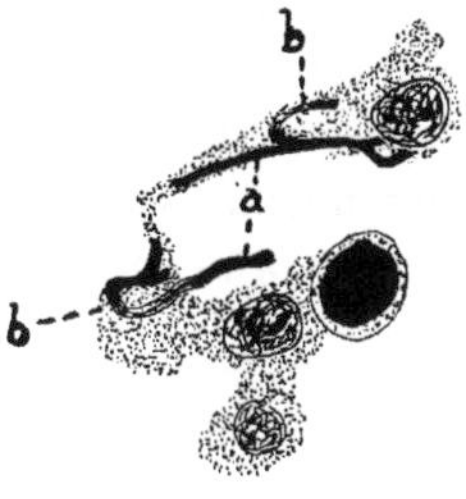

10

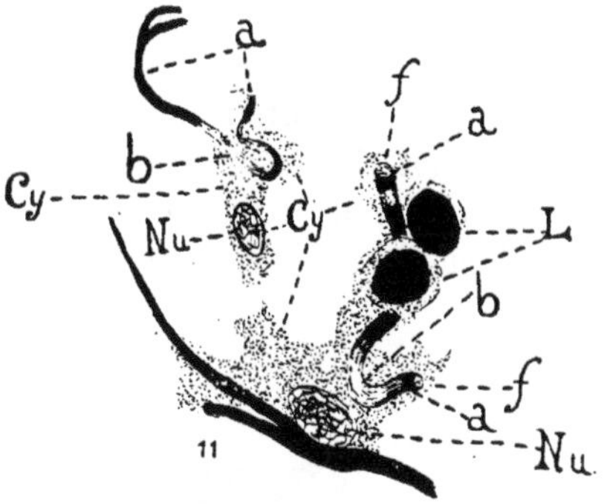

11

Fig. 10 Accurately drawn from a section of the spleen of man, showing the actual course of fibrils of blackened reticulum through the cytoplasm of fixed connective tissue cells. The parts *a* of the fibrils end by making a sharp turn which passes out of the plane of the section. The loops formed at *b* are shaded light, and in the section they lie below the level of the cytoplasm as readily demonstrated by change of focus. The black portions, *a*, are above the level of the cytoplasm. Bielschowsky stain. Camera lucida; occ. 1, obj. $\frac{1}{12}$ hom. im.

Fig. 11 Areas similar to those shown in fig. 10, but drawn from a section of a lymphatic gland of man. Similar appearances were very numerous in this section. *a* and *b*, as in the preceding figure; *cy*, cytoplasm; *f* cut ends of fibrils; *L*, lymphocytes; *Nu*, nucleus of a fixed connective tissue cell; at the top of the figure a fibril forms a U-shaped loop which passes through the cytoplasm of a "fixed" cell, entering from below and coming out above. Two similar fibers are also shown. Bielschowsky stain, after stained with acid fuchsin. Camera lucida; occ. 1, obj. $\frac{1}{12}$ hom. im.

Yet bearing in mind that we are dealing with a method of impregnation only, and are subject to all the limitations of such methods, one is not fully warranted in drawing inferences of chemical similarity between the mesenchymal and the reticular fibrils.

D. FIBROGLIA

The occurrence of fibrils blackening with silver in the mesenchymal cells suggests a possible identity with the fibroglia fibrils of Mallory ('03, '04) for such fibrils were found by that observer to be abundant in developing connective tissue. In order to accurately compare the fibrils shown by these two special methods one must first consider the mature tissues, only thereafter the developing tissues.

In mature tissues Mallory states that fibroglia fibrils "are not very common in normal tissues except possibly in one situation and have to be hunted for with an oil immersion lens." This is certainly not the case with the argentiferous fibers which occur abundantly in a great variety of places among normal tissues and which are of sufficient size to be seen as networks among

12

Fig. 12 The basket-cells of a coil gland of the human finger-tip, darkened by haematoxylin. Haidenhain's iron-haematoxylin. Camera lucida; occ. 1, obj. $\frac{1}{12}$ hom. im.

the other fibers with very low magnification. Again the staining reactions of the two sets of fibrils are different. Mallory describes the basement membranes as the "one situation" where fibroglia fibrils are common in normal tissues, and the subepithelial basket-cells of the sweat glands—regarded by Benda ('93, '94) as muscle cells—as the place where the largest fibroglia fibers occur. As these last fibers can be easily located they form a definite unit for comparison. With Mallory's stain they are red; with iron haematoxylin they blacken when the stain is not too much extracted (fig. 12). Both of these reactions are characteristic for fibroglia, and, as McGill ('08) has shown, they are also characteristic for myoglia fibrils. But with silver impregnation these fibers are not in the least blackened, nor is the thin layer of collaginous tissue upon which they rest. It would therefore appear that the basement membrane of the sweat glands, unlike most other basement membranes, contains no reticulum but is formed by collaginous fibers together with the peculiar basket-cells, be they fibroglia or muscle. If one compares in the same way the reticulum of lymphoid tissues one arrives at similar conclusions as to the non-identity of 'reticulum' (viz. those fibers which blacken with the Bielschowsky method) and fibroglia. It would therefore appear that in the mature tissues there is no identity between fibroglia and reticulum nor for the same reasons can there be between fibroglia and the fibers which blacken with Bielschowsky's stain. These last are identical with certain fibers which are colored blue by Mallory's stain.

SUMMARY

Briefly summing up we find that the Bielschowsky stain applied to the connective tissues of mature individuals exerts a selective

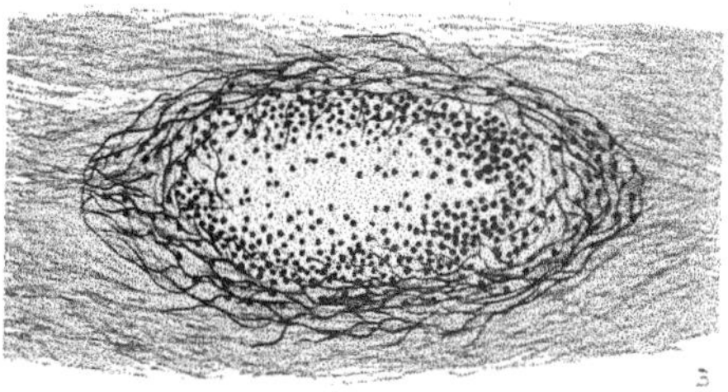

Fig. 13 A small lymphatic nodule from the submucosa of the human esophagus, showing the 'perifollicular plexus' of Ciaccio sharply defined, but with reticular fibers intertwining with the collaginous fibers.

The collaginous tissue is drawn free-hand, the reticulum by camera lucida; occ. 1, obj. $\frac{1}{6}$.

action, blackening certain fibers which are certainly not identical with either elastic or fibroglia fibers, which in many cases certainly are identical with the fibers of reticulum, and which in some cases show a certain tendency suggesting possible transitions between reticular and collaginous fibers. The typical collaginous fibers do not blacken, but take on a characteristic golden brown color; nevertheless, in certain locations and under certain conditions some fibers which we have been accustomed to regard as collaginous, e.g., within dense connective tissue bundles or in embryonic mesencyhme, do blacken somewhat, though never so typically nor with such clear and sharp definition as do the fibers of "reticulum" in mature tissues. The further elucidation of this atypical reaction of the collaginous fibers must be sought in the histogenesis of the connective tissues. For the present we may safely consider the black reaction of fibers of mature connective tissue to the Bielschowsky stain to be distinctive of "reticulum" in all satisfactory preparations, viz. those in which the collaginous fibers assume a golden brown tint.

When such tissues as nerve, muscle and embryonic mesenchyme are excluded the Bielschowsky method serves as a well-nigh specific stain for the reticulum of Mall.

BIBLIOGRAPHY

ALAGNA, G. 1908 Anat. Anz., Bd. 32, p. 178.
BALABIO, R. 1908 Anat. Anz., Bd. 33, p. 135.
BENDA 1893–1894 Dermatol. Zeitschr., Bd. I, p. 94, (Quoted by Mallory).
BIELSCHOWSKY, M. AND POLLACK, B. 1904 Neurol. Centralbl., vol. 23, p. 387.
CESA-BIANCHI, D. 1908 Anat. Anz., Bd. 32, p. 41.
1908 Internat. Monatschr. f. Anat. u. Physiol., Bd. 25, p. 1.
CIACCIO, C. 1907 Anat. Anz., Bd. 31, p. 594.
FAVARO, G. 1909 Internat. Monatschr. f. Anat. u. Physiol., Bd. 26, p. 301.
FERGUSON, J. S. 1905 Normal histology and microscopical anatomy, New York and London, pp. 666–7.
FÜRBRINGER, M. 1909 Gegenbaur, Lehrb. d. Anat., 8 auf., Bd. 1.
LEVI, G. 1907 Monitore zool. ital., 18, 290.
McGILL, C. 1908 Internat. Monatschr. f. Anat. u. Physiol., Bd. 25, 90.
MALL, F. 1896 J. Hop. Hosp. Rep., vol. 1, p. 171
1902 Amer. Jour. Anat., vol. 1, p. 329.
MALLORY, F. 1903–1904 Jour. Med. Research, vol. x, p. 334.
MARESCH, R. 1905 Centralbl. f. allg. Pathol., Bd. 14, 641.
SIEGFRIED, M. 1892 Habilitationsschrift, Leipzig, (quoted by Mall).
1902 J. of Physiol., 28, 319.
STUDNIČKA 1903 Anat. Hefte, Bd. 21, 279.
1906 Zeitschr. f. wis. Mik., Bd. 23, 414.
WOGLOM, W. H. 1909–10 Proc. N. Y. Path. Soc., vol. 9, p. 146.

[Reprinted from BIOLOGICAL BULLETIN, Vol. XXI., No. 4. September, 1911.

A PRELIMINARY NOTE ON THE RELATION OF NORMAL LIVING CELLS TO THE EXISTING THEORIES OF THE HISTOGENESIS OF CONNECTIVE TISSUE.

JEREMIAH S. FERGUSON, M.S., M.D.,

ASSISTANT PROFESSOR OF HISTOLOGY, CORNELL UNIVERSITY MEDICAL COLLEGE, NEW YORK CITY.

The connective issue of the adult is in the embryo derived from the mesenchyme. The same is true of certain other tissues, endothelium, muscle, and cartilage, bone and dentine, if these last be not included within the scope of the term connective tissue. It is in this narrower sense that the term connective tissue is herein used.

In considering the origin of connective tissue one has to take into account two distinct elements, cells and fibers. Of the cells there are at various stages in the embryonic mesenchyme three distinct types: (*a*) the wandering cells, viz., leucocytes, which are more directly concerned with the processes of hæmatopoiesis; (*b*) those cells which give rise to the true wandering connective tissue cells of the mature tissues, the mast cells, plasma cells, etc., whose history is more or less closely concerned with that of the embryonic leucocytes; (*c*) those typical connective tissue cells which are related to the fibers and which are commonly known as the "fixed" connective tissue cells. It is without doubt only the last type which is concerned with the origin of the connective tissue fibers, hence the present note deals only with this type of cell.

At least two types of fibers have also to be considered as arising in the connective tissue mesenchyme, the elastic and the collaginous. We are here concerned only with the latter type, for the reason that the elastic fibers arise at a later period; thus the origin of the connective tissue fibers concerns primarily the collaginous, or so-called "white fibers." It is unnecessary to distinguish at the early stage under consideration between such varieties of collaginous fibers as "reticulum" and "fibroglia," for the former undoubtedly arises in exactly the same manner as the ordinary "white fibers," and of the histogenesis of the làtter little or nothing is known if, indeed, it can be considered as an entity distinct from the other connective tissue fibers.

The numerous theories which have been proposed to account for the origin of the connective tissue fibers may be reduced to three chief divisions:

1. The embryonic connective tissue or mesenchymal cells become directly transformed by elongation into connective tissue fibers.

2. The fibers arise in the ground substance between the cells either by transformation of that ground substance or as a secretion from the adjacent cells.

3. The fibers arise within the cells either as distinct fibers (Schwann, 1839), as granules which fuse to form fibers (Spuler, 1896; Lavini, 1909), as an epicellular protoplasmic fibrous layer (Lwoff, 1889), or as an ectoplasm about the cell (Hansen, 1899) or forming the syncytium (Mall, 1902).

The theory of direct transformation by elongation of cells into fibers may well be abandoned and, except it be construed along the lines of the various theories of intracellular origin, it is worthy only of passing notice. Among other things the existence of an overwhelming number of fibers relative to the number of cells present in connective tissue argues against direct transformation, and modern microscopical methods are not able to detect the phases of such transformation either in normal growing tissues or in the process of wound repair. For several decades the results of observations have been in support of either the intercellular or the intracellular group of theories.

To Henle (1841) we owe the theory of the intercellular (extracellular) origin of connective tissue fibers, the fibers being supposed to arise from the intercellular ground substance, not directly from the cells. Latterly this theory has received but little support unless it be from the observations of Merkel (1895), who found that in the umbilical cord fibers appear first to arise in locations relatively remote from the cells. Later the cells wandered into these fibrous areas, and at a still later period the fibers had so increased in numbers that they necessarily laid in relation to the cells.

The weak point in Merkel's deduction appears to be the disregard of the locomotive properties of mesenchymal or young connective tissue cells. If these cells, except for the incidents of mitosis, are stationary, then Merkel's observations would render conclusive evidence to support the theory of the extracellular origin of fibers. But that such cells possess at least a limited power of motion simulating the amœboid character has long been known. An adequate theory must account for the activity of these cells. Merkel's theory presupposes them to be stationary till fibers have appeared and only later to acquire amoeboid characters.

In searching for a tissue which would offer opportunity for the study of the activities of connective tissue cells in the living

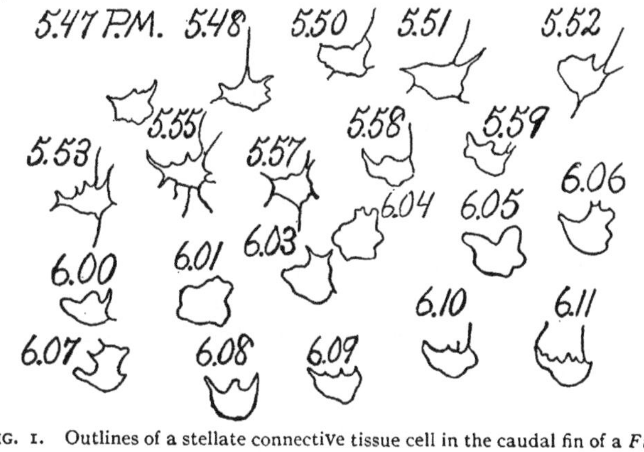

FIG. 1. Outlines of a stellate connective tissue cell in the caudal fin of a *Fundulus* embryo 4.5 mm. long (one day after hatching). The figures record the moment of completion of each drawing, the whole observation extending over a period of 24 minutes. The fish was afterward returned to water and swam for two days before being killed for further histological use. The cell observed corresponds in appearance with that shown by W. Flemming (*Arch. f. Anat.*, 1897, Fig. 9. Plate VI.) in the mesentery of a *Salamander* larva. It shows active amœboid motion and extensive changes of form. Its locomotion was not recorded, it was quite limited. × 800, camera lucida.

animal under normal conditions, I have found that the median fin of young fish embryos affords a subject for study which throws some light on Merkel's observations.

In a *Fundulus* embryo of 6 mm. total length—several days after hatching—I find in the median fin between the double layer of cutaneous epithelium the first signs of the dermal fin-rays. Blood-vessels have not entered, and between the fin-rays are few if any connective tissue fibers, certainly none in the distal portion of the fin. nvading the base of the fin is a mass of mesenchymal cells, mostly of the round cell type. Scattered through the fin are, here and there, isolated connective tissue cells in very limited numbers, distributed over an area representing the proximal one half or two thirds of the fin, and forming a sort of skirmish line, as it were, in advance of the army of round cells. These stellate cells can often be distinctly seen in the living animal. Later, connective tissue cells in abundance wander into the fin and fibers appear. But it cannot be said that the fibers appear in advance of the cells, for the "skirmish line" precedes the appearance of fibers, these advance cells becoming later intermingled with those of the subsequent invasion.

The advance cells, like all other stellate connective tissue cells in the fins of embryo fishes, I find to be actively amœboid, capable not only of motion but to some extent of locomotion. Since these cells are travelling through an area in which fibers are only just beginning to appear, the location of early fibers at points relatively distant from the cell in "fixed" tissue at once loses its significance and thus robs the theory of the extracellular origin of fibers of one of its strongest supports. My observations may be readily verified on any living, free-swimming *Fundulus* embryo within the first week after hatching.

Connective tissue fibers may be supposed to arise as a product of secretion of the mesenchymal cells, the extracellular deposition of fibers thus occurring under the influence of the cells. This is but a slight deviation from the strict extracellular origin presupposed by the theory of Henle and Kölliker. Such secretion must appear either in fluid form or as granules. In either case it is difficult to conceive why the secretion should take on a linear form if the cells remain stationary, and still more difficult if the cells are in motion, unless that motion be extremely slow and in one direction only.

By observation of the fins of embryo fish, chiefly *Fundulus*, I have been able to observe that the spindle cells of embryonic connective tissue possess locomotion and move largely in one direction, but their motion and locomotion is extremely active, and so far as I can observe in the living animal they leave behind no trail of secretion, certainly no observable granules: still more important, the fibers appear to have been formed in great numbers prior to the appearance of the type of spindle cells in any considerable proportion relative to the other types of cells, round and stellate, already present in the primitive connective tissue. Hence I cannot conceive of the spindle cells as producing fibers by direct secretion.

But it is at the time of the predominance of stellate cells that the fibers make their first appearance in the fish's fin, as is likewise the case in the tissues studied by other observers. These stellate cells rapidly change their form, throwing out and retracting their processes in rapid succession, as can be seen in any living, free-swimming *Fundulus* embryo under 25 mm. total length. Fig. 1 shows the outline of such an early connective tissue cell in the caudal fin, its changes of form at intervals for a period of twenty-four minutes being accurately depicted with the aid of the camera lucida. It is impossible to both watch and draw all the changes in form occurring during this period, but sufficient are shown to demonstrate very active amœboid motion. By observing these cells in relation to a relatively fixed object, *e. g.*, a chromatophore, or a joint in an adjacent fin-ray, I have been able to detect in them considerable locomotion. I find that, unlike the spindle cells, the stellate cells appear to travel equally well in all directions. Neither following their locomotion nor in trail of a retracted process do I find any indication of secreted granules, nor of anything other than the most delicate, preformed fibers. One would suspect that if new fibers were being formed by a process of secretion they would appear within twenty-four minutes after such a cell or its process had occupied a given position in which fiber should reasonably be expected to be formed. Though I have repeated the experiment many times I have failed to observe such evidence of secretion, and these results do not appear to be consistent with a theory of direct secretion of fibers by the primitive connective tissue cells.

The intracellular theories of the origin of connective tissue fibers may be divided into two classes, those in which the fibers are presumed to arise in the peripheral portion of the cell, as proposed by Lwoff (1889), or in a modified peripheral portion of the cellular or syncytial protoplasm, ectoplasm or exoplasm, as proposed by Hansen (1899) and afterwards by Mall (1902), and those in which the fibers are presumed to have a direct intracellular origin as observed by Spuler (1896) and later by Livini (1909).

The "circumcellular" origin of fibers proposed by Lwoff, since it must take into account the separation of fibers from the cells as well as their origin from intracellular granules, does not fundamentally differ from the later ectoplasmic modifications

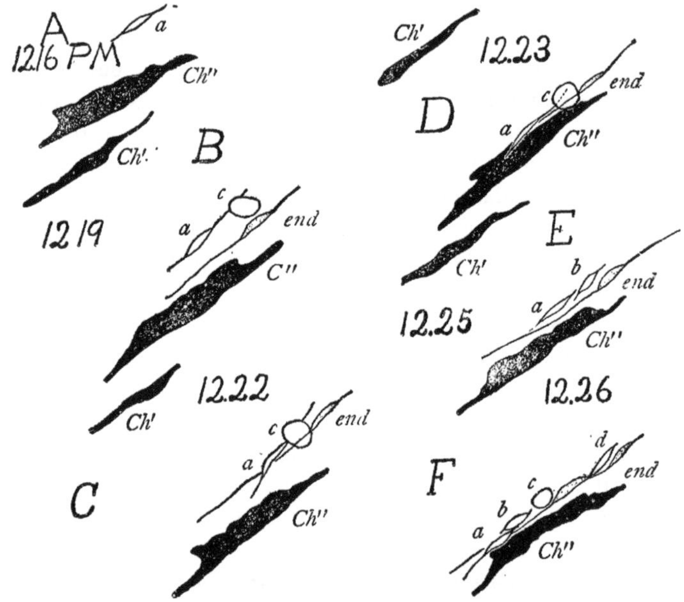

FIG. 2. A record of the locomotion of a spindle connective tissue cell, *a*, for a period of 10 minutes, adjacent chromatophores bordering on the fin-rays on either side being taken as the relatively fixed points. The cell was found to have covered a distance of about 50μ, a rate of 1μ in every 12 seconds. *a–d*, connective tissue cells; *c*, a round cell, the others spindle shaped; *Ch'*, *Ch''*, two chromatophores; *end*, endothelium of an afferent blood-vessel.

of Hansen and Mall. These later authors, basing their investigation on "fixed" material, presuppose a constant but gradual recession of the endoplasmic area with coincident changes in the exoplasm in order to account for the transference of fibers from a position within the cell, in the ectoplasm, to an extracellular position. This hypothesis is entirely harmonious with the appearances to be observed in "fixed" tissue with the methods used. However, it is open to two objections, viz., the observations of Spuler, confirmed in a casual and not wholly satisfactory way by Livini, that fibers actually lie within the cells, and not always at the periphery, as Hansen's and Mall's theories presuppose, but even in direct contact with the nucleus; and, secondly, if the cells are in active amœboid motion, with locomotion as well, as is certainly the case in the fins of the embryo *Fundulus*, then one can hardly conceive that the more or less fixed syncytium described by Mall, whose observations can be readily verified by any of the usual histological methods applied to the tissue used, is sufficiently stationary to permit the retraction or shrinkage of endoplasm, thus leaving the endoplasmic fibrils outside the bounds of the cellular protoplasm, for according to Mall's theory true cells are lost, being replaced by nucleated endoplasmic areas anastomosing with their neighbors to form a continuous syncytial net.

I have many times observed the fins of *Fundulus* embryos between the lengths of 4.5 mm. (at the time of hatching) and 25 mm., the pectoral and caudal fins being usually selected, and in no single instance have I observed either a stellate or a spindle cell which did not exhibit changes of form and locomotion, the latter being sometimes very limited, at other times, as shown in Fig. 2, quite extensive. Hence the ultimate theory of connective tissue histogenesis, it would seem, must take into account the relative activity of the mesodermic cells. So far as I appreciate them, none of the theories thus far advanced fully meet all of the observed conditions. Further studies of both living and "fixed" tissue may further elucidate the true conditions involved in the histogenic process.

The fins of living fish embryos offer for the development of connective tissue a most desirable subject for study, for in the subcutaneous tissue between the dermal fin-rays one finds a definite connective tissue area, in which, with the exception of the somewhat related chromatophores, there is no other structure to confuse the picture. This area is bordered on either side by the dermal fin-ray, beneath which an efferent blood-vessel descends, and alongside of which the return or afferent vessel ascends the fin; it is covered only by the pavement cells of the dermal epithelium, whose outlines are sharply defined, lying at a more superficial focal plane than the connective tissue. The chromatophores are easily recognized both by their color, black to yellow, and by their peculiar branching form, so that even those relatively or almost entirely devoid of color can be easily differentiated from the smaller colorless connective tissue cells. Moreover, the chromatophores are prone to lie in immediate relation to the fin-rays and to the blood-vessels.

In conclusion I desire to emphasize the opinion that in the study of histogenesis of connective tissue we must take into consideration the conditions surrounding the living cells and draw conclusions from "fixed" tissue only in the light of our knowledge of the living.

To furnish conclusive results living cells must be observed under normal conditions and not merely under those most unusual, or even pathological, conditions which surround the observation of growing tissues either in the mesentery studied by Flemming and others, open to the criticism of the presence of inflammatory changes, or in the tissue cultures by the more recent method of Harrison, Carrel and Burrows, which is surrounded by the necessity for the interpretation of results arising under entirely new surroundings and conditions as yet but little understood. Both in the mesentery and in tissue cultures movements of the connective tissue cells and even locomotion have been observed. I now add to these former results the record of the observation of motion and locomotion in cells under wholly normal conditions, in animals which underwent no operation other than in some instances the administration of chloretone, and which remained alive, active and normal for some time subsequent to the period of observation. For example, the particular fish which furnished the subject for Fig. 1 was returned to sea water and remained actively swimming for two days; he could have been kept much longer.

For the opportunity of pursuing this study I am indebted to the Marine Biological Laboratory at Woods Hole, Mass.

Reprinted from THE ANATOMICAL RECORD, VOL. 5, NO. 12
December, 1911

ON THE STROMA OF THE PROSTATE GLAND, WITH SPECIAL REFERENCE TO ITS CONNECTIVE TISSUE FIBERS

JEREMIAH S. FERGUSON

From Cornell University Medical College, New York City

FIVE FIGURES

The prostate gland consists essentially of secreting ducts and alveoli embedded in a stroma of connective tissue containing much smooth muscle. The smooth muscle is divisible into three portions: (1) the inner, which surrounds the urethra and consists of longitudinal and circular fibers; (2) the outer, which encapsulates the gland and is likewise disposed in longitudinal and circular bundles; and (3) that intermediate muscle which lies between the inner and outer divisions and whose fibers form interlacing bundles having a general trend from without inward.

Between the bundles of the intermediate muscle the glandular tissue is disposed in more or less slender conical lobules beginning with a broad base formed by dilated acini in the outer portion of the gland and tapering to an apex from which a duct leads inward to finally open on the surface of the prostatic urethra in the region of the colliculus seminalis.

The volume of smooth muscle in the framework or stroma is considerable; roughly, it forms nearly one-half of the tissue which is interposed between the glandular alveoli. Its fibers are coarse, and are easily observed in all preparations. They are specially distinct in sections which have been stained according to the iron-haematoxylin method of Heidenhain.

The muscle is embedded in a delicate network of connective tissue fibers which, unlike the usual arrangement, penetrate the muscle bundles and form between their individual fibers a close but delicate mesh.

more abundant in the narrow septa between the lobular acini. In this location the meshes are frequently occupied by the capillary vessels which occur with great frequency close beneath the glandular epithelium (fig. 2).

It is usually stated that the acini of the prostate have no membrana propria or basement membrane. Kölliker[6] states that beneath the prostatic epithelium a membrana propria is entirely wanting. Oppel[7] says that careful examination with the best immersion systems failed to reveal any structure in the layer beneath the epithelium. Sections of the prostate prepared with the Bielschowsky method show beneath the glandular epithelium of nearly all of the acini a very definite membranous layer, consisting of closely interlaced β-collaginous fibers, upon which the epithelium directly rests (fig. 3, *b.m.*). On examining such sections, even with very low magnification (e.g., × 60) one is immediately impressed with the prominence of the membrane. In haematoxylin-eosin sections this membrana propria, for such we may call it, can scarcely be recognized; with iron-haematoxylin it can scarcely be discerned and is not sharply differentiated; with the Mallory connective tissue stain it can be distinguished but lacks prominence because of the absence of color differentiation from the α-collaginous fibers.

The fibers of the β-collaginous membrane of the mucosa are continuous with the similar fibers which form the coarser network of 'white fibers' between the walls of the glandular acini, and whose meshes form the peculiar oval areas already described, such areas as are characteristic of the β-collaginous fibers wherever found, and to which attention has already been drawn by Ciaccio[8] in his description of the 'perifollicular plexus' about the nodules of lymphoid organs. This peculiar conformation appears to be due to a tendency of these fibers to pursue a spiral course. I have observed this tendency not only in the prostate and in lymphatic organs, but elsewhere, e.g., among the collaginous fibers of the derma, and among the fibers of elastic tissue (liga-

[6] Kölliker Gewebelehre, 6te Aufl., Bd. 3, S. 470.
[7] Oppel, Lehrb. d. vergleich. mik. Anat., Teil 4, S. 397.
[8] Ciaccio, Anat. Anz., 1907, vol. 31, p. 594.

mentum nuchae, arterial tunica intima, etc.); in this last location their spiral character is most apparent (fig. 4).

The β-collaginous fibers closely invest the blood-vessels, the capillaries as well as the adventitia of the arterioles and venules. Oppel[9] has shown an excellent figure of the relation of these fibers to the capillary vessels but without directing attention to their peculiar character. The capillaries are enclosed within the oval meshes formed by these peculiar fibers. The endothelium rests directly upon the fibers.

The disposition of the β-collaginous fibers within the muscular tissue of the prostate is also peculiar and characteristic. The larger bundles of smooth muscle are not always penetrated throughout, but in the smaller bundles the β-fibers form delicate, rounded oval meshes, whose fibers are intricately interlaced, and within the spaces of which are the individual muscle cells. Even if a smooth muscle fiber lies isolated within the prostatic stroma it will usually be found surrounded by the interlacing β-fibers by which it will be inclosed within an oval space (fig. 5). The muscle fibers are not, however, over closely grasped by the enveloping β-fibers, there being usually a narrow interval between muscle and collaginous fiber.

SUMMARY

The method of Bielschowsky discloses within the prostatic stroma, in addition to smooth muscle and elastic fibers, *two types of collaginous fibers*, the one being of the wavy variety usually designated as typical 'white fibers' though here they are of rather unusual delicacy, and the second a type which we have called the β-collaginous and which in successful Bielschowsky preparations are characterized by an opaque black color reaction; these last form a membrana propria supporting the glandular epithelium, and an interlacing network throughout the intergland-ular tissue. They closely invest the blood vessels and penetrate between the individual fibers of the smooth muscle bundles. That they are not elastic fibers is amply demonstrated by the

[9] Oppel, loc. cit.

fact that they do not react to the special stains for elastic tissue, and by the additional fact that elastic fibers in known locations, e.g., in the tunica intima of the arteries, or in the ligamentum nuchae, remain colorless when treated in accordance with the Bielschowsky method.[10]

The above study was partially conducted on sections which had been prepared under my direction by Mr. I. Rosen, first year student in the Cornell University Medical College, for whose assistance I am greatly indebted.

PLATE 1

EXPLANATION OF FIGURES

1 α-and β-collaginous fibers from the stroma of the human prostate gland. The β-fibers are coarse and black, the α-fibers much finer and of a golden brown color in the original preparation. Bielschowsky stain. Drawn with Edinger projection apparatus, × 900, reduced one-fourth in reproduction.

2 The epithelium and intervening stroma of adjacent prostatic alveoli. *β*, β-collaginous fibers; *bl. ves.*, blood-vessels; *ep.*, secretory epithelium of the prostate. Bielschowsky stain. Drawn with Edinger projection apparatus, × 270, reduced one-fourth in reproduction.

3 Several alveoli and the intervening stroma of the prostate gland. The basement membrane is prominent, even with relatively low magnification. *b.m.*, basement membrane *ep*, epithelium of the prostatic alveoli. Bielschowsky stain. Drawn with Edinger projection apparatus, × 200, reduced one-third in reproduction.

4 Isolated β-collaginous fibers from the tunica intima of the human aorta, showing their typically spiral course. Bielschowsky stain. Drawn with Edinger projection apparatus, × 270, reduced one-fourth in reproduction.

5 The β-collaginous fibers of a muscle bundle of the prostatic stroma. At the right of the fibers the smooth muscle fibers were transversely, at the left of the figure they were longitudinally cut. The muscle fibers are not represented; each mesh at the right of the figure enclosed a single muscle fiber, and a single muscle fiber occupied each interstice between adjacent β-collaginous fibers at the left. α-collaginous fibers, indicated by the finer and lighter lines are found between muscle bundles and about the coarser β-fibers. *L.S.*, portion longitudinally cut; *T.S.*, portion transversely cut. Bielschowsky stain. Drawn with Edinger projection apparatus, × 270, reduced one-fourth in reproduction.

[10] Ferguson, Am. Jour. Anat., vol. 12, no. 3, 1911.

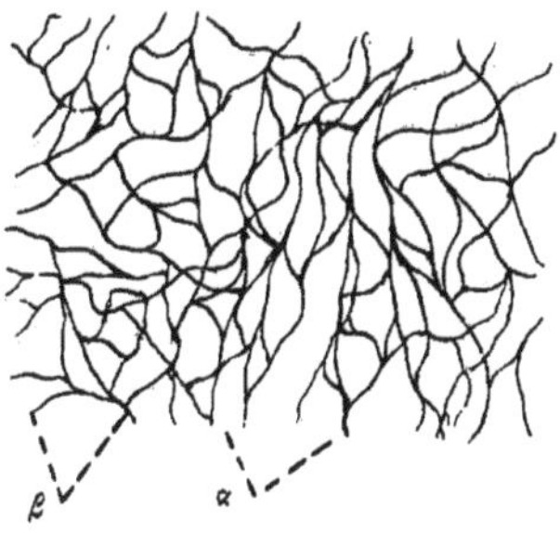

5

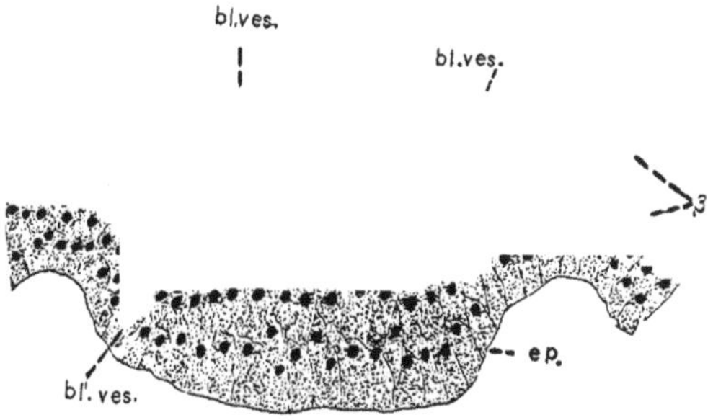

2

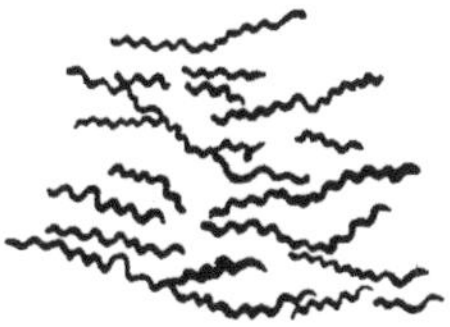

4

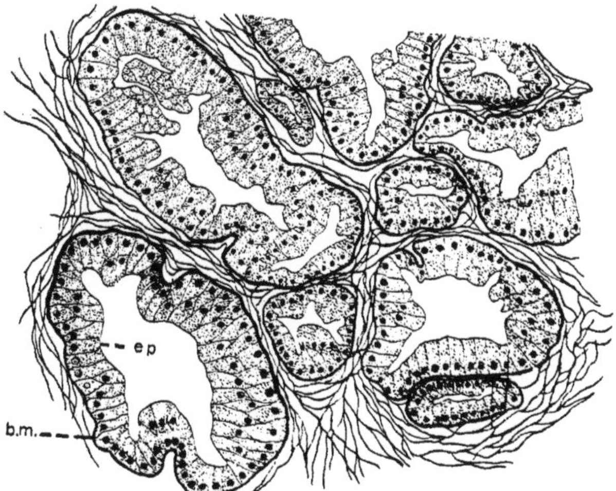

3

THE TEACHING OF VISCERAL ANATOMY, OR ORGANOLOGY

JEREMIAH S. FERGUSON, M.S., M.D.
Assistant Professor of Histology, Cornell University Medical College
NEW YORK

Medicine as a science is yet in its infancy; its composite development, its evolution, forms a steadily increasing curve, but each of its component parts form irregular curves with frequent rises, falls and stationary levels, when historically considered. The close of the last century marked the completion of a period of rapid development in surgical science, during which it was, perhaps, eminently desirable that the medical student should view anatomy with exaggerated reference to its surgical importance.

The beginning of the present century offers promise of an equally important development on the side of internal medicine. The former necessity is now superseded by the latter, and for the internist an accurate and extensive familiarity with visceral anatomy is of fundamental importance.

If there was a time when students could be allowed to devote weeks to the careful dissection of an arm or a leg and then for want of time pass cursorily over the abdominal contents and perhaps fail entirely to open the thorax, cranium or spinal canal, that time is past: the demands of the internist in the present generation call not only for careful visceral work in the dissecting-room but for an amplification of that course by such methods as shall result in accurate conceptions of living structure, which in turn can result only from the familiarity obtained through extensive first-hand acquaintance with the tissues themselves. It is for the satisfaction of these demands that courses modeled along the lines which I am about to indicate must be offered to the medical student of to-day.

The form, position and external relations of the organs are best observed in the "fixed" material of the dissecting-room, and if only recently prepared or carefully preserved tissue is used, the conception in the mind of the student will be well formed. But it is essential that the tissue be perfectly preserved and that the dissection shall be so rapid and careful as to allow no opportunity for excessive evaporation with coincident changes of color, texture and size. Even at its best, preserved dissecting-room material will show considerable changes in color and some hardening of texture. Material which has lain for weeks on the table during the dissection of the limbs and body wall, unless it has received the best of care, is not entirely satisfactory for the proper study of the viscera. To obtain the best results from the standpoint of organology a body freshly prepared by injection should be brought directly from the preparation room, its paries at once opened, as at autopsy, and its viscera carefully surveyed, as rapidly as is consistent with thoroughness, first for their position and general relations and then for form and structure. Such organs as are not badly altered by disease should then be removed, their connections to paries and to adjacent organs and especially their vascular connections being carefully noted; then their surface markings are to be carefully observed and their architecture thoroughly studied. Even at the loss of some material the finer details of dissection involving the topographic anatomy of the organ will be better pursued on other material.

But even at its best, dissecting-room material only partially fulfils the requirements of an accurate conception of the living viscera. Accurate conceptions and exact knowledge of any matter are obtained only by intimate first-hand acquaintance with the material. Abundance of opportunity for handling anatomic material should be afforded the student; no source of supply should be neglected. A course in organology is therefore incomplete unless it utilizes an abundance of fresh tissue, which only the autopsy room can supply. Since the nearest approach to the actual living organs obtainable for anatomic study is found in the human tissues at autopsy, and in the fresh tissues of the lower animals, both of these sources should be extensively utilized for the purposes of organology.

The recently dead animal presents organs, to all intents and purposes, in a living condition. In certain cases their cells can be transplanted and caused to grow; indeed, they may even be cultivated extraneously as has been demonstrated by R. G. Harrison and others.

The fact that the lower mammalian organs differ anatomically from the human is not an objection to their utilization, for, indeed, the very differences may often be used to broaden the conception of human structure. The fact that the sheep's kidney is lobulated, or that the kidney of the horse or dog has but one pyramid does not detract from the value of either in demonstrating the structural relations and appearance of the living human kidney, and if properly utilized they respectively elucidate a logical conception of the markings on the surface and the contents of the sinus of the human organ. Again, the objective study of the surface markings of the adult human lung is difficult of interpretation without reference to the child, the infant, and the lower animals.

But while lower animal tissues may be of great value in the demonstration of certain structural features they cannot be relied on as material for any great portion of such a course as is under discussion. Human anatomy must be primarily taught from human material. The handling and study of unaltered human material from the autopsy room offers the most valuable instruction to the anatomic student.

The utilization of this source of material cannot be left till such time as the student may pursue his work in the autopsy room under the instruction of the pathologist, for there he will be concerned, not with anatomic details, but with pathologic lesions, and his facility in pathologic study will largely depend on the accuracy and fulness of his anatomic conceptions, and on his familiarity with the structure and appearance of the normal rather than diseased viscera.

The result of a pedagogic experience of some years has convinced me of the fundamental importance of anatomic instruction based on fresh human tissue. A student of anatomy cannot see too much material of this kind; he should use it for most careful study.

Much normal tissue is removed at autopsy by the pathologist; with proper cooperation it should find its way, while still fresh, to the department of anatomy. A course in organology based on such material, amplified by the use of lower mammalian and other animal tissues, by museum specimens carefully prepared to preserve form, color, and to show the vital appearance, as nearly as possible, of certain special features of anatomic structure, and the whole broadened by coincident demonstrations on dissecting-room material which has been carefully and freshly prepared with organs fixed *in situ,* offers to the student of anatomy a well-nigh ideal opportunity for developing an accurate conception of the human body as it actually exists in a functionally active condition.

That such a course is utopian is not true. The fact that just such a course has been offered for the past three sessions in the laboratories of the Cornell University Medical College demonstrates its feasibility. The essentials are some time and energy on the part of the corps of anatomic instruction, and a very little friendly cooperation on the part of the pathologists. The time demanded from the student is little, because what is consumed in actual work renders so rich a return in improved knowledge of structure and of general, fundamental, anatomic characters that the work of the dissecting-room, and especially of the microscopic laboratory may thereby be very materially abbreviated. For this reason such a course in organology is best taught in connection with microscopic anatomy and the work so correlated that the gross anatomy of the organ immediately precedes its histology.

The consideration of the value of a course in organology on the above lines directs attention at once to certain phases of the relative value of the didactic and laboratory types of instruction. The value of the former type depends on the assumption that known facts are more readily obtained by the novice through the recorded observations of his predecessors, while the laboratory type presupposes that an accurate and utilizable knowledge must result only from personal contact with the object of study.

For this last type of work a progressively increasing value is represented by the book, the model, and the actual material. Given unlimited time and unlimited material, a trained student has little need of other sources of information. But for the great bulk of medical students time in the laboratory is not unlimited; hence the intensity of the laboratory work is greatly enhanced if accompanied by ample opportunity for reading and reflection in study hours. In the course which I have hitherto offered, the work of the student is also greatly facilitated by the use of a brief syllabus which indicates the names of structures of anatomic interest, without comment or description, arranged in the order in which they are uncovered.

With such opportunity it is remarkable how rapidly an accurate systematic conception of an organ may be acquired. In the laboratories of the University of Chicago it has been found that the study of the human brain is not only simplified but accelerated by a preliminary examination of the less complex nervous system of the dogfish, and the course is therefore begun with the dissection of the selachian nervous system. A true conception of the general fundaments, the gross divisions of structure as it were, renders the more intricate and complicated system so much more readily mastered as to result in an actual saving of time to the student as well as in a more utilitarian knowledge of the subject.

Likewise, in my laboratory at the Cornell University Medical College the survey of the fundaments of structure represented by gross organology as outlined above rendered possible an actual shortening of the time allotted to microscopic anatomy, and, far more important, I believe that the student better appreciates the full structure of the organs so studied. It is a matter of observation that he turns from his anatomy to his physiology better prepared to apply his anatomic knowledge to the deduction and explanation of function than was possible under other conditions. When one has studied in the gross the bundle of His and its ramifications, seen it in the fresh tissue, as well as in the dissected specimen and the microscopic section, and all have been properly correlated by coincident study, he has a preparation for the consideration of heart-block which cannot be obtained by the isolated work of the dissecting-room or the histologic laboratory. When those courses have been correlated by the study of the conduction system in fresh tissue and the transference of bits of known location to the stage of the microscope for verification of structure, the student thus prepared is in full possession of the anatomic mechanism, so far as it concerns the bundle in question, on which the phenomenon of heart-block largely depends.

A course in organology based on fresh material is, therefore, closely related to the work of the microscopic laboratory, and for him who pursues it the utmost of value is only obtainable when the facilities of the autopsy, the dissecting-room, and the microscopic laboratory have been coincidently placed at his disposal. With the organization of broad-minded departments of anatomy with gross and microscopic work provided for under the same head, an arrangement now generally adopted in leading medical schools at home and abroad, such correlation of work appears entirely feasible and eminently desirable.

330 West Twenty-Eighth Street.

Reprinted from The Journal of the American Medical Association
May 27, 1911, Vol. LVI, pp. 1544-1546

ERRATA

132, first word, eighth paragraph, for Candy read Gandy.

134, end of fourteenth line, for to read of.

136, fifteenth line, for Doudena read Duodena.

1, explanation of figure 6, for "This natural-size sketch represents enal mucosa," etc., read "This sketch represents the duodenal etc.

The Anatomical Record, Vol. 5, No. 3, March, 1911.

Reprinted from THE ANATOMICAL RECORD. VOL. 5, NO. 3.
March, 1911.

DUODENAL DIVERTICULA IN MAN

WESLEY M. BALDWIN

Cornell University Medical College

TWENTY-SIX FIGURES

Generally speaking diverticula of the duodenum are rare in comparison to the number of diverticula described in the literature as occurring elsewhere in the intestine. These have been divided by most authors into two classes, congenital and acquired, which are terms sufficiently explanatory in themselves. There are also two sub-groups known as 'true' and 'false,' the basis of this division being entirely anatomical. A 'true' acquired diverticulum presents in its walls all coats of the normal intestine, *i.e.*, mucosa, submucosa, muscularis, and peritoneum (fig. 1). The presence or absence of the last named constituent depends, however, upon the site of the diverticulum, namely, whether arising from the mesenteric border of the gut and thus projecting between the layers of the mesentery or whether situated along the convex free surface of the intestine. On the other hand, in the 'false' variety of acquired diverticula the muscularis is wanting, the walls, then, being formed by mucosa and submucosa alone. Accordingly, one might readily conceive of this form as being a hernia of the mucosa with the submucosa through a rent in the muscular layers of the intestine, and this conception has been verified in some instances by actual findings of diverticula in which the muscularis ceased abruptly at the orifice of the sac (fig. 2).

This Cornell series of duodenal diverticula is remarkable for two reasons: first, that so large a number as fifteen should be discovered among a series of only 105 duodena, and, secondly, because of the fact that the series includes the largest diverticulum yet mentioned.

For purposes of convenience in the matter of description of these specimens, I have divided them into the following classes based upon their gross anatomical form. In fig. 3, *a*, represents a funnel-shaped diverticulum, *b*, a tubular or cylindrical form, and *c*, a spherical or globular diverticulum.

Further, with the idea of making this paper a complete presentation of the subject of duodenal diverticula, I have collected the descriptions of the specimens heretofore reported in the literature and herewith present them in abstracted form.

The Cornell series

Specimen 1. (Figs. 4 and 5) This specimen (no. 353) was removed from the body of a white male, age 65, weight 150 (estimated), who had died of senility. Six centimeters caudal to the major duodenal papilla a wide-mouthed, funnel-shaped diverticulum opened upon the cephalic surface of the third portion of the duodenum. The sac extended cephalad a distance of 1.0 cm. lying dorsal to the head of the pancreas but did not penetrate it. The fundus measured 2.0 cm. transversely by 0.7 cm. in the dorso-ventral direction. The structure of the sac-wall was similar to that of the duodenum, all of the coats, however, being reduced in thickness. There was no relation of the fundus to any of the large vessels or to the ducts of the pancreas.

Specimen 2. (Figs. 6 and 7) The history of this specimen was not available (s795). In the descending portion of the duodenum there was a large diverticulum with a circular orifice 1.5 cm. in diameter. The fundus, globular in shape, extended medially a distance of 2.5 cm., lay dorsal to the head of the pancreas and possessed an approximate diameter of 2.0 cm. The bile and main pancreatic ducts traversed the ventral wall of the sac, opening at the ventro-caudal angle of its orifice. The minor papilla was not related to the diverticulum. The mucosal lining presented several folds, but the structure of the wall differed in no particular from that of the duodenum excepting that all the layers were thinner and the glands less numerous.

Specimen 3. (Fig. 8) White female, age 82, 5 feet, 2 inches tall, weight 140 (estimated), died of dysentery (no. 287). In this

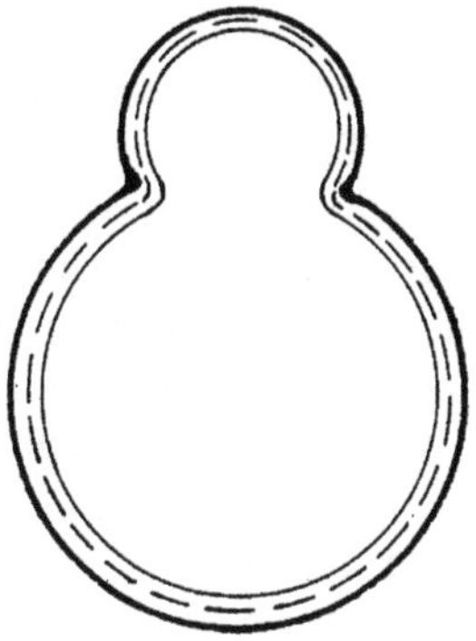

1

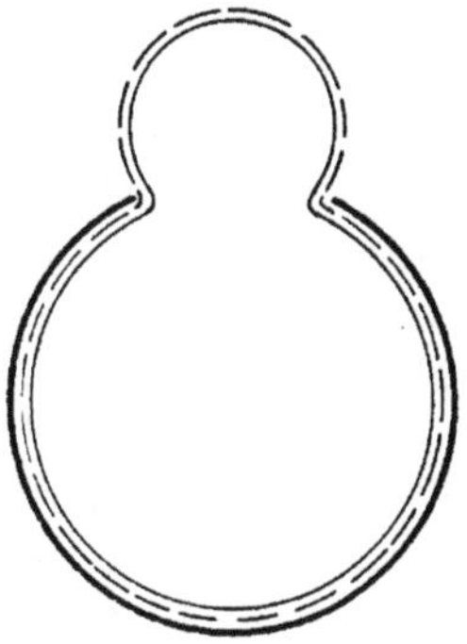

2

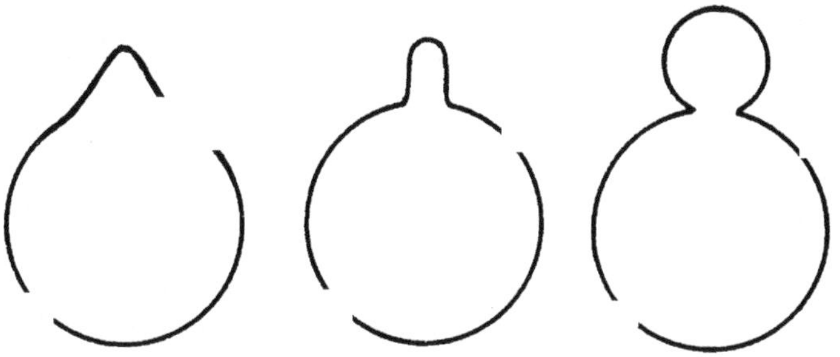

A B C

3

Fig. 1 In this diagram the intestine is represented in cross-section with a 'true' acquired diverticulum opening into it. The thin inner line represents mucosa, the interrupted line, submucosa, while the heavy line represents the intestinal muscularis. As is to be seen, all three layers are carried uninterruptedly over the fundus of the sac.

Fig. 2 In this diagram a 'false' acquired diverticulum is figured opening into the intestine. The muscularis, represented by the heavy outer line, ends abruptly at the mouth of the diverticulum, the mucosa and submucosa, however, represented by the thin inner line and the broken line, respectively, constitute the wall of the diverticulum.

subject a diverticulum of unusual size projected into the head of the pancreas from the medial wall of the descending portion of the duodenum. Its communication with the duodenum measured 4.0 cm. in the long axis of the intestine by 2.5 cm. transversely, being somewhat quadrilateral in shape. The fundus, loosely attached to the substance of the pancreas which encompassed it, presented proportions approximately equal to those of the head of the gland, *i.e.*, it measured 6.0 cm. cephalo-caudally, 3.0 cm. dorso-ventrally, and 4.0 cm. transversely. Within the sac a prominent fold of mucosa 0.9 cm. broad and 2.2 cm. long passed cephalo-caudally across the dorsal wall. The bile duct, traversing this fold at some distance from its free edge, terminated at the major duodenal papilla 1.0 cm. within the sac at the junction of its dorsal and caudal walls. The main pancreatic duct passed dorsal to the diverticulum to open at the major papilla with the bile duct. Upon the duodenal mucosa, cephalad and ventral to the orifice and about 1.0 cm. from it, lay the minor duodenal papilla with the opening of the accessory pancreatic duct, the latter traversing the pancreatic substance of the head of the gland ventral to the fundus of the diverticulum. The mucosa of the diverticulum was smooth with but few villi, and devoid of intestinal valves and of intestinal glands. A single thin layer of muscularis existed over the fundus and became somewhat thickened at the orifice of the sac. Numerous large blood vessels were to be seen distributed along the parietes.

Specimen 4. (Figs. 9 and 10) This specimen (no. 433) was removed from the body of a white female, age 53, weight 98, diagnosis 'heart disease.' Lying directly cephalad and dorsal to the major duodenal papilla in the second or descending portion of the duodenum an orifice, measuring 0.7 cm. transversely and 0.5 cm. in the long axis of the intestine, communicated with a spherical diverticulum. The fundus of this sac, with dimensions somewhat more ample than those of the orifice, lay dorsal to the head of the pancreas, to which it was adherent by loose connective tissue fibrils, and extended medially a distance of 0.5 cm. The sac bore no immediate relation to the orifice of the accessory pancreatic duct, but both the main pancreatic and bile ducts traversed its ventro-caudal wall. No sphincter-like arrangement of the duodenal muscularis existed at the orifice of the sac. All of the elements of the duodenal wall were represented in the parietes of the diverticulum.

Specimen 5. (Fig. 11) Male, age 55, white, weight 140 pounds (estimated), height 5 feet 7 inches, diagnosis cerebral hemorrhage (no. 377). The medial wall of the descending portion of the duodenum presented a quadrilateral orifice with sharp and well-defined edges measuring 2.5 cm. in the long axis of the intestine and 1.5 cm. transversely. This orifice opened into a large spherical diverticulum which projected medially a distance of 2.5 cm. into the substance of the head of the pancreas midway between its dorsal and ventral surfaces. The fundus measured 3.0 cm. cephalo-caudally by 1.0 cm. dorso-ventrally. The bile duct in company with the main pancreatic duct traversed the dorsal wall of the sac, both opening at the major papilla which was situated 0.7 cm. within the orifice upon the dorsal wall of the diverticulum. A prominent fold of mucosa bearing this papilla upon its crest ran caudally across the dorsal wall. The minor papilla was located upon the duodenal mucosa 1.2 cm. cephalad and ventral to the diverticulum. Microscopically the fundus showed a thin muscularis completely investing the sac. No sphincter was present. The villi were more scattered over the fundus.

Specimen 6. (Figs. 12 and 13) The data in this case (s503x) were not obtainable. At the common orifice of the bile and main pancreatic ducts a globular diverticulum 1.5 cm. in depth and 1.2 cm. in the other dimensions, lying dorsal to the head of the pancreas, opened into the duodenum through an irregularly quadrilateral orifice measuring 1.1 cm. in the long axis of the duodenum and 0.4 cm. transversely. The ducts lay along the caudal wall of the sac with their orifice at its caudal lip. With the exception of the one-layered structure of the muscularis and the thinness of the several coats in general, the constituents of the wall of the sac did not differ from those of the duodenum. There was no sphincter at the orifice.

Specimen 7. (Fig. 14) Female, age 61, white, height 5 feet 6 inches, weight 150 (estimated), death from sepsis (no. 341). The figure represents a small, shallow, funnel-shaped diverticulum upon the medial wall of the second portion of the duodenum presenting in its depth the conjoined orifice of the bile and main pancreatic ducts. The structure of the parietes of the depression was precisely similar to that of the duodenal wall. This diverticulum presented an orifice 0.4 cm. in diameter and a depth of but 0.2 cm.

Specimen 8. (Figs. 15 and 16) Male, age 60, white, height 5 feet 8 inches, weight 160 (estimated), diagnosis endocarditis (no. 330). In this specimen a large diverticulum communicated through a comparatively small orifice with the duodenum 4.5 cm. caudal to the major duodenal papilla. The body of the diverticulum lay dorsal to the caudal portion of the head of the pancreas and measured 3.0 cm. in a cephalo-caudal direction, 2.0 cm. transversely, and 0.7 cm. in the dorso-ventral axis. The orifice was irregularly oval, the long axis, parallel to that of the duodenum, being 1.0 cm. while the transverse axis measured 0.5 cm. There existed at the orifice a sphincter-like ring of muscularis. The walls of the sac were somewhat attenuated especially over the fundus. The mucosa presented numerous folds representing intestinal valves but did not differ in other respects from that of the duodenum.

Specimen 9. (Figs. 17 and 18) Male, age 47, white, height 5 feet 6 inches, weight 140 (estimated), diagnosis 'consumption' (no. 363). At the summit of the duodeno-jejunal flexure a wide-mouthed, funnel-shaped diverticulum extended cephalad a distance of 1.0 cm. and lay dorsal to the head of the pancreas. Neither of the pancreatic ducts bore any relation to the diverticulum. The muscularis did not present a sphincter-like arrangement at the orifice, nor did the structure of the walls differ in any respect from that of the duodenal wall. No large vessels were in association with the fundus of the sac.

Specimen 10. (Figs. 19 and 20) Male, age unknown, white, height 5 feet 8 inches, weight 140 (estimated), diagnosis chronic bronchitis (no. 305). This cylindrical diverticulum lay directly cephalad to the major duodenal papilla, the common bile duct passing along its ventro-caudal wall. The orifice, irregularly round, measured approximately 0.4 cm. There was no sphincter. The diverticulum passed medially and cephalically a distance of 1.0 cm. adjacent to the common bile duct and lying wholly behind the head of the pancreas. The minor duodenal papilla and the accessory pancreatic duct had no immediate relation to the diverticulum. The walls presented the usual elements of the intestinal wall, however, somewhat thinned.

Specimen 11. (Fig. 21) Male, age 59, white, height 5 feet, weight 160 (estimated), death from 'heart disease' (no. 308). This shallow diverticulum extended only 0.4 cm. from the cephalic wall of the transverse portion of the duodenum towards but not into the head of the pancreas. It lay 6.5 cm. caudal to the orifice of the bile duct and its orifice measured 0.5 cm. in diameter.

Specimens 12 and 13. (Fig. 22) The data in this specimen (s503) could not be obtained. There were two small, cylindrical diverticula situated side by side on the same transverse plane immediately cephalad to the minor papilla but with no immediate relation to the accessory pancreatic duct. Each measured 0.3 cm. in width by 0.4 cm. in depth. The fundus of each was buried in the tissue of the head of the pancreas. Their walls presented a normal mucosa and submucosa; the circular muscularis was thickened at the orifice to form a sphincter, the muscularis then became thinner and formed a single stratum over the fundus.

Specimen 14. (Figs. 23 and 24) The data in the history of this specimen (s804) could not be ascertained. Located upon the cephalic surface of the transverse portion of the duodenum 6.0 cm. caudal to the major duodenal papilla a small tubular diverticulum 1.0 cm. deep lay dorsal to the head of the pancreas. This diverticulum measured 0.5 cm. in transverse diameter and 0.3 cm. dorso-ventrally. The orifice measured only 0.2 cm. in the long axis of the duodenum and but 0.1 cm. transversely. The muscular stratum covering the fundus of the diverticulum was very thin; in other respects, however, the wall did not differ from that of the duodenum. Most of the muscle fibres covering the fundus were derived from the circular layer of the intestine. The ducts of the pancreas, of course, had no relation to the fundus of the sac.

Specimen 15. (Figs. 25 and 26) Male, age 56, white, height 5 feet 7 in., weight 140 (estimated), death from dysentery (no. 365). Five centimeters caudal to the major duodenal papilla upon the cephalic surface of the third portion of the duodenum a large funnel-shaped diverticulum projected towards the dorsal surface of the head of the pancreas. The side walls sloped gently towards the fundus of the sac, whose maximum depth approximated 1.0 cm. At the orifice, where the dimensions were greatest, the long diameter measured 2.5 cm. in the direction of the long axis of the duodenum with the transverse dimension 2.0 cm. The walls of the sac possessed all the coats of the duodenum, the muscularis being represented, however, by only a thin stratum of muscle fibres. The ducts of the pancreas bore no relation to the sac.

The essential facts of the Cornell series may be briefly stated as follows. One duodenum presented two diverticula which were located immediately cephalad to the minor duodenal papilla. Six of the diverticula were in the pars inferior duodeni while the other seven lay in immediate relation to the major duodenal papilla. All of these diverticula were situated upon the left or concave side of the duodenum extending towards the pancreas. Four projected directly into the gland, eight lay behind, while the other three were caudal to the head of the gland. None of the diverticula presented any evidences of inflammatory conditions. All of them belonged to the 'true' variety.

Following are the descriptions collected from the literature

While these abstracts have been made as brief as possible an effort has been made to keep closely to the exact wording of the original in the matter of the size, shape, etc., of the diverticula.

The first duodenal diverticulum mentioned in the literature was described by Morgagni in 1761. This specimen, which was obtained from the body of a man who had died of apoplexy, was situated two fingerbreadths caudal to the pylorus, the orifice being large enough to admit a finger. The sac exhibited no traces of pathological changes.

Rahn (1796) found a duodenal diverticulum in the emaciated body of a woman 22 years old who had died of chronic emesis. This sac-shaped diverticulum was closely related to the pylorus and presented a mucosal fold at its orifice not unlike that of the pylorus. The condition of gastroptosis was present in this cadaver.

Fleischmann ('15) reported the following specimens:

Specimen 1. Male, 64, had used brandy freely, killed. The common bile duct opened into a bladder-like diverticulum of the duodenum. Likewise the pancreatic duct opened into a similar adjacent but smaller diverticulum. The orifice of the larger diverticulum was devoid of intestinal valves. The walls of these sacs were thin and lacking of muscularis save at the orifice where a sphincter-like arrangement existed.

Specimen 2. Male, old age. The common bile duct opened into a bladder-shaped diverticulum of the duodenum. Adjacent there were three similar diverticula with lengths of a quarter, a half, and a full inch respectively. Two of these were bladder-shaped; the third, and largest, was constricted and pointed. The orifice of each presented a sphincter of muscularis and a crescentric fold of mucosa. The intestinal valves of the neighborhood were deficient.

Specimen 3. Male, 28, drowned. The ductus pancreaticus and the ductus choledochus opened separately each through a small duodenal diverticulum. Lying near these diverticula there was a third of the size of a pigeon's egg and, towards the first portion of the duodenum, still another smaller though similar diverticulum.

Albers ('44) mentioned one diverticulum located in the horizontal portion of the duodenum. It was scarcely one inch long and presented a contracted orifice with a marked fold of mucosa.

Barth ('51) reported in a female, aged 60, a duodenal diverticulum, composed of all of the coats of the intestine, penetrating the head of the pancreas and of a size sufficient to admit the little finger. This diverticulum seemed independent of the pancreatic ducts.

Habershon ('57) remarked having seen in Guy's museum a duodenal diverticulum near the orifice of the bile duct, but gave no further description of the specimen.

Roth ('72) reported the following series of specimens:

Specimen 1. Male, workman, 50, dead from a blow on the head. In the descending portion of the duodenum, 3 cm. cephalad to the orifice of the common bile duct and 0.5 cm. cephalad to the minor papilla, there was a finger-like pouch 1.5 cm. long running caudally and medially into the head of the pancreas. The fundus was covered ventrally by a thin layer of pancreatic tissue with a thicker layer dorsally. The muscularis existed only at the orifice in the form of a sphincter; elsewhere the walls were constituted by mucosa and a thin layer of connective tissue.

Specimen 2. Male, 69, emaciated. At the level of the orifice of the ductus choledochus, *i.e.*, in this specimen, the junction of the pars descendens and pars transversus inferior duodeni, a cylindrical diverticulum communicated with the duodenum through an orifice as large as a pea and extended 1.5 cm. along the bile duct into the head of the pancreas. Its walls were composed of thin mucosa and a thin layer of connective tissue. A second diverticulum, similar in form, direction and extent of long axis, structure, and in relations, existed 1.5 cm. cephalad to it. These diverticula were separated from each other by a thin layer of pancreatic tissue and by the common bile and pancreatic ducts. A prominent fold of mucosa limited the constricted orifices. The muscle fibres of the duodenum evidenced fatty degeneration.

Specimen 3. Male, 67, emaciated. Three centimeters cephalad to the orifice of the common bile duct a diverticulum of the duodenum of the size of a walnut penetrated the head of the pancreas. This rounded diverticulum was composed of mucosa. Its orifice was of the size of a lentil.

Specimen 4. Male, 58, emaciated. In the descending portion of the duodenum there was a diverticulum of the size of a hazelnut.

Specimen 5. Female, 49, emaciated. In the descending portion of the duodenum two thin-walled diverticula, one smaller and medial to the orifice of the ductus choledochus, the other, more cephalad and as large as a cherry, projected towards the head of the pancreas. The larger communicated with the duodenum through an orifice sufficient to admit the little finger.

Schüppel ('76) reported that he found a duodenal diverticulum present in seven instances among 45 bodies studied. At Kiel, however, among 200 bodies he found but one diverticulum.

Schirmer ('93) mentioned two specimens. The first, 2.2 cm. cephalad to the orifice of the common bile duct, was 2.5 cm. deep and 1.3 cm. broad; through the second, of the size of a pea, the accessory pancreatic duct communicated with the duodenum.

Under the heading "Large Pseudo-Diverticulum of the Duodenum," Pilcher ('94) described a pathological curiosity into which the duodenum opened and from which the jejunum passed. The sac presented mucosa in only one small strip. Elsewhere the parietes were devoid of epithelial lining and exhibited inflammatory changes.

Good ('94) collected five specimens. Two of these received mere mention, one being from the body of a female of twenty-seven who had died of cholera nostras.

Specimen 1. Female, 77. Lateral and cephalad to the orifice of the common bile duct a duodenal diverticulum of the size of a plum extended into the head of the pancreas. The muscularis, deficient over the fundus of the sac, existed at the constricted orifice in the form of a sphincter.

Specimen 2. Female, 54, died of carcinoma of the ovary with metastatic involvement of the liver and peritoneum. One and a half centimeters cephalad and to the right of the orifice of the bile duct a thin-walled cylindrical diverticulum 2.2 cm. deep communicated with the duodenum through an orifice 0.9 cm. in diameter. The bile duct passed ventral and to the right of the sac while a layer of pancreatic tissue 0.5 cm. thick lay dorsal to it. The parietes contained no muscle fibres.

Specimen 3. Female, 51. A spherical diverticulum 1.8 cm. in diameter opened into the duodenum 3.5 cm. cephalad and 1.4 cm. medial to the orifice of the bile duct and 10.0 cm. caudal to the pylorus. The fundus projected medially a distance of 2.5 cm. into the substance of the head of the pancreas, which separated it from the bile duct. No muscle fibres were present in the wall of the sac nor did the mucosa present any folds.

Seippel ('95) reported a cylindrical diverticulum of the duodenum at the level of the major duodenal papilla, 5.0 cm. in diameter and extending about 3.0 cm. dorsally and cephalically, the fundus being surrounded by pancreatic tissue. Microscopically the wall of the sac revealed a mucosa, somewhat thinned, a muscularis, and a muscularis mucosae. No other muscle fibres were prolonged over the fundus of the sac, the muscularis being represented there by a thin layer of connective tissue.

Hansemann ('96) reported a specimen of multiple diverticula in a man of 85 who had died of pneumonia. He had been fat. The intestine contained about 400 diverticula varying in size from a hemp-seed to a pigeon's egg. Several isolated examples were in the duodenum, but no description of these diverticula was given.

Helly ('98) saw in a duodenum two diverticula, one located above each papilla. The caudal of the two, corresponding to the major papilla extended into the head of the pancreas.

Charpy (98) reported one specimen taken from the body of an aged woman. The main pancreatic duct was small and terminated blindly at a cul-de-sac of the duodenal wall. In another specimen a small diverticulum caudal to the orifice of the bile duct opened into the ampulla of Vater but was not associated with either duct.

Letulle ('99) described two specimens. The first, of a depth of from 1.5 cm. to 1.6 cm., was composed of mucosa, with a few intestinal glands and no duodenal glands, of muscularis mucosae and of connective tissue. The muscularis of the duodenum ended abruptly at the orifice of the diverticulum. In the other specimen five pouches were grouped about the common orifice of the main pancreatic and common bile ducts, four being cephalad and one caudal. These pouches extended medially towards the head of the pancreas, but did not penetrate it.

Nattan-Larrier ('99) described a duodenum in which the common bile duct emptied into a small cul-de-sac containing also the orifice of the main pancreatic duct.

Marie ('99) says that he found in the body of a man of 45 a duodenum which presented two diverticula. The larger extended about 5.0 cm. medially, dorsal to the head of the pancreas and communicated with the duodenum by a constricted orifice 2.0 cm. in diameter. Both the bile and the pancreatic ducts opened at the mouth of the sac. The other smaller but similar diverticulum, 1.5 cm. deep, lay more cephalad and opened into the duodenum through a rounded orifice 1.0 cm. in diameter. The mucous lining of these diverticula was thin, smooth, devoid of villi, and of duodenal glands, there remaining but little else than a layer of cylindrical epithelium. The muscularis of the duodenum stopped abruptly at the orifice of the sac which it surrounded in the form of a ring.

Jach ('99) made the following report:

Specimen 1. Female, 64. On either side of the major duodenal papilla a diverticulum extended into the substance of the head of the pancreas with the bile duct lying between them. The larger or medial of these two was as large as an egg, globular in shape and possessed a somewhat constricted orifice sufficiently ample, however, to admit two fingers. The pancreatic duct traversed its caudal wall. The smaller, lateral diverticulum was as large as a walnut. Four centimeters caudal to the major papilla two diverticula of the size of a cherry and with orifices admitting the little finger, lay about 2.0 cm. apart, one being caudal and medial to the other. Their dorsal surfaces were not covered with pancreatic substance.

Specimen 2. Immediately caudal to the pylorus a cylindrical diverticulum with a mouth large enough to admit the thumb extended 3.0 cm. caudally, dorsally, and medially from the first portion of the duodenum. Because of its cephalad position this diverticulum had no relation either to the major duodenal papilla or to the head of the pancreas.

Specimen 3. Male, 58, carcinoma of the rectum. In the first portion of the duodenum 2.0 cm. from the pylorus there was an obliquely placed cicatrix. Between this and the pylorus an orifice large enough to admit one finger opened into a spherical diverticulum 1.0 cm. deep, the fundus of which passed cephalad to the pylorus. Its dorsal surface was not covered with pancreatic tissue.

Rolleston and Fenton ('00) described three specimens:

Specimen 1. Female, emaciated, hepatic cirrhosis. Two adjacent duodenal pouches immediately cephalad to the orifices of the bile and pancreatic ducts opened into the duodenum through separate but similar orifices about 0.6 cm. in diameter and extended medially into the head of the pancreas. The bile duct traversed the fold of mucosa separating the diverticula and opened into the duodenum apart from the main pancreatic duct.

Specimen 2. This presented a pouch about 1.8 cm. cephalad to the major papilla, 1.25 cm. in diameter and buried to a depth of about 1.8 cm. in the head of the pancreas.

Specimen 3. This pouch was similar to specimen 2 and lay 1.25 cm. cephalad to the major papilla, approximately 0.6 cm. deep and 0.6 cm. in breadth. The walls of all of these sacs consisted of merely mucosa enveloped in loosely arranged connective tissue.

Gandy ('00) reported one specimen. Cause of death, strangulated hernia. Nine centimetres caudal to the major papilla a saccular diverticulum opened into the dorso-cephalic surface of the third portion of the duodenum through a circular orifice 1.0 cm. or 1.2 cm. in diameter. The depth of the diverticulum was 1.5 cm. the fundus being located at the caudal border of the pancreas. The wall of the sac was composed of a thin layer of mucosa devoid of folds and of a thin layer of connective tissue.

LeRoy ('01) mentioned one specimen. Female, 64, emaciated. This exhibited a cancer of the major duodenal papilla in association with a diverticulum. Dorsal to the papilla a spherical diverticulum 1.1 cm. in diameter and 1.0 cm. deep opened into the duodenum through a circular orifice 0.9 cm. in diameter limited by a mucosal fold. The wall of the sac was composed of a much-thinned mucosa and a layer of connective tissue. The muscularis spread out in a thin layer over the fundus of the sac, which did not penetrate the pancreas.

Hodenpyl ('01) presented two cases of multiple spurious diverticula of the intestine. In the first, the duodenum and the upper portion of the jejunum were the seat of a number of thin-walled cysts varying in size from that of a pea to that of an egg. Apparently no diverticula were found in the duodenum of the second specimen.

Keith ('03) figured a case of ptosis of the pancreas from the London Hospital Medical College Museum. Traction upon the common bile duct had apparently been the factor in producing a diverticulum of the duodenum at the major papilla. No further description was given.

Dorrance ('08) mentioned a diverticulum 3.5 cm. caudal to the orifice of the bile duct. The cavity measured 3.0 cm. in depth by 5.0 cm. in circumference. The walls were composed of thin mucosa and of a very thin layer of circular muscle.

Bassett ('08) reported a specimen in a male, white, aged 78, who had died of epithelioma of the lower lip. On the medial wall of the duodenum 10.0 cm. caudal to the pylorus a diverticulum extended 2.0 cm. cephalically and medially along the medial surface of the bile duct into the head of the pancreas. The orifice, somewhat smaller than the fundus of the sac, measured 2.0 cm. in the long axis of the gut and 1.0 cm. transversely. The bile duct traversed the lateral lip, opening finally at the caudolateral angle of the orifice. The mucous lining of the pouch was atrophic, degenerated, and infiltrated in places with small round cells, the duodenal glands being absent. The muscularis ended abruptly at the orifice; over the fundus the parietes consisted of mucosa alone.

Another specimen. Male, white, 78, miliary tuberculosis. This sac-like diverticulum penetrated the head of the pancreas from the median wall of the duodenum 10.0 cm. caudal to the pylorus, 3.0cm. cephalad and 2.0 cm. medial to the opening of the bile duct. The rounded orifice measured 1.5 cm. in the long axis of the duodenum by 1.0 cm. transversely. Mucosa and submucosa, much folded and containing isolated bits of pancreatic tissue, composed the wall of the sac. The muscularis of the duodenum ended abruptly at the orifice of the diverticulum.

Jackson ('08) described a diverticulum of unusual proportions springing from the transverse portion of the duodenum in a male, aged 50, who had died of pneumonia. The fundus of the sac lay dorsal to the head of the pancreas and extended cephalically a distance of 3.5 cm., measuring 3.0 cm. transversely by 2.0 cm. ventro-dorsally, and communicated with the cephalic wall of the duodenum through an orifice about 5.0 cm. in diameter. There was no relation of the sac to the ducts of the pancreas. The wall of the diverticulum consisted of mucosa, a greatly hypertrophied muscularis mucosae, and a rudimentary muscularis.

Summing up, then, we may say that exclusive of the Hansemann and Hodenpyl specimens, above mentioned, and of the

Cornell series, there are reports of only 67 specimens of duodenal diverticula available in the literature. We believe, however, that this does not represent the frequency with which these anomalies occur in the dissecting room. Many elude observation, others are noted but not reported, and again some are mentioned in papers under irrelevant headings and consequently are with difficulty collected from the literature. Schroeder, commenting upon the Fleischmann series in 1854, remarked that there were in the Fleischmann museum so large a number of duodenal diverticula as to lead one to draw the conclusion that these anomalies were found more frequently in the duodenum than in the remainder of the intestine. Apparently, this completed series, however, was never described. In this connection it is of interest to note that the sac in several of these specimens lay ventral to the head to the pancreas, a condition rarely seen by other observers. Schüppel, as noted before, found seven instances among 45 specimens examined, but in Kiel among 200 bodies examined with this end in view, he found only one diverticulum. Fairland and Calder both wrote of instances of duplication of the duodenum which might be considered as diverticula very much elongated.

Much has been written concerning the etiology of these various forms of diverticula. There has been little question over the causative factors which have produced the Meckel's diverticulum of the adult, but the so-called 'acquired' diverticula have been the occasion of much controversy. 'False' diverticula, which belong to the class of 'acquired' diverticula, possess no musculature in their walls. These have been regarded by Seippel either as a hernia of the mucosa between the muscle fibres of the intestinal wall or as a localized bulging of the intestinal wall with a subsequent atrophy of that portion of the muscularis overlying the fundus of the sac thus produced.

As causative factors in the instance of 'true' and of 'false' 'acquired' diverticula, fatty degeneration of the muscularis, atrophy of the pancreas, and ptosis of the duodenum have been advanced by Roth, Jach, Keith, and others. Instances of the production of diverticula through traction upon the duodenal wall have been cited by Hansemann, Birsch-Hirschfeld, Klebs, and others. Increase in pressure from intestinal contents, liquid or gaseous, has been cited and diverticula have been experimentally produced in animal cadavers from these causes by Klebs, Edel, Heschl, Good and Hanau, and Hansemann. These authors noted, further, that in the instance of experimentally produced herniae of the mucosa, the seat of hernia was almost invariably at the point of entrance or emergence of the blood vessels, particularly the veins. Since these results were more readily obtained in old marasmic subjects, the conditions found in the majority of diverticula reported were thus apparently verified as, at least, contributing factors. Chlumsky experimented upon the living animal bowel and found, contrary to the results previously obtained with dead intestines, that the rupture generally occurred at some point opposite the mesentery. The traction effect of an accessory pancreas has received considerable attention and several observers have reported instances where apparently this condition was the sole etiological factor, *i.e.*, Zenker, Neumann, Weichselbaum, Nauwerck, Hansemann, and others. Fleischmann says that the production of diverticula at the major duodenal papilla is favored by reason of the weakening of the duodenal wall at the point of passage of the common bile and pancreatic ducts through the muscularis.

In a paper upon the accessory pancreatic duct and the minor duodenal papilla soon to appear, the author remarks that in 60 instances among 100 specimens of duodenum and pancreas examined, there existed at the major duodenal papilla a distinct hollowing of the duodenal wall, the papilla with the orifices of the common bile and main pancreatic ducts in these cases being situated in this depression. This condition is suggestive either of a persistence of the original diverticulum from which the liver and the ventral anlage of the pancreas developed, or illustrates the traction effect of the ducts upon the duodenal wall. Possibly, then, these specimens of so-called 'true' 'acquired' diverticula are in reality "congenital" to the extent that they represent a development dating back to the time of hollowing of the duodenal buds in the formation of the ducts of the liver and pancreas.

In a majority of the specimens reported in the literature, the descriptions are so meagre that a complete analysis of the series cannot be made. The following interesting facts, however, may be presented. This tabulation includes the Cornell series.

Total number of duodenal diverticula.................................83

In addition to this number Hansemann reported one duodenum with several diverticula and Hodenpyl two duodena, one with several diverticula. In the other, the latter did not state clearly whether there were any duodenal diverticula.

Eliminating the Hansemann and the Hodenpyl specimens from the series and considering that the 8 in the Schüppel series were single, *i.e.*, one diverticulum in each duodenum, we may, with the restrictions mentioned above classify the diverticula as follows:

Duodena presenting:

One diverticulum...............48 Two diverticula............ .. 9
Three diverticula............... 0 Four diverticula................ 3
Five diverticula.................... 1
Total number of duodena (including Hansemann's and Hodenpyl's)... 63

Sex

Males.............21 Females................14 Not given.......... 28

Age

Available in 30 instances Youngest......22 (female) Oldest.. ...85 (male)
Two others were reported as 'aged.'

Condition of cadaver

'Emaciated'....... 7 'Lean'................ 5 'Medium'........... 5

Constituents of wall of diverticulum

With muscularis...20 Without muscularis...16 No report...........47

Position of fundus of diverticulum

Ventral to pancreas head....... Several of Fleischmann series (see above)
Penetrating pancreas head..17
Dorsal to pancreas head.... 14
The remainder projected towards but not into the head of the pancreas.

Location of diverticula

Convex side of duodenum		0
Concave (pancreas) side of duodenum		72
Specimens not described in full	Schüppel	8
	Letulle	1
	Hansemann	(?)
	Hodenpyl	(?)
	Good	2
Near pylorus		5
At minor papilla		9
With accessory duct opening into diverticulum		1
Near major papilla		12
At major papilla		24
Bile duct opening into diverticulum		4
Main pancreatic duct opening into diverticulum		3
(In one of these the duct was occluded).		
Bile and pancreatic ducts together opening into diverticulum		4
Diverticulum in third portion of duodenum		12

In conclusion I may say that it is a genuine pleasure for me hereby to express my sincere appreciation of the helpful advice given by Professors Gage and Kerr in the preparation of this paper and also of the numerous and great courtesies shown by the departments of Anatomy and of Histology and Embryology.

BIBLIOGRAPHY

ALBERS 1844 Atlas Abt. 4. Taf. 21. Figs. 9 und 2. Erläuterungen zum Atlas 4. S. 262.

BARTH 1851 Bull. de la Soc. Anat., vol 26, p. 90.

BASSETT 1907–8 Duodenal diverticula; with especial reference to diverticula associated with the pancreatic and biliary tracts. Tr. Chicago Path. Soc., 7, p. 83.

BIRCH-HIRSCHFELD 1895 Lehrbuch der Path. Anat., 4 Aufl., S. 656.

CALDER 1733 Medical essays and observations, reprint, Edinburgh, 1, p. 205.

CHARPY 1898 Variétés et Anomalies des Canaux Pancréatiques. Journ. de l' Anat., p. 720.

CHLUMSKY 1899 Ueber versch. Methoden d. Darmvereinigung. Beitr. z. klin. Chir., Band 15.

DORRANCE 1908 A diverticulum of the duodenum. Univ. of Penn. Med. Bull. April.

EDEL 1894 Ueber erworbene Darmdivertikel. Virchow's Archiv. Bd. 138, S. 347.

FAIRLAND 1879 Congenital malformation of bowel; Amussat's operation. Brit. Med. Jour., 1, p. 851.

FLEISCHMANN 1815 Leichenöffnungen, p. I, Erlangen.

GANDY 1900 Bull. de la Soc. Anat. de Paris, Année 75, p. 691.

GOOD 1894 Casuistische Beiträge zur Kenntniss der Divertikelbildungen u. s. w. Inaug. Diss., Univ. Zürich, p. 47.

HABERSHON 1857 Observations on the diseases of the alimentary canal. London, p. 145.

HANAU 1896 Bemerkungen zu der Mittheilung von Hansemann "Ueber die Entstehung falscher Darmdivertikel" in diesem Archiv, Bd. 144. Hft. 2, s.400, Virchow's Archives, vol. 145.

HANSEMANN 1896 Ueber die Enstehung falscher Darmdivertikel. Archiv für path. Anat., Bd. 144, Hft. 2, S. 400.

HELLY 1898 Beitrag zur Anatomie des Pankreas und seiner Ausführungsgänge Archiv für mik. Anat., p. 773.

HESCHL 1880 Wien med. Wochenschr., no. 1, u. 2, spec. S. 5.

HODENPYL 1901 Two cases of multiple spurious diverticula of the intestine. Proc. N. Y. Path. Soc. (1899–0), 182.

JACH 1899 Ueber Duodenaldivertikel. Diss., Kiel.

JACKSON 1908 An unusual duodenal diverticulum. Jour. of Anat. and Physiol., vol. 42.

Keith 1903 On the Nature and Anatomy of Enteroptosis (Glénard's disease). Lancet, London, vol. 1, p. 640.

Klebs 1869 Handbuch d. path. Anat., Bd. 1. S. 271, Berlin.

1899 Die allgemeine Pathologie, theil 2, S. 100, Jena.

LeRoy 1901 Divertikel prévatérien congénital et cancer de l'ampoule de Vater déterminant une obstruction biliare. Jour. des Sciences Med. d. Lille, 2, p. 593.

Letulle 1899 Malformations duodenalés; diverticules péri-vatériens. La Presse Medical, p. 13.

Marie 1899 Bull. et Mem. Soc. Anat. de Paris, p. 982.

Morgagni 1761 De Causs. et sed. morb. Epist. 34. par., 17.

Nattan-Larrier 1899 Malformations multiple, retrecissement du duodénum, dilatation de l'oesophage, communication interventriculaire. Bull. et Mem. Soc. Anat., Paris, p. 981.

Nauwerck 1893 Ein Nebenpankreas. Ziegler's Beiträge, Bd. 12, S. 28.

Neumann 1870 Archiv d. Heilkunde, Bd. 11, S. 200.

Pilcher 1894 Large pseudo-diverticulum of the duodenum. Annals of Surgery, vol. 20.

Rahn 1796 Scirrhosi pancreat. diagnos. obs. 14. Göttingae.

Rolleston and Fenton 1900–'01 Two anomalous forms of duodenal pouches. Jour. Anat. and Physiol., vol. 35, p. 110.

Roth 1872 Ueber Divertikelbildung am Duodenum. Archiv. für Path. Anat., 41, p. 197.

Schirmer 1893 Beitrag zur Geschichte und Anatomie der Pankreas. Inaug. Diss., Basel, p. 58.

Schroeder 1854 Ueber Divertikelbildung am Darmcanale. Inaug. Dis., Erlang.

Schüppel 1876 Ziemmsen Handbuch der spec. Path. und Ther.

Seippel 1895 Ueber erworbene Darmdivertikel. Inaug. Diss., Univ. Zürich, p. 21.

Weichselbaum 1883 Bericht d. K. K. Krankenanstadt, Rudolph-Stift, Wien. 4.

Zenker 1861 Nebenpankreas in der Darmwand. Virchow's Archives, Bd. 21, S. 369.

PLATE 1

EXPLANATION OF FIGURES

4 This sketch, made from the actual specimen, represents the duodenal mucosa with the mouth of the diverticulum. The duodenum was opened along that border opposite to the orifice of the diverticulum.

5 In this diagram the diverticulum shown in fig. 4 is represented as seen from the dorsum. The fundus is represented projecting cephalically from the third portion of the duodenum and lying dorsal to the head of the pancreas.

6 This ~~natural size~~ sketch represents the doudenal mucosa with the orifice of the diverticulum. The duodenum was opened opposite to the site of the diverticulum. Note the close relation of the orifice to the major duodenal papilla.

7 In this diagram the dorsal surface of the head of the pancreas is shown encircled by the duodenum. The fundus of the sac represented in fig. 6 is seen projecting in close proximity to the bile duct and dorsal to the head of the pancreas.

8 In this sketch the mucosa lining the interior of the diverticulum can be seen. Just within the orifice and running cephalo-caudally across the dorsal wall of the sac a prominent fold of mucosa is seen presenting on its summit caudally the major papilla. This diverticulum is the largest yet described.

9 In this sketch the close relation of the orifice of the diverticulum to the major duodenal papilla is represented.

10 This diagram presents the relation of the fundus of the diverticulum, represented in fig. 9, to the dorsal surface of the head of the pancreas and especially to the bile and pancreatic ducts.

REFERENCE NUMBERS

1 Orifice of diverticulum.
2 Duodenum (dorsal aspect).
3 Head of pancreas (dorsal aspect).
4 Diverticulum.
5 Major duodenal papilla.
6 Ductus choledochus.
7 Mucosal fold.
8 Ductus pancreaticus.

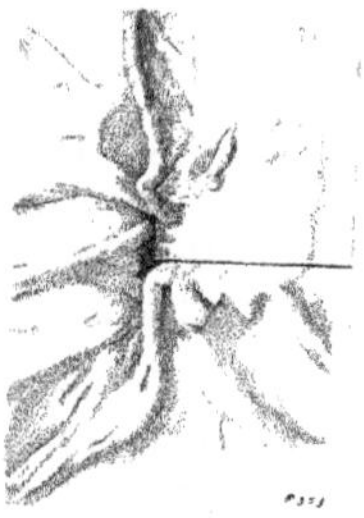

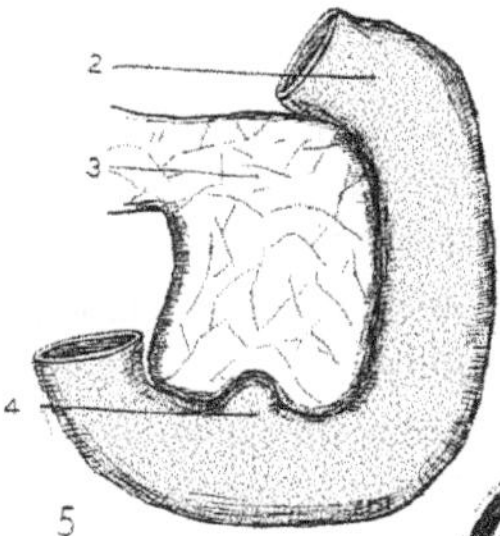

5

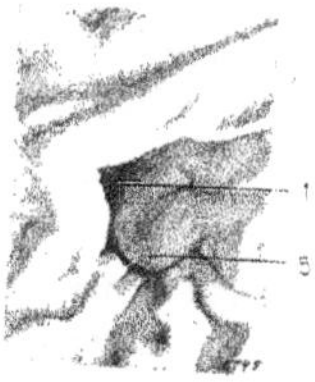

6

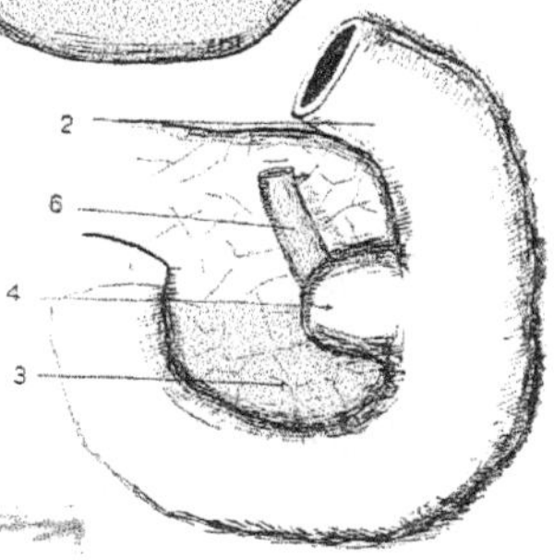

7

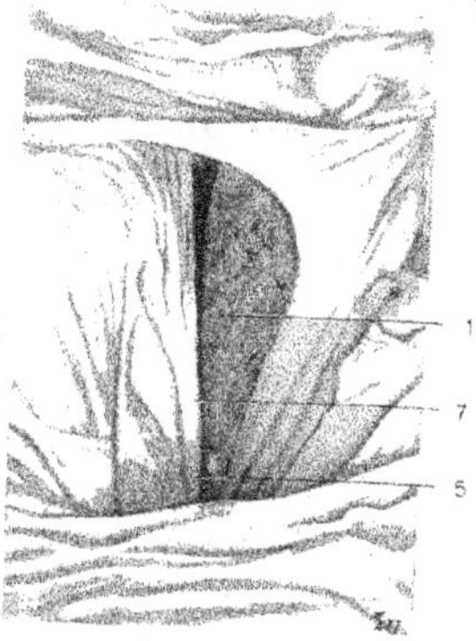

8

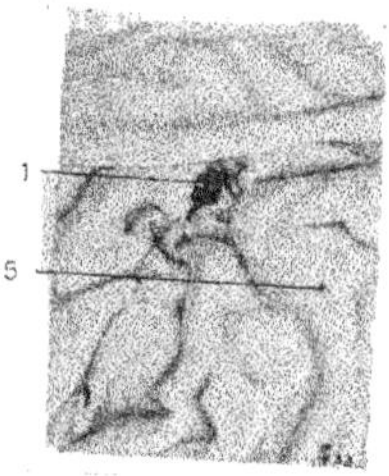

9

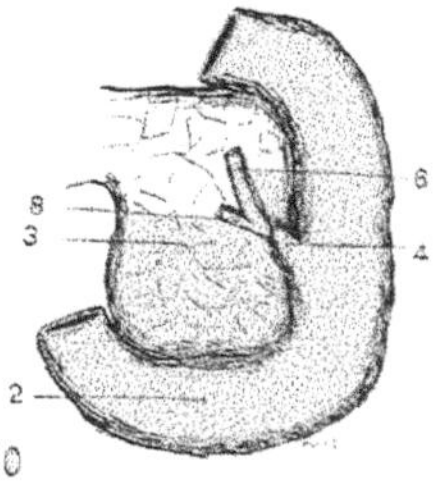

10

PLATE 2

EXPLANATION OF FIGURES

11 This sketch resembles in general appearance fig. 8. The fold of mucosa traversing the dorsal wall of the sac bearing the major duodenal papilla is well shown. These structures lie just within the orifice of the diverticulum.

12 This sketch shows the close relation of the orifice of the diverticulum to the major duodenal papilla.

13 This diagram shows the relation of the diverticulum, represented in fig. 12. to the dorsal surface of the head of the pancreas and to the bile and main pancreatic ducts.

14 In this sketch the common orifice of the diverticulum, the bile, and the main pancreatic duct is shown upon the mucosa of the second portion of the duodenum.

15 This sketch shows the orifice of a diverticulum situated in the third portion of the duodenum.

16 In this diagram the fundus of the diverticulum, represented in fig. 15, is shown in relation to the dorsal surface of the head of the pancreas. There is, of course, no relation to the bile or pancreatic ducts.

REFERENCE NUMBERS

1 Orifice of diverticulum.
2 Duodenum (dorsal aspect).
3 Head of pancreas (dorsal aspect).
4 Diverticulum.
5 Major duodenal papilla.
6 Ductus choledochus.
8 Ductus pancreaticus.

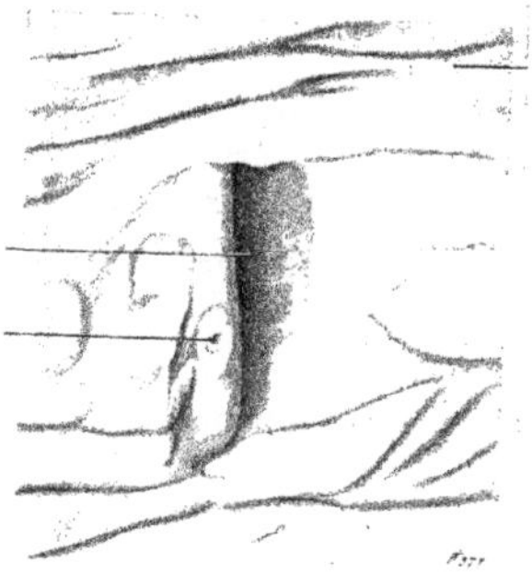

11

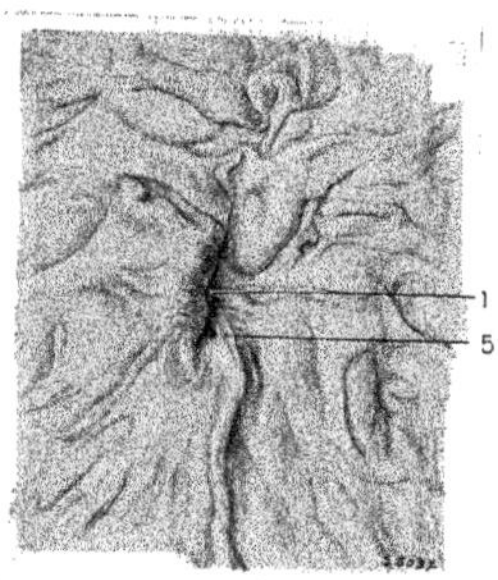

12

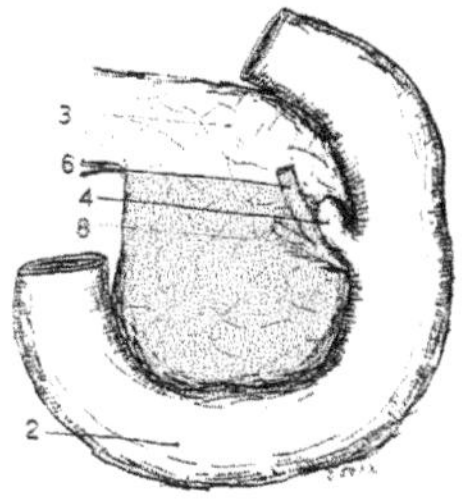

13

14

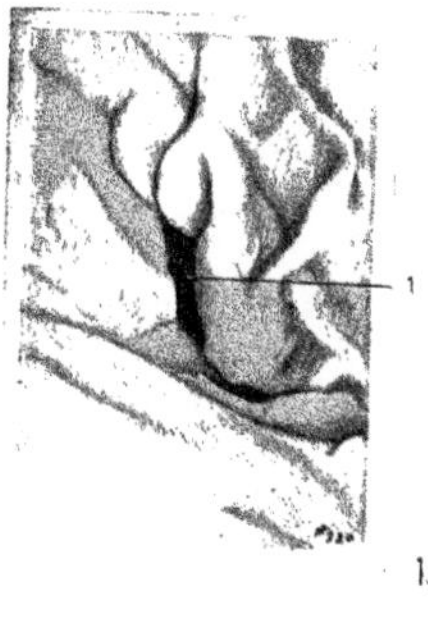

15

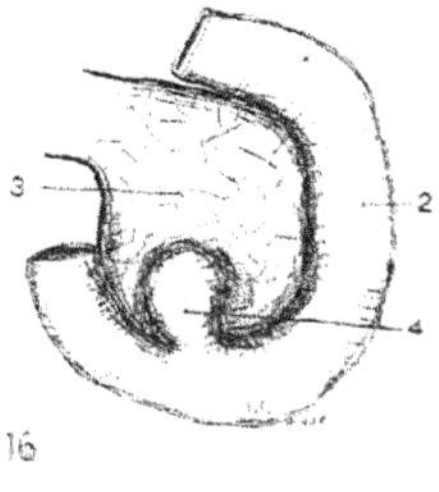

16

PLATE 3

EXPLANATION OF FIGURES

17 This sketch represents the duodenal mucosa exposed to show the orifice of the diverticulum.

18 In this diagram the head of the pancreas encircled by the duodenum (fig. 17) is represented as viewed from the dorsum. The fundus of the diverticulum located at the duodenal-jejunal flexure is shown in relation to the pancreas.

19 In this sketch the orifice of the diverticulum lies immediately cephalic to the major duodenal papilla.

20 This diagram represents the relation of the fundus of the diverticulum, shown in fig. 19, to the bile and main pancreatic ducts and to the dorsal surface of the head of the pancreas.

21 In this sketch the orifice of the diverticulum is shown upon the mucosa of the third portion of the duodenum.

REFERENCE NUMBERS

1 Orifice of diverticulum.
2 Duodenum (dorsal aspect).
3 Head of pancreas (dorsal aspect).
4 Diverticulum.
5 Major duodenal papilla.
6 Ductus choledochus
8 Ductus pancreaticus.

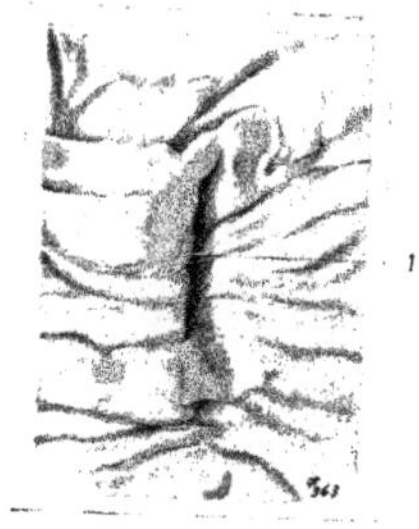

17

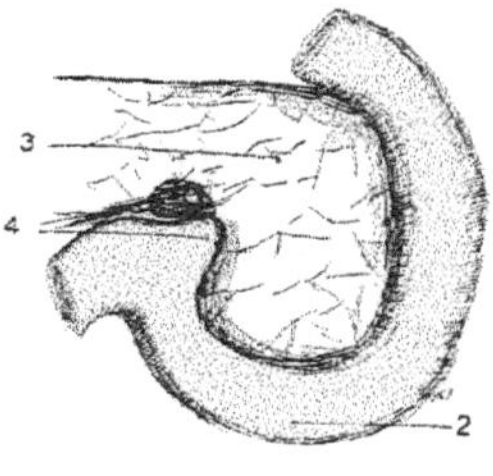

18

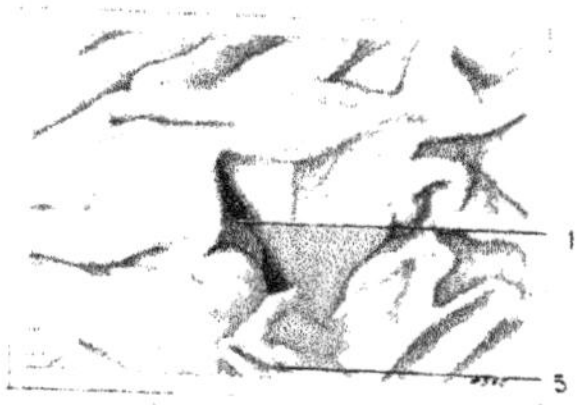

19

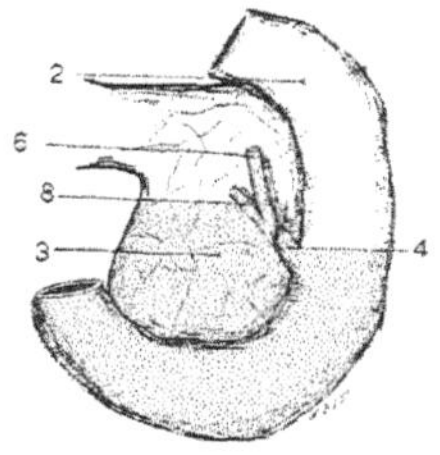

20

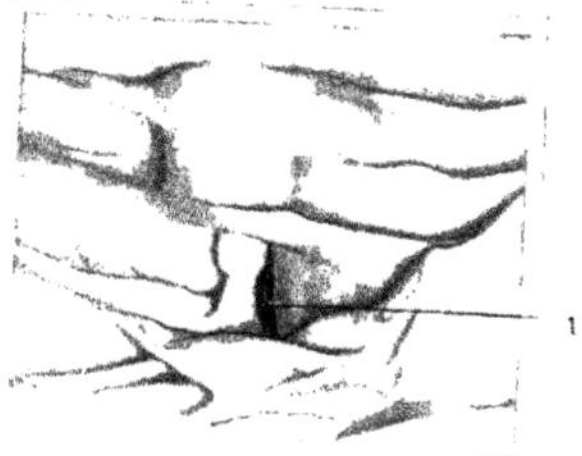

21

PLATE 4

EXPLANATION OF FIGURES

22 In this sketch the relation of the small tubular diverticula to the minor duodenal papilla is seen.

23 This sketch represents the mucosa of the third portion of the duodenum with the orifice of the diverticulum.

24 In this diagram the fundus of the diverticulum, represented in fig. 23, is shown as seen from the dorsum. The fundus extends cephalically lying dorsal to the head of the pancreas but it has no immediate relation either to the ducts of the pancreas or to the bile duct.

25 In this sketch a large diverticulum is represented opening into the third portion of the duodenum.

26 This diagram represents the fundus of the diverticulum, shown in fig. 25, as seen from the dorsum. It lies dorsal to the head of the pancreas but has no relation to the ducts of the gland or to the bile duct.

REFERENCE NUMBERS

1 Orifice of diverticulum.
2 Duodenum (dorsal aspect).
3 Head of pancreas (dorsal aspect).
4 Diverticulum.
5 Minor duodenal papilla.

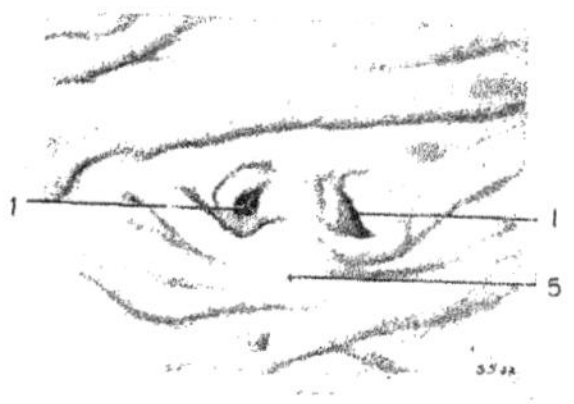

22

23

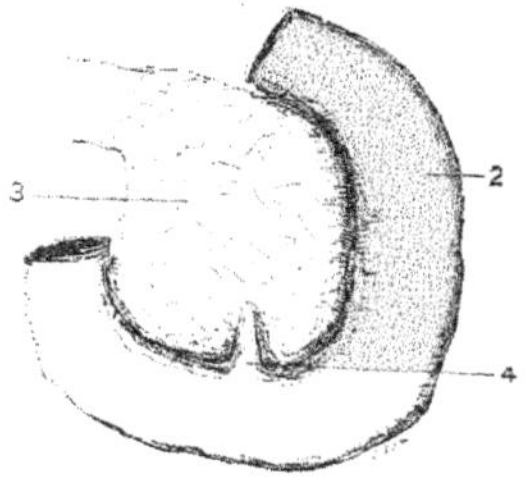

24

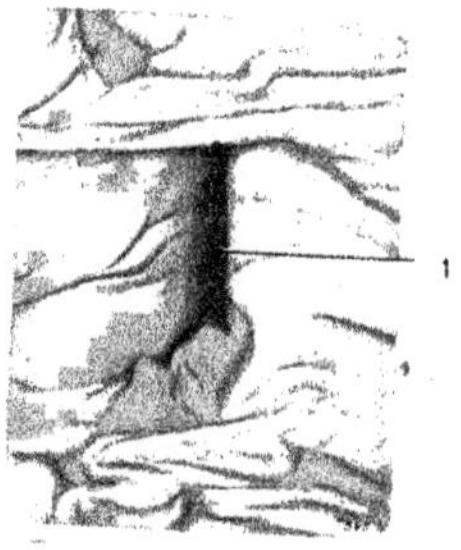

25

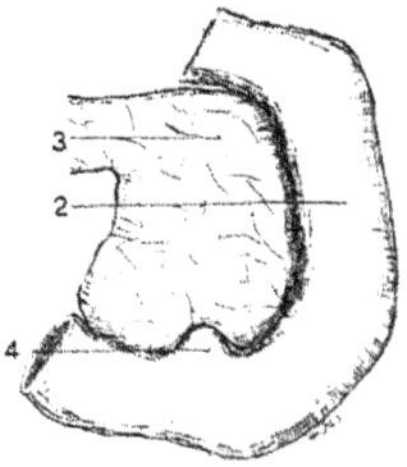

26

Reprinted from THE JOURNAL OF ANATOMICAL RECORD, VOL. 5, NO. 5.
May, 1911

THE PANCREATIC DUCTS IN MAN, TOGETHER WITH A STUDY OF THE MICROSCOPICAL STRUCTURE OF THE MINOR DUODENAL PAPILLA

W. M. BALDWIN

Cornell University Medical College

TWELVE FIGURES

As ordinarily described in the text-books, there exist in the substance of the pancreas, two ducts; one, the larger and more constant, called the 'pancreatic duct' or 'duct of Wirsung,' and the other, smaller and comparatively inconstant, the 'accessory pancreatic duct' or 'duct of Santorini.' The main duct, beginning in the tail of the pancreas, courses from left to right through the body and neck to the head of the gland where it bends caudally, after receiving the accessory duct, and, traversing the head of the gland, perforates the duodenal wall to empty into the duodenum in company with the common bile duct; occasionally, however, apart from it. The accessory duct, on the other hand, is confined to the cephalo-ventral segment of the head which it traverses from its point of junction with the main duct to the minor duodenal papilla where it either empties into the duodenum or terminates blindly. This minor papilla in the duodenal mucosa bears a cephalo-ventral relation to the major papilla containing the ampulla of Vater and lies about 1.8 cm. from it. The one noticeable feature of the anatomical descriptions of these parts is the discordance of opinion concerning the terminal relations of the accessory duct.

In the year 1641, Moritz Hoffmann discovered the duct of the pancreas while working on a rooster and showed his findings to Wirsung, who the following year dissected the duct in the pancreas of a human body. In a letter to Jean Riolan, Jr., Professor of Anatomy in Paris, Wirsung gave to the world the first account of his important discovery.

Wirsung had the duct reproduced on a copper plate from which but few copies were struck off. According to Choulant only two copies are known to be preserved. Schirmer (1893) saw one in the University of Strassburg and had a photolithographic reproduction of it made.

To Jo. Dominici Santorini belongs the credit for the first description of the accessory pancreatic duct and for the first representation approximating accuracy of the arrangement of the ducts in the adult human pancreas. He called attention to the existence of two papillæ in the duodenal mucosa and figured them in table 12 of his published work. In this connection, also, mention should justly be made of the name of Regner de Graaf, who previously had reported that, contrary to what had been the prevailing opinion, the pancreas might present two or even three ducts.

A complete list of the workers upon this particular problem in connection with the pancreas is lengthy; it includes such names as Vesling, who reported two ducts, apparently in lower animals, Thomas Bartholinus, Bernard Swalwe, G. Blasius, Johannes von Muralt, and Christianus Ludovicus Welschius. Now that the identity of the ducts was established, investigators began to report anomalous conditions of these passages; as Albrecht von Haller, Tiedemann, Mayer, and M. Bécourt. J. F. Meckel's was a significant statement in explanation of the causative factors involved in the production of the numerous anomalous conditions observed, *i. e.*, that atrophy of the duodenal end of the accessory duct was the developmental rule. A further list of workers at this period includes such names as Huschke, Jean Cruveilhier, and Sappey. Since the time of Claude Bernard, who in 1846 revived interest in the accessory duct, which had apparently been neglected, much attention has been given to the relation of the accessory duct to the main duct and to the duodenum. An incomplete list of the investigators thus engaged with the number of specimens studied is as follows: Bécourt, 32; Verneuil, about 20; Henle, Sappey, 17; Hamburger, 'mehr als 50'; Schirmer, 105; Schieffer, 10; Helly, 50; Charpy, 30; Letulle, 21; Opie, 100.

A study of the different methods employed by the investigators in their efforts to ascertain the condition of patency or occlusion of one or both ends of the accessory duct is of two-fold interest, because it illustrates the ingenuity of the workers, and, secondly, gives a probable explanation of the inharmonious results of their work.

For example, Claude Bernard used injections of metallic mercury which he forced into the main duct; Sappey, likewise working with mercury, ligated the ampulla and injected through the common bile duct. Schirmer, however, availing himself of Henle's objection to the use of mercury as an injection fluid for reason of its liability to burst through what might be a natural barrier at the blind duodenal end of the accessory duct, had recourse to the ingenious method of blowing air at a low pressure through the duodenal orifice of the main duct while the whole gland was submerged in water. Charpy used the injection method. His fluids were alcohol and some coagulable fluids, followed in some instances by the air injection method. Taking his cue from Charpy's comment upon Schirmer's method in which the former said that it was well to dislodge by friction the mucous which might obstruct the ducts, Helly went one step farther towards accuracy by subjecting the minor papilla to a microscopical examination, after he had injected the ducts with a gelatin mass. His findings substantiated the previously advanced objections. Several times the injection mass broke down a natural barrier thus giving rise to the erroneous conclusion, had the microscopical examination not followed, that the channel in life had been patent. On the other hand, Helly found that several times a small accumulation of mucous was sufficient to completely block the accessory duct.

This present work comprises a study of one hundred specimens of adult human pancreas removed, with the exception of four derived from autopsies, from the bodies used in the regular dissecting courses in anatomy in the Cornell University Medical College at Ithaca, New York. The bodies had been embalmed with a mixture of equal parts of carbolic acid, glycerin, and 95 per cent ethyl alcohol. The ages of the individuals ranged from 21 to 95 years. There were 57 males and 21 females plus a series of twenty-two specimens from which the identification tags had been lost and consequently all data. Death in no instance had been caused by pathological processes localized either in the pancreas or in the duodenum.

The method followed in the examination of the specimens was as follows: the ductus pancreaticus was located by gross dissection, with the aid of a lens magnifying two diameters, in the neck of the gland where the pancreatic tissue overlies the superior mesenteric vessels. At this level the duct approaches the dorsal surface of the gland and is readily found usually about midway between the cephalic and caudal borders at a depth of 2 or 3 mm. in the gland substance. Once located, the duct was easily traced both towards the tail of the gland and towards the duodenum. In both of these regions it was found to lie nearer the dorsal than the ventral surface of the gland.

The junction with the accessory duct was most quickly reached by working along the ventral surface of the main duct beginning at the neck and proceeding towards the duodenum. In those anomalous instances where this duct could not be located by this method, the duodenum was opened along its right border and the position of the minor papilla ascertained. Then, using this as a guide, the accessory duct was sought for in the glandular tissue cephalad to the level of the papilla. In those instances where no junction of the accessory duct and the main duct could be readily ascertained upon gross dissection, a ligature was passed around the duodenal end of the accessory duct at the point of perforation of the duodenal wall, and the main duct injected with a stain, either aqueous eosin or methylene blue. Regurgitation of the fluid into the accessory duct evidenced the presence of a communication between the two ducts. However, in no instance was the accessory duct injected with air or any fluid as a means of ascertaining the condition of patency of its duodenal termination.

The relation of the ducts to each other being thus established, the minor papilla entire, including the adjacent duodenal wall and a small portion of pancreatic tissue, was imbedded in paraffin, sectioned in series, and stained with the hæmatoxylin and eosin method. In four instances the accessory duct was of such a large calibre as to be readily followed through the papilla by gross dissection. In these few specimens the papilla was not sectioned and studied microscopically.

The relation of the main pancreatic duct to the termination of the bile duct was studied by gross dissection, while by slitting open both ducts their part in the formation of the major duodenal papilla was ascertained. The major papilla, however, was not sectioned or studied under the microscope.

In an investigation of this nature covering so much ground and productive of so many data it has seemed wise to present the facts of the problem in the topical order herewith listed.

1. The duodenal mucosa; its papillæ and intestinal valves.
2. The main pancreatic duct; course, tributaries, and drainage.
3. The duodenal termination of the pancreatic duct in the major papilla and its relation to the bile duct.
4. The accessory pancreatic duct; course, tributaries, and drainage.
5. The minor papilla; relation to the accessory duct and microscopical structure.

6. The relation of the main pancreatic duct to the common bile duct at the duodenal wall.
7. The bile duct and the major papilla.

1. The duodenal mucosa; its papillæ and intestinal valves

Attention was given to a study of the arrangement of the intestinal valves. These were exposed by laying open the duodenum along its convex border. The minor and major papillæ were present in every instance of this series of one hundred specimens. Locating the major papilla presented but little difficulty. On the other hand, it was occasionally only as the result of the most careful search that the minor papilla could be identified. It lay cephalad and on a plane ventral to the major papilla in ninety of the specimens. In eight instances the two papillæ were on the same vertical plane, the minor papilla being cephalad. Finally, in the two remaining specimens the minor papilla lay upon the same transverse plane with the other papilla but ventral to it. The fact is worthy of special mention that in no instance did the minor papilla occupy a position either caudal to the major papilla or dorsal to it. Separating these papillæ, the average distance, measured from center to center, was 2.0 cm. The shortest distance observed was 0.9 cm., and the longest 3.5 cm., and the mean distance 2.1 cm.

One specimen presented three papillæ; the minor papilla occupying the usual position relative to the major papilla and 2.3 cm. from it. The accessory papilla lay 1.0 cm. directly cephalad to the minor papilla. This third papilla had no pancreatic duct opening through it.

Notwithstanding the apparently hap-hazard and chance disposition of the smaller and incomplete mucosal folds in the vicinity of the papillæ, there could be identified in these specimens a marked conformity of the larger intestinal folds or valves to a fixed and entirely characteristic arrangement. In order that a more intelligible description might be made of these valves, I have divided them into two classes, *i. e.*, 'primary' and 'secondary.'

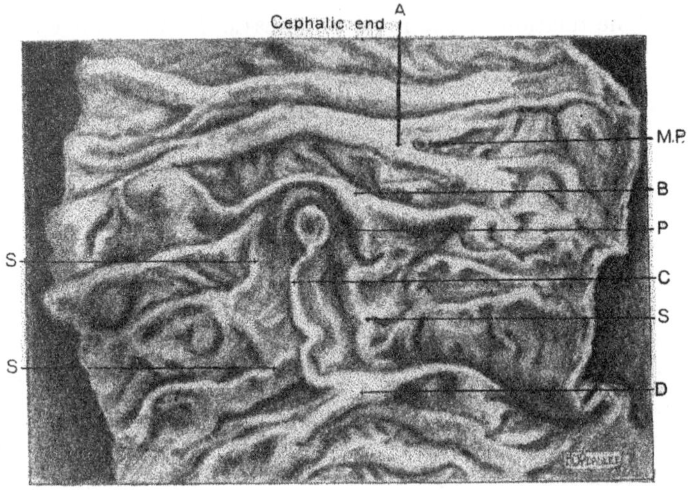

Fig. 1 (natural size) represents the typical distribution of the 'primary' and 'secondary' folds of duodenal mucosa in the region of the two papillæ in the descending portion of the duodenum. *M.P.* Minor papilla. *P.* Depression containing the major papilla with the orifices of the bile and the pancreatic duct. *C.* Plica longitudinalis duodeni. *A,B,D.* 'Primary' folds. *S,S,S,* 'Secondary' folds.

The basis of this classification is dependent entirely upon the size and constancy of the folds (fig. 1).

The minor papilla (*M.P.*) occupies a position upon a prominent 'primary' transverse fold or valve (*A*) and often at its bifurcation as represented in the drawing. It lies not on the ridge or crest of the valve but within the angle of bifurcation on the side of the fold. About 0.5 cm. caudal to this, a second, also 'primary,' fold (*B*) traverses the duodenal wall. Beginning at this second valve a prominent 'primary' longitudinal fold (*C*) proceeds caudally at a right angle and in the direction of the long axis of the duodenum. This is the 'plica longitudinalis duodeni' of the textbooks. Upon its summit and close to its cephalic extremity, in fact, overlapped by the fold indicated at (*B*), lies the major papilla presenting the orifices of the bile and pancreatic ducts. This plica passes caudally uninterruptedly across two or three 'primary' and 'secondary' (*S*) folds to terminate in another 'primary' transverse fold (*D*) located at the junction of the descending and transverse portions of the duodenum. Scattered among these 'primary' plicæ are many small 'secondary' folds (*S*), which have no definite or constant arrangement, which branch often, and which join or fuse with the larger folds at varying angles.

In addition to these features the observation was made upon thirty of the specimens that if the valves were disregarded, ironed out, so to speak, and merely the general contour of the duodenal wall considered, at the level of the major papilla a distinct bulging or hollowing of the wall towards the head of the pancreas was demonstrable. This feature was superadded to the constant dorso-ventral curvature of the duodenal wall and existed as a distinct entity. Furthermore, in sixty of the one hundred specimens, the major papilla occupied a small, localized, but, nevertheless, distinct pitting of the duodenal wall (*P*). This pitting was produced by simply a localized exaggeration of the general hollowing of the medial wall of the duodenum mentioned above. To the mind of the author, however, this hollowing was suggestive of a possible persistence of the original diverticulum from which in the embryo both the liver and the ventral pancreas developed.

2. The main pancreatic duct; course, tributaries, and drainage

In order clearly to understand the arrangement of the ducts and their accessory features in the adult, our consideration must be turned to the embryology of the pancreas, which has engaged the attention of many workers, including the names of His, Phisalix, Zimmermann, Felix, Hamburger, Janosik, Jankelowitz, Swaen, Helly, Völker, Kollmann, Ingalls, and Thyng. According to the results of these investigators the human pancreas develops from the duodenum at the level of the hepatic diverticulum from two buds or anlages, one ventral, and the other dorsal, the former being in association with the hepatic diverticulum.

Diverse views are entertained at present concerning the duplicity of the ventral anlage, some maintaining that the bud is single from the first while others hold that at the beginning it consists of two lateral halves which subsequently fuse. The author has recently published a paper descriptive of an unusual form of adult pancreas which possibly exemplifies a persistence of the earlier embryological condition of the primitive anlages.

Concerning the dorsal anlage we may say that contrary to what had been previously demonstrated by Hamburger, Felix, and Janosik, Thyng's studies seem to prove that "the dorsal pancreas arises from the intestine distinctly anterior to the hepatic diverticulum." From the dorsal bud the cephalic portion of the head and all of the neck, body, and tail of the pancreas develop, the ductus pancreatis dorsalis draining these portions. The ventral pancreas

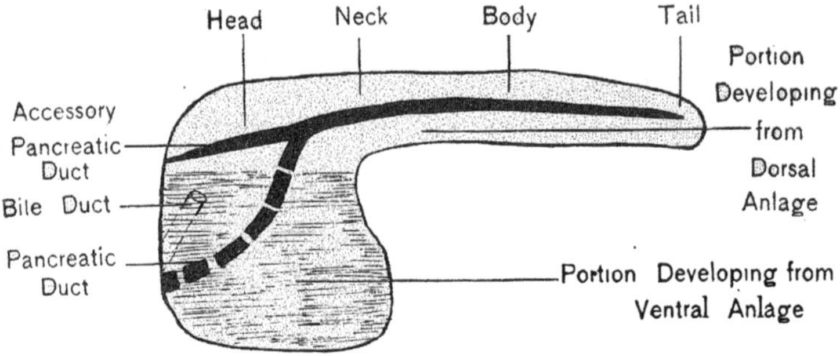

2

enclosing its duct, the ductus pancreatis ventralis, forms ultimately the caudal portion of the head of the gland (fig. 2).

The accompanying figure (2) represents diagrammatically the parts of the gland derived from the two anlages. The clear portion traversed by the heavy unbroken line is developed from the dorsal anlage, while the shaded portion is derived from the ventral anlage. The terms suggested by Thyng 'ductus pancreatis ventralis' and 'ductus pancreatis dorsalis' are particularly applicable from the standpoint of their embryological relations.

As development progresses, however, the ducts unite as is shown in the sketch, the duct of the dorsal anlage then undergoing a certain degree of atrophy at its duodenal end to thus produce the adult arrangement (see also fig. 3).

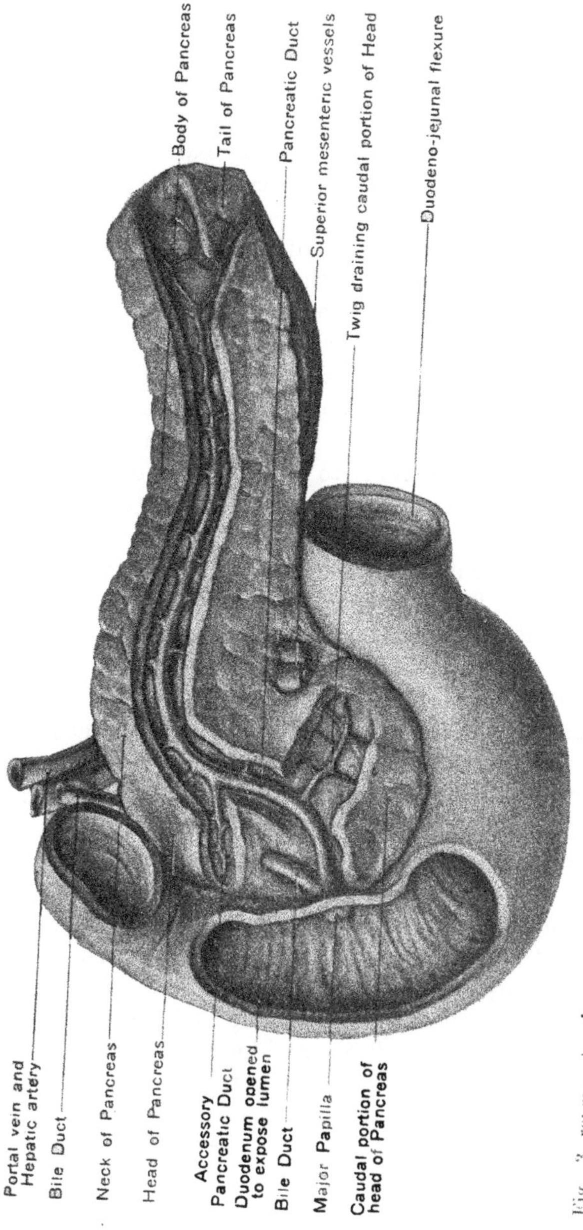

Fig. 3 represents the ventral surface of the pancreas and duodenum, with a portion of the glandular substance removed to show the disposition of the ducts as ascertained by this investigation. A window has been cut in the duodenum to expose the major papilla.

By keeping these facts of embryology in mind the anatomical findings of this investigation have a much clearer interpretation. The main duct was observed to begin in the tail of the gland through the convergence of a number of small duct radicles. It could not be demonstrated that these conformed to any particular arrangement. Pursuing a more or less tortuous course, the duct passed thence from left to right, traversing the glandular substance of the body of the pancreas and approximating the dorsal rather than the ventral surface of the gland. Furthermore, the duct lay nearer the cephalic than the caudal border of the body.

Upon arriving at the head of the gland, the main duct inclined somewhat abruptly caudally and dorsally with the convexity towards the right and approached the dorsal surface of the head of the gland. Reaching the level of the major duodenal papilla, the duct now ran almost horizontally to the right to join with the caudal aspect of the bile duct and empty with it into the major papilla (fig. 3).

The tributaries of the main duct in the body of the gland were observed to join that duct almost invariably at right angles and also to alternate with tributaries of the opposite side in the level at which they joined the main duct. These same features in turn characterized somewhat less noticeably, however, the radicles of these tributaries of the main duct. Only in the head of the gland was the conformity to these rules departed from. Here the tributaries were, occasionally, of some irregularity, first, in the angle of junction, and secondly, in the arrangement of the radicles.

Here also there existed one large unpaired trunk quite variable in appearance but well represented in fig. 3. This is the chief channel of drainage of the small lobe of the head (lobe of Winslow) which lies dorsal to the superior mesenteric vessels. Winslow in 1732, and later Charpy, called attention to its constancy. In four or five of the specimens dissected the last two tributaries joined this duct at the same level, and in such a manner and of such proportions as to appear to form a third pancreatic duct traversing the caudal segment of the head parallel to the main pancreatic duct. In no instances, however, was there any observable direct communication between these twigs and the duodenum.

The main duct drained the whole of the body, tail, and neck of the gland, and in addition to these parts, in 66 per cent, or fifty of the seventy-six specimens studied for this purpose, the dorsal half of the head with nearly the whole of the caudal portion of the ventral half (fig. 2). This restricted the accessory duct to the ventro-cephalic portion in the immediate neighborhood of the minor papilla and to a small portion of the ventro-caudal segment. The main duct usually drained the whole of the region of the head adjacent to the neck. In four specimens the accessory duct drained the whole of the cephalic half of the head. In three instances the main duct was restricted to the dorsal and caudal portion of the head. In one other example the accessory duct drained nearly the whole of the head of the gland.

The outside diameter of the main duct, in those specimens with a normal arrangement of ducts (88), taken a few millimeters before it emptied into the duodenum and with the duct flattened out, averaged 3.25 mm. The mean diameter was 3.0 mm. In the body of the gland the duct averaged 3.0 mm. The smallest main duct measured 1.5 mm. and the largest 4.5 mm.

There were three specimens in the series which presented a rather unusual arrangement of ducts as represented in fig. 4. In these instances the main duct, descending towards the right into the caudal portion of the head, described a 'loop,' as shown in the figure, before finally proceeding horizontally to the major papilla. The accessory duct in these instances occupied its usual ventral and cephalic position and joined the main duct before the beginning of the 'loop.' In no instances was the main duct duplicated in the body of the pancreas as described by Bernard nor was there found the spiral disposition of the pancreatic duct as described by Hyrtl and figured in his Corrosion Anatomie.

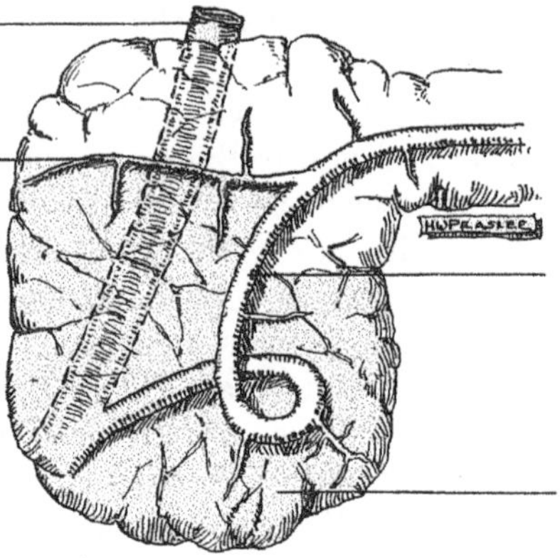

Fig. 4 A rough, schematic sketch of the ventral surface of the head of the pancreas showing the typical arrangement of the pancreatic duct in those specimens in which the 'loop' disposition prevailed.

3. The duodenal termination of the pancreatic duct in the major papilla and its relation to the bile duct

Ever since Bidloo first noted the papilla common to both the bile and the pancreatic duct, the relation of these two ducts to each other in the ampulla has been the subject of considerable investigation. Bernard and Laguesse each mentioned one specimen in which the main pancreatic duct opened into the duodenum apart from the orifice of the bile duct. Bécourt also recorded another instance. Schirmer reported twenty-two specimens (about 47 per cent) among forty-seven investigated in which a mucosal fold separated the orifices of the ducts in such a manner that a true ampulla did not exist. Opie examined one hundred specimens. In eleven instances no ampulla was present, the two ducts entering the duodenum separately. In the remaining cases the ampulla varied in length from less than 1 mm. to 11 mm., while in only thirty specimens did this measurement equal or exceed 5 mm. The ampullary orifice had an average diameter of 2.5 mm. Among twenty-one specimens which Letulle studied in only six was there a true ampulla.

The main pancreatic duct in this series of one hundred specimens approached the caudal aspect of the ductus choledochus to fuse with its wall before penetrating the duodenal wall (fig. 3). In two instances the main duct emptied into the caudal aspect of the bile duct at 1.3 cm. and 0.7 cm. respectively from the duodenal wall. Upon opening the ducts it was a noticeable fact that, notwithstanding the apparent fusion of the walls outside of the duodenum, the lumina did not unite until the duodenal wall had been perforated.

Fig. 5 represents diagrammatically the two classes into which the specimens reported and those studied in this investigation

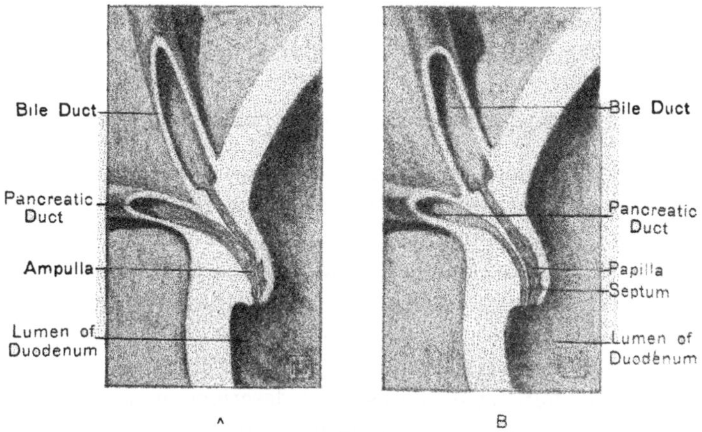

5

seem to fall. In *A* the walls of the two ducts are seen to fuse at the level of the duodenal wall. The lumina, on the other hand, do not fuse until the papilla has been entered. The thin mucous septum is shown separating the two ducts for a distance of at least one half of the papilla where the true ampulla can then be said to begin. Fig. 5, *B* represents the other general appearance noted, *i. e.*, complete isolation of the two ducts. The figure also gives a fairly good representation of the foliated appearance of the mucosa observed in the ampulla and mentioned at an earlier date by Bernard.

The distance from the mouth of the major papilla to the point of junction of the two ducts in the ampulla averaged 4.8 mm. (mean 4.0 mm.) in the ninety specimens dissected. In twenty of the specimens (about 22 per cent), there could be found no junction of the ducts, each opening side by side separately into the duodenum through the major papilla. This appearance is represented in fig. 5, *B*. A true ampulla was not present in these cases. In two specimens the distance observed was 0.5 mm. In twelve instances the partition was only 2.0 mm. from the mouth of the ampulla.

In but one pancreas was the duodenal end of the main duct occluded (fig. 6). The duct in this instance was a mere impervious twig which opened neither into the bile nor the accessory duct. The accessory duct drained the whole gland.

4. The accessory pancreatic duct; course, tributaries, and drainage

The accessory duct was found to be present in each of seventy-six specimens examined with that object in view. It was located entirely within the substance of the cephalo-ventral segment of the head (fig. 3), and pursued an arched course towards the duodenum. In no instance did it occupy a position wholly caudal to the main duct.

Invariably the accessory duct lay upon a plane ventral to that of the main duct. Two curves were described in its passage to the duodenum; the first of these, more pronounced and with its concavity cephalad, occupied the duct end, while the other the shorter of the two, was situated at the duodenal end with its concavity looking caudad. This condition, present in forty-two (64 per cent) of sixty-six specimens, is not clearly enough represented in fig. 3. In twenty-one specimens (31 per cent) the duct described a wide curvature with its concavity cephalad. Leaving the main duct it proceeded into the caudal portion of the head of the gland, then, turning to pass ventral to the main duct, emptied into the minor papilla. This appearance is represented in fig. 8. In the three remaining specimens (5 per cent) the usual curvatures of the duct were reversed, *i. e.*, a caudal concavity in the duct half with

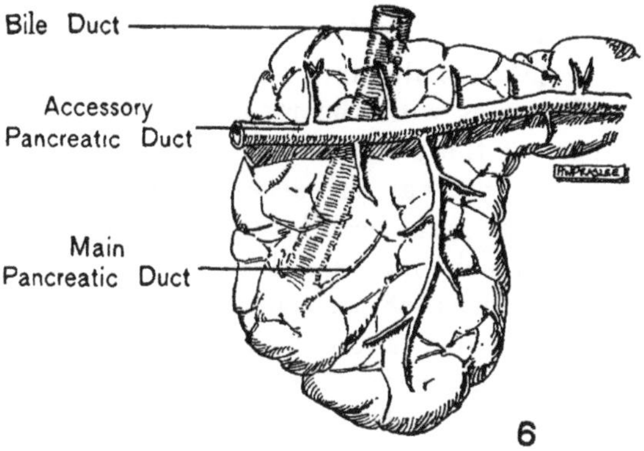

6

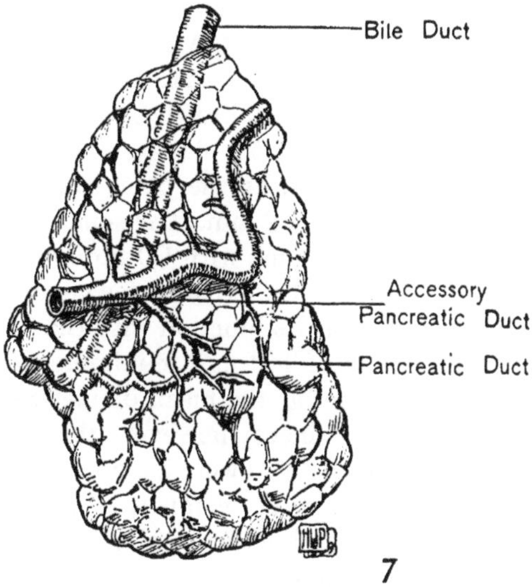

7

Fig. 7 In this schematic sketch the ventral surface of the head of the pancreas is represented. Here the ducts are 'inverted,' *i. e.*, the accessory duct conveys most of the drainage from the neck, body, and tail of the gland into the duodenum. The main duct, occupying its usual caudal and dorsal position, is inferior in size to the accessory duct but joins the bile duct to empty with it through the major papilla into the duodenum.

a cephalic concavity at the duodenal end. Apart from the twenty-one specimens above noted, in forty-five (69 per cent) the duct was restricted to the cephalic and ventral segment of the head.

Charpy's work agrees with the results of this investigation regarding the part of the gland drained by the accessory duct. Opie, however, thought that the accessory duct drained "the anterior and lower part of the head" restricting for the main duct a smaller mass of parenchyma 'behind the larger lobe.'

In fifty-eight specimens (88 per cent) the duct approached the duodenum with diminishing calibre; in six specimens (9 per cent) the duodenal end was larger (fig. 7) while in two (3 per cent) both ends were of the same size. These figures, as would be expected, were compiled from those specimens in which the duct united with the main duct in the usual manner, namely, from sixty-six specimens. In ten other specimens where no demonstrable junction was present, the accessory duct naturally approached the duodenum with an augmenting calibre (figs. 9 and 10).

The outside diameter of the flattened accessory duct in these sixty-six specimens, taken at the point where it perforated the duodenal wall, averaged 1.2 mm. The smallest observed was 0.75 mm. and the largest 2.0 mm., with 1.0 mm. as the mean diameter of this end of the duct. Under the same conditions the other end of the accessory duct at its junction with the main duct measured 1.75 mm. with limits of 1.0 mm. minimum and 3.0 mm. maximum and with 1.5 mm. as the mean diameter.

In three other specimens, however, the maximum diameters observed at the duodenal end were 2.5 mm., 3.0 mm., and 3.5 mm. respectively, but these were instances of inversion of the ducts, *i. e.*, the main duct was inferior in size to the accessory duct as represented in figs. 7 and 9.

These facts bear out Meckel's statement that in the fœtus the two pancreatic ducts possess the same calibre, but as development progresses the accessory duct undergoes a natural atrophy at its duodenal end. This fact was also noted by Bernard and verified still later by Schieffer upon five human fœtuses from 7.5 to 9 months of age.

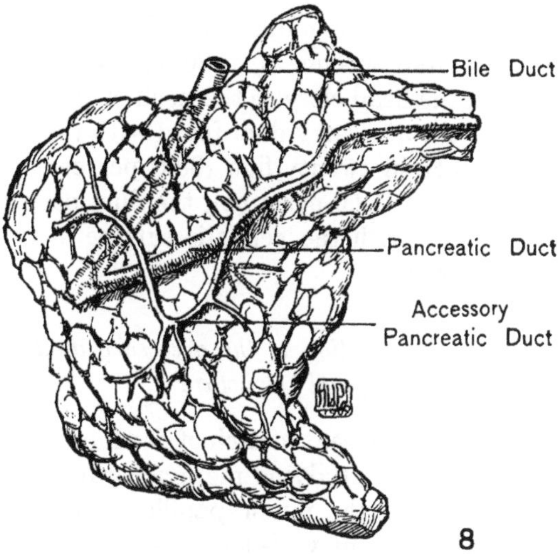

8

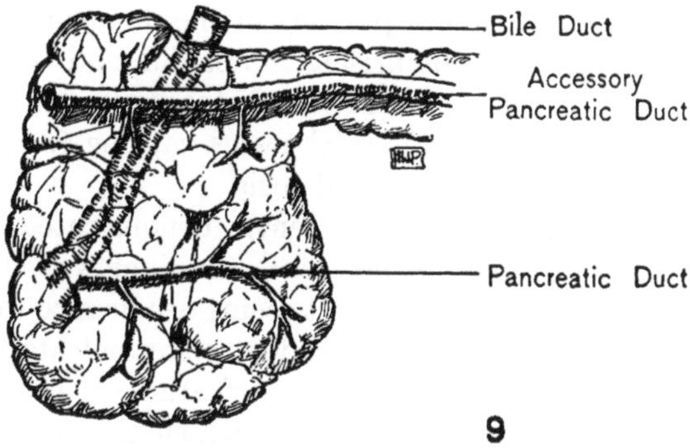

9

Fig. 8 This sketch represents the ventral surface of the head of the pancreas showing the accessory duct passing through the caudal portion of the head.

Fig. 9 represents a condition present in five specimens of the seventy-six dissected (6.5 per cent). The ducts do not unite. The accessory duct, larger than the main duct, drains the whole of the body, tail, neck, and cephalic half of the head. This is a persistence of the embryonic arrangement of ducts. To these ducts the terms ductus pancreatis ventralis for the main duct and ductus pancreatis dorsalis for the accessory duct are particularly applicable.

In ten of seventy-six specimens (13.2 per cent) the accessory duct failed to join with the main duct (figs. 9 and 10). The junction in the other sixty-six specimens, or 86.8 per cent of cases, was found invariably in the head close to the neck of the gland (fig. 3). Among these latter the accessory duct fused with the ventral surface of the main duct in twenty-five (38.0 per cent), with the caudal surface in twelve (19.0 per cent), the duct passing ventral to the main duct; and with the cephalic surface in twenty-nine specimens (43.0 per cent).

5. The minor papilla; relation to the accessory duct and microscopical structure

The minor papilla was present in each of one hundred specimens examined. As a means of studying more accurately the relation of the accessory duct to the minor papilla, forty-six out of a total series of fifty specimens were subjected to a microscopical examination without first having been injected as a means of ascertaining the condition of patency of the duodenal end of the duct. A block of tissue comprising the papilla and the duodenal wall with the adjacent pancreatic substance was imbedded in paraffin, sectioned in series in thicknesses varying from 12 to 40μ and stained with hæmatoxylin and eosin.

Forty-one specimens (82 per cent) demonstrated a patent accessory duct. In five (10 per cent) the duct was closed, terminating blindly at the papilla. It seems needless to say that in these last specimens the accessory duct communicated with the main duct through an ample orifice. A feature especially worthy of mention was the abrupt manner in which the accessory duct in these five instances became constricted from an ample lumen to one of capillary dimensions and then terminated abruptly at the papilla. This abrupt dwarfing of the duct was no exceptional feature confined to these five isolated specimens. It was the rule rather than the exception. In brief, as was frequently verified, amplitude of calibre was no criterion of patency. In the four remaining specimens of the series of fifty selected (8 per cent), the patency of the accessory duct was so manifest as to be demonstrable upon gross dissection. In these, therefore, no microscopical examination was made. This gives, then, among fifty specimens examined, a total of five (10 per cent) which did not communicate with the duodenum. Further, in six other specimens the accessory duct did not unite with the main duct, giving, therefore, a total percentage of practical importance of eleven specimens out of fifty (22 per cent) in which fluid could not pass from the main duct into the duodenum through the accessory duct.

The shape of the papilla was uniformly rounded or conical with a diameter averaging 2 mm. and an aperture quite variable,

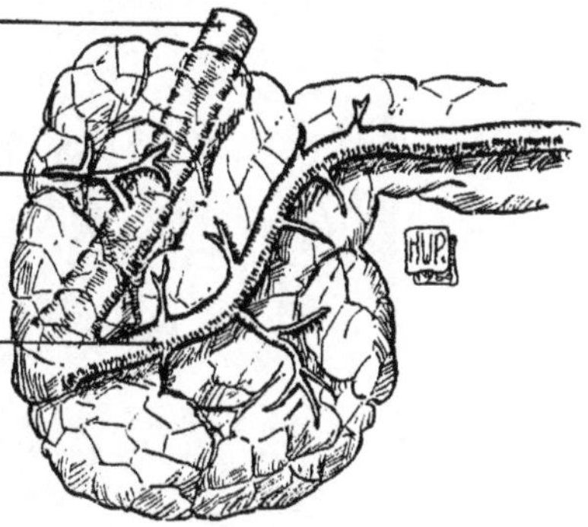

Fig. 10 shows an arrangement found in five specimens (6.5 per cent) of the seventy-six dissected. The accessory duct is isolated and smaller than the main duct. It drains but a small region in the immediate neighborhood of the minor papilla, through which it opens into the duodenum.

most often not visible to the unaided eye. The epithelial covering did not differ in appearance from that found in the rest of the duodenum (fig. 11, *E*). The mass of the papilla was composed of a core (*C.C.*) This core, imbedded in the mucosa and submucosa of the gut and extending obliquely from the muscularis (*M*) to the epithelial covering of the papilla (*E*), consisted of a supporting framework of dense connective tissue, and appeared as a constant factor in the structure of the papilla. It was present when the accessory duct failed, indeed, its size, which contributed largely to the proportions of the papilla, seemed less referable to the presence of the accessory duct than to the

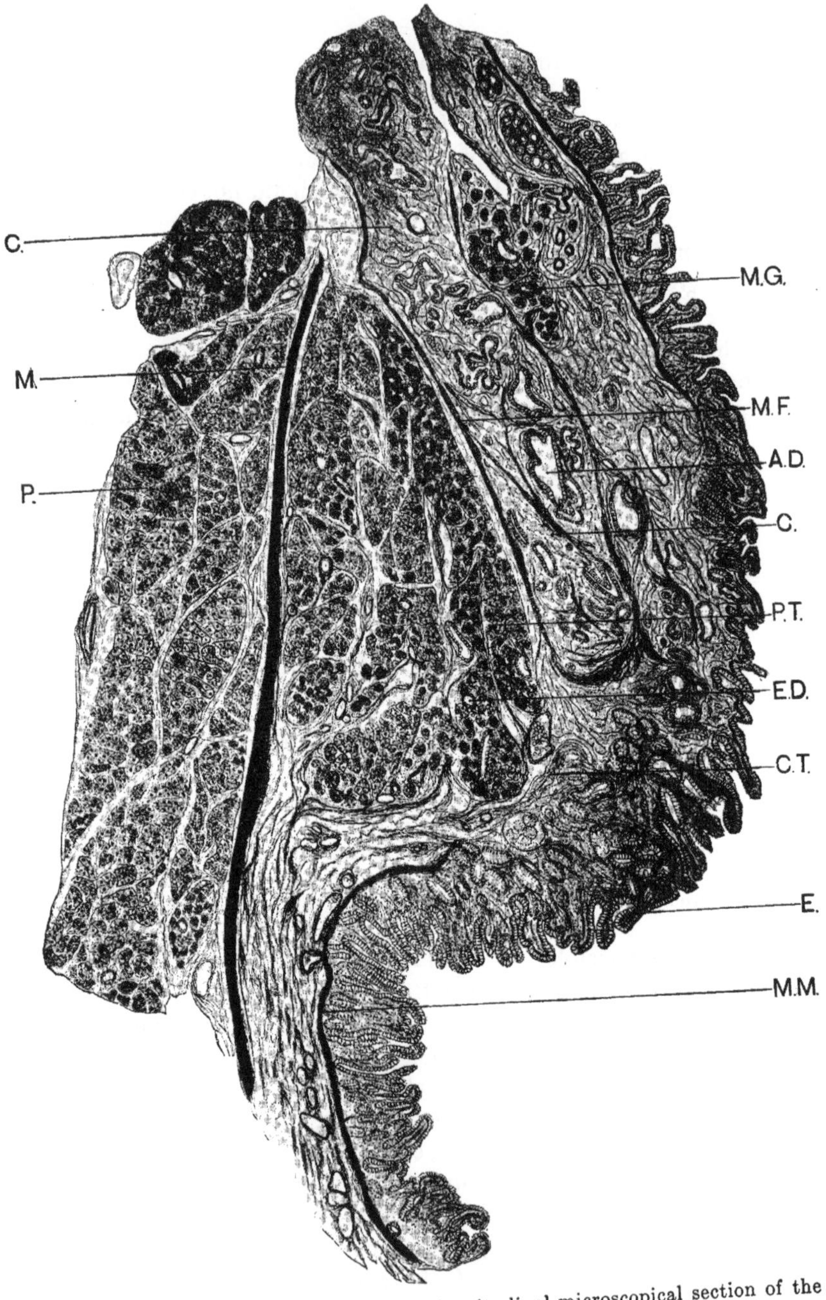

Fig. 11 (magnified 18 diameters) A longitudinal microscopical section of the minor duodenal papilla showing the passage of the accessory duct through the 'core' of connective tissue.

quantity of mucous glandular tissue enclosed within its stroma. The prominent features, then, of the papilla were this core of dense connective tissue containing many smooth muscle fibres (*M.F.*) and enclosing much mucous glandular tissue, with the accessory duct (*A.D.*) traversing the middle of its substance. Its whole appearance was strongly suggestive, however, of functional regression remindful of Meckel's observation regarding the developmental atrophy of the duodenal end of the accessory duct.

The accessory duct (*A.D.*) passed directly from the pancreatic tissue (*P*) of the head of the gland, which accompanied it up to the duodenal wall, through both layers of muscular tissue (*M*). Entering immediately into the substance of the core, it passed through the middle of its stroma to open finally into the duodenum. At the level of perforation of the duodenal wall, it underwent an abrupt caudal bending. The angle of this flexure, as previously noted by Helly, varied from 20° to 30°. In thirty-seven of the fifty specimens (74 per cent) the duct passed in a caudoventral direction through the papilla; in six specimens (12 per cent) it curved caudodorsally; and in the remaining seven specimens (14 per cent) horizontally ventral. Occasionally it was noted that the fibres of the muscularis formed a sphincter-like ring around the duct at the level where it perforated the duodenal wall.

There was no difficulty experienced in tracing the accessory duct through the pancreatic tissue which accompanied it up to the duodenal wall. The lumen was direct, uniform, and either gradually enlarging or diminishing in calibre. Once that the muscularis was perforated, however, the appearance of the duct was transformed to a remarkable extent. The lumen now became tortuous and irregular, dilating and narrowing, and, at times, branching to reunite farther along in its course. To add to the complexity of this arrangement the association with the duct in the core of numerous mucous glands, whose individual or combined ducts either opened directly into the accessory duct or independently into the intestine, rendered the tracing of the lumen of the duct particularly difficult.

The amplitude of lumen of the accessory duct as it approached the papilla offered no trustworthy suggestion of its condition of patency or occlusion in the core. Oftentimes a duct with the largest calibre dwarfed instantly to capillary dimensions upon entering the core. On the other hand, occasionally the smaller and most unpromising ducts traversed the core with a direct, unsinuous, and even enlarging lumen. In six specimens (12 per cent) the lumen of the accessory duct gradually increased in size as it traversed the papilla towards the epithelial covering. In ten specimens (20 per cent) the lumen sustained a pronounced diminution in calibre, while in the remaining thirty-four cases (68 per cent) the duct was so tortuous and irregular as to make it impossible to say whether there was an actual increase or a reduction in size.

In the five instances in which the accessory duct did not communicate with the duodenum, the duct was found to have perforated the muscularis. In the core immediately adjacent to the muscular coat, however, it suddenly underwent a diminution in calibre and terminated blindly in the connective tissue of the stroma of the core (*C*). In some of these specimens the strands of connective tissue separating the duct lumen from that of the adjacent mucous glands were so delicate that it seemed possible and, indeed, quite probable that they could be broken down by an injection mass even under the lightest pressure, thus giving rise to erroneous conclusions as to the condition of patency of the duct. Thus was confirmed both Henle's and Helly's objections to injection methods.

In those instances where the duct was comparatively ample and the lumen could be followed with little difficulty through the papilla, the mucous glandular, and pancreatic tissues played only subsidiary parts in the formation of the core and but little muscular tissue was discernible. In the majority of instances, however, the duct, being constricted, formed but a small part of the core. In these specimens the connective tissue and muscular stroma were very prominent. In these, too, aside from the epithelial lining of the lumen, there was no distinct wall to the duct. The true wall ceased where the duct perforated the muscular coat of the duodenum.

The duodenal orifice in the instance of the small ducts appeared like that of an intestinal gland, the lumen proper of the duct, indeed, seemingly, opening into the fundus of the tubule and the side walls not differing from those of the usual intestinal gland. Larger ducts, however, opened through what appeared to be several fused tubules. This orifice occurred either upon the caudal slope or upon the summit of the papilla, seldom upon its cephalic aspect.

Many glands (*M.G.*), which from general appearances seemed to be mucous in character, were found associated with the accessory duct. The characteristic staining of mucous glands could not, however, be obtained in the instance of these glands owing, doubtless, to their imperfect fixation. Kölliker noted the occurrence of these glands in the walls of the larger ducts of the pancreas apart from the papillæ.

These glands occurred in large, irregular, spherical groups situated either with the accessory duct within the core or immediately outlying it in the loose connective tissue of the papilla. The ducts of those glands situated within the core opened through irregular channels into the accessory duct. Those located near the epithelial extremity of the core imbedded in the connective tissue of the papilla opened directly upon the surface of the duodenal mucosa, while those farther removed from the epithelium emptied by longer channels, either into the accessory duct or upon the surface of the mucosa. The presence or absence of the accessory duct did not seem to influence the number of these glands so much as might be expected. In the five specimens of occlusion of the duct, they opened either directly upon the surface of the mucosa or indirectly through a lengthy, tortuous channel which occupied the usual position of the accessory duct in the core. When the duct was very large and patent the mucous glands were fewer in number and much more scattered. No instances were found where the glands were entirely absent.

In confirmation of Helly's and of Opie's earlier observations, small masses of pancreatic tissue (*P.T.*) were found in two situations, first, within the core close to the duodenal muscularis; secondly, in the loose connective tissue of the papilla usually upon the caudal aspect of the core. This pancreatic tissue differed from the tissue of the pancreas itself only in the distribution of the supporting connective tissue, the latter occurring in thick, well-marked septa isolating the lobules and acini from each other. The ducts from the acini united into larger trunks which emptied either directly into the accessory duct or independently upon the epithelial surface of the duodenum.

The unstriped muscle tissue (*M.F.*) contributed largely to the thickness of the septa of the core. The fibres were scattered either parallel to the long axis of the core or were disposed circularly around some of the tubules. The amount of muscular tissue was greater towards the muscularis side of the papilla but few fibres reaching the level of the intestinal glands. There could not be observed any relation between the condition of patency or occlusion of the accessory duct and the number of these fibres.

Claude Bernard thought that the papilla was contractile.

The relation of the muscularis mucosæ (*M.M.*) of the duodenal wall to the tissue of the core could not be ascertained in every instance. In some specimens it was continuous with the muscular tissue of the core. In many others, however, it could be clearly seen that there was no continuity of this structure with that in the core.

The following pages present a tabular compilation of all of the work which has been done upon the features presented in this problem.

The minor papilla

	PRESENT	ABSENT
Schirmer	103	1
Helly	47	3
Charpy	29	1
Verneuil	20	0
Baldwin	100	0

Thus it will be seen from the above that in about 98 per cent of specimens the minor papilla is present.

But four specimens of three papillæ have been reported:

Schirmer	1
Letulle	1
Rollestin and Fenton	1
Baldwin	1

None of these accessory papillæ have been studied microscopically. The papillæ are of possible interest in two ways, first, because of the occasional bifid character of the ventral anlage, and, secondly, because of the occasional appearance of a third duct in the caudal region of the head.

The accessory duct

	PRESENT	ABSENT
Schirmer	101	3
Charpy	29	1
Helly	50	0
Verneuil	20	0
Santorini	?	0
Bernard	?	0
Hamburger	50+	0
Sappey	17	0
Opie	100	0
Baldwin	76	0
About	443	4

Complete absence of the accessory duct according to these figures seems to be a rare anomaly since it occurs in less than 1 per cent of specimens.

Condition of accessory duct at duodenal end. With microscopical method

	PATENT	CLOSED
Helly	40	10
Baldwin	45	5
Total	85	15

With injection method

Schirmer	85	19
Charpy	9	21
Opie	79	21
Verneuil	20	0
Sappey	16	1
Total	209	62

According to the microscopical method, 15 per cent of ducts are occluded; with the injection method about 23 per cent are closed at the duodenal end.

Relation of main to accessory duct

	JUNCTION	NO JUNCTION
Opie	90	10
Duval	?	1
Helly	48	2
Charpy	28	2
Schirmer	97	7
Verneuil	20	0
Baldwin	66	10
Total	349	32

According to these figures a junction between the main duct and the accessory duct is to be expected in over 90 per cent of specimens.

The accessory duct is larger than the main duct

	SPECIMENS EXAMINED	INVERSION
Schirmer	104	3
Charpy	30	3
Bernard		1
Morel and Duval	?	1
Opie	100	11
Bimar	?	1
Moyse	?	1
Baldwin	76	3
Total	310	24

Distance between major and minor papillae

	NUMBER EXAMINED	AVERAGE	LIMIT
		cm.	*cm.*
Bernard	?	3.5	(2.0–4.0)
Schirmer	104	2.5–3.5	
Letulle	21	1.8	(1.0–3.5)
Baldwin	100	2.0	(0.9–3.5)

6. *The relation of the main pancreatic duct to the common bile duct at the duodenal wall*

Schirmer mentioned eleven instances among a series of forty-seven specimens in which the pancreatic duct opened into the bile duct and also fourteen instances in the same series in which the bile duct opened into the pancreatic duct. In these cases the single conjoined duct was the only one entering the ampulla of the papilla. Verneuil seemed to believe that usually the pancreatic duct received the bile duct and that, accordingly, the ampulla of the papilla belonged to the pancreatic duct. The main duct did not fail in any of the ninety specimens of my series, in one, however, it was occluded at its duodenal end. Helly saw one instance where the main duct was absent; Schirmer, four; Cruveilhier, one; Charpy, one. (See topic 3, page 208.)

Occasionally the common bile duct opens into the duodenum in company with the accessory duct. No such instance was found in my series. Schirmer mentions five. Tiedemann mentions one case where both pancreatic ducts emptied separately into the duodenum apart from the common bile duct.

	THE DUCTS JOIN TO FORM AN AMPULLA	NO JUNCTION
Bernard	?	1
Laguesse	?	1
Schirmer	25	22
Opie	89	11
Letulle	9	12
Baldwin	70	20
Total	193	67

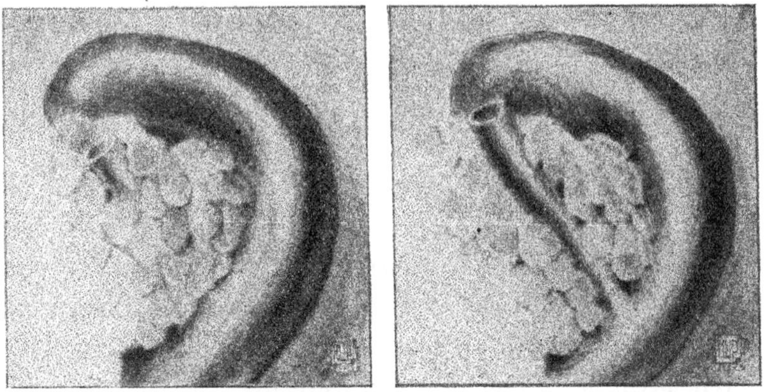

A B

Fig. 12 (dorsal view) represents the two conditions of the bile duct. In *A* the duct passes through the tissue of the head of the pancreas. In *B* the duct grooves the head of the gland but is not entirely surrounded by pancreatic tissue.

In about 25.8 per cent of specimens the ducts open separately into the duodenum. In 74.2 per cent the ducts have a common ampulla.

7. *The bile duct and the major papilla*

As an unavoidable adjunct to this study of the ducts of the pancreas the relations of the terminal or pancreatic portion of the bile duct were considered in this series of one hundred specimens. The duct ran invariably caudally towards the median surface of the second portion of the duodenum lying dorsal to the head of the pancreas and producing a furrow upon that surface. In no instance did it pass, as was observed by Helly, in a groove between the duodenum and the pancreas. In 80 per cent of the specimens the pancreatic tissue completely surrounded the duct for a distance varying from 0.5 cm. to 5.0 cm. In 5 per cent of specimens the duct received a partial investment without being entirely enclosed by glandular tissue, while in the remaining 15 per cent of specimens the bile duct grooved but was not covered by the tissue of the head.

The lumen of the bile duct underwent a marked contraction at the duodenal wall before its junction with the main pancreatic duct (fig. 5). Cephalad to this level a distinct bulging or ampulla was noticeable. The difference in calibre between these two adjacent portions was less appreciable upon the external surface of the duct than upon the internal. The outside diameter of the bile duct at the level of its perforation of the duodenal wall was 5.4 mm. in the one hundred specimens, that of the ampulla of the duct averaged 6.4 mm. The largest bile duct observed measured 15.0 mm. and the smallest 3.0 mm. This gives 6.0 mm. as the mean diameter of this duct. These measurements are outside diameters taken with the duct flattened out.

Letulle and Nattan-Larrier reported nineteen specimens in which the common bile duct traversed the head of the gland, often only a thin strip of pancreatic tissue separating the duct from the duodenum. Usually the glandular tissue extended a distance of only 2 or 4 cm. along the duct wall. The bile duct underwent a diminution in size in the last centimeter averaging 8 to 9 mm. in diameter. Several were from 12 to 14 mm.

O. Wyss found five specimens among twenty-two in which the terminal portion of the common bile duct penetrated the head of the gland. Helly studied forty specimens, in about half of which the duct lay in a canal of pancreatic tissue.

Judging from the results of these investigations, we should expect to find the terminal portion of the bile duct imbedded in pancreatic tissue in about 65 per cent of specimens.

Because of the nature of the material used in this investigation it was found impossible to use the whole series of one hundred specimens in the several portions of this problem. As many of them as were suitable, were utilized, however, with the result that a smaller number of specimens had to be reported upon in many of the essential features of the problem. This accounts, therefore, for the somewhat confusing use of varying numbers of specimens.

In conclusion I wish to express my sincere appreciation of the valuable advice and assistance given by Professor Gage, Dr. Kerr, and by Dr. Kingsbury in the preparation of this paper and for the numerous courtesies shown by their departments.

BIBLIOGRAPHY

BALDWIN, W. M. 1910 An adult human pancreas showing an embryological condition. Anat. Rec., vol. 4, no. 1, pp. 21–22.

1910 A specimen of annular pancreas, Anat. Rec., vol. 4, no. 8, pp. 299–304.

Duodenal diverticula in man. Anat. Rec., vol. 5, no. 3, pp. 121–141.

BARTHOLINUS, TH. 1651 Anatomia reformata.

BÉCOURT, M. 1830 Recherches sur le pancréas.

BERNARD, CL. 1856 Mémoire sur le pancréas.

BIDLOO, GOVERT 1685 Anatomia humani corporis.

BIMAR 1887 Conduits anormaux du pancréas. Gaz. hebdom. de Montpellier.

BLASIUS, G. 1677 Zootomia s. anatomia hominis.

BRACHET 1897 Sur le développement du foie. Anat. Anzeiger. B. 13, N. 23, S. 621–636.

CHARPY, A. 1898 Variétés et anomalies des canaux pancréatiques. Journ. de l'Anat. et de la Phys., p. 720.

CHOULANT, J. B. 1852 Geschichte der anatomischen Abbildungen.

CRUVEILHIER, J. 1833 Traité d'anatomie descriptive.

FELIX, W. 1892 Zur Leber und Pancreas Entwickelung. Arch. f. Anat.

GEGENBAUR, C. 1890 Anatomie des Menschen.

DE GRAAF, REGNER 1671 Tractatus anat. med.

VON HALLER, A. 1764 Elementa physiologiae corp. hum.

HAMBURGER 1892 Zur Entwickelung der Bauchspeicheldrüsen. Anat. Anzeiger.

HELLY, K. 1898 Beiträge zur Anatomie des Pankreas und seiner Ausführungsgänge. Arch. f. mikr. Anat.

1901 Zur Pankreasentwickelung der Säugethiere. Arch. f. mikr. Anat. u. Entwickelungsgesch.

1904 Zur Frage der primären Lagebeziehungen bei der Pankreasanlagen des Menschen. Arch. f. mikr. Anat.

HENLE, J. 1866 Eingeweidelehre.

HIS, W 1885 Anatomie menschlicher Embryonen. Bd. 3, Leipzig.

HUSCHKE 1845 Traité de splanchnologie et des organs des sens. Traduction de Jourdan. Paris.

HYRTL, J. 1873 Die Corrosions—Anatomie und ihre Ergebnisse. Wien.

INGALLS, N. W. 1907 Beschreibung eines menschlichen Embryos von 4.9 Mon. Archiv. für mikros. Anat. u. Entwickelungesch. 70, 506–576.

JANKELOWITZ, A. 1895 Ein junger menschlicher Embryo und die Entwicklung des Pankreas bei demselben. Arch. f. mikr. Anat.

JANOSIK 1895 Le Pancréas et la Rate. Bibliogr. anatom.

JOUBIN 1895 Développement des canaux pancréatiques. Th. de Lille.

KOLLMANN, J. 1897 Handatlas der Entwickelungsgeschichte des Menschen.

KÖLLIKER 1889 Handbuch der Gewebelehre des Menschen, 6 Aufl.

LAGUESSE 1894 Le pancréas apres les travaux recents. Journ. de l'Anat.

LETULLE 1898 Arch. des Sciences médicales.

MAYER 1819 Journal complémentaire des Sciences médicales, t. 3, p. 283.

MECKEL 1812–1816 Handbuch der path. Anatomie und Anatomie comparée, t. 3.

MILNE-EDWARDS 1860 Leçons sur la Physiologie et l'Anatomie comparées.

MOREL ET DUVAL 1883 Manuel de l'anatomiste.

VON MURALT, J. 1677 Vademecum anatomicum...

NATTAN-LARRIER 1898 Bull. Soc. Anat., Paris.

NOTHNAGEL 1898 Handbuch der speciellen Pathologie und Therapie. (Oser. Pankreas)

OPIE, E.L. 1903 Anatomy of the Pancreas. Johns Hopkins Hospital Bulletin, vol. 14.

Disease of the Pancreas, Its Cause and Nature. 1903. Second edition, 1910.

PHISALIX 1888 Étude d'un embryon humain de 10 mm. Arch. de Zoolog. Exp. et Générale.

RIOLANUS, J. 1649 Opera anatomica, p. 811.

ROLLESTIN AND FENTON 1900–1 Jour. Anat. Physiol., vol. 35, p. 110.

SANTORINI, J. D. 1775 Anatomici summi septemdecim tabulae quas edit. Michael Girardi.

SAPPEY 1873 Traité d'anatomie.

SCHIEFFER 1894 Du Pancréas dans la série animale. Th. de Montpellier.

SCHIRMER, A. M. 1893 Beitrag zur Geschichte und Anatomie des Pankreas. Inaugural Dissertation. Basel.

STOSS 1891 Zur Entwickelungsgeschichte des Pancreas. Anat. Anzeiger.

SWAEN, A. 1897 Recherches sur le développement du foie, du tube digestif, de l'arrière-cavité du péritoine et du mésentère. Jour. de l'Anat. et de la Physiol.

SWALWE 1668 Pancreas pancrene adornante.

THYNG, F. W. 1908 Models of the pancreas in embryos of the pig, rabbit, cat, and man. Am. Jour. Anat., vol. 7, no. 4.

TIEDEMANN 1819 Journal complémentaire des Sciences médicales, t. 4, p. 330.

TIEDEMANN AND GMELIN 1826 Verdauung nach Versuchen.

VERNEUIL 1851 Mémoire sur quelques points de l'anatomie du pancréas. Gaz. médic. de Paris.

VESLING, J. 1647 u. 1666 Syntagma anatomicum.

VÖLKER, O. 1903 Über die Verlagerung des dorsalen Pankreas beim menschen. Arch. f. mikr. Anat.

WELSCHIUS, L. C. 1698 Tabulae anatomicae.

WINSLOW 1732 Exposition anatomique.

WIRSÜNG, C. 1642 Figura ductus cuiusdam....

WYSS, O. 1866 Zur Aetiologie des Stauungsikterus. Virchow's Archiv.

ZIMMERMANN 1889 Rekonstruktionen eines menschlichen Embryos von 7 mm. Anat. Anz.

Reprinted from THE ANATOMICAL RECORD, VOL. 6, NO. 3
March, 1912

A METHOD OF FURNISHING A CONTINUOUS SUPPLY OF NEW MEDIUM TO A TISSUE CULTURE IN VITRO[1]

MONTROSE T. BURROWS

From the Anatomical Department, Cornell University Medical College, New York

ONE FIGURE

The method devised by Harrison for the cultivation of tissues in vitro has demonstrated that many varieties of tissues may survive and grow for a period of time outside the animal organism. Harrison originally devised this method for studying conditions surrounding the early differentiation of the nerve fiber and for this purpose it has answered admirably as it probably will for the differentiation of many other tissues. During the last year and a half this method with its modifications has been used by many workers as a means of studying the effects of salts and various animal extracts and fluids upon growth. The experiments of this kind have shown that absolute and constant differences, expressed in terms of rate and extent of growth are not obtained when identical physical conditions surround in each case the growing tissue unless the media used in these experiments are of widely different composition and contain substances which act promptly on the cell (cytotoxines, etc.). Such a result is to be expected in cultures where the period of great activity is immediate and endures for only a short period of time. The tissue vitality is undoubtedly sufficient to allow a considerable activity even in a medium which would ultimately bring about cellular death. Again, the chemical composition of the medium necessarily changes continuously throughout the period of cellular activity. In such cultures no sustained rate of growth is to be expected even if an ideal medium be employed. It was necessary, therefore, to improve the technic so that growth might

[1] Read before the American Association of Anatomists, December 27, 1911, at Princeton, N. J.

be continued over a longer period and render the growth more nearly comparable to cellular activities in the animal body. In this way problems of cellular metabolism and problems dealing with tissue growth and differentiation in various media may be attacked.

The first step toward such improvements has resulted in a method which supplies the tissue continuously with a known quantity of new media and at the same time removes the waste products without in any way disturbing the growing cells. The fresh medium is carried by means of a cotton wick from a reservoir at one end of a slide through a culture chamber and into a receiving reservoir situated at a lower level at the opposite end of the slide (fig. 1). In the culture chamber the wick is teased apart into its individual fibers which adhere to the surface of the cover glass and thus simulate a capillary system. The tissue is placed in this open network of cotton fibers and held there by a drop of coagulated plasma. The culture medium passes slowly along the wick through the culture and collects in the receiving chamber. The medium about the tissue is continuously changed by this means.

The supplying chamber (fig. 1, *a*) is blown from glass. It consists of two compartments, the horizontal or reservoir, and the vertical or wick chamber. These two chambers are connected by a glass tube, which comes from the bottom of the reservoir up and over to enter the upper end of the wick chamber through a small capillary point. The reservoir is open to the outside by a long vertical tube. The wick chamber is connected with the culture chamber (fig. 1, *b*) by two tubes. One tube carries the wick; the other acts as an air tube which equalizes the pressure on the two ends of the wick.

The rim of the culture chamber *b*, is made of cork. The central cavity of the cork is closed by a cover glass above and a long glass slide below. The tubes of the supplying chamber and receiving chamber enter through holes cut horizontally through this cork rim (see figure). All parts are carefully sealed together with paraffin. The wick coming from the supplying chamber is spread out on the under surface of the cover glass and here the culture is planted among the cotton fibrils.

The receiving chamber, fig. 1, *c*, is made of glass, and consists of a horizontal reservoir situated below the level of the slide. One tube connects this reservoir with the culture chamber and another tube opens to the exterior as shown in the figure. The wick passes from the cover glass through the glass tube to end above the surface of the liquid in the receiving chamber. The wick does not completely fill the tube. The arrangement allows free air communication between the receiving chamber and the culture chamber. The receiving chamber receives air by its communication

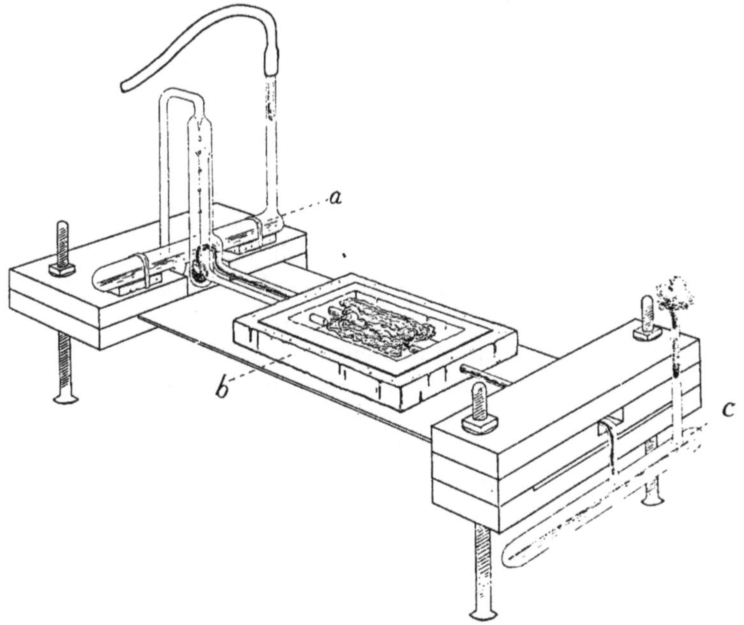

Fig. 1 Complete culture apparatus with rubber tube which connects with air pressure; *a*, supplying chamber; *b*, culture chamber; *c*, receiving chamber.

with the exterior through the vertical tube which is plugged with loose sterile cotton.

Blocks of wood held together with bolts are fastened by friction rigidly to the glass slide. A stand for the apparatus and firm supports for the glass chambers are thus formed. The entire apparatus is compact and strong and the contained culture may be examined under the microscope.

The glass chambers with the wick inserted are sterilized in the autoclave. The cork is sterilized in hot paraffin. The cover glass and the slide are sterilized by dry heat. To set up the apparatus the cork is removed from the hot paraffin and drained, the chamber tubes are inserted into it and the cover glass sealed over its surface. The wick is teased apart on the inner surface of the cover glass and the tissue planted. The apparatus is then immediately placed on the slide and all connections sealed with paraffin, and the chambers are now fastened to the wooden supports.

The reservoir of the supplying chamber is filled with the liquid medium and its open tube is plugged with dry sterile cotton. The open end of this vertical glass tube is connected with the pressure apparatus. The fluid is driven at a constant rate by air pressure from the reservoir into the wick chamber.

To refill or empty the chambers the vertical tubes are flammed, the cotton removed and the fluid entered or removed by sterile pipettes. Freshly sterilized cotton plugs are inserted and the apparatus again connected with the air pressure. Great care must always be taken to keep all parts of the apparatus and the media free from bacterial contamination.

Growth in such an apparatus has been tested with embryonic chick tissues in a medium of blood serum prepared from adult chickens. Growth of such tissues is vigorous and can be maintained for a considerable period of time. Hearts of embryo chicks and small pieces of heart muscle can be kept beating with great regularity. Small bits of ventricle beat actively until all muscle fibers have wandered apart and are lost in dense connective tissue outgrowths.

With this apparatus I have been able to test the effects of various media on the growing cells during any period of their activity. The growth is first established in control media. The media to be tested is then added and its effects recorded. In this manner the effects of many media may be tested on the growing cells without in any way disturbing or changing their physical surroundings. Changes in the chemical composition of the media can be tested by a comparison of the original media with samples obtained from the receiving chamber.

[From The Johns Hopkins Hospital Bulletin, Vol. XXII, No. 242, May, 1911.]

THE RELATIONSHIP BETWEEN THE NORMAL AND PATHOLOGICAL THYROID GLAND OF FISH.

By J. F. Gudernatsch.

(From the Department of Anatomy, Cornell University Medical College, New York City.)

Recent investigations of the thyroid gland of Teleosts have [152] revealed a great many facts that may be of value to the comparative pathologist. The thyroid gland of these fish attracts at present a good deal of attention in cancer research, since it is often liable to cancerous degeneration, especially in artificially reared trout and salmon.

The normal anatomy of the gland was briefly described in a paper read before the meeting of the American Association of Cancer Research, November 27, 1909. * It was especially emphasized that one of the most striking features of the thyroid in bony fish is the absence of a connective tissue capsule, such as exists in other vertebrates. Later this fact was again pointed out by Marine and Lenhart. It will readily be seen that the absence of a capsule makes it rather difficult to define the normal extension of the gland. It would perhaps be better not to use the term "thyroid gland" at all in this group of animals, since physiologically isopotent units (follicles) are not so arranged as to form a closed organ, but are distributed over a wide area (Fig. 2 and Plate 1). This distribution varies not only with the species, but also with the individual, and is dependent entirely on mechanical influences, mainly pressure from the sides of the surrounding tissues, and pull enacted by the growing connective tissue fibers and [153] blood and lymph vessels. The latter force works chiefly in the

* An extensive analysis of the normal conditions of the Teleost thyroid will be found in Jour. of Morphology, V. 21, Suppl., 1911.

[153] early development of the gland, when thyroid cells are carried off from the main point of growth to distant regions, where they form new centers of multiplication.

The thyroid gland develops around the stem of the ventral aorta (Fig. 1), in many species the main bulk lying between the branches to the first and second gill arches. This locality is filled with connective tissue and fat, and is enclosed dorsally by cartilages or bones and ventrally by muscles. Thus the region available for the thyroid tissue is rather limited, and therefore the follicles tend to fill every space that is offered by the surrounding structures. Not uncommonly follicles are found far from the center of thyroid development, invading muscles (Plate 1, *C*) or creeping into the crevices which exist

[152]

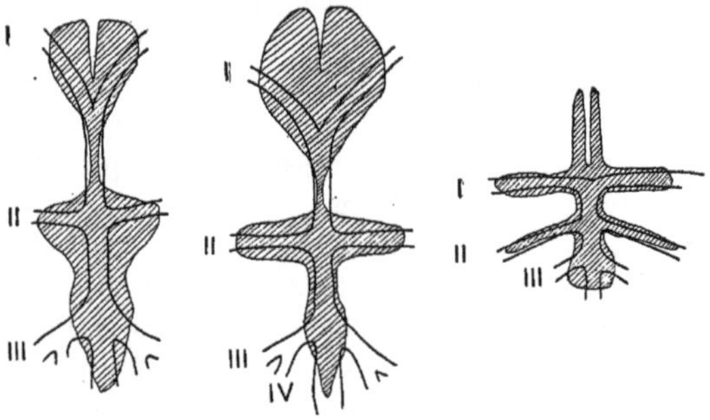

Fig. 1.—These diagrammatic drawings show the expansion of the thyroid gland in the three species: Oncorhynchus, Salvelinus and Opsanus. *I*, aortic bifurcation; *II*, *III* and *IV*, the second, third and fourth branchial arteries.

between the osseous lamellæ of the bones in this region (Fig. 2 and Plate 1, *A*).

The thyroid gland of the Teleosts is thus a rather indefinite organ in its shape, having the tendency to lose its unity and break into numerous small parts. In some species, the trout and others, this tendency manifests itself most strikingly, so that thyroid follicles are found even far out in the gill arches along the gill filaments (Plate 1, *D*).

The spreading apart of the thyroid follicles over a wide area and the invasion of neighboring tissues are a normal feature and of no pathological significance. By such an in-

vasion the surrounding structures are not destroyed. Often [153]
the term "invasion" is even incorrect. Thus Marine and Lenhart's statement that the normal follicles invade the bones is not appropriate, since they do not invade true bone or cartilage tissue, but merely the spaces that are present *between* the osseous or cartilaginous lamellæ (Fig. 2 and Plate 1, *A*). In their paper Fig. 6 demonstrates this fact definitely, although it is supposed to show a true invasion. On the other

[154]

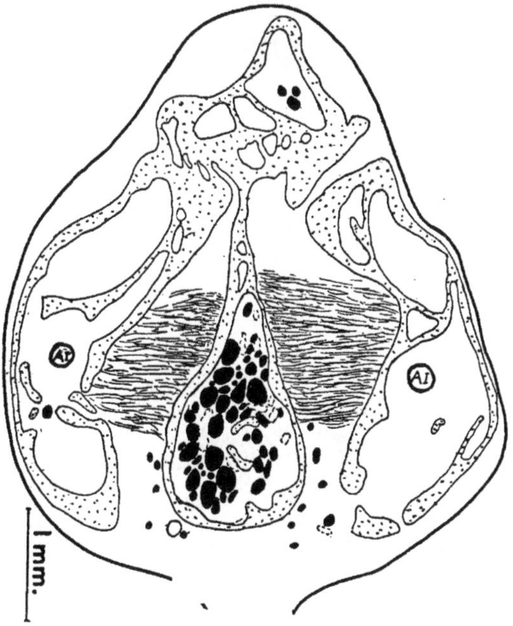

FIG. 2.—Section through the thyroid gland of Brevoortia, anterior to the aortic bifurcation. *AI*, first branchial arteries.

hand, Gaylord was able to show specimens in which thyroid [154]
tissue, belonging to a diseased gland, had actually invaded or infiltrated true cartilaginous tissue. The latter invasion, of course, is never seen in the anatomy of the normal gland, but is a strictly pathological feature. Whether or not it is due to a cancerous growth of the gland, may still be an open question. Strong evidence seems, however, to point in that direction.

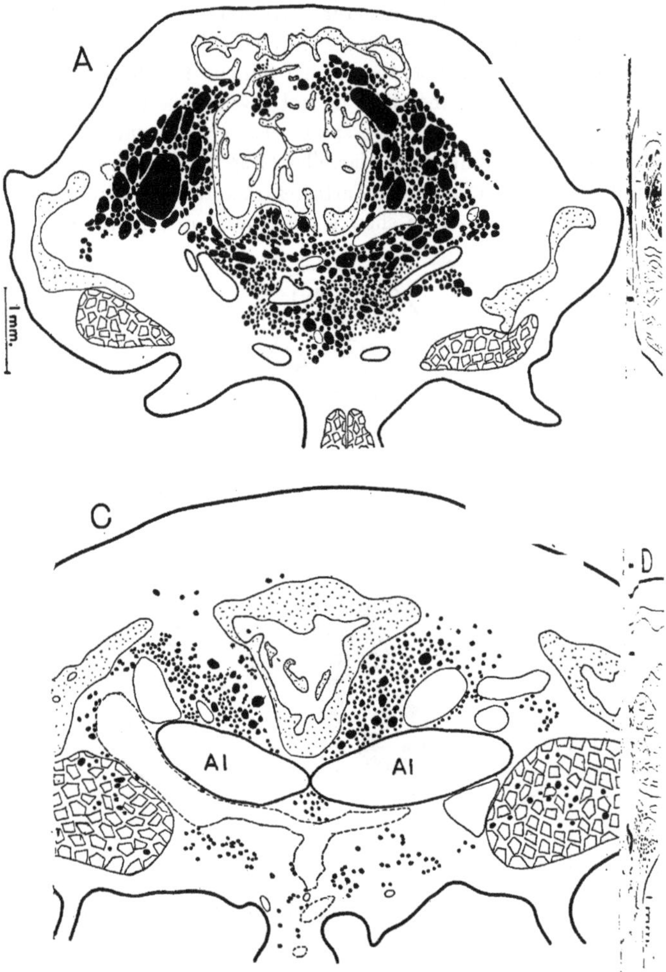

PLATE I.—*A* and *B*, sections through the thyroid gland of a
bifurcation; *B*, near the second branchial arteries, the region f
of Salvelinus. *C*, in the aortic bifurcation; *D*, near the secon b
(Thyroid follicles in all figures shown in solid black. Traw
parts stippled. Arteries in heavy lines. Veins in light lines
branchial arteries.)

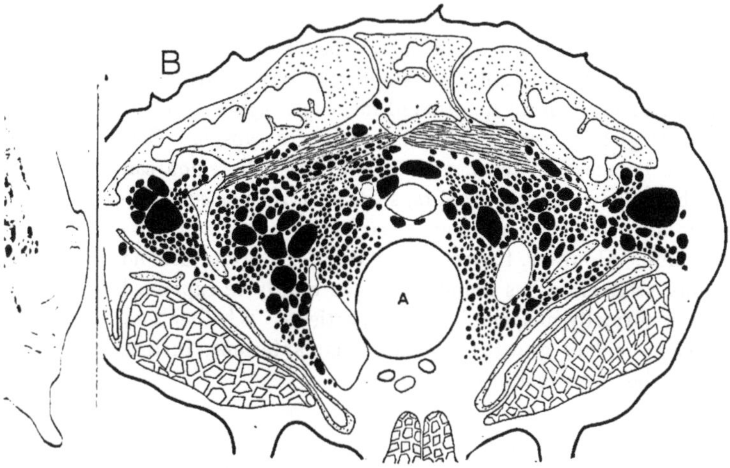

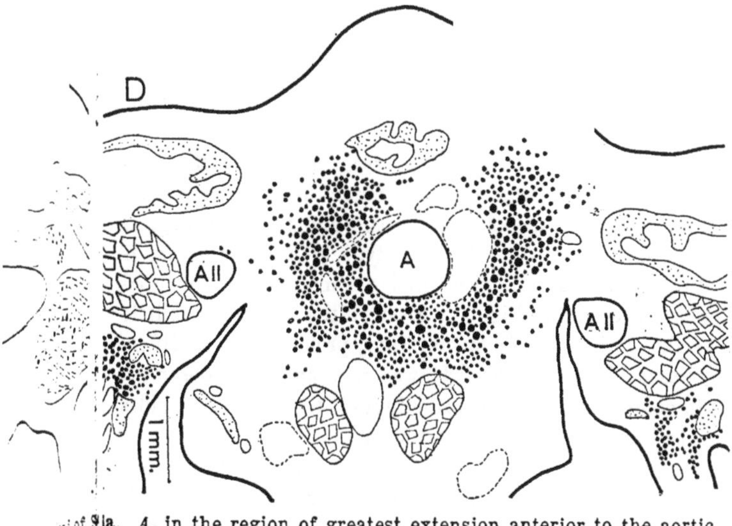

...of ...la. *A*, in the region of greatest extension anterior to the aortic ...n ...reatest extension. *C* and *D*, sections through the thyroid gland ...nd ...nchial arteries.
...Trans...se muscles lined. Longitudinal muscles in polygons. Skeletal ...lines. ...ymph sinuses in broken lines. *A*, ventral aorta, *AI* and *AII*,

[154] Should the so-called thyroid carcinoma of brook trout be a cancerous growth and not a mere hyperplasia, as Marine and Lenhart believe, then the question of metastasis again demands the pathologist to keep in mind the lack of a capsule. Certainly no detached nodules in or around the gill region can safely be called secondary tumors, since such misplaced structures in all probability are merely parts of the primarily diseased gland. However, tumors on the tip of the jaw or around the anus, as Gaylord has found them, can hardly be explained as due to normally misplaced thyroid particles. Whether they are true secondary growths or simply implantations, further experimental-investigations may show.

Histologically the thyroid gland of the Teleosts offers a great many interesting peculiarities. The size of the follicles varies between very wide limits. Aside from the fact that in young embryos it is naturally very small, the size of follicles cannot be taken as a reliable indication of the age of the fish. It seems much more probable that it is a sign of the age of the individual follicle.

The follicular epithelium varies from an almost flat to a very high, columnar type. It is different in each individual and may somewhat depend on the age and the physiological condition of the animal. Yet the type of the epithelium is not always uniform in all the follicles, sometimes very marked differences are found (Fig. 3). Even in the individual follicle the height of the epithelium may vary, probably due to different pressure from outside. It was now and then observed, that in oblong follicles the epithelium on the two longer sides would be lower than on the shorter ones.

The colloid material is sometimes present in all the follicles, in other glands it may appear in some only, in still others it may be entirely lacking. It, again, is no definite sign of the age of the animal, although it may be somewhat dependent on the age of the fish, its sex (egg-carrying females, for instance) and other inherent factors that need further investigation. The colloid certainly is a sign of the physiological state of the individual follicle, yet we do not understand it well enough to interpret our observations in a definite manner. There can be no doubt that all the follicles of a gland are not in the same state of physiological activity. Otherwise it cannot be explained why (Hürthle's) colloid-forming cells appear in some follicles only, sometimes in a group of neighboring follicles, so that we can easily distinguish "colloid zones" from non-colloid-forming parts of the gland. [154]

In interpreting their results after iodin treatment of the thyroid gland of the pike Marine and Lenhart lay great stress on the histological appearance of the treated and not treated

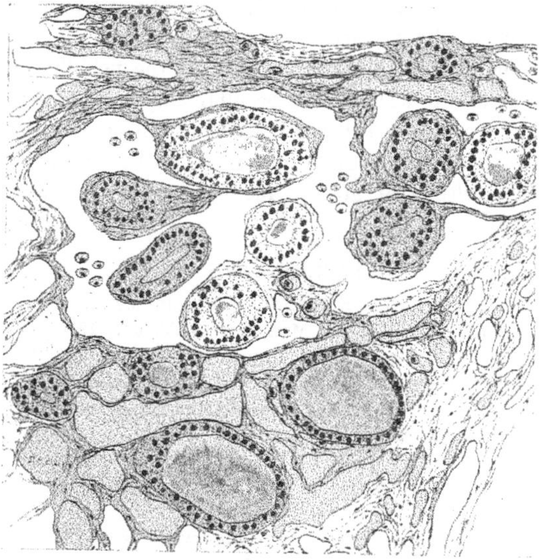

FIG. 3.—Section through the thyroid gland of Salvelinus fontinalis. Note the different heights of the follicular epithelium. Dia. 1:300.

glands. Yet from the above discussion it seems obvious that the anatomy of the gland as well as its histology, as far as the type of epithelium and the colloid formation are concerned, makes it rather difficult for the microscopist to distinguish a normal from a hyperplastic, and the latter in turn from a "reverted" thyroid gland in these fish.

The presence or absence of the colloid material has no

[154] significance whatsoever. If all glands, the follicles of which
spread out far from the main bulk even into the gill region,
are highly hyperplastic, then, according to Marine and Len-
[155] hart, the colloid material should be nearly or entirely absent
in them. Yet it is present throughout the gland.

The type of the epithelium is also a perfectly unreliable guide in regarding a gland as hyperplastic. When, of course, the epithelium shows marked foldings and protuberances into the lumen, the hyperplastic condition is evident.

Further studies on the carcinoma of the fish thyroid will have to take into account the peculiar anatomy and histology of this organ in the Teleosts. Many conditions which might be regarded as pathological may prove to be normal as soon as our knowledge of all the factors involved is sufficiently broadened.

BIBLIOGRAPHY.

Gaylord, H. R.: An Epidemic of Carcinoma of the Thyroid Gland Among Fish. J. Am. Med. Ass., LIV, 227, 1910.

Gudernatsch, J. F.: The Structure, Distribution and Variation of the Thyroid Gland in Fish. J. Am. Med. Ass., LIV, 227, 1910.

Gudernatsch, J. F.: The Thyroid Gland of the Teleosts. J. of Morphology, XXI, 709, 1911.

Marine, D., and Lenhart, C. H.: On the Occurrence of Goitre (Active Thyroid Hyperplasia) in Fish. Johns Hopkins Hosp. Bull., XXI, 95, 1910.

Marine, D., and Lenhart, C. H.: Observations and Experiments on the so-called Thyroid Carcinoma of Brook Trout (Salvelinus fontinalis) and Its Relation to Ordinary Goitre. J. Exper. Med., XII, 311, 1910.

Reprinted from The American Journal of Anatomy, Vol. 11, No. 3
March, 1911.

HERMAPHRODITISMUS VERUS IN MAN

J. F. GUDERNATSCH

From the Department of Embryology, Cornell University Medical College, New York City

SEVEN FIGURES

THREE PLATES

Hermaphroditismus verus is of such rare occurrence and so eminently important in our knowledge of the development of the genital organs, that it would seem worth while to add a new case to the few so far recorded. Numerous instances of supposedly true hermaphroditism have been described, but only in rare instances have they stood a critical consideration.

The microscopic diagnosis in many cases has been incorrect, particularly in cases where the normal structure of the tissues had been altered by neoplasms so that their identification was almost impossible. Some instances of true hermaphroditism have been reported in which histological examination was neglected; yet without a microscopic investigation the correct interpretation of malformations of this kind is at least doubtful.

The 'ovotestis' to be described in this article was taken from an individual forty years old who came to the hospital to be operated upon for tumor of the right inguinal region. In the left inguinal canal a similar, but somewhat smaller nodule was detectable.

The external genitals were of the female type; labia majora and minora were well developed and an introitus vaginæ was present. The noticeably peculiar feature was the extremely enlarged clitoris with the opening of the urethra on its ventral side, so that the organ offered rather the aspect of a hypospadic penis than that of a clitoris.

No uterus was present and the vagina ended blindly (atresia vaginæ). During the course of embryonic development, therefore, the greater part of the Müllerian ducts had been lost and the formation of the genital apparatus must have inclined toward the male type. A prostate-like organ was felt attached to the urethra, yet its real nature remains doubtful, since no microscopic examination could be made. In other cases in which both a vagina and prostate have been found they were seen to be connected with one another, but in the present instance no detectable communication existed between the vagina and the supposed prostate.

The distribution of the hair on the body was that typical of the female. The pelvis was wide, but mammary glands were not developed, and the larynx was externally that of the male. The secondary sexual characters were not so decidedly of the male type that there arose any doubt about the sex of the individual, although the knowledge of the malformation existed. The individual was believed by herself and her associates to be a woman.

The individual has never menstruated, sexual intercourse has never taken place, and libido sexualis is not present. Her psychic disposition is that of a woman and she earns a living as a cook.

The tumor in the left inguinal region was left in place since it did not cause inconvenience and that in the right channel was removed. The nodule extirpated had about the form of a testicle with attached epididymis. It measured 6 cm. in length and 5 cm. in width and thickness. Histologically it proved to be male genital tissue, with the exception of a small nodule which exhibits a structure very similar to that of an ovary. Dr. James Ewing diagnosed this region of the specimen as true ovarian tissue. Others who were asked to examine the sections agreed with Dr. Ewing in considering the tissue ovarian.[1]

[1] I wish to thank Prof. A. Kohn, the well known histologist in the German University of Prague, for his careful examination of the sections and for the many suggestions he made in regard to their interpretation. The preparations were also demonstrated before the Eighth International Zoölogical Congress in Graz, and the structure was interpreted by all as ovarian.

THE TESTICULAR STRUCTURE

The mass of the testicle is surrounded by an extremely broad tunica albuginea, the elements of which are arranged in a somewhat undulating manner and contain elastic fibres with but few nuclei and vessel. As might be expected the testicular tissue proper (fig. 2) is not of the appearance of the normal male sexual gland since the organ developed under very abnormal conditions. It resembles the well known pathological condition of a degenerating testicle, and the hyaline type of degeneration is typical of the kryptorchic mammalian testicle retained in the inguinal canal. The sections through the contorted tubules vary much in size, the smaller ones are circular in shape, the larger ones more or less oval and some a little bent or even S-shaped owing to the plane of section. The average diameter of the tubuli contorti is much smaller than normal. The epithelium which lines the tubules shows only one row of basal cells, and these are propably all of the Sertoli's cell type being rather large, triangular or cubical in shape with clear cytoplasm and large nuclei (fig. 2, *ct*). From their surfaces protoplasmic processes project into the lumen. Germ cells seem to be entirely absent since there are no indications of spermatogonia, spermatids or spermatozoa. It is not improbable that in the younger years of the individual anlagen of germ cells were present in the tubules, but failed to undergo further development on account of the abnormal physiological conditions. It must be assumed that formerly germ cells existed, for without them the development of male sexual glands is hardly conceivable.

The cells of Sertoli show different stages of degeneration as is indicated by the different densities and staining abilities of the nuclei. In some cells the nuclei are shrunken and lie in a nuclear cavity. This degeneration of the epithelial cells is probably due to the above mentioned hyaline degeneration of the wall of the seminal tubules. The inner layers of the tunica propria upon which the epithelial cells lie are swollen, while the constituent cell nuclei disappear entirely, forming a hyaline mass which lies as a band between the follicular wall and the epithelium (fig. 2, *ct*). This swollen band of cells pushes the epithelium towards the lumen which thus becomes more and more occluded and often with gradual dissolution of the cells becomes entirely obliterated. The hyaline band exhibits a somewhat fibre-like structure with processes from it extending between the epithelial cells. It varies in thickness and in some places is entirely missing, and along with this variation in thickness the degeneration of the epithelial cells presents a regional distribution. In some regions the epithelial cell nuclei stain in a normal manner and are situated towards the periphery, in other regions the cell limits are indistinct and the nuclei lie near the lumen. This accords with former observations, and Finotti states that in the kryptorchic testicle, even when the individual is not a hermaphrodite and the germinal epithelium in places develops spermatozoa, the degeneration is not of uniform degree in all regions.

Wherever there is a membrana propria left in the form of a thin layer of spindle-like connective tissue cells, elastic fibres are usually present.

The testicle under consideration offers still another peculiarity. The interstitial tissue is enormously increased so that the seminal tubules are in places pushed far apart (fig. 2, *i*). Connective tissue fibres are comparatively scarce in this tissue.

The interstitial cells are normal in appearance, the majority being of a triangular or polygonal shape. The large nucleus (there are occasionally two nuclei) possesses one or more nucleoli and is usually excentrically situated. The cytoplasm is somewhat denser than that of ordinary connective tissue cells, it is finely granular and often contains in some regions a very fine brown pigment. Of the so-called Reinke's crystals nothing could be detected.

The interstitial cells are usually arranged either in small irregular groups or narrow streaks, often, however, they are united in large compact nests and grow so excessively that they actually invade the tunica albuginea. There is no relationship between these cells and the blood vessels as is often claimed.

This striking increase in the number of interstitial cells is a feature well known to the human embryologist as well as to the pathologist. During the fourth month of embryonal development these cells constitute about two-thirds of the parenchyma of the testicle. Their number is also very much increased in the kryptorchic testicle. Whether the large amount of interstitial tissue in the present case is an embryonal condition due to arrested development, or simply a secondary pathological feature, is difficult to say. The latter, however, seems to be more probable since there are no indications, except perhaps the entire lack of germinal epithelium, that the gland did not develop normally. It probably degenerated later on account of its unusual position. Finotti claims that the gland does not degenerate for such a reason, but on account of an early predisposition to do so.

The entire accessory system of the male genital gland, rete testis, ductuli efferentes, ductus epididymis and vas deferens, are present. The globus major is, as far as the arrangement of the efferent ducts is concerned, rather well developed. The epithelium, however, is very degenerate in places (fig. 4), though towards the duct of the epididymis it approaches a normal condition. The epithelium in some tubules is of the low cubical type without foldings (fig. 3), while in others it shows the normal projections into the lumen with alternating columnar and cubical cells.

The structure of the epididymis resembles in parts that of the epoöphoron and since both organs are derived from the Wolffian body, it is not impossible that in a true hermaphrodite we might find them somewhat mixed.

The ductus epididymidis is normally developed. The muscular coat of the efferent ducts increases in thickness as they approach the epididymis. There are numerous elastic fibres among the muscle cells; if these, as Stöhr states, do not appear until puberty is reached we must conclude that the entire efferent system reached a mature state independently of the testicle. This is also emphasized by the fact that the better developed parts are those farthest removed from the testicle. The duct of the epididymis, for instance, has a normal, highly columnar, ciliated epithelium, which shows no signs of degeneration. In the lumen is seen cell detritus, a finely granular mass and numerous concrements. The muscular coat of the spermatic cord is much increased.

The sclerotic blood vessels, as everywhere seen in the sections, are typical of the kryptorchic testicle.

THE OVARIAN STRUCTURE

The female portion of the genital gland is rather small, the rudimentary ovary being a little nodule only 3 mm. in length and 2 mm. in width and thickness. It is enclosed within a cyst in the tunica between the testicle and the head of the epididymis (fig. 1, *o*). The typical ovarian stroma is easily recognized by the arrangement of the spindle-shaped connective tissue cells (fig. 5). This structure is nowhere else to be found in the human body. Cortical and medullary portions are distinguishable. The former consists of dense connective tissue rich in cells, and traversed by small blood vessels. The slender cells sometimes resembling smooth muscle fibres, are arranged either in strands or twirls. In the central portion of the ovarian body the connective tissue contains fewer nuclei and its elements are arranged in broader streaks. The blood vessels are large and bent.

The entire nodule is surrounded by a single-layered, cubical or cylindrical epithelium (fig. 5) which although rather primitive shows here and there slight cellular differentiation. Some cells are larger and broader and their nuclei are large and more circular and contain less chromatin than the neighboring cylindrical cells (figs. 6, 7, *p*). These cells are very probably primordial ova, yet a definite diagnosis cannot be made. However the decision that the body is ovarian in structure is sufficiently warranted by the typical stroma with its surface epithelium.

The ovary remained in an early stage of development as is indicated by the rather high columnar cells in certain regions (fig. 6). A migration of primordial ova into the stroma and the formation of Graafian follicles has not taken place, probably due to the abnormal conditions of development. This accords with earlier reports on the subject which state that in all cases of hermaphroditism, whether true or spurious feminine, the epithelial part of the ovary is below the normal in development. Various transitions have been described from almost mature ovaries to those containing only primordial follicles or even empty follicles.

The primitive and rather small female portion found in this hermaphroditic genital gland indicates perhaps that in many cases of spurious hermaphroditism traces of ovarian tissue might be found, provided the entire testicular tissue be thoroughly searched, so far, however, this has never been done.

That both types of germinal tissue are in a hypoplastic condition is explained by Halban in the following way, the impulse for development, which normally is concentrated upon one system, in cases of hermaphroditism is called to act upon two systems and thus is insufficient to force either to the normal degree of development.

In the present case male and female tissues are found in close contact in one gland, but an intermixing of the two kinds of tissue does not exist, and therefore the term "ovotestis," as is often used for this kind of gland, does not seem to be appropriate. In the true ovotestis of invertebrates male and female sexual cells are produced by one glandular structure.

The male part of an 'ovotestis' is as a rule considerably larger and further developed than the female. Kopsch and Szymonovicz have observed this in all hermaphroditic vertebrates and it likewise holds for the present case. This individual, however, shows many more external female characters than one should expect when considering the large male part of the genital gland. If the interstitial cells of the testis are really responsible for the accessory sexual characters then the person in question should show a typical male condition. The body of the individual is not a perfect woman, yet the male characters are not so outspoken that she could be called a man-woman. This fact is surprising from the study of the genital gland tissue, as far as this could be investigated, but it must be remembered that the actual amount of ovarian tissue the individual really carried in its body is not known. As has been mentioned above, a nodule existed in the left inguinal canal similar to that described from the right, this in all probability may have also been genital gland tissue and it might have contained a preponderance of ovarian material.

Such is not improbable since in malformations of this kind a great difference between left and right genital glands has often been observed. Salén, for instance, describes a case of true hermaphroditism, in which the right side contained an "ovotestis," while on the left a perfect ovarium was found. Lilienfeld states that in all cases of disturbed development of the genital region the female type is predominant on the left side. In this individual a large ovary may have been present on one or both sides higher up in the abdominal cavity, or there may have been more ovarian tissue present during an earlier period of life than can now be detected. This latter suggestion would account for the development of the strong female characters of the individual. Kermauner states in Schwalbe's "Die Missbildungen des Menschen und der Tiere" that "whether the entire defect may be regarded as primary aplasia, or later involution, cannot always be decided. In no case can it be entirely denied that microscopic remnants of ovarian tissue, perhaps transformed beyond recognition, may be located somewhere in the abdominal cavity."

Whatever circumstances were responsible for the strong female characters of this person, it is interesting that along with them the male sexual apparatus developed to the perfection here described. The cavities and the concrements of the epididymis would indicate that a secretory function was performed by the epithelium.

This case of true hermaphroditism recalls the old theory of Waldeyer according to which there is a bisexual anlage of the genital gland, as opposed to Lenhossek's idea of an indifferent anlage. Instances in which male and female genital tissue are found next to each other speak at least for Waldeyer's view that the ovary develops from a different region of the genital ridge from that of the testicle even though they may not entirely support his theory of hermaphroditism. In all the cases of true hermaphroditism the ovary occupies the same relative position to the testicle. It seems strange that there should always be a sharp distinction between the two kinds of tissue and never an undefined mixing of both elements (true ovotestis) as might be expected, if all cells of the germinal epithelium could produce either male or female tissue.

Every embryo has the anlagen of the efferent ducts for the expulsion of both male and female products of the genital glands, which indicates that the male and female sexual apparatus are rather distinct from one another, thus it does not seem impossible that two distinct regions of the germinal epithelium might exist next to one another, one giving off the male, the other the female primordial cells. Why is the Müllerian duct always laid down in the male if there are no female tendencies in the undifferentiated embryo? It seems that in every embryo there is a trace of a female tendency. Some authors, Benda for instance, go so far as to claim "that the primary anlage of the entire sexual system of the vertebrates must be regarded as female." Waldeyer's view, however, may be correct, that the Müllerian duct alone is the primary efferent duct for the genital glands in both sexes. He believes that in its function it corresponds to the primitive opening for the products of the gonads in the lowest vertebrates, the porus abdominalis. This scarcely seems possible, since in some fish adbominal pores and efferent genital ducts exist side by side and thus the second do not replace the first. Benda on the basis of Waldeyer's view may be justified in his conclusion that at no time do both ducts, Wolffian and Müllerian, exist as parallel genital ducts. Yet this is not entirely true, for the Wolffian duct, even in early periods, when it certainly serves as the mesonephric duct, must possess the potential faculty of developing into a male genital duct. This faculty is possessed by the duct in all embryos, whether the further development is male or female.

In the resulting female the possession of the quality to develop the mesonephric duct into a male sperm duct seems likely, if not proven, by the fact that the remnants of the Wolffian body, the Epoöphoron and Paroöphoron, closely resembling the structure of some parts of the epididymis, are found to exist.

The present case is interesting in still another direction. The sister of this hermaphrodite also shows irregularities in the formation of the external genital organs. Unfortunately only the testimony of laymen could be secured regarding this. In man a hereditary tendency towards hermaphroditism has never been scientifically proven, though several cases have been supported by laymen. Reuter, however, found among three pigs of one litter one true and two spurious hermaphrodites, and in a later litter from the same sow a pseudo-hermaphrodite occurred.

From this investigation the sex of the individual remains undetermined. According to Virchow it is an 'individuum utriusque generis.' According to Klebs the condition is a typical hermaphroditismus verus. Klebs regards an individual as a true hermaphrodite, when the genital glands of both sexes are united in it. The physiological state is of no importance, simply the anatomical fact. An anatomical hermaphroditism seems to be all we can expect to find in vertebrates. The physiological hermaphroditism, as is normally the case in invertebrates, may hardly be looked for in man. In the higher vertebrates the persistence of both genital glands, is looked upon as an imperfect development and under such circumstances it is only natural that their physiological faculty should be reduced.

Our knowledge of the etiology of these malformations is almost nil, since in general our conceptions of the principles involved in the development of sex are still rather vague. It remains for the anatomist and embryologist, perhaps for the experimenter, to bring about a deeper understanding of these abnormalities. So long as the interest in them is reserved for pathologists and clinicians, and the malformations of the external genitals remain the only thing of interest, the literature regarding such anomalies will be as Benda states "overcrowded with sensational reports," with no exact investigation of the anatomical and histological features.

Hermaphroditism in the sense that separate testicles and ovaries are found has not been demonstrated in man, nor even in other mammals beyond doubt. Yet there are four cases of the so-called ovotestis on record, two of these with neoplastic changes in the male portion. The present is therefore the fifth recorded case of true hermaphroditism in man.

LITERATURE CITED

BENDA, C. 1895 Hermaphroditismus und Missbildungen mit Verwischung des Geschlechtscharakters. Ergebn, d. allg. Path., vol. 2, p. 627.

BORN, G. 1894 Die Entwickelung der Geschlechtsdrüsen. Erg. An. u. Entw., vol. 4, p. 592.

CORBY, H. 1905 Removal of a tumor from a hermaphrodite. Brit. Med. J., vol. 2, p. 710.

FIBIGER, J. 1905 Beiträge zur Kenntnis des weiblichen Scheinzwittertums. Virchow's Arch. f. path. An., vol. 181, p. 1.

FINOTTI, E. 1897 Zur Pathologie und Therapie des Leistenhodens nebst einigen Bemerkungen über die grossen Zwischenzellen des Hodens. Arch. f. klin. Chir., vol. 55, p. 120.

HALBAN, J. 1903 Die Entstehung der Geschlechtscharaktere. Arch. f. Gynaek. vol. 70, p. 205.

HANSEMANN, D. 1895 Über die grossen Zwischenzellen des Hodens. Virchow's Arch. f. path. Anat., vol. 142, p. 538.

HIRSCHFELD, M. 1905 Ein Fall von irrtümlicher Geschlechtsbestimmung. Monatschr. f. Harnkr. u. sex. Hyg., vo.. 2, p. 53.

1905 Ein seltener Fall von Hermaphroditismus. Monatsschr. f. Harnkr. &. sex. Hyg., vol. 2, p. 202.

HOFMEISTER 1872 Untersuchungen über die Zwischensubstanz im Hoden der Säugetiere. Sitzungsb. Akad. Wiss. Wien, vol. 65.

JANOSIK, J. 1887 Bemerkungen über die Entwicklung des Genitalsystems. Sitzungsber. Akad. Wiss. Wien, vol. 99, 3. Abt., p. 260.

KOPSCH, FR. u. SZYMONOWIEZ, L. 1896 Ein Fall von Hermaphroditismus verus bilateralis beim Schweine, nebst Bemerkungen über die Entstehung der Geschlechtsdrüsen aus dem Keimepithel. An. Anz., vol. 12, p. 129.

LUKSCH, F. 1900 Über einen neuen Fall von weit entwickeltem Hermaphroditismus spurius masculinus internus. Ztschr. f. Heilk., Abt. f. Path., vol. 21, p. 215.

MEIXNER, K. 1905 Zur Frage des Hermaphroditismus verus. Ztschr. f. Heilk., Abt. f. prakt. Anat., vol. 26, p. 318.

MIHALKOWICZ, G. 1885 Untersuchungen über die Entwicklung des Harn- und Geschlechtsapparates bei Amnioten. Intern. Monatsschr. f. An. u. Phys., vol. 2, p. 1.

NEUGEBAUER, F. 1908 Hermaphroditismus beim Menschen. Leipzig.

PHILIPPS, J. 1887 Four cases of spurious hermaphroditism in one family. Transact. Obst. Soc. London, vol 28, p. 158.

PICK, L. 1905 Über Adenome der männlichen und weiblichen Keimdrüse bei Hermaphroditismus verus und spurius. Berl. klin. Woch., vol. 43, p. 502.

1905 Über Neubildungen am Genitale bei Zwittern. Arch. f. Gynaek., vol. 76, p. 191.

PLATO, J. 1897 Die interstitiellen Zellen des Hodens und ihre physiologische Bedeutung. Arch. f. mikr. Anat., vol. 48, p. 281.

REINKE, Fr. 1896 Beiträge zur Histologie des Menschen. Arch. f. mikr. Anat., 1896, vol. 47, p. 34.

REIZENSTEIN, A 1905 Über Pseudohermaphroditismus masculinus. Münchn. med. Woch., vol. 52, p. 1517.

SALÈN, E. 1899 Ein Fall von Hermaphroditismus verus unilateralis beim Menschen. Verh. Deutsch. Path. Ges., vol. 2, p. 241.

SCHICKELE, G. 1906 Adenoma tubulare ovarii (testiculare). Hegar's Beitr. z. Geburtsh. u. Gynaek., vol. 11, p. 263.

SCHWALBE, E. 1906 Die Morphologie der Missbildungen des Menschen und der Tiere, β. Jena.

SIMON, W. Hermaphroditismus verus. Virchows' Arch. f. path. An., vol. 172, p. 1.

SPANGARO, S. 1900 Über die histologischen Veränderungen des Hodens und des Samenleiters von Geburt an bis zum Greisenalter. Anat. Hefte, vol. 18.

TOURNEUX, F. 1904 Hermaphroditisme de la glande génitale chex la taupe femelle adulte et localisation des cellules interstitielles dans le segment spermatique. Comp. rend. de l'assoc. des anat., Toulouse, p. 49.

UNGER, E. 1905 Beitrag zur Lehre vom Hermaphroditismus. Berl. klin. Woch., vol. 42, p. 499.

WALDEYER, W. 1870 Eierstock und Ei. Leipzig.

PLATE 1

EXPLANATION OF FIGURES

1 General view of the relative positions of testicle, *t*, ovary, *o*; and epididymis, *e*. Dia. 1:17.

2 Testicular tissue, showing the hyaline wall of the convoluted tubules, *ct*; and the large masses of interstitial cells, *i*. Dia. 1:90.

PLATE 2

EXPLANATION OF FIGURES

3 and 4 Sections through different parts of the epididymis. The tissue in fig. 3 resembles somewhat a parovarian structure. Dia. 1:90.

PLATE 3

EXPLANATION OF FIGURES

5 Ovarian tissue. Dia. 1:50.

6 and 7 Germinal epithelium of the ovary; *p*, somewhat differentiated cells, probably primordial ova. Dia. 1:200.

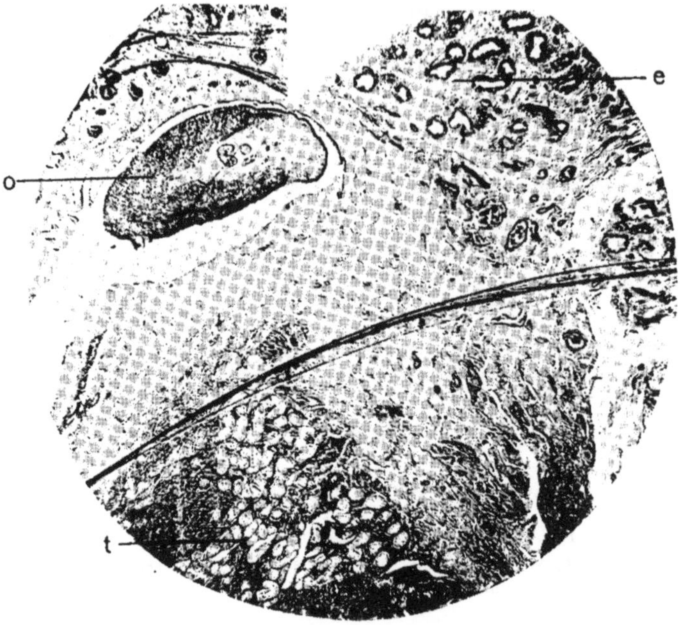

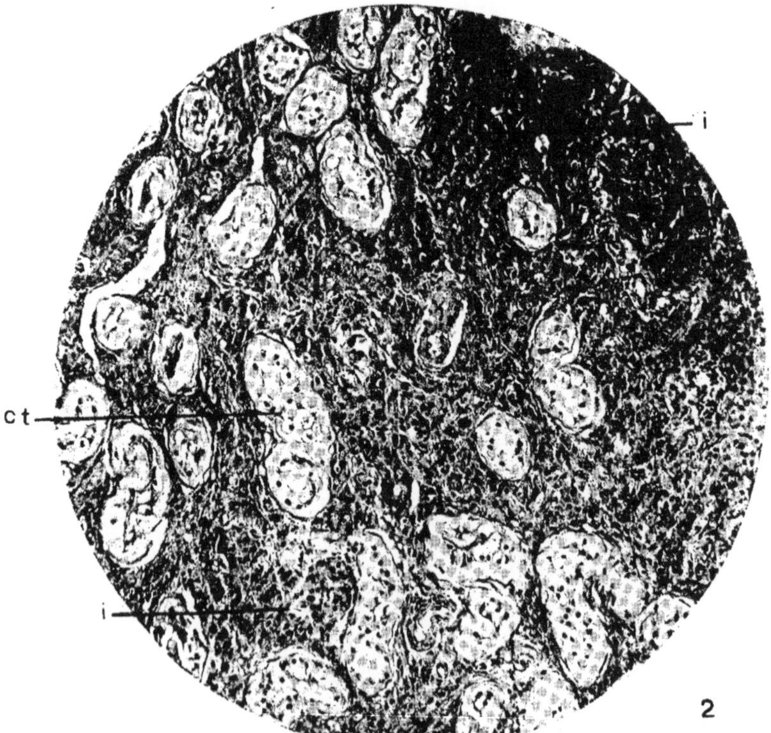

2

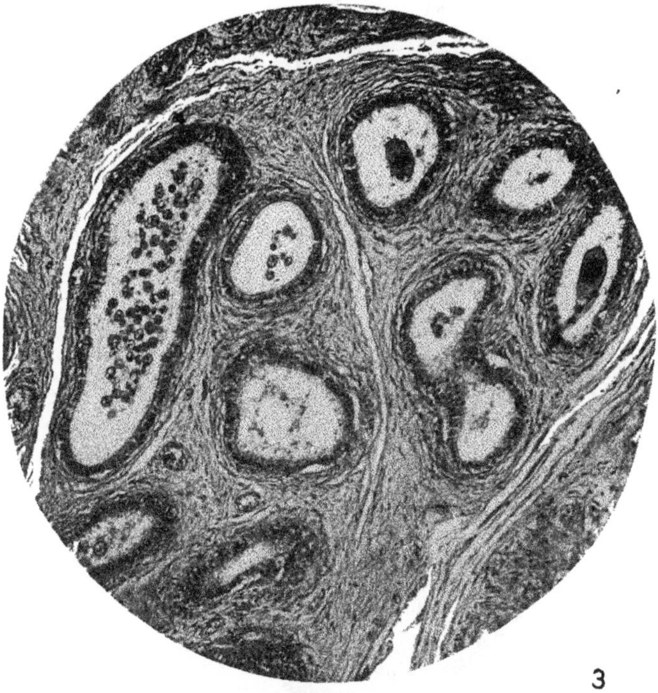

3

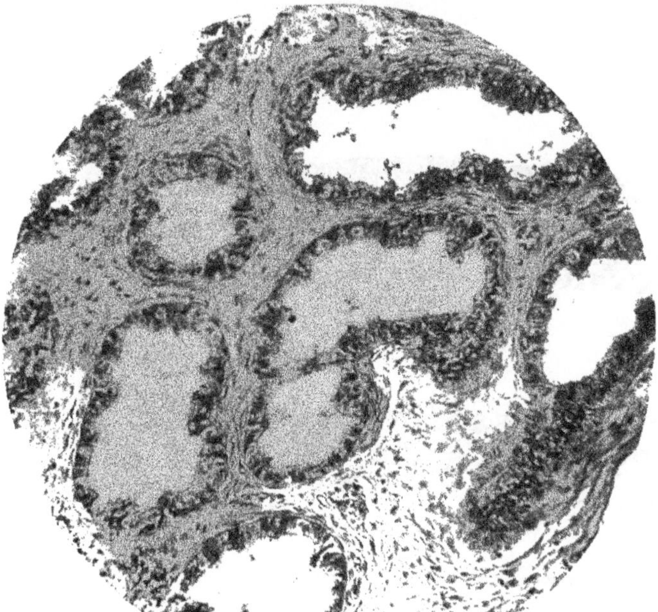

4

5

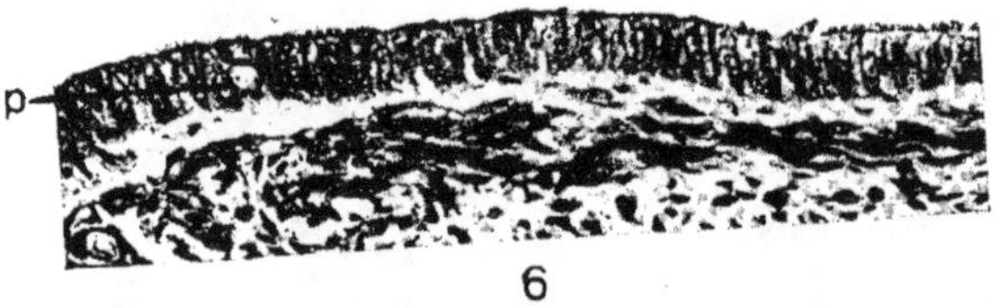

6

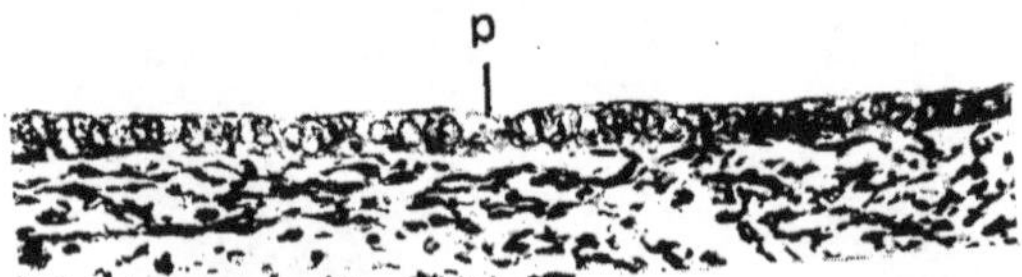

7

Abdruck aus den
Verhandlungen des VIII. Internationalen Zoologen-Kongresses zu Graz vom 15.—20. August 1910.

Ein Fall von Hermaphroditismus verus hominis[1]).

Von J. F. Gudernatsch (New York City).

Eine 40 Jahre alte Frau war wegen eines Tumors in der rechten Leistengegend, der ihr seit einiger Zeit Beschwerden verursacht hatte, auf die chirurgische Klinik gebracht worden. Bei Betastung konnte an der bezeichneten Stelle eine etwa hühnereigroße Geschwulst, in der linken Gegend eine ähnliche, etwas kleinere, ermittelt werden.

Die äußeren Genitalien boten das Bild des Weiblichen dar. Labia majora und minora waren gut ausgebildet, ein introitus vaginae vorhanden. Auffällig war nur die übermäßig große Ausbildung der Clitoris und die Ausmündung der Harnröhre an der Unterseite des Clitorisschaftes, so daß man eher das Bild eines hypospadischen Penis vor sich hatte. Bei innerer Untersuchung der Genitalregion wurde dagegen konstatiert, daß kein Uterus vorhanden war, sondern die Vagina blind endigte; wohl aber ist ein prostata-ähnlicher Körper tastbar. Es ist somit der größte Teil der Müllerschen Gänge verloren gegangen, die Ausbildung der Genitalorgane hat sich also in der für das männliche Geschlecht geltenden Richtung bewegt. Ob übrigens der der Harnröhre anlagernde Körper Prostata war, ist sehr fraglich, da die histologische Diagnose fehlt; denn eigentlich müßten ja Scheide und Prostata miteinander in Zusammenhang stehen.

Weiterer Befund: Die Behaarung des Mons veneris weiblich, sonst am Körper keine Behaarung, das Becken breit, Milchdrüsen nicht entwickelt, Kehlkopf eher männlich, im Gesicht ein leichter Bartanflug, wie er etwa bei Frauen nach der Menopause auftritt, Kopfhaar lang. Der Gesamthabitus und die Bildung der äußeren Geschlechtsorgane sind nicht so männlich, daß über das Geschlecht des Individuums Zweifel aufgetaucht wären. Es wurde von sich selbst und seiner Umgebung für ein Weib gehalten.

---

[1]) Die ausführliche Arbeit ist erschienen in „The American Journal of Anatomy, Vol. XI."

Die Kranke hat niemals menstruiert, Beischlafversuche haben keine stattgefunden, Libido war nie vorhanden. Die Psyche ist weiblich; Patientin verdient als Köchin ihren Lebensunterhalt.

Die Geschwulst der linken Inguinalgegend wurde vorläufig belassen, da sie keine Beschwerden verursachte, die der rechten wurde entfernt. Das exstirpierte Gebilde hatte annähernd die Form eines Hodens mit Nebenhoden. Daß man es mit einem solchen tatsächlich zu tun habe, bestätigte die histologische Diagnose. Die Frau wurde deshalb als männlicher Scheinzwitter bezeichnet. Erst als mehr Schnitte aus einer ganz zufällig gewählten zweiten Partie des Tumors angefertigt wurden, fand man ein kleines Gebilde, das unter dem Mikroskop eine der des Ovariums sehr ähnliche Struktur erkennen ließ. Wegen der Seltenheit eines derartigen Befundes und seiner großen entwicklungsgeschichtlichen Bedeutung wurden die Schnitte auch mehreren Fachleuten, unter anderen Herrn Professor Alfred Kohn in Prag, zur Begutachtung vorgelegt. Die Struktur wurde als unzweifelhaft ovariell bezeichnet.

Schnitte durch die Hodensubstanz zeigen kein normales Gewebe, was ja erklärlich ist, da die Keimdrüse unter ganz abnormalen Verhältnissen zur Entwicklung gelangt ist. Sie bietet das jedem Pathologen wohlbekannte Bild des degenerierenden Hodens dar, und zwar ist diese Art der Degeneration, die hyaline, für den im Leistenkanal retinierten, kryptorchischen Säugerhoden typisch. Man sieht Durchschnitte durch die gewundenen Samenkanälchen in Menge, aber von dem dieselben unter normalen Verhältnissen auskleidenden Epithel findet sich nur eine Reihe von basalen, wohl ausschließlich Sertolischen Zellen. Der Hode begann also zuerst eine normale Entwicklung, erreichte eine verhältnismäßig hohe Ausbildung und ist erst später entartet. Die hyaline Degeneration der Membrana propria und die Degeneration des Samenkanälchenepithels ist übrigens nicht an allen Stellen gleich weit vorgeschritten. Letzterer Befund steht im Einklang mit früheren Angaben. So hat schon Finotti gezeigt, daß beim Leistenhoden nicht immer alle Partien gleichmäßig degenerieren.

Noch eine andere Eigentümlichkeit zeigen die Schnitte durch den Hoden. Das Hodenzwischengewebe ist außerordentlich mächtig entwickelt, so zwar, daß die interstitiellen Zellen oft zu ganzen Nestern vereinigt sind. Auch das ist eine typische Erscheinung, einerseits beim etwa vier Monate alten Fötus, wo diese Zellen nach Nagel zwei Drittel des Parenchyms ausmachen, andererseits beim kryptorchischen, degenerierenden Hoden. Welcher von den beiden Fällen hier zutrifft, ob der Reichtum an Zwischengewebe ein embryonaler Zustand ist oder ein sekundär pathologischer, ist schwer zu entscheiden. Das letztere hat die größere Wahrscheinlichkeit für sich; denn es ergibt sich kein

Grund gegen die Annahme, daß sich der Hode anfangs normal entwickelt hat und erst später infolge der ungewöhnlichen Lagerungsverhältnisse degeneriert ist.

Auch die sklerotischen Gefäße, wie sie in den Schnitten zu sehen sind, sind für den im Leistenkanal retinierten Hoden typisch.

Die übrigen Teile des Hodens, Rete, Epididymis, Ausführungsgänge, sind vorhanden, alle natürlich mehr oder weniger pathologisch verändert. Nebenhoden ist ziemlich typisch gebaut, ähnelt aber gleichzeitig auch sehr der Struktur des Parovariums. Beide Gebilde sind ja Abkömmlinge des Wolffschen Körpers und als solche homologe Organe.

Den weiblichen Anteil am Aufbau der Keimdrüse stellt ein sehr kleines, etwa bohnenförmiges Knötchen dar. Am Schnitt kann man an der Anordnung der spindelförmigen Bindegewebszellen sofort das typische ovarielle Stroma erkennen, wie es sonst in keinem Organ des Körpers gefunden wird. Man sieht genau so wie in der Rindensubstanz des normalen Ovariums kleinere Gefäße in einem dichten, kernreichen Bindegewebe, dessen meist schlanke Elemente hie und da eine wirbelartige Anordnung erkennen lassen und teilweise glatten Muskelzellen ähneln. In der zentralen Partie des Gebildes haben wir ähnlich wie im Mark des Ovariums große, gewundene Gefäße; das Bindegewebe ist viel ärmer an Kernen als in der Rindenschicht und in gröberen Lagen angeordnet. Das Gebilde ist an seiner Oberfläche von einem einreihigen Epithel bekleidet, wie wir es vom Ovarium her kennen. Bei näherer Durchsicht der Schnitte bemerkt man hie und da eine Differenzierung in diesem Epithel; einzelne Zellen sind etwas größer und breiter als die ihnen benachbarten Zylinderzellen, sehen etwas gequollen aus, und die Kerne in ihnen sind ärmer an Chromatin als die in den übrigen Zellen. Man dürfte kaum fehl gehen, diese Zellen als Anlagen von Ureiern anzusehen. — Die Diagnose „ovarielles Gebilde“ läßt sich schon durch den typischen Bau des Stromas mit dem umgebenden Epithel begründen. Allerdings muß gesagt werden, daß das Organ auf einer äußerst frühen Entwicklungsstufe stehen geblieben ist, worauf auch die hohen Zylinderzellen des Keimepithels hindeuten. Zur Ausbildung eines Follikelapparates, Einwandern von Ureiern in das Stroma und Anlage von Graafschen Follikeln ist es nicht mehr gekommen. Dies stimmt wiederum völlig mit den bisherigen Befunden auf diesem Gebiete überein. Es sind alle bisher bekannten Fälle von Hermaphroditismus, seien sie echte oder unechte, dadurch ausgezeichnet, daß der epitheliale Teil des Ovariums immer hinter der Norm, gewöhnlich stark, zurückbleibt, so daß von nahezu reifen Ovarien bis zu solchen, die nur Primärfollikel, ja selbst leere Follikel zeigten, alle Übergänge gefunden wurden.

Aus den anatomischen und histologischen Befunden geht hervor, daß der männliche Anteil am Aufbau dieses Ovotestis bedeutend größer und auch in der Entwicklung weiter vorgeschritten ist als der weibliche. Es soll übrigens, wie Kopsch und Szymonowicz feststellten, auch bei niederen Säugetieren, falls ein Ovotestis sich findet, der ovarielle Anteil immer kleiner sein als der testikuläre. Und doch sind in unserem Falle die sekundären Sexualcharaktere durchaus nicht absolut nach der männlichen Seite ausschlaggebend, obwohl man bei dem großen Reichtum an interstitiellem Gewebe gerade das Gegenteil erwarten sollte. Die Patientin ist zwar kein absolutes Weib, doch sind die männlichen Anzeichen nicht einmal so stark, daß man sie als Mannweib bezeichnen könnte. Sie gilt ja übrigens für ihre Umgebung als Weib. Dieser Umstand scheint mit den Befunden an der Keimdrüse in Widerspruch zu stehen. Doch wissen wir leider nicht, wieviel ovarielles Gewebe die Frau tatsächlich in ihrem Körper besitzt. Wie eingangs erwähnt, findet sich bei der Patientin in der linken Inguinalgegend ein ähnlich knotiges Gebilde, wie das aus der rechten entfernte. Es ist mit größter Wahrscheinlichkeit anzunehmen, daß es sich auch dort um Keimgewebe handelt, der Charakter desselben aber ließe sich nur durch eine genaue histologische Untersuchung feststellen. Daß tiefgreifende Unterschiede zwischen rechter und linker Keimdrüse bestehen, wurde schon öfters beobachtet. So fand sich z. B. bei dem Fall von Hermaphroditismus verus, den Salén beschreibt, linkerseits ein vollständiges Ovarium, rechts eine Zwitterdrüse. Es soll nach Lilienfeld überhaupt bei menschlichen Zwittern auf der linken Seite das weibliche Geschlecht überwiegen. Es kann aber auch anfänglich mehr weibliches Keimgewebe vorhanden gewesen sein als später gefunden werden kann; „denn", sagt Kermauner in Schwalbes „Die Mißbildungen des Menschen und der Tiere", „ob der vollständige Defekt auf primäre Aplasie oder nachträgliche Involution zurückzuführen ist, ist nicht immer zu entscheiden; in keinem Falle kann man es ausschließen, daß mikroskopische Reste des Ovariums, vielleicht zur Unkenntlichkeit verändert, irgendwo an der Bauchwand liegen geblieben sind".

In dem hier beschriebenen Falle scheint es unwahrscheinlich, daß der Knoten, der auf der linken Seite getastet werden kann, nicht auch zum größten Teile männliches Keimgewebe enthält; denn es dürfte ja ein Descensus versucht worden sein. Unbedingt aber muß die Frau ihrem, wenn auch nicht ausgesprochen, so doch ziemlich stark weiblichen Typus nach zu schließen, viel mehr weibliches Keimgewebe in sich haben, als wir in dem kleinen Knoten auf der rechten Seite gefunden haben. Es ist sehr wahrscheinlich, daß auf der einen oder der anderen oder auf beiden Seiten sich neben dem Leistenhoden noch Ovarien höher oben in der Bauchhöhle vorfinden. Das sind natürlich Mutmaßungen, die nur durch eine genaue Autopsie und mikroskopische Diagnose auf ihre Richtigkeit geprüft werden können. Sicher läßt sich nur sagen, daß die Person viel weibliche Charaktere aufweist, und es erscheint dann interessant, daß trotz dieses Umstandes der männliche Keimapparat sich zu einer derartigen Höhe entwickeln konnte, wie die Präparate es zeigen. Es ist sogar, wie man aus den Hohlräumen und Konkrementen im Nebenhoden schließen muß, unbedingt zu einer sekretorischen Tätigkeit gekommen.

Der Fall ruft die Anschauung Waldeyers von der ursprünglich zweigeschlechtlichen Anlage des Keimbezirkes gegenüber der neueren Lenhosseks von der anfangs indifferenten ins Gedächtnis zurück. Beide Anlagen stehen, man könnte sagen, im labilen Gleichgewicht zueinander, bis die eine aus unbekannten Gründen die Oberhand gewinnt und die andere dann unterdrückt wird. In diesem Falle wäre eben die schwächere Anlage doch noch bis zu einem gewissen Grade zum Durchbruch gekommen.

Noch in einer anderen Hinsicht ist der demonstrierte Fall interessant. Die Schwester dieses Hermaphroditen zeigt nämlich ebenfalls Unregelmäßigkeiten im Bau der äußeren Geschlechtsorgane. Für den Menschen ist bisher Heredität oder verwandtschaftliche Beziehung beim Hermaphroditismus noch nicht wissenschaftlich einwandfrei nachgewiesen worden, wohl aber fand Reuter bei einem Wurf von drei Schweinen einen echten und zwei Pseudohermaphroditen, und in einem späteren Wurf desselben Mutterschweines wiederum einen Pseudohermaphroditen.

Ich bin geneigt, die Patientin als echten Zwitter anzusehen und stütze mich dabei auf die Einteilung von Klebs, der von einem echten Zwitter spricht, wenn die Keimdrüsen beider Geschlechter in einem Individuum angelegt sind. Die physiologische Leistungsfähigkeit kommt dabei nicht in Betracht. Hermaphroditismus in dem Sinne, daß Hoden und Eierstöcke beiderseits getrennt vorkommen, ist beim Menschen noch nicht gefunden worden. Hingegen sind vier Fälle von sogenanntem Ovotestis — zwei allerdings mit neoplastischer Veränderung des männlichen Anteiles — unbestreitbar festgestellt worden, denen sich der vorliegende Fall als fünfter anschließt.

[Reprinted from Biological Bulletin, Vol. XXII., No 3 February, 1912.]

THE OSMOTIC AND SURFACE TENSION PHENOMENA OF LIVING ELEMENTS AND THEIR PHYSIOLOGICAL SIGNIFICANCE.[1]

J. F. McCLENDON.

Contents.

Preface.

This paper formed the basis for two lectures given before the class in physiology at Woods Hole, July 7 and 8, 1911, although owing to limited time, some parts were omitted. Since then there has appeared a second edition of Höber's "Physikalische Chemie der Zelle und Gewebe," which reviews much of the literature considered in this paper. However, owing to an entirely different mode of presentation, it is hoped that the present treatment of the subject might be helpful to many general readers, some of whom would not read Höber's book.

[1] From the Embryological Laboratory of Cornell University Medical College, New York City.

I am indebted to several persons for suggestions, especially to Dr. Ralph Lillie[1] and Professor B. M. Duggar.

I. Introduction.

The object of this paper is to bring the "vital" phenomena, as far as possible, within the scope of physics and chemistry, and not to elucidate physical and chemical processes. It should therefore be borne in mind that the osmotic phenomena of "dead" systems are not all satisfactorily explained.

The Vant Hoff-Arrhenius theory of osmosis concerns itself with the number of particles, molecules and ions, in solution, and is applicable to dilute solutions, in which the total volume of the dissolved particles is negligible. However, in more concentrated solutions, the volume of the dissolved particles is of the same importance as the volume of the molecules in gases, as expressed in Van der Waal's equation. Also the dissolved particles bind molecules of the solvent and so reduce the volume of the free solvent.

That the molecules and ions of a dissolved substance bind some molecules of the solvent, follows from the work of Jones and his collaborators.[2] Compare also the work of Pickering.[3] Jones concludes that the larger the number of molecules of water of crystallization, the greater the hydrating power of a substance in aqueous solution. The number of molecules of water bound by one molecule of the solute usually increases with dilution up to a certain point (the boundary between concentrated and dilute solutions, beyond which there is no heat of dilution). The bond between ions and the solvent is also indicated by the phenomenon known as "electrical transference." If an electrolyte and a non-electrolyte be dissolved in water and an electric current passed through the solution, water will be carried along with the ions to the electrodes.

With these corrections, the Vant Hoff-Arrhenius theory accounts for osmotic pressure, but does not show why many substances exert no osmotic pressure, in other words, why no

[1] Cf. this journal, 1909, XVII., 188.

[2] "Hydrates in Aqueous Solution," Pub. No. 8, Carnegie Ins. Wash., 1907.

[3] Whetam, "The Theory of Solution," 1902, Cambridge, p. 170.

membranes have been found that are impermeable to them. Overton supposed that the substance, in order to diffuse, must dissolve in the membrane. Kahlenberg and others consider a solution as a chemical combination between solute and solvent, and osmosis as a series of chemical reactions between the membrane and the two solutions, continuing until equilibrium is established. The essential points in the theory are: that the membrane is not a molecule sieve, but a substance with specific properties, and the chemical characters of the membrane and of the dissolved substances affect osmosis.

Willard Gibbs found that the more a solute lowers the surface tension of a solution, the more it tends to pass out of the solution, *i. e.*, by osmosis, or if this is prevented, to collect at the surface of the solution. This law has been extensively investigated and confirmed by I. Traube. For instance, in general, lipoid-soluble substances lower the surface tension of water and tend to diffuse out of it, whereas electrolytes slightly raise the surface tension of water and attract water from the adjacent phase. Osmosis may occur in opposite directions simultaneously. Gibbs and Traube state that the greatest osmotic flow is from the solution of lower surface tension to that of the higher, but this is not generally accepted. Osmosis consists of two distinct processes, from one solution to the membrane, and from the membrane to the second solution.

In case the membrane consists of two or more chemically different membranes placed one on another, osmosis consists of a series of steps; and Hamburger[1] made double membranes through which certain substances diffuse more rapidly in one direction than in the other.

Traube calls the bond between solute and solvent the "attraction pressure." In general, attraction pressure of ions increases with valence. The less the attraction pressure of the solute, the more it lowers the surface tension and tends to pass out of the solution. The presence of one solute lowers the attraction pressure of another in the same solution, and the greater the attraction pressure of a solute the more it lowers that of another. We might express this idea by saying that one substance takes

[1] *Biochem. Zeit.*, 1908, XI., 443.

part of the solvent away from the second and increases the concentration of the second substance. This may explain the effect of a harmless substance in increasing the toxicity of a poison. Schnerlen[1] observed that a solution of phenol below the threshold of toxicity for certain bacteria is rendered toxic by adding NaCl. Stockard showed that the toxicity of pure solutions of salts on fish eggs is increased by the addition of sugar, although the total osmotic pressure of the mixture is less than that of the normal medium.[2]

Just as Traube's precipitation membranes are absolutely impermeable to certain substances, so do living cells show this selective permeability. For instance, the vacuole fluid or cell sap of certain plant cells contains colored substances which do not diffuse into the protoplasm surrounding the vacuoles. If a cell be placed in a solution of the pigment, the protoplasm remains colorless. If the protoplasm be squeezed out of the cell into a solution of the pigment, it does not invariably become stained. However, if the cell is injured in certain ways, or dies from any cause, the pigment diffuses out of the vacuoles into the protoplasm and thence into the surrounding medium. We might conclude that the protoplasm in general is impermeable to the color, but at death it becomes permeable. On the other hand, Pfeffer[3] gives evidence for the existence of a mechanical membrane on the surface of the cell and lining the vacuoles. De Vries[4] placed cells into 10 per cent. KNO_3 solution colored with eosin. The plasma membrane and granular plasm died and stained long before any dye entered the vacuoles. However, the granular plasm may have absorbed all the dye, thus preventing its entrance for some time, without the necessity of any resistance of the vacuole membrane. Since protoplasm may be squeezed out in the form of droplets and still appears to be surrounded by membranes, Pfeffer concluded that the membrane was formed by the contact of the protoplasm with the medium

[1] *Arch. exp. Path.*, 1896, XXXVII., 84.

[2] However the NaCl in Schnerlen's and sugar in Stockard's experiment may have increased the permeability to the toxic substances, as discussed in later chapters.

[3] "Pflanzenphysiologie."

[4] *Jahrb. wiss. Bot.*, 1885, XVI., 465.

or with cell sap. He supposed these membranes to be the semipermeable parts of the cell, and that they became altered at death. Pfeffer called this membrane on the cell surface the "plasma membrane."

Whereas the nuclear membrane and certain vacuole membranes are semipermeable, these are lacking in erythrocytes, which are therefore good objects for testing the question whether the protoplasm in general, or merely its surface, is semipermeable. Höber[1] by two very ingenious but complicated methods, one based on dielectric capacity, determined the electric conductivity of the interior of the erythrocyte without rupture of the plasma membrane. Since the conductivity of the interior (about that of a .2 per cent. NaCl solution) was found to be many times greater than that of the erythrocyte as a whole, the membrane must be relatively impermeable to ions. There is much other, but less direct, evidence that the semipermeability resides in the plasma membrane, namely: the rapidity of change in permeability of certain cells, the sudden increase in permeability of a cell after swelling to a certain size (due presumably to rupture of the plasma membrane), the ease with which mild mechanical treatment increases the permeability, and the localization of electric polarization at the cell surface.

Quincke[2] supposed these membranes to be of a fatty nature. This idea was carried further by Overton, who considered the plasma membrane to be composed, not of neutral fats, but of substances of the class which are called "lipoids," which included non-saponifying ether soluble extracts of organs, *i. e.*, cholesterin, lecithin, cuorin, and cerebrin. He found[3] that all basic dyes were easily absorbed by living cells, but not most of the sulphonic acid dyes. This corresponded to their solubility in melted cholesterin, or solutions of lecithin and cholesterin, or particles of lecithin, protagon or cerebrin. His argument is somewhat weakened, however, by the fact that cholesterin decomposes on melting, and that if lecithin is allowed to absorb water its solvent power changes.

[1] *Arch. f. d. ges. Physiol.*, 1910, CXXXIII., 237, and Eighth Internat. Physiol. Congress, Vienna, 1910.

[2] *Sitzber. d. Kon. Preuss. Akad. d. Wissensch. zu Berlin*, 1888, Bd. XXXIV.

[3] *Jahrb. wiss. Bot.*, 1900, XXXIV., 669.

Many of Overton's critics do not distinguish between lipoids proper and a host of ether-soluble substances which are also called lipoids, and of the data which they present we will consider only that on lipoids proper. Ruhland[1] found that certain dyes stain plant cells but are not soluble in solutions of cholesterin (and vice versa). Robertson[2] observed that methyl green freed from methyl violet was insoluble in a nearly saturated solution of lecithin in benzol, whereas it stained living cells. Höber[3] obtained Ruhland's results, when using certain animal cells, but found that certain nephric tubule cells absorb all dyes that are not suspension colloids.

Faure-Fremiet, Mayer and Schaeffer[4] state that pure cholesterin does not stain with any dyes (contrary to Overton), malachite green (considered lipoid-insoluble by Ruhland and Höber) stains lecithin, and Bismarck brown (considered lipoid-insoluble by Ruhland) stains lecithin, cholesterin-oleate and cerebrin. A mere trace of free fatty acid greatly affects the behavior of lipoids toward stains.

Mathews[5] considers the absorption of dyes by cells as a chemical process. Since basic dyes combine with albumin in alkaline solution, lipoids in the membrane are not necessary for the absorption of such dyes.

Traube objected to Overton's hypothesis on the ground that Overton's plasmolytic series is the same as found by Brown, who used the membrane of the barley grain,[6] and the same as the series of the attraction pressures of the substances in water. But Traube admits in his later papers that the chemical character of the membrane affects osmosis.

We may conclude that, although the plasma membrane of some cells may be lipoid in character, this has not been proven, but, in general, it is more permeable the more the diffusing substance lowers the surface tension of water.

[1] *Jahrb. wiss. Bot.*, 1908, XLVI., 1, and *Ber. Deutsch. bot. Gesellsch.*, 1909, XXVI., 772.

[2] *Jour. Bio. Chem.*, 1908, LV., 1.

[3] *Biochem. Zeit.*, 1909, XX., 55.

[4] *Arch. d'Anat. Mic.*, 1910, XII., 19.

[5] *Jour. Pharmacol. and Exp. Ther.*, 1910, II., 201.

[6] But this is not true of all seed coats. Atkins, *Sci. Proc. Roy. Dublin Soc.*, XII., n. s., No. 4, p. 35, observed that the membranes of the bean seed are freely permeable, semipermeable plasma membranes arising only after germination.

Nathanson[1] supposed the plasma membrane to be a mosaic of lipoids and "protoplasm," but it is evident that if the lipoid portion is not continuous, it can not make the cell impermeable to any substance.

Czapek[2] states that lipoid solvents cause cytolysis when the surface tension of the solution is reduced to .68, and concludes from this that the plasma membrane contains glycerine tri-oleate since its emulsion reduces the surface tension of water to this figure.

The diffusion of water-soluble substances through swollen-plates, "gels" or "sols" of gelatine, varies inversely with the viscosity (Arrhenius). The great hysteresis of gelatine gels is taken advantage of to show that diffusion depends on the viscosity and not on the per cent. of gelatine, at the same temperature.[3]

The absorption of water by a gelatine plate increases its permeability, and the temperature and therefore the presence of substances which affect this swelling of gelatine affect its permeability. Impregnation of colloidal membranes with bile salts, alcohol, ether, acetone or sugar changes (usually increases) their permeability. The effects of substances on the rate of diffusion through gelatine plates, and on their swelling (viscosity) and melting point are not always quite parallel.[4]

In case the substance added to the membrane is removable, the change in permeability becomes reversible, which is true in regard to many of the substances mentioned above. Changes in non-living membranes are usually only partially reversible or are irreversible. Denaturalization of a colloid membrane by heat, heavy metals, or other coagulative agents which induce chemical changes in the membrane, or the addition of substances which cannot be removed, produce irreversible changes in permeability.

That the permeability of the membranes in living tissue is increased at death is proven by a host of observations. The electric conductivity increases enormously at death. Contained

[1] *Jahrb. wiss. Bot.*, 1903, XXXVIII., 284; 1904, XXXIV., 601, and XL., 403.

[2] *Ber. deutsch. bot. Gesell.*, 1910, 28, 480.

[3] Zangger, Asher & Spiro's *Ergeb. der Physiol.*, 1908, VII., 99.

[4] Zangger, *loc. cit.*

substances diffuse out, substances in the medium (fixing fluids, stains, etc.) diffuse in. There is a more general mixing of tissue substances. Enzymes come in contact with proteids and autolysis results.

Certain substances are known to increase the permeability of membranes in tissues of the body. Thus ether, chloroform, etc., increase the penetration of fixing fluids, and the exit of contained substances, and the mixing of tissue substances. In this way they increase autolysis.

II. Osmotic Phenomena in Plants.

It is evident that water, salts, carbon dioxide and oxygen can, at least occasionally, penetrate plant cells, as otherwise no growth could occur. In case of the higher plants, the same is true of sugars and other bodies.[1] Janse[2] found that so much KNO_3 is absorbed by *Spirogyra* cells in 10 minutes, that it may be easily detected microchemically with diphenylamin-sulphuric acid.

Osterhout[3] grew seeds of *Dianthus barbatus* in distilled water. The rate of growth during the several days of observation was normal. In nature, calcium oxalate crystals are found in the root hairs, but are not formed in the distilled water cultures, showing that the Ca comes from the medium. If placed in calcium solutions, crystals became large enough to see with the polarizing microscope in four hours, showing permeability to Ca.[4]

Nathanson[5] found that nitrates and other substances entered the cell. Ruhland also observed penetration of salts.

Traube-Mengarini and Scala[6] conclude that salts enter plant cells only through the partition walls. At these places there appears an "acid reaction" (bluing of methyl violet). They

[1] See Laurent in Livingstone, "The Rôle of Diffusion and Osmotic Pressure in Plants," 1903, p. 67.

[2] *Versl. en Medeel. der Konikl. Akad. van afdeel. Naturs.*, 3. Reeks, IV. part, 1888, p. 333.

[3] *Zeits. f. physik. Chem.*, 1909, LXX., 408.

[4] But compare von Mayenberg, *Jahrb. f. wiss. Bot.*, XXXVI., 381, who found little penetration of salts into fungous hyphæ. And see Demoussy, *Comptes Rendus*, CXXVII., 970.

[5] *Jahrb. wiss. Bot.*, XXXVIII., 284; XXXIX., 601; XL., 403.

[6] *Biochem. Zeit.*, 1909, XVII., 443.

interpret this as showing that the anion of the salt unites with an H ion of an amino group, forming a free acid, and the kation of the salt unites with the protoplasm. It appears to me that the basis of this conclusion is very slight.

Permeability may be investigated by a study of plasmolysis, which consists in the shrinkage of the surface protoplasm away from the cellulose cell wall, due to the osmotic pressure of the hypertonic solution of a dissolved substance which does not penetrate. A regaining of turgor by the cell while in the hypertonic solution indicates slow penetration of the substance. The plasmolytic method was originated by Nageli, who also noted that a shrinkage resembling plasmolysis but accompanied by outward diffusion of dissolved substances, occurs at death or severe injury to the cell.[1]

The plant cell is surrounded by an elastic cell wall. The internal osmotic pressure may be divided into three resultants: that causing rounding up of the cell is called turgor, that resulting in stretching of the cell wall is sometimes distinguished as turgescence, and that resisting the surface tension of the cell, "central pressure."

The plasmolytic experiments of DeVries[2] and others[3] are interpreted by them as indicating a selective impermeability of the plasma membrane to neutral salts.

In the plasmolytic experiments of Overton[4] all salts plasmolyzed permanently. Non-electrolytes fell in four groups, thus: Cane sugar, dextrose, manit, glycocoll > urea, glucerin > ethylene-alcohol, acetamid > methyl-alcohol, acetonitril, ethyl-alcohol, phenol, aniline, isobutyl-alcohol, isoamyl-alcohol, methyl acetate, ethyl acetate, butyl aldehyde, acetone, acetaldoxim. Diffusion of substances of homologous series increased with molecular weight.

Overton ascertained the permeability of plant cells to alkaloids

[1] "Pflanzenphysiol. Untersuchungen," 1885.

[2] *Zeit. physikal. Chem.*, 1888, II., 415; 1889, III., 103.

[3] Cf. Livingstone, "The Rôle of Diffusion and Osmotic Pressure in Plants." Chicago, 1903; Janse, *Bot. Centlb.*, 1887, XXXII., 21; Duggar, *Trans. Acad. Sc. St. Louis*, 1906, XVI., 473.

[4] *Vierteljahrschrift der Naturforschers. Gesell. in Zurich*, XLIV., 88; *Jahr. wiss. Bot.*, 1900, XXXIV., 669.

by their precipitation of the tannic acid in the cell sap. Most alkaloids penetrate rapidly, but only in the form of the free (undissociated) base produced by hydrolysis. Hence the penetration (precipitation and toxic effect) may be prevented by adding a little acid to the medium.

Pfeffer had shown that methylene blue is precipitated by tannic acid in the cell sap of certain plants.

Some discussion has arisen as to whether the mechanism of the entrance of dyes into plant cells is similar to that of alkaloids. Overton showed that lipoid soluble basic dyes penetrate easily. He at first supposed that only the free color base (undissociated) is able to penetrate the cell.[1] Overton found, however, that triphenylmethane and chinonimid dyes disprove his assumption, showing that it is at least not general. This question was taken up again by Harvey[2] who found that neutral red or methylene blue, which stain *Elodea* leaves in tap water, do not do so if just enough acid be added to the water to prevent any free color base from forming.

He observed that, although these dyes are not precipitated in the cell sap of this plant, they become more concentrated in the cell sap than in the medium. Neutral red is bright red in the cell sap, indicating that the reaction is acid (no free color base is present). He supposes that the absence of any of the dye in the form of the free color base prevents it from diffusing out of the cell, hence it becomes more concentrated within than without.

In using the plasmolytic method, if a cell does not recover from plasmolysis in a solution of a salt, it is said to be impermeable to that salt. However, the cell may recover, but may be killed by penetration of the salt, and shrink again. It is possible that Overton and others failed in some cases to note this transient recovery. Contrary to Overton, Osterhout[3] found *Spirogyra* permeable to alkali-salts and alkaline earth salts, but more

[1] In this connection it is interesting to note that Robertson observed that free color bases, and to a less extent free color acids, are much more soluble in fats than are their salts. This is what we should expect, since the salts dissociate in water, and ions are insoluble in fats.

[2] *Science*, 1910, n.s., XXXII., 565.

[3] *Science*, 1911, n. s., XXXIV., 187; XXXV., 112.

easily to Na than to Ca. It is plasmolyzed by $.2M$ $CaCl_2$ and not by the isosmotic $.29M$ NaCl but by $.38M$ NaCl. $.195M$ $CaCl_2$ and $.375M$ NaCl just failed to plasmolyze. On mixing 100 c.c. $.375M$ NaCl with 10 c.c. $.195M$ $CaCl_2$, thus decreasing the osmotic pressure of the former, marked plasmolysis occurred. This indicates that Ca decreases the permeability to Na.[1] From further work by the same author, not yet published, it appears that Na increases and Ca decreases the permeability of certain marine plants. Also Fluri[2] obtained increase in permeability by salts of aluminium, yttrium and lanthanum.

DeVries plasmolyzed cells of *Tradescantia*, containing blue cell sap, with 4 per cent. KNO_3 solution, then added nitric acid until the color changed to red. The acid made the cells permeable to KNO_3 for they regained their turgor and finally burst. This explains the easy penetration of acids into cells. Pfeffer[3] found that if red beet cells, petals of *Pulmonaria*, stamen hairs of *Tradescantia* and other anthocyan-containing cells are placed in extremely dilute HCl or H_2SO_4, they suddenly turn red, indicating immediate penetration of the acid. If allowed to remain but a short time, the cells are not killed, and the color change is reversed on returning the tissues to acid-free water.

I have repeated these experiments, using cells of red beet, red cabbage and red nectar glands of *Vicia faba*, and find that mineral acids penetrate, but that (the lipoid soluble) acetic acid penetrates much more rapidly and also more easily alters the plasma membrane, causing pigment to diffuse out, if not cautiously applied. Alkalis also penetrate, but (the lipoid soluble) ammonia penetrates much more rapidly than the others. Ammonia does not so easily increase the permeability to the pigment as does acetic acid.

Ruhland[4] after staining root hairs of *Trianea*, etc., with the indicators, methyl orange and neutral red, found that mineral acids as well as lipoid soluble acids penetrated.

[1] The work of Kearney, Report 71, U. S. Dept. of Agriculture, indicates that Ca prevents the plasmolytic and toxic effect of Mg. but this is "false plasmolysis" following death.

[2] *Flora*, 1908, XCIX., 81.

[3] "Osmotische Untersuchungen," Leipzig, 1877, p. 135.

[4] *Jahrb. wiss. Bot.*, 1908, XLVI., 1.

One defect in the plasmolytic method is the fact that the cellulose cell wall, if not very thick, is elastic, and a slightly hypertonic solution may cause the cell to decrease in volume without pressing the protoplasm away from the cell wall. This source of error may be eliminated by substituting calculations of the volume of the cells (as necessary for animal cells) for observations on plasmolysis.

It is well known that movement, and in many cases increase in size of plants is due to changes in turgor of the cells. If we exclude the turgor changes in aerial plants produced by variations in the ratio of the water supply to the transpiration, turgor changes may be due to changes in the osmotic pressure of the external medium, or of the cell sap (due to metabolic changes) or to changes in the permeability of the plasma membrane. Lepeschkin[1] has confirmed Pfeffer in showing that changes in permeability of stipule cells accompany (or immediately precede) changes in turgor. By chemical analysis of the medium he has shown that an outward diffusion of dissolved substances, from the cells, accompanies loss of turgor, and by plasmolytic experiments, that the permeability to certain substances increases.

It is interesting to note the force that may be exerted by such changes in turgor. From measurements of the pull of a stamen hair of *Cynara scolumus* or *Centaurea jacea* on loss of turgor following stimulation, it seems not improbable that the change in turgor amounts to 2–4 atmospheres (Höber). This also indicates the strength of the cell wall necessary to prevent rupture of the plasma membrane. The osmotic pressure of the juices pressed out of plants varies from 3.5–9 atmospheres.[2] The pressing out of the juices causes an error due to chemical changes; on the other hand, in taking the freezing point or pieces of plant tissues, an error arises from lowering of the freezing point by the walls of the capillary spaces. Müller-Thurgau[3] found the Δ (corrected freezing point lowering) of plant tissues $= .8–3.1°$. Many plants respond to light by definite movements, produced

[1] *Ber. deutsch. bot. Gesell.*, XXVI. (*a*), 725.

[2] DeVries, *Pringsheime Jahrbucher wiss. Bot.*, 1884, XIV., 427; Pantanelli, *ibid.*, 1904, XL., 303.

[3] *Landwirtschaftl. Jahrb.*, 1886, XV., 490.

by turgor changes in certain of their cells. Trondle[1] found that light produced changes in permeability of these cells.

Changes in permeability may not only affect the turgor, but also the assimilation and excretion, and consequently the metabolism and growth of the cells. Chapin[2] observed that CO_2 in certain doses is a stimulant to the growth, not only of green plants but also of moulds. As only a few saprophytes can decompose CO_2, it is not probable that its effect is nutritive. A similar stimulating action of ether and various salts, even such toxic ones as those of zinc, was previously known. These salts probably stimulate without penetrating the cells, since Zn, for instance, is not a constituent of protoplasm.[3] This leads one to suppose that the initial effect of all of these substances is on the surface, changing the permeability of the cells.

Wächter[4] found that potassium decreases the permeability of onion cells. Sugar diffused out of sections of *Allium cepa* placed in distilled water or hypotonic sugar solutions, but a trace of potassium salt entirely prohibited the diffusion. When the K was removed the diffusion recommenced.

Czapek[5] determined increase in permeability by the exosmosis of tannin in cells of *Echeveria* leaves. Various monovalent alcohols and ketones, ether, ethyl urethan, di and tri acetin, Na-oleate, oleic acid, lecithin and cholesterin all just caused exosmosis of tannin in concentrations (aqueous solutions) which had a surface tension of about 0.68. It would appear therefore that these substances, chiefly of the class of indifferent narcotics, alter the cells if they diffuse into them, or diffuse into certain structures such as the cell lipoids or the plasma membrane. It seems more reasonable to suppose that the plasma membrane is the structure affected, and the more the substance lowers the surface tension of water, the more it diffuses into the plasma membrane. When this membrane is altered, it allows escape of tannin. Some substances such as chloral hydrate are effective

[1] *Jahrb. f. wiss. Bot.*, 1910, XLVIII., 171.

[2] *Flora*, 1902, XC., 348.

[3] Cf. Loeb, "Dynamics of Living Matter," pp. 73, 74.

[4] *Jahrb. wiss. Bot.*, 1905, XLI., 165.

[5] "Über eine Methode zur direkten Bestimmung der Oberflächenspannung der Plasmahaut von Pflanzenzellen," Jena, G. Fischer, 1911.

in less concentration, and probably affect the cell chemically as well as physically.

Mineral acids caused exosmosis of tannin when the concentration just exceeded 1/6,400 normal, and the effect is probably due to H ions. At this same concentration Kahlenberg and True[1] found the growth of seedlings of *Lupinus albus* to cease. It appears, therefore, that this cessation of growth is due to increased permeability, causing decreased turgor of the cells.

Changes in permeability may also affect secretion (excretion). The addition or formation of alcohol or acetates causes yeast and other fungi to secrete (excrete) for a short time, various substances, especially enzymes which do not come out in a culture medium lacking the reagent.[2] It appears that the alcohol or acetates increase the permeability of the fungi to these substances.

My own experiments[3] indicate that pure $MgCl_2$ solutions increase the permeability of yeast. A certain per cent. of yeast and dextrose in .3 molecular $MgCl_2$ eliminated CO_2 more rapidly than .5*M* NaCl or .325*M* $CaCl_2$, all which have about the same freezing points. Also, the CO_2 elimination was more rapid in the magnesium solution than in a solution of the same concentration of $MgCl_2$ with either of the other salts in addition, or in a solution containing NaCl and $CaCl_2$ in the same concentrations as in their respective pure solutions, or in a solution of all three salts, or in tap or distilled water. In order to determine whether the magnesium entered the cells I took two equal masses of compressed yeast and agitated one in H_2O and the other in a molecular solution of $MgCl_2$ for 5 hours, then washed each rapidly in H_2O by means of the centrifuge. The ash of the magnesium culture = .048 gram, that of the control = .0466 gram. Evidently the Mg did not enter the yeast to any great extent, and probably acted on the surface, increasing the permeability.

Ewart[4] observed that after placing plant tissue in 2 per cent. HCl and washing in water its electric conductivity (ionic permeability) was increased. If one portion of the plant is stimulated, the stimulus may be transmitted to other portions. In

[1] Kahlenberg and True, *Botanical Gazette*, 1896, XXII., p. 81.

[2] Zangger, "Asher and Spiro's Ergeb. d. Physiol.," 1908, VII., 144.

[3] McClendon, *Am. Jour. Physiol.*, 1910, XXVII., p. 265.

[4] "Protoplasmic Streaming in Plants," Oxford, 1903, p. 96.

this way increase in electric conductivity was produced by stimulation of a point outside the path of the current.

Whereas many plants are very sensitive to sudden and extreme changes in osmotic pressure, Osterhout[1] found that certain marine algæ thrived when subjected daily to a change from fresh water, to sea water evaporated down until it crystallized out, and vice versa. He does not state whether these algæ survive extreme plasmolysis, or whether they are so easily permeable to salts as not to be plasmolyzed by the saturated sea water or burst by the fresh water.

For regulation to slight changes in the osmotic pressure of the medium, a change in size of the cell altering the turgescence, or tension of the cell wall, is sufficient.

If *Tradescantia* cells are placed in a hypotonic solution, they begin to swell. But soon crystals of calcium oxalate are formed in the cell sap, and in this way the turgor, due chiefly to oxalic acid, is reduced.[2] It would be interesting to know what is the source of the Ca. Was it previously in combination with proteids?

The accommodation to a hypertonic medium takes place, according to van Rysselberghe, partly through absorption of substances of the medium and partly through metabolic production of osmotic substances, chiefly the transformation of starch into oxalic acid.[3]

III. Bio-electrical Phenomena.

1. *In Plants.*

Change in permeability of the plasma membrane to ions would necessarily cause electrical change due to its influence on the migration of ions. These electrical changes actually occur, and may be easily studied.

Stimulation or wounding in plants is accompanied by an electronegative variation of the affected surface. This negative region spreads in all directions over the surface, but the rate of

[1] Univ. of Cal. Pub., Bot., 1906, II., 227.

[2] Van Rysselberghe, Mém. d. l'Acad. royale de Belgique, 1899, LVIII., 1.

[3] Compare von Mayenberg *Jahrb. f. wiss. Bot.*, XXXVI., 381.

propagation[1] is much slower than the similar process in muscle or nerve.[2]

Pfeffer[3] supposed that the plasma membrane is normally permeable to ions of only one sign. Since the normal cell surface is positive in relation to the cell interior (cut surface) we may conclude that the plasma membrane is normally more permeable to kations (less permeable to anions). Just as the negative variation of wounding is due to the removal or rupture of the plasma membrane, so the negative variation of stimulation would, on the membrane hypothesis, be due to increase in permeability of the plasma membrane to the confined anions.

An alternative hypothesis is that these electrical changes result from changes in metabolic activity. The production of an electrolyte whose anion and kation have very different speeds of migration (such as an acid or alkali) would cause electrical changes. But how are we to account for changes in metabolic activity? There exists varied evidence for changes in permeability, and it is simpler to assume that changes in metabolic activity and electrical changes are both the result of changes in permeability.

Kunkel[4] tried to explain the vital electrical phenomena as the result of the movement of fluids in the vessels of the tissues, but bio-electrical changes may occur without such movement of fluids (Burdon-Sanderson).

Kunkel observed in 1882[5] that the movement of the leaf of *Mimosa pudica* is accompanied by an "action current," or negative variation of one surface of the pulvinus. Similar results on *Dionæa* leaves were obtained by Munk[6] and specially studied by Burdon-Sanderson.[7] It was stated above that Lepeschkin had shown that the turgor changes in plants were accompanied or immediately preceded by changes in permeability to certain substances. The electrical phenomena suggest that the turgor

[1] Which is in *mimosa* 600–1,000 times as fast as the geotropic impulse in a root.

[2] Fitting, "Asher and Spiro's Ergeb. d. Physiol.," 1906, V., 155.

[3] "Pflanzenphysiologie."

[4] *Arch. f. d. ges. Physiol.*, 1881, XXV., 342.

[5] See Winterstein's "Handbuch der vergleichenden Physiologie," III. (2), 2, p. 214.

[6] *Arch. f. Anat. u. Physiol.*, 1876, XXX., 167.

[7] *Proc. Roy. Soc. London*, 1877, XXV., 441; *Philos. Trans.*, 1888, CLXXIX., 417.

change is accompanied (or immediately preceded) by increase in permeability of the plasma membrane to anions. Burdon-Sanderson states that, whereas the movement resulting from turgor change begins 2.5 seconds after stimulation, the negative variation reaches its maximum 1 second after stimulation. This may be due to the mechanical inertia, or the time required for the diffusion of substances.

It was stated in the preceding chapter that light changes the permeability of the plasma membrane, and Waller[1] found corresponding electrical changes due to light, but not always in the same direction in different plants. This inconstancy in direction is probably due to the fact that light not only influences the permeability, but also the assimilation, and changes in assimilation produce electric changes. This is supported by the fact that Querton[2] found that assimilation as well as electric change is most affected by the longer light rays.

2. *In Muscle and Nerve.*[3]

Ostwald[4] proposed the hypothesis that the electric phenomena of muscle, nerve and the electric organs of fish (which may reach several hundred volts) are produced with the aid of semipermeable membranes. The alternative theory of Hermann, which would account for the current of injury by assuming the production of some electrolyte (alkali?) in the wounded region, whose anions and kations have very different speeds, seems less probably to be the correct one.

According to the "membrane theory," the muscle or nerve element is surrounded by a semipermeable membrane allowing easier passage to kations than to anions. The kations passing through the membrane are held back by the negative field produced by the confined anions, but owing to their kinetic energy, the kations pass out far enough to give the outside of the cell surface a positive charge. Therefore any portion of the surface that is made freely permeable to anions becomes electronegative

[1] *Jour. of Physiol.*, 1899–'00, XXV., 18.

[2] "Contribution à l'étude du mode de la production de l'électricité dans etres vivantes," *Travaux de l'Institut Solvay*, 1902, V.

[3] Cf. R. Lillie, *Amer. Jour. Physiol.*, 1911, XXVIII., 197.

[4] *Zeit. physik. Chem.*, 1890, VI., 71.

in relation to the remainder of the surface. This negative variation may be produced by artificially removing or altering a portion of the membrane (producing the current of injury) or as the result of normal stimulation, making it permeable to anions (action current).

Bernstein resorted to mathematical proof of this hypothesis. We will not here go into details, but the gist of the matter is that if the process were as we have imagined it, the electromotive force of the current of injury, or action current, should be proportional to the absolute temperature. He found this to be true for temperatures between 0° and 18°, but between 18° and 32° the E.M.F. was found to be too small. The muscle was not permanently injured by exposure to the higher temperatures for the length of time necessary for the experiments. Bernstein explained this discrepancy by the further assumption that at the higher temperatures the plasma membrane became slightly more permeable to anions.[1]

Since the muscle contains a higher per cent. of potassium than the blood plasma or lymph, it might be supposed that K ions passed outward through the plasma membrane and gave the surface of the muscle element the positive charge. But if this were the case, the current of injury should be reversed by placing the muscle in a solution containing potassium in greater concentration than in the muscle. This reversal, however, was shown by Höber not to occur. Since lactic and carbonic acids are produced by muscle and diffuse out in increased amount on contraction, one might suppose H ions to give the muscle surface the positive charge. This is only a guess (and a poor one, since undissociated molecules of CO_2 and lactic acid are lipoid-soluble) but may be convenient until some better one is proposed. Perhaps the carbonic acid combines with amphoteric proteids, which

[1] This is similar to the conclusion reached by Biataszewicz, *Bull. d. l'Acad. d' Sc. d. Cracovie, Sc. Math. e. Nat.*, Oct., 1908, p. 783, in regard to the unfertilized frog's egg. In order to explain his observation that the rate of swelling in tap water increased 5 times for every 10° rise in temperature, he assumed that heat increased the permeability to H_2O. This would seem to be the simplest explanation, provided the swelling were not due to chemical production of osmotic substances: and since the Δ of the ripe ovarian egg is .48° but is reduced to .045° after oviposition, *Biochem. Zeit.*, 1909, XXII., 390, much if not all of the swelling is probably due to the initial osmotic pressure of the egg interior.

then set free H^+ and HCO_3^- ions, thus increasing the ionization and therefore reducing the number of undissociated molecules, which can escape.[1]

Since Osterhout showed that certain electrolytes may alter the permeability of cells, we might expect to find, on the membrane hypothesis, an effect of salts on the electric polarization of muscle. Höber[2] observed that a portion of the surface of a muscle treated with certain salts, KCl for instance, becomes electro-negative (more permeable to anions) whereas a portion treated with NaI or LiCl becomes positive (still less permeable to anions than is the normal unstimulated muscle). The order of effectiveness of the ions is as follows: $Li < Na < Cs < NH_4 < Rb < K$ and $CNS < NO_3 < I < Br < Cl <$ valerianate, butyrate, propionate, acetate, formate $< SO_4$, tartrate. Similar ionic series were found by Overton, R. Lillie, Schwartz, Mathews, Grützner, Höber, and Mayer in the effect of salts on the functional activity of muscle, nerve and cilia, but the exact relation of these phenomena to permeability is not understood in every case. Pure solutions of salts of alkali metals may "inhibit" muscle by *increasing* permeability, but salts of alkali earth metals are said to "inhibit" by *decreasing* permeability. . Mayer says that the effect of salts on cilia is the reverse of that of muscle, but the relation of this to permeability is not known. Since ions affect the aggregation state of hydrophile colloids in the same or exactly reversed order, and the kation series is found in no other known physico-chemical phenomena, it might be supposed that the semipermeable membranes of muscle are colloidal.

It seems probable that sugar solutions inhibit the activity of muscle by increasing the permeability, but since sugar is not an electrolyte this question cannot be tested by electric methods.

A negative variation of muscle may also be produced by the so-called "hæmolytic" substances, but is irreversible, whereas that produced by salts may be reversible. In this connection it

[1] Roaf, *Q. J. Exper. Physiol.*, 1910, III., 171, supposed the anion to be protein; however it has not been shown that proteids, or even amino acids diffuse out on stimulation. I do not see that the speculation of Galeotti, *Zeit. f. Allgem. Physiol.*, 1907, VI., 99, is at all explanatory.

[2] *Loc. cit.* and *Pflüger's Arch.*, 1910, CXXXIV., 311.

is interesting to note that Overton[1] found the permeability of muscle to be similar to that of plant cells.

It might appear to the reader that the membrane theory is merely wild speculation. What proof have we that on injury or during contraction the muscle is more permeable to any ion?

DuBois Reymond[2] and Hermann[3] explained the fact that living muscle has a greater electric resistance than dead muscle on the hypothesis that the resistance of living muscle is due to the presence of membranes, which become more permeable at death. They demonstrated the resistance of muscle tissue to the passage of ions by the fact that electric polarization occurs in muscle tissue on the pasage of an electric current. It seems to me that Kodis[4] and Galeotti[5] take a step backward, in attributing the decreased resistance of dead muscle to the liberation of ions. Galeotti tried to support his view by determinations of the freezing points of the living and dead muscle, but found on the contrary that the change in electric conductivity of the muscle did not correspond to the change in the osmotic pressure.

Du Bois Reymond[6] observed that the electric conductivity of muscle changes on (during?) contraction and Galeotti[7] found it to be greater on strong contraction than on weak contraction, and least on fatigue-exhaustion or cold-anæsthesia. However, the duration of a contraction is momentary (about 1/5 second for frog's muscle) and it is not clear that these investigators measured the conductivity accurately during such a brief period, in fact they probably measured it after contraction. Therefore I decided to repeat these experiments, using a method by which I could measure the conductivity during the actual contraction period, as well as in the unstimulated condition.[8]

[1] *Pflüger's Arch.*, 1902, XCII., 115.

[2] "Untersuchungen über thierische Electricität," 1849.

[3] *Pflüger's Arch.*, 1872, V., 223, VI., 313.

[4] *Am. Jour. Physiol.*, 1901, V., 267.

[5] *Zeit. f. Biol.*, n. f., 1902, XXV., 289; 1903, XXVII., 65.

[6] *Loc. cit.*

[7] *Loc. cit.*

[8] McClendon, *American Journal of Physiology*, 1912, XXIX., 302.

Experimental.

Platinum electrodes, platinized with platinic chloride containing a little lead acetate, and of a form similar to those designed by Galeotti, were used. Galeotti stimulated the muscle through the same electrodes used in measuring the electric conductivity, by switching on a different electric current. Though it were possible to throw a switch quickly enough to have the current for measurement of conductivity pass through the muscle during contraction, it would be necessary to use a string galvanometer to take the reading, and this method would probably not be very accurate. A more accurate method is that of Kohlrausch, in which a rapidly alternating current reduces polarization at the electrodes and in the tissue, but it is necessary to throw the muscle into tetanus in order to have time for the reading. I accomplished this by using the same current for stimulation and measurement of conductivity. A very small induction coil was fitted with a rheostat in the primary. Another rheostat in the secondary could be thrown out of the circuit by a switch. By adjusting the rheostats, a current strong enough to be distinctly heard in the telephone, yet too weak to stimulate the muscle, was obtained. By switching the resistance out of the secondary circuit, the current could immediately be increased so as to throw the muscle into tetanus. Since the Wheatstone bridge was used, the difference in current strengths had no direct effect on the readings. The conductivity increased from 6 to 28 per cent. (being usually about 15 per cent.) on stimulation.

We have, then, evidence for the increase in permeability of muscle to ions during contraction, but what relation has this to the mechanism of the contractile process? It has been suggested by D'Arsonval, Quincke, Imbert, Bernstein, Galeotti and others that the increased permeability to ions causes a disappearance of the normal electrical polarization of the elements, whose surface tension consequently increases, causing them to round up (shorten). But what are the elements concerned? It would be confusing to assume them to be the fibers, as then the function of the complicated internal structure would be unexplained. They are probably not the sarcous elements (portions of fiber between 2 Z-lines) as the rounding up of these elements would elongate the muscle. And even though contraction were produced by inequality in surface tension, as assumed by Macallum[1] the total surface change would be so small as not to account for the energy liberated in contraction. In order to avoid this last difficulty Bernstein made use of hypothetical ellipsoids. These were surrounded by elastic material to account for elongation of the muscle.[2]

The great differences of potential (several hundred volts) that may be produced by the electric organs of fish, is achieved by the arrangement of the modified muscle plates in series. All of the plates have the nerve termination on the same side. On stimulation of the nerve, each plate becomes negative, first on the nerve termination side, and thus the negative side of one plate touches the positive side of the next plate. In this way the direction of the current may be determined by studying the anatomy of the innervation. This rule, discovered by Pacini, finds an exception only in *Malopterurus*, whose electric organ is supposed by Fritsch to be derived, not from muscle but from skin glands.

The electric fish are *relatively* immune to electric currents passed through the medium. This is not merely an apparent immunity due to the fish being out of the path of the current, or the current being short circuited by sea water (in case of marine fish). I have received severe shocks from a torpedo that was entirely submerged in sea water.

3. *Amœboid Movement.*[3]

The normal unstimulated surface of plant and animal tissues is electro-positive in relation to the cut or injured surface of the cells. We have given reasons for assuming that this indicates greater permeability of the plasma membrane to kations than to anions, the latter accumulating in the cell interior, gives it a negative charge.

There are two reasons for believing that this is true also of the *Amœba:*

[1] *Science*, n. s., 1910, XXXII., 822.

[2] Meigs., *Am. Jour. Physiol.*, 1910, XXVI., 191, supposes the rounding up of muscle elements due to increased turgor.

[3] McClendon, *Arch. f. d. ges. Physiol.*, 1911, CXL., 271.

1. If a weak electric current is passed through water in which an *Amœba* is suspended, it is carried passively toward the anode, indicating that it has a negative charge. This charge may be due to confined anions.

2. If a stronger electric current is passed through an *Amœba*, it begins to disintegrate first at that surface nearest the anode. The disintegration is probably due to the accumulation of ions retarded by the plasma membrane. The ions in the medium are free to pass around the *Amœba*, but the contained ions must pass the plasma membrane in order to migrate to the electrodes. Since the disintegration is toward the anode, it is probably due to anions which cannot get out of the *Amœba*. Since no corresponding disintegration begins toward the kathode, the plasma membrane is probably more permeable to kations.

The surface tension of the *Amœba* is very low, and apparently increases on strong stimulation (indicated by rounding up of the *Amœba*). We saw that stimulation in plant and muscle cells caused increased permeability to ions, and consequently disappearance of the normal electrical polarization, and thereby causing increased surface tension. We might conclude therefore that the low surface tension of the *Amœba* is caused by electric polarization, due to the production of some metabolic electrolyte whose anions cannot escape; and that strong stimulation causes increased permeability and hence disappearance of the electrical polarization.

This would explain all negative tropisms of the *Amœba*. The surface tension of the portion most strongly stimulated is increased, and the *Amœba* flows away from the stimulus.

In order to explain positive tropisms we would have to make another assumption. If the stimulus did not act directly on the plasma membrane, but penetrated the *Amœba* and acted on the protoplasm, and increased the production of the metabolic product producing polarization of the plasma membrane, it would thereby decrease the surface tension. The local decrease in surface tension would cause the *Amœba* to flow toward the source of the stimulus, just as the quicksilver drop in dilute HNO_3 flows toward potassium bichromate in Bernstein's experiment.

All stimuli producing positive tropism would then have to penetrate to a greater or less distance into the *Amœba*. But the same stimulus thus acting on the interior might, in greater intensity, affect also the plasma membrane, increasing its permeability and changing the positive to negative tropism. Such a change of the sign of tropism has been observed.

Soap lowers the surface tension of fats and lipoids, and Quincke, Bütschli, Loeb, Robertson and others supposed that lowering of the surface tension of living cells might be due to soap. However, I found that soap always causes negative tropism in *Amœba*, probably because it increases the permeability of the plasma membrane.

4. *The Propagation of the Bio-electric Changes.*

On the hypothesis, that the electric phenomena in muscle and nerve, as well as other animal and also plant tissues, is due to change in permeability to ions, we might hope to explain the wave-like propagation of these changes. Since extraneous electric currents "stimulate" all tissues (presumably by increasing permeability) thus causing them to produce additional electric phenomena, it seems natural that these latter would be self-propagating. It is probably the negative variation of nerve which stimulates the muscle, and the negative variation of the portion of the muscle fiber adjoining the nerve ending, which stimulates the adjacent portions of the muscle. Nernst found mathematical proof that electric stimulation is due to change in ionic concentration at the semipermeable membranes.

I have found evidence that the negative variation (current of injury) in plants, may strongly affect adjacent cells. If an electric current of suitable density is passed through plant or animal tissue, negatively charged colloids in the protoplasm migrate toward the anode. I have observed this movement in living cells, and the resulting displaced bodies in histological sections. In certain cases there may be some doubt whether the colloids moved toward the anode, or water toward the kathode, but in others, easily distinguishable bodies such as chromatin granules or threads moved toward the anode.

If the tip of a root be cut off we observe a negative variation of the cut surface. This produces an electric current through the medium and surrounding tissue. The fact that the current actually passes through adjacent cells is shown by a displacement of their contained colloids, identical in appearance with the displacement produced by the currents used in the above experiments. Nêmec[1] apparently observed these changes but did not correctly describe or interpret them.

The fact that an electric current on increase (make) stimulates muscle at the kathode, and the fact that the muscle surface is normally positive in relation to the interior (cut surface), probably indicates that stimulation is produced by a rapid depolarization of the muscle surface.

If this reasoning be applied to an individual contractile element, we may assume that the current causes kations to leave the outer surface of the membrane, and other kations to be attracted toward the inner side of the membrane, and thus the polarization disappears or may even be reversed. Just how this causes an increase in permeability of the membrane is a matter which we will leave to the future for discussion.

It has been supposed that the stimulated region acts as kathode to the adjacent portions, and these in turn act as kathodes to the next portions and so the stimulus is propagated.

Stimulation of a part of the surface, causing it to become more permeable to ions, depolarizes the adjacent parts of the surface owing to the fact that confined anions migrate through the permeable region and neutralize the charges of the kations on adjacent parts of the impermeable region (see Fig. 1). For this reason the increase in permeability is propagated.

This explanation of the phenomenon in a single element holds for a tissue made up of many elements provided these are in contact, as illustrated by the accompanying Fig. 2. This is probably the mechanism of propagation of the negative variation (and "stimulus") in many plant and animal tissues.

This mechanism accounts for the movement of the negative variation over a surface. But it may be possible for this electric change to jump from one element, to another not touching it. The observations on the current of injury, cited above, show that

[1] "Reizleitung u .d. reizleitenden Strukturen b. d. Pflanzen," Jena, 1901.

increased permeability of part of a tissue surface, may cause electric currents to flow through cells some distance from the wound. These currents probably stimulate the cells through which they pass, which in turn become permeable and produce electric currents. This explains the propagation of stimuli

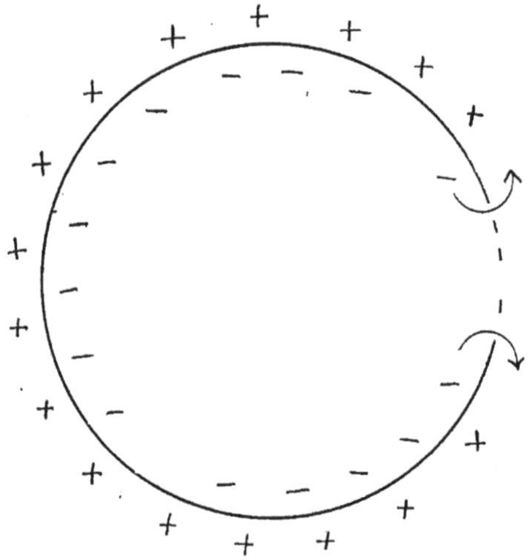

FIG. 1.

Anions represented by minus sign, kations represented by plus sign. Arrows denote the direction of migration of ions. The large circle represents the plasma membrane, the dotted line denoting the permeable and the continuous line, the impermeable portion.

through loose tissues, and the structural changes, as observed by Nêmec.

The rate of propagation of the "wound stimulus" is very slow, whereas that of propagation of the "stimulus" (negative variation) in sensitive plants is more rapid, and that of the nerve impulse still more rapid. We have not, however, sufficient data to show whether this is a mathematical objection to the hypothesis.

The streaming movements in plants may be stopped by a strong stimulus or "shock." This stimulus is usually propagated in one or more directions. Ewart[1] states that the rate of propagation at 18° in a single elongated cell of *Nitella* is 1–20 mm.

[1] *Loc. cit.*

per sec., but where it has to pass cell walls .001–.03 mm. per sec. However, the stoppage of the streaming was his criterion of the presence of the stimulus, and probably the banking of the stream

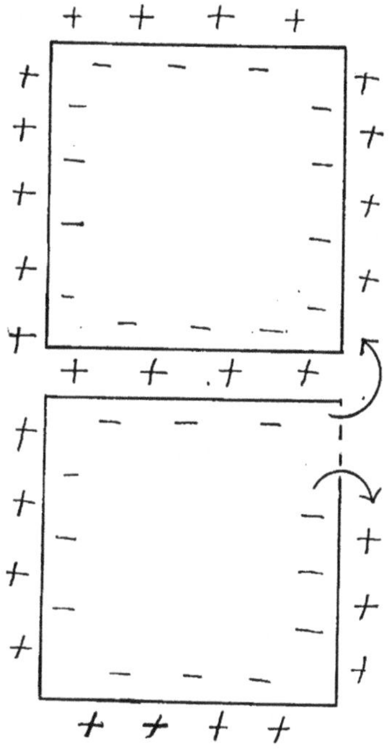

FIG. 2.

The squares represent the plasma membranes of adjacent cells. For further explanation see Fig. 1.

at one point, soon stopped the whole stream thus simulating the propagation of the stimulus.

IV. Narcosis.

If stimulation consists in increase in permeability, we should expect anæsthetics to prevent this change. The object of this chapter is to present evidence that may support or refute such a hypothesis.

Overton observed that warm- and cold-blooded vertebrates, insects and entomostraca, require practically the same concentration of the anæsthetic for narcosis. Certain groups of worms require double, and protozoa and plants six times this concentration. We might conclude from this that nerves (and especially medullated nerves?) are more susceptible to narcosis than are other cells. All groups of worms contain nerves, but Loeb has shown that certain worms may perform coördinated movements after the nerves are cut, hence the higher concentration of the narcotic required to quiet them. However it should be remembered that *over-stimulation* causes rounding up and quiescence of *Amœba* and muscle may be paralyzed by increasing the permeability. The growth of plants is increased by a certain concentration of ether and retarded by a greater concentration. It may be that true narcosis (decreased permeability) of protozoa and plants cannot be produced by such substances as ether, etc.

Vertebrate nerve tissues are rich in lipoids (which have similar solubilities to neutral fats) and it is therefore significant that Overton and also Meyer[1] found that the partition coefficient of anæsthetic between olive oil and water corresponds to its anæsthetic power. Meyer[2] showed further, that with change of temperature, the change in the partition coefficient between oil and water, and the anæsthetic power of the substance were parallel. Pohl, Frantz, Gréhaut, and Archangelsky found that chloroform, ether, alcohol, chloral-hydrate or acetone, became more concentrated in the central nervous system than in other tissues. This is probably due to the absorption of the narcotic by the lipoids (especially the immense mass of myelin) in the nerve tissues.

If it could be proven that the plasma membrane consists of lipoids, this solubility of narcotics might be considered direct evidence for or against the permeability hypothesis, but lacking such proof we must first attack the subject from another side.

Höber[3] observed that ethyl-methane, phenyl-methane, chloral-hydrate, chloroform and hypnon, in *low concentration* prevent the production by salts, of the current of injury on muscle. He showed that in *lethal doses* on the contrary these narcotics do

[1] *Arch. exp. Path. u. Pharm.*, 1889, XLII., 109.

[2] *Arch. exp. Path. u. Pharm.*, 1901, XLVI., 338.

[3] *Pflüger's Arch.*, 1907, CXX., 492, 501, 508. Cf. R. Lillie, *Am. Jour. Physiol.*, 1912, XXIX., 373.

not prevent but even *produce* a current of injury, in this way explaining data which might otherwise seem to contradict the first statement. Galeotti and Cristina[1] observed that ether, ethyl-chlorid, and chloroform produce a current of injury on frog's muscle.

We may conclude, then, that anæsthetics, in the concentration producing narcosis, so change the plasma membrane as to prevent salts from making it permeable to anions. This is probably also true of nerve, since Höber found that ethyl-methane in low concentration prevented the sensitizing of nerve with K_2SO_4.

Höber has attempted to connect these facts with the lipoid solubility of narcotics. Moore and Roaf[2] had observed that *small quantities* of such narcotics as chloroform, alcohol, ether, or benzol, precipitated lipoids extracted from organs and suspended in water. But Höber and Gordon[3] found that colloidal solutions of lecithin were not precipitated, but were made transparent by ether or chloroform in *high concentration*. Similarly, Goldschmidt and Pribram[4] observed that lecithin suspended in NaCl solution, which is dissolved by chloral hydrate, methane, or cocaine, in high concentration, is precipitated by them in low concentration. On the other hand, Koch and McLean[5] state that chloral, hypnon, acetone, or pure ether, do not change the size of colloidal particles of lecithin (*i. e.*, make them easier or more difficult to salt out). Calugareanu[6] explains the mechanism of the precipitation of lipoids by anæsthetics by the increase in size of the particles due to absorption of the anæsthetic.

Thus there seems to be a parallel difference between the action of low and high concentrations of anæsthetics on muscle and nerve, and the action of the same on lipoid suspensions, but this does not hold true for all cases. Moore and Roaf[7] conclude that anæsthetics are bound, not only by lipoids, but also by proteids,

[1] *Arch. allg. Physiol.*, 1910, X., 1.
[2] *Proc. Roy. Soc. London*, 1904, LXXIII., 382; 1906, LXXVII., 86.
[3] *Hofmeisters Beitrage*, 1904, V., 432.
[4] *Zeit. f. exper. Path. u. Ther.*, 1909, VI., 1.
[5] *Jour. Pharm. and Exp. Ther.*, 1910, II., 249.
[6] *Biochem. Zeit.*, 1910, XXIX., 96.
[7] *Loc. cit.*

and their charactersitic action on the permeability of the living cell may be due to their action on proteids. In other words, the plasma membrane may be entirely proteid.

It is well known that during narcosis little or no oxygen is absorbed by nerve tissue. Verworn and his pupils assumed that the narcotic directly suppressed oxidation. On the other hand Mansfeld[1] supposed that the narcotic dissolving in a lipoid plasma membrane made it less permeable to oxygen. It would be more in harmony with the phenomena considered in previous chapters, to suppose that the narcotic in low concentration decreased the permeability of the plasma membrane to the anions and molecules of some acid end product of oxidation, and thus stopped the combustion. An objection to this hypothesis is made by Warburg[2] who found that phenylurethan, which only slightly reduces oxidation in certain cells, fertilized eggs, delayed cell division enormously. With greater concentration of the narcotic, oxidation was greatly reduced.

V. Osmotic Properties of the Blood Corpuscles.

Hamburger and Bubonavik[3] have concluded that the erythrocytes are permeable to K, Na, Ca and Mg. However, the opposite conclusion was reached by previous workers.

Gyrn's,[4] Hedin,[5] Traube[6] and others observed that the erythrocytes are relatively impermeable to neutral salts (exc. NH_4 salts) amino acids, various sugars and hexite, slowly permeable to erythrite, more permeable to glycerine, and easily permeable to monovalent alcohols, aldehydes, ketones, esters, ether, and urea. In general, it may be said that the erythrocyte is permeable to lipoid-soluble substances or those that lower the surface tension of water. Such substances (for instance, ether) become more concentrated in the corpuscle than in the serum. Saponin becomes 120, and ammonia 880 times more concentrated in corpuscle than in serum.[7]

[1] *Pflüger's Arch.*, 1909, CXXIX., 69.

[2] *Zeit. physiol. Chem.*, LXVI., 305.

[3] *Arch. internat. de Physiol.*, 1910, X., 1.

[4] *Pflüger's Arch.*, 1896, LXIII., 86, and *Koninkl. Akad. von Wetensch. Amsterdam*, 1910, p. 347.

[5] *Pflüger's Arch.*, 1897, LXVIII., 229; 1898, LXX., 525.

[6] *Biochem. Zeit.*, 1908, X., 371.

[7] Arrhenius, *Biochem. Zeit.*, 1908, XI., 161.

The erythrocytes are practically impermeable to ions. Stewart[1] observed that they offered a great resistance to the electric current. It is difficult to remove all of the serum from a mass of erythrocytes, but Bugarsky and Tangl, working independently of Stewart, obtained sediments of corpuscles having a conductivity of only 1/50 that of the serum. This indicates that the corpuscles are practically impermeable to both classes of ions, for if permeable to ions of one sign, they would probably not be such good insulators. The electric conductivity of the ash (made up to equal volume) of the corpuscles is about that of the serum, although the osmotic pressure of the solution of ash of the latter is greater.[2]

Hence an increase in electric conductivity of the corpuscles (as will be considered below) indicates increased permeability to ions. After the corpuscle becomes permeable to ions, further increase in conductivity might be due to liberation of ions from combinations with colloids in the interior. However many ions, for instance PO_4, cannot be liberated without incineration or other rigorous treatment. Increase in conductivity of the blood by laking agents has been proven to be chiefly due to increased permeability of the corpuscles, since the conductivity of the serum never shows so great an increase on the addition of the laking agent, and is usually diminished (by the hæmoglobin) if the corpuscles are present.

The portion of the normal corpuscle presenting the greatest resistance to the electric current is the surface layer, since Höber[3] observed that the conductivity of the interior of the corpuscle (determined by its dielectric value) is many times greater than that of the corpuscle as a whole. Peskind[4] caused bubbles of nitrogen to form within the corpuscle and observed that they were retained by a superficial membrane. This may be the membrane which resists the electric current.

The chemical composition of the corpuscle is supposed to bear some relation to its permeability. Aside from the hæmoglobin, and the rather low water content (60 per cent.) the corpuscle

[1] *Science*, Jan. 22, 1897.

[2] Moore and Roaf, *Biochem. Jour.*, III., 155.

[3] *Pflüger's Arch.*, 1910, CXXXIII., 237.

[4] *Am. Jour. Physiol.*, VIII.

is composed of lecithin and cholesterin with a little nucleo-proteid. It is probable that these lipoids are chemically different in different species of animals, since Lefmann[1] observed that the lipoids of erythrocytes of the same species are not toxic, whereas those of another species may be very toxic.

The distribution of these substances in the corpuscle has not been ascertained. Pascucci[2] supposed the corpuscle to be a bag of proteid impregnated with lecithin and cholesterin and filled with hæmoglobin. He found that artificial lecithin-cholesterin membranes were made more permeable to hæmoglobin by the laking agents, saponin, solanin and tetanus or cobra poison. Dantwitz and Landsteiner suppose the lecithin to be in combination with protein.

Hoppe-Seyler assumed the hæmoglobin to be in combination with lecithin in the corpuscle, and Bang[3] has shown that lipoids may be fixed by hæmoglobin. It seems evident that there does not exist an aqueous solution of hæmoglobin within the corpuscle, since hæmoglobin crystals may be made to form in *Necturus* corpuscles without extraction of water. Furthermore, Traube and Goldenthal[4] find that hæmoglobin has a hæmolytic action, and unless there exists some body within the corpuscle which antagonizes this action (as serum does) a hæmoglobin solution could not be retained by the corpuscle. Probably all of the so-called "stroma" constituents, not in combination with the hæmoglobin, form the plasma membrane of the corpuscle.

Under certain conditions, the hæmoglobin comes out of the corpuscles, and the blood is said to be laked. Laking of "fixed" corpuscles occurs only after the removal of the fixing reagent. Thus, sublimate-fixed corpuscles may be laked by substances which combine with mercury, such as potassium iodide, sodium hyposulphite or even serum proteids. The fact that they may be laked by heating in water is probably because the nucleo-histone is not fixed by sublimate. This process is prevented by hypertonic NaCl solution, presumably on account of its power to precipitate nucleo-histone (Stewart). Formaldehyde-fixed corpuscles may

[1] *Beiträge chem. Physiol. u. Path.*, XI., 255.
[2] *Hofmeister's Beiträge*, 1905, VI., 543, 552.
[3] *Ergeb. d. Physiol.*, 1907, VI., 152.
[4] *Biochem. Zeit.*, 1908, X., 390.

be laked by ammoniacal water, at a temperature which must be higher, the more thoroughly they have been fixed. Ammonia combines with formaldehyde.

Stewart[1] supposes that the hæmoglobin must be liberated from some compound before the blood can be laked. We cannot say that the corpuscle is always permeable to hæmoglobin from within outward. However the corpuscle probably is impermeable to it from without inward, since it does not take up hæmoglobin from a solution, and after the blood is laked the serum contains hæmoglobin in greater concentration than the "ghosts" do.

At any rate, permeability to hæmoglobin appears to be independent of permeability to salts, since Rollett[2] found that laking by condenser discharges may set free the hæmoglobin without the corpuscle becoming permeable to ions. Stewart[3] concluded that the same is true of laking with sodium taurocholate (even after considering the depressing action of hæmoglobin on the conductivity).

Stewart[4] and others had already shown that blood laked by minimal applications of such laking agents as freezing and thawing, heating (to 60°), foreign serum, and autolysis (spontaneous laking) cause but a slight increase in the permeability to ions, whereas the continued application of some of these agents, or especially such violent reagents as distilled water and saponin, cause a marked increase in electric conductivity. On the other hand, if saponin is added to defibrinated blood at 0°, the conductivity of the corpuscles to ions begins to increase before any hæmoglobin escapes from the corpuscles.

The liberation of the hæmoglobin by some laking agents may be due to the direct action of the reagent in breaking up the compound in which the blood pigment exists, but is probably sometimes a secondary effect, following increase in permeability to electrolytes.

It has been shown that many laking agents, lipoid solvents, saponin unsaturated fatty acids, soaps, and hæmolysins (containing lipase) are such as would alter lipoids physically or

[1] *Jour. Pharm. and Exper. Therapeutics*, 1909, I., 49.

[2] *Pflüger's Arch.*, 1900, LXXXII., 199.

[3] *Am. Jour. Physiol.*, X.

[4] *Jour. Physiol.*, 1899, XXIV., 211.

chemically, whereas pressure, trituration, shaking, heat, condensor discharges, freezing and thawing, water, drying and moistening, salts (including bile salts), acids and alkalis, might act also on proteids.

Since any treatment which causes great swelling[1] of the corpuscle leads to loss of hæmoglobin, it is probable that stretching or breaking of the surface film increases its permeability. But laking may occur without swelling, and even crenated corpuscles may be laked by sodium taurocholate.

Höber[2] observed that the relative action of ions in favoring hæmolysis is: salicylate>benzoate>I>NO_3, Br>Cl>SO_4 and K>Rb>Cs>Na, Li. Since this is the order in which they affect the aggregation state of colloids, their action is probably on the aggregation state of the colloids of the corpuscle (proteids or lipoids or their combinations).

The permeability of formaldehyde-fixed corpuscles to ions, is greatly increased by extraction of the lipoids with ether, or by treatment with substances such as saponin, which act on lipoids. Since the proteids have been thoroughly fixed, it is evident that they play no part in this process, though they may do so in the non-fixed corpuscles.

The relation of lipoids outside of the corpuscles to hæmolysis has been extensively investigated, and cannot be fully treated here. Willstätter found that cholesterin combines with one of the saponins, destroying its hæmolytic power. Iscovesco[3] concludes that cholesterin combines with soap, and prevents its toxic action.

Changes in permeability of the corpuscles to ions were studied chemically before the application of the electrolytic method. Hamburger[4] and Limbeck[5] observed that when CO_2 is passed through blood, chlorine passes from serum into corpuscles and the alkalescence of the serum is increased. On the other hand, the distribution of sodium and potassium is not changed.[6]

[1] Roaf, *Q. J. Exper. Physiol.*, III., 75, supposes this swelling to be due to ionization and hence increased osmotic pressure of hæmoglobin.

[2] *Biochem. Zeit.*, 1908, XIV., 209, and *loc. cit.*

[3] *Comptes Rendus, Soc. Biol.*, 1910, LXIX., 566.

[4] *Zeit. f. Biol.*, 1891, XXVIII., 405.

[5] *Arch. exp. Path.*, 1895, XXXV., 309.

[6] Gürber, *Sitzungsber. physik.-med. Ges. Wurzburg*, 1895.

Koeppe[1] and Höber[2] explain this process in the following manner: The lipoid-soluble CO_2 enters the corpuscle, and by reacting with alkali albuminates in the protoplasm, gives off more anions than it does in the serum. During the presence of CO_2, the corpuscle is permeable to anions, and the $CO_3^{=}$ or HCO_3^- ions pass back into the serum, being exchanged for Cl^- ions to equalize the electrical potential. Sodium bicarbonate being more alkalescent than sodium chloride, the titratable alkalinity of the serum is increased.

This explanation is supported by the following facts: When CO_2 is passed through a suspension of erythrocytes in cane sugar solution the latter does not become alkaline. If CO_2 is passed through a mass of centrifuged erythocytes, which are then added to physiological salt solution, the latter becomes more alkaline than the serum in Hamburger's experiment. Any sodium salt may be substituted for serum, and its anions will pass into the corpuscles.[3] Also the number of ionic valences passing into the corpuscle is constant, *i. e.*, if sulphate is used only half as many ions enter the corpuscles as when chloride or nitrate is used. The process is reversed by removal of the CO_2.

This same phenomenon has been observed in leucocytes by van der Schroeff.

There seems to be some relation between hæmolysis and agglutination of the corpuscles. Arrhenius[4] supposed that agglutination by acids is due to the coagulation of the proteids of the envelope. However, since agglutination is followed by precipitation, it seems probable that the loss of the negative electric charge which tends to keep the corpuscle in suspension and causes it to repel every other corpuscle, is partly responsible for the phenomena.

The fact that water-laking is preceded by agglutination might be explained if we assume that increase in permeability to ions leads to loss of electric charge. The charge may be due to the charges on the colloids of the corpuscle or to semi-permeability to ions. The corpuscle is very poorly permeable to ions, but may

[1] *Pflüger's Arch.*, 1897, LXVII., 189.

[2] *Pflüger's Arch.*, 1904, CII., 196.

[3] Hamburger and van Lier, *Engelmann's Arch.*, 1902, 492.

[4] *Biochem. Zeit.*, 1907, VI., 358.

be slightly more permeable to some one ion than to others. If this ion were more concentrated in the plasma or in the corpuscle, the latter would become electrically charged, and a general increase in ionic permeability would lead to a reduction or loss of this charge. The loss of charge would favor their coming in contact with one another and their precipitation, but their cohesion is probably due to some other change, possibly the exit of adhesive substances, on increase in permeability.

VI. Absorption and Secretion.

1. *Absorption through the Gut.*

If a live vertebrate intestine be filled with one portion of a physiological NaCl solution, and suspended in another portion of the same solution, fluid will pass through the wall of the gut from within outward. Cohnheim[1] found that holothurian gut behaves in the same way toward sea water, and the absorption stops if the gut is injured with chloroform or sodium fluoride.

It might be supposed that the hydrostatic pressure produced by the contraction of the musculature, is the driving force of absorption, but on the contrary, Reid[2] found that the wall of the rabbit's intestine behaved in the same way when used as a diaphragm.

Salt is absorbed by an intestine filled with a very hypotonic solution of it, and water may be absorbed when the solution is very hypertonic.

Blood salts enter the intestine when it is injured by an extremely hypertonic solution, or sodium fluoride, chinin or arsenic.

Grape sugar and sodium iodide may pass from without inwards through the wall of a normal holothurian intestine.

Traube[3] claims that absorption is explained by his observation that the surface tension of the contents of the gut is less than that of the blood, but this does not apply to the experiments in which an identical solution was placed on each surface of the wall of the gut. Traube[4] found that the addition of a substance

[1] *Zeit. physiol. Chem.*, 1901, XXXIII., 9.

[2] *Jour. Physiol.*, 1901, XXVI., 436.

[3] *Pflüger's Arch.*, 1904, CV., 559. Cf. Iscovesco, *Comptes Rendus, Soc. Biol.*, 1911, LXXI., 637.

[4] *Biochem. Zeit.*, 1910, XXIV., 323.

lowering the surface tension increased the absorption of $NaCl$ by the gut.

Absorption is probably due to irreciprocal permeability of the wall of the gut. Hamburger showed that dead gut and even artificial membranes showed irreciprocal permeability to certain substances. These artificial membranes were of different composition on their opposite surfaces (parchment paper-chrome albumin, or parchment paper-collodion) and he assumed that the wall of the gut is composed of two osmotically different layers. In reality there may be more than two such layers, and the plasma membranes of the individual cells of the gut may show irreciprocal permeability.

Traube[1] showed that the rate of absorption of a substance by living gut is usually greater the more it lowers the surface tension of water. The order of ions is: $Cl > Br > I > NO_3 > SO_4$, HPO_4 and K, Na > Ca, Mg. The order of non-electrolytes, according to Katzenellenbogen[2] is: glycocoll <urea<acetone, mannit<erythrite<glycerine<acetamid, methylalcohol, propylalcohol, amylalcohol.

The rate of absorption through dead ox gut according to Hedin[3] is: $Br > NO_3 > Cl > SO_4$ and K > Rb > Na > Li > Mg and mannit <erythrite < glycerine < urethan < glycocoll < amylenhydrate < glycol < urea < propylalcohol < isobutylalcohol < methylalcohol, ethylalcohol.

The action of poisons on absorption may be due to the alteration of the plasma membranes of the individual cells. Mayerhofer and Stein[4] state that even sugar in certain concentrations increased the permeability of the gut.

2. *Osmotic Relation of Aquatic Animals to the Medium.*

Fredericq found that the salt content of the body fluids of marine invertebrates is about the same as that of sea water. Henri and Lalou[5] showed that the osmotic exchange between cœlom fluid of sea urchins and holothurians and medium is chiefly

[1] *Pflüger's Arch.*, CXXXII.
[2] *Pflüger's Arch.*, CXIV., 522.
[3] *Pflüger's Arch.*, 1899, LXXVIII., 205.
[4] *Biochem. Zeit.*, 1910, XXVII., 376.
[5] Winterstein, II. (2), 2.

water. If the sea water was diluted with ¼ vol. of isotonic cane sugar solution, the salt content of the cœlom fluid is very little lowered in 4 hours, and only traces of sugar appear in it. The result is the same with isotonic urea (which easily penetrates most plasma membranes). But the salt content of the blood of elasmobranchs and teleosts is about half that of the sea.

Botazzi and his colleagues observed that the osmotic pressure of the blood of elasmobranchs is about equal to that of the medium, the salts in the blood being supplemented by organic substances, chiefly urea, of which there is 2–3 per cent.

If elasmobranchs are placed in concentrated sea water, the osmotic pressure of the blood rises, but the ratio of urea to salts remains the same. G. G. Scott found that changes in the density and osmotic pressure of the blood of elasmobranchs accompany changes in the salt content of the medium.

However, in marine teleosts as well as all fresh-water animals which have been studied in this respect, both salinity and osmotic pressure of the body fluids are very different from that of the medium.

The osmotic pressure of the blood of marine teleosts is about half that of the sea, but in fresh-water teleosts it is still less (but much greater than the fresh water). This indicates that there must be a change in the osmotic pressure of the blood as the fish ascends a river. Greene[1] observed that it took salmon 30–40 days to pass the brackish water, in which time they were acclimatized to fresh water. After being in fresh water 8–12 weeks, the osmotic pressure of the blood was reduced only 17.6 per cent. This reduction may be partly accounted for by the absorption of the osmotic substances in the blood by the sexual glands. In harmony with this view is the fact that the osmotic pressure of the blood of the female was reduced much more than that of the male. One salmon, that was very weak and probably dying, showed 32 per cent. decrease in Δ of blood. Sumner[2] observed that changes in weight and salt content of marine teleosts accompany, but are not proportional to changes in the medium.

[1] U. S. B. F., 1904, XXIV., 445; 1909, XXIX., 129; *Jour. Exp. Zoöl.*, 1910, IX.

[2] Bull. U. S. B. F., 1905, XXV., 53, and *Am. Jour. Physiol.*, 1907, XIX., 61.

Overton observed that if the cloaca and mouth of a frog in fresh water are closed, the frog constantly increases in weight. This can be prevented by the addition of .7 per cent. NaCl to the medium. In a hypotonic solution water is constantly absorbed by the skin and excreted by the kidneys. Fischer's[1] experiment, in which ligature of the leg of a frog caused great swelling below the ligature is probably to be explained by the fact that water was absorbed by the skin but could not reach the kidneys, since the blood circulation was stopped. In regard to Fischer's explanation, compare the results of Sidbury and Gies.[2] Sumner concluded that in the fish, the gills are the chief seat of osmotic exchange.

It appears, therefore, that osmosis occurs through the integument (including gills), kidneys and gut simultaneously, and since the contents of the gut and kidney tubules are not the same as the medium, we should not expect an osmotic equilibrium between the body fluids and the medium. Furthermore, all three of these membranes may show irreciprocal permeability.

Fresh-water fish and non-migratory marine fish are killed by great changes in the medium, even though it be very gradual.

Bert maintained that if fresh-water fish are placed in sea water, the gill capillaries contract and become blocked by the distorted corpuscles. In naked-skinned fishes, not only the gills are affected, but water may be lost from the tissues.

Bert and Sumner both agree that the salts in sea water cannot be replaced by any other substance, without causing the death of certain marine fishes. Mosso[3] claimed that when sharks are placed in fresh water, the gill capillaries become so blocked with laked corpuscles that physiological salt solution could not be forced through them. He observed that the differences in the resistance of certain fish to changes in the salt content of the medium, corresponded to differences in the resistance of their blood cells to the hæmolytic action of such changes. Sumner,[4] however, states that this blocking of gill capillaries does not occur in sharks or marine teleosts in fresh water.

[1] Fischer, M. H., "Œdema," J. Wiley & Sons, 1910.

[2] *Soc. Exper. Bio. and Medicine*, 1911, VII., 104.

[3] *Biol. Centlb.*, 1890, X., 570.

[4] *Proc. Seventh Internat. Zoöl. Congress*, Boston, 1907.

Sumner showed that as the fish becomes enfeebled by the abnormal medium, it becomes more permeable to salts.[1] Whether the direct action of the abnormal medium, or the blocking of the gill capillaries, produce the increase in permeability, has not been experimentally tested. However, the gills themselves would not be *asphyxiated* by blocking of their capillaries, and it seems probable that the change in permeability is due to the direct action of the medium.

We may conclude therefore that the death of the fish results from the osmotic exchange. This may be sufficient to cause death while the fish still maintains its normal semi-permeability, or death may occur only after increase in permeability, due to the direct action of the medium on the osmotic membranes.

A similar increase in permeability may explain Wo. Ostwald's observations on fresh-water *Gammarus* in pure salt solutions.[2] He found that the ratio of the rapidity of death to the concentration is about constant up to a certain point, above which it is much greater. This critical concentration has nothing to do with the osmotic pressure, since it is different for different salts. Perhaps at this concentration the salt made the membranes more permeable.

Schücking[3] found that nicotine and strychnine made the skin of *Aplysia* more permeable to salts. Since cocain retarded shrinkage in hypertonic solution, he supposed that the hydrostatic pressure produced by the muscles aided shrinkage. However the hydrostatic pressure is probably very small, and the effect might have been due chiefly to an increase in permeability to salts, produced by the cocain.

3. *Secretion of Lymph and Tissue Juice.*

Höber supposes the raising of the osmotic pressure by the katabolism of the tissues, causes fluid to be drawn out of the bloodvessels, and states that the lymph in the thoracic duct has a greater osmotic pressure than the blood.

Traube states that the surface tension of transudates and

[1] Cf. Greene, above.

[2] *Pflüger's Arch.*, 1905, CVI., 568.

[3] *Arch. Anat. Physiol.*, Physiol. Abt., 1902, 533.

exudates is always greater than that of the blood. He cites a case in which a transudate was caused to be absorbed by injecting into it a substance which decreased its surface tension.

4. *Excretion.*

Milk and bile have about the same osmotic pressure as the blood, but urine is almost dry in some animals; it is usually hypertonic in man but may be hypotonic.

Traube maintains that the surface tension of the normal urine is always greater than that of the blood, and that this is the driving force in excretion.

However, Höber and others suppose that the substances to be excreted may be formed into solid bodies in the tubule cells, and thrown out into the lumen.

If lipoid-insoluble dyes are fed to frogs, granules in the cells of certain segments of the kidney tubule are stained with them. The dye is not first excreted by the glomeruli and then absorbed from the lumen by the tubule cells, for if the vena Jacobsoni, which supplies the tubules, is ligatured, no staining occurs, although the renal arteries still supply the glomeruli.

The stained granules in the tubule cells are thrown out into the lumen and pass into the bladder. These granules usually dissolve to form a slimy substance in the urine, but some of them may remain intact.

The circulation in mammalian kidneys cannot be controlled in the same way, but after intravenous injection of a certain lipoid-insoluble dye, no stain may be detected in the walls of the glomeruli, although the tubule cells are stained. The stain in the lumen does not appear above the level of the stained tubule cells. In the excretion of carmine, it may be found in granules in the tubule cells and lumen, similar to those found in frog's kidneys.

It has been supposed that urea is excreted by collecting in these granules and passing out with them, but it would be even simpler to assume that some substance is excreted into the lumen, which combines with urea and so lowers the concentration of that in solution, thus accelerating its excretion.

The chief recommendation for the granules is their valve-like action, which would account for the secretion of urine against a concentration gradient, but a simpler mechanism of such a process is shown in Hamburger's double membranes.

The blood pressure may aid in the secretion of the water of the urine, which is eliminated chiefly through the glomeruli, but its insignificance in the elimination of urea is shown by the fact that after increasing the volume (and therefore pressure) of rabbit's blood 70 per cent. by transfusion, the urea elimination was not or only very slightly increased.

VII. Cell Division.

Various hypotheses as to the cause of cell division have been advanced by the morphologists. Hertwig, supposed that when the ratio of nucleus to cytoplasm is less than normal, the cell will divide.[1] Gerassimow[2] subjected cells of *Spirogyra* to low temperatures and other abnormal conditions and obtained an increased amount of chromatin in some of them. These cells did not divide until the ratio of nucleus to cytoplasm was as great as at the time of division of a normal cell.

I found that chromatin is not necessary for cell division.[3] After extracting the chromosomes from the starfish egg, I caused it to divide. In this case the ratio of nucleus to cytoplasm was zero; however the cell did not continue to divide indefinitely.

There is no easy method of determining the ratio of nucleus to cytoplasm. Some cells contain large vacuoles whose contents are not considered as cytoplasm. Eggs contain fat drops and granules compounded of protein and lipoids. These are not considered as cytoplasm by all investigators. If the granules and oil are included as cytoplasm, the ratio of nucleus to cytoplasm is very small, and yet the egg cell does not divide unless "stimulated" by the sperm or some other means.

R. Lillie[4] observed that chemical substances, which in low concentration cause the *Arbacia* egg to divide, in high concentration cause outward diffusion of the red pigment (echinochrome) and compared this to the laking of erythrocytes.

[1] He is not confirmed by Conklin, *Jour. Exper. Zoöl.*, 1912, XII., 1.

[2] *Bull. Soc. Imp. Nat.*, Moskau, 1904, No. 1.

[3] McClendon, *Arch. f. Entwicklungsmech.*, 1908, XXVI, 662.

[4] Biol. Bull., 1909, XVII., 188.

This is made more striking by the fact, mentioned first by Loeb, that hæmolytic agents are effective in artificial parthenogenesis. R. Lillie observed that pure solutions of sodium salts caused the egg to divide, the order of effectiveness of anions being $Cl < Br < ClO_3 < NO_3 < CNS < I$. He also found that these salts could be inhibited by others ($CaCl_2$, $MgCl_2$), as is characteristic of the antagonistic effects of salts in physiological phenomena, and the precipitation of colloids.

I found that the sea urchin's egg contains fatty substances, and relatively large amounts of lecithin probably in combination with proteids. I found that *Toxopneustes* eggs freed from the jelly-like coverings, contained about 10 per cent. lecithin (alcohol extract ppt. with acetone) and about 2 per cent. of an extract soluble in alcohol or acetone and containing rosettes of fat-like crystals. This extract blackened strongly with osmic tetroxide and effervesced on adding dry Na-carbonate in water, then emulsified, probably it contained unsaturated fatty acid.

According to a private communication by Mathews, the egg of the starfish contains lecithin and an unsaturated fatty acid, but no cholesterin. In this last characteristic it differs markedly from the erythrocyte. There is no way of determining whether these substances enter into the composition of the plasma membrane, but the facts are presented in order to indicate the possibilities.

We have seen that the exit of hæmoglobin is probably not due to increased permeability to this substance. It is possible that the same is true of echinochrome. I found that the echinochrome in the egg shows a continuous spectrum, whereas that extracted in various ways shows characteristic bands. It may possibly be held by chemical combination in the egg.

However I found other evidence for increase in permeability of the sea urchin's egg coincident with beginning development:[1]

1. Fertilized eggs are caused to shrink more quickly than unfertilized eggs, with isotonic sugar solution. Presumably the fertilized eggs are more permeable to the substances exerting the internal osmotic pressure.

2. The electric conductivity of the egg increases about ¼ when

[1] McClendon, *Amer. Jour. Physiol.*, 1910, XXVII., 240.

it is fertilized or made parthenogenetic with acetic acid, indicating increased permeability to ions.

Lyon and Shackell[1] and Harvey[2] observed that methylene blue and neutral red enter fertilized eggs more quickly than unfertilized eggs. Harvey supposed that only the free color base (undissociated) entered, since the addition of a little acid to the sea water prevented the staining of the eggs.

Mathews[3] considered the penetration of stains into the egg as a chemical process (the stain forming a salt combination with the lecithin or proteins of the egg surface).

Harvey observed, further, that NaOH penetrates fertilized more easily than unfertilized eggs, but the eggs are killed by the alkali.

The fact that the unfertilized frog's egg continues to swell *for a long time* in water (Biataszewitz) whereas the osmotic pressure of the fertilized frog's egg is *quickly* reduced to equal that of the medium (Backmann and Runnström) indicates increase in permeability to osmotic substances on fertilization. In this connection it is interesting to note that Bataillon,[4] Brachet, and myself[5] caused the unfertilized frog's egg to rotate normally and segment merely by pricking it.

It has been supposed by various observers that the "formation" of the fertilization membrane in very closely related to the segmentation of the egg. Loeb observed that the sea urchin's egg may develop without the formation of a fertilization membrane, and I have confirmed this observation, and shown that it is very probably wrong to suppose that this is a case of failure in "pushing out" of the membrane. Apparently "membrane formation" is not essential for the segmentation of the egg, although by furnishing protection it may insure the development of the embryo.

Loeb postulated that an osmotically active colloid exists in the unfertilized egg, but is so covered with lipoids that it does not absorb water until it is squeezed out or otherwise exposed

[1] *Science*, 1910, XXXII., 250.
[2] *Ibid.*, p. 565.
[3] *Jour. Pharmacol. and Exp. Ther.*, 1910, II., 201.
[4] *Arch. Zool. Expér.*, 1910 (5), VI., 101.
[5] McClendon, *Amer. Jour. Physiol.*, 1912, XXIX., 298.

at the surface of the egg, at the beginning of development (when it fills the so-called "perivitelline space"). I observed that this substance bears a positive charge (is basic) since it migrates toward the kathode when an electric current is passed through sea water containing the fertilized egg.

The unfertilized egg is imbedded in a mass of jelly which is probably mucin. This jelly bears a negative charge (is acid) since it combines with color bases.

When the positively charged colloid is exposed at the surface (on increase in permeability) and comes in contact with the negatively charged jelly, the two mutually precipitate at their surface of contact, thus forming the fertilization membrane. But if all of the jelly is washed off of the egg before the latter is caused to develop, no fertilization membrane is formed (as I have observed) because no two oppositely charged colloids are brought in contact, but the basic colloid may with difficulty be seen as a refractive layer, which has been mistaken for a poorly developed "fertilization membrane."

The observation of Lyon[1] makes it appear that catalase comes out of fertilized more quickly than unfertilized eggs, probably due to increased permeability.

Lyon observed that CO_2 came out of fertilized more quickly than unfertilized eggs, and O. Warburg, Loeb and myself[2] observed that oxygen is absorbed more rapidly by the former. We might ask: Does increased permeability allow increased oxidation, or is increased oxidation the primary cause of the increased respiration?

The permeability change is the simplest explanation, and in what other way could oxidation be increased? Loeb supposed the sperm carried an oxidase into the egg.[3] But no addition of oxidase is concerned in artificial parthenogenesis, and Loeb assumed that the oxidase (or other enzyme, kinase?) is held in the egg periphery and cannot penetrate the egg interior until the permeability is increased.

In addition to oxygen, oxidase, and escape of CO_2, hydroxyl

[1] *Am. Jour. Physiol.*, 1909, IV., 199.

[2] McClendon and Mitchell, *Jour. Biol. Chem.*, 1912, X., 459.

[3] In this connection it is interesting to note that Masing, *Zeit. physiol. Chem.*, 1910, LXVI., 265, failed to find more iron in sperm than in sea water.

ions are necessary for the rapid oxidation of the sea urchin egg (Loeb), and Harvey showed that the unfertilized egg is practically impermeable to OH ions of low concentration. The increased permeability allows hydroxyl ions in the sea water to penetrate the egg, as shown by Harvey, and, since the sea is always alkaline, this may explain the increased oxidation.

Asters always develop in the egg before segmentation. In the normal egg these have some relation to the division of the nucleus, but even if a nucleus is not present, I have observed that the cytoplasm constricts along a line on the surface farthest removed from the centers of the asters.

The constriction of the cytoplasm is probably due to a band of increased surface tension (or to decreased surface tension at the poles). This might be caused by local increase in permeability to ions, causing decreased polarization, at the equator (or increased polarization at the poles, due to increased production of the polarizing electrolyte in the asters).

The same reasons that were given for assuming that the surface of the *Amœba* is electrically polarized, hold good for the egg. The first change is probably a general increase in surface tension, indicated by rounding up of the egg. Later this may become localized from internal causes and result in cleavage.

Hyde[1] observed local changes in electric polarization of *Fundulus* eggs during cleavage, indicating that surface tension changes and cleavage are due to this cause.

It has been objected that the segmentation of the egg is not a typical case of cell division, since the egg cell is "wound up" and ready for some "stimulus" to set it going, whereas tissue cells must "grow" or "rest" after each division before dividing again.

It may be true that growth is prerequisite to division, but this cannot be formulated quantitatively. In the spore-formation of certain organisms, a cell may divide in a relatively short time into myriads of almost ultra-microscopic cells.

Hertwig may be right, in general, in assuming that the relative growth of nucleus and cytoplasm influences division, but the difficulties in proving this have been indicated, and this cannot

[1] *Am. Jour. Physiol.*, XII., 241.

be expressed in chemical terms. It is generally supposed that nucleic acid is a more abundant constituent of the nucleus than of the cytoplasm, but much evidence has appeared for believing that it is often present in considerable quantities in the cytoplasm. Loeb supposed that the segmentation of the sea urchin egg is accompanied by an "autocatalytic" synthesis of nucleic acid, since the nuclei increased in number. But Masing[1] and more recently Shackell[2] by chemical analysis found as much nucleic acid in the unsegmented egg or 1-cell stage as in the blastula stage.

There is some indirect evidence that increase in permeability may cause an increased division rate of tissue cells. Though cell growth may influence division, it is probable that permeability influences growth.

Various "stimuli" cause increased proliferation of cells of the germinal layer of the skin. It is commonly known that mechanical stimuli increase growth of the skin.

Bernhard Fisher observed that Sudan III. or Scharlack R[3] cause increased proliferation of the epidermis. When the dye is injected under the skin of a rabbit the skin grows toward the dye.

Furst[4] found that gradual increase of temperature caused a corresponding increase in proliferation of tissue cells (due to increased chemical reaction and inflammation of the tissue). But when a certain temperature was reached a sudden jump in the increase in proliferation was observed without a corresponding increase in inflammation. This is similar to the phenomenon seen in unfertilized eggs, where a rise in temperature beyond a certain point causes segmentation.

It has also been observed that electrical stimulation may cause increased proliferation of tissue cells.

All of these changes (electrical, thermal, or mechanical stimulation, or treatment with lipoid soluble substances) cause in-

[1] *Zeit. physiol. Chem.*, 1910, LXXVII., 161.

[2] *Science*, 1911, n. s., XXXIV., 573.

[3] Which are practically insoluble in water but soluble in fats and lipoids and, as I have observed, slightly in lipoid-protein combinations.

[4] See v. Dungern u. Werner, "Das Wesen Bösartigen Geschwülste," Leipzig, 1907, p. 65.

creased permeability and segmentation of the sea urchin's egg. Therefore, from analogy, we may conclude that increase in permeability may cause tissue cells to divide.

The "wound stimulus" to regeneration of tissue may also cause increased permeability of the cells.

In a preceding chapter it was shown that the "current of injury" produced by the negative electric potential of a wounded surface is common to animal and plant tissues. The wounded cell acts as an electric generator and a current flows through neighboring cells.

I observed that if a current is passed through living tissue, which is subsequently fixed and stained, basophile substances will be found displaced toward the anode. In sections of tissue adjacent to a wound the extent of the current is indicated by the displacement of basophile granules. The current affects first the cells in contact with the wounded cells, then extends in some directions more than others. Electric currents ("currents of growth") continue for many days after the wound has healed.

Since electric currents cause sea-urchin eggs and tissue cells to divide and proliferate, probably these bio-electric currents constitute the so-called "formative stimulus" of regeneration.

Embryonic cells, cells of germinal regions, and cancer cells are distinguished by their great power of proliferation, or rapid division. It is probable that the plasma membranes of these cells are more permeable than those of other tissue cells in the same medium or under the same conditions.

Cancers have been produced by the action of X-rays (electric pulsations) on the skin. The cells in the skin were so changed that they proliferated more rapidly. Similarly, electric changes have been observed to start the egg cell to rapid proliferation. There is probably some irreversible change in the permeability of these cells, which does not, however, make the plasma membrane incapable of subsequent reversible changes in permeability (*i. e.*, the change is unlike what occurs at death of the cell).

The suggestion that cancer cells are more permeable than tissue cells in general may possibly be of therapeutic importance. Loeb has shown that fertilized eggs are more sensitive than unfertilized eggs to various toxic substances (probably because these substances enter the fertilized eggs more easily). The same explanation may possibly be applied to the effect of sugar on certain living cells. The unfertilized eggs of the frog, *petromyzon*, sea urchin and annelid have been caused to segment, by placing them in sugar solutions. Mayerhofer and Stein[1] observed that sugar in certain concentrations increased the permeability of the gut to certain salts, and in this condition the gut was more easily injured by the diffusion of substances.

Similarly Stockard observed that sugar increased the toxicity of pure solutions of salts on the *Fundulus* egg. Morgan and Stockard[2] showed that this was not due to the inversion of sugar or to the osmotic pressure, and supposed that the sugar might combine chemically with the salt. It seems probable that the sugar increased the permeability to salt. The fact that sugar in fresh water is toxic whereas the same amount of sugar in the normal medium (sea water) is not toxic or less toxic, indicates that the salts within the *Fundulus* egg are the same as those outside (in sea water), and increase in permeability to them does not lead to diffusion while they remain in sea water, but diffusion takes place in fresh water.[3]

If it be shown that cancer cells are more permeable, substances may be found which kill cancer cells more easily than tissue cells as explained below.

Whereas a certain increase in permeability of the cell seems to cause division, a very great increase in permeability causes death (hæmolysis, cytolysis, bacteriolysis). It has been shown that certain lysins are specific for certain cells, probably because the plasma membranes of these cells differ chemically.

The fertilized egg is more easily cytolyzed than the unfertilized egg by certain substances. It therefore appears that the more permeable the cell is in the beginning, the more easily is the permeability brought to the point which causes cytolysis.

Hence it is probable that certain substances may be found by which cancer cells can be more easily cytolyzed than normal tissue cells.

[1] *Biochem. Zeit.*, 1910, XXVII., 376.

[2] BIOL. BULL., 1907, XIII., 272.

[3] In the absence of sugar I have shown that no diffusion takes place in fresh water. *Amer. Jour. Physiol.*, 1912, XXIX., 295.

It has been shown that narcosis is accompanied by decreased permeability. On the other hand, certain forms of inhibition of muscle are accompanied by an increase in permeability. May certain cells be inhibited in proliferation by an increase in permeability, too great for cell division but not great enough for cytolysis? The great oxidation rate in eggs inhibited in cleavage by very hypertonic solutions as determined by Warburg, seem to indicate this.

It has been shown that certain tissue cells inhibit the proliferation of others. In the healing of wounds, the epidermis inhibits the growth of connective tissue. If a wound remains uncovered by epidermis for a relatively long time, processes of connective tissue may grow outward, but this is prevented by the growth or transplantation of epidermis over the wound.

Perhaps the proliferation of the connective tissue is due to abnormal "stimuli" (bio-electric currents, diffusion of substances) such as cause proliferation in regenerating tissue generally. The presence of epidermis over the wound might protect the connective tissue from these "stimuli."

The foregoing facts and the speculations based on them may not be of far-reaching importance in themselves, but they suggest lines of research, which if followed, it is hoped, will add a great deal to cell physiology and pathology and be an aid to the understanding of many problems in therapeutics.

Reprinted from the American Journal of Physiology.
Vol. XXIX. — January 1, 1912. — No. III.

AN ATTEMPT TOWARD THE PHYSICAL CHEMISTRY OF THE PRODUCTION OF ONE-EYED MONSTROSITIES.

BY J. F. McCLENDON.

[*From the Embryological Laboratory of Cornell University Medical College, New York City and the U. S. Bureau of Fisheries Laboratory, Woods Hole, Mass.*]

EXPERIMENTAL teratology is one of the oldest branches of the physiology of development. However, the majority of the experiments have been performed on difficult material, such as the eggs of frogs, birds, and mammals, and with uncertain results. Stockard[1] observed that cyclopia, or the numerical defect in the eyes, may be easily produced in embryos of the marine fish, Fundulus heteroclitus. He found the defects to arise, apparently, from the retardation in the growth of the brain region lying between the Anlagen of the two optic vesicles. Probably this region is more sensitive than any other to toxic substances.

No one has recorded cyclopia in Fundulus embryos kept under normal conditions. A number of monstrosities develop from the eggs of fish, especially those, such as the trout, which require a very high oxygen tension for their development. When the temperature rises, and consequently the oxygen tension falls, many abnormal forms appear. However, I know of no observations, except my own, on the repeated natural occurrence of cyclopia in fish. During two successive seasons I found about one tenth of one per cent of the smelt embryos, at Cold Spring Harbor, Long Island, developed in streams, or in hatching jars of the New York State Fish Hatchery, to be cyclopic. Stockard observed abnormalities in Fundulus embryos developed in the fresh water at this place.

I have attempted to find the cause of these abnormalities, but have

[1] STOCKARD: Journal of experimental zoölogy, 1907, iv, p. 165; 1909, vi. p. 286; American journal of anatomy, 1910, x, p. 369.

not completed this work. In fact, only one clue has been found, and that is the CO_2 content of the water. The fresh water at Cold Spring Harbor comes from deep artesian wells and springs, and is heavily charged with carbonic acid. Dr. R. A. Gortner found more than ten times as much CO_2 in some samples of water that had just emerged as was found in pond water. Ordinary distillation did not change the CO_2 content. It seems probable that the carbonic acid is the cause of cyclopia in these smelt embryos, but I have found no proof of this assumption. The eggs are deposited naturally in large masses, with spaces between the eggs. Those on the surface of the masses develop faster, due probably to larger amount of oxygen or light, or smaller amount of CO_2. But in the hatching jars the eggs are separated, and all are exposed to the same conditions.

I produced cyclopia in Fundulus embryos with a number of salts and with volatile anæsthetics. The solutions might be very hypertonic or very hypotonic. On the one hand, distilled water containing only traces of dissolved substances or substances to which the egg is freely permeable, were effective, while, on the other hand, 3.4 volumes of sea water concentrated to one volume were likewise so. It seems, therefore, that osmotic pressure is not primarily the cause of cyclopia.

However, the eggs decreased in volume when placed in theoretically isomotic solutions producing cyclopia, tending to show that these solutions increased the permeability of the egg to the substances producing the internal osmotic pressure, or turgor. We need only to suppose that the cells between the eye Anlagen are more easily affected than other cells of the embryo, to explain the action of the solutions. The decrease in turgor would retard growth in the same way as a hypertonic solution prevents development.

It is well known that great increase in permeability means death, and that death is always accompanied by increased permeability. In my experiments the concentration of salts or anæsthetics causing cyclopia was always just slightly below the lethal dose (Table I). The eggs were placed in the solutions at the two-cell stage, and usually remained in these solutions until the eyes appeared. Those apparently cyclopic were removed to sea water and observed more minutely at a later date, then finally made into histological preparations.

In one case cyclopia was observed in distilled water containing minute traces of heavy metals,[2] and it will be observed from Table I that hypotonic solutions of NaCl and LiCl produced the same result

TABLE I.

Substance.	Dist. water solutions.		Sea water solutions.	
	Lethal dose.	Cyclopic dose.	Lethal dose.	Cyclopic dose.
KCl	.....	...	.8–.9 mol.	.....
NaCl	.4–.5 mol.	.3–.36 mol.	1.2–1.6 "	1.04–1.36 mol.
LiCl	.2–.25 "	.2 "	.66 "	.33–.66 "
$MgCl_2$	.4–.5 "	...	.4–.5 "	.3–.38 "
$CaCl_2$	.2–.3 "	...	.2–.3 "	.....
NaOH	.006–.01 mol.	.006 "	$>$.01 "	.....
HCl	[1].00015-.0002 mol.	...	.0025 "	.....
Cane sugar . . .	1. molecular	...	$>$.5 molecular	.....
Dextrose	.....	...	1.–2. "	.....
Methyl alcohol . .	.....	...	$<$1.8 vol. %	.....
Ethyl "	3.15–4.75 vol. %	3 vol. %	3.5–4.75 vol. %	2–3.66 vol. %
Amyl "	.....	...	.2 "	.1–.15 "
Acetone	.....	...	3.2 "	1.6–2.4 "
Phenol	.....	...	$<$1 "	.....
Ether	.....	...	1/6 saturated	1/8 saturated

[1] At about the same concentration KAHLENBERG and TRUE, Journal of the American Medical Association, July 18, 1896, p. 138, found the growth of certain plants to cease and showed that this was due to H^+ ions.

as the traces of metal. The addition to sea water of the chlorides of Na, Li, or Mg, thus raising the osmotic pressure, produced cyclopia.

[2] The traces of metal came from the still, and the experiment was designed to test the effect of water distilled in metal. The distilled water used in all other experiments was re-distilled in a fused quartz condensing tube, and there was no cork, rubber, or other organic material in the apparatus.

Stockard[3] obtained one cyclops monster by adding $MnCl_2$ to sea water.

It is interesting to note that merely the alteration of the relative concentration of the salts already present in sea water produces cyclopia.

In order that these salts should act on the region between the eye Anlagen after the neural tube has formed, they must first penetrate the embryo. According to Overton and others, living cells are relatively impermeable to electrolytes. Therefore I determined the permeability of the Fundulus egg under the conditions of the experiments in which cyclopia was produced. The eggs were impermeable to Cl ions, but permeable to the kations. By a kationic exchange, the salt content of the interior could be qualitatively but not quantitatively altered (see Appendix II).

Anæsthetics belong to the class of substances which freely penetrate living cells. They also change the permeability of cells to other substances. Czapek[4] observed that many anæsthetics increase the permeability of plant cells to the contained tannin, in "iso-capillary" concentrations. By iso-capillary he designates solutions which have the same surface tension. The anæsthetics which behave in this way are the so-called indifferent anæsthetics, *i. e.*, substances which alter the cells physically but not, or to a small extent, chemically, and whose effects are wholly reversible unless the concentration producing narcosis is far exceeded.

In order to determine whether this rule applied to cyclopia, I made iso-capillary solutions of various anæsthetics and observed their relative effects. Some difficulties were met in determining the surface tension of the solutions. According to I. Traube, in homologous series of anæsthetics, a molecule of each member is three times more powerful in lowering the surface tension than a molecule of the preceding member. However, Traube's rule appears to be only approximate, and difficulties are met in determining the molecular concentration of the solutions. Therefore I measured the surface tension directly with Traube's stalagmometer, making corrections for specific gravity. The apparatus used by Czapek was tried but given up.

Since the surface tension was measured directly, there was no

[3] STOCKARD: Journal of experimental zoölogy, 1907, iv, p. 187.

[4] CZAPEK: Über eine Methode zur direkten Bestimmung der Oberflächenspannung der Plasmahaut von Pflanzenzellen, Jena, G. Fisher, 1911.

necessity of having the substances absolutely pure, although the character of impurities would be significant. I used absolute ethyl alcohol and pure ether, acetone and phenol, but the other substances were not pure.

Large glass bottles were partially filled with the solutions, a relatively small number of eggs in the two-cell stage added, and the

TABLE II.

Iso-capillary solutions.	6 per cent methyl alcohol.	3.36 per cent ethyl.	.2 per cent amyl.	3.2 per cent acetone.	1/6 saturated ether.	1 per cent phenol.
Undiluted . .	All dead	Nearly all dead, cy.	All dead	All dead	All dead	All dead.
Diluted to three fourths.	" "	Cy.	Monop.	Cy	Cy.	" "
Diluted to one half	" "	Cy.	Cy.	Cy.	...	"
Diluted to three tenths.	" "	...	...	..	...	" "

bottles tightly corked and set aside for forty-eight hours; the eggs were then transferred to sea water. In the tables, *Cy.* signifies the production of true cyclopia (1 median eye), *Monop.* indicates monopthalmia asymmetrica (one lateral eye normal, the other vestigial or absent).

Table II shows that cyclopia-producing solutions of ethyl and amyl alcohol, ether[5] and acetone, are iso-capillary, also the lethal doses are iso-capillary. Ethyl and amyl alcohol and ether have been classed as indifferent anæsthetics. Acetone may be considered as almost indifferent chemically, at least to many cells.

All of the solutions of methyl alcohol or phenol that were used killed the eggs. These substances probably act chemically on the eggs. Czapek found methyl alcohol to fall into the class of indifferent anæsthetics as regards plant cells. It may be objected that the toxic action of my sample was due to impurities, but evidently a contamination with amyl alcohol and acetone could not be responsible for the effect,

[5] My results with ether are not as exact as with the other substances. The high vapor tension of ether prevented exact measurements of surface tension of its solution. If the drops were in contact with circulating air, the concentration of ether in the surface of the drop was lowered. If the drops fell into an enclosed space, the ether vapor in the air altered the surface tension of the drop.

since these substances behave as indifferent anæsthetics, and their action is additative.

Feré[6] found the teratogenic power of alcohols injected into hen's eggs, measured in volumes per cent, to fall in the following series, ethyl < prophyl, butyl < amyl. Since, according to Traube, each member of this series lowers the surface tension of water three times more powerfully than the one preceding, we may conclude that they were equally effective in "iso-capillary" concentrations. However, the data of Feré are too inadequate for any final conclusions.

It is interesting to note that Child[7] produced cyclopia in flat-worms, Planaria, with anæsthetics.

Appendix I (Tables).

KCl in sea water:

Molecular concentration	.3 to .8	.9
Result	50 per cent dead.	All dead.

Pure NaCl:

Molecular concentration	.2	.3	.35	.4
Result	All hatched.	None hatched, monop.	Monop.[1]	Cy.[2]
Molecular concentration	.4	.5–1		
Result	Nearly all dead.	All dead.		

NaCl in sea water:

Molecular concentration	.9	1.	1.04	1.12
Result	All normal.	59 per cent dead.	Cy.	Cy.
Molecular concentration	1.2	1.28	1.36	1.6–2.0
Result	60 per cent dead, cy.	Monop.	Cy., monop.	All dead.

Pure LiCl:

Molecular concentration	.2	.25	.3–.4
Result	Cy.	Nearly all dead.	All dead.

LiCl in sea water:

Molecular concentration	.33	.4	.5	.6	.66
Result	All normal.	Monop.	Monop.	Monop.	Monop.

[1] Monop. = monopthalmia asymmetrica. [2] Cy. = cyclopia.

[6] Feré: Cinquantenaire de la Société de Biologie, 1899, p. 360. Reviewed in Ballentyne: Antenatal pathology and hygiene, Edinburgh, Green & Co., 1904, ii, p. 210.

[7] Child: Biological bulletin, 1909, xvi, p. 277; 1911, xx, p. 309.

Pure $MgCl_2$:

Molecular concentration . .	.3	.4	.5	
Result	75 per cent dead.	95 per cent dead.	All dead.	

$MgCl_2$ in sea water:

Molecular concentration . .	2.8	3	.32	.33
Result	53 per cent dead.	57 per cent dead, monop.	63 p. c. dead, monop, cy.	Cy.
Molecular concentration . .	.34	.36	.38	.4
Result	Monop, cy.	83 per cent dead, cy.	88 per cent dead, monop.	97 per cent dead.

$CaCl_2$ in sea water:

Molecular concentration . .	.3
Result	All dead.

Pure $CaCl_2$:

Molecular concentration . .	.2	.3
Result	30 per cent dead.	All dead.

Volumes of sea water concentrated to one volume:

1.9	2.2	2.5	2.65	2.8	2.95	4.
50 per cent dead.	42 per cent dead.	28 per cent dead, monop.	Cy.	62 per cent dead, monop., cy.	Monop.	All dead.

Appendix II (Permeability of the Egg).

It was thought desirable to know something of the composition, especially the salt content, of the egg, but the analysis was not completed. After remaining one hour in sea water, 44.318 gm. of unfertilized eggs were wiped dry with filter paper. After desiccation to constant weight, the dry substance was found to be 20 per cent of the total. The dry substance contained:

Alcohol-ether extract	12.2	per cent.
Insoluble ash	5.47	" "
Soluble ash	3.18	" "

The chlorides in the soluble ash = 4.82 c.c. of a normal solution. Therefore the chloride content of the water in the eggs (35.576 c.c.) was about .135 normal, or only .27 as much as in the sea water, which contained from .49 to .51 normal.

An estimate of the magnesium content may be made from the following, although the sample was too small for accurate results. Ten gm. eggs contained 2 gm. dry substance. The ash analyzed for Mg by the method of W. Gibbs gave .0336 gm. magnesium pyrophosphate. From this the magnesium content in the 8 c.c. of water in the eggs may be calculated as .03 molecular, or one half as much as in sea water.

From these preliminary ash analyses we might infer that the eggs, or certain regions within them, are impermeable to Cl and Mg ions and $MgCl_2$. However, it appeared to me that this might be determined by dialysis, especially since these eggs will live in distilled water.

In the following experiments the water was re-distilled in quartz. The eggs were fertilized and washed in many changes of water for about two hours. They were then placed in a graduate, and water or salt solution added to make 100 or 200 c.c., and soaked for a certain length of time. Three fourths of the total volume was decanted, evaporated, charred, and analyzed. The eggs were placed in a second volume of the solution and the process repeated.

Vol. of eggs.	Solution.	Hour.	T. Cl.[1]	Hour.	T. Cl.	Hour.	T. Cl.	Hour.	T. Cl.	Hour.	T. Cl.
12	H_2O	2d	None	16th	None	41	None	..	..	..	..
12	$M/4MgSO_4$	2d	None	"	"	"	"	..	..	..	..
12	H_2O	2d	None	"	"	"	"	68	None	113	None
12	Nitrates	2d	None	"	"	"	"	"	"	"	"

[1] T. Cl. = chlorine by titration.

The above table shows the egg to be impermeable to chlorine ions. The nitrates were of the same kations as are contained in sea water.

The above table shows a kationic exchange, Na in the medium being exchanged for Mg in the eggs. No Mg came out of the eggs in distilled water because the Mg^+ ions were held back by the electric charge of the Cl^- ions in the egg. The ash analysis given above suggests that the egg is normally impermeable to Mg ions, and that the pure NaCl solution, or Na+K, may increase its permeability to Mg.

In this connection it must be emphasized that it is the plasma membrane or surface of the protoplasm of the egg that shows this impermeability, and not the relatively firm egg shell, or chorion. I have demonstrated the permeability of the egg shell, or chorion, of fish eggs to several salts by microchemical methods. Loeb [8] suggests

[8] LOEB: Biochemische Zeitschrift, 1911, xxxvi, 275.

that change in permeability of the egg to salts may be due to the fact that the micropyle is at times open and at others closed. Since the micropyle is a hole in the egg shell, its stoppage cannot prevent the diffusion of salts, as the shell is permeable to them.

Vol. of eggs.	Solution.	Hour.	Mg. precip. in milligrams.	Hour.	Mg. precip.	Hour.	Mg. precip.
17	H_2O	19th	None	33	None	129	None
17	N/10 NaCl	"	.6[1]	"	Dense cloud	"	Dense cloud
10	H_2O	29	None	..	...	..	...
10	N/10 NaCl	"	Dense cloud	..	...	..	...
20	N/10 NaCl	90	None	..	...	..	...
20	" "	"	Dense cloud	..	...	..	...

[1] If we can rely on these small quantities, this represents .005 of the total milligrams in the eggs.

Appendix III ("Lithium Embryos").

Lithium produces a large per cent of embryos with enlarged segmentation cavities, which were observed and described by Stockard,[9] who called them "lithium embryos." I have confirmed this observation, and found similar embryos in pure solutions of .38 to .5 molecular NaCl, and in sea water solutions of ether, acetone, molecular dextrose, and .3 mol. $CaCl_2$, but in these latter the per cent of the "lithium embryos" is less than in solutions of LiCl. These embryos were studied alive, then prepared histologically and studied as serial sections. Those produced by lithium did not differ from the others.

Apparently the fluid from the segmentation cavity migrates forward at a later period and comes to lie in front of the embryo, thus producing a vesicle in the pericardial region. However it is possible that these two collections of fluid have different origins.

[9] Stockard: Journal of experimental zoölogy, 1906, iii, p. 99.

Reprinted from the American Journal of Physiology.
Vol. XXIX. — January 1, 1912. — No. III.

THE INCREASED PERMEABILITY OF STRIATED MUSCLE TO IONS DURING CONTRACTION.

By J. F. McCLENDON.

[*From the Embryological Laboratory of Cornell University Medical College, New York City.*]

ACCORDING to the membrane theory of Bernstein,[1] the muscle fibre is surrounded by a plasma membrane or surface film, allowing easier exit to one or more classes of kations than to the corresponding anions. The kations of some electrolyte which is more concentrated within than without come through the surface film and give the surface a positive electric charge. Destruction or alteration of this film or membrane causes a negative variation (the affective surface being less positive than the normal surface).

It has been suggested that potassium ions, which are more concentrated within the muscle than in the blood plasma, give the muscle the positive charge. However, one should then expect a reversal of the electric effects by placing muscle in an isotonic solution of potassium salts. This was shown by Höber[2] and Overton[3] not to occur. R. Lillie[4] suggests that lactic and carbonic acids are the electrolytes in question, and the H ions give the muscle surface the positive charge. The difficulty with this view lies in the fact that muscle, as shown by Overton, is in general permeable to substances soluble in many oils. It seems, therefore, that this question is not settled.

If the negative variation or "action current" of muscle is due to increased permeability to any ions, we should expect increased electric conductivity on contraction. It has long been known that the electric

[1] Bernstein: Archiv für die gesammte Physiologie, 1902, xcii, p. 521.

[2] Höber: Archiv für die gesammte Physiologie, 1905, cvi, p. 607.

[3] Overton: Sitzungsberichte der physikalisch-medizinische Gesellschaft, Würzburg, 1905, p. 2.

[4] Lillie: This journal, 1909, xxiv, p. 14; 1911, xxviii, p. 197.

conductivity of muscle increases at death. Du Bois Reymond[5] explained the electric resistance of living muscle by the presence of membranes, which become altered (permeable) at death. He showed that muscle becomes polarized on the passage of an electric current. It seems to me that Kodis[6] and Galaeotti[7] take a step backward in attributing the increased conductivity of dead muscle to the liberation of ions. Galaeotti tried to support his view by determinations of the freezing points of living and dead muscle, but found, on the contrary, that the change in electric conductivity did not correspond to the change in the osmotic pressure.

I have not found in the literature a clear statement that the electric conductivity of muscle has ever been measured, actually, during contraction. The contraction period (about one fifth of a second for frog's muscle) is too short for an accurate measurement of conductivity to be made. Therefore I decided to measure the conductivity during tetanus.

Experiments.

Platinum electrodes similar to those designed by Galaeotti were used. These were "platinized" with a solution of platinic chloride containing a trace of lead acetate. Since the muscle was in contact with the electrodes, there was danger of rubbing off some of the platinum black. I found that the black adhered more strongly if the electrodes were previously roughened by coating with platinum black and then heating in a flame.

The method of Kohlrausch was used to measure the conductivity. The smallest sized induction coil made for such experiments was fitted with a rheostat in the primary, and only just enough current was used as is required to work the interrupter. A second rheostat was inserted in the secondary, and the current could be reduced until it was not felt when passed through my tongue.

The muscles of the frog's thigh were used. They were placed between, and with the fibres parallel to, the electrodes, and the latter pressed together until the muscle bulged out on all sides. The elec-

[5] Du Bois Reymond: Untersuchungen über tierische Electricität, 1849.

[6] Kodis: This journal, 1901, v, p. 267.

[7] Galaeotti: Zeitschrift für Biologie, 1902, n. F. xxv, p. 289, and 1903, xxvii, p. 65.

trodes were then fixed rigidly in position and the preparation placed in a moist, constant temperature chamber.

The conductivity of the muscle was measured with too little current to cause contraction. Then, by cutting out resistance in the secondary circuit, enough current was passed to throw the muscle into tetanus and a second reading made. In all cases the conductivity was greater during contraction.

It might be objected that the heating effect of this current would change the conductivity, but control experiments on liver tissue and the tissues of certain plants, in which no change in conductivity occurred, showed this not to be the case.

Since the muscle at all times entirely filled the space between the electrodes and extended out on all sides, a change in the form of the entire muscle would not appreciably alter the conductivity. However, it is possible that the change in form of the muscle fibres might slightly alter the conductivity, but it is improbable that it would account for the large differences observed.

If the change in conductivity were due to metabolic activity in the muscle, we would expect different results depending on whether the muscle was measured first in the stimulated or unstimulated condition, but no such difference was found. If the change in conductivity is due to chemical change, the latter must be completely and instantaneously reversible.

Experiments were made both in the spring and the autumn, and a long series of measurements were made on each muscle. The increase in conductivity was greatest in fresh preparations and decreased as the muscle became fatigued, varying from 28 to 6 per cent.

Discussion of Results.

The increased conductivity of muscle during contraction may be interpreted as demonstrating the increase in permeability of some structures within the muscle to anions (since the muscle appears already permeable to certain kations). According to the membrane theory, this causes a reduction of the electrical polarization of these structures, thus causing increased surface tension and contraction.

In order that the contraction be due to increased surface tension of any structures, the latter must during relaxation be elongate in the axis of the muscle. We have such a structure in the anisotropic segment of an ultimate fibril. It is interesting to note that Duesberg [8] finds these segments to arise from fat or lipoid-containing bodies known as chondriosomes, and the presence of fat would account for the high surface tension between them and the sarcoplasm that is necessary for contraction.

Bernstein calculated the size of the structures that would be compatible with the force of contraction, and concluded that it must be smaller than any structures seen in histological preparations of muscle. He therefore postulated hypothetical ellipsoids as the elements in question.

It is possible that surface tension changes are aided by osmotic pressure,[9] since the movements of plants are due to osmotic changes following changes in permeability. However, the small size of the muscle elements makes it impossible to apply botanical methods to it.

[8] DUESBERG: Archiv für Zellforschung, 1910, iv, p. 602.

[9] *Cf.* MEIGS: This journal, 1910, xxvi, p. 191.

Reprinted from the American Journal of Physiology.
Vol. XXIX. — January 1, 1912. — No. III.

DYNAMICS OF CELL DIVISION. — III. ARTIFICIAL PARTHENOGENESIS IN VERTEBRATES.

By J. F. McCLENDON.

[*From the Embryological Laboratory of Cornell University Medical College, New York City; and the Station for Experimental Evolution of the Carnegie Institution of Washington, at Cold Spring Harbor, Long Island.*]

I. By Mechanical Stimulation.

EGGS of the wood frog, Rana sylvatica, and the tree frog, Hyla pickeringii, were caused to segment by momentary pressure. Cutting the mass of eggs from the uterus to pieces with scissors caused a few eggs to die and a few others to segment. Compression was a very unreliable method for causing segmentation, and much more uniform results were obtained by pricking with a fine needle.[1]

The females were washed with alcohol followed by a strong stream of water. The eggs were carefully taken from the uterus so as to avoid pressure and placed in shallow dishes. In some experiments the eggs were pricked immediately and then covered with water. In others the eggs were first covered with water and the jelly allowed to swell slightly, though the eggs remained adherent to the bottom and were thus held in position. The glass dish was placed under a Zeiss binocular microscope and the eggs lightly pricked with the finest sewing needle.

The extent of the puncture had to be regulated with great care in order to obtain good results. It seemed necessary merely to touch the surface of the egg in order to start development, although a puncture that did not result in a large extra-ovate was permissible. In my operations on thousands of eggs, some failed to be reached by the

[1] Bataillon: Comptes rendus de la Société de Biologie, 1911, lxx, p. 562, had previously caused European frog's eggs to segment by pricking with platinum or glass needles. He claims to have developed a few to the metamorphosis. He interprets Guyer's results on injecting lymph into frog's eggs as parthenogenesis due to pricking, but thinks the serum entering the wound may be essential.

needle, whereas some others were pricked too deeply and died. The jelly offered considerable resistance, and a sudden stab with the needle seemed less injurious than a slow motion.

Eggs which were pricked just sufficiently, rotated in the normal manner and segmented. The first cleavage was regular in a small per cent, and showed varying degrees of irregularity in the remainder. No regular later cleavages were observed, so that it is impossible to state how many cytoplasmic cleavages occurred, though divisions of the nuclei continued for a few days, not, however, at the normal rate.

In some cases the first cleavage furrow passed through the point of puncture, but since it did not do so in all cases, further study would be necessary to analyze the localization factors.

A control was always kept, care being taken not to subject any of the eggs to pressure. None of these eggs segmented, but it was noted in some cases that they finally became wrinkled and apparently decreased in volume. According to Biataszewicz,[2] the unfertilized frog's egg swells continuously. The water which I used came from an artesian well, and contained over ten times as much CO_2 as was found in pond water,[3] and it is possible that the CO_2 injured the eggs. In harmony with this view is the fact that there was a large mortality among fertilized eggs developing in it.

II. By Electrical Stimulation.

Eggs of the wood frog, Rana sylvatica, the tree frogs Hyla pickeringii and Hyla versicolor, and the toad, Bufo lentiginosus, were caused to segment by electric stimulation. The character of segmentation was similar to that in pricked eggs. A high per cent could be caused to segment. The segmentation rate seemed to be about normal, for instance the first cleavage in Hyla pickeringii occurred from two to three hours after electric stimulation, at 21° C. But in order to avoid the slightest possibility of contamination with sperm, no males or fertilized eggs were allowed in the laboratory during the experiments, and the rate of development was not directly compared with the normal.

[2] Biataszewicz: Bulletin de l'Académie des Sciences de Cracovie, 1908, p. 783.

[3] From analysis by R. A. Gortner.

The females were washed with alcohol followed by a strong stream of water. The eggs were carefully taken from the uterus, without pressure, placed in glass dishes and covered with water. An alternating current of 110 volts and 60 cycles was passed through the water from platinum or carbon electrodes held from 2.5 to 15 cm. apart. The use of an alternating current prevented appreciable polarization and chemical changes at the electrodes, but in order to avoid the possibility of the slightest chemical influence, the water was poured off of the eggs immediately after removal of the electrodes and fresh water added. Eggs very close to the electrodes were necessarily more affected by the current than those farther removed, but in comparing experiments I have regarded only those eggs midway between the electrodes. The position of the electrodes was marked, and the eggs remained adhering to the glass in their original positions.

An instantaneous exposure to the current was sufficient and prolonged exposure injurious, although the resistance of the eggs to injury varied with the species. Having the electrodes 5 cm. apart, the injurious effect was seen in Hyla pickeringii eggs after five seconds, but in toad's eggs only after twenty to one hundred and twenty seconds' exposure.

The majority of the experiments were performed in artesian water. The fact that the use of water re-distilled from barium hydrate gave as good results, indicates that the CO_2 of the artesian water was not essential to parthenogenesis. The addition of 1/50, 1/20, 1/10, or even 1/5 vols. of sea water did not prevent cleavage. Also the following pure salts and combinations were permissible: NaCl, $CaCl_2$, $MgCl_2$, Na_2CO_3, CaOH, chlorides of Na & Ca, Ca & Mg, Na & Ca & Mg.

A large number of eggs of each species were removed at various periods after stimulation and prepared histologically in serial sections. The following account of the internal changes in eggs of the wood frog applies to all except for details as to size, amount of pigment, etc.

The female pro-nucleus, in passing from the surface at the animal pole, toward (but not reaching) the centre of the egg, leaves a pigment track similar to that left by the sperm nucleus in fertilized eggs.[4] I have no doubt that this pigment track is present in normal develop-

[4] In Hyla pickeringii, where there is little pigment, the track is hardly noticeable.

ment, but has not been observed because the first cleavage furrow cuts through it longitudinally, as it is being formed, and obliterates it. In parthenogenetic eggs the furrow is usually not through the axis of symmetry, in which the pigment track lies, and therefore usually does not obliterate it.

Although the first cleavage is not usually through the axis of symmetry, it is often near this axis, and therefore very nearly regular. Sometimes a large cell is cut off from a small one. In the majority of cases, however, two and sometimes more furrows appear simultaneously. This is not due to a corresponding increase in number of nuclei, as sections show but one nucleus, which is the undivided female pro-nucleus. The metaphase of the first nuclear division is reached long after the first cleavage furrow begins to cut through the egg.

The first cleavage usually does not completely cut through the vegetative pole. Later cleavage furrows sometimes appear, but are likewise never completed.

Cleavage of the cytoplasm is brought to a standstill, and the nuclear changes continue. The nucleus divides repeatedly, filling the egg with daughter nuclei. Finally the egg becomes vacuolated and is dead within a few days.

In a preliminary note [5] I have mentioned certain probabilities in regard to the theoretical significance of these results. However, since more data are desirable, I will not continue the discussion here, but hope to do so at a later date.

[5] McClendon: Science, 1911, N. S. xxxiii, p. 629.

Reprinted from The Journal of Biological Chemistry, Vol. X, No. 6, 1912

HOW DO ISOTONIC SODIUM CHLORIDE SOLUTION AND OTHER PARTHENOGENIC AGENTS INCREASE OXIDATION IN THE SEA URCHIN'S EGG?

By J. F. McCLENDON and P. H. MITCHELL.

(*From the Embryological Laboratory of Cornell University Medical College, New York City, the Physiological Laboratory of Brown University, Providence, R. I., and the U. S. Bureau of Fisheries, Woods Hole, Mass.*)

(Received for publication, November 4, 1911.)

Loeb has shown that ^{-}OH ions favor development.[1] Our own experiments (Table 6) as well as those of O. Warburg[2] demonstrate that increase in the alkalinity of the medium increases oxidation in fertilized eggs.

According to Warburg, the egg is impermeable to ^{-}OH ions or fixed alkalies, because, although eggs stained with neutral red are changed to yellow by NH_3 of a concentration which increases the oxidation only one-tenth, similar eggs are not changed to yellow by fixed alkalies of a concentration at which oxidation is greatly increased.

These facts are, however, capable of another interpretation. The egg is filled with lipoid particles which take up the neutral red to such an extent as to render the surrounding protoplasm and sea water colorless. The ^{-}OH ions cannot freely enter the lipoids in order to change the color of the neutral red. On the other hand ammonia is lipoid-soluble and can enter.

It might be objected that the ammonia cannot react with the dye in the non-aqueous medium owing to the suppression of ionization, but whether the dye is driven out of the lipoids or changed to yellow *in situ*, the fact remains that it does become yellow.

Harvey[3] observed that the addition of but a small quantity of alkali to sea water containing fertilized eggs stained with neutral

[1] Loeb: *Chemische Entwicklungserregung des tierischen Eies*. Berlin, 1909.
[2] Warburg: *Zeitschr. f. physiol. Chem.*, lx, p. 305, 1910.
[3] Harvey: *Journ. of Exp. Zoology*, x, p. 507, 1911.

red, changed the eggs to yellow. However, the eggs were injured by the alkali and probably their permeability was increased. It is possible that alkalies or ^{-}OH ions in any concentration enter the fertilized eggs but must be present in sufficient concentration to set free ammonia or change the lipoids in order to affect the dye.

We may assume, then, unless more conclusive evidence indicates the contrary, that the ^{-}OH ions increase oxidation after penetrating the egg.

One of us has shown that unfertilized sea urchin's eggs are poorly permeable to salts and their ions, but become more permeable after fertilization or the initiation of parthenogenetic development.[1] Not only is the permeability increased but the oxidation rate is increased. Warburg observed that oxidation increases from five to seven times on fertilization (compare our Table 5, experiment III). He found also a large increase after the initiation of parthenogenetic development caused by hypertonic sea water,[2] fatty acid, alkalies or traces of the heavy metals,[3] silver or copper. Our own experiments, given below, confirm and extend these findings. There appears then to be some relation between permeability and oxidation, and the present paper is an attempt to determine what this relation is.

The living cell may be compared to a furnace, and R. Lillie[4] advanced the view that increase in permeability opens the draughts so to speak, allowing the escape of carbonic acid, and hence oxidation is increased. He supposes that the accumulation of carbonic acid and perhaps other end-products checks the oxidation, and increase in permeability to carbonic acid allows oxidation to proceed. The difficulty with this hypothesis lies in the fact that living cells have been shown by Overton and others to be freely permeable to substances which are easily soluble in fats and oils, or especially in lecithin and cholesterin. Carbon dioxide is soluble in oils and probably enters cells easily, at least there is evidence to show that red blood corpuscles are freely permeable to this gas. Fatty oils are permeable only to the undissociated molecules and not to the ions. Since the proportion of ions of carbonic

[1] McClendon: *Amer. Journ. of Physiol.*, xxvii, p. 240, 1910.
[2] Warburg: *Zeitschr. f. physiol. Chem.*, lxii, p. 1, 1908.
[3] Warburg: *Ibid.*, lxvi, p. 305, 1910.
[4] Lillie: *Biol. Bull.*, xvii, p. 188, 1909.

acid would ordinarily be small, what conditions in the egg might favor ionization of CO_2?

Not all of the alkali metals in the egg are combined with mineral acids; some are combined with proteins. This has been shown by the senior author with electric conductivity measurements of hens egg yolk given in the accompanying curves. The continuous line represents the conductivity of yolk freed from protein granules, and the dotted line, yolk containing an excess of protein

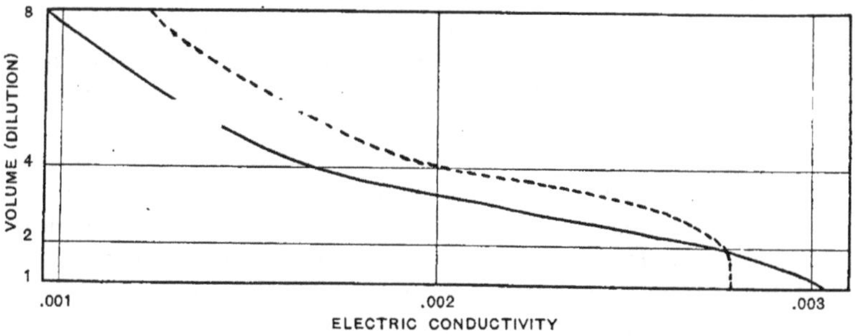

granules, separated by the centrifuge. The granules impede the current as shown by the fact that the granule-containing yolk is a poorer conductor than granule-free yolk. But on dilution, the granule-containing yolk becomes the better conductor. Therefore ions are set free from the granules on dilution.

The carbonic acid formed within the egg would react with the alkali albuminates with the formation of alkali carbonates and bicarbonates, which, notwithstanding hydrolysis, would liberate a considerable quantity of carbonic acid anions.[1] The dissociated carbonic acid, being unable to escape from the unfertilized egg would lower the ^-OH ion concentration and thus reduce oxidation. As the undissociated molecules of carbonic acid escape, more are formed by the slow oxidation in the egg.

On fertilization, the permeability to the anions of carbonic acid is greatly increased. They migrate out of the egg, and negative ions enter the egg to take their place. Since the ^-OH ions of the sea water are the fastest negative ions, they enter the egg and

[1] At about molecular concentration the equivalent electric conductivity of NaCl is 76; of Na_2CO_3, 45, at 18°.

increase oxidation. However, if some ^{-}Cl or $^{-}SO_4$ ions entered the egg they would tend to decrease the ^{-}OH ion concentration within the egg. But the senior author has shown that the fertilized *Fundulus* egg is impermeable to ^{-}Cl ions, for when placed in distilled water, or in solutions of nitrates or sulphates, practically no chlorine comes out of the eggs.[1] We may assume, therefore, that carbonic acid accumulates in the unfertilized egg until the reaction is neutral or slightly acid. But on fertilization, the permeability to carbonic acid anions is increased and the concentration of this acid is diminished so that the reaction is neutral or slightly alkaline. The increased alkalinity is the cause of the increased oxidation. We will now see to what extent our experiments bear out this assumption.

We observed that oxidation is about doubled when the egg is made parthenogenetic with carbonated sea water (Table 5, experiment II) or alkaline isotonic sodium chloride (Table 3, experiments I and II). In some cases the eggs, being physiologically different from those in other experiments, did not show the morphological signs of development in this solution, and oxidation was not doubled (Table 2: Table 3, experiment III; cf. Table 5, experiment I). The eggs begin development while in the alkaline solution, but eggs made parthenogenetic by treatment with neutral or acid solutions (neutral sodium chloride or carbonated sea water) begin development only when transferred to natural sea water or other alkaline solution. This bears out our hypothesis, for if the increased permeability remains after the egg is returned to an alkaline medium, a chance is given for an increase in alkalinity in the interior, which was lacking in the non-alkaline solution.

Certain facts may seem to contradict our assumption, but probably they merely limit its application:

1. A slight increase in ^{-}OH ions may cause even the unfertilized egg to absorb more oxygen (Table 2) and a greater increase causes it to develop. This does not necessarily show permeability of the unfertilized egg to hydroxyl ions. The increased alkalinity slowly causes an increased permeability of the egg and thus leads to parthenogenesis, but the degree of alkalinity of the medium necessary to induce development of the unfertilized egg is far greater

[1] These experiments will be published later in the *American Journal of Physiology*.

than that necessary for the development of the fertilized egg or the egg already made parthenogenetic.

2. The fertilized egg of *Arbacia punctulata* (but not of some other sea urchins) may develop in a natural medium, as Loeb observed and which we have confirmed. In other words, a hydroxyl ion concentration in the medium, greater than that of distilled water, is not necessary for development of this egg made freely permeable to carbonic acid. However this fact does not set a limit to the alkalinity of the egg interior. The egg probably contains more Na than Cl ions, and if it be impermeable to Na or Cl, the escape of carbonic acid might cause the egg interior to become alkaline or at least neutral. Eggs made parthenogenetic in some ways (neutral sodium chloride, for instance) do not develop unless transferred to an alkaline medium, but this may be due to the possibility that these parthenogenetic eggs are not quite as permeable as are fertilized eggs. The same may be inferred from oxidation measurements. Neutral sodium chloride causes the unfertilized egg to absorb more oxygen than it does in sea water (Table 4) but the increase is slight, and morphological development does not commence. If the eggs are then transferred to sea water or other alkaline solution, some of them may develop.

It appears therefore that increase in permeability is a gradual process. Although some eggs are so permeable as to be able to develop in an neutral medium others are less permeable and do not develop, or develop only in an alkaline medium. By treating eggs with parthenogenetic agents in various concentrations or for various lengths of time we may induce various degrees of permeability. Even fertilized eggs may be made more permeable by treatment with parthenogenic agents, and a corresponding increase in oxidation may be observed (Table 6). In these experiments the oxidation of the eggs in sea water was measured about ninety minutes after fertilization: they were then placed in isotonic, alkaline sodium chloride solution, in which the oxidation increased one-half, when returned to sea water the oxidation fell below its previous level in the same medium. According to Loeb, this indicates death of some of the eggs (20 per cent).

The experiments just described explain the discrepancy between the results of Warburg and those of Loeb. Warburg[1] found that

[1] Warburg: *loc. cit.*

the oxidation of the *fertilized egg* in isotonic sodium chloride solutions containing a trace of sodium cyanide, is much greater than in sea water containing the same concentration of sodium cyanide. Loeb[1] confirmed this determination, but observed further that if the cyanide is omitted (from both) no increased oxidation in sodium chloride solution occurs. The cyanogen in both sea water and sodium chloride solution depresses oxidation. Since sodium cyanide liberates ^{-}OH ions, we may conclude that the increase in oxidation in the sodium chloride solution used by Warburg was due to the increased penetration of hydroxyl ions, following increase of permeability.

In our experiments no cyanide was used, and the alkalinity of the sodium chloride solution was not greater than sea water, yet oxidation was increased. In Loeb's experiment the tendency of increased permeability to increase oxidation was counteracted by the effect of lower alkalinity, which decreases oxidations.

Alkaline sodium chloride solution also favors oxidation in eggs that have reached later stages of development, morula or blastula (Table 7). In this experiment, the rate of oxidation in sea water was rising gradually (see next section) before the eggs were placed in the alkaline soduim chloride solution, but in the latter a sudden increase of more than 50 per cent was observed.

MATERIALS AND METHODS.

The eggs of the sea urchins, *Arbacia punctulata*, were used. The animals were washed in a strong stream of fresh water and opened with precautions against introducing spermatozoa among the eggs. The ovaries were removed and placed in the first solution to be used, sea water or neutral van't Hoff's solution. The mass was strained through bolting cloth of such a grade as to allow but one egg to pass through one mesh at a time. The eggs were repeatedly precipitated by gravity in fresh portions of the solution in order to remove coelomic fluid cells (elaeocytes), and transferred with a small quantity of fluid to the determination flask.

The rate of oxidation in the various solutions was measured by comparison of the dissolved oxygen in the solution before and after the eggs had been suspended in it during a definite period.

[1] Loeb and Wasteneys: *Biochem. Zeitschr.*, xxviii, p. 340, 1910.

Winkler's thiosulphate method of oxygen determination (iodometric), as described in Treadwell's Quantitative Analysis, was used.

From 3 to 7 cc. of eggs were used in each experiment, but the actual volume of the eggs was not measured until after the oxygen determinations.

We tried a number of methods for filling the determination flask and sample bottles without an uncertain loss or gain of oxygen. Loeb collected the water in the sample bottle under petroleum. Although petroleum absorbs five times as much oxygen as water does, the oil would tend to reduce currents adjacent to the air-water surface, and thus reduce oxygen exchange. Using paraffin oil, we found that it was extremely difficult to prevent a little oil from sticking within the flask, and abandoned the method. Perhaps kerosene would have worked better, yet the quantity of kerosene that would dissolve in the water might vitiate the experiments.

Since the sea water and the solutions were shaken up and saturated with air at the given temperature before beginning the experiment, the control sample might have been taken under air, without change. But after loss of oxygen in the determination flask, a gain in oxygen would result from such treatment. By introducing the solution through a tube passing through a doubly perforated stopper, and extending to the bottom of the sample bottle, the exposed surface of the water was made as small and quiet as possible. By maintaining a constant rate of flow the error could be made to bear an approximately constant ratio to the oxidation, no matter whether the sample bottle was filled with air or some other gas. We tried the effect of introducing the sample under air, and also under hydrogen, and decided that the latter method was preferable for oxygen-low samples. In order to make all errors fall in the same direction we also collected the oxygen-high samples under hydrogen.

The experiments were so regulated that the oxygen content of the determination flask at the end of the exposure would not fall very low. However, Warburg failed to observe a decrease in the rate of oxidation in low oxygen concentration.

The water was forced out of the determination flask rapidly into the sample bottle by hydrogen under pressure. The determination flask held 332 cc., the two sample bottles 152 and 142.6 cc. respectively.

The determination flask was placed in a thermostat which was kept 2° above the temperature which the air had reached at the beginning of the experiment. In most cases the time of exposure was one hour. During the first half-hour the eggs were distributed throughout the solution once every five minutes by rotating and rocking the flask, during the last half hour they were allowed to settle to the bottom.

Although the majority of the eggs settle to the bottom in ten minutes, at the end of one-half hour there are always a few which, on account of swelling or fragmentation, have failed to precipitate. To prevent these going over into the sample bottle and causing an error due to absorption of iodine, Loeb placed filter paper over the outlet. We found that a relatively hard filter paper was necessary to retain all fragments of eggs, and that this interfered with the rapid transfer of the solution. The error due to eggs in suspension is negligible, especially since it was practically constant in all of our experiments. For instance: a sample bottle filled with water contained 8.09 parts per million of oxygen, while water from the same jar run into a sample bottle in which had been placed about one-hundred times as many eggs as the water from the determination flask, contained 7.85 parts per million, an error of 0.24 parts per million due to eggs in suspension. As in our experiments, differences of 0.1—3.0 parts per million were obtained, an error of 0.024 parts per million would not reverse the results.

When the eggs were fertilized or placed in parthenogenic solutions they lost sufficient red pigment (McMunn's Echinochrome) to color the water a straw yellow. In order to ascertain whether this organic matter would vitiate the results, we took two sample bottles, into one of which was placed a mass of elaeocytes containing about fifty times as much echinochrome as is lost from the eggs in one experiment, and syphoned into each tap water from the same jar. Tap water causes these cells to liberate their pigment. The bottle containing water only, was found to hold 6.85 parts per million of oxygen, while that contaminated by elaeocytes was titrated as 6.36 parts per million of oxygen. We repeated this using three sample bottles. Two of these filled with water gave

6.86 and 6.91 parts per million respectively and the third contaminated with elaeocytes liberating about one hundred times as much pigment as is liberated in the experiments with eggs, gave 6.09 parts per million as the titration, showing an error of 0.8 parts per million. Probably this loss was due chiefly to the broken up cells, but if due entirely to the pigment, the error in our experiments would be only .008 parts per million.

In experiment 3, the eggs were divided into two equal portions and placed simultaneously in two determination flasks of equal capacity. Therefore the eggs in the two solutions were in the same stage of ripeness. In each of the other experiments the eggs were all placed in one flask and treated successively with the various solutions. In case of fertilized eggs Warburg observed[1] that the oxidation rate rose steeply from fertilization to the 2-cell stage then gradually to the 64-cell stage. In order to determine whether this would be a great source of error on our experiments, we measured the oxygen used by a mass of fertilized eggs in sea-water during successive periods in the first six hours of development, and found it to vary from 7.93 to 9.76 tenths of a milligram (Table 7). Within the duration of the majority of our experiments, however, the variation was only from 7.93 to 8.96 tenths of a milligram or 11.5 per cent, and would be less between successive exposures. This possible source of error would not reverse the results of the majority of our experiments.

In making solutions of the same alkalinity as the sea water, a colorimetric method was used, with phenolphthalein as indicator. At the end of the season, the sea water was diluted with heavy rains and failed to color phenolphthalein. The eggs behaved abnormally, though whether this was due to the decrease in alkalinity or salinity of the sea water or to some other cause was not determined.

In making the eggs parthenogenetic with carbon dioxide, they were placed with a small quantity of sea water in a "sparklet syphon" and charged under slight pressure for about one minute. At the end of five minutes they were poured into a large volume of sea water and this was syphoned off and fresh sea water added.

[1] Warburg: *Zeitschr. f. physiol. Chem.*, lx, p. 443, 1909 and *loc. cit.*

RESULTS OF EXPERIMENTS.

I. Experiments with Unfertilized Eggs.

TABLE 1.

Oxidation in neutral van't Hoff's solution contrasted with that in neutral NaCl solution.

EXPERIMENT	SOLUTION USED	VOLUME OF EGGS	TEMPERATURE THERMOSTAT	DURATION OF STIRRING	DURATION OF SETTLING	OXYGEN USED	REMARKS
		cc.	°*C*	*min.*	*min.*	$\frac{mg}{10}$	
First period..	Neutral van't Hoff	4	24	30–	30	2.49	
Second period	Neutral $\frac{M}{2}$ NaCl	4	24	30	30	5.47	No "fertilization membrane" formed.

TABLE 2.

Oxidation in neutral van't Hoff's solution contrasted with that in sea water and alkaline NaCl solution.

EXPERIMENT	SOLUTION USED	VOLUME OF EGGS	TEMPERATURE THERMOSTAT	DURATION OF STIRRING	DURATION OF SETTLING	OXYGEN USED	REMARKS
		cc.	°*C*	*min.*	*min.*	$\frac{mg}{10}$	
First period..	Neutral van't Hoff.		24	30	30	2.29	
Second period	Sea water		24	30	30	2.65	
Third period	$\frac{M}{2}$ NaCl +−OH ions		24	30	30	3.12	"Fertilization membranes" in very small per cent.
Fourth period	Sea water		24	30	30	3.48	

TABLE 3.

Oxidation in sea water contrasted with that in alkaline NaCl solution.

EXPERIMENT	SOLUTION USED	VOLUME OF EGGS	TEMPERATURE THERMOSTAT	DURATION OF STIRRING	DURATION OF SETTLING	OXYGEN USED	REMARKS
		cc.	°C	min.	min.	$\frac{mg}{10}$	
I. One-half of eggs.....	Sea water		23	30	30	4.80	
Second half of eggs.....	$\frac{M}{2}$ NaCl + $^-$OH ions		23	30	30	6.90	"Fertilization membrane" formed.
II. First period......	Sea water		24.5	35	25	3.38	
Second period	$\frac{M}{2}$ NaCl + $^-$OH ions		24.5	35	25	7.80	"Fertilization membrane" formed.
III. First period......	Sea water	4.5	24	30	30	6.04	
Second period	$\frac{M}{2}$ NaCl + $^-$OH ions	4.5	24	30	30	6.47	"Fertilization membranes" in very small per cent.

TABLE 4.

Oxidation in sea water contrasted with that in neutral NaCl.

EXPERIMENT	SOLUTION USED	VOLUME OF EGGS	TEMPERATURE THERMOSTAT	DURATION OF STIRRING	DURATION OF SETTLING	OXYGEN USED	REMARKS
		cc.	°C	min.	min.	$\frac{mg}{10}$	
I. First period......	Sea water	5	23	30	30	2.65	
Second period	Neutral $\frac{M}{2}$ NaCl	5	23	30	30	3.60	No "fertilization membrane" formed.
II. First period....	Sea water	2.3	23.5	30	30	1.22	
Second period	Neutral $\frac{M}{2}$ NaCl	2.3	23.5	30	30	1.66	Ibid.

TABLE 5.

Effect of CO_2-parthenogenesis on oxidation. Effect of fertilization.

EXPERIMENT	SOLUTION USED	VOLUME OF EGGS	TEMPERATURE THERMOSTAT	DURATION OF STIRRING	DURATION OF SETTLING	OXYGEN USED	REMARKS
		cc.	°*C*	*min.*	*min.*	$\frac{mg}{10}$	
I. First period.....	Sea water	4.5	23.5	30	30	4.58	
After CO_2 treatment..	Sea water	4.5	23.5	30	30	6.00	"Fertilization membranes" in small per cent.
II. First period......	Sea water	4.5	24	30	30	4.11	
After CO_2 treatment..	Sea water	4.5	24	30	30	9.91	Good membrane formed.
III. First period......	Sea water		25	30	30	2.52	
After fertilization......	Sea water	5	25	30	30	12.94	98 per cent of eggs segmented.

SUMMARY.

1. The presence of ^-OH ions in the medium, increases the rate of oxidation in fertilized eggs of the sea urchin.

2. The oxidation rate of unfertilized eggs is increased by fertilization or any treatment which causes them to develop parthenogenetically.

In 1 and 2 we merely confirm and extend the observations of Warburg.

3. Since it was shown by the senior author that fertilization or parthenogenesis means increased ionic permeability of this egg, and that the *Fundulus* egg, even after fertilization, is impermeable to ^-Cl ions, the increase in permeability probably applies so far as the anions are concerned, to ^-OH and $^-HCO_3$ or $^-CO_3$ ions. The carbonic acid anions are more concentrated within, and their outward diffusion would cause a potential gradient which would pull other

II. *Experiments with Fertilized Eggs.*

TABLE 6.

Oxidation in neutral van't Hoff's solution contrasted with that in sea water and alkaline NaCl.

EXPERIMENT	SOLUTION USED	VOLUME OF EGGS	TEMPERATURE THERMOSTAT	DURATION OF STIRRING	DURATION OF SETTLING	OXYGEN USED	REMARKS
		cc.	°C	min.	min.	mgr./10	
I. First period......	Neutral van't Hoff	3.6	23	30	30	6.47	Contained some sea water.
Second period	Sea water	3.6	23	30	30	7.43	
Third period......	$\frac{M}{2}$ NaCl + $^-$OH ions	3.6	23	30	30	12 28	
Fourth period	Sea water	3.6	23	30	30	5.97	Very few eggs dead.
II. First period......	Neutral van't Hoff	4	23	30	30	3.91	Eggs washed in van't Hoff sol. and fertilized in it.
Second period	Sea water	4	23	30	30	7.17	
Third period	$\frac{M}{2}$ NaCl + $^-$OH ions	4	23	30	30	7.53	
Fourth period	Sea water	4	23	30	30	6.27	

anions, hence $^-$OH ions, into the egg, thereby increasing the internal concentration of $^-$OH ions. This increase in $^-$OH ions probably causes the increased oxidation.

4. The increase of $^-$OH ions in the medium causes even in unfertilized eggs, an increased oxidation. This is not interpreted as indicating that the unfertilized egg is normally permeable to $^-$OH ions, but that increased alkalinity causes increased permeability.

5. Increase in permeability is a gradual process. Beginning with the relatively impermeable unfertilized egg, and denoting degree of permeability by numerals, we have the following series.

TABLE 7.

Effect of alkaline NaCl on oxidation six hours after fertilization.

EXPERIMENT	SOLUTION USED	VOLUME OF EGGS	TEMPERATURE THERMOSTAT	DURATION OF STIRRING	DURATION OF SETTLING	OXYGEN USED	REMARKS
		cc.	°C	min.	min.	$\frac{mg}{10}$	
First period..	Sea water	7	22	20	30	8.90	Thirteen minutes after fertilization.
Second period	Sea water	7	22	20	30	8.96	2-cell stage.
Third period	Sea water	7	22	20	30	8.33	4-cell stage.
Fourth period	Sea water	7	22	20	30	7.93	16-cell stage.
Fifth period..	Sea water	7	22	20	30	8.13	32-cell stage.
Sixth period..	Sea water	7	22	20	30	9.76	Many retarded.
Seventh period.....	$\frac{M}{2}$ NaCl + $^{-}$OH ions	7	22	20	30	13.77	Many blastulae.
Eighth period	Sea water	7	22	20	30	8.98	All alive.

I. slightly increased oxidation, II. greater increase in oxidation rate, and imperfect "fertilization membrane" formation, III. oxidation still further increased, membrane formation perfect, followed by segmentation of the egg, IV, *ditto* except that oxidation is still further increased, and the eggs die sooner or later if oxidation is not reduced, V. oxidation enormous, membrane formation but no segmentation, premature death.

It is supposed that the primary effect of many toxic substances is an abnormal increase in the permeability of the egg, and fertilized eggs are more susceptible because they are already more permeable than unfertilized eggs.

A Note on the Dynamics of Cell Division.

A Reply to Robertson

by

J. F. McClendon.

(From the Embryological Laboratory of Cornell University Medical College, New York City.)

With 2 figures in text.

Eingegangen am 15. November 1911.

In a recent paper ROBERTSON[1]) attempts to strengthen his hypothesis of cell division put forward previously, which is as follows: In the polar synthesis of nuclein from lecithin, cholin is formed as a by-product, and diffusing in all directions, reaches a maximum concentration at the equator, where it (or a soap formed from it) diminishes the surface tension and leads to constriction of the cell. This hypothesis is erroneous for two reasons:

First: There occurs no appreciable synthesis of nuclein during cleavage. MASING[2]) and recently SHACKELL[3]) have shown that there is as much nuclein in the egg at the 2 cell stage as during the blastula stage.

Second: The surface tension is not diminished, but is relatively increased at the equator as shown in a model of cell division described below. Following the preliminary experiments of GAD, QUINCKE[4]) observed that if a drop of rancid olive oil in water is touched with an alkaline solution, even with such a slightly alkaline

[1]) ROBERTSON, T. B., Arch. f. Entw.-Mech. 1911. XXXII. S. 308.

[2]) MASING, Zeitschr. Physiol. Chem. 1910. LXXVII. S. 161.

[3]) SHACKELL, Science. 1911. N. S. XXXIV. p. 573.

[4]) QUINCKE, Sitzungsber. d. kgl. preuß. Akad. d. Wiss. zu Berlin. 1880. XXXIV. S. 791.

solution as egg white, the surface touched decreases in tension, spreads and is pushed forward. He explained that this was due to soap formation. He supposed that living cells were covered with a thin layer of oil, and movements were due to local formation of soap.

These experiments were continued by Bütschli[1]) who filled the drop of olive oil with globules of an alkaline solution. When one of these globules burst, the alkaline solution spread over the adjacent surface of the oil drop and locally lowered its surface tension, causing a protuberance. Bütschli supposed the movements of Amoeba to be of similar origin.

Bernstein[2]) described surface tension movements in drops of mercury, which are now well known to the majority of biologists. Similar movements were seen by Jennings[3]) in drops of clove oil.

Fig. 1.

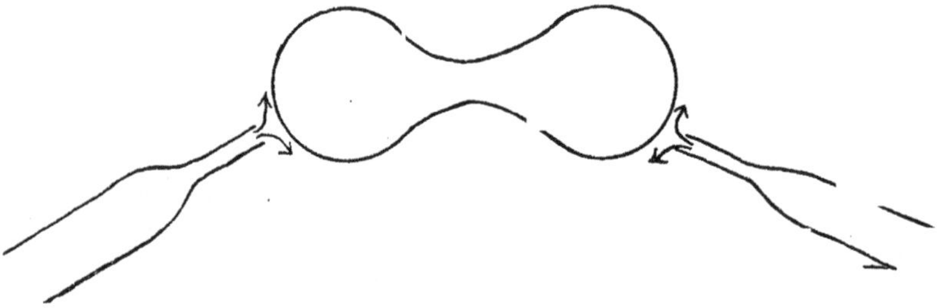

In all of the above mentioned experiments a decrease in surface tension is shown to cause a protrusion of the surface. Robertson, however, claims that a decrease of surface tension causes a receding of the surface, and when the decrease is along an equator, the drop is cut in two. He used a drop of 2 parts chloroform and 3 parts rancid olive oil and immersed it in water. When a thread soaked in n/10 NaOH was laid across the drop, the latter was cut in two, as he supposes, by the decrease in surface tension following soap formation. Contrary to Robertson I have failed utterly to obtain a division of the drop by this method. When the NaOH has diffused over the whole surface, a flattening of the drop occurs, but a constriction never takes place.

The result is very different when the NaOH is applied to the

[1]) Bütschli, Quart. Journ. Micr. Sc. 1890. XXXI. p. 99.

[2]) Bernstein, Arch. f. d. ges. Physiol. 1900. LXXX. S. 628.

[3]) Jennings, Journ. Appli. Micr. and Lab. Methods. 1902. V. p. 1597.

poles of the drop. Fig. 1 represents a drop of rancid oil and chloroform submerged in water. n/10 NaOH is applied to the poles by means of the two pipettes. If the NaOH is applied to the two poles at exactly the same time and rate, the drop constricts as in Fig. 1, and divides in two.

The details of this process are shown in Fig. 2. When the surface tension is decreased at the poles, these surfaces stretch because the tension is overbalanced by the greater surface tension at the equator. This causes a surface movement from the poles to the equator as shown by the dark arrows. The tension of the surface of the drop causes hydrostatic pressure in the interior. The surface tension, being diminished at the poles, is overbalanced by the hydrostatic pressure and the surface is protruded. This causes internal currents from the equator toward the poles of the drop, as shown by the dashed arrows.

Fig. 2.

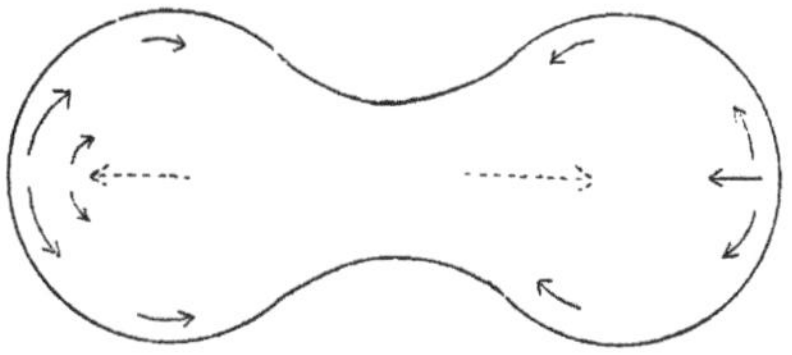

This model of cell division simulates the division of an Amoeba more than it does that of an egg or tissue cell. In these latter, the presence of a membrane or adjacent cells prevent the daughter cells from moving apart and the dumb bell shape is not formed.

The mechanics of cell division may be illustrated by a more tangible model. A rubber balloon is inflated with air and attains a spherical shape. The rubber may represent the surface film, and with uniform thickness of rubber we obtain uniform tension and spherical shape. If the equator of the balloon is re-inforced with a rubber band, the tension along the equator is increased and the balloon is constricted equatorially. The rubber band may be cemented to the balloon before it is inflated or the balloon originally made thicker along the equator, and in every case the result is the same.

The division of a mercury drop, due to decrease in surface tension at the poles following electric polarization in these regions, as observed by CHRISTIANSEN[1]) is described in FREUNDLICH's Capillar-

[1]) CHRISTIANSEN, Drud. Ann. 1903. XII. p. 1072.

chemie, p. 212—5. The mercury drop in KNO_3 solution becomes positively charged. Where the positive current enters, the drop becomes less positive, i. e. the polarization decreases, and the surface tension increases. At the opposite pole the surface tension decreases. As the current increases, the pole at which the positive current enters, passes the neutral point and becomes negative, leading to a decrease in surface tension. We have then the surface tension decreased at one pole more than at the other. One pole is strongly positive and the other slightly negative, and the neutral region forms a band around the drop nearer to the negative pole than to the positive. This neutral band is the region of greatest surface tension. The drop finally divides near the neutral region.

The reason the drop does not divide exactly in the center of the neutral region, or region of greatest surface tension, may be that surface currents hinder such a division. However, this is a detail which need not be essential to our main topic, since the consistency of mercury is very different from that of protoplasm.

Summary.

If a drop of rancid oil and chloroform is immersed in water and n/10 NaOH solution is allowed to diffuse against two opposite poles of the drop at the same time and rate, the drop constricts and divides in two along the equator.

This division is due to decrease in surface tension following soap formation at the poles, or relative increase in surface tension at the equator. Contrary to Robertson, a decrease in surface tension along the equator does not divide a drop.

Zusammenfassung.

Gibt man einen Tropfen ranziges Öl mit Chloroform in Wasser und läßt $^1/_{10}$-Normal-Natronlauge gegen zwei entgegengesetzte Pole des Tropfens gleichzeitig und gleichmäßig diffundieren, so zieht sich der Tropfen zusammen und unterliegt einer Zweiteilung entlang seinem Äquator.

Diese Teilung ist bedingt durch eine Verminderung der Oberflächenspannung, welche auf die Seifenbildung an den Polen folgt, oder durch eine relative Vermehrung der Oberflächenspannung am Äquator. Im Gegensatz zu Robertson führt eine Abnahme der Oberflächenspannung entlang dem Äquator nicht zur Tropfenteilung.

(Übersetzt von **W. Gebhardt.**)

(From the Histological Laboratory of Cornell University Medical College, New York City.)

Ein Versuch, amöboide Bewegung als Folgeerscheinung des wechselnden elektrischen Polarisationszustandes der Plasmahaut zu erklären.

Von

J. F. McClendon.

(Mit 4 Textfiguren.)

Nach der Ansicht vieler Autoren werden amöboide Bewegungen durch Änderungen der Spannung der Oberfläche hervorgerufen. Dieser Spannungszustand der Oberfläche kann entweder eine echte Oberflächenspannung oder, nach Rhumbler[1]) zuweilen eine Spannung des Ektoplasmas als Folge von Wasserverlust („Gelatinierungsspannung oder Gelatinierungsdruck") sein. In dieser Mitteilung soll nur von echter Oberflächenspannung die Rede sein.

Lässt man Kaliumbichromat gegen einen Tropfen Quecksilber in verdünnter Salpetersäure diffundieren, dann sendet der Quecksilbertropfen von der Stelle, die zuerst mit dem K_2CrO_4 in Berührung tritt, einen amöboiden Fortsatz aus. Es findet nämlich unter Verminderung der Oberflächenspannung eine Bewegung der äusseren Schichten von dem Bichromat weg und eine solche der inneren gegen das letztere zu statt. Ähnliche Strömungen wurden bei der Bewegung vieler Amöben beobachtet. Hier verursachen die Rückströmungen an der Oberfläche und die Vorwärtsströmungen im Inneren naturgemäss einen allmählichen Austausch von Ekto- und Endoplasma. Dieser Vorgang, den Rhumbler den „Ektoendoplasma-

1) L. Rhumbler, Zur Theorie der Oberflächenkräfte der Amöben. Zeitschr. f. wissensch. Zool. Bd. 83. S. 1. 1905.

prozess nennt, ist schon vor Jahren von Wallich und Montgomery[1]) (1979, 1881) beobachtet worden.

Nach Jennings[2]) findet ein derartiges Abfliessen der vorgeschobenen Oberfläche nicht bei allen Amöben statt. Doch liesse sich dessen Ausbleiben in solchen Fällen durch die Annahme einer zäheren Plasmahaut und Alveolarwandsubstanz des Protoplasmas genügend erklären. Das oberflächliche Abfliessen wäre dann teilweise auf die einzelnen Alveolen beschränkt. Rhumbler jedoch erklärt dieses Ausbleiben der oberflächlichen Strömungen auf Grund von Berthold's Theorie dahin, dass die Fortbewegung der Amöbe durch einseitiges Anhaften an der Unterlage hervorgerufen wird.

Zweifellos hat die Plasmahaut einen Einfluss auf den Mechanismus der Bewegung. Nach Quincke ist dieselbe ein Fett-, nach Overton ein lipoides Häutchen, doch ist ihre wirkliche Zusammensetzung noch nicht sichergestellt. Sie ist in Wasser unlöslich, jedoch für Wasser und andere, namentlich lipoidlösliche Substanzen durchlässig. Das Cytoplasma lebender Zellen oder Eier enthält Lipoideiweissverbindungen, die in Wasser nicht löslich sind, sich aber bei Berührung mit Wasser unter Freiwerden von Lipoiden zu zersetzen scheinen [McClendon[3]), 1910]. Nach dem Gibbs'schen Prinzip sammeln sich diese Lipoide an der Oberfläche und bilden eine Schicht, die eine weitere Zersetzung der Lipoideiweisverbindungen verhindert. Dieses lipoide Häutchen ist unter gewöhnlichen Umständen von ultramikroskopischer Dicke; wird hingegen dem Wasser etwas Alkohol zugefügt, so kann unter Umständen ein verhältnismässig dickes Häutchen auf der Oberfläche (z. B. von Hühnereidotter) geformt werden.

Dass die Plasmahaut der Amöbe für Anionen weniger durchlässig zu sein scheint als für Kationen, ist eine Vermutung, die durch zwei Tatsachen nahegelegt wird:

---

1) E. Montgomery, Elementary Functions and Primitive Organization of Protoplasm. St. Thos. Hos. Repts., London 1878, n. s. IX. 1878. — Zur Lehre von der Muskelkontraktion. II. Die amöboide Bewegung. Pflüger's Arch. Bd. 25 S. 499.

2) H. S. Jennings, Contributions to the Study of the Behavior of Lower Organisms. VI. Publ. Carnegie Instit. Washington Nr. 16 p. 129. 1904.

3) J. F. McClendon, Dynamics of Cell Division. II. Americ. Journ. of Physiol. vol. 27 p. 240. 1910.

1. Wird durch einen Wassertropfen, in dem eine Amöbe schwebt, ein schwacher elektrischer Strom geleitet, so wandert das Tier passiv gegen die Anode.

2. Wird ein starker elektrischer Strom durch das Wasser geleitet, dann beginnt der Zerfall der Amöbe am anodalen Pol.

Der zerstörende Einfluss eines elektrischen Stromes auf lebendes Gewebe ist vielleicht die Folge einer Anhäufung von Ionen, die in ihrer freien Bewegung durch gewisse Gewebsstrukturen behindert werden. Da nun an der Amöbe die Zerstörungserscheinungen zuerst an der Oberfläche auftreten, so dürfte die Plasmahaut für eine Behinderung der Ionenwanderung verantwortlich zu machen sein; die Ionen würden sich, da sie die Plasmahaut nicht leicht passieren können, hinter derselben anhäufen, und diese Anhäufung würde die sichtbaren Erscheinungen des Zerfalles hervorrufen.

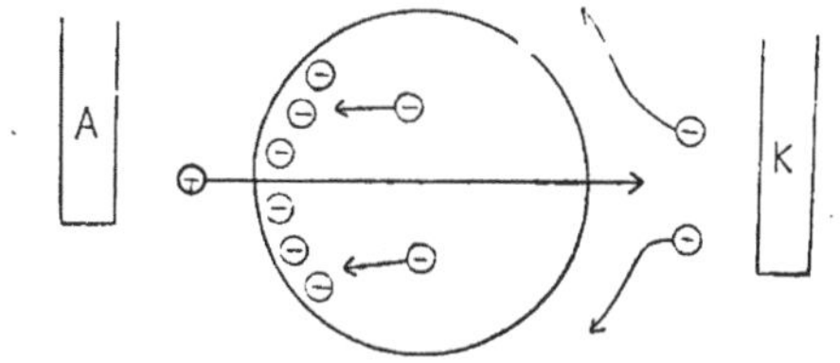

Fig. 1. Schematische Darstellung des Einflusses eines elektrischen Stromes auf eine Amöbe. *A* = Anode. *K* = Kathode. Der grosse Kreis stellt die Plasmahaut dar, die kleinen Kreise Ionen mit ihrer entsprechenden Ladung. Die Pfeile deuten die Bewegungsrichtung der Ionen an.

Der Zerfall der oberflächlichen Schichten tritt nicht an beiden Polen der Amöbe gleichzeitig auf, sondern zunächst nur an der Anode, so dass es den Anschein hat, als ob die Plasmahaut der Durchwanderung der Anionen einen grösseren Widerstand entgegensetzt als der der Kationen. Die Fig. 1 stellt diese Annahme schematisch dar. Der grosse Kreis repräsentiert die Plasmahaut, die kleineren die Ionen, deren Ladung durch + oder — bezeichnet ist. Die Bewegungsrichtung der Ionen ist durch Pfeile angedeutet. Die positiven Ionen scheinen die Amöbe ungehindert zu passieren, während die negativen ausserhalb des Zelleibes denselben leicht umgehen können. Jene negativen Ionen jedoch, die im Protoplasma eingesperrt sind, scheinen die Oberhaut nicht passieren zu können, stauen sich daher hinter derselben und rufen dann die gewissen Auflösungserscheinungen hervor.

Wird jedoch ein Strom, der zu schwach ist, um das Plasma zu zerstören, durch das Wasser geleitet, dann wandert die Amöbe passiv gegen die Anode, eine Beobachtung, die Hirschfeld's[1] (1909) Annahme einer positiven Ladung derselben nicht bestätigt. Diese passive Wanderung der Amöbe könnte in folgender Weise erklärt werden. Im Innern derselben wird ununterbrochen ein Elektrolyt erzeugt, dessen Anionen die Plasmahaut nicht passieren können. Diese geben der Amöbe eine negative Ladung, so dass das Tier, wenn ein Strom durch das Wasser geleitet wird, gegen die Anode gezogen wird.

Da unter allen Elektrolyten, die unter solchen Umständen erzeugt werden könnten, die Kohlensäure wohl der ausgiebigste ist, dürfte deren Gegenwart den oben gestellten Forderungen genügen.

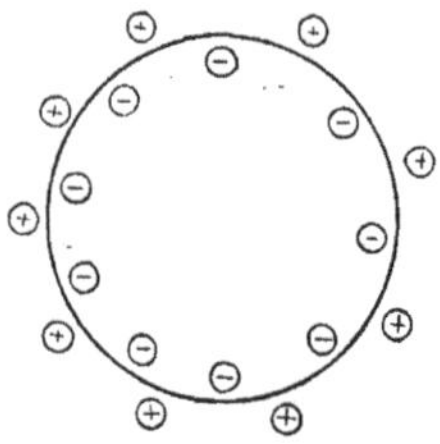

Fig. 2. Schematische Darstellung des elektrischen Polarisationszustandes der Amöbenoberfläche als Resultat einer relativen Undurchlässigkeit der Plasmahaut für die Anionen der Kohlensäure. Zeichen wie in Fig. 1.

Es ist allerdings wahr, dass das elektrolytische Dissoziationsvermögen der Kohlensäure sehr gering ist. Dies würde dann eben eine starke Konzentration derselben nötig machen, die möglicherweise in der Zelle existiert. Wir können nun annehmen, dass die H^+-Ionen aus der Amöbe herauswandern können, während die HCO_3^-- und $CO_3^{=}$-Ionen innerhalb derselben zurückgehalten werden und ihr die nötige negative Ladung geben. Diese Scheidung der Ionen würde noch durch die grössere Schnelligkeit, mit der die H^+-Ionen durch das Plasma wandern, gefördert. Die Anionen im Inneren und jene Kationen, die durch sie auf der Oberfläche zurückgehalten werden (Fig. 2), bewirken eine Polarisation der letzteren, ähnlich dem Lippmannschen Phänomen im Kapillarelektrometer. Infolge dieser Polarisation muss die Oberflächenspannung eine geringe sein, wie es tatsächlich

1) L. Hirschfeld, Ein Versuch, einige Lebenserscheinungen der Amöben physikalisch-chemisch zu erklären. Zeitschr. f. allgem. Physiol. Bd. 9 S. 529. 1909.

die Beobachtung lehrt. Obzwar die Oberfläche einer Amöbe möglicherweise aus Lipoiden besteht, besitzt das Tier eine viel geringere Oberflächenspannung als ein Öltropfen im Wasser.

Unter solchen Umständen würden amöboide Bewegungen durch örtliche Veränderungen der Oberflächenspannung erzeugt, die ihrerseits eine Folge der wechselnden Oberflächenpolarisation sind. Während die $CO_3^{=}$-Ionen vielleicht nicht leicht passieren können und die Kationen unter ihrem Einflusse zum Teil zurückgehalten werden, könnten andererseits die nicht dissoziierten Moleküle als CO_2 auswandern, da CO_2 in Lipoiden löslich ist. Der Polarisationszustand könnte dann nur so lange aufrechterhalten werden, als CO_2 innerhalb der Amöbe in gleichem Maasse erzeugt wird, als sie auswandert. Würde an irgendeiner Stelle die Kohlensäurebildung steigen oder fallen, dann würde die gleichzeitige Polarisationsveränderung des nächstliegenden Teiles der Oberfläche spontane Bewegung erzeugen. Die Kohlensäurebildung wird vielleicht zeitweise durch äussere Bedingungen beeinflusst, z. B. durch Substanzen, die von anderen Organismen in das Wasser abgegeben werden. Eine Erhöhung der CO_2-Bildung würde dann positiven Tropismus hervorbringen.

Andererseits würde eine Erhöhung der Durchlässigkeit der Plasmahaut eine Herabsetzung der Polarisation derselben zur Folge haben. In einer früheren Arbeit (1910) habe ich eine Reihe von Agentien aufgezählt, die die Durchlässigkeit der Plasmahaut für Anionen erhöhen. Dieselben Agentien bewirken auch in der Amöbe eine Erhöhung der Oberflächenspannung, die sich in einer grösseren Abrundung des Körpers äussert. Wird die Amöbe nur auf einer Seite von diesen Agentien beeinflusst, dann zeigt sie negativen Tropismus. Negativer Tropismus kann dadurch erklärt werden, dass die ihn bewirkenden mechanischen, thermischen, chemischen u. dgl. Einflüsse die Plasmahaut an der der Einwirkung nächstgelegenen Stelle auflockern; der daraus resultierende Verlust an Polarisation unter gleichzeitiger Erhöhung der Oberflächenspannung veranlasst die Amöbe, sich von dem schädlichen Einfluss zurückzuziehen.

Der Teil der Oberfläche, an dem die Plasmahaut verändert worden ist, bleibt aber nicht depolarisiert, da er seine frühere Undurchlässigkeit zurückerhält, sobald sich die Amöbe dem schädigenden Einfluss entzogen hat.

Die Polarisation ist nicht der einzige Faktor, der die Oberflächenspannung zu verändern vermag. Es ist z. B. eine bekannte Tatsache, dass Seifen die Oberflächenspannung an Lipoiden vermindern. Wenn nun die Plasmahaut aus Lipoiden besteht, sollte man [mit Bütschli, J. Loeb, Robertson, Michaelis u. a.] vermuten, dass Seifen eine Herabminderung der Oberflächenspannung an der lebenden Zelle bewirken; doch habe ich das gerade Gegenteil beobachtet. Eine Kapillarröhre wurde mit fester Seife oder Seifenlösungen von verschiedener Konzentration gefüllt und mittels der „mechanischen Hand" [McClendon[1]), 1909] so weit in Wasser eingetaucht, bis ihre feine Öffnung ganz in die Nähe einer grossen Amoeba proteus kam. Das Tier zeigte immer deutlichen negativen Chemotropismus, in dem es rasch von der Seife, die aus der Kapillarröhre heraussickerte, wegwanderte. Manchmal war die Bewegung nicht rasch genug, um das Leben des Tieres zu retten. Leider konnte der genaue Zeitpunkt, an dem die ersten Seifenmoleküle die Amöbe erreichten, nicht bestimmt werden. So liess sich daher auch nicht feststellen, ob ein kurzdauernder Vorstoss des Tieres gegen die Seifenlösung zu, wie er manchmal beobachtet wurde, als eine „Reiz-" oder spontane Bewegung aufzufassen sei. Auf jeden Fall können wir schliessen, dass der hauptsächlichste Einfluss, den die Seife auf die Amöbe ausübt, geeignet ist, die Durchlässigkeit der Oberfläche für Anionen zu erhöhen, da das Tier negativen Tropismus zeigt.

In einer kürzlich erschienenen interessanten Arbeit über die Rolle der verschiedenen Kolloidalzustände der Oberfläche der Amöben bei Nahrungsaufnahme kam Rhumbler[2]) (1910) zu dem Schlusse, dass die Annahme einer festen, elastischen und relativ dicken Oberflächenschicht notwendig ist, um jene Form der Nahrungsaufnahme zu erklären, dass er „Circumvallation" nennt. Er vermutet, dass „Circumvallation" durch eine örtliche Auflösung der halbfesten Ober-

---

1) J. F. McClendon, Protozoan Studies I. Reactions of Amoeba Proteus to Minutely Localized Stimuli. Journ. of exper. Zool. vol. 6 p. 265. — Autoreferat über vorstehende Abhandlung in Arch. f. Entwicklungsmechanik Bd. 37 S. 323. 1909. — The Reaction of Amoeba to Stimuli of Small Area. Americ. Journ. Physiol. vol. 21. 1908. Proc. Americ. Physiol. Soc. 1907 p. 13.

2) L. Rhumbler, Die verschiedenartigen Nahrungsaufnahmen bei Amöben als Folge verschiedener Kolloidalzustände ihrer Oberflächen. Arch. f. Entwicklungsmechanik Bd. 30 S. 194. 1910.

flächenschicht durch einen von der Beute ausgeübten „Reiz" eingeleitet wird [1]).

Es gibt jedoch eine Art der Nahrungsaufnahme, die der „Circumvallation" zwar verwandt, aber in folgender Weise durch Veränderungen der Oberflächenspannung erklärlich ist: wenn z. B. eine

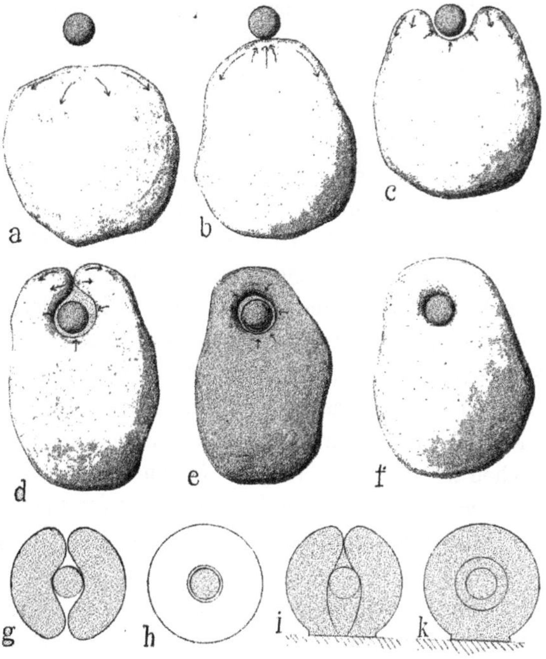

Fig. 3. Schematische Darstellung der Nahrungsaufnahme einer Amöbe. Die Pfeile deuten die Richtung an, in der eine Erhöhung der Oberflächenspannung erfolgt. Der kleine Kreis stellt eine Algenzelle dar. *g—k* = Querschnitte durch den Plasmaring der Figuren *e—f*.

Amoeba proteus sich in der Nähe einer Algenzelle befindet (Fig. 3*a*), dann bewirken auf Grund meiner früheren Voraussetzungen

---

1) Es ist zweifellos, dass eine feste Oberfläche die Nahrungsaufnahme der Amöbe nicht verhindert. Schaudinn (1899) (Generationswechsel von Trichosphaerium Sieboldi. Anhang d. Abhandl. Berliner Akad. d. Wissensch. 1899 S. 1) beobachtete die Aufnahme von fester Nahrung bei Trichosphaerium Sieboldi, dessen Körper von einer Gallerthülle umgeben ist, in die zahlreiche radiär gestellte Stäbchen, $MgCO_3$, eingebettet sind. Eine zum mindest halbstarre Oberflächenschicht ist, wie Rhumbler beweist, sogar notwendig, um die Nahrungsaufnahme durch „Invagination" zu ermöglichen.

Stoffe, die aus der Zelle ausgeschieden werden, eine Steigerung der CO_2-Erzeugung im nächstliegenden Teil des Amöbenleibes; denn die Amöbe breitet sich gegen die Alge aus. Manchmal rollt sie die letztere sogar etwas vor sich her (Fig. 3*b*). Die Berührung mit der Alge oder die stärkere Konzentration der reizenden Substanzen verursacht eine lokale Erhöhung der Durchlässigkeit und der Spannung eines Teiles der Plasmahaut, welcher das Feld verminderter Spannung durchschneidet (Fig. 3*c*). Die seitlichen Partien schieben sich nun weiter nach vorne, bis sie die Beute ganz umgeben (Fig. 3*d*) und sich dann vor derselben vereinigen (Fig. 3*e*). Die Beute ist dann von einem Protoplasmaring eingeschlossen. Die Oberflächenspannung bestrebt nun eine Verkleinerung des letzteren. Manchmal kann der dabei von allen Seiten einwirkende Druck ein Herauspressen der Beute nach oben zur Folge haben, so dass die Nahrungsaufnahme misslingt. Unter günstigen Bedingungen hingegen schliesst sich der Ring über und unter dem Fremdkörper, wie es im Querschnitt in Fig. 3*g* und *h* dargestellt ist; gleichzeitig wird auch etwas Wasser mit der Beute eingeschlossen, und zwar dürfte, wenn die Amöbe an der Unterlage anhaftet, etwas mehr Wasser eingeschlossen werden.

Diese Methode der Nahrungsaufnahme nimmt eine Zwischenstellung zwischen Rhumbler's Klassen „Circumfluenz" und „Circumvallation" ein, indem die Amöbe in Berührung mit der Beute kommt, zugleich aber Wasser mit der Nahrung aufnimmt.

Versuche, die ich an Seeigeleiern unternommen habe, unterstützen meine Ausführungen. Dass Eier amöboide Bewegungen ausführen können, ist eine bekannte Tatsache. Prowazek[1]) (1903) beobachtete an Seeigeleiern, die in die Körperflüssigkeit einer Annelide gelegt wurden, aktive Bewegungen; ein Ei nahm eine „Eläocyte" in sich auf.

Wenn ein allmählich stärker werdender elektrischer Strom durch eine isotonische Zuckerlösung geleitet wird, die Seeigeleier enthält, so muss man eine Erhöhung der Polarisation in dem der Anode nächstgelegenen Teile der Plasmahaut erwarten, da die Anionen, die hinter der Membran zurückgehalten werden, eine Anhäufung der Kationen an der Aussenseite hervorrufen (Fig. 4*a*). Die Anionen

1) Prowazek, Studien zur Biologie der Zelle. Zeitschr. f. allg. Physiol. Bd. 2 S. 385. 1903.

des Wassers können das Ei umgehen, während die Kationen es durchwandern. Die Erhöhung der Polarisation muss eine Herabsetzung der Oberflächenspannung und damit eine Ausdehnung der Eioberfläche gegen die Anode zu veranlassen. Zahlreiche Beobachtungen bestätigten diese Vermutung, indem das Ei unter den gestellten Bedingungen jedesmal ein oder mehrere Pseudopodien gegen die Anode ausstreckte. In einigen Fällen wurde auch ein Abfliessen der oberflächlichen Schichten bemerkt.

Das beschriebene Phänomen war von sehr kurzer Dauer, da das vorgestreckte Cytoplasma bald zu desintegrieren begann und osmo-

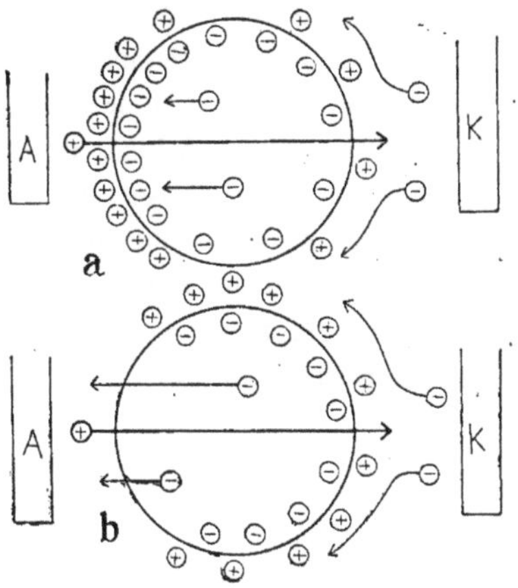

Fig. 4. Schematischer Versuch, den Tropismus der Amöbe gegen die Kathode zu erklären. Zeichen wie in Fig. 1. (In der Amöbe äussert sich Galvanotropismus an der Kathode, Kataphorese an der Anode.)

tische Erscheinungen die Bewegungen auf Grund der Spannungsdifferenzen verhüllten. Der Zersetzungsprozess wurde wahrscheinlich durch die Anhäufung der Anionen eingeleitet.

Während in einem schwachen Strom das Seeigelei Pseudopodien gegen die Anode ausstreckt, zeigt die Amöbe Tropismus nach der Kathode. Ein starker Strom aber verursacht in beiden Formen eine Zersetzung am anodalen Pol. Diese Tatsache scheint anzudeuten, dass die Verminderung der Oberflächenspannung und der Zerfall des Plasmas im Seeigelei durch ähnliche, in der Amöbe aber durch entgegengesetzte Faktoren hervorgerufen werden. Doch kann dieser anscheinende Widerspruch dadurch erklärt werden, dass auch in der Amöbe die Polarisation des anodalen Poles steigt; bevor jedoch genügend Zeit verstreicht, um eine Bewegung auszuführen, wird die Plasmahaut zerstört, so dass der Polarisationsgrad dieser Stelle unter das Normale der übrigen Oberfläche sinkt (Fig. 4*b*). Der Schlusseffekt des Stromes wäre somit verminderte Polarisation der anodalen Gegend, und die Amöbe wandert gegen die Kathode.

Die erhöhte Durchlässigkeit eines Teiles der Plasmahaut würde die Polarisation der gesamten Oberfläche etwas herabsetzen, da die Anionen jenen weniger dichten Teil durchwandern könnten. Doch wird in der Amöbe ununterbrochen CO_2 erzeugt und wahrscheinlich um so energischer, je schneller sie weggeschafft wird, wie es ja bei allen Endprodukten chemischer Vorgänge der Fall ist. Aus diesem Grunde dürfte die Polarisation des nicht desintegrierenden Teiles der Oberfläche sehr wenig, wenn überhaupt, herabgesetzt werden.

Der Potentialabfall zwischen den beiden Seiten jenes Teiles der Plasmahaut, der der Kathode zunächst liegt, ist entgegengesetzt dem Potentialabfall, der an jener Stelle von den Elektroden erzeugt würde (Fig. 4*b*). Man muss daher annehmen, dass die normale Polarisation der Plasmahaut einem Potentialabfall entspricht, der stärker ist als der des umgebenden Wassers, da sonst der erstere durch den letzteren unterdrückt würde.

PUBLICATIONS

OF

Cornell University

MEDICAL COLLEGE

STUDIES

FROM THE

Department of Anatomy

VOLUME III

1912

NEW YORK CITY

CONTENTS

Being reprints of studies published in 1912.

---

9. THE RELATION OF MUSCLE FIBRILLÆ TO TENDON FIBRILLÆ IN VOLUNTARY STRIPED MUSCLES OF VERTEBRATES. By **W. M. Baldwin.**
Morphologisches Jahrbuch, Vol XLV, 249-266 and 1 plate.

10 MUSCLE FIBRES AND MUSCLE CELLS OF THE ADULT WHITE MOUSE HEART. By **W. M. Baldwin.**
Anatomischer Anzeiger, Vol. 42, 177-181 and 2 figures.

11 DIE ENTWICKLUNG DER FASERN DER ZONULA ZINNII IM AUGE DER WEISSEN MAUS NACH DER GEBURT.
By **W. M. Baldwin.**
Archiv f. mikroskop. Anatomie, Vol LXXX, 274-305 and 2 plates.

12. RHYTHMICAL ACTIVITY OF ISOLATED HEART MUSCLE CELLS IN VITRO. By **Montrose T. Burrows.**
Science, N. S., Vol XXXVI, 90-92.

13. RHYTHMISCHE KONTRAKTIONEN DER ISOLIERTEN HERZMUSKELZELLE AUSSERHALB DES ORGANISMUS.
By **Montrose T. Burrows.**
Münchener medizinischen Wochenschrift, No. 27, 1473-1475 and 2 figures.

14. THE EFFECTS OF ALKALOIDS ON THE DEVELOPMENT OF FISH (FUNDULUS) EGGS. By **J. F. McClendon.**
Am. Jour. of Physiology, Vol. XXXI, 131-140 and 9 figures.

15. ECHINOCHROMS, A RED SUBSTANCE IN SEA UDCHINS.
By **J. F. McClendon.**
Jour. of Biological Chemistry, Vol. XI, 435-441.

16. PREPARATION OF MATERIAL FOR HISTOLOGY AND EMBRYOLOGY, WITH AN APPENDIX ON THE ARTERIES AND VEINS IN A 30 MM. PIG EMBRYO.
By **J. F. McClendon.**
Anatomical Record. Vol. VII, 51-61 and 3 figures

AN EXPERIMENTAL STUDY OF RACIAL DEGENERATION IN MAMMALS TREATED WITH ALCOHOL *

CHARLES R. STOCKARD, PH.D.
NEW YORK

It is recognized, by most observers who have studied the subject, that alcohol may play an important rôle in the causation of monstrosities and of structural defects predisposing to later disease. This view is based largely on observations on defective human beings, and the probability of its truth is sufficiently established to warrant further careful experimental analysis.

The quality of an offspring depends on two factors, the perfection of the germ cells from which it arises and the nature of the environment in which it develops. Diseased and weakened germ-cells give rise to a defective individual under all circumstances, while perfect germ-cells produce a perfect offspring *only* when the embryo develops in a normal or favorable environment. These facts may be readily demonstrated in lower vertebrates in which the development of the egg is outside the mother's body. The egg or spermatozoon in such cases may easily be chemically modified or injured before fertilization, and the embryo itself may be affected in various ways during its development by subjecting it to unusual surroundings, either physical or chemical. In other animals, such as mammals, in which the embryo develops internally, the proposition likewise holds true. In these animals, however, the problem is more difficult to completely analyze. The reactions of the parental body, the secondary conditions induced by the experimental treatment and other sources of error should be fully considered in determining whether an effect shown by the offspring is directly due to the applied stimulus or to secondary conditions. In the lower vertebrates it has been shown that given doses of certain substances induce definite developmental defects. The defects are directly due to the treatment. Is it possible by the addition of certain chemicals to the mammalian body to obtain similar definite changes in either the germ-cells or the developing embryo?

In the present paper I shall endeavor to show that alcohol does act directly on the germ-cells of mammals to a sufficient degree to render them incapable of producing normal offspring, and further, that similar treatment administered to the pregnant female may likewise act directly on the developing embryo so as to modify its resulting structure.

---

*From the Anatomical Laboratory, Cornell University Medical College.
*Manuscript submitted for publication May 10, 1912.

consider in a somewhat critical manner the literature pertaining to the actions of alcohol and other substances on the reproductive glands and developing embryos of man and lower animals.

DISCUSSION OF LITERATURE

There is an abundant literature relating to the effects of alcohol on the offspring, though little of it is scientifically reliable. I have attempted to select those cases which seem most trustworthy. Since we are more interested in the general problem of the effects of parental poisoning on the germ-cells and the embryo in mammals I have also collected the works relating to injurious substances other than alcohol. The observations and statistics on human beings in various countries are reliable only in so far as they may be substantiated and borne out by controlled experiments on lower animals. Yet in the light of animal experiments many of these human records become of surprising interest, although few if any of them may be accepted entirely as they stand.[1]

EFFECT ON THE MALE GERM-CELLS

It is a well known and universally accepted fact that alcohol does cause changes and degeneration in many of the body tissues of man. The question naturally presents itself, How, then, can the reproductive tissues escape? Nicloux and Renault have shown that alcohol has a decided affinity for the reproductive glands. In the testicular tissues and the seminal fluid an amount of alcohol is soon present which almost equals that in the blood of a person having recently taken alcohol. The proportion of alcohol in the testis as compared with that in the blood was as 2 to 3, and in the ovary of female mammals as 3 to 5. The genital glands show as great an affinity for this substance as does the nervous system. From these observations it must necessarily follow that alcohol may act on the ripe spermatozoon shortly before the time when it fertilizes the egg, and since an affected spermatozoon gives rise to a defective individual we have a probable explanation for many of the recorded defects attributed to drunkenness at the time of conception. A male, even for the first time, in a state of acute intoxication, is probably more apt to beget an abnormal offspring by fertilizing an egg at this particular period than is a non-intoxicated male although a frequent user of alcohol. The experimental data on the sensitiveness of the spermatozoon and the observations on the presence of alcohol in the seminal fluid warrant this statement.

1. Most of the literature is devoted to considerations of disease and insanity statistics and the family records of degenerates. The data are often collected in a careless fashion so that the *actual observations* are not always scientifically correct though the records are carefully and fully computed.

Lippich claims to have observed ninety-seven children resulting from such conceptions. Only fourteen of these were without noticeable defects. Eighty-three of them showed various abnormal conditions, twenty-eight were scrofulous,[2] three had "weak lungs," three showed different atrophic conditions, one watery brain, four were feeble-minded, etc. Others have made similar observations. Sullivan reported seven cases of drunkenness during conception which are fairly authentic. Six of the offspring died in convulsions after a few months, and the seventh was still-born.

Thus one finds proof by Nicloux and Renault that alcohol does reach the reproductive glands and, therefore, may affect the egg or sperm-cell, and observations seem to indicate that this effect expresses itself in the condition of the resulting offspring. Experiments on lower animals support the probability. When the perfectly normal spermatozoa of frogs are treated with *x*-ray or radium, Bardeen and O. Hertwig have shown that normal eggs fertilized by such spermatozoa all develop abnormally. Todde found that the offspring from alcoholized roosters were not quite normal and that the roosters did not succeed as well as normally in fertilizing eggs.

Combemale, 1888, was the first to experiment on the influence of alcohol on the mammalian offspring. He treated a dog for eight months with absinthe (11 gr. per day per kilo of animal weight) and paired this alcoholized dog with a normal bitch. Twelve young resulted; two were born dead, three died within fourteen days and the others died between thirty-two and sixty-seven days of intestinal catarrh, tuberculosis, etc. In a second experiment both parents were mated while normal, then the female was made drunk for twenty-three days (2.75 to 5 gr. absinthe of 72 per cent. per day per kilo). Of six young three were still-born, two had normal bodies though of weak intelligence, while one moved slowly and was very stupid. The last individual, a female. was later paired with a normal intelligent non-alcoholic dog. She gave only three young; one was deformed, club-footed with abnormal teeth. the second had a patent ductus arteriosis and died after fourteen days. while the third was poorly muscled in the hinder parts and died a few hours after birth. Thus the effects in the second generation are as pronounced as in the first although neither parent had themselves received any alcohol. The only criticism against Combemale's experiments is that an insufficient number of animals was used. Dogs often give defective pups and these may have been from poor stock, though such an interpretation is really not probable, and his results are supported by subsequent workers.

---

2. Imbault, F.: Contribution a l'étude de la fréquence de la tuberculose chez les alcooliques. Thèse de Paris, 1901. Imbault found that tuberculosis was about as common among the children of alcoholic parentage as among those of tuberculous parents.

Hodge, in 1897, obtained similar results. From one pair of alcoholic dogs he obtained twenty-three pups, eight were deformed and nine were born dead, while only four lived. In a control set forty-one individuals lived, four were deformed and there were no still-births.

Laitinen treated rabbits and guinea-pigs with various doses of alcohol and studied chiefly the changes in body conditions as to resistance against disease toxins, etc. He has also recorded observations on the offspring produced by these animals during the experiment. He is apparently more interested in the problem of the misuse of alcohol than in the scientific study of the influence of injurious substances on the offspring and in his enthusiasm to prove the point with extremely small doses of alcohol he fails to fully consider both sides of his own tables.

He used daily doses of alcohol as small as 0.1 c.c. per kilo of animal weight. This would amount to a small glass (200 c.c.) of beer per day for an adult man. His tables on careful study fail to show that so little alcohol actually does injure the offspring of the treated animals.

With alcoholized rabbits Laitinen finds that only 38.71 per cent. of the young live, while 61.29 per cent. are still-born or die shortly after birth. In the control, however, only 45.83 per cent. lived, while 54.17 per cent., more than half, were still-born or died shortly after birth. The animals were kept all together in a general cage and the pregnant females were only separated shortly before the young reached term. This is scarcely an approved method in breeding experiments, and the fact that young rabbits are so delicate and are born in a rather poorly developed state makes their careful handling necessary. The fact that more than half of the control young die, 54.17 per cent., would indicate the danger of drawing conclusions from a death-rate only 7 per cent. higher among the offspring of the treated animals.

The case of the guinea-pigs is also indifferent, 78.26 per cent. of the control young lived, while 21.75 per cent., or a little more than one-fifth, of them died. The large majority of the young of treated parents also lived, 63.24 per cent., while 37.76 per cent. died. In both sets more of the young lived than died. Guinea-pigs are easily reared and are born in a well-developed condition. On the other hand, in both of the rabbit sets more of the young died than lived.

Results which I shall record below show that larger doses of alcohol do produce definite effects on the offspring. My experiments have been performed in a different manner and from another point of view. The primary object has been to regulate or control the type of development in mammals in a definite fashion as I had succeeded in doing with lower vertebrates. In these experiments it will be demonstrated that an alcoholized male guinea-pig almost invariably begets a defective offspring even when bred to a vigorous normal female.

Rösch was the first to study the reproductive glands of alcoholics, in 1837, and found degeneration of the testicles. Lancereaux described a parenchymatous degeneration of the seminal canals. Simmonds (1898) found azoospermia in 60 per cent. of cases of chronic alcoholism; 5 per cent. of these men were sterile. Kyrle reported three cases of total atrophy of the testicular parenchyma in which death had resulted from cirrhosis of the liver due to alcohol. Kyrle attributed the atrophy of the testicle to the cirrhosis of the liver and not to chronic alcoholism.

Bertholet (1909) made an extensive examination of the influence of alcohol on the histological structure of the germ glands, more particularly on the testicles of chronic alcoholics. He found testicular atrophy in alcoholics with no cirrhosis of the liver. Bertholet observed partial atrophy of the testicles in the majority of seventy-five chronic alcoholics. These men died between the ages of 24 and 57 years, the greatest mortality being between 30 and 50 years. In thirty-seven cases, excluding syphilitics, a microscopical examination showed a more or less diffuse atrophy of the testicular parenchyma and a sclerosis of the interstitial connective tissue. The testicles were small and hard. The canals were greatly reduced in size and their lumina obliterated. Spermatogonia were atrophic. It was generally impossible to differentiate spermatocytes or spermatids. There were no dividing cells and no spermatozoa. The thick basal membrane of the canals was formed of connective tissue lamellæ with concentrated spindle cells. These conditions with slight variations were found in twenty-four cases. Such atrophic structures were already present in a drinker only 29 years old. In four cases of cirrhosis of the liver the testicular atrophy had not progressed very far and spermatozoa were still present. In five cases the microscopical conditions were less marked.

While these appearances of the basal membrane may also be observed in non-alcoholics, the extreme conditions of atrophy of the testicles were only found in alcoholics. Observing the testicles of non-alcoholics that had died of various chronic illnesses such as tuberculosis, no atrophy of the testicles or thickening of the membrana propria was found. Two such old men of 70 and 91 years still possessed spermatozoa in the canals. Bertholet concludes that the atrophy he has observed cannot be due to old age, but is due to the hurtful effects of chronic alcoholism on the reproductive glands.

Bertholet has also reported an atrophy of the ovary and ova in female alcoholics. Weichselbaum has confirmed the observations of Bertholet at his institute in Vienna.

Bertholet's observations are most important and his drawings bear out his statements. On the other hand, it is certain that the chronic alcoholic is not so often rendered sterile as his study might lead one to believe. Judging from the statistics it is not rare to find alcoholics with large families. My experiments on animals may not be of sufficient duration at the present time, yet I have male guinea-pigs that have been almost intoxicated on alcohol once per day for six days a week extending over a period of nineteen months. These animals are still splendid breeders. Nineteen months of a guinea-pig's existence is proportionally equal to a good fraction of a human life. Many of these animals have been killed and their testicles examined microscopically and found to be normal. In some cases where a male had failed to succeed in impregnating the female for several times, he was partially castrated, one testicle being taken out. In this case the testicle was found to be normal and the same male has since given offspring by other females. Ovaries have been examined in a similar way, and in no individual has the alcohol treatment caused a visible structural change in the reproductive glands. The actual physiological proof of the efficiency of the organs is shown by the ability of all animals to reproduce. The important point which I shall show in the following pages is that although there is no visible structural change in the germ-cells, nevertheless, they have been modified chemically to an extent sufficient to cause them to give rise to defective embryos or weakened individuals which die shortly after birth.

Schweighofer has recorded an interesting individual case. A normal woman married a normal man and had three sound children. The husband died and she married a drunkard and gave birth to three other children; one of these became a drunkard, one had infantilism, while the third was a social degenerate and drunkard. The first two of these children contracted tuberculosis, which had never before been in the family. The woman married a third time and by this sober husband she again produced sound children. This is an important human experiment. The female was first tested with a normal male and gave normal offspring; when mated with an alcoholic male the progeny were defective as a result of his poisoned condition. She was again tested with a normal male and found to be still capable of giving sound offspring. A number of such cases are on record.

Schweighofer states from a mass of observations that the offspring of drunkards, themselves of good sound families, show much degeneracy and defective conditions.

Other substances than alcohol seem to act directly on the germ-cells of mammals. Constantine Paul long ago pointed out that the children of people working in lead were often defective. He made the interesting observation that when the father alone was employed in such work his children were affected by it.

All of the above experiments and observations refer more particularly to the action of injurious substances on the germ-cells of the male parent.

This is the crucial proof of an effect on the germ cells. The case of the female is complex, since the substance may produce a germinal defect by acting on the egg, or it may also directly affect the developing embryo and thus act as an environmental influence on development.

THE FEMALE GERM-CELLS AND THE DEVELOPING EMBRYO

Herbst's classical lithium experiments show the influence of salt solutions on developing eggs. The experiments of J. Loeb on fish embryos, those of Morgan on the frog and my experiments on fish all show the marked influence of inorganic salts and organic compounds on the development of the embryo. I showed that alcohol caused all known

TABLE 1.—EFFECTS OF WORKING IN LEAD

	No. of Cases	No. of Pregnancies	Abortions, Premature Labor, Still-Births	Living Births	Remarks
Females showing lead poisoning symptoms..	4	15	13	2	One of the living children died in 24 hours
Females working in type foundry; previously had normal pregnancies	5	36	29	7	Four died in first year
Female in type foundry; five pregnancies	1	5	5	0	....
Females working intermittently; while there	3	3	3	0	After being away for some time had healthy children
Females with blue line on gums, only sign of poisoning	6	29	21	8	....
Male alone exposed......	?	32	12	20	8 died first year, 4 second year, 5 third year
Total................	..	120	83	37	22 died under three years

gross abnormalities of the brain in fish embryos and also gave all possible abnormal conditions of the eyes. Other substances such as ether, chloroform, chlorbutanol (chloretone), etc., also had a peculiar affinity for the developing central nervous system. These substances also act physiologically on the central nervous system of the adult.

Constantine Paul not only showed the injurious effects of lead on the paternal germ-cells, but also recorded instructive data regarding the offspring of women working in lead. More recent observers have pointed out the frequency of idiocy and other defects among the children of lead workers. Adami has tabulated the findings of Constantine Paul as shown in Table 1.

Forel states that acute alcoholic intoxication affects not only the brain, but, as Nicloux has shown, the alcohol passes quickly to the cells of the testicle or ovary and Bertholet's observations confirm this. A conception which takes place while the cells are in this poisoned state often results in a feeble-minded or degenerate child. The facts furnished by experiments on the eggs and spermatozoa of lower animals lend the strongest support to this idea and there is no experimental evidence that can be interpreted as opposed to Forel's statement.

Chronic alcoholics who consume daily certain amounts of alcohol slowly injure their germ cells. By intensive use of alcohol these cells may actually be killed or caused to atrophy. This, however, is the extreme case and before reaching a state of atrophy the cells pass through various grades of defectiveness. The stages may show no anatomic changes, but their physiologic state is indicated by the defective individuals to which they give rise in development.

Bezzola found that in Switzerland, in the years 1880 to 1890, there were 8,190 idiots. Most of the idiots were born in wine districts, and the season for the maximum birth of such children was nine months after the great national feasts, indicating, possibly, that idiots were conceived during the period of heaviest drinking. Schweighofer found the same relationship between the season for the greatest number of still-births and the feast seasons in Austria.

Martin studied the family histories of eighty-three epileptic girls in the Salpetrière (Paris). Sixty had alcoholic parents, while of the other twenty-three alcoholism was doubtful or absent in their parents. The sixty girls from alcoholic parentage had 244 sisters, of them 132, or 54.1 per cent., were dead; forty-eight, or 19.7 per cent., had had spasms during childhood.

Studying the direct ancestry of 370 insane people, Jenny Koller (in 1895) found that there were twice as many drinkers as were found in the direct ancestry of 370 sound people selected at random. Others have recorded similar observations.

Karl Pearson and Miss Elderson studied statistically 3,000 school children in England. They concluded that the children of alcoholics were often heavier than those of sober parents, they were also less diseased, had little epilepsy and tuberculosis and are actually cleverer in school. They found, however, a greater mortality among the children of alcoholics, especially of female drinkers, and concluded that only the stronger children lived, and therefore, their quality was good.

These studies have been widely criticised, and are probably not based on very thorough biological observations. They consider, in the first place, only school children. It is not known whether the parents were drunkards at the time of, or previous to the conception. The degenerate offspring of alcoholics could not enter school. The results would doubtless have been quite different if the inmates of an institution for defective children had been studied. The great body of evidence from anatomic studies of the reproductive glands of alcoholics, the animal experiments and disease records are all opposed to Pearson's conclusions.

The most valuable study that I have been able to find on the influence of alcohol on the human offspring is that of Sullivan in 1899.

Sullivan emphasizes the point that while much effort has been made to record alcoholism in the ancestry of degenerates, the important study must be made on degeneracy in the descendants of well-observed alcoholics. He studied the alcoholics among the female population of the Liverpool prison and as far as possible chose cases of alcoholism that were unaccompanied by disease or other degenerate factors.

Localization of alcoholic lesions in the body are not well worked out, yet it is unquestionable that in the criminal, as in insane alcoholics, the nervous manifestations of the intoxication occur with notable frequency, while non-nervous disorders are rare or secondary. Of these alcoholic females, thirty-one had had one or more attacks of alcoholic delirium, twenty-four had occasional hallucinations, suicidal impulses; disorders of cutaneous sensibility, and cramp in the extremities was noted in a considerable number of cases. In these patients tissues other than the nervous, so far as examination of the patients themselves could show, were comparatively immune to the poison of alcohol, and this was also true of their alcoholic relatives.

There were 100 women in the series Sullivan observed, and twenty of these gave details of female relatives of drunken habits who had children. To these 120 females were born 600 children, of whom 265, or 44.2 per cent., lived over two years; 335, or 55.8 per cent., died under 2 years or were still-born. Twenty-one of the women observed gave records of sober relatives, sisters or daughters married to sober men. The twenty-one drunken females had 125 children, sixty-nine, or 55.2 per cent., died under 2 years; the twenty-eight sober females had 138 children, and thirty-three, or 23.9 per cent., of them died under 2 years. The death-rate of children from the drunken mothers was nearly two and one-half times greater than that of the children of their near-blood relatives who were non-alcoholic. The alcoholics, however, are poor mothers and take little care of their children: this fact might possibly account for the entire difference, though such a deduction is extremely improbable.

The progressive births in the alcoholic family show interesting records. In eighty cases the number of children reached or exceeded three.

The tabulation shows an increasingly poor condition. The records of two individual cases may be mentioned by way of illustration:

idiot, sixth still-born, and finally an abortion.

Case 10: First child survived to adult life, second died of infection as child, two infants then died in convulsions in first few months, then a still-birth.

These records stand in interesting contrast with those known for syphilitic mothers in which each conception seems to be more and more nearly successful until a weak offspring is born, and finally such a mother may give birth to an apparently normal child. The syphilitic is gradually becoming less diseased and is overcoming the toxic condition as time goes on, while these alcoholic women are on the contrary becoming more and more saturated with the poison, and for this reason each succeeding birth is more decidedly defective.

TABLE 2.—SHOWING PERCENTAGE OF STILL-BORN AND CHILDREN WHO DIED IN AN ALCOHOLIC FAMILY

	Cases	Died or Still-Born, Per cent.	Still-Born, Per cent.
First born	80	33.7	6.2
Second born	80	50.0	11.2
Third born	80	52.6	7.6
Fourth and fifth born	111	65.7	10.8
Sixth to tenth born	93	72.0	17.2

The records were worse for women who had begun drinking some time previous to the first conception. In thirty-one cases they had been drinking for at least two years before the first pregnancy. Of 118 children born to these, seventy-four were still-born or died in infancy, giving 62.7 per cent. as compared with a death-rate of 54.1 per cent. for the others of the series.

In only thirty-nine of the cases were the women's parents sober people, yet the records of the offspring from these women were equally as bad as those from the sixty-one mothers who had alcoholic parents. This is a significant fact, since it indicates most strongly that the defective children are due to the direct effect of alcoholism and not to other degenerate conditions. Sullivan recorded seven known cases of conception during a state of drunkenness; six of the children died in convulsions in a few months, while the seventh was still-born.

Another observation by Sullivan which indicates that the alcohol as such is the cause of defectiveness was the fact that mothers imprisoned during pregnancy gave birth to a better child since the drinking was stopped.

Sixty per cent. of the children of all these mothers died in convulsions. This is a common manner of death for the offspring from the alcoholic mammals I have studied.

Kende found that of twenty-one families in which the father and mother both drank, ten were childless, while of the twenty-four children in the other eleven families, sixteen died early and only three were entirely normal. In eighteen families in which only the father drank, but three children in twenty-one were entirely sound, while there were many abortions and several cases of sterility.

There are numerous statistical facts showing a large percentage of alcoholics in the ancestry of prostitutes, degenerates and other inferior classes. All of the studies seem to show that alcoholism and the degenerate condition tend to occur in the same family, and Sullivan seems to control the case by showing that in some instances, at least, alcohol is the cause of degeneracy.

The real, crucial proof of the direct action of alcohol must come, however, from experiments on lower animals, where the sources of error may be entirely controlled.

Adami states: "The general belief (and we regard it as well founded) is that the children of the sot are as a body of lowered intelligence and vitality with unstable self-control." He recognizes the great difficulty of statistically proving this in man, since alcoholism is so often the accompaniment of weakness and hereditary taint, and may not be the primary cause of the condition in many families. With animals, however, the experimenter is enabled to prove that alcohol does induce a primarily degenerate condition.

One could continue to enumerate records showing the effect of alcoholism on the human offspring, yet a sufficient number of studies have been considered to show how strongly indicative the evidence is that alcohol is really the direct cause of defects in many cases. There is also little doubt that alcoholism is sometimes acquired by perfectly normal human beings, and when the tissues of such people become affected by alcohol they no doubt give rise to defective and abnormal offspring.

It is, however, an undeniable fact that alcoholism in man is very frequently an accompaniment of various degenerate conditions, and these conditions are oftentimes within themselves sufficient to account for further degeneration in the offspring. We shall, therefore, consider more fully at this point the evidence furnished by animal experimentation.

ANIMAL EXPERIMENTS

As stated above, the problem is broader than the subject of alcoholism. If it is shown that any toxic substance can act on the germ cells or developing embryo in such a manner as to change or modify its development, it necessarily follows that alcohol may induce a more or less equivalent condition, since it is definitely known to act on all animal tissues. I shall, therefore, mention the experiments with alcohol in particular, and at the same time consider other of the striking examples of environmental effects on the developing eggs of lower animals.

H. E. Ziegler treated sea-urchin's eggs with ethyl-alcohol. A 1 per cent. solution in sea-water delayed development, a 2 per cent. solution also delayed development and caused abnormal embryos, while a 4 per cent. solution prevented all development. The peculiarly typical larvæ of the sea-urchin which Herbst induced by the addition of lithium salts to sea-water have been mentioned above. Herbst's experiments furnish a striking example of a characteristic response on the part of the developing organism to a definite chemical treatment. Morgan obtained similarly definite results by treating frog's eggs with lithium, and I have shown a somewhat comparable response for the fish's egg. Other salts may give the same types of larvæ, as occurred in these cases, as McClendon has shown for the fish, and as I previously pointed out in several of my studies on the cyclopean defect. Yet with certain doses of given substances one gets greater numbers of the same defect than with any other treatment. It is not surprising that a few individuals of any one deformed type may occur in a number of different solutions. The important fact is, that with a particular treatment one is able to obtain on all occasions a large number of embryos exhibiting a perfectly clear-cut, definite defect.

Ridge got decided results by treating the eggs of the blue bottle-fly and frogs with alcohol. In solutions of 1/100 per cent. alcohol in water the development was slow. In 1/20 per cent. solutions development proceeded for only a short time and the eggs died. In one per cent. alcohol only one or two eggs started.

Ovize made an interesting observation on the influence of alcoholic fumes on developing hen's eggs. An incubator containing 160 eggs was in a cellar in which wine and brandy were being distilled. Seventy-eight chickens hatched; of these twenty-five were deformed and forty died during the first three or four days. Of the number unhatched, one-third were deformed, and 3 to 4 per cent. had only developed a short way.[3]

Féré has experimented extensively with the influence of alcohol on the developing hen's egg. Alcohol was injected into the albumen in some experiments, while in others the eggs were placed under bell jars and exposed to the fumes of evaporating alcohol. Enough of the fumes penetrated the shell and entered the egg to affect the subsequent development of the embryos. When eggs were placed in the incubator after such

---

3. These results by Ovize were taken from Forel's review.

treatment they developed more slowly than the control and a large number of malformed embryos resulted. The abnormalities were variable, yet many had defective nervous systems and a number of the embryos exhibited eye defects. Féré made no attempt to analyze the cause of the different types of deformities, and in fact he paid little attention to the structure of the defects. Yet he showed most decidedly that alcohol fumes do affect the developing embryo, as one might have inferred from the preceding observations made by Ovize.

I have repeated Féré's experiments at some length during the past two years and can confirm his results. My object has been to regulate the treatment in such a manner as to get definite types of defects with certain intensities of treatment. Up to the present time I have only partially succeeded in doing this, though in several experiments the delay in development and the general type of the defects has been rather constant. This treatment of hen's eggs with alcoholic fumes is one of the most convincing and easily performed demonstrations of the influence of alcohol on development.

Féré also experimented with hen's eggs to show the influence of differences in temperature during incubation and many other physical and chemical factors. All unusual conditions affected the development of the embryo. Féré also developed hen's eggs in glass dishes after removing them from the shell. Preyer and Loisel had previously done similar experiments, but they carried the embryo for only a day or so, while Féré succeeded in keeping the egg developing for six days. Some of these embryos develop abnormally.

I have recorded a number of experiments on fishes' eggs which show the decided effects of alcohol and a large series of other substances on embryonic development. Alcohol and various anesthetics showed a peculiar affinity for the developing nervous system and organs of special sense. In many cases other organs and parts of the embryos were apparently normal. Many of the deformed individuals hatched and lived for some time, swimming about and feeding in a typical fashion.

With alcohol solutions of given strength definite defects were induced. In some experiments dozens of embryos with typical brain and eye defects occurred, while few or no other types of deformities existed. *The experimenter has the power in these cases to predict with at least a limited degree of certainty the type of deformity which will result from a definite intensity of a particular treatment.* Embryonic development in such cases may really be regulated or controlled.

We have already considered a number of experiments on mammals which show that alcohol and other injurious substances affect the quality of the offspring. In treating mammals the case is not so simple as in treating the eggs of lower vertebrates which develop outside the parent's body. The effects in mammals may not be due directly to the substance used, but rather indirectly to the changed conditions the substances have induced in the body of the parent. It is important for this reason to know whether certain substances come in direct contact with the germ cells of the individual. As before mentioned, Nicloux and Renault have shown that alcohol may be readily found in the seminal fluid of a man shortly after drinking it. Thus the spermatozoa may come to float or swim in a weak solution of alcohol.

In the case of female mammals, Nicloux has carefully demonstrated the passage of alcohol from the blood of the mother into the tissues of the embryo. The following tabulation readily shows the results of his experiments on dogs and guinea-pigs:

TABLE 3.—PASSAGE OF ALCOHOL FROM THE MOTHER TO THE FETUS

		Amt. of Abs. Alc. Inject. per Kilo of Animal Weight, c.c.	Time of Absorption; Animal Killed, Hours	Amt. Alc. per 100 c.c. Maternal Blood, c.c.	Amt. Alc. per 100 c.c. Fetal Blood, c.c.	Amt. Alc. per 100 gm. Mother's Liver, gm.	Amt. Alc. per 100 gm. Fetal Tissue, c.c.
1.	Guinea-pig	5	5/6	0.36	0.31	....	.....
2.	Guinea-pig	5	1	0.47	0.35	.....	.....
3.	Guinea-pig	2	1	0.20	....	0.10	0.12
4.	Guinea-pig	1	1	0.13	....	0.081	0.086
5.	Guinea-pig	0.5	1¼	0.045	....	0.015	0.02
6.	Dog	3	1½	0.37	0.37	0.26	0.26

After a short period of time the amount of alcohol in the blood of the fetus is about equal to that in the blood of the mother, while there is really more alcohol in a given weight of the tissues of the fetus than is to be found in an equal weight of liver tissues from the mother.

The reality of the passage of alcohol from the mother to the fetus demonstrates the possibility of the intoxication of the fetus. Therefore, nervous disorders, anesthesia, etc., of the late fetus may result as a consequence of alcohol in the blood, while the developing embryo or early fetus will show the effects by an abnormal formation of the nervous system.

Thus the results of the experiments of Mairet, Combemale and Hodge on dogs are readily explained as the direct influence of alcohol on the paternal germ cells in the case of the treated male, or on the developing fetus within the body of the alcoholic mother. The great number of human records briefly referred to above are also readily interpreted as the result of direct alcoholic action on the germ cells and the developing embryo.

The experiments on mammals do not then really differ greatly from those on the lower vertebrates where the externally developing eggs are placed directly in various unusual solutions, since the egg or embryo although within the mother's body is readily bathed or impregnated by the alcohol contained within the mother's blood.

The only experiments with alcohol on lower mammals which do not fall completely in line with the above records are those recently recorded by Nice. He has fed mice on alcohol. Each day 2 c.c. of 35 per cent. alcohol was added to crackers and milk and placed as food for each mouse. Instead of drinking water the mice could drink 35 per cent. alcohol from a syphon which prevented evaporation. Animals treated in this way gained in weight over the control. The offspring from these alcoholic mice excelled all the other mice in growth, even when they themselves were fed alcohol. The young grew faster, however, when not given alcohol. Nice treated other mice with tobacco fumes, nicotin and caffein. The fecundity of the alcohol, nicotin and caffein mice was greater than the control while those treated with tobacco fumes had almost twice as many young as the control. The mortality of the offspring from the treated mice was, however, greater than from the control. None of the control young died, while 17.3 per cent. of the nicotin young and 11.1 per cent. of the alcoholic young died soon after birth. There was only one abortion, no still-births and none of the young were deformed.

Mice may possibly be peculiarly resistant to these drugs, though I should rather think that in the case of alcohol, at least, the animals received too little to give a pronounced effect, though it was sufficient to cause a certain fatality among the young. Weak alcohol mixed with crackers and milk no doubt rapidly evaporates. The animals possibly waited until a certain amount of the alcohol had disappeared before they ate their food, and, of course, the amount of alcohol they took instead of drinking water was very small. Mice may easily be kept on a cracker and milk diet without ever receiving water. One cannot deny, however, that the mice did receive enough alcohol to cause them to fatten more rapidly than the control, and probably to cause the death of some of their offspring.

Carrière has shown that when guinea-pigs are inoculated with various soluble products of the tubercle bacillus for several months that the number of offspring is diminished. He sometimes observed the death of the fetus or premature death of the young, while many of the living young had feeble constitutions. The action was produced when either parent was impregnated with the poison.

Mating together two inoculated animals gave 52 per cent still-born young; 28 per cent. of the living young died under sixteen days and only 20 per cent. of the young survived. When the female alone was inoculated 26.9 per cent. of the offspring were still-born, 34.6 per cent. died under sixteen days and 38.4 per cent. of the young survived. The matings with the male alone inoculated gave 16.6 per cent. still-born, 10 per cent. dying under sixteen days and 73 per cent. of the offspring survived. Thus the effect of the toxin is shown on the germ cells of both sexes.

Lustig's experiments of inoculating fowls with abrin gave results parallel to those recorded by Carrière. The offspring were less resistant to inoculations of abrin, just as the guinea-pigs were to the tuberculosis extracts when compared with control animals of the same age.

Mall has clearly shown in his monograph on the causes of human monstrosities that poor nutrition and abnormal environment are most potent factors. Only 7 per cent. of the uterine pregnancies examined gave monsters, while 96 per cent. of the tubal pregnancies produced abnormal embryos.

Ballantyne has presented in his "Antenatal Pathology" a most comprehensive consideration of the part played by abnormal environment and disease in the causation of monstrosities and developmental defects in general.

The effects of malnutrition or poor environment on the developing embryo is splendidly illustrated by the case of monochorial twins when one becomes more vigorous and pumps blood from the other through the anastomoses between their placental or umbilical vessels. In such cases one of the twins may fail to develop certain parts and may actually lack a heart, the heart of the superior embryo pumping blood through both the bodies. The various degrees of the degenerate or parasitic twin is thus produced. One individual falls behind in development and may finally actually be included within the body of the more vigorous twin. Double monsters may occur in which one individual is almost perfect, while the smaller monster is attached to some part of its body.

I have given this somewhat extensive survey of the literature in order to show that an abundance of evidence exists at the present time to indicate that the course of embryonic development may be readily modified. It is also clearly shown that the germ cells of various animals may be directly affected by different chemical treatments to such a degree that they give rise to defective individuals. The experiments of O. Hertwig and Morgan on the chemical production of spina bifida in large numbers of tadpoles and my experiments on the constant production of typical cyclopian monsters by subjecting developing fish eggs to definite chemical treatments strongly indicate that the manner of embryonic development may be definitely regulated. This exact regulation or control of development is the important goal of experimental teratology.

The problem is now in its beginning, since the actual influence of various treatments is known to be expressed in the resulting type of embryonic development.

The studies on alcoholism in mammals have failed to produce any convincing evidence of the specific actions of this poison. Yet the statistical studies on defective human beings would indicate that alcohol had a special affinity for the developing nervous system. My experiments on the influence of alcohol on the developing fish embryo demonstrated that alcohol did have a specific affinity for the central nervous system, and caused the brains of these embryos to exhibit numerous deformities, while the organs of special sense were also affected.

METHOD AND RESULT

The experiments here recorded have been undertaken in order to ascertain whether alcohol did exert a marked influence on the germ cells and developing embryos of mammals, and, if possible to demonstrate the nature and mode of action of this influence. I have used alcohol as an agent, since it may be given to guinea-pigs without greatly disturbing their normal physiological processes, and so does not produce marked conditions which might secondarily affect the results. Alcohol may remain as such in the blood and tissues of a mammal, and so may act directly just as it would when added to the sea-water in which fishes' eggs were developing. I have studied its effects experimentally on the eggs of lower vertebrates and am familiar with the defects it produces in these animals. It is an active substance and, therefore, for these many reasons lends itself admirably to experimental use.

The experiments have been conducted on guinea-pigs, since they breed fairly rapidly and rear their young without much difficulty in the laboratory. Strong healthy stock has been chosen and the animals have been carefully handled. All have remained in vigorous health and most of them have increased in size and fattened during the progress of the experiment. The males and females have been kept carefully separated and individual pairs mated from time to time.

The animals are first tested by normal matings and found to produce normal offspring. The alcoholic treatment is then begun on a given number of individuals and males and females mated in different combinations according to whether they are alcoholics or normal. An alcoholic male is mated with a normal female, the paternal test. This is the crucial test for influence on the germ cells, as here the defective offspring must be due to the chemically modified spermatozoon from which it arose, since the egg, and the mother in which the embryo developed, were both normal.

Normal untreated males are paired with alcoholic females, the maternal test. Here the defective offspring may be due either to a modified ovum or to the fact that it developed in a mother with alcoholic blood, therefore supplying an unfavorable developmental environment. Lastly, its condition may be due to both of these causes. The mammalian mother has two chances to injure an offspring, either by producing a defective egg, or secondly by supplying an unfavorable or diseased environment in which the embryo must develop.

The final combination is the mating of alcoholic individuals. This, of course, offers the greatest chance for defective offspring.

Alcohol is administered to the guinea-pigs by inhalation. At first it was given with the food, but the animals did not relish it, and therefore took less food. It was then given by stomach tube, but this method

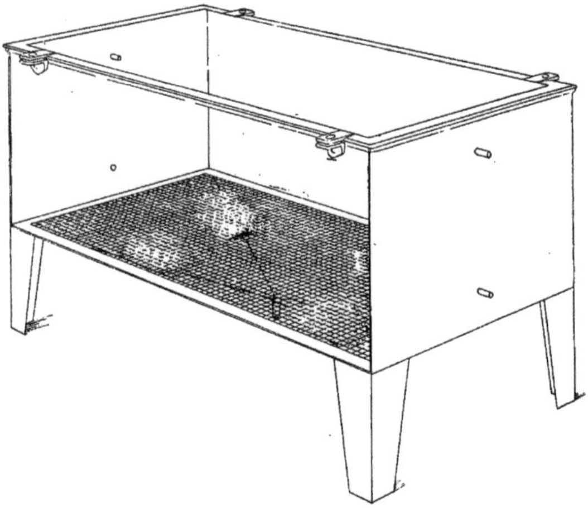

Fig. 1.—Tank for alcohol treatment. Animals are placed on the wire screen in the closed tank and inhale the fumes of alcohol evaporating from the cotton below the screen.

so upset the animals that the results might have been modified by their poor bodily condition and the bad state of their stomachs. The inhalation method is entirely satisfactory, the guinea-pigs thrive and usually gain in weight during the experiment, they have good appetites and are in all respects apparently normal. The only indication of the effects of the treatment is shown by the quality of offspring they produce.

The apparatus used for giving the alcohol consists of an air-tight copper tank 36 inches long by 18 inches wide and 12 inches deep, with a sloping bottom draining to the center. Over this bottom is placed a wire screen and below the screen cotton soaked with 95 per cent. alcohol is spread (Fig. 1). The tank is closed and allowed to stand until the atmosphere within is saturated with alcoholic fumes. A ventilation system is so arranged that a given quantity of alcohol fumes may be driven through the tank in a given time, but it has not seemed advisable to use this device, as the degree of intoxication is a better index to the physiological response of the animals, since their resistance to a given amount of the fumes is changeable. The guinea-pigs, three or four at a time, are placed on the wire screen above the evaporating alcohol, the tank is again closed and the animals are allowed to remain until they begin to show signs of intoxication, though they are never completely intoxicated. They usually inhale the fumes for about an hour. The animals are treated in this way for six days per week and some have now been treated over a period of about nineteen months. None of the effects

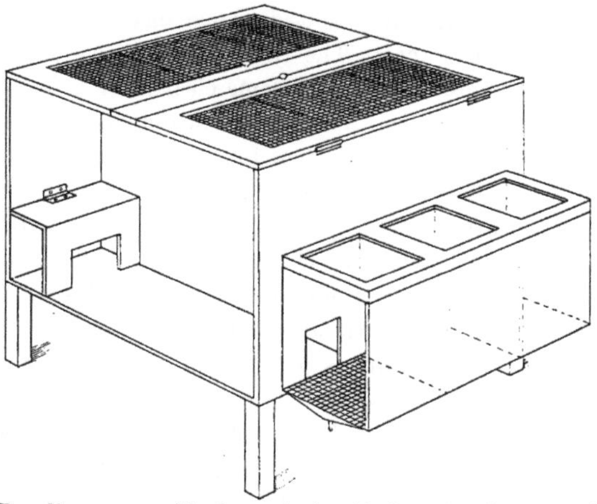

Fig. 2.—Breeding cage with fume tank attachment. Pregnant females are kept in this cage and may be driven through the drop door into the fume tank. Handling pregnant animals during the treatment is thus avoided.

are due to want of air, since the same number of guinea-pigs may remain for hours in this closed tank without showing any signs of discomfort when there are no fumes present.

In order to avoid handling the females during late pregnancy, a special treating cage is devised for them. An ordinary box run with a covered nest in which the animal lives is connected by a drop door with a metal-lined tank having a similar screen arrangement to that described for the general treatment tank (Fig. 2). The pregnant animal may be driven daily into the tank and thus treated with alcohol fumes throughout her pregnancy without having to be handled or moved about in any way that would tend to disturb the developing fetus.

During the vapor treatment the animals usually react in a manner quite similar to their behavior in weak fumes of ether or chloroform.

The majority of them sit quite motionless and sniff their noses for a time and then become somewhat drowsy. A few individuals, however, are excited by the treatment and run about the tank, becoming sexually excited, and many often fight other animals savagely. One of the males fights and bites so vigorously while taking the fumes that he has to be treated separately from all others. The fumes then have a different influence on the behavior of different individuals in much the same way that alcoholic intoxication expresses itself differently on different human beings.

During the first few weeks of the treatment the fumes cause the eyes to water so that tears run over the face. The nose and mouth also become moist and the animals sniff almost constantly. The fumes are very irritating to the mucous membranes at first. The cornea becomes irritated and finally opaque in some instances, so that the eye takes on a white appearance. The tissues seem, however, to develop a resistance to the fumes. The eyes become clear after a few months and never again become opaque. The nasal mucosa also ceases to secrete excessively unless the animal is left in for an unusually long time.

Many of the guinea-pigs have been killed after treatments of different duration up to fifteen months, and all of their viscera carefully examined. In no case have I found any changed structures due to the alcoholic treatment. The lungs, liver, stomach, intestines, kidneys, reproductive glands, brain and all other parts appear perfectly normal. The general health and behavior of the animals also indicate that they are in good condition. As before mentioned, several animals have been partially castrated during the experiment. One of the reproductive glands was removed and examined microscopically. In all cases the germ cells, ova or spermatozoa, as the case may be, were found to exhibit perfectly normal structure. One cannot claim, therefore, that this treatment is excessively severe or greater in proportional amount than the alcohol a human being often takes. The fact is that these animals have never been completely intoxicated, but receive only enough alcohol six times per week to affect their nervous states. They may be compared to a toper who drinks daily but never becomes really drunk.

While the bodies of these animals display no direct effects of the alcohol, the conditions of the offspring to which they give rise show most strikingly the effects of the alcoholic treatment. The results of mating the alcoholized guinea-pigs are summarized in Table 4.

Fifty-five matings of treated animals have been made. Forty-two of these have now reached full term and are recorded. Thirteen matings are not yet due. From the forty-two matings only seven young survived, and six of these are still living, five of which are runts, though their parents were unusually large, strong animals (Figs. 4 and 5).

The conditions of the animals in the mating pairs are shown in the first column of the table and the total results of the matings are indicated in the following columns. The first horizontal line gives the records when alcoholic males are paired with normal females. Twenty-four such matings were made. Fourteen of these gave negative results, or resulted in early abortions. Many embryos were aborted during very young stages, and some of these were deformed, though they were generally in such poor condition after being cast out into the cages that little could be learned from them. They were partially or completely eaten by the mother in most cases. The males were always kept for a number of days with the females during favorable periods, and conception should have occurred in all cases, as it did in the control matings.

TABLE 4.—EFFECTS OF ALCOHOL ON OFFSPRING OF GUINEA-PIGS

Condition of Animal	No. of Matings	No Result or Early Abortion	Still-Born Litters	No. Still-Born Young	Living Litters	Young Dying Soon After Birth	Surviving Young
Alcoholic male by normal female	24	14	5	8	5	7	5*
Normal male by alcoholic female	4	1	0	0	3	3 (*a*)	2†
Alcoholic male by alcoholic female	14	10	3	6	1	1 (*b*)	0
Summary	42	25	8	14	9	11	7‡
Normal male by normal female—Control	9§	0	0	0	9	0	17

*Four survivors in one litter, and one was a member of a litter of three, the other two died immediately after birth. (*a*) Premature. (*b*) Sixth day.

†One lived to become pregnant with two young *in utero*, one deformed. Fig. 3. Other survivor normal, the mother was not treated until after first two or three weeks of pregnancy.

‡Of thirty-two young born only seven have survived.

§One other non-alcoholic mating was made from which two young resulted: they died after the second and fourth days, respectively, and the mother died two days later; her diseased condition no doubt affected the suckling young. They have for this reason not been included in the normal control.

Only ten of the twenty-four matings resulted in conceptions which ran the full term. Half of these, or five, were still-born litters. There were three still-born litters of two young each and two of one individual each. Most of these were slightly premature, their eyes being closed and the hair sparse on the bodies. (A normal guinea-pig at birth is well covered with a hairy coat, its eyes are open and it very quickly begins to run about actively.)

Five litters of living young were born. One litter consisted of only one young. a weak individual that grew very little and died after six weeks. Two litters contained two young each. The members of one of these litters died during the first and fourth weeks, having been weak and small since birth. Both of those in the other litter were in a similarly feeble condition and died before the first month. One litter contained three young; two of these died immediately after birth; the other one is still alive, though small for its age. The fifth litter contained four young, all of which are runts, though their parents were unusually large animals (Figs. 4 and 5). *Thus out of twenty-four full-term young, of which only twelve were born alive, but five individuals have survived, and these are unusually small and very shy and excitable animals.*

It is a point of some interest that all of the young animals that died showed various nervous disturbances, having epileptic-like seizures, and in every case died in a state of convulsion. This is commonly the fate of feeble and nervously defective children.

The important fact in the above case is that only the father was alcoholic, the mother being a normally vigorous animal. *This experiment clearly demonstrates that the paternal germ cells may be modified by chemical treatment to such a degree that the male will beget abnormal offspring even though he mate with a vigorous female.* A reconsideration of the figures in the first line of the table shows really how decidedly the injured spermatozoon expresses itself in the fate of the egg with which it combines.

The second line of the table shows the results of matings between alcoholized females and normal males. These matings might be expected to give more marked results than the pervious ones, since in the treated females not only the germ cells may be affected, but the developing embryo itself may be injured by the presence of alcohol in the blood of the mother. Nicloux has shown that alcohol may pass directly from maternal blood into the embryonic tissues of a guinea-pig. The spermatozoon, however, is probably a more sensitive structure than the egg and is easily injured or killed by slightly abnormal conditions. It might possibly be that when such a specialized cell swam for even a short period of time in seminal fluid containing a trace of alcohol its chemical nature would be so decidedly disturbed as to render it incapable of inducing normal development after impregnating the egg. At any rate the few cases at present available seem to indicate that the effect on the offspring is equally as great when it is produced by an alcoholic father as by an alcoholic mother.

There are only four matings between alcoholized females and normal males. One of these gave a negative result or was possibiy aborted very early. Three living litters were born. One of these consisted of three premature young, which died shortly after birth. The remaining two litters each contained only one young, but these two animals survived. One of these guinea-pigs was born after the mother had been treated for three and one-half months. The offspring was weak and small for several months after birth, but finally recovered and developed into a normal animal. This guinea-pig was mated with an alcoholic male and became pregnant. Unfortunately, she was killed by accident, and on examination her uterus was found to contain two embryos, 33 and 32 mm. in length. One of these embryos was deformed and showed very decidedly

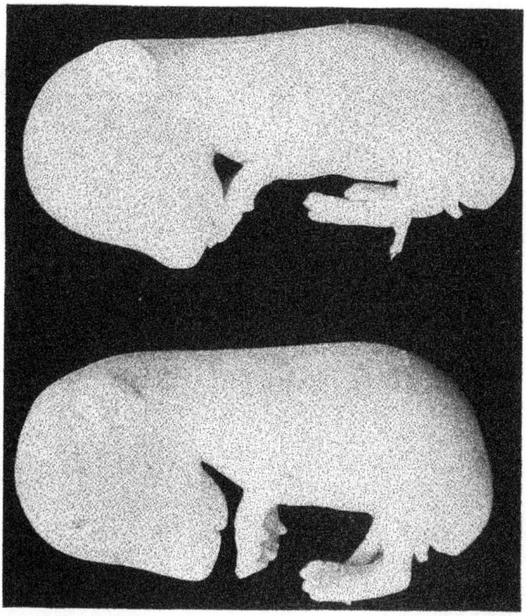

Fig. 3.—Two embryos 32 and 33 mm. in length, taken from a female that had an alcoholic mother and was mated with alcoholic male. The upper fetus has deformed hind legs and a poorly developed posterior part of the body: lower fetus is normal.

degenerate and feebly developed hind legs. The posterior end of its body was also poorly formed. This condition is readily seen in Figure 3, a photograph of the two embryos. The abnormal one has small hind legs, and one of them is badly folded under its body. This is of interest, since all of the affected offspring of alcoholic guinea-pigs are weak in their hind extremities and drag their legs. Yet none were so modified as to show a noticeable structural defect except this embryo, which had one alcoholic grandmother and an alcoholic father.

The only other survivor from an alcoholic mother is strong and full grown for its age. The mother had been treated for only two and one-half months when the offspring was born, so that she was normal during the first two or three weeks of pregnancy. No doubt the early stages of development are more easily modified to produce significant defects than are the later. This question is being more fully tested on guinea-pigs with experiments now in progress. I have shown, however, in

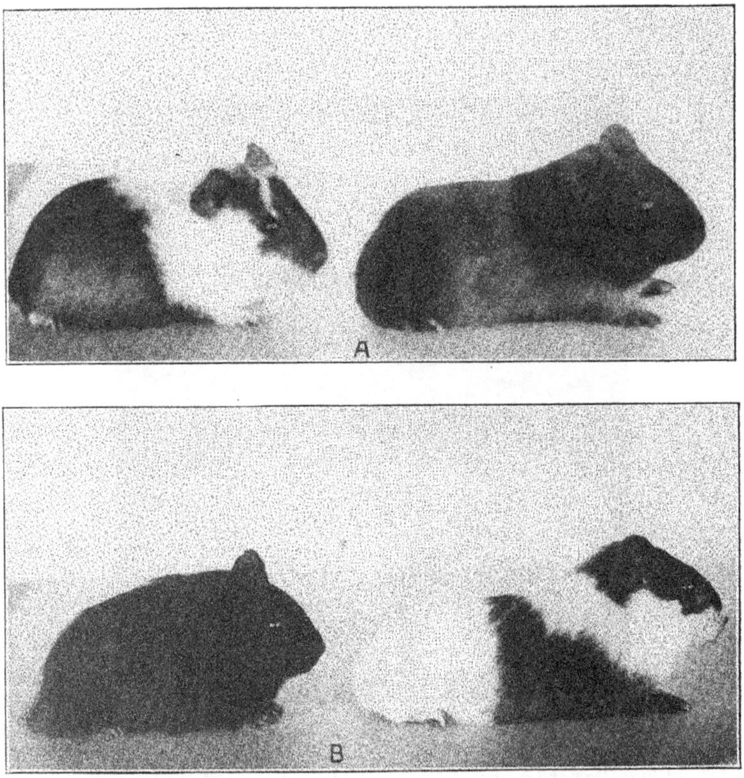

Fig. 4.—A. The animal on the left is a runt from a large alcoholic male and a large normal female; weighs 134 gm. The animal on the right, from normal parents, is larger although 1 month younger and weighs 147 gm.

B. The guinea-pig on the left is a runt, weighing 132 gm. from an alcoholic father; on the right a normal guinea-pig twice as large though only 10 days older, weighs 221 gm.

treating fish eggs that the period at which the treatment is applied is a most important factor in determining the type of defect or modification which will result. Certain salts, different strengths of magnesium chlorid, for example, which give pronounced effects when added to the sea-water containing eggs in early developmental stages, may really be ineffective after the eggs have developed beyond these stages. In the case under consideration the offspring might not have fared so well if the alcoholic treatment had been started on the mother a few weeks before conception, instead of three weeks after her pregnancy had begun. This with other points shall be more completely analyzed in future communications on these experiments.

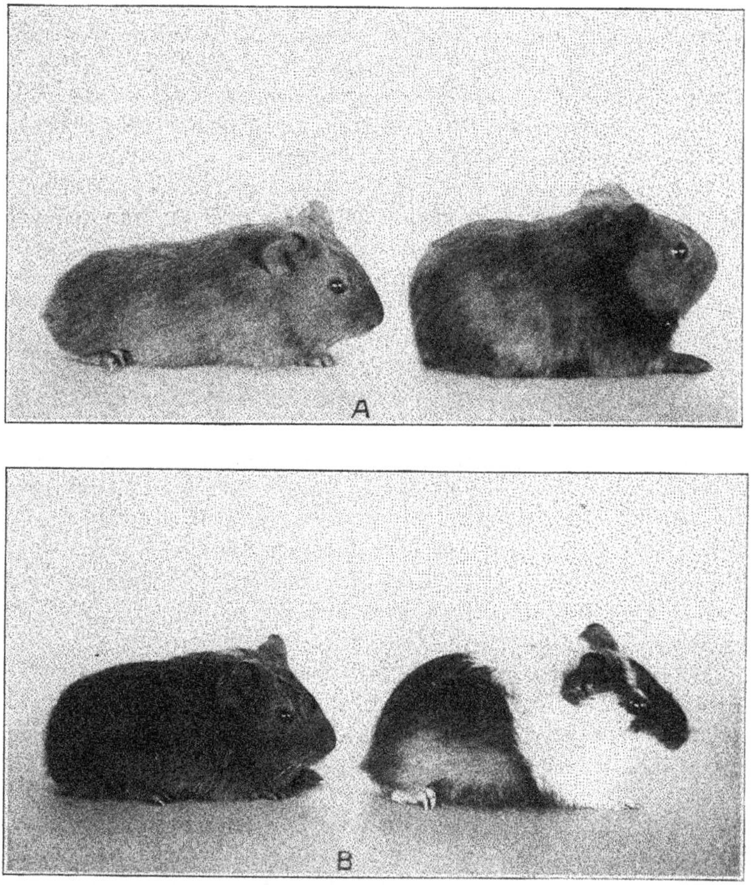

Fig. 5.—A. Two guinea-pigs from alcoholic fathers, the left one 1 month and 10 days younger than the runt on the right.
B. The left animal is the same as above, the right another of the same runt litter.

The four matings of alcoholic females and normal males resulted in three living litters in all of five individuals. Three of the young were premature and died shortly after birth, while two young survived.

Finally, we may consider the results of pairing two alcoholized individuals. The third line of the table summarizes these results. As might have been anticipated, this type of mating has given the highest fatality of all.

Ten out of a total of fourteen matings have given no offspring or early abortions, which were in many cases eaten by the mother. Three still-born litters have been produced, each consisting of two young. *Only one living litter was born from the fourteen matings in which both parents were alcoholic, and this litter consisted of but one weak individual which died in convulsions on the sixth day after birth.* This is indeed a decided effect of alcohol on the offspring when one compares it with nine control matings, all of which gave living litters containing a total of seventeen individuals, all surviving.

Two other young were produced by non-alcoholic parents and died on the second and fourth days after birth. They have not been included in the control since the mother died two days later in a diseased condition. No doubt the poor state of the mother had much to do with the fate of the suckling young. She was an animal that had only been in the experiment for a short time, and is one of the very few that have contracted disease or died during the nineteen months of the work. This might possibly go to show the influence of a diseased mother on the offspring.

The fourth line of Table 1 gives a summary of the experiments. There have been forty-two full-term matings, twenty-five of which gave no results or early abortions; eight still-born litters have occurred, consisting of fourteen individuals; only nine living litters have been born, 21 per cent. of the matings. These contained eighteen young, and but seven of this number have survived and five of these survivors are unusually small (Figs. 4, 5).

The bottom line of the table shows nine control matings. All have given living litters containing a total of seventeen young, all of them surviving. The two young that died, as stated above, were from a dying mother and not included in the control.

Records of the successive matings of ten of the female guinea-pigs are shown in Table 5. The varying ways in which the same individual has responded in different matings is noticeable. Number 10, an alcoholic female, first mated with an alcoholic male, gave one young which died on the sixth day after birth. On being remated with the same male, No. 10, gave no result. When mated with another alcoholic male, gave no result. She mated again after several months with the first male and on being killed was found to contain one embryo *in utero* about 2 weeks old.

Female 15, a normal guinea-pig, shows an instructive record. She was mated with an alcoholic male and gave birth to two still-born young.

of convulsions within four weeks after birth. She was then mated with a normal male as a control and gave one vigorous normal offspring which survived.

TABLE 5.—RESULTS OF SUCCESSIVE MATINGS OF TEN FEMALES

Animal	First Mating	Second Mating	Third Mating	Fourth Mating
No. 10 Alc.	Alc. male 4, 1 young died in 6 days	Alc. male 4 0	Alc. male 6 0	Alc. male 4, 1 embryo *in utero* 2 weeks after
No. 12 Alc.	Alc. male 5 0	Alc. male 5 0	Alc. male 4 0	....
No. 11 Alc.	Alc. male 6 0	Alc. male 6 0	Alc. male 5; 2 premat. still-born	Alc. male 4 0
No. 13 Nor.	Alc. male 5 1 still-born	Alc. male 5 0	Alc. male 4 0	....
No. 17 Nor.	Etherized male 1 0	Etherized male 1 0	....	....
No. 18 Nor.	Alc. male 5 0	Alc. male 5 0	Alc. male 6 0	....
No. 7 Nor.	Etherized male 2; 2 premat. still-born	Etherized male 2 0	Etherized male 2 0	....
No. 14 Nor.	Etherized male 3 0	Etherized male 3 0	....	....
No. 19 Nor.	Alc. male 4 0	Alc. male 6; 1 still-born	Alc. male 6 0	Alc. male 5: 4 small, active, only one-half size, but living
No. 15 Nor.	Alc. male 6; 2 still-born	Alc. male 5 0	Alc. male 5; 2 died fourth week of convulsions	Nor. male: 1 normal vigorous young

The other records are easily understood.

These experiments have suggested many questions still to be solved, some of which are now being tested, such as the length of time necessary to treat an animal before the resulting offspring is affected, whether this time is equally long for both sexes, and what amount of individual variation may exist. An important point to ascertain is whether the effects of the alcohol treatment are permanent, or does the animal recover after a time and again become capable of giving normal offspring. One of the most valuable problems is to regulate the treatment in such a manner as to induce a definite type of defect with a given kind or degree of treatment. The structure or morphology of the monsters and defective offspring which occur is to be carefully studied. Many other points might readily be suggested.

Definite and well-controlled experiments with alcohol and other substances on the mammalian offspring have not been sufficiently studied. The work is really in its beginning, and while there is much evidence to show that various toxic agents do affect and modify the offspring, facts are badly needed to demonstrate the regularity and manner of this modification. The present experiments seem to me to prove in a convincing way that alcohol may readily affect the offspring through either parent, and that this effect is almost fatal to the existence of the offspring when the parents have been treated with even fairly large doses of alcohol. Many of the cases seem to indicate further, that the tissues of the nervous system of the offspring are particularly sensitive in their responses to the induced conditions.

My assistant, Miss Craig, has aided me greatly throughout almost the entire progress of these experiments. Last year during my absence abroad she assumed entire control of the animals, and I am indebted to her for this efficient assistance.

SUMMARY

Guinea-pigs have been treated with alcohol in order to test the influence of such treatment on their offspring. Male and female animals are given alcohol by an inhalation method until they begin to show signs of intoxication, though they are never completely intoxicated. They are treated for about an hour at the time, six days per week. The treatment in some of the cases has now extended over a period of nineteen months. The animals may be said to be in a state of chronic alcoholism.

Fifty-five matings of the alcoholized animals have been made, forty-two of which have reached full term and are recorded.

From these forty-two matings only seven young animals have survived, and five of them are unusually small, though their parents were large, vigorous guinea-pigs. The following combinations were made:

1. Alcoholic males were mated to normal females. This is the paternal test, and is the really crucial proof of the influence of alcohol on the germ cells, since the defective offspring in this case must be due to the modified spermatozoa, or male germ cells, from which they arise. Twenty-four matings of this type were made, fourteen of which gave no result or very early abortions; five still-born litters were produced, consisting of eight individuals in all, and five living litters containing twelve young. Seven of these twelve died soon after birth, and only five have survived. Four of the survivors are from one litter and the fifth is the only living member of a litter of three.

2. Normal males were mated with alcoholic females. This is the maternal test. In such cases the alcohol may affect the offspring in two ways—by modifying the germ cells of the mother or acting directly on

the developing embryo *in utero*. Only four such matings were tried. One gave no offspring; three living litters were born, one consisting of three premature young that died at birth, while the other two litters consisted each of one young, which have survived. The alcoholic treatment in one of the last cases was only begun after the mother had been pregnant for about three weeks.

3. Alcoholic males were mated to alcoholic females. This is the most severe test, both parents being alcoholic. Fourteen such matings gave in ten cases no offspring, or very early abortions. Three still-born litters were produced, consisting in all of six individuals, while only one living young was born. *This single offspring from the fourteen matings died in convulsions on the sixth day after birth.*

The young that have died in the experiment showed nervous disorders, many having epileptic-like seizures, and all died in convulsions.

Nine control matings in the same group of animals have given nine living litters, consisting in all of seventeen individuals, all of which have survived and are large, vigorous animals for their ages. Two young from non-alcoholic parents died, but this mother also died two days later. Her diseased condition doubtless affected the suckling young.

Forty-two matings of alcoholic guinea-pigs have given only eighteen young born alive, and of these only seven, five of which are runts, survived for more than a few weeks, while nine control matings have given seventeen young, all of which have survived and are normal, vigorous individuals. These facts convincingly demonstrate the detrimental effects of alcohol on the parental germ cells and the developing offspring.

REFERENCES

Adami, J. G.: The Principles of Pathology, Vol. I. New York. Lea and Febiger, 1908.

Ballantyne: Antenatal Pathology, Edinburgh, Green and Son. 1902.

Bertholet, E.: Ueber Atrophie des Hoden bei chronischem Alkoholismus. Centralbl. f. allg. Path., 1909, xx, 1062.

Elderton, E., and Pearson, K.: A First Study of the Influence of Parental Alcoholism on the Physique and Ability of the Offspring. Eugenics Lab. Memoir. 1910, x, Dulan, London.

Féré, C.: Influence du repos. sur les effets de l'exposition préalable aux vapeurs d'alcool avant l'incubation de l'oeuf de poule. Compt. rend Soc. de biol.. 1899. li: Note sur la resistance de l'embryon de poulet aux traumatismes de l'oeuf. Jour. anat. et de physiol., 1897, p. 264. Remarques sur l'incubations des oeufs de poule privés de leur coquille. Compt. rend. Soc. de biol.. 1900. lii.

Forel, A.: Alkohol und Keimzellen (blastophthorische Entartung). München. med. Wchnschr., Dec. 5, 1911, lviii. 2596.

Herbst, C.: Experimentelle Untersuchungen. u.s.w.. Ztschr. f. wissensch. Zool.. 1892, iv; Mitt. a. d. Zool. Staz. zu Naepel. 1893: Arch. f. Entwicklungsmechn. d. Organ., 1896, iv.

Hertwig, O.: Urmund und Spina bifida. Eine vergleichende morphologische. teratologische Studie an missgebildeten Froscheiern. Arch. f. mik. Anat.. 1892. xxxix, 353-503.

Hodge, C. F.: The Influence of Alcohol on Growth and Development. In Physiological Aspects of the Liquor Problem. by Billings. ed. 1. p. 359. Houghton. Mifflin Co., New York. 1903.

Hoppe, H.: Die Tatsachen über den Alkohol. Ed. 3, Berlin, 1904.

Hunt, Reid: Studies in Experimental Alcoholism. U. S. Pub. Health Bull. No. 33, 1907.

Kyrle: Bericht über Verhandlungen der XIII Tagung der Deutschen pathologischen Gesellschaft in Leipzig. Centralbl. f. Path. u. path. Anat., xx, No. 77, 1909.

Laitinen, T.: Ueber den Einfluss des Alkohols auf die Widerstandsfähigkeit des menschlichen und tierischen Organismus mit besonderer Berücksichtigung der Vererbung. Tr. Kong. Inter. X Alkoholismus, Budapest, 1905; Ueber die Einwirkung der Kleinsten Alkoholmengen auf die Widerstandsfähigkeit des tierischen Organismus mit besonderer Berücksichtigung der Nachkommenschaft. Ztschr. f. Hygiene, 1908, lviii, 139.

Loeb, J.: Investigations in Physiological Morphology. III. Experiments on Cleavage. Jour. Morph., 1892, vii, 253; Ueber die Entwicklung von Fischembryonen ohne Kreislauf. Pflüger's Arch. f. d. ges. Physiol., 1893, liv, 525; Ueber die relative Empfindlichkeit von Fischembryonen gegen Sauerstoffmangel und Wasserentziehung in verschiedenen Entwicklungsstadien. Pflüger's Arch. f. d. ges. Physiol., 1894, lv, 530; Studies in General Physiology. Two volumes, Univ. of Chicago Press, 1905.

Lustig, A.: Ist die für Gifte erwordene Immunität übertragbar von Eltern auf die Nachkommenschaft? Centralbl. f. Pathol., 1904, xv, 210.

Mairet and Combemale: Influence dégéneration de l'alcool sur la descendance. Compt. rend. Acad. d. Sc., 1888, cvi, 667.

Mall, F. P.: A Study of the Causes underlying the Origin of Human Monsters, Jour. Morph., 1908, xix, 1-361.

McClendon, J. F.: An Attempt Towards the Physical Chemistry of the Production of One-Eyed Monstrosities. Am. Jour. Physiol., 1912, xxix, 289.

Morgan, T. H.: The Relation Between Normal and Abnormal Development of the Embryo of the Frog, as determined by the Effects of Lithium Chlorid in Solution. Arch. f. Entwicklingsmech., 1903, xvi; The Relation Between Normal and Abnormal Development of the Embryo of the Frog. Ibid., 1902-1905, xv-xix.

Nice, L. B.: Comparative Studies on the Effects of Alcohol, Nicotin, Tobacco Smoke and Caffeine on White Mice. I. Effects on Reproduction and Growth. Jour. Exper. Zool., 1912, vii, 133.

Nicloux: Passage de l'Alcool ingéré de la mère au foetus, etc. L'Obstetrique, 1900, xcix.

Paul, Constantin: Étude sur l'intoxication lente par les préparations de plomb, de son influence sur le produit de la conception. Arch. gén. de méd., 1860, xv, 513.

Pearson, K., and Elderton, E.: A Second Study of the Influence of Parental Alcoholism on the Physique and Ability of the Offspring. A Reply to Medical Critics of the First Memoir. Eugenics Lab. Memoir. 13, Dulan, London, 1910.

Preyer, W.: Physiologie spéciale de l'embryon. Trad. franc., 1887, p. 16.

Simmonds, H.: Ueber die Ursache der Azoospermie. Vortr. im Aerztl. Verein zu Hamburg, June, 1898; Berl. klin. Wchnschr., 1898, No. 36, p. 806.

Stockard, C. R.: The Development of Fundulus heteroclitus in Solutions of Lithium Chlorid, etc. Jour. Exper. Zool., 1906, iii, 99; The Influence of External Factors, Chemical and Physical, on the Development of Fundulus heteroclitus. Jour. Exp. Zool., 1907, iv, 165; The Artificial Production of a Single Median Cyclopean Eye in the Fish Embryo by Means of Solutions of Magnesium Chlorid. Arch. f. Entwicklungsmech., 1907, xxiii, 249; 1909, vi, 285; The Origin of Certain Types of Monsters. Am. Jour. Obst., 1909; lix, No. 4; The Independent Origin and Development of the Crystalline Lens. Am. Jour. Anat., 1910, x, 393; The Influence of Alcohol and Other Anesthetics on Embryonic Development. Am. Jour. Anat., 1910, x, 369.

Sullivan, W. C.: A Note on the Influence of Maternal Inebriety on the Offspring. Jour. Ment. Sc., 1899, xlv, 489.

Todde, C.: L'azione dell' alcool sullo svillupo e sulla funzione dei testicoli. Riv. sper. di Freniatria, 1910, xxxvi, No. 3, p. 491.

Ziegler, H. E.: Ueber die Einwirkung des Alkohols auf die Entwicklung der Seeigel. Biol. Zentralbl., June, 1903, xxiii, 448.

Reprinted from The Archives of Internal Medicine,
October, 1912, pp. 369-398

IS THE CONTROL OF EMBRYONIC DEVELOPMENT A PRACTICAL PROBLEM?

By CHARLES R. STOCKARD.

(*Read April 19, 1912.*)

Under favorable natural conditions two normal parents should, and usually do, produce a vigorous normal offspring. When, however, the conditions of development are modified or if in the second place the parents are not entirely normal the offspring is usually more or less defective. I shall attempt to show that the proper development of the offspring is dependent upon two main factors, first the physical qualities of the parental germ cells, and second the environment in which the embryo develops.

One is at first sight apt to think that deformities and defects are rare among men and other animals; but closer observation will show that the really structurally perfect individual is rather exceptional. Gross anatomical defects or monstrosities are frequently found among all animals, while lesser defects of minor importance are to be observed in a majority of individuals. These defects often cause no inconvenience, and indeed, we may be ignorant of their presence, since they are generally internal. Yet many apparently normal individuals sooner or later suffer or may actually die from some hidden developmental imperfection. The well-known congenital defects of the heart and other parts of the vascular system, digestive tract, etc., as well as the numerous developmental arrests in various parts of the body constantly remind the observer of the great loss in ability and energy that the race suffers as a result of faulty development.

These defects in construction must be considered a disease which causes the death of about 23 per cent. of the human race before or shortly after the time of birth (Sullivan's studies and French statistics), and handicaps a certain proportion of the survivors through-

Reprinted from Proceedings American Philosophical Society, Vol. li, 1912.

out their lives. We carefully study and use all known precautions to protect ourselves against post-natal infections and diseases, and much interest and time is given to combating the causes, yet little is said and scarcely anything done towards a control of development, or the hygienic protection of the developing individual.

This is really a morphological problem and is as truly a part of the fight against disease as is the treatment of abnormal physiological processes. It is not all of morphology to describe and study the detail of bodily structure, but its important task is to understand and analyze that structure, and if possible control and regulate its formation: and thus, if properly developed its goal is to relieve the race of its great structural disease—a disease which affects more individuals than any other one malady of man.

To most persons the above task seems at first thought a futile undertaking, and any one suggesting such control or preventive treatment might be interpreted as indulging in fanciful speculation. Yet the data available from the studies of defective persons in different countries of the world, and the experimental evidence furnished by work on lower animals makes the correction or prevention of developmental defects seem even today a problem to be practically handled to a slight degree at least.

To proceed as with any other disease, we must first ascertain the cause of these conditions, as the possibility of a cure depends upon the nature of the cause.

Are monstrosities and defective development due to some innate change within the germ cells of the parent, thus being incurable, as many former workers would have us believe? Or, are they due to changes produced in the germ cells by the action of some unusual condition in the body of either the male or female parent, or finally may they not be due to an unusual environment acting upon the developing embryo itself? In both of the latter cases the conditions are open to regulation or control. These questions may only be solved experimentally and the experiments have proven that the great majority of monsters are due to the action of unusual conditions upon either the parental germ cells or the developing embryo. There may be some changes of form or variations in animals which are due to innate changes in the germ-plasm but even these when fully understood may possibly be shown to result indirectly from some change in the chemical surroundings.

First to consider the modifications induced in the developing egg or embryo by a strange chemical environment. It has been found for the eggs of a number of animals that develop normally in sea-water that when certain chemicals are added to their environment they develop into various unusual forms.

I experimented for several years on fish's eggs and found that on adding any one of a large series of salts to the sea-water that the eggs developed abnormally and gave rise to a great number of monstrous individuals. The types of the monstrosities were variable, and the same kind of monster often resulted from different treatments. This was to be expected, but the important problem was to produce some definite type of monster in great numbers with any given treatment. This I finally succeeded in doing and in some experiments got as many as 90 per cent. typical cyclopean or monophthalmic monsters. These types of monsters first occurred in solutions of $MgCl_2$ in sea-water. In such solutions as many as 50 in 100 eggs formed one-eyed cyclopean embryos. Since Mg has the power to inhibit activity in animals and so acts as an anæsthetic I determined to try the action of a number of such substances on the developing eggs to ascertain whether they might also inhibit the lateral migration of eye parts. Alcohol, ether, chloroform, chloreton, etc., were employed and cyclopean monsters resulted from eggs developing in all of these substances. Alcohol gave the most decided effects and inhibited the normal production of eyes in almost all cases. All of these anæsthetics act more particularly upon the central nervous system of the adult and it is important to find that the development of the nervous system is also especially affected by them. In alcohol solutions the embryos showed almost every gross abnormality of the brain which is known to occur, and the spinal cord was often defective.

I have repeated the experiments of Féré with hen's eggs and find that when these eggs are exposed to fumes of alcohol many abnormal chicks result. When hen's eggs are placed in closed dishes over evaporating 95 per cent. alcohol enough of the fumes penetrate the shell and enter the contents of the egg to cause the developing chick to form abnormally.

McClendon has lately found that an excess of CO_2 and other substances also cause cyclopia and brain abnormalities. Many other workers have shown the effects of the environment on the developing egg.

It is, therefore, proven that the experimenter has the power to take an egg which would normally give rise to a perfect animal and by proper treatment he may cause it to form a typically abnormal individual. The monster may in many cases be able to survive and move about. No one can question that in these experiments the unfavorable environment modifies the form of the resulting individual.

Does this also occur in embryos developing in the mother's body? Children are born which exhibit the same types of deformities as those described above. Syphilitic mothers usually abort or give birth to abnormal children and there is much evidence to indicate that an alcoholic mother is more apt to produce an abnormal child than is a non-alcoholic mother.

Tubal pregnancies are common among women with venereal diseases and in such cases the embryo must necessarily develop under abnormal environment, having a poor surface for placental attachment in a region not adapted to the conditions of pregnancy. The conditions for embryonic nutrition are poor. Mall has found that while only 7 per cent. of uterine pregnancies in his records contained pathological embryos, that 96 per cent. of the embryos in tubal pregnancies were pathological, only 2 in 46 specimens being normal. This is strongly indicative of an abnormal environment as the cause of abnormalities. If these monsters were due to inherent tendencies in the germ cells one should not expect more abnormal tubal than uterine embryos.

Among lower mammals it has been shown that dogs fed on alcohol produce deformed and otherwise defective pups. I am now conducting a series of experiments with guinea pigs which show that a female treated with alcohol during her pregnancy will often abort or produce defective young, while the control animals are giving birth to normal young. Many more cases could be cited if time permitted.

Experiments on lower animals, therefore, show and human statistics seem to indicate that the cause of structural disease is often an abnormal developmental environment. To prevent such a disease the developmental conditions must be controlled and rendered as nearly normal as possible.

The second consideration is whether abnormal chemical environment may act on the parental germ cells in such a manner as to cause them to change and become incapable of giving rise to a normal individual. It is well known that certain disease toxins such as that of syphilis and substances such as alcohol and lead effect various body tissues so as to render them unfit for normal physiological activity. It is, therefore, only logical to suppose that the same or similar substances may effect the germ cells and so derange their chemical constitutions as to cause them to give rise to offspring of peculiar structure and qualities.

Bertholet has found that alcohol has a particular affinity for the reproductive glands just as it does for the nervous system. In examining the structure of the testicles from a large number of chronic alcoholics it was shown that spermatozoa were absent entirely or degenerate in form (azoospermy) in a majority of the cases. It is doubtless true that the ability of the spermatozoa to accomplish normal fertilization would be affected long before any definite structural change could be observed.

The crucial case is the treatment of the male in such a way as to render his spermatozoa unable to produce a normal development when combined with a healthy egg from a normal female. In this case the action must of necessity be on the germ cell only and not on both the egg and embryo as it might be in treating a female mammal.

It must be recognized that an individual owes its structure and character to the peculiar chemical constitution of the germ cells from which it arises. The germ cells of two species of animals are probably as different chemically as the animals are morphologically. Therefore, if the chemical nature of the germ cells is disturbed or injured by the action of poisons in the animal's blood they will probably show this injury in the type of individual to which they give rise.

Constantine Paul long ago found in studying 88 cases of pregnancy among women lead workers that 71 resulted in abortion, premature labour, or stillbirth while only 17 children were born alive and of these five died within the first year. Several of these women later produced healthy children after leaving this work. (This indicates that when the cause is known for defective development the cure may often be established by its removal.) Lead not only effects the developing fœtus but also acts directly upon the germ cells as is shown in the case of men working in lead while their wives were not exposed to the poison. Many of the offspring from such fathers are aborted and the children born are epileptic, feebleminded or generally defective.

To return to the results furnished by the guninea pig experiments referred to above—I have chosen healthy individuals and treated them daily with the fumes of 95 per cent. alcohol to about the point of intoxication. Feeding alcohol and giving it by stomach tube was first tried, both of these methods were unsatisfactory as the guinea pigs did not take alcoholic food in sufficient quantity and the stomach tube disturbed the animals to such a degree that I feared the experimental result might be vitiated even though it could be partially controlled. The inhalation method is perfectly satisfactory; the animals are placed in a copper tank having a screen floor which holds them above the evaporating alcohol. The alcohol is breathed directly into the lungs and affects the animals readily, in much the same manner as weak treatments of ether or chloroform would. The animals are thus put into a condition of chronic alcoholism, being almost intoxicated six times per week. Many of these guinea pigs have been killed and their lungs, liver and other organs examined and found to be perfectly normal so far as their appearance goes. The conjunctiva over the eyes is very often affected by the fumes, during the beginning of the treatment the eyes often become white, this is transitory in most instances and the eyes finally clear again and remain in a normal condition from then on. Most of the specimens have fattened under the alcohol treatment.

The matings have been made in such a fashion as to test several questions. First, alcoholic males are mated with normal females, paternal influence, the crucial test for the effect upon the germ cells. Second, alcoholic females are paired with normal males, the maternal influence plus the direct action on the developing embryo. Lastly, alcoholic males and females are paired.

The results of 40 such matings are shown in Table I. The decided effects of the alcoholic treatment are seen when the records are compared with those of the normal guinea pigs.

TABLE I.

MATINGS OF ALCOHOLIZED GUINEA-PIGS.

Condition of Animal.	Number of Matings.	No Result or Early Abortion.	Stillborn Litters.	Number of Still-Born.	Living Litters.	Early Deaths.	Surviving Young.
Alc. male × nor. female	24	14	5	8	5	5	5 1 + 4
Nor. male × alc. female	2	1	0	0	1	0	1 Preg. 2 in utero, 1 deformed.
Alc. male × alc. female	14	10	3	5	1	1 died 6th day	0
Summary	40	25	8	13	7	6	6 6 in 25
Nor. male × nor. female. *Control.*	8	0	0	0	8	0	15 15 in 15

In the 24 cases in which normal females were mated with alcoholic males, 14 gave negative results. Some of these probably aborted early as the parents were all fertile and the female is apt to eat the young before they have been observed when they are born prematurely. Five of the matings gave stillborn young, in some cases they were born a little before term. Litters were born alive but the young died soon after showing many nervous symptoms, such as epileptic-like seizures, and all died in convulsions. Only two litters consisted of normal offspring and these young, five in all, seem healthy though unusually small. It is thus seen that in 24 matings of *normal* females with alcoholic males only two gave normal results. Whereas in the control animals all matings have resulted in the production of normal offspring.

Only two matings were made between normal males and alcoholic females. One of these gave no result or was possibly aborted very early and lost, while the other mating produced one female offspring that lived to become pregnant by an alcoholic male. This last mentioned female was killed by accident, two embryos were found *in utero* one of which was deformed.

Fourteen matings were made between alcoholic males and females. Ten gave no result or aborted very early and were eaten, while four cases showed the following records. One young was born weak and died in convulsions on the sixth day after birth. Two cases of premature births of dead young. One female had young *in utero* when killed.

TABLE II.

Successive Matings of Ten Females.

Animal.	1st Mating.	2d Mating.	3d Mating.	4th Mating.
No. 10 alc.	Alc. male 4 = 1, young died 6th day.	Alc. male 4 = 0.	Alc. male 6 = 0.	Alc. male 4 = 1, embryo in utero 2nd week.
No. 12 alc.	Alc. male 5 = 0.	Alc. male 5 = 0.	Alc. male 4 = 0.	
No. 11 alc.	Alc. male 6 = 0.	Alc. male 6 = 0.	Alc. male 5 = 2, premature, still-born.	Alc. male 4 = 0.
No. 13 nor.	Alc. male 5 = 1, stillborn.	Alc. male 5 = 0.	Alc. male 4 = 0.	
No. 17 nor.	Eth. male 1 = 0.	Eth. male 1 = 0.		
No. 18 nor.	Alc. male 5 = 0.	Alc. male 5 = 0.	Alc. male 6 = 0.	
No. 7 nor.	Eth. male 2 = 2, premature, stillborn.	Eth. male 2 = 0.	Eth. male 2 = 0.	
No. 14 nor.	Eth. male 3 = 0.	Eth. male 3 = 0.		
No. 19 nor.	Alc. male 4 = 0.	Alc. male 6 = 1, stillborn.	Alc. male 6 = 0.	Alc. male 5 = 4, small but active.
No. 15 nor.	Alc. male 6 = 2, stillborn.	Alc. male 5 = 0.	Alc. male 5 = 2, died 4th week.	Nor. male = 1, normal young.

These results stand in marked contrast to the records of the control, which show all normal conceptions and normal offspring.

The second table shows the results of successive matings in ten of the females. The varying success of the conceptions in the same individual are striking.

Nice has quite recently recorded a similar series of experiments with alcohol on mice. Alcohol was given to the mice in their food. Nice finds that while there was a certain fatality among the offspring from alcoholic parents as compared with those from normal parents, where there was no fatality, yet nevertheless the offspring of alcoholic parents actually grew faster than those from the control. This may indicate that alcohol is not equally poisonous in its effects upon all animals, as might really be expected. The germ cells of mice may be more or less immune to the action of alcohol. It is well known that the action of alcohol is different in its effects on individuals from different human families. Some alcoholics show chiefly nervous disorders, hallucinations, delirium, etc., while others may have no nervous symptoms but exhibit various derangements of the digestive glands, kidneys, etc., or may have a fatty degeneration of almost all organs.

Finally it may be concluded that the experimental evidence goes to show that the development of an offspring may be modified by either treating the parents so as to affect their germ cells or by subjecting the developing embryo itself to unusual or injurious conditions.

The causes of many congenital defects are therefore known. It is possible to control embryonic development to such an extent as to produce abnormal structures. May not the proposition be reversed and unfavorable environments be treated in such a manner as to render them favorable to normal development? Diseased mothers may in some cases, at least, be made fit for the function of reproduction.

The regulation of structural disease becomes then a problem of morphology and hygiene. It is most important, and must precede, or go before, the selective mating of human beings or the eugenics movement. The most intellectual will rarely submit to direction in choosing a mate, yet every productive pair will welcome any possible means of improving the quality of their offspring.

While preventive measures are being used to protect the postnatal life of the individual, why not guard as far as possible its prenatal development?

CORNELL UNIVERSITY MEDICAL COLLEGE,
DEPARTMENT OF ANATOMY,
NEW YORK CITY.

An Experimental Study of the Influence of Alcohol on the Germ Cells and the Developing Embryos of Mammals.

By

Charles R. Stockard and **Dorothy M. Craig.**

(Anatomical Laboratory, Cornell University Medical College, New York City, U. S. A.)

Eingegangen am 25. Juni 1912.

The eggs of lower vertebrates which develop outside the maternal body may be induced to give rise to abnormal or monstrous forms when subjected to various unusual conditions. By regulating the chemical environment in which the eggs develop definitely typical defects may be produced in great numbers. The same type of defect may be caused by a variety of substances. This fact was shown by STOCKARD '10 in studies on cyclopia and also recently by McCLENDON in similar experiments. The occurrence of the same defect under various conditions might be expected and has often been observed by numerous workers.

The important fact, however, is that under certain conditions a particular defect can be made to occur in a large number of the embryos.

In some of the alcohol experiments (STOCKARD '10) on fish's eggs dozens of embryos with typical brain and eye defects occurred while few or no other types of deformities existed. The experimenter has the power in these cases to predict, with at least a limited degree of certainty, the type of deformity which will result from a definite intensity of a certain treatment. Embryonic development in such cases may be partially regulated or controlled.

BARDEEN '07 showed that when normal eggs of the toad are fertilized by spermatozoa which have been modified by treatment with the Roentgen Rays the eggs develop abnormally. O. Hertwig '11 has since recorded similar results by treating the spermatozoa of frogs with radium. These experiments show convincingly the influence of the male germ cell on the development and structure of the embryo. Thus in the lower vertebrates it has been possible to modify development by treating either the egg or spermatozoon.

The question now arises whether the germ cells and developing embryos of mammals may be similarly affected. The conditions in mammals are more complex since the embryo develops internally, and the substances administered to the body of the parent may induce secondary effects which will modify or confuse the results. Should a substance be secured that will act directly upon the germ cells or the tissues of the developing embryo within the parental body, then one might expect to regulate the action of this substance in much the same manner as in experimenting upon eggs developing in sea-water. From a number of recent observations it would seem that alcohol is just such a substance as is required.

It is an accepted fact that alcohol may cause changes and degeneration in many of the body tissues. How then can the reproductive tissues escape, or may they be affected without seriously injuring other tissues? Bertholet, 1909, made an extensive examination of the influences of alcohol upon the histological structure of the reproductive glands and found much degeneration and atrophy in the testicular parenchyma of chronic alcoholics, both young and old. Nicloux has shown that alcohol has a marked affinity for the reproductive glands and that alcohol may occur, as such, in the testicular tissues and in the seminal fluid of mammals. The ripe spermatozoa may, therefore, be bathed in a weak solution of alcohol shortly before fertilizing the egg, and if affected by the alcohol the spermatozoon may cause the developing eggs to give rise to a defective or deformed individual.

Nicloux has also demonstrated the passage of alcohol from the blood of the mammalian mother into the tissues of the embryo (guinea-pigs and dogs). After a short period of time the amount of alcohol in the blood of the foetus is about equal to that in the blood of the mother, while there is really more alcohol in a given weight of foetal tissue than is to be found in an equal weight of liver tissue from the mother. The reality of the passage of alcohol from the mother to the foetus demonstrates the possibility of the intoxication of the foetus.

The experiments on mammals do not then really differ greatly from those on the lower vertebrates, where the externally developing eggs may be placed directly in various unusual solutions; since the egg or embryo, although within the mother's body, is readily bathed or impregnated by the alcohol contained within the mother's blood.

The few experiments upon the influence of alcohol on the mammalian offspring are not at all conclusive and are somewhat contradictory.

Combemale and Hodge studied the effects of alcohol upon the offspring of dogs and recorded injurious effects, though their experiments were performed on very limited numbers of individuals.

Quite recently Nice has recorded experiments which seem to show that alcohol exerts a very slight effect upon the offspring of mice. He has given small doses of alcohol mixed with the food and also supplied 35 % alcohol instead of water for the animals to drink. Much of the alcohol probably evaporated from the food before it was eaten. Animals treated in this way gained in weight over the control, and their offspring excelled the control in rate of growth. The fecundity of the treated mice was greater than that of the control but there was also a greater mortality of the offspring from the treated parents. None of the control young died while 11.1 % of the alcoholic young died soon after birth. There were no abortions, no still-births and none of the young were deformed.

Mice may possibly be peculiarly resistant to alcohol though we should rather think that they received too little to give a pronounced effect, yet it was sufficient to cause a certain fatality among the offspring.

The few studies on mammals have failed to produce convincing evidence of the specific actions of alcohol. Yet the statistical data from observations on defective human beings would indicate that alcohol had a special affinity for the developing nervous system[1]). The experiments upon the influence of alcohol on the developing fish embryo (Stockard '10) demonstrated that alcohol did have a specific affinity for the central nervous system and caused the brains of these embryos to exhibit numerous deformities while the organs of special sense were also affected.

---

1) The senior author has given a somewhat extensive review of the literature pertaining to this subject in the Archives of Internal Medicine. 1912.

Method and Material.

The experiments here recorded were undertaken in order to ascertain whether alcohol did exert a marked influence on the germ cells and developing embryos of mammals, and if possible to demonstrate the nature and mode of action of this influence. Alcohol was employed as an agent since it may be given to guinea-pigs without greatly disturbing their normal physiological processes and so does not produce marked conditions which would secondarily effect the results. As before mentioned alcohol may remain as such in the blood and tissues of a mammal and may thus act directly upon the embryo just as it would when added to the sea-water in which fish's eggs were developing. It is an active substance, and, therefore, lends itself admirably to experimental use.

The experiments have been conducted on guinea-pigs since they breed fairly rapidly and rear their young without much difficulty in the laboratory. Strong healthy stock has been chosen and the animals have been carefully handled. All have remained in vigorous health and most of them have increased in size and fattened during the progress of the experiment. The males and females have been kept separated and individual pairs were mated from time to time.

The animals are first tested by normal matings and found to produce normal offspring. The alcoholic treatment is then begun on a given number of individuals and the males and females mated in different combinations according to whether they are alcoholics or normal. An alcoholic male is mated with a normal female, the paternal test, this is the crucial test of the influence upon the germ cells as here the defective offspring must be due to the chemically modified spermatozoon from which it arises, since the egg and the mother in which the embryo develops are both normal.

Normal untreated males are paired with alcoholic females, maternal test, in this case the defective offspring may be due either to a modified ovum or to the fact that it developed in a mother with alcoholic blood, therefore supplying an unfavorable developmental environment. Lastly, its condition may be due to both of these causes. The mammalian mother has two chances to injure an offspring, either by producing a defective egg or secondly by supplying an unfavorable or diseased environment in which the embryo develops.

The final combination is the mating of alcoholic individuals, this of course offers the greatest chance for defective offspring.

Alcohol is administered to the animals by inhalation. It was first given with the food, but the animals did not relish it, and therefore took less food. It was then given by stomach tube but this method so upset the animals that the results might have been modified by their poor bodily conditions and the bad state of their stomachs. The inhalation method is entirely satisfactory, the guinea-pigs thrive and usually gain in weight during the experiment, they have good appetites and are in all respects apparently normal. The only indication of the effects of the treatment is shown by the quality of offspring they produce.

The apparatus used for giving the alcohol consists of an air tight copper tank 36 inches long by 18 inches wide and 12 inches deep with a sloping bottom draining to the center. Over this bottom is placed a wire screen and below the screen cotton soaked with 95 % alcohol is spread. The tank is closed and allowed to stand until the atmosphere within is saturated with alcoholic fumes. A ventilation system is so arranged that a given quantity of alcohol fumes may be driven through the tank in a given time, but it has not seemed advisable to use this device as the degree of intoxication is a better index to the physiological response of the animals. Their resistance to the fumes is changeable. The guinea-pigs, three or four at the time, are placed on the wire screen above the evaporating alcohol, the tank is again closed and the animals are allowed to remain until they begin to show signs of intoxication, though they are never completely intoxicated. They usually inhale the fumes for about one hour. The animals are treated in this way for six days per week and some have now been treated over a period of about nineteen months. None of the effects are due to want of air since the same number of guinea-pigs may remain for hours in this closed tank without showing any signs of discomfort when there are no fumes present.

In order to avoid handling the females during late pregnancy a special treating cage is devised for them. An ordinary box run with a covered nest in which the animal lives is connected by a drop door with a metal lined tank having a similar screen arrangement to that described for the general treatment tank. The pregnant animal may be driven daily into the tank and thus treated with alcohol fumes throughout her pregnancy without having to be handled or moved about in any way that would tend to disturb the developing foetus.

Results.

During the vapor treatment the animals usually react in a manner quite similar to their behavior in weak fumes of ether or chloroform. The majority of them sit quite motionless and sniff their noses for a time and then become somewhat drowsy. A few individuals, however, are excited by the treatment and run about the tank becoming sexually excited and may often fight other animals savagely. One of the males fights and bites so vigorously while taking the fumes that he has to be treated separately from all others. The fumes thus have a different influence upon the behavior of different individuals in much the same way that alcoholic intoxication expresses itself differently on different human beings.

During the first few weeks of the treatment the fumes cause the eyes to water so that tears run over the face. The nose and mouth also become moist and the animals sniff almost constantly. Alcoholic fumes are very irritating to the mucous membranes at first. The conjunctiva of the eye becomes irritated and finally opaque in some instances, so that the eye takes on a white appearance. The tissues seem, however, to develop a resistance to the fumes. The eyes become clear after a few months and never again become opaque. The nasal mucosa also ceases to secrete excessively unless the animal is left in the tank for an unusually long time.

Many of the guinea-pigs have been killed after treatments of different duration up to fifteen months and all of their viscera carefully examined and the reproductive glands studied microscopically. In no case have we found any changed structures due to the alcoholic treatment. The lungs, liver, stomach, intestines, kidneys, reproductive glands, brain and all other parts appear perfectly normal. The general health and behavior of the animals also indicate that they are in good condition. As before mentioned several animals have been partially castrated during the experiment. One of the reproductive glands was removed and examined microscopically. In all cases the germ cells, ova or spermatozoa, were found to exhibit perfectly normal structure. One cannot claim, therefore, that this treatment is excessively severe or greater in proportional amount than the alcohol a human being often takes. The matter of fact is that these animals have never been completely intoxicated but receive only enough alcohol six times per week to affect their nervous states.

They may be compared to a toper who drinks daily but never becomes really drunk.

While the bodies of these animals show no direct effects of the alcohol, the conditions of the offspring to which they give rise exhibit most strikingly the effects of the alcoholic treatment. The results of mating the alcoholized guinea-pigs are summarized in Table I.

Table I.

Effects of Alcohol on the Offspring of Guinea-Pigs.

Condition of animal	No. of matings	No result or early abortion	Still-born litters	No. still-born young	Living litters	Young dying soon after birth	Surviving young
Alcoholic male by normal female .	24	14	5	8	5	7	5[1]
Normal male by alcoholic female	4	1	0	0	3	3 premat.	2[2]
Alcoholic male by alcoholic female	14	10	3	6	1	1 6th day	0
Summary	42	25	8	14	9	11	7[3]
Normal male by normal female . Control	9[4]	0	0	0	9	0	17

Fifty-five matings of treated animals have been made. Forty-two of these have now reached full term and are recorded. Thirteen matings are not yet due. From the forty-two matings only seven young survived, and six of these are still living, five of which are small for their ages though their parents were unusually large strong animals.

[1]) Four survivors in one litter, and one was a member of a litter of three, the other two died immediately after birth.

[2]) One lived to become pregnant with two young in utero, one deformed. Other survivor normal, the mother was not treated until after first two or three weeks of pregnancy.

[3]) Of thirty-two young born only seven have survived.

[4]) One other non-alcoholic mating was made from which two young resulted, they died after the second and fourth days respectively and the mother died two days later, her diseased condition no doubt affected the suckling young. They have for this reason not been included in the normal control.

The conditions of the animals in the mating pairs are shown in the first column of the table and the total results of the mating are indicated in the following columns.

The first horizontal line gives the records when alcoholic males are paired with normal females. Twenty-four such matings were made. Fourteen of these gave negative results or resulted in early abortions. Many embryos were aborted during very young stages and some of these were deformed, though they were generally in such poor conditions after being cast out into the cages that little could be learned from them. They were partially or completely eaten by the mother in most cases. The males were always kept for a number of days with the females during favorable periods and conception should have occurred in all cases, as it did in the control matings. Only ten of the twenty-four matings resulted in conceptions which ran the full term. Half of these, or five, were still-born litters. There were three still-born litters of two young each and two of one individual each. Most of these were slightly premature, their eyes being closed and the hair sparce on their bodies. (A normal guinea-pig at birth is well covered with a hairy coat, its eyes are open and it very quickly begins to run about actively.)

Five litters of living young were born. One litter consisted of only one young, a weak individual that grew very little and died after six weeks. Two litters contained two young each. The members of one of these litters died during the first and fourth weeks having been weak and small since birth. Both of those in the other litter were in a similarly feeble condition and died before the first month. One litter contained three young, two of these died immediately after birth; the other one is still alive though small for its age. The fifth litter contained four young all of which are runty though their parents were unusually large animals. Thus out of twenty full-term young of which only twelve were born alive but five individuals have survived and these are unusually small and very shy and excitable animals.

It is a point of some interest that all of the young animals that died showed various nervous disturbances having epileptic-like seizures and in every case died in a state of convulsion.

The important fact in the above cases is that only the father was alcoholic, the mother being a normal vigorous animal. This experiment clearly demonstrates that the paternal germ cells of mammals may be modified by chemical treatment to such a degree that the male will beget abnormal offspring even though he mate with a vigorous female. A reconsideration of the figures in the first line of the table shows really how decidedly the injured spermatozoon expresses itself in the fate of the egg with which it combines.

The second line of the table shows the results of matings between alcoholized females and normal males. These matings might be expected to give more marked results than the previous ones, since in the treated females not only the germ cells may be affected but the developing embryo itself may be injured by the presence of alcohol in the blood of the mother. NICLOUX has shown that alcohol may pass directly from the maternal blood into the embryonic tissues of a guinea-pig.

The spermatozoon, however, is probably a more sensitive structure than the egg and is easily injured or killed by slightly abnormal conditions. It might possible be that when such specialized cells swam for even a short time in seminal fluid containing a trace of alcohol that their chemical nature would be so decidedly disturbed as to render them incapable of inducing normal development after impregnating the eggs. At any rate the few cases at present available seem to indicate that the effect on the offspring is equally as great when it is produced by an alcoholic father as by an alcoholic mother.

There are only four matings between alcoholized females and normal males. One of these gave a negative result or was possibly aborted very early. Three living litters were born. One litter consisted of three premature young which died shortly after birth. The remaining two litters each contained only one young but these two animals survived. One of these guinea-pigs was born after the mother had been treated for three and one half months. The offspring was weak and small for several months after birth but finally recovered and developed into a normal animal. This animal, a female, was mated with an alcoholic male and became pregnant. Unfortunately she was killed by accident and on examination her uterus was found to contain two embryos of 33 and 32 mm. in length. One of these embryos was deformed and showed very decidedly degenerate and feebly developed hind legs. The posterior end of its body was also poorly formed. This is of interest since all of the affected offspring of alcoholic guinea-pigs are weak in their hind extremities and drag their legs. Yet none were so modified as to show a noticeable structural defect except this embryo, which had one alcoholic grandmother and an alcoholic father.

The only other survivor from an alcoholic mother is strong and full grown for its age. The mother had been treated for only one and one-half months when the offspring was born so that she was normal during the first two or three weeks of pregnancy. No doubt the early stages of development are more easily modified to produce significant defects than are the later. This question is being more fully tested on guinea-pigs with experiments now in progress. STOCKARD has shown, however, in treating fish's eggs that the period at which the treatment is applied is a most important factor in determining the type of defect or modification which will result. Certain salts, different strengths of magnesium chloride for example, which give pronounced effects when added to the sea-water containing eggs in early developmental stages may really be ineffective after the eggs have developed beyond these stages. In the case under consideration the offspring might not have fared so well if the alcoholic treatment had been started on the mother a few weeks before conception, instead of three weeks after her pregnancy had begun. This with other points shall be more completely analyzed in future communications on these experiments.

The four matings of alcoholic females and normal males resulted in three living litters consisting in all of five individuals. Three of the young were premature and died shortly after birth while two young survived.

Finally, we may consider the results of pairing two alcoholized individuals. The third line of the table summarizes these data. As might have been anticipated this type of mating has given the highest fatality of all.

Ten out of a total of fourteen matings have given no offspring or early abortions which were in many cases eaten by the mother. Three still-born litters have been produced each consisting of two young. Only one living litter was born from the fourteen matings in which both parents were alcoholic and this litter consisted of but one weak individual which died in convulsions on the sixth day after birth. This is indeed a decided effect of alcohol upon the offspring when one considers the nine control matings, all of which gave living litters containing a total of seventeen individuals all surviving.

Two young which were included in the control and died, should not really be counted. They died four days after birth and the mother died two days later in a diseased condition. No doubt the poor state of the mother had much to do with the fate of the suckling young. She was an animal that had only been in the experiment for a short time (this was her first mating) and is one of the very few that have contracted disease or died during the nineteen months of the work.

The fourth line of Table I gives a summary of the experiments. There have been forty-two full-term matings, twenty-five of which gave no offspring or early abortions, eight still-born litters have occurred consisting of fourteen individuals, only nine living litters have been born, 21 % of the matings, these contained eighteen young and but seven of this number have survived and five of these survivors are runts or small for their ages.

The bottom line of the table shows nine control matings, all have given living litters containing a total of seventeen young, all of them surviving. Two young that died, as stated above, were from a dying mother and are not included in the control.

Records of successive matings of ten of the female guinea-pigs are shown in Table II. The varying ways in which the same individual has responded in different matings is noticeable. Number 10, an alcoholic female, first mated with an alcoholic male gave one young which died on the sixth day after birth. On being remated with the same male No. 10 gave no offspring. Then mated with another alcoholic male gave no offspring. She mated again after several months with the first male and on being killed was found to contain one embryo in utero about two weeks old.

Female No. 15, a normal guinea-pig, shows an instructive record. She was mated with an alcoholic male and gave birth to two still-born young. Mated to another alcoholic male and gave negative result. Remated with the second male and gave two young both of which died of convulsions in four weeks after birth. She was then mated with a normal male as a control and gave one vigorous normal offspring which survived.

The other records are easily understood.

These experiments have suggested many questions still to be solved, some of which are now being tested. The length of time necessary to treat an animal before the resulting offspring is affected, whether this time is equally long for both sexes, and what amount of individual variation may exist. An important point to ascertain

Table II.

Successive Matings of 10 Females.

Animal	1st Mating	2nd Mating	3rd Mating	4th Mating
No. 10 Alc.	Alc. male 4, 1 young died in 6 days	Alc. male 4 0	Alc. male 6 0	Alc. male 4, 1 embryo in utero, 2 weeks after
No. 12 Alc.	Alc. male 5 0	Alc. male 5 0	Alc. male 4 0	
No. 11 Alc.	Alc. male 6 0	Alc. male 6 0	Alc. male 5 2 premat. still-born	Alc. male 4 0
No. 13 Nor.	Alc. male 5 1 still-born	Alc. male 5 0	Alc. male 4 0	
No. 17 Nor.	Etherized male 1 0	Etherized male 1 0		
No. 18 Nor.	Alc. male 5 0	Alc. male 5 0	Alc. male 6 0	
No. 7 Nor.	Etherized male 2 2 premat. still-born	Etherized male 2 0	Etherized male 2 0	
No. 14 Nor.	Etherized male 3 0	Etherized male 3 0		
No. 19 Nor.	Alc. male 4 0	Alc. male 6 1 still-born	Alc. male 6 0	Alc. male 5 4 small, active only 1/2 size, but living
No. 15 Nor.	Alc. male 6 2 still-born	Alc. male 5 0	Alc. male 5 2 died 4th wk. convulsions	Nor. male 1 normal vigorous young

is whether the effects of the alcoholic treatment are permanent, or does the animal recover after a time and again become capable of giving normal offspring. One of the most valuable problems is to regulate the treatment in such a manner as to induce a definite type of defect with a given kind or degree of treatment. The structure or morphology of the monsters and defective offspring which occur is to be carefully studied. Many other points might readily be suggested.

Definite and well controlled experiments with alcohol and other substances on the mammalian offspring have not been sufficiently studied. The work is really in its beginning, and while there is much evidence to show that various toxic agents do effect and modify the offspring, facts are badly needed to demonstrate the regularity and manner of this modification. The present experiments seem to us to demonstrate in a convincing way that alcohol may readily effect the offspring through either parent, and that this effect is almost fatal to the existance of the offspring when the parents have been treated with even fairly large doses of alcohol. Many of the cases seem to indicate further, that the tissues of the nervous system in the offspring are particularly sensitive in their responses to the induced conditions.

Summary.

Guinea-pigs have been treated with alcohol in order to test the influence of such treatment on their offspring. Male and female animals are given alcohol by an inhalation method until they begin to show signs of intoxication, though they are never completely intoxicated. They are treated for about one hour at the time, six days per week. The treatment in some of the cases has now extended over a period of nineteen months. The animals may be said to be in a state of chronic alcoholism.

Fifty-five matings of the alcoholized animals have been made, forty-two of which have reached full term and are recorded.

From these forty-two matings only seven young animals have survived, and five of them are unusually small though their parents are large vigorous guinea-pigs.

The following combinations were made:

1) Alcoholic males were mated to normal females. This is the paternal test, and is the really crucial proof of the influence of alcohol on the germ cells, since the defective offspring in this case must be due to the modified spermatozoa, or male germ cells, from which they arose. Twenty-four matings of this type were made, fourteen of which gave no offspring or very early abortions; five still-born litters were produced consisting of eight individuals in all, and five living litters containing twelve young. Seven of these twelve died soon after birth and only five have survived. Four of the survivors are from one litter and the fifth is the only living member of a litter of three.

2) Normal males were mated to alcoholic females. This is the maternal test, in such cases the alcohol may affect the offspring in two ways, by modifying the germ cells of the mother or by acting upon the developing embryo in utero. Only four such matings were tried. One gave no offspring, three living litters were born, one consisting of three premature young that died at birth, while the other two consisted each of one young which has survived. The alcoholic treatment in one of the last cases was only begun after the mother had been pregnant for about three weeks.

3) Alcoholic males were mated to alcoholic females. This is the most severe test both parents being alcoholic. Fourteen such matings gave in ten cases no offspring or very early abortions. Three still-born litters were produced consisting in all of six individuals, while only one living young was born. This single offspring from the fourteen matings died in convulsions on the sixth day after birth.

The young that have died in the experiment showed nervous disorders many having epileptic-like seizures and all died in convulsions.

Nine control matings of the same group of animals have given nine living litters consisting in all of seventeen individuals, all of which have survived and are large vigorous specimens for their ages.

Fourty-two matings of alcoholic guinea-pigs have given only eighteen young born alive and of these only seven, five of which are runts, survived for more than a few weeks, while nine control matings gave seventeen young all of which have survived and are normal vigorous individuals. These facts convincingly demonstrate the detrimental effects of alcohol upon the parental germ cells and the developing offspring.

Zusammenfassung.

Meerschweinchen wurden mit Alkohol behandelt, um den Einfluß einer solchen Behandlung auf ihre Nachkommenschaft zu erweisen. Männliche und weibliche Tiere bekommen Alkohol mittels einer Inhalationsmethode, bis sie Zeichen von Intoxication aufweisen, doch werden sie niemals völlig betrunken gemacht. Sie werden jedesmal etwa 1 Stunde lang an 6 Tagen der Woche behandelt. Die bisherige Behandlung erstreckt sich in einigen der Fälle auf eine Dauer von 19 Monaten. Die Tiere befinden sich sozusagen in einem Zustande von chronischem Alkoholismus.

Es wurden 55 Paarungen der alkoholisierten Tiere vorgenommen, von denen 42 bis zum vollen Schwangerschaftsablauf kamen und hier benutzt sind. Von diesen 42 Paarungen sind nur 7 junge Tiere am Leben geblieben, und 5 von diesen sind ungewöhnlich klein, obgleich ihre Eltern große, kräftige Meerschweinchen sind.

Die folgenden Kombinationen wurden versucht:

1) Alkoholische Männchen wurden mit normalen Weibchen gepaart. Das ist also die Prüfung des väterlichen Einflusses und stellt den entscheidenden Versuch betreffs des Einflusses des Alkohols auf die Keimzellen dar, da in diesem Falle die beeinflußten Spermatozoen oder männlichen Keimzellen die Ursache der defekten Nachkommenschaft sein müssen, von denen sie ihren Ausgang nimmt. 24 Paarungen dieses Typus wurden veranstaltet, von denen 14 überhaupt keine Nachkommenschaft oder sehr frühzeitige Aborte ergaben. 5 totgeborene Sätze enthielten alles in allem 8 Individuen und 5 lebende Sätze ergaben 12 Junge. 7 von diesen 12 starben bald nach der Geburt und nur 5 sind am Leben geblieben. 4 der Überlebenden stammen aus einem Wurf und das 5. ist das einzig Überlebende aus einem Wurf von 3 Individuen.

2) Normale Männchen wurden mit alkoholischen Weibchen gepaart. Das stellt also die Untersuchung des Einflusses der Mutter dar. In solchen Fällen kann der Alkohol die Nachkommenschaft auf zweierlei Weise affizieren: durch eine Einwirkung auf die mütterlichen Keimzellen oder eine Einwirkung auf den sich im Uterus entwickelnden Embryo. Nur 4 solche Paarungen gelangten zur Untersuchung: 1 ergab keine Nachkommenschaft, lebende Sätze wurden 3 geworfen, von denen einer aus 3 frühgeborenen Jungen bestand, welche bei der Geburt starben. Die andern Sätze bestanden aus je 1 Jungen, das leben blieb. Die Alkoholbehandlung in einem dieser letzten Fälle setzte erst ein, als die Mutter bereits seit ungefähr 3 Wochen trächtig war.

3) Alkoholische Männchen wurden mit alkoholischen Weibchen gepaart. Das ist der stärkste Versuch, da ja beide Eltern alkoholisch sind. 14 solche Paarungen ergaben in 10 Fällen keine Nachkommenschaft oder sehr frühzeitige Aborte. 3 tote Würfe wurden hervorgebracht mit im ganzen 6 Individuen, und nur ein einziges lebendes Junge geboren. Dieser einzige Nachkomme von den 14 Paarungen starb in Konvulsionen am 6. Tage nach der Geburt.

Die während der Versuche gestorbenen Jungen zeigten nervöse Störungen, manche hatten epilepsieähnliche Zufälle und alle starben in Krämpfen.

9 Kontrollpaarungen von derselben Tiergruppe ergaben 7 lebende Würfe, die im ganzen aus 17 Individuen bestanden, die sämtlich am Leben blieben und für ihr Alter kräftige Exemplare darstellen.

42 Paarungen alkoholisierter Meerschweinchen haben nur 18 lebende Junge ergeben, und von diesen lebten nur 7, darunter 5 Kümmerlinge, länger als einige Wochen, während 9 Kontrollpaarungen 17 Junge ergaben, welche alle am Leben blieben und normale, kraftvolle Individuen sind. Diese Tatsachen demonstrieren überzeugend den schädigenden Einfluß des Alkohols auf die elterlichen Keimzellen und die sich entwickelnde Nachkommenschaft.

(Übersetzt von **W. Gebhardt.**)

Literature cited.

BARDEEN, C. R., Abnormal Development of Toad Ova Fertilized by Spermatozoa Exposed to the Roentgen Rays. Journ. Exp. Zool. Vol. 4. pp. 1—44. 1907.

BERTHOLET, E., Über Atrophie der Hoden bei chronischem Alkoholismus. Centralbl. f. allgem. Path. Bd. 20. S. 1062. 1909.

HERTWIG, O., Die Radiumkrankheit tierischer Keimzellen. Bonn, Cohen, 1911.

HODGE, C. F., The Influence of Alcohol on Growth and Development. In »Physiological Aspects of the Liquor Problem«. First Ed. by Billings, Houghton, Mifflin, N. Y. pp. 359—375. 1903.

MAIRET et COMBEMALE, Influence dégénération de l'alcool sur la descendance. Compt. rend. Acad. de Sc. Vol. 106. p. 667. 1888.

MCCLENDON, J. F., An Attempt towards the Physical Chemistry of the Production of One-Eyed Monstrosities. Am. Journ. Physiol. Vol. 29. pp. 289—297. 1912.

NICE, L. B., Comparative Studies on the Effects of Alcohol, Nicotine, Tobacco Smoke and Caffeine on White Mice. I. Effects on Reproduction and Growth. Journ. Exper. Zool. Vol. 12. pp. 133—152. 1912.

NICLOUX, Passage de l'alcool ingéré de la mère au foetus etc. L'Obstétrique. Vol. 97. 1900.

STOCKARD, C. R., The Influence of External Factors, Chemical and Physical, on the Development of Fundulus heteroclitus. Journ. Exp. Zool. Vol. 4. pp. 165—201. 1907.

—— The Artificial Production of a Single Median Cyclopean Eye in the Fish Embryo by means of Solutions of Magnesium Chlorid. Arch. f. Entw.-Mech. Bd. 23. S. 249—258. 1907. Journ. Exp. Zool. Vol. 6. pp. 285—337. 1909.

—— The Influence of Alcohol and Other Anæsthetics on Embryonic Development. Am. Journ. Anat. Vol. 10. pp. 369—392. 1910.

Fütterungsversuche an Amphibienlarven.

Von J. F. Gudernatsch (New York).

(Vorläufige Mitteilung.)

(Der Redaktion zugegangen am 9. Mai 1912.)

Es wurde versucht, der zurzeit im Mittelpunkt des Interesses aller Biologen stehenden Frage, welche Rolle die meisten drüsigen Organe, namentlich die Drüsen mit sogenannter „innerer Sekretion", im Haushalte des Organismus spielen, auf dem Wege der Verfütterung experimentell näher zu treten. Zu den Versuchen wurden im Jahre 1911 Quappen von Rana temporaria und esculenta verwendet, jetzt werden diese Versuche an den gleichen Tieren und an Bufo- und Tritonlarven fortgesetzt. Zur Verfütterung werden Säugetierorgane verschiedentlichen Ursprungs verwendet, vom Pferd, Rind, Schwein, Hund, Katze, Kaninchen usw. Es hat sich bis jetzt gezeigt, daß die Speziesherkunft der Organe ohne wesentliche Bedeutung für ihre Wirkungsweise ist. Verfüttert werden: Thyreoidea, Thymus, Nebenniere, Hypophyse, Hoden, Ovarium, Milz, Leber, Pankreas und Muskel. Die genannten Substanzen werden frisch kurz nach Entnahme aus dem Organismus oder, nachdem sie höchstens zwei Tage auf Eis gelegen sind, verfüttert. Im Jahre 1910 ausgeführte Versuche, den Einfluß von Extrakten obiger Substanzen und solcher verschiedener benigner und maligner Tumoren auf die Entwicklung von Fisch- und Amphibieneiern zu studieren, führten aus äußeren Gründen zu keinen befriedigenden Resultaten.

War es a priori zweifelhaft, ob von Säugern entnommene Organe bei Verfütterung an Amphibien auf deren Entwicklung einen spezifischen Einfluß ausüben könnten, so hat der Verlauf der Experimente alle Zweifel darüber verscheucht. Natürlich bleibt immer noch die Frage offen, ob diese Organe, nachdem sie den Amphibiendarm passiert haben, auf den betreffenden Organismus den gleichen Einfluß ausüben, der ihnen in ihrem Ursprungsorganismus zukommt.

Sehr deutliche Resultate ergeben die Verfütterung von Thyreoidea und Thymus, zweier Organe, die ja auch bei Wachstum und Differenzierung des sich entwickelnden Organismus eine große Rolle spielen. Werden sie verfüttert, so sind beide Organe in ihrer Wirkung auf jene Vorgänge gerade entgegengesetzt und auch der Zeitpunkt, auf dem ihr Einfluß am stärksten in die Erscheinung tritt, dürfte ein verschiedener sein. Die Thyreoidea scheint am raschesten zu wirken, je älter die Tiere bei Beginn der Fütterung sind, bei der Thymus scheint das Umgekehrte der Fall zu sein.

Wird Thyreoidea auf irgend einem Stadium gefüttert, so hört jedes Weiterwachstum der Quappen auf und die Tiere schicken sich sofort zur Metamorphose an. So konnten Quappen, die noch keine Extremitäten besaßen, innerhalb 7 Tagen dazu gebracht werden, Hinter- und Vorderbeine zu entwickeln und den Schwanz zu reduzieren. In diesem Frühjahr ist es gelungen, Quappen, die erst 16 Tage alt (16 Tage nach dem Verlassen des Eies) waren, zur Bildung der Vorderextremitäten zu bringen. Da die Thyreoideafütterung jedes Weiterwachstum unterdrückt, so sind das Resultat derselben ganz kleine (Zwerg-) Frösche. Dabei ist es ganz gleichgültig, auf welcher Altersstufe die Behandlung beginnt oder welche Nahrung vorher verfüttert wurde. Waren die Quappen zu Beginn der Fütterung sehr klein, so sind es die metamorphosierenden Tiere auch, größere Quappen aber ergeben größere Frösche.

Bei der Verfütterung von Thymus sind die Resultate gerade entgegengesetzt. Die Tiere wachsen anfangs sehr rasch, es werden große Kaulquappen erzeugt, je länger dieselben aber unter dem Einflusse der Thymus stehen, um so mehr wird die Differenzierung hinausgeschoben und die Metamorphose eventuell ganz unterdrückt. So kommt es, daß aus demselben Satz die mit Thyreoidea gefütterten Tiere sehr rasch, innerhalb 1 bis 2 Wochen, zur Metamorphose gebracht werden können, während die mit Thymus gefütterten Tiere selbst viele Wochen später, nachdem die Kontrolltiere schon längst metamorphosiert haben, zum größten Teil noch ganz undifferenzierte Quappen sind ohne Extremitäten und teilweise gar nicht zur Metamorphose kommen.

Interessant sind auch Färbungsunterschiede der verschieden behandelten Tiere, z. B. die tiefdunkle Farbe der Thymusquappen (Ausbreitung der Pigmentzellen), die auffallend helle Farbe der Nebennierequappen (Kontraktion der Pigmentzellen) usw.

Ein eingehender Bericht über die Gesamtversuche von 1911 und 1912 sowie über die histologischen Ergebnisse und die Beeinflussung der Regeneration durch die verschiedenen Substanzen wird später gegeben werden.

Druck von Rudolf M. Rohrer in Brünn.

Feeding Experiments on Tadpoles.

I. The influence of specific organs given as food on growth and differentiation.

A contribution to the knowledge of organs with internal secretion.

By

J. F. Gudernatsch,

Department of Anatomy, Cornell University Medical College, New York City.

[From the department of histology, German University of Prague. Director: Prof. ALFRED KOHN [1]).]

With plate IX.

Eingegangen am 11. Juli 1912.

During a stay in the Zoological Station of Naples, spring 1910, an attempt was made to study the influence of various organic extracts on the development of fish (*Belone*, *Gobius* etc.) and amphibian (*Rana*) eggs. The substances used were extracts from mammalian tissues, viz. thyroid, thymus, testicle, ovary, hypophysis, adrenal, pancreas, cancer of ovary, carcinoma of liver and carcinoma of rectum. Different quantities of the extracts, in corresponding degrees of concentration, were added to the sea-water, containing the recently fertilized eggs. The eggs were kept for various lengths of time — up to 20 hours — in this mixture and afterwards transferred into pure sea-water. In every case an influence upon the developing eggs was noticeable; and the disturbances of the normal development caused by the various substances were different. The difficulties met in these experiments were unusually great, partly on account of the rapid decomposition of the extracts brought from New

[1]) My best thanks are due to Prof. KOHN for permitting me to work in his laboratory, furnishing ample material and giving me out of his enormous experience valuable help and advice as well as for carefully revising the manuscript.

York — on board ship they were kept in cold storage, but there was no ice-box in the laboratory — and partly because in that season it was very difficult to get sufficient material; in two months *Belone* eggs were brought to the laboratory only twice. In spite of these inconveniences a great number of eggs were kept under observation up to the time of hatching. Yet the work could not be carried on systematically with sufficient repetitions and control experiments. Besides it was not to be expected that the influences of the various organs on development would allow of any conclusions as to the function of the respective organs, viz. thyroid gland etc. It was more likely that the disturbances of the normal development were of a general type caused perhaps by the change in the osmotic pressure of the surrounding medium etc.

To show that the influences of the various substances upon development were actually different, the following table may be given. It cannot, of course, be used for any generalizations; for it is the result of only one experiment on *Belone* eggs. The curves show the respective percentages of the living and developing eggs after a certain number of days; viz. of each 100 eggs there were on the second day still living: in the control 87, after addition of ovary exctract 85, etc. to .... carcinoma recti extract 30; at the beginning of the third day 76, 65, ... 26, etc. After the first day (each 100) a different decline of the curves is already visible. A certain percentage of the eggs, 13 of the control, died, since naturally not all eggs were equally able to develop, others were not fertilized etc. For this reason it may be that on the second day the curves rather run parallel. A striking decline of all curves with the exception of the control and thymus curves appears on the fifth day. On this day in normal development, the heart begins to beat. Many eggs which up to that time remained alive, although probably with diminished energy, seem not to have been able to survive this critical point, the starting of pulsation. From the sixth day on the curves run more or less parallel. The control shows after this time a comparatively strong decline. An explanation for this phenomenon may be that under the influence of the extracts the more feeble eggs were killed in the earlier stages of development, while in pure sea-water they were able to go on developing for some time before their vitality was exhausted.

Table I.

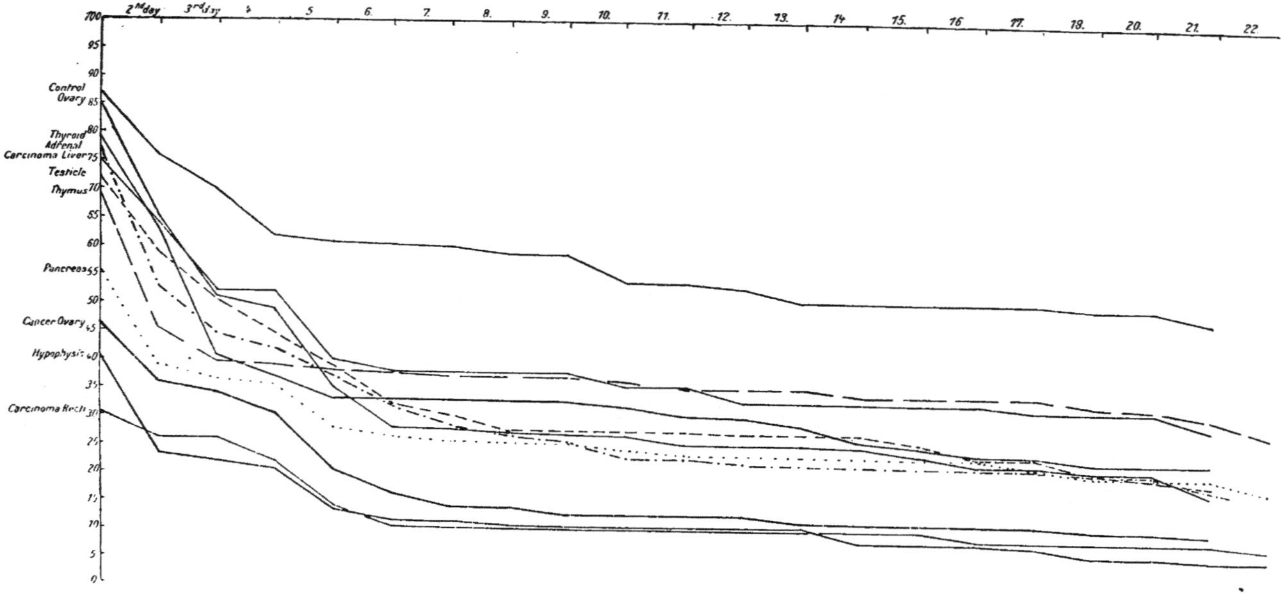

2nd day
3rd day
4
5
6.
7.
8.
9.
10.
11.
12.
13.
14
15.
16
17.
18.
20.
21.
22
100
95
90
Control
Ovary 85
Thyroid 80
Adrenal
Carcinoma Liver 75
Testicle
Thymus 70
65
60
Pancreas 55
50
Cancer Ovary 45
Hypophysis 40
35
Carcinoma Recti 30
25
20
15
10
5
0

Later on the advice of Prof. KOHN, the various organs have been applied as fresh feeding material. The tadpoles of *Rana temporaria* and *Rana esculenta* were chosen for the experiments. These feeding experiments carried on during the summer of 1911 in the histological laboratory in Prague gave some very interesting results.

The tadpoles were kept in bowls, each containing 15 to 20 individuals, and were fed three or four times a week on the different organs. *Rana temporaria* was used in two sets of different ages. As food were tried: thyroid, liver, adrenal, hypophysis, and muscle from horse, thymus from calf, testicle and ovary from dog or cat. Some organs from rabbits and pigs also were given. The origin of the organs used apparently made no difference in their action. The food was put into the water and was ravenously taken by the animals. With each experiment one group was left unfed as control to test how much nourishment the animals could take from the tap-water which in Prague is very rich on organisms. The water was changed daily, on hot days sometimes twice. This frequent change of water as well as the accumulation of the products of metabolism in the water between the changes may have exerted some influence on the development, yet the prevailing conditions were the same for all animals. Unfortunately it was found impossible to carry a constant current of air or water through the great number of dishes. It was also found impossible on account of the artificial feeding to keep plants in the dishes. The feeding was continued till shortly after the appearance of the fore legs, then the animals cease to take food for some time on account of the transformation of the jaws, and since breathing through the lungs begins the metamorphosed animals leave the water and look for fresh food.

Experiment I.

Rana temporaria in two sets of different sizes. Fig. 1 *a*—*e*, 2 *a*—*e*. Original size on May 23rd 1911 of set I 1.1—1.7 cm, of set II 1.8—2.3 cm.

The differences in size of the animals of the same set diminish somewhat with continued feeding, so that the deviations from a mean become less obvious. The following table gives the sizes of the different groups 26 days after the beginning of the experiment:

		set I	set II
May 23.	Original size, average . . .	1.4 cm	2.05 om
June 18.	Average size		
	control unfed	2.2 -	3.0
	liver	3.6 -	4.1
	muscle	3.5 -	3.9
	thymus	3.6 -	4.1
	adrenal	2.8 -	3.9
	hypophysis	2.8 -	—
	testis	2.5 -	3.5
	ovary	2.1 -	3.3 -
June 14.	thyroid	1.1 -	1.8 -

Testis and ovary could not be fed regularly on account of the material lacking, so that these animals are not much ahead of the unfed control.

The thyroid fed animals had died on June 16 as fully developed frogs, though dwarfs in size; therefore their measurement on June 14 is given, on which day the fore legs in set I were also noticeable.

From the notes the following data may be given[1]):

May 23. Beginning of the experiment.

May 30 (up to this day food had been given 4 times). All thyroid II show hind legs.

June 4. 3 thymus II show hind legs.

June 5. The differences in size between the individual groups become marked, thyroid I and II are smaller and thymus I and II bigger than the other animals. The differentiation (limbs, form of the body etc.) is most evident in thyroid.

June 6. More thymus II and some of the other groups grow hind legs.

June 7. The differences in size and differentiation have become more striking. Thymus II are the biggest animals, they have, however, retained the typical form of tadpoles. Thyroid II begin to grow fore legs and to reduce their tails, their bodies are markedly frog-like.

June 8. All thyroid I grow hind legs. Thyroid II have become typical, but very small frogs and begin to jump. Some adrenal I

[1]) For the sake of simplicity, the organ given is used throughout the paper as an index of the respective group. I and II mean: size I and II. For instance, liver I means: tadpoles, size I, fed on liver, etc. Thymusthyroid means: tadpoles, fed first on thymus, later on thyroid.

and II show a somewhat lighter color than the rest of the tadpoles.

June 14. Thyroid II begin to die off. Thyroid I grow fore legs and start to absorb their tails.

June 16. Thyroid I die off. 2 thymus II grow fore legs — i. e. 9 days later than thyroid II. The lighter coloring of the adrenal I and II becomes more evident.

June 18. Thymus I and II are the biggest tadpoles of each set. The difference in length between thymus I and II and liver I and II is not marked, yet the thymus tadpoles are broader, have stronger legs etc.

June 20. The liver show a greenish coloring. Some thymus I and liver I grow hind legs, i. e. 12 days later than thyroid I.

June 23. Thymus II seem to show a retarded differentiation being behind liver II, muscle II and adrenal II.

June 25. This difference becomes more evident. All adrenal tadpoles show a color markedly lighter than that of the rest of the groups.

June 27. Thymus I are much bigger, yet far less differentiated than liver I and muscle I.

June 29. For several days no more of the thymus II have grown fore legs, while more and more of the liver II, muscle II, adrenal II and even some of the poorly fed testicle II and ovary II have done so.

July 2. Thymus I and II are very big and their color is very dark. No more thymus II develop fore legs, while of liver II, muscle II and adrenal II there is only one in each group without fore legs. Even of the poorly fed testis II and ovary II there are only 3, respectively 5 without them. Adrenal I grow hind legs, muscle I fore legs.

July 3. All liver II, muscle II and adrenal II tadpoles have grown fore legs.

July 8. The adrenal II frogs which on July 3 had been taken out of the water and placed on sand are now just as dark as the frogs of the other groups. There is only 1 of liver I and 3 of muscle I without fore legs, while there are still 9 of thymus II and 10 of thymus I without them, though the latter tadpoles are much bigger than the former.

July 9. Hypophysis I and testis I grow hind legs, adrenal I are the largest, almost as big as thymus I.

July 11. Adrenal I grow fore legs.

July 13. There are only 2 of muscle I without fore legs, while there are still 7 of thymus I, and 5 of thymus II lacking fore legs.

July 17. Some hypophysis I and testis I grow fore legs, the last muscle I grows fore legs, while there are still 3 of thymus I and 3 of thymus II without them. Hypophysis I tadpoles gradually become rather transparent, especially their heads. They show on the right side of their body a greenish swelling beneath the skin. The green color is also seen through on the right side of their bodies, yet there is no swelling there. The tails of some are twisted in a peculiar manner.

July 21. Hypophysis I begin to die one after the other without completing their metamorphosis. There are still 3 thymus I and 3 thymus II without fore legs, their bodies assume a very irregular shape and become very broad and bloated.

July 26. The last adrenal I grows fore legs.

Aug. 3. 1 more thymus I grows fore legs, there remain 2 without fore legs.

Aug. 4. 1 more thymus I grows fore legs, there remains 1 without them, 2 thymus II die without fore legs. One only survives.

Aug. 5. The last thymus I and thymus II die without fore legs.

Aug. 17. Some of the unfed control grow hind legs.

It is evident from the data just given that the thyroid and thymus tadpoles (Fig. 1*a*—*b*, 2*a*—*b*). reacted most peculiarly to their specific foods, while the liver, muscle and adrenal animals showed a more indifferent behavior. However, the very light coloring of the adrenal was striking as compared with the very dark color of the thymus and the dark greenish one of the liver tadpoles. It is highly probable that the light color of the adrenal is not the result of the feeding with adrenal, but was merely a contraction of the pigment cells due to the contents of the chromaffine cells going into solution (adrenalin reaction). The gradually developing transparency of the hypophysis fed tadpoles must also be mentioned as well as the fact that most of them died without completing their metamorphosis. The nature of the green swelling in their abdomen can only be determined by microscopic examination.

The results of the testis and ovary feeding are inconclusive since regular feeding was impossible on account of the difficulties met with in providing the food. The behavior of the other groups, however, is characteristic, and there is also a definite control given, as the experiments were conducted on two different sets with corresponding results.

The quickest results were seen in the thyroid groups. While an increase in body size was lacking, the differentiation of the body was extremely rapid, both hind and fore legs appeared earlier than in any other group and the animals metamorphosed long before those fed on other substances. It was peculiar that every change in the body form set in almost simultaneously in all the animals of one set, so that the corresponding stages of development were reached within 24 hours or less; for instance, all had their hind or fore legs come out on the same day etc. In no other group could such a uniform development be observed. The only explanation of this can be the increased velocity of the differentiation processes. In the other groups only a few animals at first began to grow hind or fore legs, and often many weeks elapsed before all the others had reached the same stage in development. This is the natural course of events, since *a priori* not all tadpoles possessed the same vitality. The thyroid food, however, enacted such a strong accelerating influence on the body differentiation that the differences in time which existed in the development of the individual tadpoles were so reduced, that they hardly remained noticeable. The greatest difference in time, between the slowest and the most rapid differentiation of thyroid tadpoles of one set was less than one day.

The difference in time between the thyroid groups and those fed on other substances was as might be expected greater, the longer the treatment lasted. For instance, while only 5 days (May 30—June 4) lie between the appearance of the hind legs in thyroid II and thymus II, this interval in set I — I is the younger set, therefore was fed longer — is 12 days (June 8—June 20).

The precocious body differentiation of the thyroid fed tadpoles did not allow the animals to continue their growth, the result of the metamorphosis were therefore extremely small (pigmy-) frogs (Fig. 1 *a*, 2 *a*).

The feeding with thymus showed an influence on the development of the tadpoles, exactly the opposite of that caused by the thyroid diet. Its consequence was a prolonged increase in size beyond the normal, the metamorphosis, however, was much retarded or not completed at all as the animals died before that time. With this retarded development the individual differences, of course, were much emphasized. Not all individuals of one set grew their hind or fore limbs on the same day, as in the thyroid groups, but there were intervals of days and weeks between the corresponding stages in different individuals. Those tadpoles that possessed the least amount of vital energy had to be fed longest. They were, therefore, longest under the influence of the retarding food.

The later an organ develops in normal ontogeny, for instance fore legs later than hind ones, the more its appearance was postponed by the thymus and accelerated by the thyroid diet. The hind legs of thyroid II and thymus II appeared at an interval of 5 days (May 30—June 4), for the fore legs the interval was 9 days (June 7—June 16). This can also be expressed in the opposite way: the younger a tadpole is at the beginning of the feeding, the greater is the retarding influence on development by the thymus treatment and the accelerating influence by the thyroid treatment. For instance, in the appearance of the hind limbs in thyroid II and thymus II (older set) the difference in time is only 5 days (May 30—June 4), while in thyroid I and thymus I (younger set) 12 days (June 8—June 20). In this comparison those thymus fed tadpoles with quickest differentiation, about 10%, were chosen. If all the thymus tadpoles were considered, the average difference in time would be much greater; for thymus I and II, although fed regularly and abundantly, needed about two months before all had completed their development, while all the other groups had metamorphosed long before.

Thus the influence of the thymus food was such that in the beginning it caused a rapid increase in body size, going beyond the normal, while later on it postponed the metamorphosis extremely. The color of the animals became very dark during the experiment. Those tadpoles most backward in development showed a clumsy bloated shape.

Experiment II.

A group of *Rana temporaria* tadpoles, originally selected for ovary feeding, had been fed only twice, May 23 and May 25, with that substance, after this time, up to the start of experiment II, July 6, through 43 days, they had starved. The short feeding of ovary was of so little influence, that these animals differed in no respect from the unfed control tadpoles (Fig. 3). From July 6 on a part of these tadpoles were fed on thyroid, another part on thymus. The differences in the results were most evident (Fig. 6*a*, *b*) and corresponded to those of experiment I.

The diary reads as follows:

July 6. Start of the experiment. Average size 2.75 cm.
July 9. (after 3 days only) thyroid grow hind legs.
July 11. A difference in the sizes of thyroid and thymus is noticeable. 2 thymus grow hind legs.
July 12. Size of thyroid 2.6 cm, of thymus 3.2 cm. Thyroid assume frog-shape.
July 13. Thyroid grow fore legs and swim on their back.
July 14. Thyroid have completed their metamorphosis and begin to die.
July 17. Thymus are very big, entire length 3.7 cm (body 1.3 cm, tail 2.4 cm) and are very dark colored.
July 18. 2 thymus are still without hind legs. From to-day on these two will be fed on thyroid, so that the experiment now runs thus:

	Thymus.	Thymusthyroid.
July 21.		After 3 days only! Appearance of hind legs.
July 22.		Body assumes frog-shape. The dark (thymus) color has disappeared.
July 23.		Fore legs! The animals swim on their back. Length of body 0.8 cm — 0.9 cm - - tail 1,3 - — 1.3 - entire length 2.1 cm — 2.2 cm
July 24.	Not until to-day 2 of this group grow fore legs, although on July 17 they were so much further along than those in the right column.	
Aug. 5.	The last one completes its metamorphosis, 22 days later than thyroid, and 13 days later than thymusthyroid.	

Thus experiment II ends with the same results as experiment I. The feeding on thyroid causes an extremely rapid differentiation of the body with a complete suppression of growth (compare Fig. 3 and 6*a*), the feeding on thymus furthers the growth (Fig. 6*b*), but retards the differentiation. This is most strikingly seen in the subexperiment described in the right column. Those tadpoles which were backward most on July 17 metamorphose after having been fed on thyroid for only 5 days, sooner than those farthest along in development, which remained on thymus.

In this experiment the influence of the thyroid food made itself manifest after only 3 days. The reason for this might be that the animals had starved through 43 days and had thus become older without being able to develop. They were, one might say, in a condition of latent overripeness and the first application of food rapidly caused a further development.

Experiment III.

Taken alone experiment III would not allow of any conclusions, since it was done with only a few animals. Yet the results attained are absolutely the same as those of experiments I and II, and therefore furnish a confirmation of the latter.

Some control animals of experiment I which had been starving since the first feeding, May 23, through 51 days, were fed on thyroid from July 13, others on liver.

The experiment ran as follows:

	Thyroid.	Liver.
July 13.	Average size 2.4—2.6 cm	
July 18.	Hind legs appear, tail gets shorter, body assumes frog-shape	
	length of body 0.7 cm	0.9 cm
	- - tail 0.8 -	1.9 -
	entire length 1.5 cm	2.8 cm
July 19.	They swim on their back.	
July 20.	Fore legs just noticeable. They die off.	
July 25.		hind legs appear (7 days later than in thyroid!).
Aug. 16.		fore legs appear (17 days later than in thyroid!).

Experiment III again shows the extremely strong influence of the thyroid food in accelerating the development as compared with an indifferent food, as liver can be regarded. At the same time it confirms the above statement, that the differences in time between corresponding stages of development become greater the later an organ appears in normal ontogeny. The time between the appearance of the hind legs in thyroid and liver was an interval of 7 days, while the interval between the appearance of the fore legs in the two sets was 17 days.

To gain a further control of the results of experiments I—III on *Rana temporaria*, a similar set of experiments was repeated on *Rana esculenta.*

Experiment IV.

Tadpoles of *Rana esculenta* were fed in groups of 20 on thyroid, thymus and liver and one group was left unfed. For the feeding on thyroid 3 groups of different sizes were used, the smallest ones in group I, the largest ones in III. Group II as well as the liver, thymus and control groups consisted of tadpoles of the intermediate size.

The differences in size at various times of the experiment are given in the following table:

		Control	Liver	Thymus	Thyroid I	Thyroid II	Thyroid III
July 6.	Size at the start of the experiment		cm 2.5—3.0		2.1	cm 2.7	3.3
July 17.	Length of body[1])	1.0	1.2	1.4	0.7	1.0	1.2
	- - tail	1.5	1.7	1.7	1.1	1.4	1.8
	entire length	2.5	2.9	3.1	1.8	2.4	3.0
July 21.	Length of body	1.0	1.2	1.4	0.7	1.0	1.1
	- - tail	1.5	1.7	1.8	1.0	1.1	1.6
	entire length	2.5	2.9	3.2	1.7	2.1	2.7
July 31.	Length of body	1.0	1.2	1.4			
	- - tail	1.5	1.8	2.1			
	entire length	2.5	3.0	3.5			
	breadth of body	—	0.6	0.7			
Aug. 10.	Length of body		1.2	1.5			
	- - tail		1.9	2.3			
	entire length		3.1	3.8			
	breadth of body		0.7	0.8			
	- - tail		0.4	0.7			

[1]) At the beginning of the experiments only the entire length of the animals was measured. Later it was found better to determine the lengths of the body and tail separately.

		Liver	Thymus		
Aug. 16.	Length of body	1.2	1.5		
	- - tail	2.0	2.5		
	entire length	3.2	4.0		
	breadth of body	0.7	0.9		
			a	b	
Aug. 29.	Length of body	1.3	1.6	1.9	a = smallest
	- - tail	2.1	2.6	3.2	b = biggest
	entire length	3.3	4.2	5.1	
	breadth of body	0.7	0.9	1.0	

From the record of the experiment may be mentioned:

July 16. Liver are big and show a greenish color. Thymus are very big and dark. All thyroid II and III grow hind legs.

July 17. Thyroid I grow hind legs. Some thyroid II and III show buds of fore legs, thyroid II breathe very rapidly and swim on their backs. The bodies of all thyroid I—III assume frog-shape.

July 20. Thyroid II begin to die, they have typical frog-shape. Thyroid III swim on their backs.

July 21. All thyroid II are dead.

July 23. Thyroid I and III begin to die.

July 24. All are dead. Their bodies are typically frog-like.

July 31. The unfed control begin to die. The thymus lose their dark color and become lighter than liver.

Aug. 13. Liver begin to grow hind legs, 28 days later than thyroid.

Aug 15. Thymus begin to grow hind legs, 30 days later than thyroid.

Aug. 13—Sept. 15. Liver die one after the other without completing their metamorphosis.

Sept. 5—Sept. 7. Thymus die one after the other without completing their metamorphosis.

This experiment has therefore given the same results as those attained on *Rana temp.* The effect of the thyroid diet is again striking.

At present it is not clear, why the liver and thymus tadpoles in spite of the continual feeding (July 6—Sept. 7, Sept. 15) did not complete their metamorphosis, but died before. The only respect in which the *Rana esculenta* experiments differed from those on *temporaria* was the higher temperature of the water and air. The former were undertaken during the hottest period of the summer of 1911, while the latter had been completed before that time. However, it is unlikely that the rise in the temperature itself should have enacted such a retarding influence on the development of the tadpoles. One should rather expect the contrary. Although the high temperature may not be directly injurious, it may indirectly create unfavorable conditions for artificially rearing the animals. The water in which the tadpoles were kept contained a large amount of organic substances constantly undergoing decomposition much more rapidly than on cooler days. Therefore the accelerating influence of the higher temperature may well have been counteracted by this process.

Still another reason may account for the delay in development. *Rana esculenta* is less fit than *temporaria* to be reared under artificial surroundings, therefore in general less resistent to aquarium conditions. Furthermore, it sometimes happens that under apparently favorable conditions *esculenta* tadpoles do not complete their metamorphosis before the following spring. BARFURTH and TORNIER state that overfeeding may postpone the metamorphosis.

The thyroid tadpoles did not succumb to any of the above mentioned influences. This can easily be explained by the fact that the thyroid treatment did not have to last very long on account of the immensely accelerating influence of that food.

During the first half of the experiment the thymus fed animals showed the same dark pigmentation as the *temporaria* did, later on this dark color disappeared and they became even lighter than the liver fed tadpoles.

Experiment V.

The aim of this experiment was to study the influence that a sudden change in the food given would have on the development of *Rana esculenta* tadpoles. For this purpose a set of animals which had been fed on liver since July 6 was on July 21 put on thymus-, another set on thyroid diet. The same was done with thymus fed animals which were put on liver and thymus respectively. Thyroid fed tadpoles for feeding on liver and thymus unfortunately could not be used. With other animals it was tried, however, to stop the rapid progress in differentiation after thyroid diet by giving liver or thymus, but without results.

a. Liver fed tadpoles, put on thymus or thyroid diet on July 21.

		Liverthymus.	Liverthyroid.
Average size:			
July 21.	Length of body	1.2 cm	1.2 cm
	- - tail	1.7 -	1.7 -
	entire length	2.9 cm	2.9 cm

	Liverthymus.		Liverthyroid.	
July 24.			After 3 days feeding! hind legs appear.	
July 27.			Frog-shape is noticeable.	
	Length of body	1.2 cm	1.1 cm	
	- - tail	2.1 -	1.6 -	
	entire length	3.3 cm	2.7 cm	
July 31.	Length of body	1.3 cm	1.0 cm	
	- - tail	2.3 -	1.4 -	
	entire length	3.6 cm	2.4 cm	
	breadth of body	0.75 -	0.6 -, length of legs 0.3 cm.	
			Swim on the back, air vesicles in the gill region, begin to die off.	
	(continued unter c.)			

b. Thymus fed tadpoles, put on liver or thyroid diet on July 21.

	Thymusliver.		Thymusthyroid.	
July 21.	Length of body	1.4 cm	1.4 cm	
	- - tail	1.8 -	1.8 -	
	entire length	3.2 cm	3.2 cm	
July 24.			After 3 days feeding! hind legs appear.	
July 27.			Frog-shape is noticeable.	
	Length of body	1.4 cm	1.3 cm	
	- - tail	2.1 -	1.8 -	
	entire length	3.5 cm	3.1 cm	
July 30.	Length of body	1.4 cm	1.1 cm	
	- - tail	2.1 -	1.5 -	
	entire length	3.5 cm	2.6 cm	
	breadth of body	0.7 -	0,6 -, length of legs 0.6 cm.	
			Swim on the back, air vesicles in the gill region.	
Aug. 2.			Begin to die. A few have the buds of the fore legs out.	
Aug. 3.			The last ones die.	
			Length of body	1.0 cm
			- - tail	1.5 -
			entire length	2.5 cm
			breadth	0.6 -
	(continued under c.)		length of legs	0.6 -

c. Continuation of the left columns of the above tables.
A comparison of liverthymus and thymusliver.

Date	Liverthymus.		Thymusliver.
July 21.	Length of body	1.2 cm	1.4 cm
	- - tail	1.7 -	1.8 -
	entire length	2.9 cm	3.2 cm
July 27.	Length of body	1.2 cm	1.4 cm
	- - tail	2.1 -	2.1 -
	entire length	3.3 cm	3.5 cm
July 31.	Length of body	1.3 cm	1.4 cm
	- - tail	2.3 -	2.2 -
	entire length	3.6 cm	3.6 cm
	breadth of body	0.75 -	0.7 -
Aug. 5.	Length of body	1.45 cm	1.4 cm
	- - tail	2.5 -	2.3 -
	entire length	3.95 cm	3.7 cm
	breadth of body	0.8 -	0.7 -
Aug. 10.	Length of body	1.5 cm	1.4 cm
	- - tail	2.6 -	2.3 -
	entire length	4.1 cm	3.7 cm
	breadth of body	0.8 -	0.7 -

Aug. 13. Hind legs appear, 20 days later than in thyroid (compare the right columns of a and b).

Date	Liverthymus.		Thymusliver.
Aug. 16.	Length of body	1.6 cm	1.4 cm
	- - tail	2.8 -	2.3 -
	entire length	4.4 cm	3.7 cm
	breadth of body	0.8 -	0.7 -
Sept. 2.			Begin to die without developing fore legs.
Sept. 20.	Begin to die without developing fore legs.		

Experiment V shows that a thyroid diet, started even at an advanced stage of differentiation and after other substances have been fed, is able to influence the further development intensely. It seems of little importance, which substances were fed before the thyroid, except that the relative sizes are different. The liverthyroid went almost parallel with the thymusthyroid (Fig. 8*a*, *b*). Some minute differences, however, were noticeable, yet further experiments with a combined diet will have to determine their importance. Some of the thymusthyroid, for instance, showed buds of the fore legs, while the liverthyroid died before that stage showing characteristic responses to thyroid. These features as swimming on the back, formation of air bubbles in the gill region and others will be discussed later.

The liverthyroid and thymusthyroid were far ahead of the liverthymus and thymusliver and also of the liver and thymus of experiment IV.

The liverthymus and the thymusliver ran almost parallel with the exception that the liverthymus grew bigger than the thymusliver. The thymus diet, therefore, furthers growth even, when it is given at an advanced stage of differentiation, but apparently less than when given to younger animals.

The following comparison of tables IV and V is interesting: liverthymus Vc become gradually larger than liver IV, thymusliver Vc smaller than thymus IV. The thymus food thus seems to act differently at different ages, and it may be possible to find a time or stage for its optimum influence such as is also surmised for the thyroid diet.

In liver IV the hind legs appeared on August 13, in thymus IV on August 15. This difference in time is rather small and further experiments must show, whether or not it is significant. At any rate, this observation agrees with those made on *Rana temporaria*, which showed that the thymus food retarded the development. Liverthymus V and thymusliver V grew their hind legs on August 13, i. e. on the same day as liver IV. Thus the partial feeding on thymus seems not to have caused the same delay in development as the exclusive thymus diet. However, a difference of only 2 days, observed on one set of animals does not allow of conclusions.

Experiment VI.

This experiment can be regarded as a supplement to experiment V, at the same time it furnishes a further confirmation of the results of former thyroid feedings. Tadpoles that had been fed on liver and thymus 15 days longer than the corresponding groups of experiment V, thus were 15 days older, were put on the thyroid diet on August 5.

This last experiment shows that the thyroid when food given even at a very advanced stage of differentiation can cause an accelerated development. 5 days after the beginning of the experiment hind legs appear, this is still 3 and 5 days sooner than in the control animals liver IV and thymus IV. The effect of the previous feeding on different

		Liverthyroid.	Thymusthyroid.
Aug. 5.	Length of body	1.2 cm	1.4 cm
	- - tail	1.9 -	2.2 -
	entire length	3.1 cm	3.6 cm
	breadth of body	0.6 -	0.8 -
Aug. 10.	Length of body	1.2 cm	1.3 cm
	- - tail	1.8 -	2.1 -
	breadth of body	0.6 cm	0.7 cm

Hind legs appear after 5 days feeding. (The liver IV and thymus IV do not grow them until August 13 and August 15.)

	Liverthyroid	Thymusthyroid
Aug. 12.	Frog-shape is noticeable.	
Aug. 14.	Swim on the back.	
Aug. 15.	2 grow fore legs.	
Aug. 16.	These two (a, b) die, the rest grow fore legs	

	a	b	rest	
length of body	0.9	1.05	0.9 cm	1.15 cm
- - tail	1.2	1.0	1.2 -	1.7 -
entire length	2.1	2.05	2.1 cm	2.85 cm
breadth of body	0.5 cm			0.7 -

	Liverthyroid	Thymusthyroid
Aug. 17.	Begin to die off.	Swim on the back.
Aug. 18.		1 grows fore legs.
Aug. 19.	Last ones die	Last ones die.

	smallest	largest	smallest	largest
length of body	0.8 cm	0,9 cm	1.0 cm	1.1 cm
- - tail	0.6 -	0.9 -	1.2 -	1.4 -
entire length	1.4 cm	1.8 cm	2.2 cm	2.5 cm
breadth of body	0.5 -	0.5 -	0.65 -	0.7 -

substances before the thyroid diet here also manifests itself in the different sizes of the animals. During the entire experiment thymusthyroid remain bigger than liverthyroid; on the other hand liverthyroid develop quicker than thymusthyroid, which is suggested also by experiment V. If further experiments of this kind give similar results, we shall have additional evidence, that thymus food postpones the metamorphosis. In fact, at the beginning of experiment VI the liver IV must have been ahead of thymus IV, although macroscopically the difference was not evident; for liver IV grew their hind limbs on August 13 and thymus IV on August 15.

General discussion.

The most striking and at the same time unquestionable results were attained by thyroid feeding. They were the same in all experiments. The influence of the thyroid food was such that it stopped any further growth but on the contrary led to an abnormal diminution of the size in the animals treated, while simultaneously it accelerated the differentiation of the body immensely and brought it to a premature end. It was of little importance, at which stage of differentiation the thyroid diet began or which kind of food had been given before. Under all circumstances the influence of the thyroid food became noticeable in a very short time.

This influence must have been very strong, as can be concluded from two kinds of observations. First, within a very short time, 3—5 days, after the beginning of the experiments changes in the outer features of the animals were noticeable; second, the influence on all tadpoles of one group was uniform and rather parallel. While, for instance, in other groups not fed on thyroid the influence of the food became evident gradually, without abolishing the individual differences, so that the individuals of one group grew their hind legs, fore legs etc. one after the other, often at intervals of many days, the thyroid diet, on the other hand, brought all the animals of one group within a few hours, not more than 24, to the same stage of development. However, it cannot be said that the individual differences were entirely abolished. The measurable signs of these differences, the intervals between the corresponding phases of development, were greatly reduced since the entire period of development was much shortened.

One of the most peculiar features is that the time at which the feeding begins is of no importance as regards its results. The stages of development of the animals to be treated may be chosen, but always the same results will be obtained. Animals in different stages of development, others that had starved for many weeks, and still others that had before been fed on other substances were placed on thyroid diet with exactly the same results: within a few days the rapid differentiation of the body began. Thus extremely young or very old tadpoles could be forced to undergo their methamorphosis quickly. The lower and upper limit of age for the start of a successful thyroid diet will be determined later. The upper limit is probably the time shortly before completing their metamorphosis, when the tadpoles stop feeding in general. How near to the time of hatching the lower limit can be brought further experiments will show[1]). The tadpoles that were available for the experiments here recorded had been hatched for some weeks.

The second influence of the thyroid diet, the suppression of growth, is merely the consequence of the precocious development, and this in turn seems to be caused by the well known activity of the thyroid agens to stimulate metabolism. The thyroid agens accelerates the metabolism which leads to a rapid reduction of the larval organs and thus to a premature metamorphosis. As soon as thyroid food is given the differentiation of the body begins. Hand in hand with the progressing metamorphosis goes, more than in the case in normal development, a reduction of the body mass (resorption of the tail, loss of water, therefore an increasing compactness of the body etc.) The outcome of such precocious metamorphosises are then very small (pigmy) frogs. This mass reduction was especially striking in the experiments on *Rana esculenta*.

The thyroid showed still other peculiar influences on the behavior of the tadpoles. Towards the end of the metamorphosis the animals hardly moved about in the water. They were always lying quietly, generally on their backs. When disturbed they would move for a few seconds in a somewhat convulsive manner and then drop again to the bottom of the dish, while tadpoles fed on other material would swim about for a long time. The reason for this may be that the thyroid fed tadpoles always began to reduce their tail before the extremities were at all or sufficiently strongly developed. The extremities, even if fully developed, were always extremely thin, merely thread-like (Fig. 6 *a*), and could hardly be used for swimming a long time.

At one time *Rana esculenta* tadpoles of the different groups were placed in small dishes with equal quantities of water, to which equal amounts (about 5 drops) of chloroform had been added. This was done so as to be able to photograph the animals. All tadpoles remained the same length of time in the mixture. All animals survived the narcosis very well except the thyroid fed ones which died in it.

At another time *Rana temporaria* tadpoles were taken out of the water and placed on wet filter paper to photograph them. During this procedure, which of course was somewhat rough, the thyroid

[1]) In recent experiments (1912) which will be discussed in a later paper I succeeded in forcing *Rana temporaria* tadpoles to grow fore legs as early as 15 (!) days after leaving the egg.

died, while the others stood it. Thus in different ways it was seen that the thyroid fed tadpoles possessed far less resistance against noxious influences than the others, as if the thyroid food had weakened their systems enormously. One cannot, however, speak of a poisoning of their body in the true sense, since that would not have allowed the rapid progress in development.

In the tables given above several dates are mentioned at which the animals began to swim on their backs. This, too, is one of the features observed only in thyroid tadpoles. Before the animals completed their metamorphosis, about 3—4 days previous, they began lying on their backs and floated passively on the surface of the water. They breathed very heavily and rapidly. Even when disturbed and swimming actively they did not usually turn over. It seemed as if the animals were passively forced to take this peculiar position; as under the skin in the gill region there were always one or two air bubbles visible, as if during the closure of the gill opening air had been enclosed. If the animals did not die these air bubbles were usually absorbed after which the animals assumed a normal position. It was seen that the swimming on the back always began shortly before the completion of the metamorphosis and its early appearance was watched.

The influence of the thymus diet on the development of the tadpoles was as evident as that of the thyroid, but less striking. The thymus food caused an accelerated growth beyond the normal (giant tadpoles) and at the same time it retarded or completely suppressed the differentiation of the body. In doing so individual differences were very much emphasized, so that an interval of several weeks elapsed between the metamorphosis of the first and the last tadpole, while in normal development the difference amounted to days only. The strongest tadpoles or better those which at the start of the feeding had progressed most in their development were best able to keep pace with the control. Those, however, which were backward in their development at the time the thymus diet began stayed much behind the control, since they were attacked by the thymus at a less advanced stage of differentiation, and further because they remained longest on thymus diet.

The thyroid and thymus diets were thus diametrically opposite in their influences. Their relative action, however, corresponds with the views held regarding the physiological properties of these organs.

Experiments of the kind discussed in this paper may perhaps give a direction for further studies towards a rational application of thymus and thyroid preparations.

It is not the purpose of this experimental paper to discuss the extensive literature on the functional and therapeutic importance of the organs with an internal secretion. Reference is simply made to the numerous papers in which the therapeutic value of thyroid preparations for the stimulation of metabolism and ossification, and the influence of the thymus on growth in the early periods of individual life are being discussed. A list of them will be found at the end of this paper.

Liver and muscle were about equal in their action on development and did not seem to influence especially the normal progress of differentiation. Since so far they appeared to be indifferent food stuffs the tadpoles fed on liver or muscle were regarded as a control to the other feedings. However, under natural conditions the animals have a food supply quite different from a constant meat diet, yet for various reasons it was impossible to study the development of control animals on a more vegetal or mixed diet[1]).

The tadpoles fed on adrenal[2]) developed somewhat slower than those fed on liver or muscle, otherwise quite normally. The outcome of the metamorphosises were especially large and strongly developed frogs (Fig. 5*b*).

A prolonged diet of hypophysis did not force the animals to complete their methamorphosis. They all died before that stage. No conclusions can be drawn from this fact, since these tadpoles were not fed as regularly as the others on account of great difficulties encountered in providing the food[3]). The feeding on testis and ovary was also unsatisfactory for the same reason.

---

[1]) Such experiments are now being done, spring 1912, and they will be discussed in a later paper. The difference in macroscopic development between a vegetal and liver or muscle diet is slight.

[2]) Experiments on feeding adrenal cortex and medulla separately will be discussed in a later paper.

[3]) More extensive experiments on feeding the two lobes of the hypophysis will be discussed in a later paper. So far they do not confirm the above statement.

Preliminary experiments were also undertaken on *Rana esculenta* to study the influence of different diets on regenerating animals. So much can be said that of tadpoles which hat a piece of their tail amputated the thymus fed ones regenerated quickest, while the thyroid fed ones, although they did regenerate a part, showed the typical precocious metamorphosis. In one experiment the average length of the regenerated part of the tail was: in thymus 3.5 mm, in thyroid 3.2 mm, in liver 2.9 mm; in another experiment: liver 3.1 mm, thymus 4.6 mm; later liver 6 mm, thymus 9 mm. Regeneration of the tail begins even when the animals are near the point where they resorb their larval organs, otherwise the thyroid fed ones would not have regenerated. BARFURTH showed that *Rana fusca* tadpoles which metamorphosed even 2 or 3 days after the operation tried to regenerate the amputated part of their tails.

The influence of the different food stuffs on the pigmentation has been mentioned before. The animals were kept under the same conditions of light and temperature and in the same kind of dishes. The position of the dishes was changed daily in a certain rotation so that the minute differences in light and temperature were abolished as much as possible.

The liver fed tadpoles were rather dark, gradually assuming a greenish tint. The thymus fed tadpoles of *Rana temporaria* grew extremely dark with the progress of the experiments until they became almost black; those of *R. esculenta* grew dark in the beginning, later, however, they became lighter. The adrenal fed tadpoles after 3—4 weeks became extremely light in color. Those fed on hypophysis lost their pigment more and more and became almost transparent, but this may have been the consequence of the irregular feeding.

TORNIER has studied the influence of varying quantities of food on the pigmentation of *Pelobates fuscus* tadpoles and found that a minimum food ratio gives albinotic, a maximum ratio highly melanotic larvae and frogs. So the melanism of the thymus fed tadpoles may have been partly caused by an overrich diet, yet they were much darker than those in the other groups, although all were fed sufficiently well. Why the *Rana esculenta* larvae which in the first weeks of the experiments were as melanotic as the *temporaria*, later lost their dark appearance, cannot be explained at present. Very minute differences in temperature, as KAMMERER points out, may easily cause a change in pigmentation.

The very dark and the very light adrenal (cortex and medulla) tadpoles seem, roughly estimated, to have possessed equal amounts of pigment. In the thymus fed animals the pigment cells were spread out very much in a star-like manner, in the adrenal fed ones they were completely contracted. Former experiments with adrenalin would warrant the suggestion that the extract from the chromaffine cells of the medulla which dissolved in the water caused the pigment cells to contract[1].

The histogenetic processes must have been influenced very much by the different diets. The investigation of the thyroid and thymus fed material promises especially interesting results. The report on this topic will be given later.

More experiments, especially with mixed diets, are necessary to clear up all the questions concerned in this discussion. At any rate, these experiments may open a new and extensive field of work in experimental morphology, in which success is rather certain.

At present one fact alone deserves notice, that the food stuffs given fresh were able to pass the stomach without losing at least some of their specific properties. It still remains an open question, whether their action, after they have passed the intestinal canal, is entirely the same as that which they exert as functionating organs. Before this question is solved, no conclusions can be made on the rôle of these organs in the household of the body. However, so far it has been shown that a diet on thyroid substance or the application of thyroid tablets can to a certain degree substitute the normal function of the thyroid gland. — It must also be kept in mind that mammalian organs were fed to amphibians.

Summary.

A number of mammalian organs, especially those with an internal secretion, thyroid, thymus, adrenal, testis, ovary, hypophysis, liver, muscle etc. were given as food to tadpoles of *Rana temporaria* and *esculenta*. It was seen that each organ exerted a certain influence on growth and differentiation of the animals. Most striking was the

[1] Compare: Lieben, S., list of literature. Recent (1912) experiments, however, so far indicate that the feeding on adrenal cortex causes a much lighter pigmentation than an adrenal medulla diet.

influence of the thyroid food. It caused a precocious differentiation of the body, but suppressed further growth. The tadpoles began to metamorphose a few days after the first application of the thyroid and weeks before the control animals did so. The influence of the thymus was quite the opposite, especially during the first days of its application it caused a rapid growth of the animals, but postponed the final metamorphosis or suppressed it completely. The action of the other organs must be studied further before definite statements can be made. The thymus diet gave very dark, melanotic tadpoles, the adrenal diet extremely light albinos, the liver diet dark ones with a greenish tint.

Zusammenfassung.

Verschiedene Säugetierorgane, namentlich solche mit innerer Secretion. Thyreoidea, Thymus, Nebenniere, Hoden, Eierstock, Hypophyse, Leber, Muskel usw. wurden an Kaulquappen von *Rana temporaria* und *esculenta* verfüttert. Jede Fütterung übte einen andern Einfluß auf das Wachstum und die Differenzierung der Tiere aus. Äußerst auffallend war die Wirkung der Schilddrüsennahrung. Sie verursachte eine rapide Körperdifferenzierung, die zu einer vorzeitigen Metamorphose führte, wobei aber jedes Weiterwachstum unterdrückt wurde. Die Kaulquappen begannen ihre Metamorphose wenige Tage nach der ersten Schilddrüsendosis und um Wochen früher als die Kontrolltiere. Der Einfluß der Thymusnahrung war gerade entgegengesetzt. Sie bewirkte namentlich in den ersten Tagen ein schnelles Wachstum der behandelten Tiere, schob aber die Metamorphose immer weiter hinaus oder unterdrückte sie gänzlich. Der Einfluß der übrigen Organe muß noch weiter studiert werden. Die Thymusverfütterung ergab tief dunkel, fast schwarz gefärbte Quappen, die Nebenniere ganz lichte, albinotische Tiere, die Leber dunkle, mit einem Stich ins Grünliche.

List of Literature.

Barfurth, D., Versuche über die Verwandlung von Froschlarven. Arch. f. mikr. Anat. 1887. Bd. 29. S. 1.

—— Der Hunger als förderndes Prinzip in der Natur. Arch. f. mikr. Anat. 1887. Bd. 29. S. 28.

Basch, K., Beiträge zur Physiologie und Pathologie der Thymus. Jahrb. f. Kinderheilk. 1906. Bd. 64. 1908. Bd. 68.

Biedl, A., Innere Secretion. Wien-Berlin 1910.

Gudernatsch, J. F., Fütterungsversuche an Kaulquappen. Demonstr. Verh. Anat. Ges. 26. Vers. München 1912.

—— Fütterungsversuche an Kaulquappen. Vorl. Mitteil. Centralbl. f. Physiol. 1912.

Hammar, I. A., Fünfzig Jahre Thymusforschung. Ergebn. d. Anat. u. Entwicklungsgesch. 1910. Bd. 19.

KAMMERER, P., Künstlicher Melanismus bei Eidechsen. Centralbl. f. Physiol. 1906. Bd. 20. S. 261.
—— Über künstliche Tiernigrinos. Verhandl. zool.-bot. Gesellsch. Wien. 1907. Bd. 57. S. 136.
—— Die Wirkung äußerer Lebensbedingungen auf die organische Variation im Lichte der experimentellen Morphologie. Arch. f. Entw.-Mech. 1910. Bd. 30. S. 379.
LIEBEN, S., Über die Wirkung von Extrakten chromaffinen Gewebes (Adrenalin) auf die Pigmentzellen. Centralbl. f. Physiol. Bd. 20.
TORNIER, G., Nachweis über das Entstehen von Albinismus, Melanismus und Neotenie bei Fröschen. Zool. Anz. 1907. Bd. 32. S. 284.
VINCENT, SWALE, Internal Secretion and the Ductless Glands. Lancet. 1907.

Explanation of Figures.

Plate IX.

Fig. 1 *a—e*. *Rana temporaria* set I (smaller size), photographed June 11 1911. Natural size. *a* tadpoles fed on thyroid, already changing into frogs. Tail is shortening, fore legs appear. *b* tadpoles fed on thymus, *c* on liver, *d* on muscle, *e* on adrenal.

Fig. 2 *a—e*. *Rana temporaria* set II (larger size), photographed June 11 1911. Natural size. *a—e* as in Fig. 1. The thyroid fed tadpoles have all metamorphosed, the tail in some has almost disappeared, fore legs are well developed.

Fig. 3. *Rana temporaria* tadpoles that had been used as an unfed control in experiment I, thus starved till July 13. Photogr. July 13. Nat. size.

Fig. 4. The same tadpoles as in Fig. 3, photogr. 7 days later, July 20. *a* had been fed in the mean-time on thyroid, and are already metamorphosing into pigmy frogs. *b* had been fed on liver. These tadpoles do not metamorphose until 19 days later.

Fig. 5. *Rana temporaria* set I frogs. Photogr. July 25. Natural size. *a* fed from the beginning of experiment I on adrenal. *b* one of the tadpoles that originally had starved. Their size on July 6 was that of the tadpoles in Fig. 3. From July 6 to July 17 these tadpoles, 5 *b*, were fed on thymus, from July 18 on thyroid. On July 21 hind legs appeared, July 23 fore legs. Notice the small body and the much shortened tail of a frog metamorphosing under thyroid influence, while the adrenal frogs, 5 *a*, at the time of metamorphosis are large and still have their long tadpole tails.

Fig. 6. *Rana temporaria* set I, photogr. July 13 1911. Natural size. Animals that originally had starved. Their size on July 6 was that of the tadpoles in Fig. 3. From July 6 to July 13. *a* were fed on thyroid, *b* on thymus. *a* are changing into pigmy frogs, fore legs appear, tail shortens, *b* are still huge tadpoles. Compare also these thymus tadpoles with Fig. 4 *b* fed for the same time on liver.

The animals in Fig. 4—6 are all of the same age and the same original size, set I, but fed on different organs.

Fig. 7. *Rana esculenta*, photogr. August 9, nat. size. *a* fed on thymus since July 6, *b* fed on liver since July 6, *c* on thymus from July 6 to July 20, on liver since July 21, *d* fed on liver from July 6 to July 20, on thymus since July 21.

a and *c*, *d* which either entirely, *a*, or at a time of their development, *c*, *d*, were fed on thymus, are much bigger than those fed on liver only, *b*.

Fig. 8. *Rana esculenta*, photogr. August 19 1911, nat. size. The animals were dead, when being photographed, therefore the curved tails. *a* fed from July 6 to August 5 on thymus, from August 6 to August 18 on thyroid. *b* fed from July 6 to August 5 on liver, from August 6 to August 16 on thyroid.

The animals in Fig. 7 and 8 are all of the same age, but fed on different substances, and were photographed on the same day.

Anmerkung. Vorliegende Untersuchungen wurden im Sommer 1911 in meinem Institute durchgeführt. Das Manuskript war bereits im Dezember 1911 druckfertig. Eine schwere Krankheit, die mich lange zur Untätigkeit zwang, hat die Veröffentlichung verzögert.

Prag, Mai 1912. Prof. ALFRED KOHN.

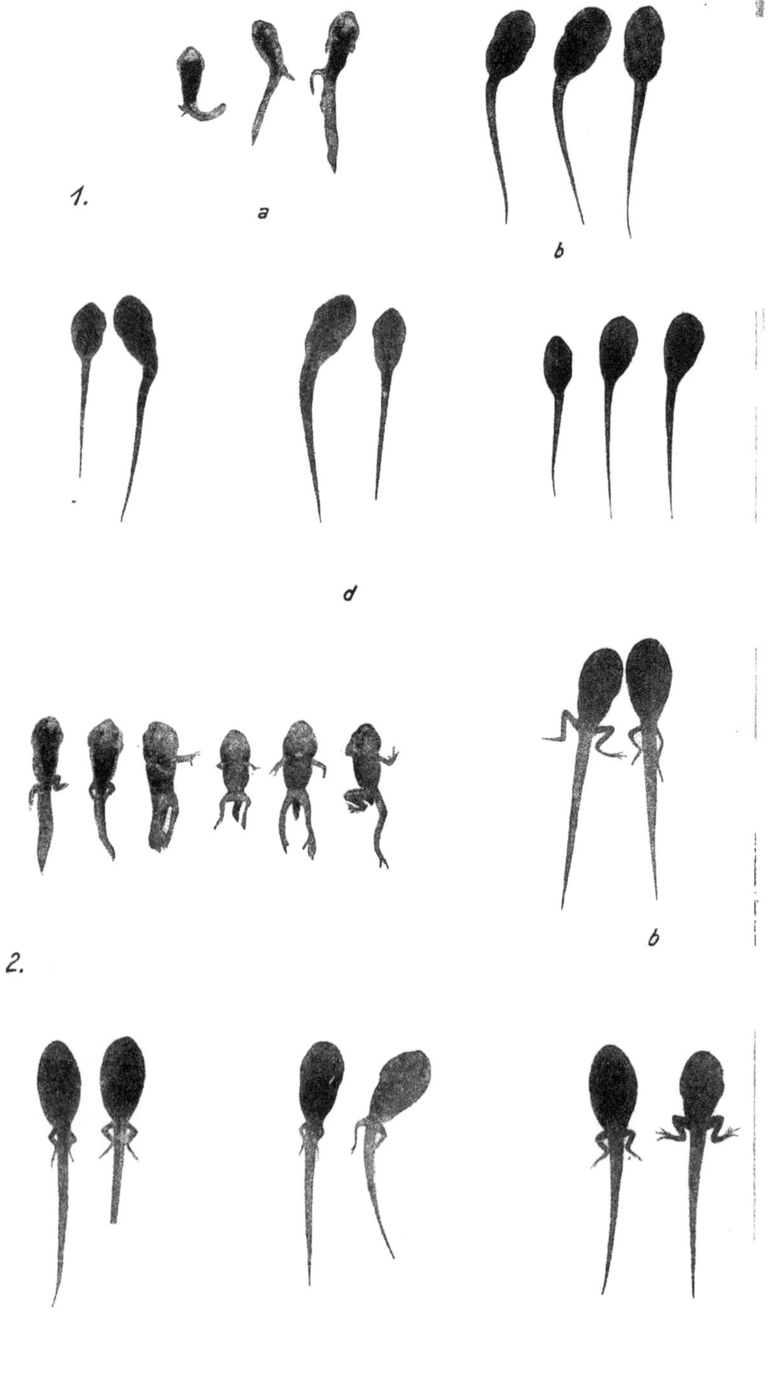
1.
a
b
d
b
2.

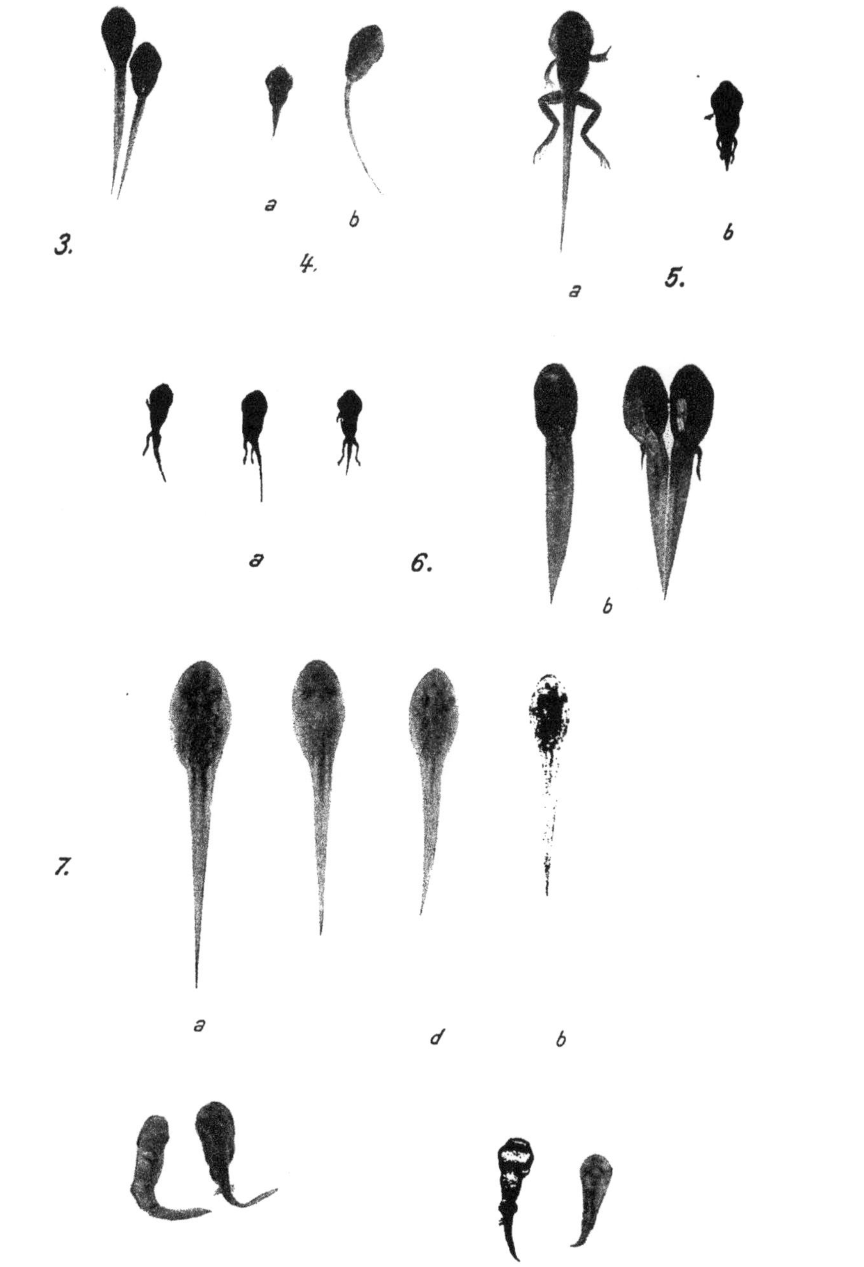
3.
a
b
4.
a
b
5.
a
6.
b
7.
a
d
b

Reprinted from THE AMERICAN JOURNAL OF ANATOMY VOL. 13, NO. 2
May 1912

THE BEHAVIOR AND RELATIONS OF LIVING CONNECTIVE TISSUE CELLS IN THE FINS OF FISH EMBRYOS WITH SPECIAL REFERENCE TO THE HISTOGENESIS OF THE COLLAGINOUS OR WHITE FIBERS

JEREMIAH S. FERGUSON

Assistant Professor of Histology
Cornell University Medical College, New York City

TEN FIGURES

In the process of connective tissue development the cells first arise, the fibers later appear. This sequence is established beyond controversy. The ontogenetic relation of cell and fiber is not, however, so thoroughly established. The theories advanced may be grouped under three heads: (1) Intra-cellular origin, the cells may transform into fibers (Schwann, Valentin, Boll, Flemming, Spuler, Livini). (2) Extra-cellular origin, the fibers arise in the intercellular substance by its fibrillation, or possibly as a secretion from the cells (Henle, Merkel, Virchow, Kölliker). (3) Epicellular origin, the fibers form in an ectoplasm at the surface of the cell (Schultze, Hansen, Golowinski, Mall).

These theories have all been primarily founded upon the results of examination of 'fixed' or 'killed' tissue, or upon the study of fresh teased tissue. Living connective tissue has been studied in the mesentery of the frog and other animals under conditions which are accompanied by marked inflammatory reaction and certain stages of the formation of exudates and of scar tissue have been thus investigated, and more recently movements of connective tissue cells have been observed in tissue cultures but so far as I know the theories of the histogenesis of connective tissue have not been examined with reference to the behavior of living cells under normal conditions.

Living connective tissue cells have been seen in tissue cultures by Harrison, Burrows, Carrel and Burrows, Margaret R. and W. H. Lewis and others, to exhibit a certain motility, and Harrison has recently emphasized the stereotropic tendency of connective tissue cells in cultures when in contact with foreign surfaces, glass, spider-web, etc. But so far as I know, the histogenesis of connective tissue fibers has not been so studied, and at best the culture method is open to some criticism on the ground that while the connective tissue cells are undoubtedly alive and active, yet they exist under very unusual, if not abnormal, conditions whose effects have not yet been subjected to complete analysis. Under these conditions the behavior of the connective tissue elements while probably similar, is not certainly in exact conformity with that of the tissue within the embryo.

In order that deductions based upon these several methods of examination be adequately controlled it appeared desirable that developing connective tissue be studied in the living animal under conditions which were in every respect normal, or which, at least, resulted in no inflammatory reaction. In mammals this endeavor is fraught with considerable difficulty owing to the size of the mammalian embryo and the depth beneath other tissues, often not transparent, at which the connective tissue lies.

During the past summer I had the opportunity, through the courtesy of the Marine Biological Laboratory at Woods Hole, of studying connective tissue in the fins of living fish embryos under conditions which were wholly normal and unaccompanied by any evidence of inflammatory reaction.

If a free swimming Fundulus embryo is placed on a hollow ground slide it will continue to swim, often actively, and its heart beat and circulation are maintained. It may be observed for some minutes and at the end of observation may be returned to the aquarium to continue an uneventful existence for hours or days thereafter. If a drop of chlorotone is added, or frequently without its addition, the fish will remain quiet for some minutes, thus permitting continued observation of connective tissue cells in his semitransparent fins. Certainly cells studied under these conditions are open to no criticism of abnormality.

The viability of the animals is unaltered for I have kept them for several days after such observation without any indication of decreased activity on their part. Even embryos which have been quieted by chlorotone, as well as those immersed for hours in a solution of Bismark brown in sea water, I have resuscitated and kept alive and in an apparently normal and usual condition for two or three days; they could easily have been kept longer had it seemed advisable.

The tissue selected for observation was in the fins of free swimming pelagic and Fundulus embryos. The embryos studied were chiefly of Fundulus and varied from 5 to 20 mm. in total length. The most favorable subjects were from the time of hatching, 5 to 6 mm., up to 12 mm. in length. The pectoral and caudal fins were usually selected as most available for observation. In such embryos the fin consists of a central frame work formed by the jointed rays, lepidotrichia, with their attached muscles, and a superficial integument of pavement epithelium with its subjacent basement membrane. The finer fin rays, actinotrichia, continue the jointed rays to the margin of the fin. The fin at this stage is very thin and the epidermis lies almost in contact with the fin rays. But between adjacent rays is an interval which lodges on either side the afferent and efferent blood vessels, bordered by chromatophores, and between them a loose mass of mesenchymal connective tissue in which the cells may be readily observed.

In embryos 5 to 6 mm. long the connective tissue in the pectoral fins consists chiefly of a mass of round cells confined to the proximal portion, and beyond this mass a distal fringe or 'skirmish line' of scattered stellate cells. In the unpaired fins, which are less advanced in their development only the scattered stellate cells are represented, the invasion of the round cell mass having not yet occurred. In later stages, as in the caudal fin of the same embryo, the zone of round cells has advanced distalward among the actinotrichia nearly to the fin margin, leaving behind between the lepidotrichia an area of more mature cells, stellate and spindle, and a few fine fibers well separated by broad spaces occupied by tissue fluids. The spindle cells and fibers preponderate in the proximal, the round cells in the distal zone of the fin. Hence, one follows the sequence of development in passing from the distal toward the proximal portion of such a fin. Older embryos show the same zones of transition but in them the formation of fibers in the proximal region is more advanced.

In mammalian tissue one finds three stages in the histogenesis of connective tissue, a primitive cellular stage, a syncytial stage, and a fibrous stage. The first is characterized by the predominance of round cells, the second by stellate, the third by spindle and lamellar cells. The same succession of cell types is present in the fins of embryo fish and there is a corresponding succession of histogenic stages. Fibers do not appear prior to the appearance of cellular processes. Fine fibers appear coincidently with stellate cells, coarse fibers and fiber bundles develop later.

In the distal portion of the fin fine fibers first appear in the round cell area coincidently with the transition from round to early stellate forms. At exactly this period I have observed the first indication of motion, the throwing out of pseudopods by the round cells, in the connective tissue cells of the living embryo. Fig. 1 shows such changes in two cells on the border of the round cell area near the posterior end of the ventral fin. There is at this time relatively little locomotion, as is shown in the figure by referring the position of the cells *a* and *b* to the relatively fixed point, a prominence on the margin of an adjacent chromatophore (*ch*).

The first appearance of fibers in the distal portions of the fins has been very properly connected by Harrison, and by Goodrich with the origin of the dermal fin rays from the 'scleroblast' cells which closely resemble the connective tissue cells and like them are of mesodermal origin. In the region of the actinotrichia in the distal portion of the fin, it is difficult to distinguish between the early forms of these coarse fibers and the true connective tissue fibers, but the actinotrichia are confined to the region of the last one or two joints of the jointed fin rays, and there they project, as Goodrich has shown, from between the two opposed dermal plates which form the distal section of the jointed fin ray. If therefore one studies a region proximal to the last section and selects the interval between the jointed rays the primitive actinotrichia are thereby excluded.

In such portions of the caudal fins of 6 mm. embryos, and in equivalent places in later stages, are typical connective tissue fibers mostly occurring as coarse longitudinal bundles with fine oblique anastomoses. Single fibers occur in the intervals of the coarser bundles. It is along these fibers and fiber bundles that the stellate and spindle cells are disposed. These cells are readily seen in the living fish, though the ease of observation is subject to much variation in different individuals and to a less extent in different portions of the same embryo.

My observations were made on living embryos immersed in sea water, some with, some without the addition of chlorotone. In some cases a few drops of a saturated solution of Bismark brown were added to the sea water in which the fish was kept, the effect of which after a time was to slightly increase the color contrast between the connective tissue cells and surrounding structures. The stain seemed almost inocuous, for fish could be kept in it for several days without apparent effect on their vitality. Many of the fish thus examined were later killed, and the fins stained and mounted in toto, or sectioned. The various cell types seen in life were readily recognizable in corresponding locations in the stained preparations.

It is in life difficult or impossible to distinguish between the spindle and lamellar types, though in 'fixed' tissue they may be morphologically distinct. In the living animal one can see a stellate or a spindle cell elongate, approach and flatten itself against a connective tissue fiber or fiber bundle, becoming sometimes so attenuated as to be scarcely distinguishable from the fiber against which it lies; it may at any time acquire increased thickness. Such a relation to a connective tissue fiber is shown by the cell *b* in fig. 2. The relationship is again exhibited by the two cells shown in fig. 3, one of which *a*, approached a small fiber bundle, became flattened against it, then rotated to the opposite side of the fiber at 9.30 A.M., and later freed itself from the contact. Its locomotion can be observed in relation to the chromatophore (*ch*) which served as a fairly fixed point. Similar cells are frequently seen flattened against the surface of fiber bundles, blood-vessels, or fin-rays, and exhibiting a slow stereotropic locomotion. Many of these cells would seem to be identical with those which in stained preparations we are accustomed to call lamellar cells.

That connective tissue cells exhibit a certain amount of motion is no new observation. It has been well known since the inflammatory reaction to injury or infection was studied in the mesentery by Arnold and others. I have observed that the extent and rapidity of the motion varies with the cell type. The round cell, or primitive type, presents relatively little motion, it being limited, so far as I have observed, to the very slow projection and retraction of minute pseudopods. Even this evidence of activity seems rather to be limited to those later phases of the cellular stage which foreshadow the transformation of the round cells to the stellate type of the succeeding stage. This transformation is indicated by the fact that the motion is more noticeable near the border of the round cell area than in its interior, and also because at the extreme margin of such a cellular area one may by careful scrutiny observe an extensive alteration from round to stellate types, some cells passing rapidly to an approximate spindle form. The type of motion exhibited by the round cells, when observable, is well shown by fig. 4, cells *a–c* being observed at the extreme margin of the round cell area, cells *d–e* just within the margin, and cells *f–g* well in the interior of the area.

While the general trend of cell change is from round to stellate to spindle cells, a change may often be observed to occur in the reverse direction, as occurred to the cell shown in fig. 5, and that in fig. 6. Such retrograde changes are less frequently observed, and the transformation is less extensive than are the progressive changes from the round to the stellate forms. The retrograde stellate phase is also more frequently of a transient character (fig. 5). Thus, a stellate cell may by retraction of its processes temporarily assume a spheroidal form but it soon again projects pseudopods and regains its stellate character. Or a typical, bipolar, spindle shaped cell may extend a third process, or even several additional processes (figs. 2, 5 and 6), but, so far as I have observed, such processes are limited in size and usually of short duration. This reverse transformation may be likened to an elastic rebound brought about by an inherent resistance to change of form reacting against an impelling force which directs the transformation from the round to the spindle type. The cell frequently balks at the change, but the general trend from round to stellate and from stellate to spindle form is inevitable.

Motion resulting in change of form is perhaps most active in the stellate type of connective tissue cell. The general trend of this motion seems to be indicated in fig. 5 *I*, in which a typical round cell selected for observation at the margin of the round cell mass in a pectoral fin of a 6 mm. embryo was seen within a period of six minutes to elongate and then to pass through successive stellate shapes to a typical spindle form. But the succession is not always so rapid. Stellate cells exhibit all sorts of morphological transformations in rapid sequence (fig. 7) and this stage of connective tissue development is of relatively more transient duration than either the preceding or the succeeding stage. Moreover, the shape of the cell is undoubtedly influenced to some extent by its surroundings and the duration of a particular stellate, spindle or lamellar shape may in some cases be thus determined.

Likewise, spindle cells undergo considerable transformations in form, the most frequent of which undoubtedly result in the lamellar shapes on the one and in the stellate on the other hand. Because of the limitations of the microscope in the delineation of the 'third dimension' it is most difficult in the colorless living tissues to differentiate between the lamellar and spindle types of cell but the evidence of fixed and stained tissues shows the lamellar to be the more mature, the spindle the earlier type, and I have observed nothing in the living tissues to indicate the contrary unless indeed it be that both types appear to be somewhat dependent on their surroundings, for as already stated these forms, in the same cell, seem to be more or less interchangeable. That spindle cells frequently and freely revert to the stellate type there is abundant evidence. There is also evidence that these cells may be capable of still further transformations than those of mere form. A syncytial stage in the development of connective tissue has long been assumed. That this stage in its most typical form presents those cell pictures which we are accustomed to regard as stellate cells is well known. It is generally recognized that this syncytial stage passes into one in which the fibers appear and the syncytium is replaced by a tissue of cells and fibers. The syncytial stage has been presumed to be preceded by a cellular stage and to those who have traced the origin of the mesoderm from the time of egg fertilization it would appear logical, even necessary, that at a sufficiently early period a cellular character must obtain, though Mall has questioned the preëxistence of this cellular condition. The transformation from the cellular to the syncytial condition has been ascribed on the basis of stained sections, to either of two processes: either the syncytium arises by incomplete division of preëxisting cells or the syncytium results from the fusion of the preëxisting cells. That some syncytia arise by incomplete cell division is very probably true. This appears specially obvious in such placental tissues as the superficial cells of the chorionic villi. I know of no convincing evidence that it does occur in the connective tissues.

Since I have been unable to observe mitotic figures in the living connective tissue cells of the fish which are under discussion I cannot offer any evidence pro or con the origin of a connective tissue syncytium by incomplete cell division. I have, however, frequently observed a phenomenon which simulates the fusion of processes of adjacent stellate cells after the manner of a typical connective tissue syncytium. In figs. 2 and 8 *II* the neighboring cells, which were at first entirely distinct and separate, were within a brief period seen to send out processes which on contact apparently fused. But of course one cannot say without subsequent fixation and staining of the identical cells, a process presenting the greatest difficulties, that the fusion was actual and complete. Even in stained sections the question is often difficult to determine. While the fusion was apparent I am not at all sure that it was actual. Not, however, in every case when cell came into contact with cell did such apparent fusion occur. This is shown in fig. 8 *I*, in which processes from the cells *a* and *b* came into contact tip to tip, yet though fusion seemed imminent it did not occur and the contour of each cell at the point of contact remained clear and distinct. Moreover it would seem that since connective tissue cells move extensively along the surfaces of the syncytium that syncytium could scarcely arise by fusion of its cells.

The spindle cells exhibit a certain stereotropism. They are prone to take their position alongside a connective tissue fiber or fiber bundle or against the surface of a blood-vessel or dermal fin-ray. When in contact with a broad surface, such as that of a blood-vessel or one of the lepidotrichia, the cells frequently assume a flattened, lamellar form. This is shown by stained sections, in which that type of cell predominates in these locations; and by the observation of living spindle cells which frequently move up to a blood-vessel or a fin-ray and then become so thinned out against the surface that they finally vanish, being in the living tissue indistinguishable from the refraction lines which surround the larger bodies. Again, the spindle cells very frequently move up to a connective tissue fiber or bundle and then elongate along the narrow filament until, as before, the cell finally appears to vanish by its extreme attenuation. Such a result was observed a moment later than the recorded observation in the case of the cell shown in fig. 5*I*.

It frequently happens that the spindle cells after such elongation again thicken to a typical spindle form, and may even throw out other processes, but in so doing, if the cell is observed in relation to some relatively fixed point, e.g., a joint of a dermal fin-ray, a chromatophore, or a blood-vessel, it will be seen that the cell has changed its relative position; it has exhibited locomotion. Locomotion is not a distinguishing character of the spindle cell; it is exhibited by the stellate cells, possibly also to a very limited extent by those round cells which are only just beginning to present pseudopod formation. But the character of the locomotion in the several types of cells differs decidedly. In the stellate type locomotion may take any direction and resembles a very active amoeboid motion, processes being extended along the surfaces of fine fibers, then either retracted or increased in size until the whole cell has come to occupy the place of the former process. Though locomotion in the stellate cells is not entirely confined to the direction of visible fiber lines, yet a projecting process of such a cell often appears to envelope or to become coincident with a fiber. In the spindle cells locomotion is always so far as I have observed, in the direction of the fiber lines: usually these cells merely slide along the surface of fibers, blood-vessels and similar structures.

I have observed that the stellate cells are more prone to lie in relation with the finer, the spindle cells with the coarser fibers; the coarser fibers in most cases, because of their size, being presumably fiber bundles rather than single fibers. This relationship is to be expected in as much as in stained preparations one finds the stellate cells present with those finer fibers which represent the earlier stages in fiber formation.

That fibers do lie without the cell in both embryonic and mature connective tissue is generally conceded. That they lie within the cell in reticular tissue, which in a way is comparable to an early or embryonic type of connective tissue, I have recently demontrated by means of the Bielschowsky stain.[1] The types of fiber development by fusion of intracellular granules described by Spuler and by Lavini though perhaps not conclusively demonstrated, at least show that certain granules which are in relation with the first appearance of fibrils do lie within the substance of the stellate, mesodermal, connective tissue cells. Moreover, I have found in embryonic tissues (fig. 9) just such appearances as I have described for reticular tissue.[2] By means of the Bielschowsky method such appearances can be shown throughout embryonic connective tissue. I have observed them in pig embryos, of various ages, in the limb buds, the head, the cervical region, and in the back throughout the whole length of the embryo from the occiput to the caudal tip, also in the umbilical cord. In many of these locations I have made similar observations on human embryos of older stages but in which the connective tissue was still actively developing. One is at a loss to explain the method by which fibers arising within the cells arrive at a location outside the cell body when these cells are in active motion. The

[1] Am. Jour. Anat., vol. 13, page 277, 1911.
[2] Loc. cit., in which see especially fig. 4 and fig. 8, pages 285 and 289.

ectoplasmic theory of Hansen does not satisfactorily account for it and its elaboration by Mall is not as specific in this particular as one might wish. These theories do not appear to fully harmonize with the relatively active motion and locomotion of the connective tissue cells which I have observed in the fins of living fish and which Harrison, Burrows and others have also to some extent recognized in tissue cultures. The cells are not sufficiently quiescent to permit of endoplasmic retraction with deposit of ectoplasmic fibers unless this retraction is rapidly performed, in which case it should be observable in the living embryo. I have in one or two cases suspected such a method of deposit but have not as yet been able to convince myself that it actually occurs; in fact I now doubt if it occurs at all.

The ectoplasmic theory presupposes that the fibers arise at the surface of the cell. This I have found to be not always the case. The clear delineation of fibers by the Bielschowsky method makes it possible to follow their course within the cell more carefully than ever before and I find that the blackened fibrils within the cell both in pig embryos (fig. 9) and in the fish's fin very frequently pass close to the nucleus, sometimes ending almost in contact with this structure, but more frequently passing by so closely as to be in actual contact with the nuclear membrane. I am aware that Golowinski using the iron haematoxylin method, demonstrated the presence of fibers at the surface of the connective tissue cells of the umbilical cord and that the apparent relation to the nucleus was explained by him as due to obliquity of section. But I have not in my preparations been able to convince myself of the adequacy of this explanation. I have found fibers to be not always at the surface of the cell, they may and frequently do penetrate entirely through the cytoplasm of the cell, as I have previously described[3] for mature reticular tissue. In the developing connective tissue, as well as in reticular tissue, such penetration of cells by the fibers is so frequent as to appear quite characteristic. It seems to me that the intimate relation of connective tissue cells and fibers in embryonic tissues can only be accounted for by tak-

[3] Loc. cit., see fig. 10, page 293.

ing cognizance of the plasticity of the connective tissue cellular cytoplasm, and also of the active motion of connective tissue cells during the period in which the fibers are being formed, so that the finer connective tissue fibers become, by the cellular activity, embedded in the plastic cytoplasm of the cells during their stereotropic locomotion. The plastic character of the cellular cytoplasm is admirably shown by the rapid changes in form of the connective tissue cells in the fins of living fish embryos.

I have already stated that the spindle cells of connective tissue in the fins of living fish undergo active locomotion. In fact this seems to be a most prominent function of the spindle cell type. Most frequently the cell glides along connective tissue fibers which often appear to be thus partially enveloped by the cytoplasm. In recording this stereotropism I am able to corroborate, for the living cells of embryo fish, the observations of Harrison on tissue cultures in which he finds that the connective tissue cells are specially prone to follow along the surface of fixed objects. Such objects in normal living subjects are most frequently the connective tissue fibers and fiber-bundles already deposited, though as previously stated, I have also observed connective tissue cells moving along the surface of the dermal fin rays and of blood-vessels. In this form of activity the cells adapt themselves more or less to the shape of the surface along which they are moving. They wrap themselves about or rotate around the finer fibers (fig. 3) and they flatten themselves against the larger objects (figs. 2, 3 and 5). In this attenuated condition they still move along the surface of fibers, often at a considerable rate of speed. One such cell I have recorded in a preliminary communication[4] was found in ten minutes to have covered a distance of 50μ, a rate of 1μ in every twelve seconds.

The striking similarity of the living, spindle, connective tissue cells to those of fixed tissue is indicated in fig. 10 which shows several such cells from a 100 mm. pig embryo. The magnification is the same as that used for the observation of the living cells. The similarity in the form of the cell and the relation of cells to fibers is apparent on comparison with the preceding figures. One

[4] Biol. Bull., vol. 21, page 272, fig. 2, 1911.

can scarcely avoid the interpretation that the cells shown in fig. 10 were at the moment of 'fixation' moving along the surface of the fiber bundles against which they lie.

The pronounced morphological relation between the connective tissue cells and fibers cannot but have an equally close functional relation. What those functional relations may be we are not now possessed of the data to fully determine.

Further studies will be necessary to fully understand the part played by the active moving connective tissue cells in the production and growth of the collaginous fibers. So far as they are now determined the essential phases of the process in which the connective tissue cells are concerned appear to be: (1) the connective tissue arises from a primitive mass of round mesenchymal cells; (2) there is a change of form from round to stellate, to spindle, and eventually to lamellar cells; (3) certain fibers seem first to appear within the cells, possibly at their surface (Hansen, Golowinski); (4) there is formed a reticulum pervading the intercellular ground substance whose fibers may be, though they not necessarily are, identical with those first arising within the cell; (5) coincident with the origin of fibers there begin amoeboid movements in the stellate and spindle cells; (6) there is an increase in size and number of fibers in the reticulum and they aggregate into bundles, synchronously with which first the stellate and later the spindle cells move along the surface of the fiber and fiber bundles.

In conclusion I desire to express my sincere thanks to the Marine Biological Laboratory at Woods Hole, Massachusetts, for the opportunities so kindly placed at my disposal.

BIBLIOGRAPHY

Arnold 1893 Arch. f. path. Anat., vol. 132, p. 502.
Boll, Franz 1872 Arch. f. mikr. Anat., vol. 8, p. 28
Burrows, M. T. 1911 Jour. Exp. Zool., vol. 10, p. 63.
Carrel, Alexis, and Burrows, M. T. 1910 Jour. Am. Med. Assoc., vol. 55, 1379.
1911 Jour. Expr. Med., vol. 13, p. 416.
Ferguson, J. S. 1911 Am. Jour. Anat., vol. 12, p. 277.
1911 Biol. Bull. vol. 21, p. 272.

FLEMMING 1891 Festschr. f. R. Virchow.
1897 Arch. f. Anat., p. 171.
1906 O. Hertwig's Handbuch. d. vergleich. u. exper. Entwickl. d. Wirbeltiere, vol. 2, p. 1.
GOLOWINSKI 1907 Anat. Hefte, vol. 33, p. 205.
GOODRICH E. S. 1904 Quart. Jour. Mic. Sc., vol. 47, p. 465.
HANSEN, F. C. C. 1899 Anat. Anz., vol. 16, p. 417.
HARRISON, R. G. 1893 Arch. f. mik. Anat., vol. 43.
1895 Anat. Anz., vol. 10, p. 138.
1910 Jour. Exp. Zool., vol. 9, p. 787.
1911 Science, vol. 34, p. 279.
HENLE, J. 1841 Allgemeine Anat., S. 197, u. 397.
v. KORFF 1907 Ergebnisse d. Anat. u. Entwickel., vol. 17, p. 247.
LEWIS, MARGARET R. AND W. H. 1911 Anat. Record, vol. 5, p. 377.
LIVINI, F. 1909 Monitore zool. Ital., vol. 20, p. 225.
MALL, F. P. 1902 Amer. Jour. Anat., vol. 1, p. 329.
MERKEL, FR. 1895 Verhandl. d. Anat. Gesellsch., vol. 9, p. 41.
SCHWANN, THEODOR 1839 Morphological Researches, trnsl., 1847.
SPULER 1897 Anat. Hefte, vol. 7, p. 115.
VALENTIN R. Wagner's Handwörterbuch der. Physiologie, vol. 1, p. 670.
VIRCHOW, RUDOLF 1852 Verhandl. d. Phys. Med. Ges., vol. 2, pp. 150, 314.
1860 Cellular Pathology, transl. from 2d German ed., p. 43.
ZIEGLER 1903 General Pathology, transl. from 10th German ed., p. 353.

PLATE 1

EXPLANATION OF FIGURES

1 From a Fundulus embryo, of 5.5 mm. total length, showing beginning amoeboid movements of two cells, *a* and *b*, on the border of the round cell area at the posterior extremity of the ventral median fin. The observation extends over a period of fifteen minutes. The last seven drawings were made without change of focus for the purpose of eliminating variation in form due to the examination of different levels. *a*, *b*, two connective tissue cells. *Ch*, prominence on the surface of a chromatophore, the body of the cell is not represented. The numerals in this and succeeding figures indicate the exact time at which each recorded drawing was completed.

2 Connective tissue cells, one of which, *a*, exhibits transformation from a spindle to a stellate type, and another, *b*, becomes flattened against a connective tissue fiber. There was an apparent anastomosis between cell *a* and the protoplasmic process *p* of an adjacent cell. *f*, connective tissue fiber; *j*, margin of a joint of a fin ray, giving a fixed point in relation to which locomotion may be determined. Other letters and numerals as in the preceding figures.

3 Two connective tissue cells exhibiting some locomotion. One of these, *a*, assumed a lamellar like relation to a fiber bundle while rotating about it. At 9.25 A.M. this cell became momentarily so thin as to almost escape observation. *Ch*, *Ch'*, chromatophores. Other letters and numerals as in the preceding figures.

1

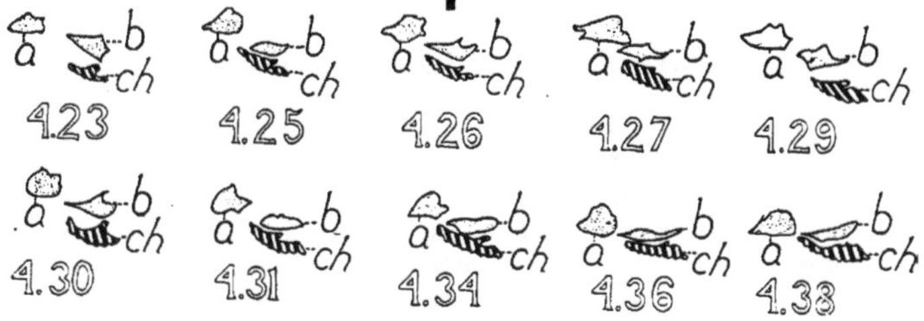
a
b
ch
4.23
4.25
4.26
4.27
4.29
4.30
4.31
4.34
4.36
4.38

2

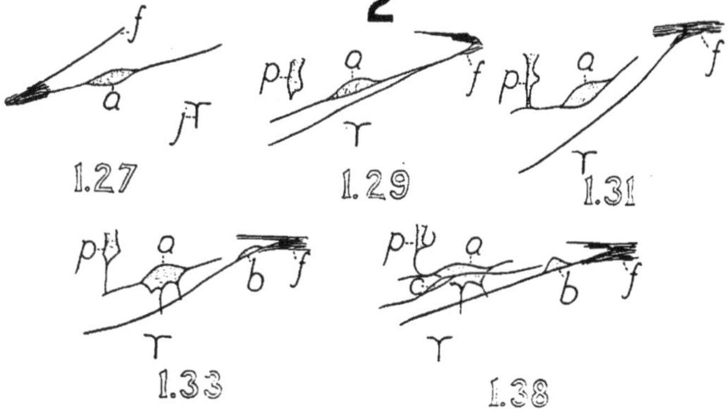
f
a
p
b
1.27
1.29
1.31
1.33
1.38

3

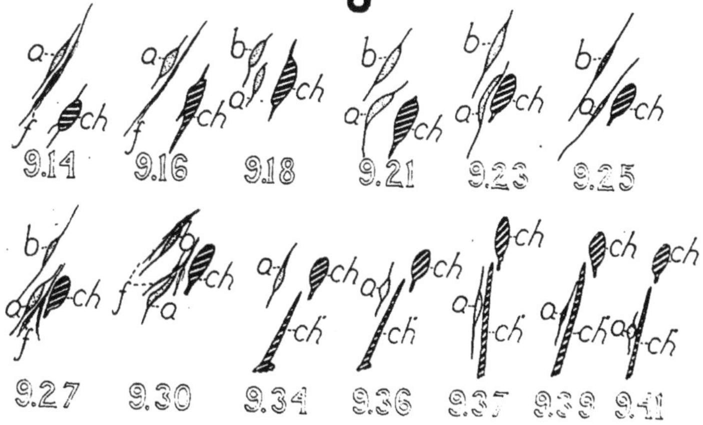
a
b
f
ch
9.14
9.16
9.18
9.21
9.23
9.25
9.27
9.30
9.34
9.36
9.37
9.39
9.41

PLATE 2

EXPLANATION OF FIGURES

4 Amoeboid motion resulting in change of form exhibited by connective tissue cells of the primitive or 'round' type. *I*, cells *a–c*, from the extreme margin; *II*, cells *d–e*, from just within the margin; and *III*, cells *f–g*, from the interior of a round cell area. *I* and *II* from the pectoral, *III* from the caudal fin. Numerals as in the preceding figures.

5 *I*, transformation of a round to a spindle cell in the pectoral fin of a Fundulus embryo 6 mm. long, 20 days after fertilization, 11 days after hatching. From 4.20 to 4.22 P.M. there was in this cell an apparently retrograde change from spindle to stellate form but at 4.24 P.M. this had been proven temporary. *II*, transition of a stellate cell to a temporary spindle form. Letters and numerals as in preceding figures.

6 Apparent retrograde change from spindle to a stellate form in a cell undergoing rather slow locomotion. The stellate phase of such cells is nearly always temporary. From the same embryo as fig. 5; *cap.*, blood-capillary. Other letters and numerals as in the preceding figures.

4

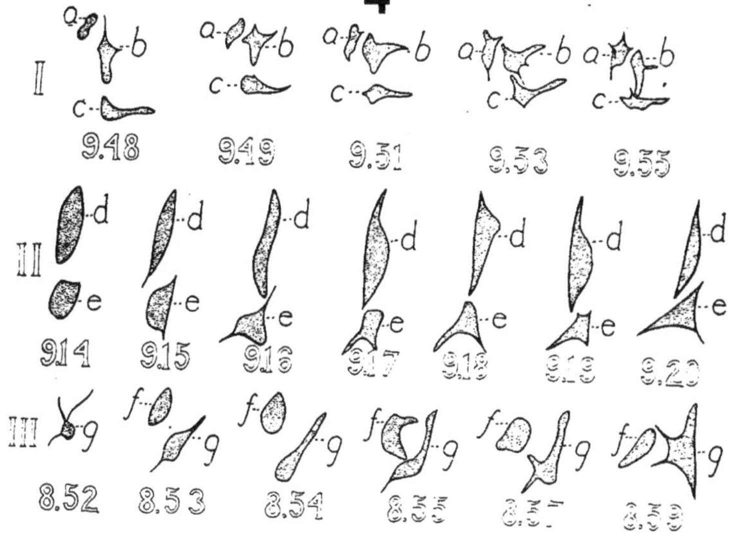
I
a
b
c
9.48
9.49
9.51
9.53
9.55
II
d
e
9.14
9.15
9.16
9.17
9.18
9.19
9.20
III
f
g
8.52
8.53
8.54
8.55
8.57
8.59

5

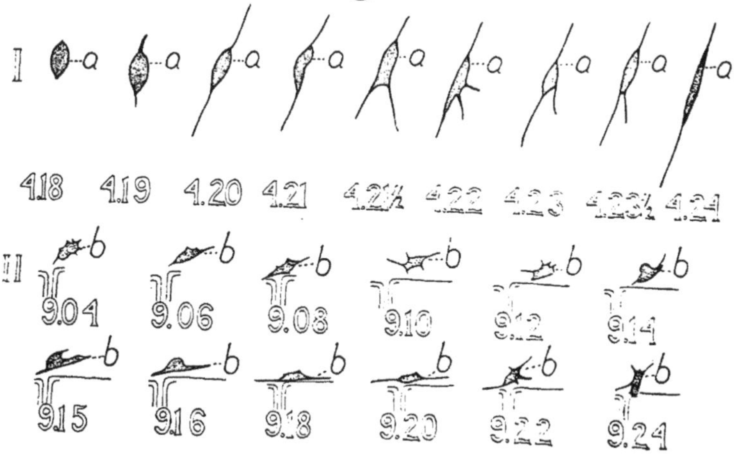
I
a
4.18
4.19
4.20
4.21
4.21½
4.22
4.23
4.23½
4.24
II
b
9.04
9.06
9.08
9.10
9.12
9.14
9.15
9.16
9.18
9.20
9.22
9.24

6

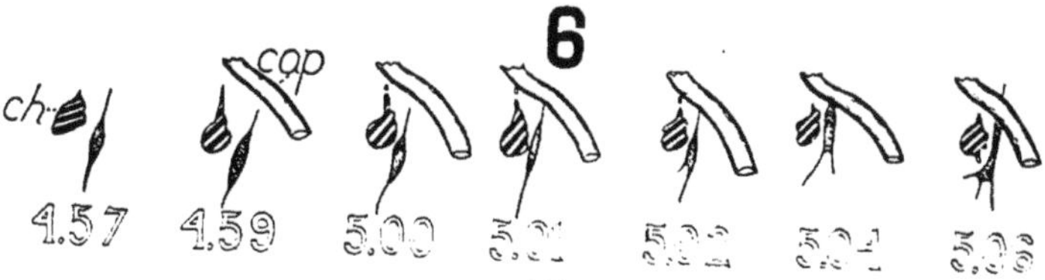
ch
cap
4.57
4.59

PLATE 3

EXPLANATION OF FIGURE

7 Stellate connective tissue cells from the fins of four embryo fish, *I–IV*. exhibiting rapid change of form. Owing to difficulties of observation it is not possible to make drawings oftener than at 1 to 3 minute intervals; hence, the actual changes of form were much more frequent than the record shows. *A-K*. nine connective tissue cells. Numerals as in the preceding figures.

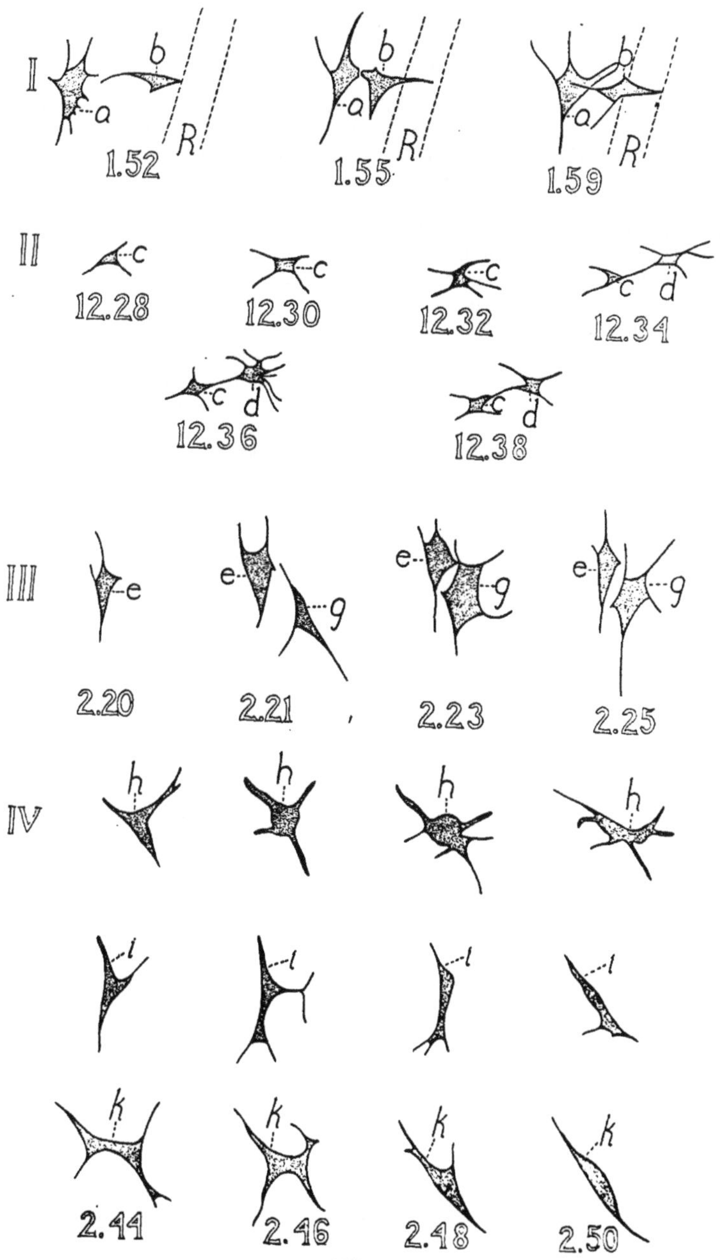
I
a
b
R
1.52
1.55
1.59
II
c
d
12.28
12.30
12.32
12.34
12.36
12.38
III
e
g
2.20
2.21
2.23
2.25
IV
h
l
k
2.44
2.46
2.48
2.50

PLATE 4

EXPLANATION OF FIGURES

8 Stellate connective tissue cells exhibiting locomotion and, on contact, apparent fusion. *I*, from the pectoral fin of a 10 mm. Fundulus embryo. The fusion between cells *a* and *b* is apparent only. *b* and *c* appear to form part of the anastomosing syncytium. *II*, from the caudal fin of a 6 mm. Fundulus embryo. Cells *e* and *g* on coming into contact at 4.44 P.M. apparently fused after the manner of the cells which form the delicate early connective tissue syncytium. Letters and numerals as in the preceding figures.

9 Stellate connective tissue cells in the subectodermal mesenchymal syncytium of a 25 mm. embryo pig. Fibrils pass through the cells very close to the nucleus. Bielschowsky stain.

10 Connective tissue cells from the praevertebral (*I*) and intermuscular (*II*) connective tissue of a 100 mm. pig embryo. The form of the cells and their contact relations to adjacent connective tissue and muscle fibers is strikingly similar to the amoeboid connective tissue cells of the living fish embryo. The appearance suggests that such cells were quite probably moving along the surfaces with which they are in contact.

8

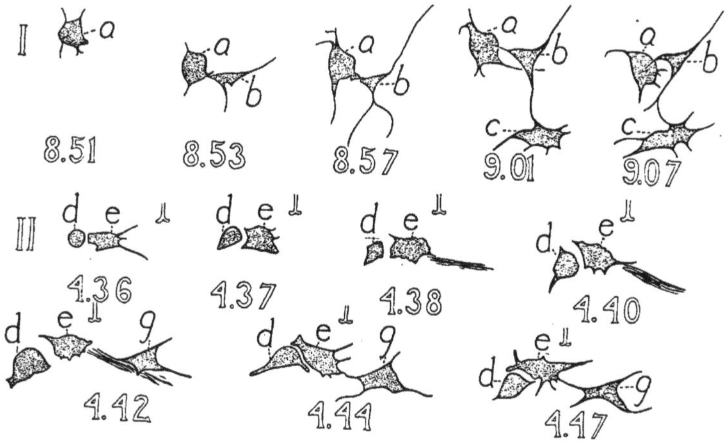
I
a
b
c
8.51
8.53
8.57
9.01
9.07
II
d
e
g
4.36
4.37
4.38
4.40
4.42
4.44
4.47

9

10

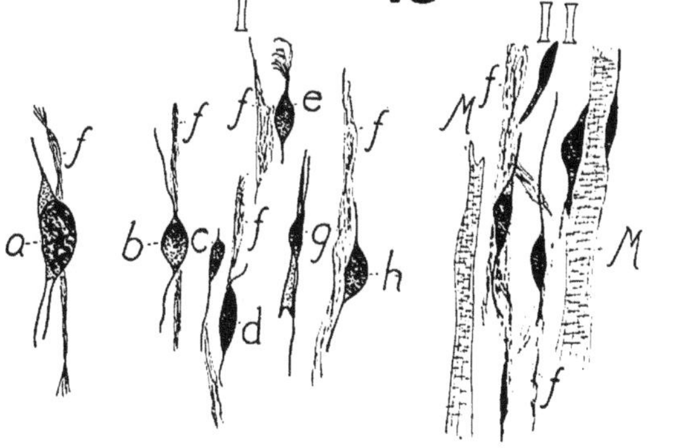
I
II
a
b
c
d
e
f
g
h
M

The Relation of Muscle Cell to Muscle Fibre in Voluntary striped Muscle.

By **W. M. Baldwin,**
Cornell University Medical College New York City.

(From the Biological Laboratory at Bonn.)

With table VII and VIII.

(Der Redaktion zugegangen am 19. Juli 1912.)

It may be stated in general terms that our present conception of a voluntary muscle fibre is that of an enormously large, multinucleated cell invested by a cell membrane or sarcolemma and containing the several elements of contractility. Or, as a recent textbook of histology phrases it, „Die quergestreifte Muskelfaser wäre also als ein vielkerniges, fadenförmiges Plasmodium mit deutlicher Zellmembran zu bezeichnen" (SOBOTTA). That this "cell membrane" invests, in addition to the muscle fibrillae and muscle nuclei, at least two entirely different forms of protoplasm,[1]) so far as their morphological appearances are concerned, there has been but little question but at the same time comparatively little mention in the literature of this feature. The one form immediately invests the nuclei and possesses the characteristics of cellular protoplasm as elsewhere observed; a granule-laden spongioplasm network with interstices of clear hyaloplasm. These general features are wanting in the other form, i. e., the sarcoplasm of the muscle fibre.

[1]) Many of the published plates of such investigators as HEIDENHAIN, HOLMGREN, NUSSBAUM, SCHNEIDER, MARCEAU, WERNER, PRENANT, etc. represent these differences most clearly.

These remarks, naturally, apply more particularly to the adult fibre. In the developing fibres the quantity of protoplasm investing the nuclei is relatively greater and much more coarsely granular. During this period it possesses apparently very intimate developmental relations to the muscle fibrillae and to the sarcoplasm since its quantity diminishes as the latter increases.

Investigators who have worked upon the structure of striped muscle have in general regarded a close functional relationship as existing between the sarcoplasm and muscle fibrillae on the one hand and between the muscle nuclei and adjacent differentiated protoplasm on the other, yet the exact morphological connection existing between the latter and the sarcoplasm has been but little examined beyond the general assumption that the two stood in intimate structural contiguity with each other.

It was with the express intention of carefully considering this relationship that I undertook the study of the muscle fibre, confining my efforts for the present to the voluntary cross-striated muscles of the tadpole, frog, chicken, calf, white mouse, gray mouse, and cat. I used for the investigation specimens of the extrinsic muscles of the eyeball, intercostales, rectus abdominalis, latissimus dorsi, adductors of the thigh, sartorius, flexors of the foot together with the caudal muscles of the tadpole. These various muscles were fixed in sublimate, Flemming's solution, Meves's solution, and in 96 % alcohol. They were imbedded in paraffin, some after Schultze's collodion and cedar oil method, and sectioned longitudinally, obliquely, and transversely in thicknesses varying from 2 to 5 μ. As stains I made use of picric acid alcohol, fuchsin S, picro-fuchsin, Gage's chloral-hematoxylin, Schultze's alcoholic-hematoxylin, eosin-xylene, and an alkaline ferric-tannate solution.

The sketches which I present, while they were made from actual specimens, are accurate, however, only so far as concerns the exact relationship of the protoplasm immediately investing the muscle nuclei to the muscle fibrillae and to the sarcoplasm of the muscle fibre. That is to say, inasmuch as the structural features of these other muscle elements did not bear directly upon the subject of the investigation, I made no effort, beyond a mere approximation of accuracy, to delineate either their structure and their relationship to each other or the ultimate structural features of the muscle nuclei. Accordingly, these figures are only to be regarded as authoritative with these two qualifications.

In most instances, I have represented only a single muscle nucleus with the adjacent muscle fibrillae and sarcoplasm. The relations shown were found to be characteristically present throughout the entire series. For the demonstration of the facts, however, I deemed it sufficient to represent only a single nucleus and the immediately adjacent muscle fibrillae. Hence these are not to be regarded as exceptional instances.

I have represented in Fig. I a portion of a fibre of an extrinsic muscle of the eyeball of a white mouse fourteen days old. At this age the eyelids are open. The fibre is invested by its sarcolemma lying upon which at one end there are several connective tissue cells and fibres. In the middle of the muscle fibre a nucleus surrounded by a considerable amount of protoplasm is shown. The characteristic features of cellular protoplasm as generally recognized are to be observed in connection with this protoplasm, i. e., strands of granule-laden spongioplasm composing a network which encloses clear hyaloplasm. The exact extent of this protoplasm in either direction is not shown, because of its passage out of the level of the oblique section. It is to be observed, however, that the structure of this protoplasm presents a marked contrast to that of the relatively uniformly and faintly staining sarcoplasm of the muscle fibre. It is further to be noted that the latter is sharply delimited from the spongioplasm network by the presence of a distinct membrane. The fibrils of spongioplasm can be observed to be attached to the internal surface of this membrane. That the existence of this membrane, as a distinct structural entity limiting the protoplasm, is not merely an appearance owing to the presence of overlying contingent muscle fibrillae is most clearly demonstrated by reason of the obliquity of the section. The relatively clear spaces intervening between the obliquely-sectioned fibrillae are observed to be bridged over by this structure. This particular feature is more clearly represented in the lower end of the section. In brief, then, we are dealing here with features which represent every characteristic of cell structure, a nucleus, cell protoplasm, with spongioplasm and hyaloplasm, and a cell membrane, all imbedded in a voluntary muscle fibre.

The clear spaces observed within the cell body immediately above and below the level of the nucleus I have regarded as artefactitious, referable as much to a retraction of the nucleus as to a corresponding shrinkage of the cellular protoplasm. At the same time attention must be called to the fact that in spite of these circumstances there is no indication in the section whatever either of a retraction of the spongioplasm from the cell wall or of the cell wall from the sarcoplasm outlying it. In many of the specimens, I have identified spongioplasm fibrillae which were attached to the nuclear membrane by one end and by the other to the cell wall.

I desire particularly to call attention to the relation of the cell wall to the nuclear membrane at the level of the nucleus. The two lie so close together as to be apparently fused. It is owing to the juxtaposition of these two structures that, when a muscle fibre presenting such a cell is studied in cross-section at the level of the nucleus, the cell membrane escapes observation altogether and the nucleus appears to lie imbedded in and in direct contact with the sarcoplasm. My own transverse sections of this same muscle and of all of the other muscles which I have studied, furnish ample evidence of this appearence, yet in those transverse sections in which, either as the result of the agents used in the preparation of the specimen or where, as the result of a tear in the muscle fibre, the nucleus was pulled away from the sarcoplasm, the presence of the cell membrane was then manifested and its existence at the level of the nucleus established. Furthermore, when transections which exhibit no traces whatever of shrinkage are studied either directly above or immediately below the level of the nucleus, the cell membrane can be seen most clearly and the fact observed, as well that it passes completely and uninterruptedly around the cell protoplasm.

In fig. II I have represented an oblique section of a similar muscle fibre presenting appearances comparable to those just described. A muscle cell occupies the middle of the fibre. Several connective tissue cells and fibrillae lie upon the sarcolemma. Very often in studying fibres of the extrinsic eye muscles of the white mouse, I have encountered specimens in which, apart from the characteristic muscle spindles, several muscle nuclei were closely associated with each other in the middle of the fibre. I have found as many as five nuclei so situated, imbedded in a protoplasmic mass which extended in the long axis of the fibre. This mass was everywhere marked off from the sarcoplasm by a membrane, similar in morphological structure and in relations to that which I have described above in connection with fig. I.

The question naturally arising in such instances was whether such a protoplasmic mass represented a single muscle cell containing several nuclei or whether the nuclei each represented distinct cells whose boundaries, because of their thinness, were not to be separately differentiated. In most instances I was not able to answer this question even in specimens studied under the best conditions of fixation and of stain. The specimen at hand furnishes, however, a suggestion of the probable solution of the difficulty. I can commit myself, however, to a positive statement regarding the general problem only so far as this one cell in particular is concerned. The protoplasm occupying the upper portion of the sketch and investing the nucleus there situated, is marked off from that found in the lower end by an obliquely-placed membrane, which, though faintly stained and somewhat indistinct, has the morphological appearance of a true cell wall. The lower mass of protoplasm contains as the next section in the series shows, also a nucleus.

The presence of such elongated masses of protoplasm is not restricted to the mouse. They can be observed as well in the extrinsic eye muscles of the three-weeks chicken. As yet, however, I have not detected these masses in the adult muscles of the thigh, leg, abdomen, or thorax of the white mouse, frog, chicken, gray mouse, or cat. In such muscles the nuclei are usually isolated from each other, while the quantity of protoplasm immediately investing them is relatively much less in amount.

Figure III represents a portion of a thigh muscle of an adult frog, showing features similar to those observed in connection with the white mouse muscles. The section was cut obliquely; consequently the muscle fibrillae cannot be traced for any considerable distance. I have represented in the sketch a single muscle cell with the adjacent fibrillae. This cell was imbedded in the muscle fibre and bore no immediate structural relationship to the sarcolemma. By reason of the obliquity of the section the internal surface of the cell membrane can be seen at the upper end of the cell. There is no question as to its presence or as to its continuity. As a distinct limiting membrane it can be traced from the vicinity of the fibrillae situated upon one side of the cell to those on the other.

Very often it is difficult to differentiate between the telophragma (Z) lines of the muscle fibrillae and the reflected cell membrane at the poles of the cell. In this particular specimen, however, this difficulty is lessened by reason of the fact that the features of cross-striation of the muscle are not distinctly marked. The cell wall, accordingly, can be seen all the more clearly. By reason also of the alcoholic hematoxylin stain used on this preparation the fibrillae are much lighter in color than the cell wall. Hence the possibility in a longitudinal section of mistaking the border of a muscle fibril for the cut edge of the cell wall is also eliminated. The cell membrane is readily demonstrable as a formed structure separated from the muscle fibrillae by a narrow interval of sarcoplasm. What is, however, a much more significant fact pointing towards the presence of the cell wall is, as the sketch shows, the presence of several strands of spongioplasm which terminate upon this cell wall. In no cell which I have studied have I ever seen a fibril of spongioplasm terminating upon a muscle fibril.

At the lower end of the cell two, sectioned, muscle fibrillae lie upon the external surface of the cell wall. These fibrillae passed out of the plane of the section in their passage around the cell body. Several published plates would lead us to infer that in some instances at least, muscle fibrillae terminated at a muscle cell. I have never observed such an instance, however, in my own specimens. In every section the fibrillae diverged from each other in order to pass uninterruptedly beyond the cell.

The remarks, which I made regarding the relation of the cell wall to the nuclear membrane opposite the center of the nucleus in connection with the mouse muscles, are to be emphasized at this place with regard to the frog muscle. The cell membrane everywhere separates the nuclear membrane from the sarcoplasm. And, likewise, in those specimens, where a shrinkage of the nucleus has taken place, its presence is most readily made out.

In some of the transverse sections of frog muscle the muscle fibrillae were torn away from each other. I have represented such a section in fig. IV. The sketch was taken at an optical level a little above the center of the nucleus. This shows a cell located in the middle of a muscle fibre. The muscle fibrillae lying above and to the left were undoubtedly torn away from a more intimate relation with the cell. From a careful study of the preparation, I concluded that the nucleus as well had been removed slightly in a direction upwards and to the left. In this direction the cell wall itself was torn. It remained intact, however, and in its proper position relative to the muscle fibrillae and sarcoplasm upon the other two sides. At the left side of the cell the edge of the cell wall can be seen to project from its attached position to the sarcoplasm into the clear space occasioned by the tear in the muscle fibre. It is further to be observed that the part projecting into this tear is unaccompanied by sarcoplasm. These features are less clearly outlined upon the right side of the cell. The specimen furnishes to my mind the most conclusive evidence as to the existence of the cell membrane as a distinct, independent, morphological structure.

A cell membrane is characteristically associated not only with those muscle nuclei which are located in the middle of a muscle fibre, but as well with peripheral-lying nuclei. I have observed the same relationship of nucleus to protoplasm, of protoplasm to cell wall, and of cell wall to sarcoplasm in connection with those nuclei which are applied to the sarcolemma. The cell membrane everywhere separates the sarcoplasm from the spongioplasm network of the cell. Upon that surface of the cell which is directly applied to the perimysium, however, the cell wall fuses with that structure and loses its identity.

In figs. V and VI are represented portions of muscle fibres removed from the thigh of an adult cat. Both of these cells lay upon the periphery of the muscle fibre in contact with the perimysium. The fibre in fig. V was sectioned longitudinally and that in fig. VI obliquely. A fold occurring at the level of the nucleus accounts for the transected fibrillae in the first. In both instances the cell membrane can be seen distinctly. It separates the cellular protoplasm containing the nucleus from the contractile elements of the muscle fibre. Upon that side of the cell which is applied to the perimysium it is fused with this structure and cannot be traced.

In the study of similar preparations presenting these features it was necessary to differentiate muscle nuclei applied directly to the internal surface of the perimysium from those connective tissue cells lying upon the external surface of that structure and from those which were, to all appearances, in the perimysium itself.[1] The fact is quite self-evident in the figures, particularly in fig. VI, that the nucleus is imbedded among the muscle fibrillae. That is to say, this relation is more intimate than would be expected were the cell situated outside of the sarcolemma but causing an indentation

[1]) This last statement requires some qualifications in view of the recent work of SCHIEFFERDECKER upon the sarcolemma. The nuclei to which I refer are so situated within what appears to be the sarcolemma that this latter structure splits apparently to enclose them. At any rate it is not possible to see any formed structure between such nuclei and the relatively clear sarcoplasm. I will comment more at length upon these special features, however, in a forth-coming paper.

of the muscle fibre. Furthermore, the morphological appearances of these cell walls are precisely similar to those noted before as characteristic of cells found imbedded among the fibrillae in the middle of a muscle fibre. Again, a difference can very readily be observed between these peripheral muscle cells and what, to all appearances, are cells located in the perimysium. The latter are small and spindle-shaped, with a compact and deeply-staining nucleus surrounded by a relatively small amount of dense and uniformly deeply-stained protoplasm.

From the study of extrinsic muscles of the eyeball of a calf, I was convinced of the presence of a cell membrane similar in relations and in its morphology to that which I had observed in the other vertebrates.

In fig. VII I have represented an entire muscle fibre cut transversely. The sketch was taken a little above the level of the center of the nucleus. At the latter level the cell membrane was not visible as a separate structure owing, as I have previously mentioned, to its contact with the nuclear outline. At the level represented, however, it is readily seen. I desire to direct particular attention to its relation to the muscle fibrillae. I have never found it in direct contact with these structures. An interval of sarcoplasm always intervenes between the two. In general terms it may be stated that this intervening layer of sarcoplasm averages in thickness to the cross diameter of a muscle fibril.

The sections of latissimus dorsi of the three-week-old chicken also show the same general features of cell protoplasm and cell wall. In fig. VIII I have represented a transection of an entire muscle fibre under the same conditions as those in fig. VII. The cell wall lies at a little distance from the adjacent fibrillae. It stains very deeply in contrast to the intervening sarcoplasm. At levels above and below that of the nucleus fibrils of spongioplasm can be traced to its internal surface where they find an attachment.

In the study of the longitudinal preparations of striped muscle fibres, there are introduced several factors which render the identification of the wall of the muscle cells particularly difficult. First among these to be mentioned, is the presence of the parallel-running muscle fibrillae. The free edge of a fibril may be very easily mistaken for the cell membrane. What is more significant, however, is the overlying or underlying of the membrane by these fibrillae thereby overshadowing the delicate cell outline. Still another difficulty is encountered at the extremities of the cells where the telophragma lines bridge over the narrowed cell protoplasm. Dependent upon this latter factor is the difficulty in the determination of the cell extent particularly enhanced. Then, lastly, scattered among the muscle fibres many connective tissue cells are found whose long axis corresponds very often to that of the muscle cells and, accordingly, to that of the muscle fibre. In their general outline, in the morphological characteristics of their protoplasm, and in that of their nucleus, as well, these cells often resemble the muscle cells very closely. Therefore, in the instance of such cells occurring upon those aspects of the muscle fibre either facing towards, or turned away from the observer it is, at times, not easy to differentiate the cells as muscle cells or as connective tissue cells. In other words to say whether the cells lie inside of the sarcolemma or outside of it.

In fig. IX I have represented a portion of a fibre from an extrinsic muscle of the eyeball of a calf. Two muscle cells, whose nuclei indent the longitudinally-cut fibre, are to be seen. The marked morphological differences between the protoplasm of these cells and the sarcoplasm of the fibre are most apparent. Notwithstanding, this fact by reason of the presence of the parallel muscle fibrillae, it is exceedingly difficult in this specimen to determine the exact limits of this protoplasm. In specimens stained with alcoholic hematoxylin and then properly extracted, however, the differentiation of cell membrane from fibrillae is rendered possible. Serving as good controls for such longitudinal sections are obliquely-sectioned preparations, for in such the cell wall can most readily be distinguished apart from the fibrillae.

In focusing down through this particular specimen, I encountered, first, the telophragma lines with the muscle fibrillae upon the upper aspect of the fibre, then the cell nucleus and protoplasm, and lastly, the same lines upon the under aspect of the fibre. Hence I was most positive that these were muscle cells within the fibre and not connective tissue cells either overlying or underlying the muscle fibre border.[1])

At this place again, I desire to point out the closeness of approximation of the nuclei to the muscle fibrillae. The latter are

[1]) I have projected the telophragma lines upon the same plane with the nucleus and cell protoplasm. They do not encircle the peripheral cells, however, nor do they traverse the cell protoplasm. I have made these points the subject of a forth-coming paper on the sarcolemma.

diverted out of their course slightly at its level. At this same level the cell wall and nuclear membrane are in contact with each other and are not separately distinguishable.

The structural relationships noted in connection with the calf's-eye muscle, I observed as well in the latissimus dorsi muscle of the three weeks chicken. Fig. X represents a portion of a fibre from this muscle. This is an alcoholic hematoxylin preparation which, notwithstanding the proximity of the muscle fibrillae, demonstrates the cell outlines very clearly. That the two cells seen were muscle cells and not connective-tissue cells located without the perimysium, I established by determining their position relative to the telophragma lines.

A muscle fibre from an extrinsic muscle of the eyeball of a white mouse is represented in fig. XI. This is a chloral-hematoxylin preparation. The outlines of the two cells upon the right of the figure are not clearly seen. Both of these cells, however, are muscle cells. In the instance of the single cell situated upon the left border of the fibre, we can note a morphological similarity to those represented in fig. IX.

In the specimens of leg, intercostal, and rectus abdominalis muscles of the adult mouse, the muscle cells are relatively less numerous, yet the cell membrane is everywhere clearly defined.

I present fig. XII for the purpose of demontrating one of the difficulties encountered in establishing the extent of the protoplasm of a cell in a longitudinal direction. The specimen is a tail muscle of a tadpole about 5,0 cm long. In either direction, as is shown, the cell protoplasm is drawn out to a pointed extremity lying between parallel-running muscle fibrillae, and occupying a space about equal in cross-diameter to that of the average interval existing between adjacent fibrillae. It is readily understood from a careful study of the figure how the elements of cross-striation, where overlying the narrow, pointed extremity of the cell, are very readily confused with that extremity. Indeed, for this reason, it is extremely difficult to determine in any given instance the true limit of the cell.

In conclusion it may be stated in general terms that the cell membrane is seen to best advantage in those oblique sections which are cut at a thickness of 2 μ and are stained with either alcoholic hematoxylin or with alkaline ferric tannate, and in such instances where the cut edge of the membrane, owing to the obliquity of the receding cell wall, is underlaid by nothing but the clear hyaloplasm of the cell. In many such instances where the staining is of the proper depth the internal surface of the membrane itself can be made out and its continuity, delicacy, and similarity to other cell walls observed. With the hematoxylin stain, the clear fibrils of spongioplasm can be seen to be attached to its internal surface. The interval of sarcoplasm existing between the muscle fibrillae and the cell membrane can also be noted. That the membrane is more than the thickened edge of the sarcoplasm seems to my mind to be established by its staining reaction, its presence as a distinct lamina separate from the sarcoplasm, in those instances where the muscle fibre has been torn, and by the attachment of fibrils of spongioplasm to its internal surface. That it is not an artefact produced by the shrinkage of protoplasm toward and upon the edge of the sarcoplasm, I believe, because it can be clearly observed in specimens where there is no indication whatever of shrinkage elsewhere in the tissues.

Concerning the query as to whether all of the nuclei with their surrounding protoplasm are invested with a cell wall, I can answer, that in all of the specimens which I have as yet studied and which were properly cut and stained, I have been able to make out the cell wall. Because of this fact, in such muscles as the intercostales which I have studied in their entirety, I consider that the muscle fibrillae are to be regarded as extracellular and not, as we have heretofore believed, intracellular. In the instance of the other voluntary striped muscles, since I have as yet observed no cell which did not possess a definite cell wall, I feel reasonably sure, though I have not as yet sectioned and studied an adult muscle as the sartorius or latissimus dorsi of the cat in its entirety, that the generalization, that the voluntary, striped-muscle fibrillae of the adult should be regarded as extracellular, is justifiable.

The apparent reason why this membrane has been overlooked by the very great number of investigators working upon muscle is referable not so much, I believe, to a fixation method fault as to a staining defect. I have, for instance, restained slides which were at first stained with eosin, fuchsin S, picro-fuchsin, and even with alcoholic hematoxylin, in which, however, the extraction of the stain from the muscle fibrillae was relatively not sufficient, and in which, accordingly, the cell membrane could not be seen, and in the same section after treatment with alcoholic hematoxylin and after a proper extraction of that stain, I have been able to define the membrane most clearly.

In order to guard against the false appearances produced by

shrinkage in stained sections, I studied many preparations of living muscle of the frog and the tadpole. In all of them the muscle cell outlines were very clearly seen. In fig. XIII I have represented a single muscle cell with the adjacent fibrillae from a thigh muscle of an adult frog. The outline of this sketch was made within ten minutes from the time that the portion of muscle was removed from the thigh of the living animal. No fixatives or other chemicals producing shinkage were employed on the specimen. The presumption is fair, I believe, that the sketch represents the condition as seen in a living muscle fibre. The outline of the cell body with its distinct cell wall is most clearly seen. The muscle fibrillae are very lightly stained with dilute methylene blue in comparison to the protoplasm of the cell body. The fibrils of spongioplasm can be observed to be attached to the cell wall. This preparation eliminates all doubt as to the presence of a definite cell wall in the living muscle and which is not produced in fixed and stained sections by artefactitious changes.[1])

Many workers, such as MALL, FLEMMING, SPALTEHOLZ, HANSEN, and others, have observed that the anlages of connective tissue fibrils are located within the genetic cell bodies. An adult fibril of such tissues is, however, either wholly extracellular or partly extracellular and partly intracellular. The names of ROLLETT, TOURNEAUX, MEVES and FERGUSON, in addition, may be mentioned in support of this latter statement.[2]) The same developmental sequence is observed in the instance of elastic tissue fibrillae. The work of HARRISON may also be cited, — „Die wichtigste Tatsache aus den vorgegangenen Beobachtungen ist, daß diese Skeletteile — die Hornfäden und Flossenstrahlen (of teleosts) — keinen intercellulären Ursprung haben, sondern das Ergebnis einer direkten Umbildung von gewissen Zellenteilen darstellen, welche von dem Mesenchym abzuleiten sind.“ The anlages of muscle fibrillae, as well, have an

---

[1]) In reviewing the literature it is not surprising to find that many of the published plates of striped muscle structure present what appears to be a muscle cell membrane. HEIDENHAIN's „Plasma und Zelle“, contains such figures, as do also the following references to invertebrate muscle, LEYDIG's „Untersuchungen zur Anatomie und Histologie der Tiere“, of 1883 and NUSSBAUM's „Anatomische Studien an californischen Cirripedien“ of 1890. It is a remarkable fact, however, that none of these authors recognized the significance of the membrane nor the extracellular character of the muscle fibrillae.

[2]) SPALTEHOLZ has recently questioned this view, holding that the fibrillae of the adult tissues are still intracellular.

intracellular position as Heidenhain, Godlewski, Meves, Marceau, Bardeen, among others have demonstrated. My own studies of the adult, voluntary, striated, muscle fibres show that the voluntary muscle fibrillae have an extracellular position. The intermediate steps in histo-myo-genesis are as yet incomplete, still with the facts at hand derived from these two extremes, a seeming parallelism can be drawn between the histogenesis of the connective and elastic tissues on the one hand and that of the voluntary striped muscle fibrillae on the other. Accordingly it would appear to be justifiable to attribute to this form of muscle fibre a position among the connective tissue group of structures and not to regard it as a purely cellular structure. My conception of a voluntary striped muscle fibre is, therefore, not that of a gigantic, multinucleated cell, but rather of a composite contractile structure enveloped by a sheath, the sarcolemma, which encloses, in addition to the elements of contractility, muscle fibrillae, sarcoplasm, etc., muscle cells, which cells present all of the recognized features of general cell structure, cell membrane, cell protoplasm, consisting of spongioplasm and hyaloplasm, and a nucleus containing nucleoli. [1])

Zusammenfassung.

Die quergestreifte Muskelsubstanz ist nur anfangs eine intracelluläre Bildung und wird später extracellular verlagert, so daß Muskelzellen und fibrilläre Substanz mit dem zugehörigen Sarkoplasma getrennte Bestandteile werden. Die Histogenese des Bindegewebes und der Muskelfasern ist identisch.

Eine quergestreifte Muskelfaser ist somit, entgegen der bisherigen Auffassung, keine vielkernige Riesenzelle, sondern von viel komplizierterem Bau. Was bis jetzt Sarkolemm genannt wurde, umschließt Muskelzellen, Sarkoplasma und Muskelfibrillen. Die

[1]) There still remain to be solved many problems involving the voluntary muscle cells and fibres. A much greater number of lower vertebrate muscles should be considered. Not only cardiac muscle but unstriped fibres in the whole vertebrate series should be reviewed in the light of the facts which I have set forth. Other pertinent questions are the relation of the nerve fibrillae to these extra-cellular fibrillae and several aspects of the histo-physiological phenomena of contraction; the relation of the lymph spaces to the fibrillae; the extent of the muscle cells in a longitudinal direction; the position of the telophragmata relative to the muscle cells; etc., etc. Accordingly, my treatise can be considered to be but little more than a preliminary communication.

Zellen haben eine Membran; ihr Protoplasma besteht aus Spongioplasma und Hyaloplasma und umschließt einen Kern mit Nukleolen. Über die Natur des Sarkolemm wird in einer folgenden Abhandlung eingehender berichtet werden.

Bibliography.

Since the recent treatises of HEIDENHAIN and of KEIBEL and MALL contain good working bibliographies dealing with the questions involved in this investigation, I have considered it superfluous to append a long literature list. In addition to some works to which I have especially referred, I have added only those articles which have appeared since the time of writing of these two books.

AIMÉ, Bandes intercalaires et bandes de contraction dans les muscles omo-hyoïdiens de la tortue. Bibliogr. Anatom., 1911.

GAGE, The Microscope. 11th Edit. Comstock Publ. Co., Ithaca, N. Y.

HARRISON, Über die Entwicklung der nicht knorpelig vorgebildeten Skelettteile in den Flossen der Teleostier. Archiv f. mikrosk. Anat., Bd. XLII, 1893.

HEIDENHAIN, Plasma und Zelle. II. Lief. Handb. d. Anat. d. Mensch. Herausg. v. K. v. BARDELEBEN, 1911.

KEIBEL und MALL, Handbuch der Entwicklungsgeschichte des Menschen. Leipzig, S. Hirzel, 1910.

LELIÈVRE et RETTERER, Des différences de structure des muscles rouges et blancs du lapin. C. R. soc. Biologie, t. LXVI, 1909.

— —, Structure de la fibre musculaire du squelette des Vertébrés. Ibid.

LEYDIG, Untersuchungen zur Anatomie und Histologie der Tiere, 1883.

LOUIS DES ARTS, Über die Muskulatur der Hirudineen. Jenaische Zeitschr. f. Naturw., Bd. XLIV, 1909.

MEVES, Archiv f. mikr. Anat., Bd. 75, 1910.

NUSSBAUM, Anatomische Studien an Californischen Cirripedien. Bonn 1890, M. Cohen & Sohn.

PRENANT, Problèmes cytologiques généraux soulevés par l'étude des cellules musculaires. Journ. de l'Anat. et de la Physiol., XLVIIIe Année, 1912, No. 3, Mai-Juin, p. 259.

SCHULTZE, Zeitschr. f. wissensch. Mikroskop. u. f. mikroskop. Technik, Bd. XXVII, 1910.

SOBOTTA, Atlas und Lehrbuch der Histologie und mikroskopischen Anatomie des Menschen. J. F. Lehmann, München 1911, S. 41.

THOMA, Über die netzförmige Anordnung der quergestreiften Muskelfasern. Arch. f. pathol. Anat., Bd. CXLI, 1908.

THULIN, Studien über den Zusammenhang granulärer, interstitieller Zellen mit den Muskelfasern. Muskelfasern mit spiralig angeordneten Säulchen. Anat. Anzeiger, Bd. XXXIII, 1908.

Explanation of figures of Plates VII—VIII.

Fig. I. This is an oblique section through an extrinsic muscle fibre of the eyeball of a white mouse fourteen days old. The preparation was stained especially deeply with Gages chloral hematoxylin. The cell membrane is everywhere distinctly seen with the attached spongioplasm fibrils. Magnification 1500 diameters.

Fig. II. This preparation is similar to the above. The core of protoplasm is, apparently, interrupted at a level slightly above the middle of the section. Magnification 1500 diameters.

Fig. III. This represents a muscle cell imbedded in a thigh muscle of an adult frog. Only the immediately adjacent muscle fibrillae are represented. The stain used was alcoholic hematoxylin, the fibrils of spongioplasm of the cell body, accordingly, lack the granular deposit of the chloral hematoxylin stain used on other preparations. The outline of the cell is most distinct. At the upper end of the sketch the internal surface of the obliquely-sectioned cell membrane with its attached fibrils of spongioplasm is to be observed. Magnification 1500 diameters.

Fig. IV. A preparation stained like the above and from the same muscle. The two groups of muscle fibrillae lying above and to the left are separated from the muscle cell by an artificial tear in the fibre. At the left side of the cell the torn edge of the cell wall can be seen to project into the clear space occasioned by the tear unaccompanied by sarcoplasm. Magnification 1500 diameters.

Fig. V. A portion of a longitudinal section of a muscle fibre of the thigh of an adult cat stained with alcoholic hematoxylin. A slight fold in the fibre occurring at the level of the nucleus explains the presence of the four transected muscle fibrillae. The muscle cell in this figure is closely applied to the internal surface of the sarcolemma. Its deeply stained protoplasm is very sharply marked of from the sarcoplasm of the muscle fibre by the cell wall. Magnification 1500 diameters.

Fig. VI. This is a portion of a thigh muscle of an adult cat sectioned obliquely and stained with alcoholic hematoxylin. The muscle cell rests against the investing sarcolemma. The distinct outline of the cell wall is not exaggerated in the sketch. The protoplasm of the cell body is very faintly stained and stands out in sharp contrast to the uniformly deep stain of the sarcoplasm. This sketch was made a little above the level of the center of the nucleus, consequently a space, occupied by cell protoplasm, intervenes between the latter and the cell wall. At the level of the nucleus similar cells show the nuclear membrane in contact with the cell membrane. Upon the sarcolemma-side of the cell the wall of the latter is fused with the sarcolemma. Magnification 1500 diameters.

Fig. VII. This is a transected extrinsic eye muscle fibre of a calf. The stain used was alcoholic hematoxylin. The sketch was made above the level of the center of the nucleus. The attachment of the cell wall to the sarcolemma is well shown. The intervening area of sarcoplasm between the cell wall and the muscle fibrillae has been carefully represented. Magnification 1000 diameters.

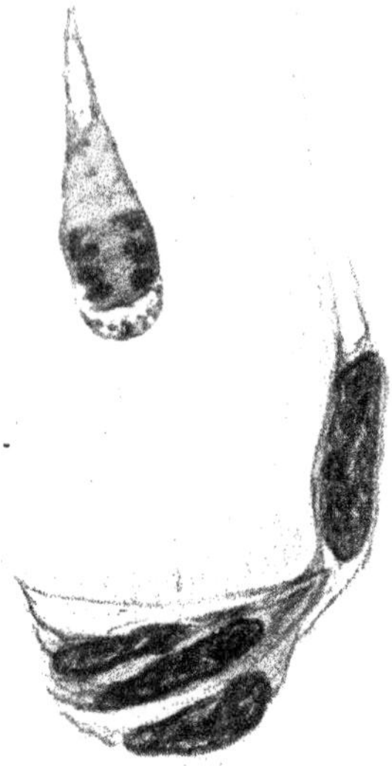

Fig. 1.

Fig. 6.

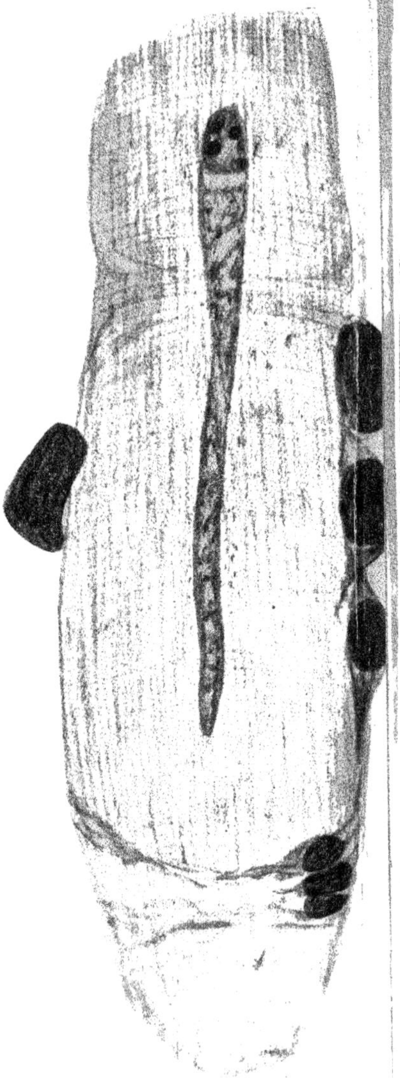

Fig. 2.

Fig.

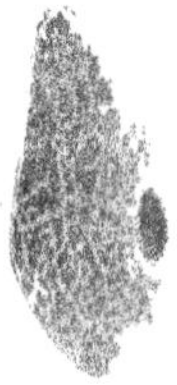

Fig.

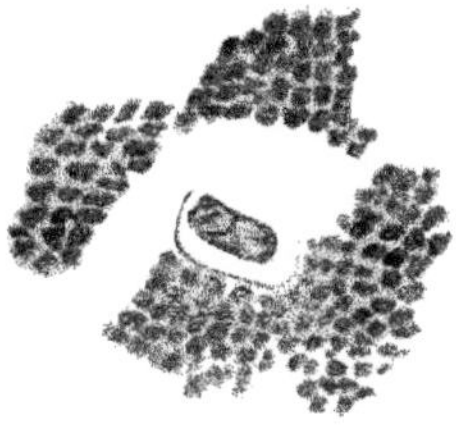

Fig. 4.

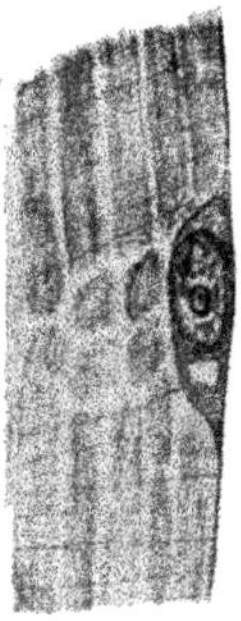

Fig. 5.

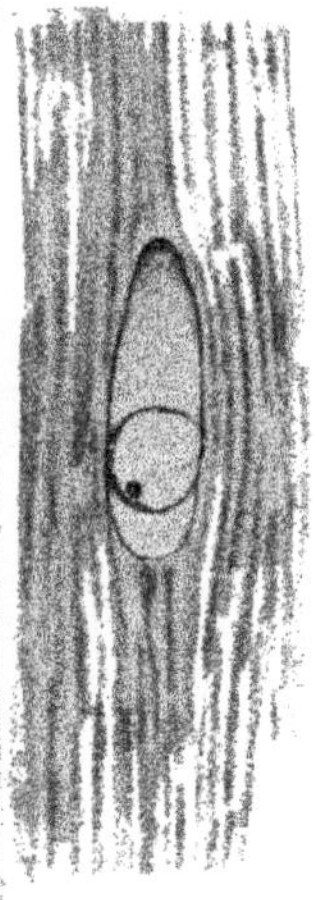

Fig. 3.

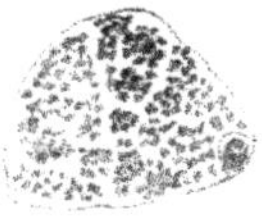

Fig. 8.

Fig. 9.

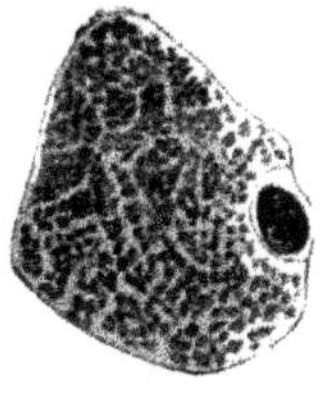

Fig. 7.

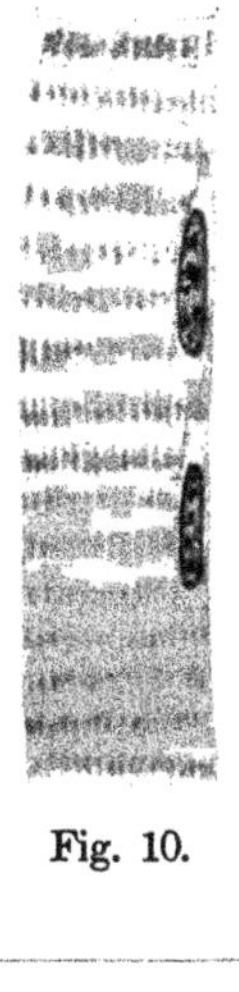

Fig. 10.

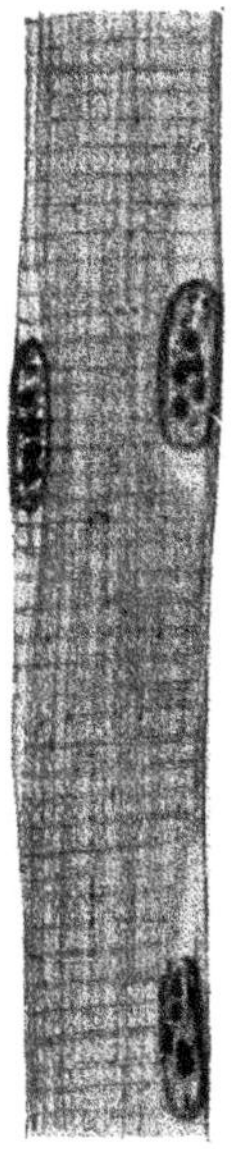

Fig. 11.

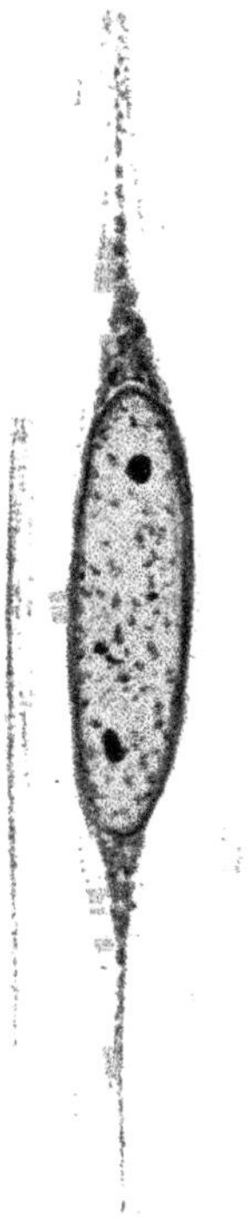

Fig. 12.

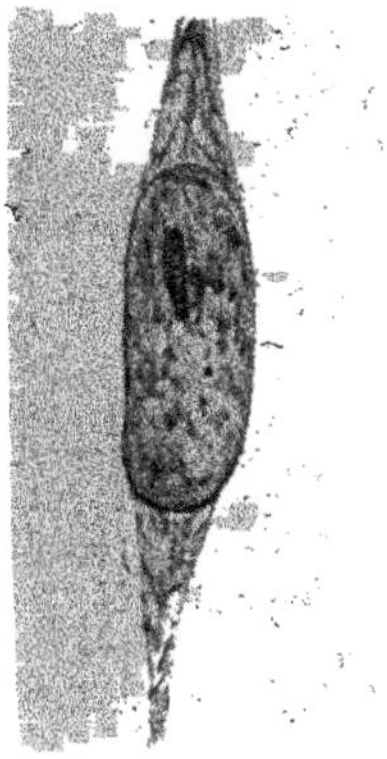

Fig. 13.

W. M. Baldwin del. Verlag von **Gustav Fischer** in **Jena.**

Fig. VIII. This represents a transected fibre of the latissimus dorsi muscle of a three-weeks old chicken. The relations are the same as those described under fig. VII. Magnification 1000 diameters.

Fig. IX. This is a portion of a fibre of an extrinsic eye muscle of a calf cut longitudinally. Two muscle cells lying between the sarcolemma and the muscle fibrillae are represented. Notwithstanding the marked contrast between the structural features of the cell protoplasm and those of the sarcoplasm, because of the overlying and parallel running fibrillae it is impossible in this preparation to identify the cell membrane. I have projected the telophragma lines upon the same level with the cell protoplasm. They do not traverse this latter structure. Stain Gages chloral hematoxylin. Magnification 1000 diameters.

Fig. X. This is a portion of a fibre of the latissimus dorsi muscle of a three weeks-old chicken represented in longitudinal section. The telophragma lines can be seen to lie over the muscle cells. The stain used was alcoholic hematoxylin. The cell outlines, therefore, are clearly represented. Magnification 1000 diameters.

Fig. XI. This figure represents a longitudinal section of a fibre of an extrinsic eye muscle of a fourteen-day old white mouse, stained with chloral hematoxylin. Three muscle cells are seen. The general morphological appearances noted in figures IX and X are to be seen here as well. A muscle fibre traverses the face of the nucleus on the left. Magnification 1000 diameters.

Fig. XII. This is a muscle cell imbedded in a caudal muscle of a tadpole about 5,0 cm long. The immediately adjacent muscle fibrillae only are represented. At either end of the spindle-shaped cell body the protoplasm is drawn out to a pointed extremity. The determination of the exact extent of this protoplasm is very difficult, notwithstanding the marked differences between its structure and that of the adjacent sarcoplasm. The stain used was alcoholic hematoxylin and the magnification 1000 diameters.

Fig. XIII. This is a portion of a fresh thigh muscle of an adult frog showing a muscle cell imbedded among the muscle fibrillae. The fibre is represented in longitudinal section. The stain used was methylene blue. Everywhere the cell outline stands out in sharp contrast to the comparatively faintly stained muscle fibrillae. This specimen being a fresh preparation and not having been treated with any fixative reagents or other chemicals save the methylene blue stain alone represents practically a living cell and muscle fibre in which no shrinkage, contraction or other artefactitious change has yet appeared. The cell was located and the sketch outlined within ten minutes from the time of removal of the specimen of muscle from the living frog. Magnification about 1500 diameters.

The relation of the Sarcolemma to the Muscle Cells of voluntary vertebrate striped Muscle Fibres and its morphological nature.

By **W. M. Baldwin,**
Cornell University Medical College New York City.

With plate IX.

(From the Biological Laboratory at Bonn.)

(Der Redaktion zugegangen am 1. August 1912.)

In a recent communication[1]) I have set forth the conception as the result of studies upon various striated muscles that the voluntary striped muscle fibre of the adult vertebrate should be regarded not as a plasmodium containing many nuclei or a giant multinucleated cell, which view has been entertained up to the present day, but, rather, as a composite contractile structure consisting of muscle fibrillae and of sarcoplasm enclosed by sarcolemma and containing, in addition, many muscle cells. Each one of the numerous nuclei imbedded in the muscle fibre presents nucleoli, is immediately surrounded by differentiated cellular protoplasm consisting of strands of spongioplasm with interstices of clear hyaloplasm, and is completely and distinctly limited from the muscle fibrillae and the sarcoplasm of the muscle fibre by a cell membrane. In other words, every feature of cell structure is to be observed. The muscle fibrillae are, accordingly, extracellular, while the sarcolemma should not be regarded as a cell membrane in the sense that it envelopes a multinucleated mass of protoplasm. In my earlier contribution, however, I left unanswered many contingent questions, such as the structural position of the sarcolemma, its

[1]) Zeitschr. f. allg. Physiol., 1912.

exact relation to the enclosed muscle cells, the relation of the telophragma lines to such cells, the longitudinal extent of the latter, etc. The purpose of my present writing is to elucidate certain of these problems; in particular, however, the morphological relation of the sarcolemma to the muscle cell walls.

In accepting the claim that the muscle fibre contained complete cells and not merely isolated nuclei, questions as to the relation of the telophragmata to such cells, i. e., whether the cell protoplasm was traversed by the uninterrupted telophragmata, were immediately presented for solution. Granting that the telophragmata are septa which completely traverse the muscle fibre (Heidenhain, Cajal, Mc Callum, Renaut, Marceau, Holmgren, etc.) then at the level of the muscle cells either the protoplasm of the latter is traversed or the telophragmata are interrupted. In the latter contingency especially in reference to the peripheral muscle cells the views entertained by M. Heidenhain, Marceau, Renaut, Holmgren, Zimmermann, that the telophragmata are directly attached to the sarcolemma must suffer some exceptions or the sarcolemma must be looked upon as surrounding each muscle cell in addition to investing the fibre as a whole. Nothing, so far as I know, has been stated definitely by the various investigators considering these problems regarding the traversal of the differentiated protoplasm investing the muscle nuclei. Indeed, the true nature of this protoplasm has been, as well, overlooked. At the level of the nuclei, however, i seems to be generally admitted, and is well represented in many of the figures of the workers whom I have cited, that the telophragmata do not traverse the muscle nuclei.

This present contribution is based upon studies of various voluntary muscles of Axolotl, chicken, white mouse, and cat. The technic and staining methods employed are the same as I described in my preliminary contribution. The secret upon which the derivation of good results depends is none other than overstaining with long prolonged extraction. In the present study I have confined my attention to those muscle cells which occupy a peripheral position in the muscle fibre lying in contact with the sarcolemma. The results present additional evidence bespeaking the new conception of the voluntary striped muscle fibre, which I have outlined above, and verify with increased emphasis the fact that the muscle fibrillae are extracellular, or intercellular, structures in the adult.

In order to present clearly and consecutively the fundamental facts ascertained by my studies, I have selected a series of white mouse preparations which are represented in figs. 1 to 4, inclusive. These are fibres from a thigh muscle of a three-fourths grown white mouse and are sketched at a magnification of 1000 diameters. The features noted in this series are to be taken not as limited to the muscles of this form of vertebrate, but, rather, as typical of the general condition found and as well presented in all of the other vertebrate muscles which I have studied.

In fig. 1 a portion of such a muscle fibre cut in its long axis is represented. The free edge of the fibre presents a single muscle cell with its relatively large nucleus. The muscle fibrillae are seen to be deviated slightly out of their course in their passage by the nucleus. Some of these fibrillae lie upon an optical plane lower than the level of the nucleus. One fibril is represented as traversing the face of the nucleus. In brief this peripheral muscle cell occupied that aspect of the muscle fibre turned at right angles to the observer. The presence of the bulgings of the sarcolemma between the attachment of the telophragmata noted upon the right aspect of the fibre may be accepted as evidence of the fact that the fibre was fixed in a condition of contraction. By reason of these bulgings the relation of the sarcolemma to the cell can be all the more readily observed. It is seen to pass across the face of the nucleus. Careful focusing demonstrated that the same relation obtained on the under aspect of the cell.

The specimen was very deeply stained with chloral hematoxylin, so deeply, in fact, that those aspects of the sarcolemma but slightly inclined towards or away from the eye could be readily observed as distinct laminae. Therefore it was possible to see at the level of the serrated edge of the sarcolemma traversing the face of the nucleus, two surfaces of sarcolemma, one upon the uppermost optical level of the fibre and the other between this and the muscle cell. Indeed, the continuity of these two surfaces could be readily observed at the serrated edge. Furthermore, the telophragmata could be followed in unbroken continuity from the uppermost aspect to the lower around the serrated edge. Between these two aspects lay the muscle fibril which I have represented and to which I have previously referred. The sarcolemma was, in fact, indented by the muscle cell and lay between the latter and the muscle fibrillae at the level of the center of the muscle cells. It did not enclose the muscle cell. At the level of the center of the muscle nucleus, because of its proximity to the nuclear membrane, its identity was obscured.

These same relations of sarcolemma, of telophragmata, and of muscle cell could be seen upon the under aspect of the fibre. As I have mentioned above, the general appearance presented was that of an indentation of the fibre by the muscle cell.

The question naturally arising as to whether the cell under consideration was a connective tissue cell overlying or underlying the muscle edge is negatived by the relations of sarcolemma which I have already outlined. Further support is contributed by a study of the relationship of the protoplasm of the extremities of the cell, (c).

In such positions the serrated border of the sarcolemma occupies the extreme edge of the muscle fibre. At such levels, too, the cell protoplasm lies among the muscle fibrillae as I have represented, in other words, is encompassed by the sarcolemma and the telophragmata.

Fig. 2 represents a portion of a muscle fibre containing a muscle cell similar to that in fig. 1 but rotated through an arc of 90^{0} so that the cell occupies the uppermost aspect of the fibre and consequently faces the observer. Many of the conditions unexplained in the foregoing figure are readily understood by a study of this sketch. There is no question as to its being a muscle cell and not a connective tissue cell. It is buried in the muscle fibre. The telophragmata are observed to mark off the bulgings of the sarcolemma upon either border of the muscle fibre. They do not, however, pass across the surface of the nucleus. Herein lie the characteristic and significant features of the specimen.

Overlying the muscle cell the serrated edge of the sarcolemma is observed to enclose a spindle-shaped space. The telophragmata proceed only up to this serrated edge. In this deeply-stained hematoxylin preparation the sarcolemma can be seen to be folded back upon itself in order to pass around the cell body. Hence the presence of the serrated interval, hence, also, the absence of the telophragmata upon the face of the cell. The sarcolemma does not traverse this uppermost aspect of the cell. It is reflected, rather, between the muscle fibrillae and the cell. In other words, it is invaginated into the muscle fibre by the cell.

I assured myself that the spindle-shaped space was not owing to the passage of the section knife through the uppermost aspect of the fibre by a most careful study of similar chloral-hematoxylin preparations which were especially deeply stained. In these the sarcolemma as a distinctly stained and homogenous membrane could be seen to be reflected upon itself at the serrated edge. There was no indication whatever of an artefactitious tear or section cut in the fibre.

In figs. 3 et 4 similar muscle fibres containing peripheral muscle cells are represented in transection. An entire muscle fibre and a portion of another are seen in figure 3. Each contains a muscle cell. That at A was encountered by the section knife at a level corresponding to A in fig. 2, hence the nuclear membrane, cell membrane, and immediately investing sarcolemma appear to be one. The nucleus seems to be imbedded in and in direct contact with the sarcoplasm of the fibre. Such appearances I have commented upon more fully in my other paper.

The cell indicated at C was transected at a level corresponding to level C in fig. 2. Nothing of the nucleus is to be seen; strands of spongioplasm, however, traverse the protoplasmic substance of the cell which is invested by cell wall plus sarcolemma.

This fact must be especially noted. According to our interpretation of the appearances presented by the cell in fig. 2 the nucleus and cell wall lie in direct contact with the perimysium of the muscle fibre, i. e., the sarcolemma does not lie upon the peripheral aspect of the cell, but, on the contrary, is reflected between it and the muscle fibrillae. The protoplasmic extremity of the cell at C, however, does not lie upon the periphery of the fibre, but passes deeper and deeper into the substance of the muscle fibre lying among the fibrillae. A narrow interval of sarcoplasm, hence, intervenes between the fused perimysium and sarcolemma encompassing the muscle fibre as a whole and the fused cell wall and sarcolemma immediately investing the cell protoplasm. Just how this sarcolemma comes to be invaginated into the muscle fibre among the fibrillae is rendered clear by a reference to diagram 5. Of these features I will speak more at length later.

The nucleus in fig. 4 has been sectioned at the level B (fig. 2), i. e., through one of its poles. At this level the nuclear membrane is separated from the cell wall and sarcolemma by an interval of cell protoplasm. It does not lie in contact with the investing membrane of the muscle fibre. The place of infolding of the sarcolemma to invest the nucleus is well represented at X. In the rest of its extent about the muscle fibre, its identity as a distinct membrane is lost because of its fusion with the investing perimysium. The place of infolding is the serrated edge observed in longitudinal sections (fig. 2) and encloses the cigar-shaped space referred to under the latter figure. Several muscle fibrillae occupy the angles of reflection. These were observed in figs. 1 and 2 to traverse the face of the nuclei. By reason of these infoldings the muscle cells really lie outside of the muscle fibre. In other words, the muscle fibrillae are intercellular. The muscle cells are merely received into a depression in the muscle fibre, lined with sarcolemma while the extremities of the cells penetrate among the fibrillae in tube-like prolongations of this same sarcolemma (c, fig. 3). The fact should be particularly noted in this specimen, since it is well demonstrated, that the sarcolemma is wanting between the muscle cell and the investing perimysium of the fibre opposite the level of the nucleus.

This interpretation of the relation of the sarcolemma to peripheral muscle cells renders reasonably clear the significance of the relation of the telophragmata to such cells. It is a significant fact that none of the figures published represent these lines as transecting the nuclei. In general they are represented as terminating abruptly in the immediate vicinity of the nuclei. Undoubtedly, because of too faint staining, both the infoldings of the sarcolemma at the edges of the nuclei and the presence of the serrated edge enclosing the cigar-shaped interval overlying the face of the nuclei have been overlooked.

The sketch 5 is a semi-diagrammatic representation of a muscle fibre cut longitudinally exactly through the middle of a peripheral muscle cell. The sketch is intended to demonstrate the relation of the sarcolemma to the cell wall. For the purpose of making this relation clearer the cell wall, C, is drawn with a narrow interval separating it from the sarcolemma, A—B. For the sake, also, of clearness the cell protoplasm with its fibrillae of spongioplasm is omitted. Consequently, the nucleus appears to lie in a clear space.

Tracing the sarcolemma, A, of the periphery of the fibre towards the nucleus, we note that at F it is reflected and passes then in contact with the cell wall, C, back to A. The angle F is the serrated edge outlining the cigar-shaped interval mentioned above. It contains, as shown, two muscle fibrillae cut obliquely which diverged out of the exact long axis of the fibre in order to proceed uninterruptedly around this interval. Between the angles of reflection of the sarcolemma, the cell wall lies in direct contact with the perimysium, D. That this last named structure contains nuclei is shown by the cell represented at E upon the left border of the fibre. Many of such spindle-shaped cells are to be found throughout the entire muscle series. But little doubt exists, then, that the perimysium is a connective tissue structure. As yet, however, I have failed to find a single nucleus in the true sarcolemma.

A portion of two muscle fibres cut longitudinally is represented in fig. 6. These are from a thigh muscle of an adult cat. By reason of a tear in the specimen the two fibres are separated from each other by an interval which is bridged, nevertheless, by several connective tissue strands. The fibre on the left contains two muscle cells united by cellular protoplasm which tunnels the peripheral portion of the muscle fibre. The relations of the crenated sarcolemma edge to the nuclei and of the perimysium to the upper muscle cell are the same as those noted in connection with the foregoing figures and, hence, require no further amplification. The lower of the two cells, however, presents several elucidating features.

The crenated border of the sarcolemma passes across the face of this nucleus. Because of the separation of the two fibres, however, the fibrillae constituting the perimysium are pulled away from the cell body and from the lower portion of the muscle fibre. One of these fibrillae is seen to be derived from a connective-tissue cell. Hence, the lower portion of the muscle fibre is naked so far as its perimysium investment is concerned. The sarcolemma alone clothes this portion of the fibre, and in it, incidentally, no cells are to be found. The removal of the perimysium has left the cell wall bare. The delicate nature of this structure, therefore, can be seen. In the extreme lower end of the sketch the cell protoplasm is observed to pass into the interior of the muscle fibre. Careful focusing and a study of the relation of the telophragmata to this protoplasmic mass proved that this cell extremity lay imbedded among the fibrillae of the muscle fibre. This specimen furnishes the most conclusive evidence upon the identity of the sarcolemma as a homogenous membrane, of the perimysium as a connective tissue structure, and of the presence of a distinct cell wall of the muscle cells.

The following four sketches are presented for the purpose of demonstrating the intimacy of contact of the sarcolemma with the perimysium. The first two figures are precisely similar to such transections which are ordinarily represented in text-books. The sketch (fig. 7) was taken from a section of thigh muscle of a white mouse three-fourths grown and demonstrates portions of four muscle fibres transected. The nucleus at A is that of a peripheral muscle cell that at B, a perimysium cell. The latter possesses a spindle-shaped body from whose extremities the perimysium passes off to encircle the muscle fibre.

Fig. 8 is a portion of a transected thigh muscle of an adult cat. The cell at A is a connective tissue cell and gives off several processes. The cell at B is similar to the cell indicated at B in the foregoing figure, a spindle-shaped perimysium cell whose processes are incorporated in that structure. In the instance of both of these perimysium cells the observation can very readily be made that, apparently, at least, the perimysium splits at the cell to enclose it, yet the internal of the two layers formed by this splitting process is so thin that, opposite the middle of the cell nucleus, its presence cannot be determined. Indeed. between the nucleus and the sarcoplasm no formed structures can be discerned, yet our studies of longitudinal sections, as we have seen above, show that the exceedingly thin sarcolemma does lie between the fibrillar, cell-containing perimysium and the sarcoplasm. Its thinness and its intimacy of contact with the perimysium are such that, only in those places where it leaves the latter to be reflected around the peripheral muscle cells, can its identity be recognized.

That this intimacy of contact with the perimysium is real and not apparent, the last two sketches demonstrate. The muscle fibre in fig. 9 is a transected trunk muscle of Axolotl and that in fig. 10 a similar section of the latissimus dorsi of a three-weeks old chicken. In both specimens, tears had been caused. As a result the perimysium was pulled away from the sarcoplasm. It is to be noted that the sarcolemma adheres to the former. The evidence for this fact must be given negatively, since is not demonstrable upon the border of the sarcoplasm imbedding the muscle fibrillae. This is seen to be rather ragged in outline such as would not be expected were the sarcolemma in situ. The appearances in fig. 10 seem to negative this latter statement, however, since the free edge of the sarcoplasm appears to possess a sharply defined thickened border, — a true sarcolemma. It must be noted in the sketch, however, that a small lappet upon the left aspect of the sarcoplasm indicates that a tear through the latter had been occasioned in the specimen. As a result a group of thirteen or fourteen fibrillae with their imbedding sarcoplasm was pulled away from the main fibril group. The significant fact demonstrated by the specimen is that in the angle of this tear the same apparently sharp edge of the sarcoplasm is observed. In other words, the edges of the torn sarcoplasm present the same distinct border as is observable elsewhere upon the surface of the sarcoplasm. Therefore, it is very probable that this seeming border is not sarcolemma but an artefactitious product. a kind of superficial increase in density of the sarcoplasm.

The remarks which I have made above and the conclusions which I have drawn, apply particularly to peripheral muscle cells. I feel sure by induction, however, that the same relation of sarcolemma to cell wall and of cell wall to cell protoplasm obtains in the instance of central-lying muscle cells. In fact, I have observed many such cells encompassed by sarcolemma. Until a new negative staining method is devised by which the cell protoplasm is stained and the muscle fibrillae unstained, however, I cannot commit myself upon the derivation of the sarcolemma immediately investing the cell wall, i. e., whether it represents a portion isolated from that enveloping the fibre or is derived from the same in a tube-like process tunneling the length of the muscle fibre. So far as I have gone with my research studies the latter appears to be the case, yet the evidence is not so coherent as to be presented in an absolutely convincing whole, by reason, as I have mentioned above, of a staining difficulty. In brief, the fundamental problem involved is the determination of the exact extent of the protoplasmic masses belonging to the imbedded cells. The difficulties present in the solution of this problem still await solution.

I have not yet observed a single centrally-located muscle cell, however, whose protoplasm was traversed by the telophragmata of the muscle fibre. I am in a position further to corroborate fully the views entertained by HEIDENHAIN, SCHIEFFERDECKER, MC CALLUM, ENDERLEIN, HOLMGREN, ZIMMERMANN, among others, that in the adult striped muscle the telophragmata are directly attached to the sarcolemma. The only interruptions occurring in this attachment are apparent rather than real and are to be noted in connection with the protoplasmic masses of the central and of the peripheral muscle cells. In such instances the telophragmata are directly attached to the sarcolemma immediately investing these muscle cells. They do not traverse either the protoplasm or the nucleus of such cells. Such observations as those of ENDERLEIN on Gastrus equi and on Hypoderma Diana, where the telophragmata traversed the protoplasmic cellular masses outlying the muscle fibrillae but enclosed within the sarcolemma, I have failed to verify in the adult muscles of the higher vertebrates which I have studied

Without entering too extensively into the history of the investigations upon the nature and relations of the sarcolemma, the following brief statements of the more pertinent facts should be made. As early as 1840 BOWMAN named the structure, which we have generally considered as sarcolemma, and described it as a structureless membrane serving as an envelope for the muscle fibres. REICHERT subsequently questioned this view of its morphological character, however, maintaining that it was a connective tissue structure and contained nuclei. These two papers marked the beginning of a half-century dispute which has lasted down to the present day. The credit belongs to LEYDIG, however, for first advancing the view of the sarcolemma of the vertebrate muscle as a cuticular structure. CALBERLA also held the same opinion. HENLE, on the contrary, found nuclei in it. CAJAL, HOCHE, HEIDENHAIN, MARCEAU, RENAUT, ZIMMERMANN, and HOLMGREN concluded, in spite of the fact that they could discerm fibrils and cells in the sarcolemma, that in cardiac muscle fibres the latter should be regarded as a cuticular thickening of the sarcoplasm. One of the most recent contributors upon the subject is SCHIEFFERDECKER. This investigator utilized the trunk musculatur of Petromyzon for his purposes. In the "central" fibres of such muscles he found the examples, long desired by investigators, where sarcolemma, devoid of perimysium, alone enveloped the fibres. In the study of such specimens he came to the conclusion that the sarcolemma of the voluntary striated muscle was a homogenous cuticular thickening and possessed neither cells nor fibrils. I, too, have studied such specimens and, as well, have arrived at the same conclusions. I differ in opinion from the latter and from MAURER, among others, however, in regarding this thickened cuticula not as a cell membrane. It is apparently a derivative of the sarcoplasm, by a process of condensation and in this respect alone resembles a cell membrane. It does not, however, encompass a multinucleated mass of protoplasm, since the sarcoplasm imbedding the muscle fibrillae, as I have demonstrated above does not contain nuclei.

My own studies have demonstrated in addition, that in the muscles which I have utilized and under the conditions which I have enumerated the sarcolemma of the adult, striped, voluntary muscle can be observed as a homogenous, fibre-less, and cell-less membrane.

The cause of the long controversy upon the subject is referable to the fact that the structure seen in ordinary transections of voluntary striped muscle fibres, as figures 7 and 8 show, is not sarcolemma but sarcolemma plus perimysium. The cells and fibres detected in it belong not to the sarcolemma but to the perimysium.

The conclusions to which I have arrived, then, as the result of my studies upon these adult voluntary striated vertebrate muscles are:

1. The sarcolemma is a thin, structure-less, homogeneous membrane completely investing the muscle fibre.

2. It contains no cells and no fibrillae.

3. It is infolded from the perimysium to invest the peripheral muscle cells.

4. At such places it separates the muscle cells from the sarcoplasm of the muscle fibre.

5. At such places, likewise, the muscle cell wall opposite the level of the nucleus lies in direct contact with the perimysium.

6. Upon its internal surface the sarcolemma is always in direct contact with the sarcoplasm.

7. The telophragmata are always directly attached to its internal surface.

8. Its external surface is fused with the perimysium.

9. The sarcolemma and the perimysium together constitute the investing envelope of the muscle fibre.

10. Only at the position of the peripheral muscle cells are these two structures separated from each other.

11. The perimysium is a connective tissue structure containing both spindle-shaped cells and fibres.

12. The sarcolemma is so thin and so closely applied to the perimysium as to escape observation in transections excepting at those levels where it is infolded to invest the peripheral muscle cells.

13. It can be seen to best advantage as a separate layer in chloral-hematoxylin preparations and at those levels in longitudinal sections where it is infolded to invest the peripheral muscle cells.

14. The telophragmata do not traverse the protoplasm of the muscle cells nor their nuclei.

All of these results are merely confirmatory of my earlier observations, i. e., — that the voluntary, striped muscle fibre of the adult vertebrate is not a giant, multi-nucleated cell, but, on the contrary, a composite contractile structure consisting of muscle fibrillae and of sarcoplasm enveloped by sarcolemma, and, in addition, numerous muscle cells. In other words, the muscle fibrillae are extracellular structures. The muscle nuclei are not directly imbedded in the sarcoplasm of the fibre but are separated from it by a layer of sarcolemma and by cell protoplasm which possesses a distinct cell wall. The terms muscle fibre and muscle cell as applied to this form of striped muscle are not synonymous. Furthermore, the muscle fibrillae are comparable to the fibrillae of adult connective tissues to the extent, at least, that both are intercellular in position.

Zusammenfassung.

Das Sarkolemm ist eine dünne, strukturlose, homogene Hülle der Muskelfaser ohne Zellen und Fibrillen. Seine äußere Oberfläche schmiegt sich dem Perimysium an und nur da, wo die Muskelzellen gelegen sind, weicht es von dem Perimysium zurück, so daß die Muskelzellen zwischen Perimysium und Sarkolemma liegen und das Sarkolemm nur das Sarkoplasm und die Muskelfibrillen umschließt.

Die Telophragmata sind mit der inneren Oberfläche des Sarkolemms verwachsen und dringen nicht in das Protoplasma der peripheren Muskelzellen ein.

Das Perimysium ist bindegewebiger Natur; es enthält Fasern und spindelförmige Zellen.

Perimysium und Sarkolemm zusammen umgeben die Muskelfaser und sind nur dort voneinander getrennt, wo die peripheren Muskelzellen gefunden werden. Das Sarkolemm ist so dünn, daß es an Schnitten der Beobachtung sich entzieht, außer an den Stellen, wo es in der Gegend der peripheren Muskelzellen von dem Perimysium sich zurückklappt und eine Mulde für diese Zellen bildet. Am besten kann das Sarkolemm an Chloral-Hämatoxylinpräparaten sichtbar gemacht werden.

Somit sind die Muskelfibrillen extrazelluläre Bildungen. Die Muskelzellen sind vom Sarkoplasma und den Muskelfibrillen durch das Sarkolemm getrennt und gemeinschaftlich mit diesen beiden Bestandteilen der Muskelfaser vom Perimysium umgeben. Die Muskelzellen haben eine deutliche Zellmembran. Die Muskelfibrillen sind den ausgebildeten Bindegewebsfibrillen vergleichbar, da beide extrazellulär gelegen sind.

Muskelfaser und Muskelzelle sind nicht synonym, soweit es sich um die hier untersuchten Arten quergestreifter Muskeln handelt.

Bibliography.

BOWMAN, Phil. Transactions, Vol. 2, 1840, Vol. I, 1841.

CALBERLA, Archiv f. mikr. Anat., Bd. XI.

CAJAL, Intern. Monatschr. 5. Bd., 1888.

ENDERLEIN, Archiv f. mikr. Anat., 55 Bd., 1900.

HEIDENHAIN, Plasma u. Zelle in Handbuch d. Anat. d. Menschen, hrsg. von v. Bardeleben, 1911.

HENLE, Allgemeine Anatomie, 1841.

HOCHE, Bibliogr. Anat., No. 3, 1897.

HOLMGREN, Anat. Anz., 31. Bd., Archiv f. mikr. Anat., 71. Bd.

LEYDIG, Untersuchungen zur Anat. u. Hist. der Tiere, 1883.

MCCALLUM, Anat. Anzeiger, 13. Bd., 1897.

MARCEAU, Bibliogr. Anatomique, 1902. — Ann. des Sc. nat., 8 série, Zoologie, T. 19, 1903. — Arch. d. Anat. micr., T. 7, 1905.

MAURER, Handb. d. Entwicklungsl., hrsg. von O. Hertwig, 3. Bd., 1. Teil, 1906.

REICHERT, Archiv f. Anat. u. Phys., 1841.

RENAUT et MOLLARD, J., Revue génér. d. hist. T. I, Fasc. 2, 1905.

SCHIEFFERDECKER, Untersuchungen über die Rumpfmuskulatur von Petromyzon fluviatilis, etc. Archiv f. mikr. Anat., Bd. 78, 1911.

ZIMMERMANN, Arch. f. mikr. Anat., 75. Bd., 1910.

Explanation of figures of Plate IX.

Fig. 1. A longitudinal section of a portion of a thigh muscle of a three-fourths grown white mouse. A peripheral muscle cell with its prominent nucleus is located upon one side of the fibre. A indicates the level of the center of the nucleus; B the extremity of the same where the nuclear membrane lies at some distance from the cell wall; C the protoplasmic extremity of the cell prolonged into the muscle fibre among the fibrillae. Magnification 1000 diameters.

Fig. 2. A surface view of a peripheral muscle cell of a muscle fibre the same as that in fig. 1. The serrated edges of the reflected sarcolemma traverse the face of the nucleus leaving a space between in which the muscle cell wall lies in direct contact with the perimysium. The letters A, B, and C correspond to those of fig. 1. Magnification 1000 diameters.

Fig. 3. A transection of two muscle fibres derived from the same source as that of fig. 1. At A a peripheral muscle cell is represented cut at the level indicated by the same letter in figs. 1 and 2. The nucleus lies in direct contact with the cell wall. At C another peripheral muscle is seen. This was encountered at the level C of figs. 1 and 2, i. e., only the protoplasm of the cell prolongation is seen. A layer of sarcolemma invests the cell wall. A narrow interval of sarcoplasm intervenes between this and the investing sarcolemma and perimysium of the muscle fibre, P. Magnification 1000 diameters.

Fig. 4. Similar to fig. 3. At B a muscle cell cut transversely at the level B of figs. 1 and 2 is seen. A narrow interval of protoplasm intervenes between the nucleus and the cell wall. At x the sarcolemma is reflected from its position in contact with the perimysium to invest the muscle cell, thereby separating it from the sarcoplasm and fibrillae of the muscle fibre. Magnification 1000 diameters.

Fig. 5. A semidiagrammatic sketch of a surface view of a portion, of an adult voluntary striped muscle fibre cut through the median longitudinal axis of a peripheral muscle cell. The cell wall, C, sarcolemma A—B, and perimysium, D, are separated from each other by a narrow interval. The protoplasm of the cell is omitted. F, the angle of infolding of the sarcolemma by which is produced the serrated edge noted in fig. 2 as traversing the face of the nucleus. Between the two angles of reflection (upper and lower in the figure) the cell wall lies in immediate contact with the perimysium. At E a spindle-shaped cell of the perimysium is shown. Magnification about 1500 diameters.

Fig. 6. A portion of two adjacent thigh muscle fibres of an adult cat cut in their long axis. Between the two fibres the interval, occasioned by handling of the tissue, is traversed by several strands of perimysium fibrils. At the lower end of this interval a connective tissue cell receives one of these fibrils. Two peripheral muscle cells occupy the right border of the muscle fibre and are connected with each other by a strand of protoplasm. The serrated edge of the sarcolemma traversing the face of the nuclei is well shown. At D the perimysium is torn away from the muscle fibre leaving the cell membrane at C bare. Magnification 1000 diameters.

Fig. 7. In this sketch a portion of four muscle fibres of a thigh muscle of a three-fourths grown white mouse is shown. The fibres are transected and demonstrate two nuclei, the one on the left being that of a peripheral muscle cell cut at the level A of figs. 1 and 2 and for this reason not demonstrating either the cell wall or the sarcolemma intervening between it and the sarcoplasm, while that on the right is a spindle-shaped perimysium cell. Between this latter cell and the enclosed sarcoplasm of the fibre the sarcolemma cannot be made out by reason of its thinness and the closeness of application to the perimysium. Magnification 1500 diameters.

Fig. 8. This represents a portion of a transected muscle of the thigh of an adult cat. A connective tissue cell with several fibres is adjacent to the perimysium which presents a spindle-shaped cell. Between the latter and the sarcoplasm the sarcolemma cannot be seen as a distinct layer.

Fig. 9. This figure is presented to show that, in a transected voluntary muscle of Axolotl where the investing envelope has been torn away by mechanical means from the fibrillae, the adhesion between the sarcolemma and the perimysium is greater than that between the former and the sarcoplasm. A bipolar perimysium cell is seen in the torn envelope. The sarcolemma, exceedingly thin and though not demonstrable upon the internal surface of the perimysium because of its intimate incorporation with the latter, is very manifestly not present upon the ragged, and irregular edge of the sarcoplasm. Magnification 1000 diameters.

Fig. 10. A transected fibre of the latissimus dorsi muscle of a three-weeks old chicken. The features demonstrated are similar to those of fig. 9. Magnification 1500 diameters.

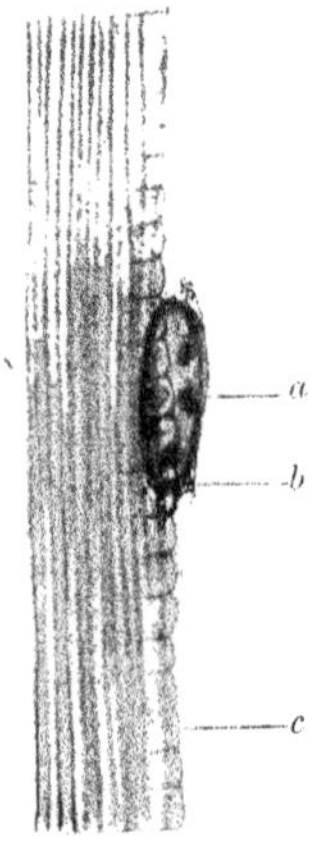

Fig. 1.

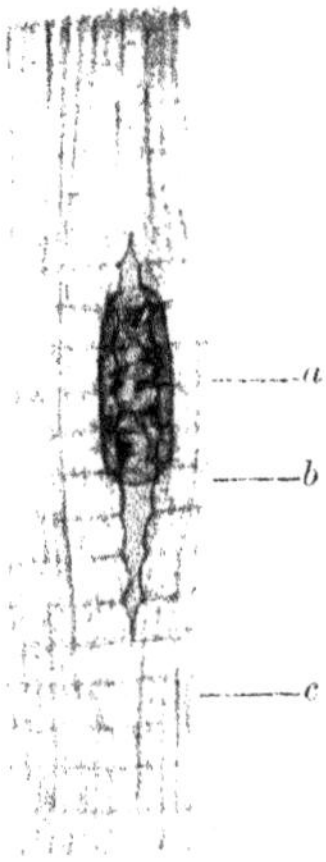

Fig. 2.

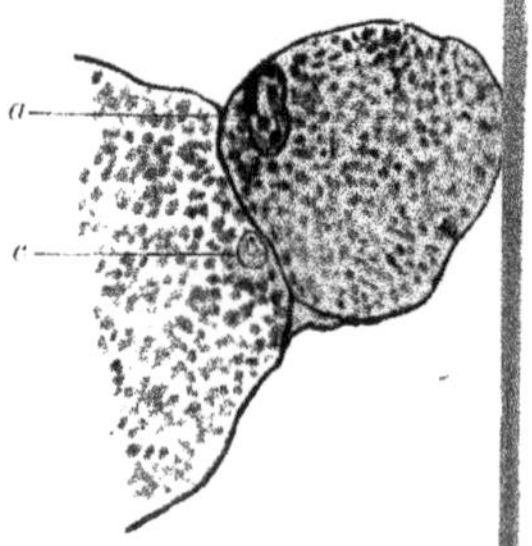

Fig. 3.

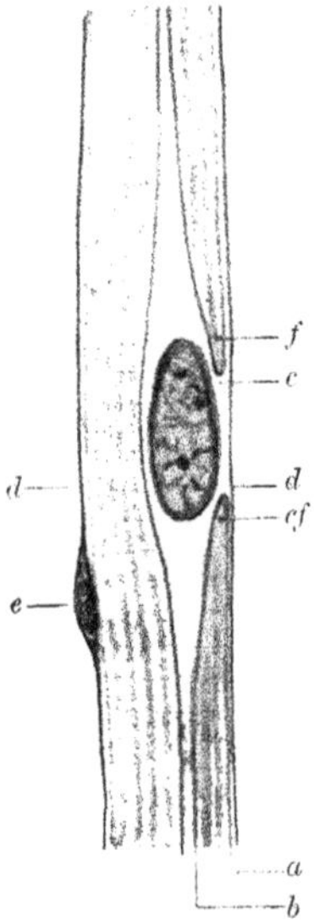

Fig. 5.

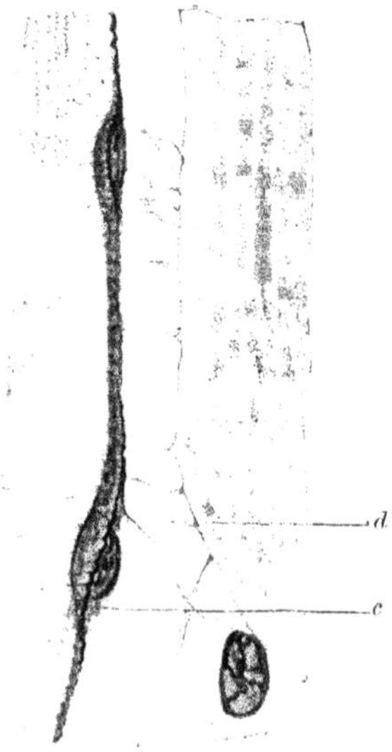

Fig. 6.

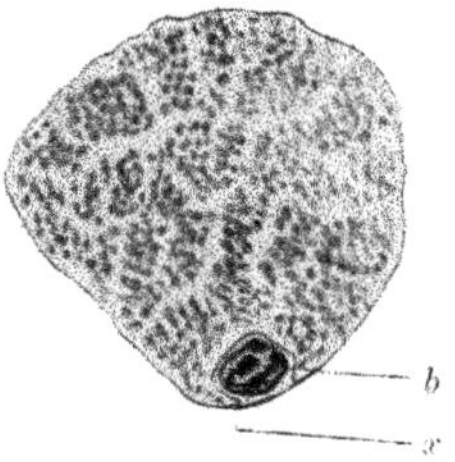

Fig. 4.

Fig. 7.

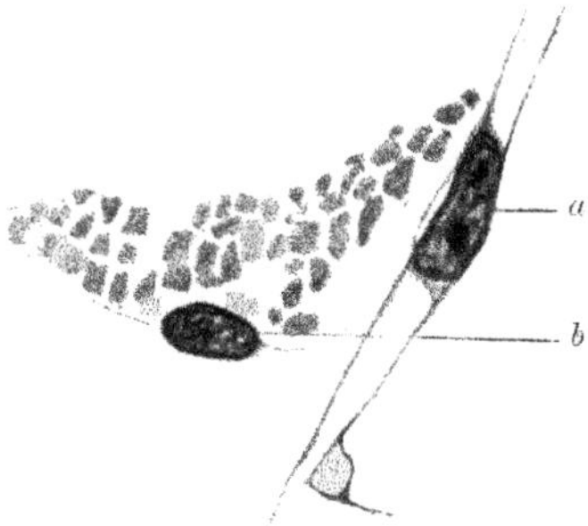

Fig. 8.

Fig. 9.

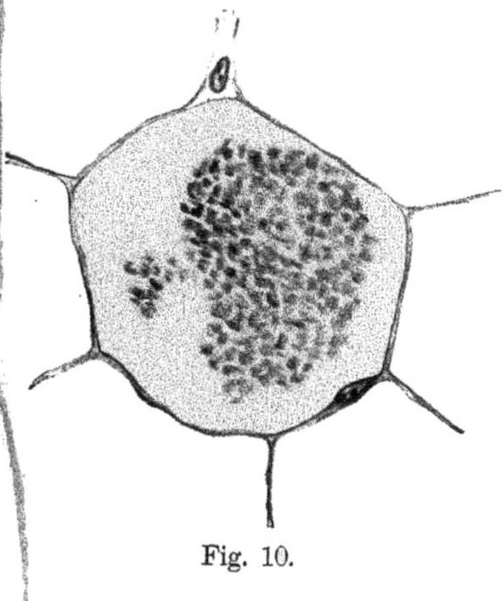

Fig. 10.

scher in Jena.

(From the Biological Laboratory at Bonn.)

The Relation of Muscle Fibrillae to Tendon Fibrillae in voluntary striped Muscles of Vertebrates.

By

W. M. Baldwin

(Dept. of Anatomy, Cornell University Medical College, New York City.)

With plate VII.

The long-contended question over the structural relationship of tendon fibrillae to muscle fibrillae has been brought to our attention by a recent contribution of O. SCHULTZE. This investigator carried out his studies on various muscles of *Hippocampus*, *Amphioxus*, several amphibia, and man, and arrived at the conclusion that these two structural features, muscle fibril and tendon fibril, were directly continuous with each other, perforating the sarcolemma. Being at the time engaged in a study of certain morphological features of striped-muscle structure, I reviewed my own preparations demonstrating these features with the express purpose of advancing our knowledge in this subject over a larger number of higher vertebrates than this author had used.

The preparations comprised sections of various muscles, such as intercostales, latissimus dorsi, rectus abdominalis, gastrocnemius, erector spinae, extrinsic muscles of the eyeball and various muscles of the thigh (together with caudal muscles) of such vertebrates as the tadpole, frog, calf, cat, white mouse, chicken, gray mouse. In addition I utilized specimens of living muscle of the frog and of the tadpole as controls to the fixed and stained preparations. The paraffin method of imbedding was employed in conjunction with SCHULTZE's excellent collodion and chloroform method of infiltration. The sections varied in thickness from 2 μ to 5 μ.

The stains used were picric acid, methylene blue, fuchsin S, and eosin, together with combinations of these and various alcoholic and aqueous solutions of hematoxylin among the latter SCHULTZE's and also GAGE's. Several of the more significant of the preparations were stained, decolorized, and then restained by another method in order not only to serve as controls to the simple stained sections but to demonstrate, in addition, the reaction of the several structures under consideration to the various methods previously employed. By means of a method of blunt dissection of the fixed and stained preparations upon the slide, several muscle fibres with their tendons were isolated from adjacent structures, by which was rendered possible a more careful and detailed study of their structural relationship.

In fig. 1 I have represented such an isolated muscle fibre with its attached tendon of an extrinsic eye muscle of a three-weeks old chicken. The instance is typical of the usual fibre-termination observed in these muscles. It can be seen that the sarcolemma at its extremity is drawn out to three distinct pointed processes which are not in any way continuous either with the sheath of the tendon or with the fibrillar components of the tendon. Furthermore, the pointed extremities of the tendon fibrillae are observed to be inserted into the recesses between the sarcolemma processes. This feature is worthy of special attention because owing to it an appearance is produced, when such an arrangement is rotated and studied in a vertical optical plane, of tendon fibrillae lying inside of the sarcolemma sheath of the muscle fibre. The individual muscle fibrillae, upon approaching the sarcolemma, lose their several features of cross-striation but can be traced, however, as slender, faintly-stained thread-like structures up to the internal surface of the sarcolemma upon which they terminate. The sarcolemma is in general much thinner than the cross-diameter of an average muscle fibril. Hence from the morphological arrangement of the parts thus effected, the muscle fibre appears to be dovetailed into the tendon with this exceedingly thin membrane as the only structure separating the muscle fibrillae from the tendon fibrillae.

Fig. 2 is another fibre from the same muscle, and demonstrates the same general features. Two muscle nuclei imbedded in granular protoplasm separate two groups of muscle fibrillae from each other. Each group, however, terminates in several pointed sarcolemma processes which are dovetailed with the fibrils of the corresponding tendon-fibril groups. Here again the muscle fibrillae, losing gradually their features of cross-striation, are, nevertheless, readily traceable up to the internal surface of the sarcolemma. There is no evidence, however, either upon morphological or upon staining-reaction grounds for the assumption that the fibrillae of muscle and tendon penetrate the sarcolemma. Neither can it be demonstrated that the sarcolemma is prolonged over the tendon or among its constituent fibril bundles. It terminates bluntly, rather, in a number of cone-shaped processes.

Fig. 3 represents the termination of an extrinsic muscle of the eyeball of a twenty-two-day-old white mouse. The observations made above in connection with the chicken muscles apply equally well to the muscles of this group in the mouse. Again, there is observed, the dovetailing of the sarcolemma processes with the tendon fibrillae, the termination of the muscle fibrillae at the sarcolemma, and the separation by the latter of the tendon fibrillae from the former.

I particularly desire to refer again to the insertion of the tendon fibrillae into the intervals between the cone-shaped prolongations of the sarcolemma through which arrangement certain appearances are produced which are exceedingly liable to be misinterpreted. This thin membrane is the only structure separating the muscle fibrillae from the tendon fibrillae. The latter pursue a course parallel to that of the former but terminate bluntly upon the external surface of the sarcolemma, whereas the muscle fibrillae run for a comparatively short distance upon the internal surface of the sarcolemma and finally lose their identity by fusing with it. Such conclusions can only be drawn from a study of those fibrillae which lie upon the same horizontal optical plane, i. e., upon that aspect of the sarcolemma which faces at right angle to the observer. When the muscle fibres and tendon fibres are cut in exactly their long axis, one must bear in mind that the uppermost and undermost aspects of the cone-shaped sarcolemma end are obliquely inclined to the vertical optical axis of the observer. This fact adds to the difficulty of interpreting the relation of the fibrillae more so than would be the case if the sarcolemma surface lay in a horizontal optical plane. Hence the solution of the question is much dependent upon the manipulation of the fine adjustment screw. Bearing in mind the fact that the sarcolemma is so thin as to be almost perfectly transparent, when studied upon these aspects of the fibres, the difficulty in determining the exact relationship of the muscle fibrillae to the overlying or underlying and parallel-running tendon fibrillae is well nigh impossible with our present optical instruments. When viewed in such vertical planes it appears as if those tendon fibrillae terminating in the sarcolemma indentations were really within the muscle fibre. Undoubtedly a neglect to take into consideration these facts has led to several erroneous conclusions such as are represented in various published figures. Accordingly, I have preferred to base my conclusions upon a careful study of those fibrillae which occupy the same horizontal optical plane. Under such circumstances the sarcolemma presents an unbroken contour. No evidence can be found bespeaking a continuity of tendon fibril with muscle fibril among these various extrinsic eye muscles of the white mouse, gray mouse, chicken, or calf.

One of the best bits of evidence upon this question is furnished by muscles of the bipenniform type. I have represented in fig. 4 a portion of a single muscle fibre of this type of an adult white mouse with its attached central tendon. As is readily seen the tendon fibrillae (*A*) lie at an angle of about 125° with the muscle fibrillae (*C*). The sarcolemma investing the fibre is considerably thickened at that end applied to the tendon. The numerous muscle fibrillae are represented half-schematically but their relation to this sarcolemma end is faithfully reproduced. Between the sarcolemma and the tendon there are to be seen several layers of connective tissue fibres and cells (*B*). I was able by means of various methods of hematoxylin staining to stain at once these three structures, central tendon, intervening connective tissue, and muscle fibrillae three different colors upon the same slide. The tendon fibrillae were yellowish-white, the connective tissue (peritendinum) reddish-brown, and the muscle fibrillae deep brown. There was absolutely no structural continuity to be seen between either the tendon fibrillae and the connective tissue fibrillae or between the latter and the muscle fibrillae. At no place could it be demonstrated that the muscle fibrillae traversed the connective tissue sheath in order to reach the central tendon, nor did they perforate the sarcolemma and turn at an angle to join the connective tissue fibrillae.

In muscles belonging to this type of structure the relation of the muscle fibres to the tendon is precisely similar to that of those muscles which are attached to bones, to the bone upon which they find their insertion. Such muscle fibres have only an indirect relation to the bone since their actual attachment is direct to the periosteum. Such is the case with these bipenniform muscles. Only

through the medium of this tendon-investing, connective tissue sheath, or peritendinum, comparable to the periosteum in the above instance, do the muscle fibres establish their connection with the tendon. The matter involved then is a consideration rather of the relation of the muscle fibrillae to the peritendinum (peritenonium) fibrillae.

A similar instance of a muscle fibre of this type with its attached tendon is represented in fig. 5 under a magnification of 1500 diameters. The peritendinum, consisting of several layers of connective tissue fibres and cells, separates the obliquely-inclined muscle fibre from the tendon. Due to a tear in the tissues an interval exists between a portion of the muscle fibre and the peritendinum and another between the latter and the tendon. In the rest of their extent however, the various structures have maintained their proper relationship to each other. The figure demonstrates five muscle fibrillae (*A*), which proceed directly up to the sarcolemma without losing their features of cross-striation, such as was the case observed with the mouse and chicken extrinsic eye muscles. The tendon end of the sarcolemma is very noticeably thickened and presents upon its internal surface several small elevations upon each of which a muscle fibril is inserted. Sections stained with picrofuchsin demonstrate that such elevations belong to the sarcolemma rather than to the muscle fibrillae. Moreover, sections stained with hematoxylin-fuchsin indicate that the sarcolemma, with these elevations has a different staining reaction from that of the peritendinum, and that both of these differ from the tendon. Morphologically there is no evidence that the sarcolemma is any other than a homogeneous structure unperforated and untraversed by any formed fibrillae. With differential staining it appears as a thickened homogeneous, unstriated, and non-fibrillar membrane. Nor is there any evidence of the peritendinum fibrillae turning at an angle to perforate it.

Another instance of the general type of muscle termination as demonstrated in figs. 1, 2, and 3 is furnished by the caudal musculature of the tadpole. A portion of a muscle fibre with its attached tendon fibrillae is represented in fig. 6. This fibre was removed from a tadpole about 5,0 cm long. The sarcolemma is very thin and is seen in the figure to be drawn out into a number of cone-shaped processes. Into each of these prolongations as many as from ten to forty muscle fibrillae enter and, without suffering any reduction in diameter or losing their features of cross-striation, proceed directly up to the internal surface of the sarcolemma with which they fuse. To each one of these cone-shaped sarcolemma processes a single tendon fibril is attached. These fibrillae vary among themselves in diameter, the average size is, however, about that of a muscle fibril, which, on the contrary, are generally uniform in size. Were the muscle fibrillae in direct continuity with the tendon fibrillae then each one of the former must be reduced greatly in diameter before becoming continuous with a tendon fibril. There is, however, no morphological evidence of such a reduction in size.

Unlike the tendon end of the sarcolemma in such muscles as are of the bipenniform type, the sarcolemma of these cone-shaped processes is not noticeably thickened. On the contrary, it is remarkable for its uniform thinness. Were it thickened, one might look therein for morphological evidence of muscle fibrillae, reduced in calibre, passing along its surface or through its substance in order to establish a conjunction with the single tendon fibril.

I have already stated that the tendon fibrillae vary in size. Such variations are not always proportionate, however, to the variations in size of the sarcolemma processes to which they are attached or to the number of muscle fibrillae therein contained. Such might be the case were a direct continuity of the two structures present. The absence of a correlation in size among these several structures might be adduced, therefore, as an additional fact arguing against the continuity of the tendon fibrillae with the muscle fibrillae.

In the figure 6 two cells are demonstrated among the tendon fibrillae. Judging from their morphological appearances and their relation to the tendon fibrillae, I have concluded that such cells were fibroblasts. In some instances I have found the tendon fibrillae traversing the cell protoplasm. In the figure the larger of the two cells gives off two delicate fibrillae each of which is attached to a pointed extremity of the sarcolemma. These are, moreover, the only fibrillae attached to these respective sarcolemma extremities. Upon the other side of the fibroblast several similarly slender fibrils stream off in the general direction of the other tendon fibrillae. I have, naturally, interpreted such fibroblastic processes as developing tendon fibrillae, and upon this interpretation have found an explanation for the disparity in size between the several fibrillae i. e. the smaller tendon fibrillae represent younger fibrillae. Hence the size of such fibrillae is in no direct wise associated with the size of the cone-shaped sarcolemma processes upon which they are inserted nor with the number of muscle fibrillae which terminate in such sarcolemma processes.

Apart from these considerations, however, the staining reactions demonstrate very clearly that the tendon fibrillae pass only up to the sarcolemma. They do not penetrate it, neither do the muscle fibrillae. The thickness of the sarcolemma everywhere separates the two structures from each other.

Further evidence is contributed by the developing intercostal muscles which I have studied, and of which I have sketched three fibres in figs. 7, 8, and 9. In figure 7 each of the three pointed extremities of the sarcolemma has a connective tissue fibril attached to its apex. The fibril on the right is seen to be derived from a distinct cell body which encloses a relatively large nucleus. The elements of cross-striation do not accompany the muscle fibrillae up to the extremity of the sarcolemma. In the vicinity of the latter these fibrillae somewhat resemble the tendon fibrillae in morphological characters, yet there is no morphological appearance to be noted from which it might be inferred that the homogeneous sarcolemma is perforated by the passage of either form of fibril. The staining of the section shows, in addition, that whereas the muscle fibrillae are thin and very faintly stained, the tendon fibrillae in these instances are relatively deeply stained and much thicker. At no place have I seen these tendon fibrillae inside of the sarcolemma as Schultze has represented in his figures.

Fig. 9 represents somewhat semidiagrammatically the typical muscle termination observed in these developing muscles. The termination of the muscle fibre as a whole is seen to be bluntly pointed, but the sarcolemma end is observed, in addition, to present upon close examination with a high power very many small cone-shaped processes to each of which a connective tissue fibril is attached. The younger the muscle fibre, the smaller are these processes and the less the numbre of muscle fibrillae terminating within each of them. In the course of development as these processes increase in size and the number of contained muscle fibrillae partakes of a proportional increment, but a single tendon fibril, notwithstanding, is found to be attached to the apex of these processes, in other words these apical fibrillae do not multiply in this type of muscle. Herein lies another significant fact denying the continuity of tendon fibrillae with muscle fibrillae.

Fig. 8 is of especial significance because it represents in the same muscle fibre the two forms of muscle ending which I have mentioned above, the one in which the fibre is inserted upon on obliquely-placed surface, as represented by the instance of the bipenniform muscles sketched in figs. 4 and 5, and the other where the insertion takes place upon a tendon extremity whose fibrillae pursue the same linear direction as those of the muscle fibrillae. At the same time this particular fibre affords an explanation of the genesis of the sarcolemma eminences or projections to which I have previously referred and figured in sketches 4 and 10 (*A*), which occur upon the internal surface of the tendon end of the sarcolemma and afford attachment to the muscle fibrillae. The right side of this sketch demonstrates three cone-shaped sarcolemma prolongations to each of which a tendon fibril is attached. These fibrils are typical of the arrangement observed in the tadpole tail. That portion of the sarcolemma end upon the left side of the figure, however, is closely applied to the obliquely-placed fibrillae of the perichondrium enclosing the cartilage of the developing rib. The component structures upon this side of the muscle fibre have the same arrangement as was seen in the bipenniform muscles. The arrangement upon the right side is to be regarded accordingly as transitional, since in the adult animal all of the muscle fibrillae are disposed as is shown upon the left side of the figure. This single muscle fibre, then, represents at once the earlier developmental condition and as well the adult condition; therefore, there can be found in it the probable explanation of the origin of those projections of the sarcolemma which are shown in fig. 10 (*A*).

In the earlier developmental stages the muscle fibres terminate in the manner represented in fig. 9 i. e., by a number of cone-shaped sarcolemma processes to each of which a connective tissue fibril is attached. In the course of development as the muscle fibres gradually approach and are applied to their definitive insertion, in such instances where this insertion is upon an obliquely inclined structure, as periosteum or peritendinum, the sarcolemma loses these cone-shaped terminal features. This takes place by a flattening of the apices of these cones and a synchronous fusion of the adjacent walls of neighboring cones. By the flattening of the terminal sarcolemma a better adhesive surface is presented to the flat surface of the structure affording attachment to the muscle. By the fusion of the adjacent cone-walls is brought about the presence in the adult of the projections of the sarcolemma which extend into the muscle fibre. They have no connection developmentally or morphologically with the tendon fibrillae, since at first there are no tendon fibrillae attached to that portion of the sarcolemma from which these projections are derived. Neither is it probable that they represent a portion of the muscle fibrillae which might have become transformed into sarcolemma-like tissue.

This genetic sequence explains as well the appearances presented by such muscles as I have represented in the first three figures where the tendon fibrillae and the muscle fibrillae occupy the same linear direction and where in the adult condition the several pointed extremities of the sarcolemma, characteristic of the younger developmental stages, are still demonstrable. The tendon fibrillae occupying the intervals between adjacent extremities assume their position at a stage in the developmental cycle later than the appearance of the definitive form of the muscle fibrillae. In other words, the muscle fibrillae are already attached to the internal surface of the sarcolemma of the intervals before the supplementary tendon fibrillae grow into these intervals and become attached to the external surface of the sarcolemma. Accordingly, we can find another argument in this fact of genesis, in addition to the one based upon morphological grounds and mentioned before, against the acceptance of the view that the tendon fibrillae effect continuity with the muscle fibrillae by perforating the sarcolemma and then coursing a considerable distance in the muscle fibre. Indeed, were structural continuity an established fact, if the chronological order of the development of these »interval« tendon fibrillae and of the muscle fibrillae were alone considered we should expect rather to find a prolongation of a portion of undifferentiated muscle fibrillae through and outside of the sarcolemma in order to meet the developing tendon fibrillae, and not vice-versa as some authors have represented. The presence of extrasarcolemmatous muscle fibrillae has never been observed among vertebrates, so far as I am aware, and in my own preparations is most positively denied by the staining reactions. The only fibrillae lying outside of the sarcolemma are those which have a connective tissue origin.

In general it may be said, then, that the voluntary striped muscles of adult vertebrates terminate in one of two general arrangements the determination of which is dependent upon the relation of the long axis of the tendon to that of the muscle fibre.

When the direction of the two coincide, the muscle fibre of the adult retains its earlier developmental features to the extent that the sarcolemma end still preserves its cone-shaped blunt projections into which the muscle fibrillae, presenting all of their features of cross-striation, enter to fuse with its internal surface. These sarcolemma projections with their intervening recesses are dovetailed into corresponding features of the tendon extremity. The second general type of muscle end is observable in those other muscles where the long axis of the tendon meets that of the muscle fibre at an angle. This form is to be regarded as a developmental derivative of a condition which in the younger specimens conformed to the type of structure which we have designated under I. So far as my own observations have led me I have not yet seen a developing muscle fibre which did not conform to the first type.

In other words, before the muscle fibre has grown far enough to reach its definitive oblique insertion it terminates in cone-shaped sarcolemma processes which give attachment to tendon fibrillae whose direction corresponds to the long axis of the muscle fibre itself.

Those muscles conforming to the bipenniform type of arrangement present the most positive evidence against a continuity of muscle fibrillae with tendon fibrillae, since in these a layer of dense connective tissue, the peritendinum, intervenes and separates the two structures from each other. The features to be considered with especial care in these muscles, therefore, are the peritendinum fibrillae and the muscle fibrillae. Continuity between the latter and the tendon fibrillae is absolutely out of the question. The presence of the greatly thickened sarcolemma end whose homogeneous nature can be very readily made out with a magnification of 1500 diameters, renders possible the most definite answer to the question of continuity so far as these muscles are concerned. In no instance is there the slightest indication of a fibrillar structure perforating this homogeneous membrane. There is surely no turning of the peritendinum fibrillae observable as must be the case were continuity to be established with the obliquely lying muscle fibrillae. We are in a position, therefore, so far as these muscles of the adult are considered, to corroborate most completely the views of Ranvier, Weismann, Kölliker and Motto-Coca, that no continuity exists between the tendon fibrillae and the muscle fibrillae.

The same conclusions may be drawn as well in the instance of those other muscles conforming to this general type where there are no peritendinum fibrillae intervening between the tendon and the obliquely-placed muscle fibres. Explaining upon developmental grounds the presence of the sarcolemma infoldings into the muscle fibre, and ascertaining through staining reactions the identity of these infoldings in the adult with the sarcolemma, and, in addition, by the same methods most positively establishing the morphological differences between such structures and the muscle and tendon fibrillae, and moreover, failing to observe any turning of the tendon fibrillae at an obtuse angle, as would be necessary to establish continuity with the muscle fibrillae, we are in a position here, as well, to deny in this type of muscle the continuity of tendon fibrillae with muscle fibrillae.

Regarding those other muscles which conform to the first type of termination, where the linear direction of the tendon fibrillae in the adult corresponds to that of the muscle fibrillae, it must be confessed that the problem is not so readily solved, the answer, however, can just as positively be given. The chief difficulty encountered is the exceeding thinness of the sarcolemma and the overlying and underlying of tendon and of muscle fibrillae in vertical optical planes upon the obliquely inclined surfaces of sarcolemma-end processes. The two kinds of fibrillae accordingly lie closer together.

The facts pointing strongly against continuity in this type of muscle are, — each sarcolemma process has attached to it but a single connective-tissue fibril, yet from ten to forty muscle fibrillae terminate in the respective process. A reduction in diameter of each muscle fibril is not demonstrable, such as must be the case if continuity with the single attached tendon fibril existed. Furthermore, there is no indication of a change in direction of the muscle fibrillae at the sarcolemma end in order to join the single apically-attached tendon fibril. The sarcolemma of the process is not thickened as one might expect to find, if the numerous muscle fibrillae turned and passed along its surface or through its substance in order to reach the tendon fibril. Again, the differences in size of the tendon fibrillae are not correlative to the variations in size of the sarcolemma processes or to the number of muscle fibrillae therein terminating. Further, the single tendon fibrillae can be traced up to a connective tissue cell body, a fact suggestive of their genesis. The staining reactions prove that such fibrillae pass only up to the sarcolemma end. In addition, in adult muscles the cross-striation proceeds directly up to the internal surface of the cone-shaped sarcolemma processes. This fact alone speaks most positively against a perforation of the sarcolemma by the tendon fibrillae and their subsequent passage through the muscle fibre to meet and fuse with the muscle fibrillae. Moreover, there are at first no tendon fibrillae attached to the sarcolemma end in the intervals between the cone-like processes, whereas the muscle fibrillae, undifferentiated in structure are almost from the very first already attached to the internal surface of the sarcolemma at such corresponding intervals.

A word should be said about the combination of fuchsin S with alcoholic-hematoxylin as recommended by several investigators. I have used these stains upon many of my own preparations, with the result that, dependent upon the relative concentration of the former constituent and the time of exposure, I was able to stain with the fuchsin not only the tendon fibrillae but also that portion of the terminal undifferentiated muscle fibrillae lying adjacent to and attached to the sarcolemma end. Indeed, by carrying the staining a little further I was able to stain the neighboring portions of the muscle fibrillae which presented all of the features of cross-striation. Hence, this combination of stains seems to be most unreliable as a criterion of morphological values.

For the developmental and morphological reasons which I have enumerated above, the assertion, that these ends of undifferentiated muscle fibrillae, by reason of their staining reactions, represent tendon fibrillae which have perforated the sarcolemma in order to join the muscle fibrillae, cannot be accepted as convincing to say nothing at all about being most positively denied. Yet some of the published figures intended to represent the continuity of these two kinds of fibrillae and stained with the same stains give exactly this same appearance.

In order to be positive that the cone-shaped extremities of the sarcolemma end were not the result of shrinkage of the tissue in the course of its preparation, I studied specimens of living muscle under high magnifications and ascertained that these processes were characteristic of the termination of the muscles conforming to the first type of ending. I found also that the presence of a uniformly rounded sarcolemma end such as has been figured by other investigators may be accepted as proof positive that the actual end of the muscle fibre has not been represented. I shall have occasion to refer to this feature in connection with fig. 11 upon a subsequent page.

I have presented fig. 10 for the purpose of elucidating certain appearances which have been misinterpreted by various investigators who have endeavored to establish a continuity between muscle and tendon fibrillae. The figure bears, as can be readily noted, a close resemblance to some of their published plates. This figure was sketched from a specimen of thigh muscle of an adult white mouse. Both the tendon fibrillae and the muscle fibrillae pursued the same linear direction. Upon the left side of the fibre the sarcolemma surface (*B*) at the end of the muscle fibre lay in a plane at exactly a right angle to that of the fibrillae, while on the right side it encountered these structures at an acute angle. These two portions of sarcolemma are uninterruptedly continuous with each other (*B*—*B*). The left side of the fibre demonstrates all of those features to which I have previously referred. The tendon fibrillae are separated from the muscle fibrillae by the thickened sarcolemma end. Upon the right side of the fibre the outline of the sarcolemma because of its obliquity is with much greater difficulty observed. Those fibrillae which occupy the same optical plane, for instance, the uppermost aspect of the section, can be seen to be separated from each other by the thin cut edge of the sarcolemma. Those other tendon fibrillae which occupy the middle of the thickness of the specimen seem to have perforated that membrane and to have extended into the muscle fibre. This appearance is naturally referable to the fact that they are attached to the under surface of obliquely inclined sarcolemma and hence underlie the uppermost muscle fibrillae. Their intense red stain, derived from the fuchsin, lends its color, too, to these adjacent overlying muscle fibrillae and, consequently, heightens the impression that a portion of them extends into the muscle fibre. A careful consideration of the left side of the sketch is sufficient, however, to eradicate all doubt of the discontinuity of the two kinds of fibrillae.

The particular criticism that I would raise against O. Schultze's work in that he has neglected to explain an appearance which is represented in almost every one of his figures, and which is of fundamental importance in our conception of the relation of the sarcoplasm to the sarcolemma. At places he has represented groups of three, four, and more muscle fibrillae which together perforate the sarcolemma and which are then prolonged as tendon fibrillae, i. e., the sarcolemma is interrupted at the point of perforation of not single muscle fibrillae but of groups of fibrillae. In other words, the intervening sarcoplasm, as well as the fibrillae, is continued beyond the limits of the sarcolemma end outside of the muscle fibre. Still no attempt is made to explain with what the sarcoplasm becomes contiuous or where it ends. Were the condition true, as the author has figured, then we should be compelled to modify our conception of the sarcolemma as a closed tube or envelope confining the semifluid sarcoplasm.

I have studied my own tadpole preparations with this particular point in mind and have represented in fig. 11 a muscle termination which may be considered as typical of the developing fibres at this age. The tadpole measured 1,5 cm.

I agree for the present with other investigators in naming the nucleus, which is seen imbedded in the granular protoplasmic mass surrounding the muscle fibrillae, a myogenetic nucleus, and also in referring to the investing membrane as sarcolemma (B). I have carried this membrane around the muscle fibre end, since such is the appearance produced when focusing down upon the fibre. It does not represent the end of the muscle fibrillae, however, as might at first appear, even in spite of their undifferentiated appearance. This fact can be ascertained by focusing to a deeper level in the fibre, then we get the appearance such as I have represented. The muscle fibrillae terminate in a number of cone-shaped processes, whose walls are formed of a delicate membrane, continuous with the sarcolemma, which bridges the ends of those fibrillae. This appearance is similar of that observed in the muscle fibre represented in fig. 6. These are the processes and this the membrane which have been overlooked. The specific remarks which I have made upon foregoing pages regarding the features to be noted in connection with fig. 6 apply equally well here and require no repetition.

The presence of this unbroken rounded contour of sarcolemma traversing the end of the muscle fibre, to which I have referred above, is readily explained, if we will imagine the muscle fibre as a whole rotated through an arc of 90°. Where the sarcolemma surface faces the eye it is almost perfectly transparent, but where, however, it lies in a vertical optical plane, its contour becomes manifest. Hence, the rounded sarcolemma-end appearance would be demonstrated again in the rotated condition of the fibre by that portion located at A in the figure. At the same time we can readily understand how the muscle fibrillae lying at a deeper level appear to have passed outside of the sarcolemma. And when through imperfect fixation or staining we overlook the exceedingly delicate, definitive, cone-shaped, sarcolemma processes, the conclusion is natural that the tendon fibrillae and muscle fibrillae are continuous because the sarcolemma has been perforated. This figure answers the question, as well, as to the relation of the sarcoplasm to the sarcolemma. Our earlier conception of the latter as a closed tube enclosing the sarcoplasm is correct. At no place does the sarcoplasm pass through the sarcolemma. At every point this thin envelope separates the specialized, semifluid sarcoplasm from the interstitial fluids, lymph, etc., of the outlying tissues.

I have not been able to carry out my studies on either *Hippocampus* or *Amphioxus*. It would be interesting to explain upon phylogenetic and ontogenetic grounds the differences in the adult morphological condition in these vertebrates from that which I have described above in higher vertebrate muscles.

As a final word upon the question of the cone-shaped processes of sarcolemma I desire to repeat with added emphasis, lest some criticisms regarding faulty fixation or shrinkage be made, that as controls to the fixed and stained preparations I studied many preparations of living muscle by vital staining in monochromatic light. Methylene blue aqueous solutions among others were used and of such a degree of concentration that a control tadpole lived for thirty-six hours in the same fluid which was used to stain the caudal muscles. The preparations which I employed were studied ten minutes after the removal of the muscle. No fixatives or dehydrating agents whatever were used. In every one of these living specimens the cone-shaped sarcolemma processes could be readily found. Therefore, they did not owe their presence to an artefactitious change in the muscle structure.

The following general conclusions, then, regarding these various voluntary striped muscles of the tadpole, white mouse, gray mouse, chicken, frog, and calf may be drawn.

1st — In the manner of termination of muscle fibres two general types may be recognized, one in which the long axes of the tendon and of the muscle fibres coincide, and the second in which they meet at an angle.

2nd — In neither of these two types are the muscle fibrillae in continuity with the tendon fibrillae.

3rd — Developing muscle fibres terminate in a number of cone-shaped processes of sarcolemma to the apex of which a tendon fibril is attached.

4th — In the adult those muscles conforming to type 1 still preserve the apical processes of their sarcolemma end.

5th — In the adult these processes of sarcolemma are dovetailed into the tendon end.

6th — The sarcolemma at the tendon end of such muscles is not markedly thickened.

7th — The central tendon of bipenniform muscles (type 2) is invested by a connective tissue sheath or peritendinum which consists of connective tissue fibres and cells, and which separates the tendon fibrillae from the muscle fibres.

8th — On muscles, conforming to the second type of structure, the sarcolemma end presents a flat surface which rests directly against the attached structure, be it peritendinum, perichondrium, or periosteum.

9th — This sarcolemma end is considerably thickened and is composed of a homogeneous substance.

10th — It presents a number of sarcolemma projections which project into the substance of the muscle fibre.

11th — These projections are derived from the fused, adjacent walls of the cone-shaped processes of the sarcolemma which were present at an earlier developmental stage of the fibre.

12th — The muscle fibrillae in adult muscles of this second type preserve their features of cross-striation up to the sarcolemma.

13th — This sarcolemma end is not perforated either by the tendon fibrillae, the peritendinum fibrillae, or the muscle fibrillae.

14th — The sarcolemma is not prolonged through the tendon, or over the tendon in either type of muscle.

15th — The sarcoplasm does not at any place in either type of muscle pass through the sarcolemma.

Literature.

Gage, S. H., The Microscope. 10th Edit. The Comstock Publ. Co., Ithaca, N. Y. 1908.

Kölliker, A., Handbuch der Gewebelehre des Menschen. Leipzig 1889.

Motta-Coca, A. and Ferlito, C., Contributo allo studio dei rapporti tra muscoli e tendini. Monitore zoologico. Vol. X. 1899.

RANVIER, L., Traité technique d'Histologie. 2nd Edit. 1889.

SCHIEFFERDECKER, P., Untersuchungen über die Rumpfmuskulatur von Petromyzon fluviatilis, etc. Archiv f. mikrosk. Anat. Bd. LXXVIII. 1911.

SCHULTZE, O., Über den direkten Zusammenhang von Muskelfibrillen und Sehnenfibrillen. Arch. f. mikrosk. Anat. Bd. LXXIX. 1912.

WEISMAN, A., Über die Verbindung der Muskelfasern mit ihren Ansatzpunkten. Zeitschr. f. rationelle Medizin. Bd. XII. 1861.

Explanation of figures.

Plate VII.

Fig. 1. An isolated muscle fibre with its attached tendon of an extrinsic muscle of the eyeball of a three-weeks-old chicken. The dovetailing of the sarcolemma end with the tendon fibrillae is characteristic of muscles belonging to this general type. The muscle fibrillae do not present the features of cross-striation at the muscle end. Magnification 1000 diameters.

Fig. 2. Similar to fig. 1 excepting that the muscle fibril groups are separated from each other by two nuclei. Magnification 1000 diameters.

Fig. 3. A muscle fibre from an extrinsic eye-muscle of a white mouse twenty-two days old. It presents the same general features as the two preceeding figures. Four tendon fibrillae (*A*), are to be noted upon the uppermost aspect of the sarcolemma end. Magnification 1000 diameters.

Fig. 4. A portion of a muscle fibre and tendon of a thigh muscle of an adult white mouse, which demonstrates the kind of muscle ending comprised under group 2. The tendon fibrillae (*A*), are separated from the muscle fibrillae (*C*), by the peritendinum (*B*), which contains four cells. Magnification 1000 diameters.

Fig. 5. Similar to the preceeding. The features of cross-striation can be traced up to the thickened sarcolemma where each is inserted upon a small raised elevation (*B*) of the latter. Magnification 1500 diameters.

Fig. 6. A portion of a muscle fibre of a caudal muscle of a tadpole about 5,0 cm long. Each cone-shaped sarcolemma process has attached to it a tendon fibril. Two of the processes derive fibrillae from a large fibroblastic cell situated among the tendon fibrillae. The original preparation from which this sketch was made demonstrates no thickening of the sarcolemma forming these processes.

Fig. 7. A developing intercostal muscle of a five-day old white mouse presenting three cone-shaped sarcolemma processes, each affording attachment to a connective tissue fibril. The fibril on the right is observed to be directly derived from a cell-body. Magnification 1000 diameters.

Fig. 8. Another developing fibre from the same muscle as the preceeding. In this fibre a more advanced stage of development is seen. Part of the muscle fibre is already attached to the perichondrium of the developing rib-cartilage. Three sarcolemma processes upon the right side of the fibre, however, have not yet reached their definitive attachment, but present connective tissue fibrillae affixed to their apex. Magnification 1000 diameters.

Fig. 9. A semi-diagrammatic sketch of a developing intercostal muscle illustrating the pointed processes of sarcolemma each with its attached tendon fibril. The features of cross-striation of the muscle fibre are wanting at the sarcolemma end. Magnification 1000 diameters.

Fig. 10. A thigh muscle fibre and tendon of an adult white mouse conforming to type 2. Upon the left side of the figure several inturned processes of sarcolemma are represented in the muscle fibre (*A*). Upon this side the sarcolemma is cut exactly transversely. Upon the right side of the figure the sarcolemma (*B*) is encountered obliquely by the section knife, consequently the more superficial of the muscle fibrillae are underlaid by the tendon fibrillae. The sarcolemma separates the two structures, still it appears as if the tendon fibrillae extended among the muscle fibrillae having perforated the sarcolemma. Magnification 1000 diameters.

Fig. 11. This is a portion of a muscle fibre from a caudal muscle of a tadpole 1,5 cm long. The presence of the richly granular protoplasmic mass imbedding a nucleus and surrounding the muscle fibrillae may be taken as an indication that the muscle fibre is in a developmental stage. In the adult state such protoplasmic masses are wanting. The membrane (*B*) investing the mass may be provisionally interpreted as the sarcolemma. The presence of the line continuous with the sarcolemma and carried across the fibre end does not represent the fibre end. I have elucidated this appearance upon a previous page (4). The sarcolemma proper is drawn out into a number of processes in which the muscle fibrillae end and upon whose apex a tendon fibril is inserted. Magnification 1500 diameters.

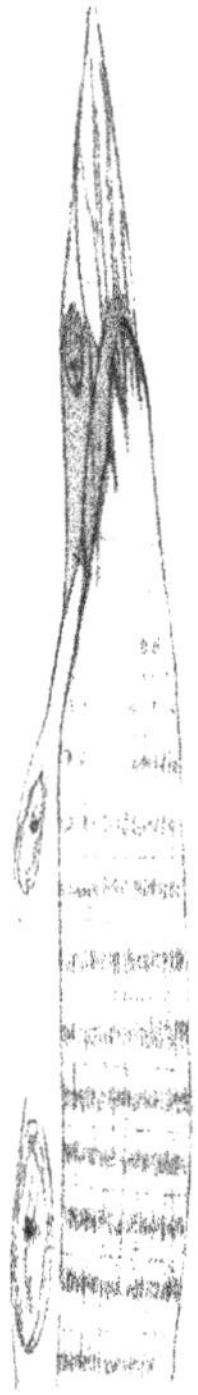

Fig. 1

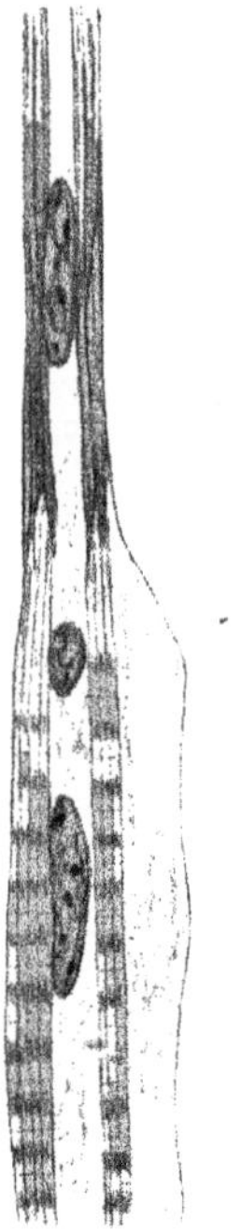

Fig. 2

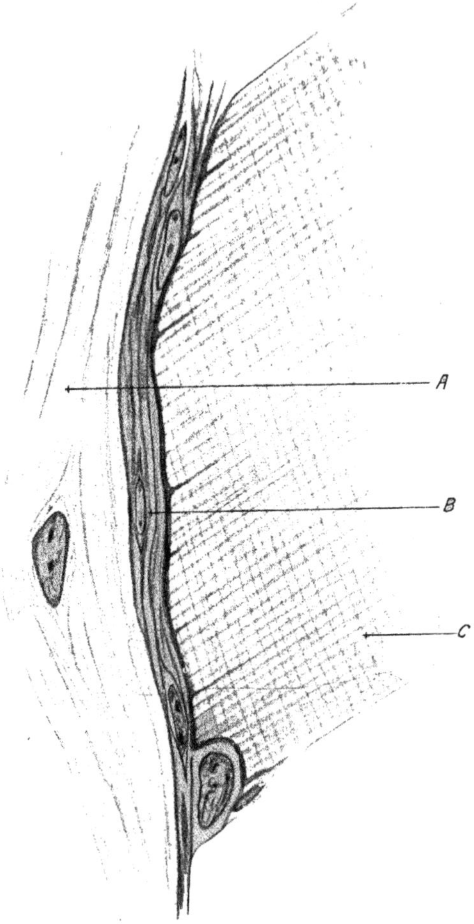

Fig. 4

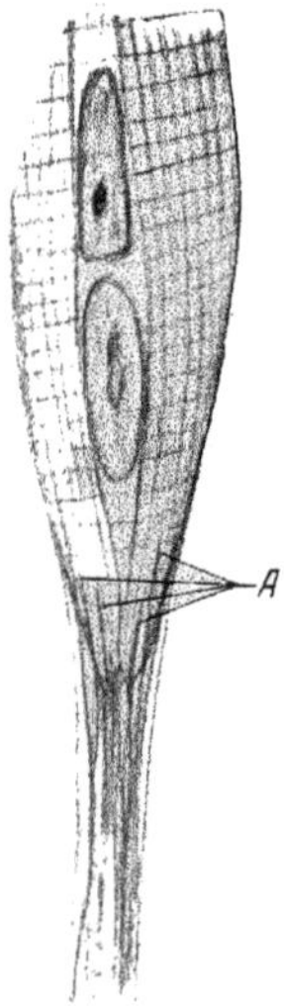

Fig. 3

Fig. 9

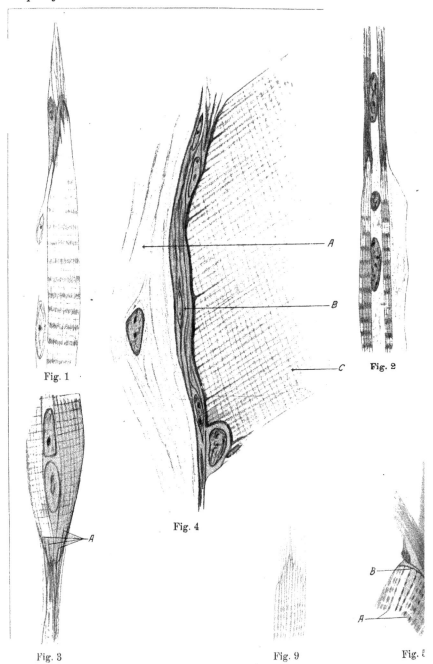

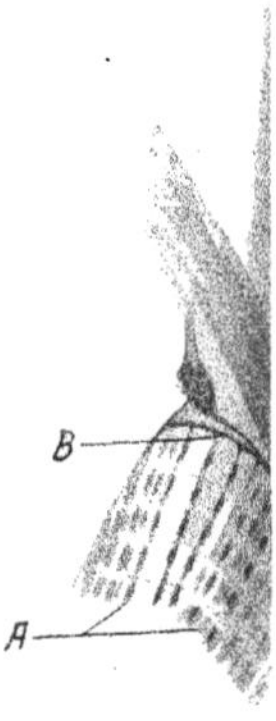

Fig. 5

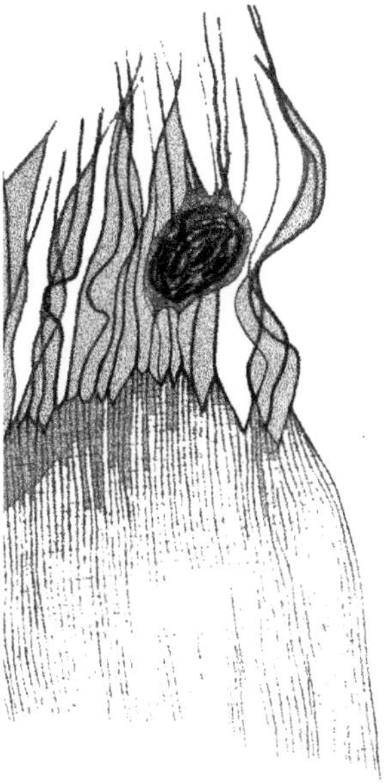

Fig. 6

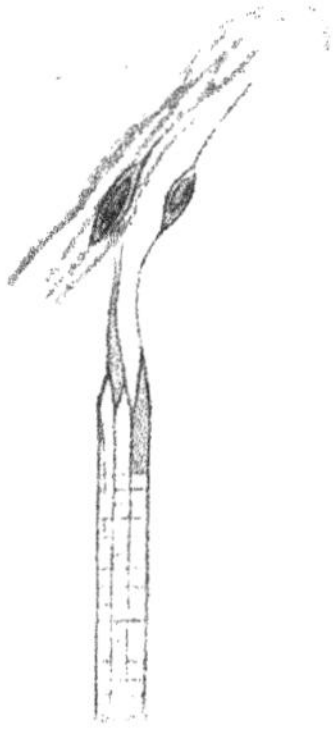

Fig. 7

Fig. 8

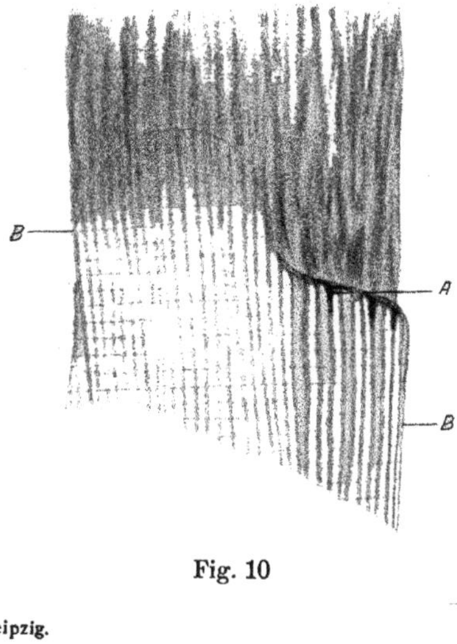

Fig. 10

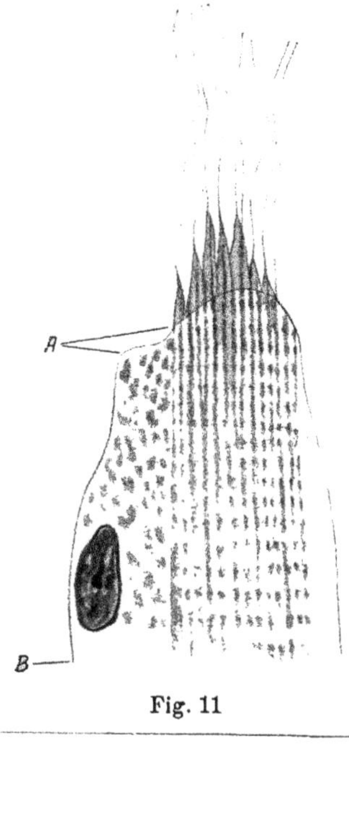

Fig. 11

Leipzig.

Abdruck aus:
Anatomischer Anzeiger.
Centralblatt für die gesamte wissenschaftliche Anatomie.
Amtliches Organ der Anatomischen Gesellschaft.
Herausgegeben von Prof. Dr. **Karl von Bardeleben** in **Jena.**
Verlag von **Gustav Fischer** in **Jena.**
42. Band, Nr. 7/8, 1912.

Muscle Fibres and Muscle Cells of the adult White Mouse Heart.

W. M. Baldwin.
(Cornell Medical College New York City.)
(From the Biological Laboratory at Bonn.)

With 2 Figures.

In two recent publications[1]) the view was advanced as the result of studies upon various voluntary striped muscles, such as latissimus dorsi, rectus abdominalis, thigh and leg muscles and extrinsic eye muscles of the chicken, cat, calf, white mouse, gray mouse, frog, and caudal muscles of the tadpole, that these muscle fibres of the adult should not be considered, as we have generally hitherto supposed, as large multinucleated cells, but rather as composite contractile structures composed of muscle fibrillae and sarcoplasm and containing muscle cells. That is to say, each one of the numerous nuclei which seem to be immediately imbedded in the sarcoplasm, represents in reality a distinct muscle cell, which presents cellular protoplasm composed of a spongioplasm network with interstices of hyaloplasm and which is completely invested by a cell membrane. This cell membrane intervenes between the protoplasm of the cell, containing the nucleus, and the imbedding sarcoplasm of the muscle fibre. In other words, since these same relations hold which all muscle cells, the muscle fibrillae and sarcoplasm are extra-cellular structures. Hence the generalization usually held, that the adult voluntary striped muscle fibre is a multinucleated cell, as stated before, is erroneous.

Still another fact bespeaking the correctness of these assertions was adduced from the studies of the sarcolemma in similar muscles and detailed in the second communication. It was found in this series that the sarcolemma was a structureless cuticula containing neither cells nor fibrils. Everywhere it stood in direct contact with the sarcoplasm of the fibre and afforded attachment to the telophragmata. Furthermore, its relation to the peripheral-lying muscle cells was such, that it was indented into the fibre by the latter, lying, therefore, between them and the sarcoplasm and not upon the peri-

1) Zeitschrift für Allgem. Physiologie (Max Verworn) 1912.

pheral fibre-aspect of these cells. The one structure found to occupy this position was the cellular and fibrillar investing perimysium of the muscle fibre. Hence such muscle cells lie outside of the sarcolemma. The latter envelope encloses only the highly specialized muscle fibrillae with the semi-fluid sacroplasm. Therefore, for these additional reasons the muscle fibrillae are to be regarded in the adult as extra- or intercellular structures.

An analogy between the histogenetic cycle of the connective tissue group and that of voluntary striped muscle can be drawn. The extremes of the cycle of the latter have been established by a number of competent observers. The intermediate steps, however, require further study. At first the myofibrillae are laid down in the genetic cell bodies. At the opposite end of the genetic course these fibrillae are extracellular. Parallel is the course of development in the connective tissue group; at first appearing as intracellular fibrillae and later being extruded from these genetic cell bodies. These facts obtain as well in the instance of cardiac musculature of the adult white mouse.

In the study of this form of striped muscle fibre the same technic was employed as that detailed in the cited papers. The sections varied in thickness from 2 μ to $3^1/_2$ μ; the stain used was alcoholic hematoxylin.

The first figure represents a longitudinal section of several muscle fibres of the ventricle. The marked morphological structural differences between the protoplasma immediately investing the nuclei and the sarcoplasm of the fibre are apparent at first observation. The cellular spongio-plasmatic network of the former is wanting in the latter. These two forms of protoplasm do not blend with each other; rather, they are sharply delimited from each other by a distinct membrane, the cell wall. With the alcoholic hematoxylin, this structure stained deeply in contrast to the neighbouring parallel-running muscle fibrillae. In focusing down through the section the uninterrupted, membrane-like nature of the cell wall could be readily noted. In contrast to this fact the slender muscle fibrillae pass into and out of focus as their level was reached and passed. Again, the features of cross-striation were not observed on the cell wall, hence the identity of the membrane apart from the muscle fibrillae was established.

The question of the longitudinal extent of such cells, as was the case with similar sections of voluntary striped muscle, is still unanswered. At the best the solution is exceedingly difficult. The delicacy of the cell wall, the overlying or underlying of it by muscle fibrillae, granulae, and telophragmata, all add to the difficulty in answering the question. The middle of the three muscle cells in fig. 1, however, possibly presents evidence upon which a correct conclusion may be drawn. The uppermost pole of this cell demonstrates what appears to be the reflected edge of the cell wall. The instance is not exceptional, since many such appearances are demonstrable throughout the entire series of sections. But the possibility of its being an obliquely-sectioned cell extremity together with the difficulties enumerated above render, with our present microscopical technic, a positive answer most injudicious.

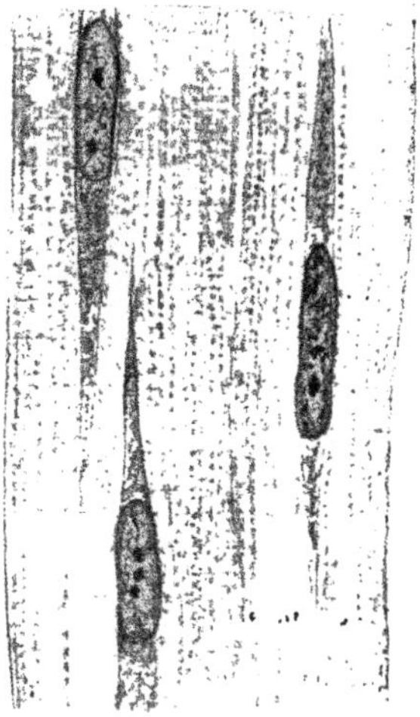

Fig. 1.

Two muscle fibres, represented as transversely sectioned, are seen in figure 2. A blood vessel occupies the angle between them. Each fibre presents a muscle cell. That on the left was encountered at the level of the middle of the nucleus; that on the right above the level of that structure. In the latter the structure of the cell protoplasm is in marked contrast to that of the sarcoplasm. The presence of a cell wall separating the two is unquestionable. The spongioplasm network of the cell is relatively heavily laden with granules. Notwithstanding, the clear fibrillae of this network can in many levels be traced directly up to the internal surface of this cell wall upon which they end. They do not at any place find an insertion upon the muscle fibrillae. A narrow interval of sarcoplasm, equal in general to the cross-diameter of an average muscle fibril, intervenes between the latter and the cell wall.

The cell on the left is of interest chiefly because it demonstrates appearances comparable to those observed in connection with similarly cut sections of voluntary striped muscles. The remarks made in the cited articles regarding such appearances apply here as well. At such levels the membrane appears to be wanting, i. e., the nucleus seems to be immediately imbedded in the sarcoplasm. Hence the view generally held that the muscle fibrillae are intracellular, being contained in a giant, multinucleated cell. Sections above or below the level of the nucleus, however, demonstrate the cell wall completely circumscribing the cell protoplasm. The inference seems to be justifiable, furthermore, that it is not wanting at the level of the nucleus. In torn preparations where the nucleus has been mechanically removed from its intimate position in relation to the sarcoplasm, and in those other preparations where the nucleus is shrunken, the presence of a distinct wall continuous with the remaining portions of the cell wall can be detected. It appears not to be an artefactitious product, since its outline is regular and definite and it is uniformly and deeply stained. Were it due to retracted and shrunken protoplasm we should expect to find it irregular in outline, varying in thickness, and differing in different levels in intensity of staining. It presents none of these artefactitious criterions. Nor does it in any portion of its extent seem to be merely the free, unthickened edge of the sarcoplasm. It possesses on the contrary a definite contour and a definite staining reaction. Owing merely to the juxtaposition of the nucleus its outline is overlooked in normal unshrunken tissues. Naturally, therefore, at such levels no spongioplasm fibrillae are attached to its internal surface.

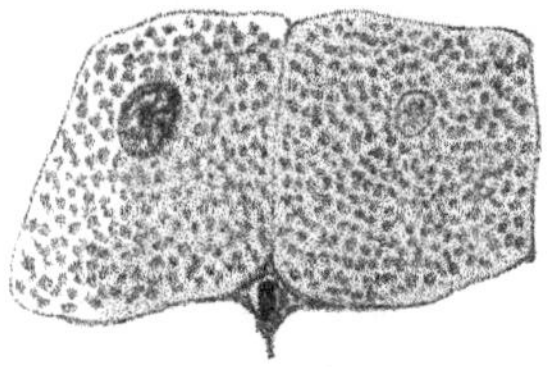

Fig. 2.

The two sketches in figure 1 and 2 are not intended to represent exceptional instances in cardiac muscle structure. Such relations were observed throughout the entire series of cardiac musculature where the conditions were favorable for sharp observation, i. e., where the parts concerned were not obscured by the overlying or underlying by granulae, muscle fibrillae, telophragmata, &c.

In the study of the telophragmata of this type of muscle no instance was observed where these lines traversed either the protoplasm or the nucleus of the muscle cells. This fact may be interpreted as of the following significance. First, it bespeaks the continuity of the cell protoplasm and nucleus as appertaining to a distinct and individual morphological and functional unit, a cell. Furthermore, granting the correctness of the observations of numerous workers that the telophragmata are always directly and uninterruptedly inserted upon the internal surface of the sarcolemma, we should look for an infolding of the sarcolemma from the periphery of the muscle fibre to invest these cells. No definite proof has been ascertained as yet that such is the instance in the cardiac fibres. The presence of an investment of sarcolemma upon the muscle cells is not, however, negatived by this fact. The telophragmata are attached, apparently, directly to the external surface of the cell membrane. This fact cannot be adduced as conclusive evidence, notwithstanding, arguing against the verified observations mentioned above. The matter demands further observation upon a greater number of vertebrates.

Much remains to be studied upon the intermediate genetic steps of histo-myogenesis. Sufficient evidence of the presence of myofibrillae as intra-cellular structures in the first developmental stages exists in the literature. In the adult stages of both forms of striped muscle, so far as concerns these particular vertebrates investigated, these muscle fibrillae are extracellular. A parallelism can be drawn, therefore, between myogenesis and developmental sequence observed and, seemingly, well established in the case of the connective tissue group of structures. At first the connective tissue and the elastic tissue fibrillae are intracellular. Later in development they are extruded from the genetic cell bodies and occupy, an intercellular position. Such is the sequence as well with the striped muscle fibrillae. In this fact we can find an additional reason, first, for grouping these striped muscle fibres, voluntary and cardiac, among the connective tissue group of structures, and secondly, for not considering them as multi- or singly-nucleated giant cells. In other words the muscle fibrillae and sarcoplasm are inter- or extracellular structures. To this extent these observations corroborate those made upon the voluntary muscles and detailed in the articles cited.

The conclusion arrived at, then, is that our conception of the cardiac muscle fibre as a cell containing fibrillae and sarcoplasm is erroneous as far as concerns the adult white mouse. The terms muscle fibre and muscle cell are not synonymous. The cuticular sarcolemma invests both the highly specialized muscle fibrillae and the sarcoplasm and, in addition, muscle cells. The latter structures present a nucleus, cell protoplasm, consisting of a spongioplasmatic network with interstices of hyaloplasm, and a cell wall. By reason of the last the cells are everywhere excluded from the sarcoplasm and the muscle fibrillae.

Aus dem biologischen Laboratorium der Universität Bonn

Die Entwicklung der Fasern der Zonula Zinnii im Auge der weissen Maus nach der Geburt.

Von

W. M. Baldwin

Instructor in Anatomy. Cornell University Medical College, New York City.

Hierzu Tafel XIV und XV.

Der Gegenstand dieser Untersuchung war die Bestimmung des Ursprungs, der Entwicklung und der endgültigen Anordnung der Zonula Zinnii - Fasern des Säugetierauges, indem ich für diesen Zweck eine doppelte Reihe von weissen Mäusen benutzte, im Alter von 12 Stunden bis einschliesslich 17 Tagen, mit anderen im Alter von 22, 27 Tagen und ausgewachsenen Exemplaren. Dazu hatte ich noch Gelegenheit, Exemplare der Katze, des Kalbes und des ausgewachsenen Menschen zu studieren. Die meisten meiner Resultate aber gründen sich auf meine Studien an weissen Mäusen.

Die verwendeten Fixiermittel waren die Flemmingsche Lösung und Sublimat. Die Einbettung geschah nach der Paraffinmethode. Alle Schnitte wurden in einer Richtung gemacht, ausstrahlend vom Mittelpunkte der Linse, und in einer Dicke von 2,5—7,5 μ. Die von mir benutzten Färbemittel waren Safranin, Orcein, Chloralhämatoxylin (Gage) und Bielschowskys Nervenfaserfärbung.

Der erste Teil meiner Arbeit behandelt die Erscheinungen bei 12 Stunden, 5, 11 und 14 Tage alten Exemplaren; im zweiten Teile soll ein fortschreitendes Bild der in der Entwicklung entstehenden Veränderungen gegeben werden, welche die besonderen Gebilde, die in den Bereich unseres Problems fallen, durchzumachen haben, bis sie zu ihrer bleibenden Gestalt und Lage gelangen: im dritten Teile endlich werde ich kurz die Ansichten anderer Forscher, die Entstehung der Fasern betreffend, anführen und nach meinen eigenen Beobachtungen die Gründe für und gegen die Aufrechterhaltung dieser Ansichten ausführlich erörtern.

Die weisse Maus, 12 Stunden alt.

Die Augenlider sind noch nicht geöffnet. Die Netzhaut ist in ihrer Entwicklung so weit vorgeschritten, dass mehrere bestimmte Schichten identifiziert werden können; aber die Schicht der Zapfen und Stäbchen ist noch nicht vorhanden. Der Glaskörperraum ist von einer Menge von Fasern derart durchzogen, dass sie ein Netzwerk bilden, welches viele Zellkörper und Blutgefässe enthält. Dieses Netzwerk reicht, wie man sehen kann, distalwärts[1]) bis zur inneren Oberfläche der Linse und dringt seitwärts von ihr in den Raum zwischen der pars ciliaris retinae und der Linse ein.

Überall, obgleich in der letzten Region weniger bemerkbar, ist auf diesem Netzwerk ein körniger Niederschlag zu beobachten, der sich stark mit Hämatoxylin färbt. Dieser Zustand verdunkelt den feinen Bau der Fasern, der Zellen und der begleitenden Blutgefässe. Die Zellen sind gross, hell und spindel- oder sternförmig und enthalten einen verhältnismässig kleinen Kern. Ihre Ausläufer bilden das körnige Netzwerk. Mehrere solche Zellen sind auf der inneren Oberfläche der limitans retinae interna zu erkennen, längs welcher ihre Fasern zusammen mit anderen, vom Netzwerk ausgehenden laufen. Diese Fasern aber verbinden sich nicht mit der limitans. Die Blutgefässe, in diesem Alter verhältnismässig sehr zahlreich, liegen auf der limitans, oder sie durchziehen das Netzwerk, werden von ihm getragen und erreichen die Linse, auf der sie ein besonders reiches Netz bilden.

Die äussere Oberfläche der limitans stellt eine scharfe Linie gegen das helle Protoplasma der darunter liegenden Netzhautzellen dar, in starkem Gegensatz zu ihrer inneren Oberfläche mit den dazukommenden, aus dem Glaskörpernetzwerk stammenden Fasern. Die limitans selbst erscheint als eine deutliche, dicke und sich dunkel färbende Membran. Bei einigen Schnitten traf ich glücklicherweise ihre flache Oberfläche, so dass ihr Bau gut zu untersuchen war. Es fehlten ihr sowohl Zellumrisse als auch Fasern. Diejenigen Fasern des Glaskörpernetzwerkes, welche sich längs ihrer inneren Oberfläche hinziehen, schienen an keiner

[1]) In dieser ganzen Arbeit habe ich die Ausdrücke „distal" und „proximal" angewendet; ersterer bedeutet an dem, oder in der Richtung auf den Hornhautpol des Augapfels hin, letzterer an oder in der Richtung zum Netzhaut- oder Funduspol hin.

Stelle mit ihr zu verschmelzen oder in ihr inneres Gefüge einzudringen. Sie ist durchweg vollständig homogen. Verfolgt man sie distalwärts, so findet man, dass sie auf der Höhe der ora serrata in mehrere Schichten und Fasern sich auflöst, von denen jede sich bis zur Spitze einer ciliaren Epithelzelle der inneren Schicht fortsetzt. Die Linsenhöhle besteht noch; aber die Zellen der durchsichtigen oder proximalen Reihe sind schon in lange Säulen ausgezogen. Die Kapsel erscheint auf beiden Seiten der Linse als eine verhältnismässig dicke fibrilläre Membran, welche ein reiches Netzwerk von anastomosierenden Blutgefässen trägt. Zu bemerken ist, dass die proximale Oberfläche der Linse schon konvexer als die distale ist. Die Hornhaut zeigt drei Schichten, von denen die mittelste viele längliche Zellen mit dunkel sich färbenden Kernen enthält. Die Regenbogenhaut hat sich soweit entwickelt, dass sie nun neben ihrer zweischichtigen Epithellage auf der proximalen Seite eine deutliche Schicht von Mesenchymgewebe zeigt, in welchem an dem pupillaren Rande der Iris die Anlage des m. sphincter iridis zu erkennen ist. Die noch unversehrte Pupillarmembran erstreckt sich durch den pupillaren Zwischenraum, indem sie sich wohl an die distale Linsenkapsel anlehnt, aber nicht mit ihr verschmilzt. Seitwärts kann sie als eine von einer einzelnen Endothelzellenlage gebildete Schicht über die distale Seite des Mesenchymlagers der Iris bis zur Verbindung dieser mit der Hornhaut verfolgt werden. Bei einigen Schnitten ist in dem Raume zwischen dem Irisrande und der Linse ein Blutgefäss zu finden, welches sich bis zur distalen Oberfläche der Linse hinzieht.

Die pars optica retinae setzt sich gegen die pars ciliaris retinae an der ora serrata ab. Schon hat die Entwicklung der Ciliarfalten begonnen, und man kann bemerken, dass sie einen Kern von Mesenchymgewebe einschliessen, welcher Blutkörperchen enthaltende Gefässe besitzt. Das Ciliarepithel ist überall aus zwei Schichten von Zylinderzellen zusammengesetzt. Eine wirkliche limitans ciliaris interna dagegen fehlt. Auf der inneren Epitheloberfläche ist nichts zu sehen als der einfache, dünne Epithelrand dieser scharf gegeneinander abgesetzten Zellen. An dieser Stelle muss noch besonders bemerkt werden, dass die Zellen der inneren Schicht des Ciliarepithels in eine Spitze oder einen Fortsatz auslaufen, und dass durch die ganze Länge der pars ciliaris retinae jede Zelle einen solchen spitzen Fortsatz besitzt. Solche Fortsätze können tatsächlich distalwärts bis zur inneren Seite der Irisbasis bemerkt werden. Ferner ist zu erwähnen, dass alle jene Epithelzellen zwischen der Spitze der definitiven Ciliarfalten und der ora serrata eine Faser haben, die sich an ihre Spitze ansetzt, dass aber in der Region distal von den erwähnten Teilen viele spitze Zellen zu bemerken sind, welche keinen Faseransatz haben. Von den distalen Verbindungen dieser Fasern und ihrer Bedeutung werde ich später ausführlicher sprechen.

Die limitans retinae interna hört plötzlich an der ora serrata auf. Sie ist nicht distalwärts über das Ciliarepithel verlängert. Die limitans retinae externa ist indessen distalwärts zwischen den beiden Schichten des Ciliarepithels zu verfolgen. Sie ist distalwärts nicht so dick, erscheint jedoch trotzdem als eine deutliche, gleichartige und ununterbrochene lamina.

Das Retziussche Bündel, das sich aus Fasern zusammensetzt, die aus den Epithelzellen unmittelbar distal von der ora serrata hervorgehen und durch den ganzen Glaskörperraum strahlen, wie leicht an Embryoschnitten zu erkennen ist, ist verschwunden und an den Schnitten der 12 Stunden alten Maus nicht mehr aufzufinden. Hier gibt es keine membrana hyaloidea, und demgemäss findet noch keine Trennung des Glaskörperraumes von dem definitiven Zonularaum statt. Viele Blutgefässe mit ihren Blutkörperchen durchziehen diesen letzten Raum. Einige davon setzen sich an die Linsenkapsel an, während andere längs des Ciliarepithels verlaufen. Die grössten dieser Gefässe aber sind in der Mitte des Raumes zu bemerken, getragen von einer dünnen ringförmigen Membran. Diese reicht proximalwärts nicht über die Ebene der ora serrata hinaus und kann distalwärts bis zu jener Stelle der distalen Linsenoberfläche verfolgt werden, wo die Gefässe, welche sie trägt, sich an die Linsenkapsel anlegen. Bei einigen Schnitten liegt sie dicht an der Linse, bei anderen in nächster Nähe des Ciliarepithels. Sie wird in diesen verschiedenen Lagen von einer Reihe von Fasern gestützt, welche an die benachbarten Teile sich anlegen.

Die Membran selbst ist dünn und färbt sich stark mit Hämatoxylin. Bei starker Vergrösserung zeigt es sich, dass sie aus einer oder zwei Schichten von Protoplasmafortsätzen gebildet ist, die aus spindelförmigen oder vierseitigen Zellen stammen.

Diese Zellen, welche in der ringförmigen, stützenden Membran vorkommen, sich gelegentlich aber auch auf den Gefasswänden vorfinden, enthalten einen runden oder ovalen Zellkern und reichlich körniges, sich dunkel färbendes Protoplasma. Gewöhnlich gibt jede Zelle zwei Fortsätze ab, durch deren Vereinigung die hier besprochene Membran gebildet wird. Gelegentlich ist indessen auch nur ein Fortsatz einer spindelförmigen Zelle zu bemerken, welche zur Linse hinübergeht; aber diese Beobachtung ist nur da zu machen, wo die Membran unterbrochen ist. Mitunter sieht man auch einen ähnlichen Fortsatz, der zum Ciliarepithel übergeht, wo er sich an die Spitze einer Epithelzelle anheftet. Solche Fortsätze sind gewöhnlich dick, mit Hämatoxylin dunkel gefärbt und besitzen einen körnigen Niederschlag, ähnlich wie er auf den Fasern des Glaskörperraumes sich findet.

Noch ein anderer Zelltypus kann im Zonularaum nachgewiesen werden. Er ist jedoch anscheinend auf diese Region beschränkt, da ich ähnliche Zellen im Glaskörperraum nicht finden kann. Es sind dies grosse, unregelmässige Zellen mit einem grossen ovalen oder unregelmässigen Kern und reichlichem Protoplasma, welches sich fast gar nicht mit Hämatoxylin färbt. Solche Zellen liegen entweder auf der Gefässhaut oder in dem leeren Raum zwischen ihr und dem Ciliarepithel. Zwischen der Linse und der Membran habe ich diese Zellen nicht gefunden. Jede Zelle ist besonders gekennzeichnet durch die grosse Zahl von Fortsätzen, die sie abgibt. Letztere sind ausserordentlich fein und hell und färben sich nur leicht mit Hämatoxylin. Auch ist auf ihnen kein körniger Niederschlag zu bemerken.

Solche fadenartigen Fortsätze ziehen sich entweder längs der stützenden, ringförmigen Membran oder gegen das Ciliarepithel hin. Dagegen kann ich keine finden, die sich nach der Linse hin erstreckten. Sie verzweigen und vereinigen sich oft, wodurch sie ein dichtes und verworrenes Netzwerk von sehr feinen Fasern bilden, welches zwischen der ora serrata und den Ciliarfortsätzen am dicksten ist. Aber nur wenige Fäserchen dieses Netzwerkes sind distal von dieser Gegend zu verfolgen. Jedoch ist leicht zu beobachten, dass jedes Fäserchen sich schliesslich an die Spitze einer Ciliarepithelzelle festheftet. Solche, die in den Zwischenräumen benachbarter Epithelzellen inserieren, kann ich nicht finden. Die Fasern, welche von den spindelförmigen

Zellen der Gefässmembran entspringen, durchziehen dieses Netzwerk, um an eine Epithelzelle zu inserieren, wie ich früher bemerkt habe; jedoch wegen ihrer Dicke, ihrer dunkeln Färbung und auch wegen des körnigen Niederschlages sind sie leicht von den zarten, hellen, keine Körner führenden Fasern des eigentlichen Netzwerkes zu unterscheiden.

Ein dritter Fasertypus endlich ist noch in dem Raume zu erkennen. Dieser Typus ist indessen nicht oft zu bemerken, sondern ist auf die Spitzen der Ciliarfortsätze beschränkt. Er ist nur bei solchen Exemplaren zu beobachten, wo die Gefässmembran in nächster Nähe jener Gebilde liegt. Es ist ein kurzer und verhältnismässig sehr breiter, schwach körniger Protoplasmafortsatz, welcher von der Ciliarzellenschicht direkt zur Membran geht, mit welcher er sich allem Anscheine nach vereinigt und so einer weiteren Untersuchung sich entzieht. Die Fortsätze dieser Art scheinen einfache Protoplasmafäden der Epithelzellen zu sein.

Die weisse Maus, 5 Tage alt.

Der Glaskörperraum zeigt in diesem Alter weniger Blutgefässe und Zellen, während der körnige Niederschlag auf den Fasern, welche den Raum durchziehen, so gut wie verschwunden ist. Die verschiedenen Schichten der Netzhaut haben sich dem Alter des Tieres entsprechend weiter entwickelt. Die lim. ret. int. endet noch auf der Ebene der ora serrata, indem sie sich in eine Anzahl von Schichten auflöst, welche zu den Epithelzellen gehen. Man kann überdies beobachten, dass jetzt mehr dieser sie zusammensetzenden Schichten vorhanden sind als früher, und dass die Intercellularsubstanz zwischen den Epithelzellen der inneren Ciliarschicht auch dicker ist. Indessen kann man keine Abgrenzung dieser Intercellularsubstanz von dem gleichartigen Bau der Schichten, welche die lim. bilden, erkennen. Die beiden Gebilde stehen in direktem Zusammenhang miteinander. Bei näherer Betrachtung scheinen die verschiedenen Lamellen sich bei der Annäherung an die Fortsätze der Spitze der Epithelzellen zu spalten, und jede Hälfte einer geteilten Lamelle verbindet sich dann sofort mit der benachbarten Intercellularsubstanz. Bei noch weiterem Verfolgen nach der Seite findet man, dass diese Intercellularsubstanz in direkte Verbindung mit der gleichartigen Substanz der lim. ret. ext. tritt und damit verschmilzt. Zu der

Feststellung ihres direkten Zusammenhanges miteinander kommt hinzu, dass diese drei Gebilde: lim. ret. int., Intercellularsubstanz der ora serrata und lim. ret. ext. auch wegen ihres Verhaltens bei der Färbung und wegen ihres morphologischen Aussehens aus derselben homogenen Substanz zu bestehen scheinen.

Die lim. ret. int. geht distalwärts nicht über das Ciliarepithel hinaus. Keine lim. ciliaris interna ist in diesem Alter vorhanden. Die lim. ret. ext. andererseits setzt sich, wie schon früher bemerkt worden ist, zwischen den beiden Schichten des Ciliarepithels ununterbrochen fort. In der Höhe der ora ist sie bemerkenswert dicker als distal dazu.

Die Linsenhöhle ist noch vorhanden. Auf ihrer Kapsel sind anscheinend nicht so viele Blutgefässe zu finden als früher. Die Pupillenmembran ist noch unversehrt.

Der Zonularaum enthält dieselben morphologischen Bestandteile, wie sie in 12 Stunden alten Exemplaren gefunden wurden. Das Netzwerk ist jedoch weniger verworren und die Maschen sind grösser. Die Gefässmembran nimmt dieselbe relative Lage in dem Zonularaum ein, ist aber dünner geworden und weniger deutlich als eine Membran zu erkennen. Sie ist indessen aus denselben spindelförmigen dunkel gefärbten Zellen und ihren Fasern zusammengesetzt wie im vorigen Stadium. Unterbrechungen sind öfters zu bemerken, und in diesen kann man mehrere zarte Fäserchen sehen, die zur Linsenkapsel hinziehen. Diese Fäserchen sind in diesem Alter oft aus den grossen hellen, unregelmässigen Zellen hervorgegangen, welche früher schon beobachtet wurden. Solche Zellen findet man noch in derselben Lage im Zonularaum, nämlich entweder auf der Membran oder in dem freien Raum zwischen ihr und dem Ciliarepithel. Daneben kann man noch beobachten, dass einige der Zellen auf dem Ciliarepithel selbst liegen. Beim Verfolgen der Netzwerkfasern, welche aus diesen unregelmässigen Zellen kommen, bis zum Epithel bemerkt man nunmehr mehrere solche Fasern bis zum Intercellularraum zwischen benachbarten Zellen hinziehen, während bei früheren Exemplaren alle diese Fasern an den spitzen Fortsätzen der Epithelzellen angeheftet waren. Es haben freilich noch nicht viele Fasern ihren apicalen Ansatz verloren; doch ist die Anzahl der Epithelzellen mit Apicalfortsätzen schon merklich verringert. Gelegentlich ist eine dickere Faser, welche aus den spindelförmigen Zellen der Membran hervorgeht, bis zum Epithel zu verfolgen, wie es schon bei den 12 Stunden alten Exemplaren bemerkt wurde. Diese Fortsätze werden indessen nur selten angetroffen und nur bei Zellen, die in der Nähe der ora serrata und lim. ret. int. liegen.

Die dickeren Protoplasmafortsätze, welche früher ohne Vermittlung eines Zellkörpers direkt vom Epithel zur Gefässmembran liefen, sind, gleichzeitig mit dem Dünnerwerden und dem teilweisen Verschwinden jenes letzteren Gebildes, nicht mehr vorhanden. In diesem Alter ist keine Faser zu beobachten, welche direkt und ohne Unterbrechung vom Ciliarepithel zur Linse geht. Indessen werden viele Fasern gefunden, die, aus einer Zelle des Zonularaumes stammend, direkt zum Epithel gehen, während ähnliche Fasern aus demselben Zellkörper in entgegengesetzter Richtung zur Linsenkapsel laufen.

Die weisse Maus, 11 Tage alt.

In diesem Alter ist die Linsenkapsel von beträchtlicher Dicke, und auf ihr verlaufen viele Blutgefässe. Diese sind auf der proximalen Oberfläche zahlreicher als auf der distalen. Die Linsenhöhle ist infolge der Vereinigung der beiden Epithelschichten, welche ihre Wände bildeten, verschwunden, aber die Pupillarmembran ist noch vorhanden.

Die Ciliarregion hat sich seit dem 5. Tage sehr entwickelt. Die Ciliarfortsätze haben sich bedeutend vergrössert, erreichen jedoch die Linse noch nicht ganz. Jeder derselben enthält einen Kern von Mesenchymgewebe, das Gefässe mit vielen Blutkörperchen umschliesst. Dicht an die äussere Oberfläche der äusseren Schicht der Ciliarepithelzellen setzt sich eine Schicht sich verzweigender verlängerter Mesenchymzellen an. Einige ihrer Fortsätze sind zwischen diesen Epithelzellen zu verfolgen, wo sie sich mit der homogenen Intercellularsubstanz verbinden und nicht weiter verfolgt werden können. In keinem Falle kann man sie durch beide Epithelschichten des Zonularaumes verfolgen.

Die äusseren Ciliarepithelzellen sind kurz, von säulen- oder würfelförmiger Gestalt und körniger als jene der inneren Schicht. Verfolgt man sie proximalwärts, so findet man, dass sie an der ora serrata in die Pigmentschicht der Netzhaut übergehen. Wie bei den jüngeren Exemplaren, kann auch bei diesem die lim. ret. ext. als lim. cil. ext. ununterbrochen distalwärts verfolgt werden. An der ora serrata ist sie jedoch beträchtlich verdickt. An keiner Stelle ist eine Verschiedenheit ihres Baues von dem homogenen Gefüge der Intercellularsubstanz festzustellen, sowohl bei der äusseren, wie bei der inneren Schicht der Epithelzellen, mit denen sie sich direkt verbindet.

Weiterhin ist zu bemerken, dass die Intercellularsubstanz zwischen jenen inneren Epithelzellen, welche in dem Gürtel zwischen den Ciliarfortsätzen und der ora serrata liegen, und ebenso derjenigen im Raume zwischen benachbarten Fortsätzen gleichmässig verdickt ist. Ein eingehendes Studium des Raumes zwischen diesen Zellen an den auf dieses folgenden Stadien bis zum 14. Tage fortschreitend, zeigt, dass das Dickenwachstum dieser Substanz durch die ganze Länge der Zwischenräume in gleichmässiger Weise stattgefunden hat, d. h. die Zunahme zeigt sich nicht zuerst an einem Ende des Raumes zwischen zwei Zellen und schreitet dann allmählich zum anderen Ende vor. Andererseits zeigt die Substanz zwischen benachbarten Epithelzellen, welche auf den Ciliarfortsätzen liegen, keine so merkliche Zunahme an Dicke.

Ich habe oben erwähnt, dass sowohl die Intercellularsubstanz zwischen den Zellen der äusseren Schicht, wie auch die zwischen den Zellen der inneren Schicht mit der lim. cil. ext. zusammenhängt; jedoch nur an sehr wenigen Stellen liegen diese Intercellularsubstanzen in derselben Ebene und bilden so eine Scheidewand, welche die ganze Dicke des Ciliarepithels durchquert. Der Bau der limitans und der der Intercellularsubstanzen scheint derselbe zu sein: ein zellen- und faserloses, homogenes Abscheidungsprodukt, das sich dunkel und gleichmässig färbt.

Die Zellen der inneren Schicht sind säulenförmig und länger als die der äusseren Schicht. Sie haben ein verhältnismässig helles Protoplasma und einen zentral gelegenen, ovalen oder unregelmässigen Kern. Mitosen sind in beiden Epithelschichten nachzuweisen.

Die inneren Ränder der inneren Epithelzellen, welche sich auf den Ciliarfortsätzen finden, liegen offenbar in derselben Ebene. Eine Anzahl sich dunkel färbender Fasern, die mehr oder weniger eng verbunden erscheinen, liegen auf diesem Epithelrande und sehen wie eine limitans ciliaris interna aus. Man kann indessen an Exemplaren mit gelegentlich kürzerer Epithelzelle, oder wo infolge der Präparation diese Fasern von der darunter liegenden Epitheloberfläche entfernt sind, leicht erkennen, dass die Ränder dieser Zellen nicht merklich verdickt sind, dass eine wirkliche limitans ciliaris interna in Wirklichkeit nicht vorhanden ist. Ferner erstreckt sich die lim. ret. int. nicht distalwärts über irgend einen Teil des Ciliarepithels hinweg.

Der Zonularaum erscheint in diesem Alter verhältmässig gross und ist vom Glaskörperraum durch die Membrana hyaloidea getrennt. Diese Membran erscheint als ein dünnes, homogenes Gebilde; sie liegt proximal zur Linse und heftet sich an das Epithel der ora serrata ganz ähnlich wie das Ende der lim. ret. int. und unmittelbar distal von ihr. Bei genauer Betrachtung sieht man, dass sie sich in eine Anzahl von Schichten auflöst, von denen jede sich an der Spitze einer Epithelzelle spaltet und sofort mit der Intercellularsubstanz verschmilzt. Auf der distalen Oberfläche der Membran und dicht bei ihrer Befestigungsstelle am Epithel sind viele unregelmässig gestaltete Zellen mit ovalen oder unregelmässigen Kernen und mit mehreren Fortsätzen zu bemerken. Einige derselben sind bis zu den Epithelzellen zu verfolgen, während andere in entgegengesetzter Richtung nach der Linse hin laufen, indem sie längs der distalen Oberfläche der Membran ziehen, mit welcher sie sich schliesslich verbinden.

Einige Blutgefässe sind in dem Raume noch vorhanden. Die Membran, welche sie bei den jüngeren Exemplaren stützte, ist jedoch als eine deutliche morphologische Einheit verschwunden.

Es ist wahr, dass die dunkel gefärbten, spindelförmigen Zellen, welche dieses Gefüge durch ihre vereinigten dicken Fortsätze früher bildeten, noch in diesem Raume vorhanden sind: aber sie sind nur noch auf die Gefässwände beschränkt. Ihre Fortsätze laufen vereinzelt längs der Gefässe und verschmelzen mit anderen ähnlichen Fortsätzen, ohne jedoch eine stützende Membran zu bilden.

Die hellen, ausläuferreichen Zellen der jüngeren Exemplare sind noch ebenso zahlreich wie früher vorhanden und nehmen relativ dieselbe Lage ein wie auf der früheren Stufe. Einige davon scheinen jetzt nur zwei Fortsätze auszusenden, von denen der eine sich zur Linse, der andere zum Epithel hinzieht. Starke Vergrösserung zeigt jedoch, dass dicht beim Epithel jede Faser sich in viele feine Fäserchen teilt, die sich alle entweder an eine Epithelzelle oder an die Intercellularsubstanz zwischen diesen Zellen heften.

Tangentialschnitte belehren darüber, dass viele dieser unregelmässig gestalteten Zellen auf der Ciliarepitheloberfläche liegen, wo sie anastomosierende Fasern abgeben, die das Augeninnere umschliessen. Diese Fasern, sowie auch jene, die ich oben schon erwähnte, und die vom Epithel zur Linse ziehen, nehmen, wenn sie dicht am Epithel der Ciliarfortsätze angeheftet sind, das Aussehen einer limitans dieser Zellen an. Keine der Epithelzellen der Ciliarfortsätze jedoch erhält Fasern. Die letzteren gehen ununterbrochen weiter, ohne mit diesen Zellen in strukturale Verbindung zu treten. Einige der nach der Linse ziehenden Fasern vereinigen sich mit einem Bipolarzellenfortsatz, der auf einer Blutgefässwand liegt; aber keine Faser tritt schliesslich mit den Endothelzellen der Gefässwände selbst in Verbindung.

Ein eingehendes Studium der Anheftungsweise der Fasern an das Ciliarepithel zeigt, dass neben den beiden oben erwähnten Arten der Verbindung, nämlich an die apicalen Fortsätze und an die Intercellularräume der Epithelien noch die folgenden hinzutreten. Manche Fasern, die an den Apicalfortsätzen der Epithelzellen inserieren, haben verschiedene Beziehung zu dem Cuticularrande dieser Zellen und zu der angrenzenden Intercellularsubstanz. In einigen Fällen ist namentlich bei den jüngeren Exemplaren häufiger zu bemerken, dass die zarten Fasern an den Apicalfortsätzen plötzlich aufhören. In anderen sind die Cuticularränder der Epithelfortsätze verdickt und ziehen sich längs der angehefteten Fasern nach der Linse hin. Wiederum kann die Verdickung der Cuticularränder auf eine Seite des Fortsatzes beschränkt sein, während in noch anderen Fällen die Cuticula auf einer Seite des Fortsatzes nur auf eine kurze Strecke der Entfernung von der Intercellularsubstanz bis zur angehefteten Faser verdickt ist. In keinem Falle aber findet man einen verdickten Cuticularrand nur auf die Ansatzstelle der Faser beschränkt und ohne Verbindung mit der benachbarten Intercellularsubstanz. Wo die Fasern in diesem Alter sich direkt mit der Intercellularsubstanz verbinden, sind sie durch diese sich dunkel färbende Masse nicht weiter zu verfolgen.

Die weisse Maus, 14 Tage alt.

In diesem Alter hat das Auge annähernd seine endgültige Beschaffenheit erreicht. Die Augenlider sind offen, und die

Pupillarmembran ist verschwunden. Im Äquator der Linse sind indessen noch einige Blutgefässe zu bemerken. Die membrana hyaloïdea. welche den Glaskörperraum vom Zonularaum trennt, ist dicker und dunkler gefärbt, aber ihre Beziehungen und Verbindungen sind dieselben wie bei den Exemplaren von 11 Tagen. Der Zonularaum ist von den zahlreichen Zonulafasern eingenommen, welche gewöhnlich direkt vom Ciliarepithel zur Linsenkapsel ziehen. Sie sind an diese entweder äquatorial, oder wie bei einigen wenigen, distal zu dieser Region angeheftet. Die meisten setzen sich jedoch an jenen Teil der Linsenkapsel an, welcher sich vom Äquator zu dem proximalen Pol erstreckt. Einige der Fasern dieser letzten Gruppe heften sich an die distale Oberfläche der membrana hyaloidea. Man kann diese Fasern eine kurze Strecke auf der Linsenkapsel verfolgen; dann vereinigen sie sich anscheinend mit ihr und sind nicht weiter zu verfolgen. Andere dagegen vereinigen sich mit den Fortsätzen aus den Bipolarzellen, welche sich auf den Gefässen längs der Linsenkapsel befinden.

Die Anheftung der Zonulafasern an die pars ciliaris retinae ist wie bei der 11 Tage alten Maus auf die Teile dieses Epithels, welche zwischen den Ciliarfortsätzen und der ora serrata liegen, und auch auf die Vertiefungen zwischen benachbarten Fortsätzen beschränkt. Keine dieser Fasern ist an dem Epithel auf den Ciliarfortsätzen befestigt. Jede Faser sitzt entweder an der Spitze eines Epithelzellenfortsatzes oder an der Intercellularsubstanz zwischen aneinander grenzenden Zellen an. Es sind jedoch auch verschiedene spitze Epithelzellen zu beobachten, die keine Faseransätze haben. Im ganzen sind weniger spitze Zellen vorhanden als bei den jüngeren Exemplaren. Zu erwähnen wäre noch, dass sie bei den älteren Exemplaren weniger oft auf den distalen Teilen des Ciliarepithels angetroffen werden, aber dass sie selbst im Alter auf den Schnitten niemals ganz fehlen.

Bei eingehendem Studium der Intercellularsubstanz ist leicht zu sehen, dass das, was als direkte Verlängerungen der Zonulafasern erscheint, sich als deutliche faserartige Bänder erkennen lässt, die von der homogenen Intercellularsubstanz, in der sie liegen, sehr verschieden sind, und welche sich dunkel färben. Sie können nach der lim. cil. ext. hin verfolgt werden, ohne diese jedoch ganz zu erreichen. Bei einigen Schnitten, wo das

Messer gerade neben der flachen Oberfläche einer Schicht Intercellularsubstanz eindrang, waren diese Fasern am besten als sich dunkel färbende Fäden zu sehen, die sich nach der äusseren Zellschicht hinziehen. In keinem Falle indessen liegen diese Fäserchen, wie Tangentialschnitte zeigen, innerhalb der Epithelzellkörper.

Nicht alle Zonulafasern gehen ununterbrochen zur Linse, da viele verlängerte, spindelförmige Zellkörper ihren Lauf unterbrechen. Diese Zellen haben einen ovalen oder unregelmässigen Kern, der von reichlichem und in manchen Fällen dunkel gefärbtem Protoplasma umgeben ist. Sie geben in der Regel zwei Fasern ab, von denen eine zur Linse, die andere zum Ciliarepithel geht; hier teilt sich jede in eine Anzahl feiner Fäserchen und erlangt eine Verbindung mit den Epithelzellen ganz ähnlich wie andere Zonulafasern, in deren Verlauf keine Zellen nachzuweisen sind. Zellen sind durch den ganzen Zonularaum zerstreut; sie liegen aber gewöhnlich näher zum Epithel als zur Linse. Einige sind dicht an den Ciliarepithelzellen zu finden und senden Fortsätze aus, die sich mit Fortsätzen aus ähnlich gelegenen Zellen vereinigen und das Innere des Augapfels umkreisen.

Gelegentlich ist auch eine Zelle auf einer Zonulafaser zu bemerken, in welcher nur die Umrisse des Kernes und des Zellkörpers vorhanden sind. Die Faserfortsätze solcher „Schattenzellen“ sind trotzdem gut erhalten und dunkel gefärbt.

Das Ciliarepithel zeigt dieselben Besonderheiten wie bei den 11 Tage alten Mäusen; ausgenommen, dass es sich in verschiedener Hinsicht in einem vorgeschritteneren Entwicklungsstadium befindet. Fortsätze von Mesenchymzellen nahe dem musc. ciliaris treten in die Intercellularsubstanz der äusseren Epithelschicht ein; aber keiner dieser Ausläufer geht hindurch bis in den Zonularaum. Die Intercellularsubstanz der inneren Epithelschicht ist an jenen Stellen, wo Zonulafasern ansetzen, im Vergleiche zu der auf den Ciliarfortsätzen verdickt.

Das Protoplasma mehrerer ciliarer Epithelzellen der äusseren Schicht dringt nach innen zu in den Raum zwischen angrenzenden, darüber liegenden Zellen der inneren Epithelschicht ein. Man findet jedoch nicht, dass die Zonulafasern mit solchen Verlangerungen zusammenhängen. Auf keinem Teile des Ciliarepithels ist eine lim. cil. int. vorhanden. Ich muss auch hier wieder wie bei den 11 Tage alten Mäusen erwähnen, dass das Aussehen einer solchen limitans dadurch hervorgerufen wird, dass mehrere Zonulafasern sich über die Oberfläche von Zellen auf den Ciliarfortsätzen hinziehen; die lim. ret. int. dagegen erstreckt sich nicht über das Ciliarepithel.

Überblicken wir nun kurz die ganze Reihe der untersuchten Mäuse, so ergibt sich folgendes Bild der aufeinanderfolgenden Entwicklungsstadien:

Vom ersten Tage der Geburt an sind die lim. ret. int. und die lim. ret. ext. als deutliche Membranen vorhanden. Die membrana limitans interna ist indes die stärkere von beiden, und ihre Stärke wird noch dadurch beträchtlich vergrössert, dass sich ihrer inneren Oberfläche zahlreiche Fasern anlegen, die von dem ursprünglichen Glaskörpergewebe herstammen. Diese Membran kann an ihrem äussersten Ende nur bis zur ora serrata verfolgt werden, die schon als die Verbindung zwischen der pars ciliaris retinae und der pars optica retinae bezeichnet wurde. Dort endigt sie, indem sie sich in verschiedene Lamellen teilt, wovon jede sich an eine Epithelzelle der ora serrata anlegt. Dagegen erstreckt sich die lim. ret. ext. von Anfang an an ihrem distalen Ende zwischen die zwei Schichten von Ciliarepithelzellen als eine ununterbrochene Membran, die limitans ciliaris externa. Die nächstfolgenden Tage hindurch nimmt die lim. retinae int. sowohl an Substanz wie auch an Anzahl der Lamellen zu, in die sie sich am äussersten Ende teilt. Die Intercellularsubstanz, mit der die limitans direkt zusammenhängt, wächst in einem entsprechenden Verhältnis an Stärke und ist schon am 6. Tag in merklichen Gegensatz zu der zwischen anderen benachbarten Zellen getreten. Ihr Zusammenhang mit der lim. ret. ext. ist ebenfalls klar ersichtlich, und es muss ferner bemerkt werden, dass letztere an dieser Stelle schon beträchtlich dicker geworden ist.

Schon am ersten Tage sind die Ciliarkörperfortsätze aufgetreten, jeder mit dem zweischichtigen Ciliarepithel und einem Kern aus Mesenchymgewebe, das Blutgefässe mit ihren Blutkörperchen enthält. Die Ciliarfortsätze wachsen allmählich an Grösse und Zahl bis zum 14. Tag, wo sie ihre grösste Entwicklung erreicht haben. Es muss bemerkt werden, dass in den späteren Entwicklungsstufen einige dieser Fortsätze verästelt sind. Ferner beginnt die Intercellularsubstanz, die zwischen gewissen Epithelzellen der inneren Schicht liegt, vom 7. Tage an sich allmählich zu verstärken. Dieses Wachstum findet in einem einheitlichen Verhältnis in der ganzen Länge der Zwischenräume angrenzender Zellen statt. Es muss noch hinzugefügt werden, dass die Teile des Epithels, wo dieses Wachstum stattfindet, sich auf die ringförmige Zone zwischen den Ciliarkörperfortsätzen und der ora serrata, sowie auf die Zwischenräume zwischen den Ciliarfortsätzen beschränken. Gleichzeitig verstärkt sich auch die lim. cil. ext., soweit sie in dieser Gegend anliegt. Eine Unterscheidung der Zellen der äusseren Ciliarschicht von denen der inneren ist von Anfang an möglich wegen ihrer morphologischen Eigentümlichkeiten. Die Zellen der inneren Schicht sind in den jüngsten untersuchten Stadien fast alle spitz, und auf jeder Spitze liegt eine Faser, die von dem den Zonularaum ausfüllenden Netzgewebe herkommt. Mit fortschreitender Entwicklung werden die spitzen Zellen, die auf den Ciliarkörperfortsätzen liegen, allmählich in Zellen umgebildet, die eine flache, gegen die Linse gerichtete Oberfläche darbieten. Diese Veränderung hat sich schon am 5. Tag vollzogen. Die spitzen Zellen sind demgemäss beschränkt auf die zwischen diesen Ciliarkörperfortsätzen liegenden Täler und auf die Zone, die sich zwischen letzteren und der ora serrata ausdehnt. Aber selbst in diesen Gegenden nehmen diese Zellen, indem sie der Verschiebung der Zonulafasern von einer apicalen Insertion zu einer intercellularen sich anpassen, an Zahl ab. Indessen verschwinden sie nie ganz aus einer Schnittserie, da sogar bei der ausgewachsenen Maus noch viele solcher Zellen gefunden werden.

Die Umbildung solcher spitzen Zellen und die Veränderung in der Insertion der Zonulafasern erfolgt gleichzeitig mit dem Wachstum der Intercellularsubstanz, die zwischen den Epithelzellen der Zonulagegend liegt. Zwischen dem 8. und dem 11. Tage kann man diese Veränderungen am besten wahrnehmen. Hinzuzufügen ist, dass durch die ganze Serie Epithelzellen nachgewiesen werden können, die einen spitzen Fortsatz haben, der aber mit keiner Zonulafaser verbunden ist.

Der Glaskörperraum ist am Anfang der Serie mit der entsprechenden Glaskörpersubstanz angefüllt, die aus einem losen Netzwerk von Fäserchen besteht. Diese kommen von verästelten oder bipolaren Zellen her, welche die zahlreichen Blutgefässe umgeben.

Von Anfang an wird auf diesen Zellen und Fasern ein körniger Niederschlag bemerkt. Keine Scheidewand trennt in den früheren Entwicklungsstufen den Glaskörperraum von dem eigentlichen Zonularaum. Die Glaskörpersubstanz wird auch in dem letzteren Raum gefunden und weist dieselben Elemente in ihrer Zusammensetzung und dieselbe allgemeine morphologische Gestalt auf. Es erfolgt sodann eine allmähliche Verminderung in der Zahl dieser Elemente bis zum 5. Tag, wo der Raum von körnigem Niederschlag frei ist. Das Netzwerk ist ebenso in diesem Alter weniger deutlich, aber die Blutgefässe und die sie tragenden Zellen und Fasern sind noch vorhanden und können sogar bis zum 27. Tag nachgewiesen werden, allerdings an Zahl bedeutend verringert.

Die Membrana hyaloidea erscheint nach dem Verschwinden der körnigen Substanz aus dem Zonularaum und hat schon am 10. Tag die morphologischen Eigentümlichkeiten des fertigen Zustandes nahezu erreicht.

Neben den ursprünglichen Bestandteilen der Glaskörpersubstanz, die man im Zonularaum findet, ist dort noch ein Zelltypus vorhanden, der nicht im Glaskörperraum vertreten ist. Es sind dies helle, grosse, schwach gefärbte Zellen mit einem grossen ovalen oder unregelmässigen Kern, reichlichem Protoplasma und einem unregelmässig gestalteten Zellkörper. Zellen von diesem Typus liegen entweder auf der Gefässmembran oder frei im Zonulagebiet. Jede Zelle gibt sehr viel feine, helle Protoplasmafortsätze ab. Diese Fortsätze verzweigen sich, anastomosieren und verschmelzen sehr oft und bilden so ein wirres Netzwerk zwischen der Blutgefässmembran und der ganzen Länge des Ciliarepithels. Die einzelnen Fasern, welche solch ein Netzwerk bilden, heften sich schliesslich an die Spitze einer Ciliarepithelzelle an.

Diese Zellen nehmen an den allmählichen regressiven Veränderungen, welchen die Bestandteile der Glaskörpersubstanz unterworfen sind, nicht teil, sondern bleiben bestehen, und nehmen womöglich mit der fortschreitenden Vergrösserung des Zonularaumes verhältnismässig an Zahl zu. Gleichzeitig aber mit der schnellen Entwicklung der Ciliarfortsätze und mit den übereinstimmenden morphologischen Veränderungen in den Epithelzellen, welche sie bedecken, wird das aus den Zellen hervorgehende

Netzwerk in seiner Lage eingeschränkt. Später heftet es sich nur an das Epithel des Strahlenkranzes und an die Vertiefungen. Zu gleicher Zeit wird es entsprechend den regressiven Veränderungen der Gefässe und der daran gelegenen Bipolarzellenfasern weniger verworren. Auch aus einem anderen Grunde noch anastomosieren, gleichzeitig mit der Vergrösserung des Querdurchmessers des Zonularaumes, die Fasern aus den hellen Zellen viel weniger oft. Sie werden beträchtlich länger und dicker, dunkler gefärbt und verschmelzen in dem Augenblick, wo sie den Zellkörper verlassen. Diese Vereinigung hat die Wirkung, solche helle Zellen bipolar erscheinen zu lassen, wobei eine Faser sich nach der Linse, die andere nach dem Epithel hinzieht. Wenn indessen der letztere Fortsatz bis zum Ciliarepithel verfolgt wird, so stellt sich heraus, dass er sich in eine Anzahl ausserordentlich feiner Fäserchen zerteilt, welche in ihrem morphologischen Charakter und ihrem Verhalten gegenüber der Färbung genau den zarten Fäserchen gleichen, welche das Netzwerk bei den jüngeren Exemplaren bildeten.

Endlich wird das Netzwerk allmählich durch die Zonulafasern ersetzt, welche eine direkte Richtung vom Epithel zur Linse einschlagen. Beinahe jede Faser ist zuerst von einem Zellkörper unterbrochen. Im vorgeschrittenen Stadium vermehrt sich die Zahl der Zonulafasern, welche keine Zellkörper tragen, allmählich. Bei den ausgewachsenen Exemplaren sind sehr wenige solcher Zellkörper zu bemerken. Aber sobald diese verschwinden, treten die „Schattenzellen“ auf. Letztere haben sehr deutliche und anscheinend gut erhaltene Zonulafaserfortsätze, jedoch nur den Umriss eines Kernes und eines Zellkörpers. Ich kann daraus nur schliessen, dass diese „Schattenzellen“ entartete, helle Zellkörper sind, in welchen die Faserfortsätze schliesslich als Zonulafasern des fertigen Auges bestehen bleiben.

Seit dem ersten Bericht von Zinn im Jahre 1775 sind von den Forschern viele und verschiedene Ansichten über den Ursprung, die Bedeutung und die letzten morphologischen Verhältnisse der Zonulafasern aufgestellt worden. Collins (1891) betrachtete sie als Fortsätze der Linsenzellen, welche sich zu den Ciliarepithelzellen erstreckten und schliesslich mit diesen vereinigten, eine Ansicht, die heute von den Gelehrten kaum anerkannt wird.

Eine Anzahl von Forschern haben die Fasern aus dem ursprünglichen Glaskörpergewebe abgeleitet.[1]) Diesen Standpunkt nehmen ein: Lieberkühn, Angelucci, Loewe, Schwalbe, Haensell, Iwanoff, Salzmann, de Waele, Retzius und von Lenhossék. Die Arbeit des zuletzt genannten Forschers wurde auch an Säugetieren, einschliesslich des Menschen, ausgeführt, gründete sich jedoch grösstenteils auf ein Studium von Hühnerembryonen vom 4. Bruttage an. Er sah die Fasern frei in der distalen Verlängerung des Glaskörperraumes zwischen der Linse und dem Ciliarepithel liegen. Sie bildeten zuerst ein verzweigtes Netzwerk, ähnlich dem ursprünglichen, anderswo zu findenden Glaskörpernetzwerk, und lösen sich, unbeeinflusst vom Zusammenhange mit Zellen des ursprünglichen Glaskörpers, der Linse oder des Ciliarepithels, allmählich in deutliche verzweigte Fasern auf, die von der Linse direkt zum Ciliarepithel laufen, wo sie sich zuletzt mit der Intercellularsubstanz jener Region verbinden. Er sah zuerst einen deutlichen Zwischenraum, welcher die Zonulafasern vom Ciliarepithel trennte, der aber später von den Fasern überbrückt wurde.

Lenhosséks Entdeckungen mögen den Ursprung dieser Fasern bei Vögeln zeigen. Meine eigenen Resultate lassen mich jedoch nach dem, was ich gefunden habe, glauben, dass die Entwicklung bei Säugetieren anders verläuft. In diesem Zusammenhang ist es interessant, dass Rabl, der an Menschen, Schafen, Schweinen, Vögeln, Selachiern und Amphibien arbeitete, Ergebnisse berichtete, welche zeigten, dass die Entwicklung der Fasern beim Hühnchen von der bei anderen Tierformen verschieden ist. Bis wir demgemäss den infolge von Schrumpfung der Präparate entstandenen Fehler ausgeschaltet haben, der in einer Anzahl veröffentlichter Zeichnungen zutage tritt, und der den von Lenhossék gesehenen Raum zwischen den Fasern und dem Epithel erklären mag, müssen wir etwas skeptisch sein gegenüber der Fähigkeit der Fasern, sich frei von jeder Zellentätigkeit so zu organisieren, arrangieren und anzusetzen, wie der Verfasser es beschrieben hat. Und wir dürfen nicht den zweiten Fehler begehen, zu folgern, dass das etwa für das Hühnchen Richtige notwendigerweise auch für Säugetiere gelten müsse.

---

[1]) Auf die Controversen über die erste Entstehung des Glaskörpers wird hier nicht eingegangen.

Zinn, Cloquet, Dessauer, Claeys, Czermak, Topolowski, Collius, Agababow, Terrien und Metzner schlossen, dass die Zonulafasern aus dem Ciliarepithel entstünden. Schoen sah sie für Protoplasmafortsätze an, die aus den inneren Epithelzellen erwüchsen. O. Schultze, Sbordane, Fischel (der am Salamanderauge arbeitete), Rabl und Addario waren derselben Meinung. Damianoff indessen betrachtete diese Fasern als ein Ausscheidungsprodukt dieser Zellen. Schoen bemerkte überdies, dass jede Epithelzelle eine Faser hergab, die durch Verbindung mit ihren Nachbarn eine wirkliche Zonulafaser bildete. Von Spee beobachtete ihren Ansatz an spitze Epithelzellen und schloss daraus, dass sie in Wirklichkeit eine Art von Cuticularprodukt dieser Zellen seien. Salzmann und von Ebner teilten diese Ansicht; letzterer erklärte dazu, er könne sehen, dass einige der Fasern in die Epithelzellen eindrängen. Kölliker behauptete ebenfalls den Epithelstandpunkt die Fasern betreffend, indem er annahm, dass sie in genetischer Beziehung zu den Fasern des Glaskörpers ständen, trotz der tatsächlichen Verschiedenheit ihrer chemischen Reaktionen.

Will man durch Ausschluss zu einem Beweise kommen, auf Grund der oben erwähnten Arbeiten, dass die Zonulafasern wirklich Auswüchse der inneren Ciliarepithelzellen darstellen, so gibt es in bezug auf den Vorgang, durch welchen sie zu ihrer endgültigen Lage gelangen, nur zwei Möglichkeiten.

Die erste ist die, dass die Verbindung zwischen den Epithelzellen und der Linse in einer früheren Periode, als diese Teile einander berührten, entstand, und dass als Resultat des Auftretens des Zonularaumes und der allmählichen Vergrösserung seines Querdurchmessers diese Fasern, welche ihre Ansatzstelle an der Linse noch behaupteten, länger und länger wurden, bis schliesslich der endgültige Zustand erreicht war.

Diese Annahme kann in zwei Hauptpunkten kritisiert werden. Erstens: So viel ich weiss, hat noch kein Autor bei Säugetieren eine Form beschrieben, in welcher selbst in den früheren Stadien die Linse und die Ciliarregion jemals in direkter Berührung miteinander gestanden hätten. Eine Schicht von Mesenchymgewebe mit Blutgefässen trennt diese beiden Gebilde von Anfang an. Zweitens (hier kann ich nur von meinen Beobachtungen an der weissen Maus sprechen): Die von mir bemerkten spitzen Zellen, welche Protoplasmafortsätze abgeben, die ununterbrochen zur Linse laufen, waren jene auf den Ciliarfortsätzen, wo später keine Zonulafasern angeheftet sind. Drittens: An jenen Flächen, wo die endgültigen Fasern angeheftet sind, habe ich auf den früheren Stufen der Entwicklung keine Protoplasmafortsätze finden können, die direkt und ununterbrochen zur Linse gehen.

Die Zonulafasern müssen sich demgemäss in einer späteren Periode, wenn ein Raum zwischen der Linse und dem Ciliarepithel vorhanden ist, entwickelt haben. Wir können unsere Aufmerksamkeit daher auf die zweite Annahme lenken.

Diese ist kurz folgende: um einen Beweis von dem Epithelursprung der Fasern richtig zu begründen hinsichtlich der unzweifelhaften Tatsache, dass der Raum, durch welchen sie laufen, schon von vielen Mesenchymzellen und Fasern, die selbst einen Epithelansatz haben, eingenommen ist, müssten wir notwendigerweise diese Fortsätze auf verschiedenen Stufen des genetischen Fortschreitens gesehen haben, wie sie aus dem Epithel hervorwachsen und zur Linse fortschreiten, bis sie sich später dort ansetzen. Ein solcher Beweis fehlt in der Arbeit der erwähnten Forscher. Demgegenüber habe ich in meinen Schnitten mehrere Fasern bemerkt, die eine kurze Strecke zur Linse hin liefen und dann plötzlich endeten. Diese Fasern waren immer von anderen begleitet, welche die ganze Strecke zur Linse durchliefen. Selbst mit den besten Linsen, die mir zu Gebote standen, und bei ungefähr 2000facher linearer Vergrösserung habe ich in keinem Falle bestimmen können, ob das Ende der Fasern das spitze Ende einer wachsenden Faser oder das durch das Messer abgeschnittene Ende einer Faser war, welche sich ein wenig unter der Ebene ihrer Nachbarn befand. Auch heben Serienschnitte trotz sorgfältigster Ausführung die Schwierigkeiten nicht auf. Das Haupthindernis ist die Orientierung dieser Fasern beim Übergang von einem Schnitte zum nächsten der Serie. Und wenn man bedenkt, dass sie in einem hellen Raume liegen, verhältnismässig weit entfernt von festen Punkten, die zur Lagebestimmung dienen könnten; wenn man auch die Leichtigkeit bedenkt, wie solche Fortsetzungen durch die Präparation verloren gehen oder verschoben werden, selbst bei den am sorgfältigsten behandelten Exemplaren: so muss man gestehen, dass hierin eine Schwierigkeit für unser Studium liegt, die bis jetzt fast unüberwindlich ist.

Überdies kann ich bei weiterer Kritik dieser zweiten Annahme hinzufügen, dass meine eigene Arbeit und die vieler anderer Forscher, über deren Ergebnisse ich später ausführlicher reden werde, zeigen, dass die Zonulafasern schliesslich einen Intercellularansatz haben und nicht an einen Apicalfortsatz auf dem Ciliarepithel ansetzen. Gerade wie diese Veränderung des Ansatzes erfolgt, und durch welche verschiedenen Vorgänge sie zustande gekommen ist, das hat keiner der Verfechter des Epithelursprungs erklären können, wenn sie auch den späteren Intercellularansatz ebenfalls bemerkt haben.

Wie können wir weiter das Vorhandensein von Zellen mit deutlichem Zellumriss und Zellkernen auf den Zonulafasern und dem Ciliarepithel erklären bei Mäusen, die beinahe voll ausgewachsen sind? Lenhossék und andere haben runde oder unregelmässige Zellen mit einem deutlichen unregelmässigen Kern in solchen Lagen beobachtet und sie für Leukocyten erklärt. Ich habe diese Zellen ebenfalls gesehen und bin zu demselben Schluss gekommen. Aber die Zellen, auf die ich mich besonders beziehe, geben Zweige ab, die zur Linse und zum Epithel laufen. Wolfrum sah solche Zellen in seinen Exemplaren. Nussbaum beobachtete sie vor ihm beim Kaninchen. Ich habe sie bei jedem Schnitt durch die ganze Reihe der weissen Mäuse gefunden und sie überdies auch im Auge des erwachsenen Menschen bemerkt.

Das Fehlen oder Vorhandensein einer wirklichen Grenzmembran auf dem freien Rande der Ciliarepithelzellen der inneren Schicht ist von mehreren Forschern als Beweis für oder gegen den Lauf der Zonulafasern zu einem tieferen Ansatz in dem Epithel dieser Region angeführt worden. Lenhossék z B. glaubte an das Vorhandensein einer wirklichen limitans ciliaris interna, deren Funktion es sei, die Zonulafasern mit dem darunterliegenden Intercellulargewebe, mit dem sie direkt zusammenhängt, in Beziehung zu bringen. Diese Membran ist nach ihm ununterbrochen. Darin fand er einen Beweis gegen das tiefere Vordringen der Zonulafasern. Czermak hielt die limitans für eine hyaline Struktur, welche mit dem Glaskörper in direktem Zusammenhange steht. Aus dieser Schicht stammten die Zonulafasern. Topolowsky bestätigt diese Ansicht. Fischel sah die limitans als die direkte distale Verlängerung der lim. ret. int. an, Salzmann und v. Ebner konnten die Zonulafasern nur bis zur limitans verfolgen. Mawas glaubte an das Vorhandensein einer limitans, stellte aber fest, dass dies kein Hindernis für das tiefere Eindringen der Zonulafasern sei, von denen einige eine Strecke weit in der darunter liegenden Intercellularsubstanz zu verfolgen seien. Der letztere Forscher betrachtete ferner die lim. cil. int. nicht als wirkliche Membran, sondern nur als ein exoplasmatisches Produkt der angrenzenden Zellen.

Ich habe oben von meinen Ergebnissen in bezug auf die lim. cil. int. gesprochen. Bei der Maus ist sie sicher auf keiner Stufe der Entwicklung vorhanden. Indessen erzeugt der Lauf der Zonulafasern und anderer Zellgewebsfasern quer über die Epitheloberflächen an mehreren Stellen das Bild einer Grenzmembran.

Von Claeys wurde ein interessanter Gedanke betreffs einiger analogen und morphologischen Eigenschaften des Ciliarepithels und der eigentlichen Retina 1886 veröffentlicht. Er nahm an, dass dieses zweischichtige Epithel ein System von Stützzellen und -fasern besässe, ähnlich den Müllerschen Fasern der Retina, und dass die Zonulafasern nur die inneren Verlängerungen solcher Fasern wären. Diese Ansicht fand später einen Verfechter in Terrien. Letzterer beschränkte jedoch die Zonulafasern nicht auf diese Stützzellen, da er einige durch die ganze Dicke der Ciliarepithelschichten bis zum äusseren Mesenchymgewebe des Ciliarkörpers verfolgen konnte. Neuerdings unterstützte Metzner diese Theorie und gab an, die Zonulafasern bis zur Scheide des m. ciliaris verfolgt zu haben.

Im Jahre 1906 ging Toufesco so weit, zu behaupten, dass die Zonulafasern aus elastischem Gewebe beständen, dass sie beide Ciliarschichten durchdrängen und so eine direkte Verbindung mit ähnlichem Gewebe in der Aderhaut des Augapfels herstellten.

Ich habe schon früher berichtet, dass ich bei einigen meiner Exemplare Mesenchymzellen bemerken konnte, die an der äusseren Oberfläche der äusseren Epithelzellschicht angeheftet waren, und die gelegentlich Fortsätze abgaben, welche nach innen zu in die Intercellularsubstanz zwischen diesen Zellen eindrangen. Ich konnte sie jedoch nur eine sehr kurze Strecke zwischen diesen Zellen verfolgen, da sie morphologische Eigenschaften und Färbbarkeit besitzen wie die homogene Substanz, in der sie liegen. Daher bin ich auch nicht imstande, die Beobachtungen dieser Autoren zu bestätigen. Wenn ich nach einigen der veröffentlichten Zeichnungen urteile, kann ich nur schliessen, dass, wie ich bei ähnlichen Erscheinungen unter dem Mikroskop sah, diese Forscher mit Schnitten arbeiteten, die sehr schräg angelegt waren.

Wenden wir unsere Aufmerksamkeit zunächst auf die Intercellularsubstanz, die in den Zonulaflächen des Ciliarepithels vorhanden ist, und welche die von vielen Forschern bemerkten Zonulafaserverlängerungen enthält, so finden wir, dass in Verbindung mit dieser Sache N. van der Stricht, Leboucq und O. van der Stricht, die über die limitans des Gehör-, des Geruchs- und des Sehepithels arbeiteten, zu dem Schlusse gelangten, diese homogenen Membranen seien nicht wirkliche Membranen, sondern nur ein strukturloser intercellularer Kitt. Bei meinen eigenen Serienschnitten kann ich, wie ich schon konstatiert habe, keine geformten Gewebsbestandteile in der limitans bemerken. (Eine Beschränkung dieser Behauptung werde ich später geben.) Ihr morphologisches Aussehen und ihr Verhalten bei der Färbung ist dem der ganzen Intercellularsubstanz des Ciliarepithels, mit der sie in direktem Zusammenhange zu stehen scheint, völlig gleich.

Wenn wir in dem Falle dieser Intercellularsubstanz annehmen, dass sie als eine Art exoplasmatischen Produkts der benachbarten Zellen gebildet ist, wobei eine Zelle nicht mehr als die andere zu ihrer Dicke beiträgt, — und es gibt in der Literatur, wie ich glaube, nichts, was dieser Ansicht widerstreitet — warum sollten wir so weit gehen, zu behaupten, dass diese lim. ext., die allem Anscheine nach aus demselben Stoff zusammengesetzt ist, mehr aus den Zellen der inneren Ciliarschicht hervorgeht, als aus denen der äusseren? Dabei dürfen wir nicht vergessen, dass die Hypothese von der Analogie der inneren Ciliarepithelzellen und derjenigen der Stützzellen der Retina noch nicht sicher gestellt ist. Wir können daher gerechterweise auch nicht vermuten, dass die lim. cil. ext. ein Derivat von Zellkörpern ist, die nach innen von ihr liegen, wie es anscheinend bei der lim. ret. ext. der Fall ist.

Mawas z. B. nimmt den Standpunkt ein, dass die lim. ext. allein von den Epithelzellen der inneren Schicht gebildet ist.

Er gibt indessen in seiner Arbeit nicht genügende Beweise für die Richtigkeit dieser Annahme. Wenn wir dann im Laufe der Entwicklung beobachten, wie die Intercellularsubstanz im Zonulagebiet allmählich an Dicke zunimmt, aber nicht bemerken, dass diese Zunahme zuerst an einem Ende eines Intercellularraumes auftritt und allmählich zum anderen fortschreitet, sondern im Gegenteil in gleichmässiger Weise auf der ganzen Länge des Zwischenraumes vor sich geht: dann haben wir keinen Grund für die Annahme, dass diese Zunahme mehr den Zellen der äusseren als denen der inneren Schicht zu verdanken ist. Wir können daher die Ansicht nicht ganz anerkennen, die von einigen Forschern, z. B. Agagobow, aufrecht erhalten wird, dass die Zonulafasern Ableitungen von den äusseren Epithelzellen seien, obgleich sie aus derselben Substanz wie die Intercellularsubstanz zu bestehen schienen. Und dies auch trotz der Tatsache, dass, wie ich schon früher bemerkt habe, oftmals eine helle Protoplasmaverlängerung äusserer Zellen sich eine kurze Strecke weit zwischen die inneren Epithelzellen einschiebt.

Diese Betrachtungen erklären indessen das faserige Aussehen der Intercellularsubstanz gewisser Regionen nicht, welches von Schultze, Wolfrum, Lenhossék und auch von Mawas bemerkt worden ist. Alle diese Forscher haben das faserige Aussehen mit den Zonulafasern in Verbindung gebracht, indem sie, ausser Lenhossék, annahmen, dass es von den Verlängerungen solcher Fasern herkomme, die in der gleichartigen Intercellularsubstanz eingebettet sind. Sie haben jedoch nicht erwähnt, ob dieses Aussehen auf die Region der Zonulaansätze beschränkt war, oder ob es als charakteristisch bezeichnet werden könnte für alle Intercellularsubstanz durch das ganze Epithel, sowohl zwischen den Zellen der äusseren, wie denen der inneren Schicht. Darin aber liegt ein wichtiger Beweisgrund.

Bei meinen eigenen Untersuchungen habe ich bemerkt, dass diese Faserung vor allem auf die Intercellularsubstanz begrenzt ist, die zwischen jenen Zellen der inneren Epithelschicht lagen, an welche sich Zonulafasern heften. Ich habe sie weder zwischen den Zellen der äusseren, noch zwischen denen der inneren Schicht gefunden, die auf den Ciliarfortsätzen liegen, wo keine Zonulafasern entspringen. Zweitens erscheinen diese Fäserchen zur Zeit der Dickenzunahme dieser Substanz in den erwähnten

Regionen. Drittens ist dieses Aussehen nur in dem Alter zu finden, nachdem die Zonulafasern ihren Ansatz von den Apicalfortsätzen der Epithelzellen in die Intercellularzwischenräume verlegt haben. Endlich sind diese Fäserchen in meinen Präparaten immer in direktem Zusammenhang mit den Zonulafasern zu finden.

Hinsichtlich der Tatsache, dass neuere Forscher die Ansicht mit Nachdruck betonen, dass für das richtige Studium der Zonulafaserverlängerungen in der Intercellularsubstanz besondere Färbemethoden nötig seien, z. B. die „Heldsche Molybdänsäure-Protoplasmafärbung“, muss ich nach meiner Erfahrung, — ich habe die Bielschowskymethode angewandt — feststellen, dass die bekannten Färbemittel wie Safranin oder Chloral-Hämatoxylin (Gage) vollständig genügen, um selbst die feinsten Fäserchen zu zeigen. Das einzige Erfordernis für ihre Behandlung besteht darin, die Schnitte sehr stark zu überfärben und dann gründlich zu wässern.

Wolfrums Ansicht, dass mehrere der Zonulafasern die Zellen der inneren Epithelschicht durchzögen, ist nach der Prüfung von Tangentialschnitten des Epithels leicht als ungenau zu erweisen, wie schon Mawas gezeigt hat. Bei meinen eigenen Exemplaren habe ich sicherlich niemals bemerkt, dass eine Zonulafaser eine Epithelzelle durchzog. Ich habe indessen die Zonulafaserverlängerungen nicht durch die Intercellularsubstanz bis zur lim. cil. ext. verfolgen können. Wolfrum jedoch konnte sie bis zu dieser Membran verfolgen, wo sie in kleinen runden Anschwellungen endeten. Die letztere Bildung habe ich ebensowenig auffinden können.

Der zuletzt erwähnte Autor war imstande, bei seinen Säugetierexemplaren Gliazellen zu finden, die von der Retina in der Gegend der ora serrata in den Zonularaum wanderten, wo sie faserähnliche Fortsätze ausschickten, die sich später in Zonulafasern auflösten. Bei meinen eigenen Untersuchungen konnte ich eine solche Wanderung von Neurogliazellen nicht bemerken. Jedoch habe ich aus Wolfrums Beschreibung dieser primitiven Zellen geschlossen, dass die hellen, unregelmässigen, vielverzweigten Zellen, die ich von der ersten Stufe an beobachtete, dieselben sind, die er als Neurogliazellen bezeichnet.

Im Jahre 1895 kam Rochon-Duvigneaud zu dem Schlusse, dass die Zonulafasern „une espèce particulière de fibres conjonctives" wären. Später entdeckte Nussbaum die Bedeutung solcher Zellen für die Entstehung fertiger Zonulafasern und erklärte sie für Bindegewebszellen und -fasern, die sich sekundär mit der Linsenkapsel und dem Ciliarepithel verbinden. Meine eigenen Schlüsse, die sich auf die Ergebnisse, welche ich in dieser Arbeit niedergelegt habe, gründen, besonders aber der Mangel eines Beweises in meinen Präparaten, die Arbeit Wolfrums bestätigen zu können, lassen mich eine Ansicht annehmen, ähnlich der Nussbaums, dass nämlich die Zonulafasern aus Mesenchymzellen hervorgehen und daher als Mesenchymfasern betrachtet werden sollten.

Die Schlüsse, zu denen ich gelangt bin, sind folgende:

I. Bei der weissen Maus haben die Zonulafasern sich aus Mesenchymzellen entwickelt.

II. Diese Fasern sind zuerst an die Apicalfortsätze der Zellen der inneren Ciliarepithelschicht angeheftet.

III. Später wechseln diese Zonulafasern ihren Ansatz und dringen in die Intercellularsubstanz ein, die zwischen den Zellen der inneren Ciliarepithelschicht liegt.

IV. Im fertigen Auge durchziehen die Zonulafasern die Intercellularsubstanz nach der limitans ciliaris externa hin; aber sie enden plötzlich, ehe sie dieses Gebilde erreichen.

V. Die Zonulafasern endigen nur an jenem Teile des Ciliarepithels, welches in den Tälern zweier benachbarter Ciliarfortsätze und zwischen den Ciliarfortsätzen und der ora serrata liegt.

Literaturverzeichnis.

1. Addario, C.: Sulla matrice del vitreo nell' occhio umano et degli animali. Riforma med. no. 17, 1901.
2. Derselbe: Sulla struttura del vitreo embryonale et dei neonati, sulla matrice del vitreo et sull' origine della Zonula. Pavia, p. 75, Tav. IX, 1902.
3. Derselbe: La matrice ciliare delle fibrille del vitreo, etc. Archivio di Ottalmologia, p. 206, 1904.
4. Agababow, A: Untersuchungen über die Natur der Zonula ciliaris. Arch. f. mikr. Anat., Bd. L, p. 563—588, 1897.
5. Angelucci, Ant.: Über Entwicklung und Bau des vorderen Uvealtractus der Vertebraten. Arch. f. mikr. Anat., Bd. XIX, 1881.
6. Derselbe: Physiologie générale de l'oeil, fonctions nutritives. Encyclopédie franc. d'Ophtalmologie, t. II, p. 2—59, 1905.
7. Berger, E.: Bemerkungen zur Zonulafrage. Arch. f. Ophth., Bd. XXXI, p. 3, 1886.
8. Derselbe: Anatomie normale et pathologique de l'oeil. Masson, éd. Paris 1893.
9. Derselbe: Historische Bemerkungen zur Anatomie der Ora serrata retinae. Arch. f. Augenhlk., Bd. XXXII, p. 288, 1896.
10. Derselbe: Anatomie générale de l'oeil, in Encycl. franc. d'Ophtalmol., t. I, p. 326—373, Paris, Doin, éd. 1903.
11. Claeys, G.: De la région ciliaire de la rétine et de la zonule de Zinn. Bull. Acad. Roy. de Méd. de Belgique, 3 série, t. XX, p. 1301, 1886.
12. Derselbe: De la région ciliaire de la rétine et de la zonule de Zinn. Arch. de Biologie, t. VIII, p. 623, 1888.
13. Cloquet, J.: Anatomie de l'Homme, t. III, p. 345—347, 1828.
14. Collins, E. J.: The glands of the ciliary body in the human eye. Trans. Ophth. Soc., vol. XI, p. 53, 1890—1891.
15. Derselbe: The glands of the ciliary body, a reply to some recent criticisms concerning them. Ophth. Review, vol. XV, 1896.
16. Czermak, W.: Zur Zonulafrage. Arch. f. Ophth., Bd. XXXI, 1, 1885.
17. Damianoff, G.: Recherches histologiques sur la cristalloïde et sur la zonule de Zinn. Thèse de médecine, Montpellier 1900.
18. Dessauer, E.: Zur Zonulafrage. Klin. Monatsbl., p. 94, 1883.
19. v. Ebner: In Köllikers Handbuch der Gewebelehre des Menschen, t. III, 2 partie, 1902.
20. Fischel: Die Regeneration der Linse. Anat. Hefte, Bd. XIV, 1900.
21. Derselbe: Weitere Mitteilungen über die Regeneration der Linse. Arch. f. Entwicklungsmech., Bd. XV, p. 1, 1902.
22. Gage, S. H.: The Microscope, an introduction to microscopic methods and to histology. 10th edit., Comstock Publ. Co., Ithaca, N. Y., 1908.
23. Haensell, P.: Recherches sur la structure et l'histogénèse du corps vitré normal et pathologique. Thèse de Paris 1888.
24. Iwanoff, H.: The vitreous humour. Strickers Handb., english ed. Sydenham Society, p. 345—356, London 1873.

25. Iwanoff, H. und J. Arnold: Mikrosk. Anat. des Uvealtractus und der Linse. Graefe-Saemisch Handbuch, Erster Band, Cap. III, p. 265—320. Leipzig, W. Engelmann, 1874.
26. Kessler, L.: Untersuchungen über die Entwicklung des Auges, angestellt am Hühnchen und Triton. Inaug.-Dissert., Dorpat 1871.
27. Derselbe: Zur Entwicklung des Auges der Wirbeltiere. Leipzig, 1877.
28. Kölliker, A.: Heidelberger Anatomenkongress. Anat. Anz., Ergänz.-Heft, p. 49, 1903.
29. Derselbe: Die Bedeutung und Entwicklung des Glaskörpers. Zeitschr. f. wissenschaft. Zoolog., LXXVI, I, 1904.
30. Leboucq, G.: Contribution à l'étude de l'histogénèse de la rétine chez les Mammifères. Arch. d'anat. micros.. pl. XVII und XIX, t. X, fasc. III und IV, p. 556—606, 1909.
31. v. Lenhossék, M.: Die Entwicklung des Glaskörpers. Leipzig, F. C. W. Vogel, 1903.
32. Derselbe: Die Entwicklung und Bedeutung der Zonulafasern, nach Untersuchungen am Hühnchen. Arch. f. mikr. Anat., Bd. 77, p. 280, 1911.
33. Lieberkühn, N.: Über das Auge des Wirbeltierembryos. Schriften d. Gesellsch. z. Bef. d. ges. Naturwissensch. zu Marburg. Kassel, 1872.
34. Derselbe: Beiträge zur Anatomie des embryonalen Auges. Arch. f. Anat. und Entwicklungsgesch., 1879.
35. Mawas, J.: Sur la structure de la rétine ciliaire. C. R. Acad. des Sciences, 14 décembre, 1908.
36. Derselbe: Recherches sur l'origine et la signification histologique des fibres de la zonule de Zinn. C. R. de l'Assoc. des Anat., Réunion de Marseille, avril 1908, p. 73—78.
37. Derselbe: Note sur l'origine des fibres de la zonule de Zinn. C. R. Soc. de Biol., 13 juin 1908, t. LXIII, p. 1029—1030.
38. Derselbe: La structure de la rétine ciliaire et la sécrétion de l'humeur aqueuse. C. R. Assoc. des Anat., 11 réunion à Nancy, avril 1909, p. 280—285.
39. Derselbe: La sécrétion de l'humeur aqueuse et la structure de la rétine ciliaire à l'état normal et pathologique. Soc. d'ophthal. de Lyon, 1 Mars 1909, et Lyon méd., 1909.
40. Metzner, R.: Kurze Notiz über Beobachtungen an dem Ciliarkörper und dem Strahlenbändchen des Tierauges. Verh. Naturf. Ges., Basel. Bd. XVI, p. 481—492, 1903.
41. Nussbaum, M.: Entwicklungsgeschichte des menschlichen Auges. Graefe-Saemisch, Handb. d. ges. Augenheilk., 2. Aufl., 1900.
42. Onfray: vidé Rochon-Duvigneaud et Onfray.
43. van Pée, P.: Recherches sur l'origine du corps vitré. Archives de Biol., T. XIX, p. 317—385, 1903.
44. Rabl, C.: Über den Bau und die Entwicklung der Linse. III. Teil. Die Linse der Säugetiere, Rückblick und Schluss. Zeitschr. f. wissensch. Zool., Bd. LXVIII, p. 29, 1898.
45. Derselbe: Zeitschr. f. wissensch. Zool., Bd. LVII, 1899.
46. Derselbe: Zur Glaskörperfrage. Anat. Anz., Nr. 25, 1903.

47. Retzius, G.: Über den Bau des Glaskörpers und der Zonula Zinnii in dem Auge des Menschen und einiger Tiere. Biol. Untersuch., neue Folge, VI, no. 9, p. 67—87, pl. XXVIII—XXXII, 1894.
48. Rochon-Duvigneaud et Onfray: Expériences preparatoires à la recherche des variations de concentration des liquides intra-oculaires et de leur influence sur le tension de l'oeil. Soc. d'Opht. de Paris, 5 juillet 1904.
49. Salzmann, M.: Die Zonula ciliaris und ihr Verhältnis zur Umgebung. Eine anatomische Studie. Wien 1900.
50. Sbordone, A.: Sull' origine delle fibre della zonula di Zinn. Ophthalmologica, vol. I, fasc. I, p. 68—83, Table V, 1910.
51. Schoen, W.: Zonula und Ora serrata. Anat. Anz., p. 360, 1895.
52. Derselbe: L'occommodation dans l'oeil humain. Arch. d'ophtal., p. 81, 1901.
53. Schultze, O.: Zur Entwicklungsgesch. des Gefäßsystems im Säugetierauge. Festschr. f. Kölliker, 1892.
54. Derselbe: Mikroskopische Anatomie der Linse und des Strahlenbändchens. Graefe-Saemisch, Handbuch d. ges. Augenheilk., 2. Aufl., Leipzig 1900, Lief. XVII.
55. Schwalbe: Lehrbuch der Anat. des Auges, 1887.
56. v. Spee, F. Graf: Über den Bau der Zonulafasern und ihre Anordnung im menschlichen Auge. Verh. der anat. Gesellsch., 16. Vers., Halle, p. 236—241, 1902.
57. Van der Stricht, N.: L'histogénèse des parties constituantes du neuroepithelium acoustique, des taches et des crètes acoustiques et de l'organe de Corti. Arch. de Biologie, t. XXIII, fasc. IV, p. 541, 1906.
58. Van der Stricht, O.: Le neuro-épithélium olfactif et sa membrane limitante interne. Mémoires de l'Ac. Roy. de Belgique, t. II, fasc. II, p. 1—45, pl. I u. II, 1909.
59. Terrien, F.: Recherches sur la structure de la rétine ciliaire et l'origine des fibres de la zonule de Zinn. Thèse médecine, Paris 1898.
60. Derselbe: La structure de la Rétina ciliaire. Arch. d'opht., Bd. XVIII, 1898.
61. Derselbe: Mode d'insertion des fibres zonulaires sur le cristallin et le rapport de ces fibres entre elles. Arch. d'opht., t. XIX, p. 250—257, 1899.
62. Topolanski: Über Bau der Zonula und Umgebung nebst Bemerkungen über das albinotische Auge. V. Graefes Arch. f. Ophth., Bd. XXXVII, no. 1, p. 28—61, pl. I—III, 1891.
63. Toufesco, Sophie: Sur le cristallin normal et pathologique. Thèse medecine, Paris 1906, auch Annal. d'Oculist, Août, p. 101, 1906.
64. de Waele, H.: Recherches sur l'anatomie comparée de l'oeil des vertébrés. Journ. int. d'Anat. et Physiol., t. XIX, p. 1, auch Internation. Monatsschr. f. Anat. u. Physiol., Bd. XIX, H. 1 und 2, 1901.
65. Wolfrum, M.: Über Ursprung und Ansatz der Zonulafasern im menschlichen Auge. Graefes Archiv für Ophth., Bd. LXIX, no. 1, p. 148—171, 1908.
66. Zinn, J. G.: Descriptio anatomica oculi humani iconibus illustrata (Cap. IV. De humore vitreo, p. 23 et suiv.), apud videam Abrami Vandenhoeck, Gottingae 1775.

Erklärung der Abbildungen auf Tafel XIV und XV.

Fig. 1. Gegend der Zonula im Auge einer 12 Stunden alten weissen Maus. A = distale Partie der Linse; B = ihre Kapsel; C = ein Stück Retina, innen begrenzt von der Mm. limitans interna. Vom Ciliarepithel ist nur der proximale Teil abgebildet, der in Beziehung zur Entwicklung der Zonulafasern (D) steht; E und F = Blutgefässe; G = Stützmembran eines Blutgefässes aus Fortsätzen einer Mesenchymzelle gebildet; den stark gefärbten Zellfortsätzen liegen Körnchen auf. Einige dieser Fortsätze können bis zu den apicalen Fortsätzen der inneren Lage der Ciliarepithelien (H), verfolgt werden. J = grosse, unregelmässig gestaltete und helle Mesenchymzelle mit zahlreichen feinen, hellen Fortsätzen. Dieser Zellentypus ist auf die Gegend der Zonula beschränkt und kommt im Glaskörper nicht vor. Aus solchen Zellen entspringen die Zonulafasern des erwachsenen Tieres. Bei der 12 Stunden alten Maus erreichen, wie Fig. 1 zeigt, die Fortsätze dieser Zellen die Linse noch nicht. Ihre zahlreichen Fibrillen bilden ein dichtes Netzwerk auf dem proximalen Abschnitt des Ciliarepithelium, woran sich die Fibrillen schliesslich festheften. An jedem zugespitzten Fortsatz einer Epithelzelle sitzt eine Fibrille. Diejenigen epithelialen Zellen, welche distal zu dem von den Fortsätzen der hellen Mesenchymzellen gebildeten Netzwerk liegen und späterhin die Ciliarfortsätze decken, verlieren ihre apicalen Fortsätze und sind demgemäss beim Erwachsenen mit der Zonula nicht mehr verbunden. Die Epithelien der ora serrata (K) dagegen und die nächsten distalen liegen im Gebiet der Zonula des Erwachsenen. Es gelang mir festzustellen, dass bei der 12 Stunden alten Maus im Bereich der Zonula die Hauptmasse des fibrillären Netzwerks aus den Fortsätzen der hellen Zellen und nicht von solchen der Epithelzellen gebildet wird. Hervorgehoben zu werden verdient auch, dass die Intercellularsubstanz der Epithelien an der Zonula nicht dicker ist als distal davon, und dass um diese Zeit keine Faser des Netzwerks mit der Intercellularsubstanz der Epithelien verbunden ist. Vergr. 500.

Fig. 2. Aus mehreren Schnitten zusammengesetzte Ansicht der Zonulagegend einer 10 Tage alten weissen Maus. A = distales und B = proximales Epithel eines Teiles der Linse mit ihrer Kapsel C. D = Membrana hyaloidea als eine dünne Membran die Zonulagegend vom Glaskörper trennend und von der proximalen Partie der Linse bis zur ora serrata sich erstreckend (E). Die Retina (F) hat mehrere Zellenlagen und ist innen von der Membr. limitans interna (G) begrenzt. H = Stäbchen und Zapfen, wohl entwickelt. Die Membr. limitans externa der Retina (J) ist distal bis zwischen die beiden Epithellagen des Ciliarkörpers zu verfolgen, wo sie zur Membr. limitans ciliaris externa (K) wird. Die Membr. limitans interna retinae hängt mit der Intercellularsubstanz der Epithelien an der ora serrata zusammen, direkt proximal von der Anheftung der Membr. hyaloidea. L = ein Stück Iris, M = ein Ciliarfortsatz. Vom Epithel der Ciliarfortsätze entspringen keine Zonulafasern; solche (N), die mit dem Epithel der Zonulagegend zusammenhängen, streichen über die freie Fläche des Epithels der Fortsätze hin. Die Zonulafasern sind halbschematisch eingezeichnet und finden sich am reichlichsten proximal vom Linsenäquator (O). In der Zonulagegend sind zwei Blutgefässe getroffen; das eine enthält im Inneren verschiedene Blutkörperchen und eine Mesenchymzelle (P) auf seiner Wand. Die Fortsätze dieser Mesenchymzelle sind bis an das Ciliarepithel zu verfolgen; viele Zonulafasern gehen in die Intercellularsubstanz des Epithels, distal haben noch verschiedene Epithelzellen zapfenartige Fortsätze. Vergr. 250.

Fig. 3. Nach einem Präparat von einer 14 Tage alten weissen Maus mit offener Lidspalte genau kopiert. A = Retina, B = Ciliarepithel. (Die Retina ist im Schnitt nach vorn verschoben.) C = ora serrata; D = Membr. limitans interna; E = Membrana hyaloidea; die beiden Membranen zerfallen in mehrere Lamellen; jede derselben hängt mit der Intercellularsubstanz der inneren Lage der Ciliarepithelien zusammen. Die Intercellularsubstanz ist an dieser Stelle dicker als sonstwo in der Zonulagegend. F = Membrana limitans externa retinae an der ora serrata verdickt; G = Membrana limitans externa ciliaris mit der vorigen zusammenhängend und die beiden Epithellagen trennend. Eine ächte Membrana limitans interna ciliaris ist an diesem Präparat nicht nachweisbar. Das Epithel der äusseren Lage ist deutlich in Form, Grösse und Granulierung von dem der inneren Lage verschieden. H = Pigmentzellen der Retina. J = Teil eines Ciliarfortsatzes; auf seinem Epithel liegt eine der grossen Mesenchymzellen, wie sie um diese Zeit in grösserer Zahl sich finden. Die Zelle (K) hat zwei Fortsätze; der eine zieht zur Linse und ist kurz abgeschnitten, der andere zerfasert sich auf der Oberfläche des Ciliarfortsatzes. Eine andere grosse Mesenchymzelle (L) liegt im Zonulabezirk auf der Membrana hyaloidea; der eine ihrer verzweigten Fortsätze endet auf der Oberfläche des Epithels der ora serrata. Zwischen diesen beiden Zellen liegen die Zonulafasern; viele derselben gehen in diesem Stadium zwischen die Epithelzellen. Im Vergleich zu dem Stadium von 12 Stunden ist zu bemerken, dass bei dem 14 Tage alten Tier die Zahl der in eine kurze Spitze ausgezogenen Epithelzellen bedeutend abgenommen hat. Zuweilen sieht es aus, als wenn einige Zonulafasern in der Nähe der ora serrata durch Epithelzellen bis zur Membrana limitans externa reichten. Untersucht man solche Stellen aber sorgfältig genug, so stellt sich heraus: dass alle diese Zonulafasern auf der dem Beobachter zugewandten Seite der Zellen in die Intercellularsubstanz eingebettet sind. Keine Zonulafaser geht durch eine Zelle und keine reicht bis an die Membrana limitans externa ciliaris. Vergr. 500.

Fig. 4. Von einer 27 Tage alten weissen Maus. Ciliarepithel zwischen ora serrata (A) und äusserem Rand der Iris (B). C = der mesenchymatische Kern eines Ciliarfortsatzes mit Blutgefäss. Die musivischen Epithelzellen (D) deuten an, dass der Ciliarfortsatz etwas schräg getroffen ist. E = verzweigte Fortsätze tiefer gelegener Mesenchymzellen, die in die Intercellularsubstanz der äusseren ciliaren Epithellage übergehen. An keiner Stelle kann ein Übergang dieser Fasern in Zonulafasern nachgewiesen werden. Die Abbildung zeigt deutlich, dass die Intercellularsubstanz im inneren Zellenlager des Ciliarfortsatzes nicht verdickt ist, im scharfen Gegensatz zum eigentlichen Zonulagebiet nahe der ora serrata. Man findet am Innenrande der Epithelien keine Spitzen mehr wie früher; alle Zonulafasern heften sich an die epitheliale Intercellularsubstanz an. Diejenigen Fasern, welche scheinbar vom Ciliarkörper selbst entspringen, können proximal über den darunter gelegenen epithelialen Rand nach dem Zonulagebiet an der ora serrata verfolgt werden. Diese Fasern erscheinen unter der Form einer Membrana limitans ciliaris interna. Vergr. 500.

1

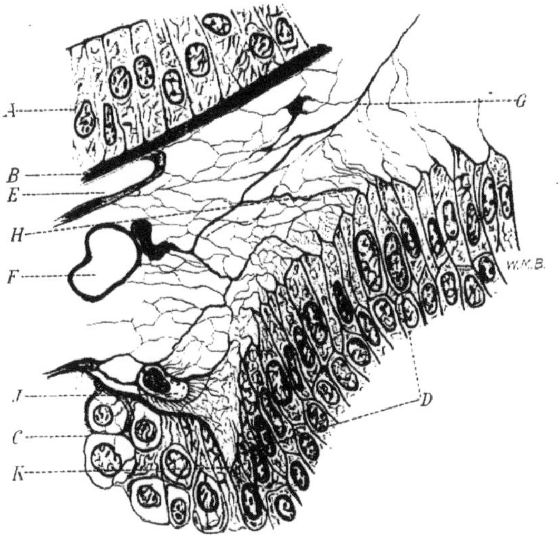

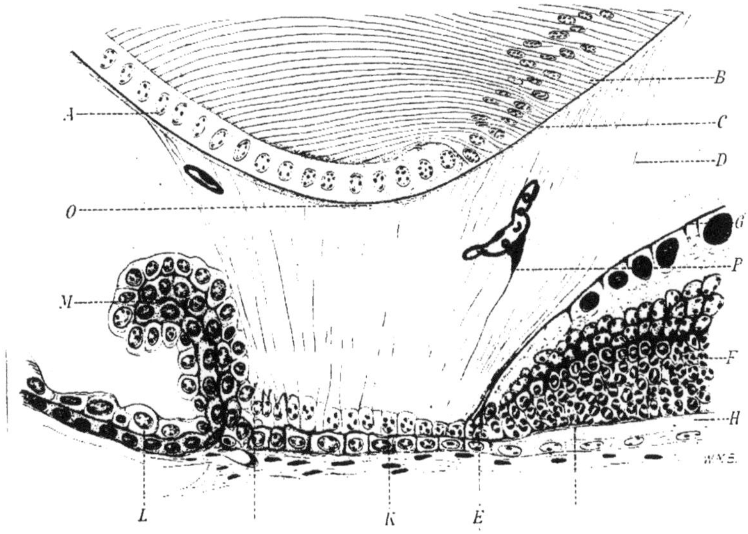

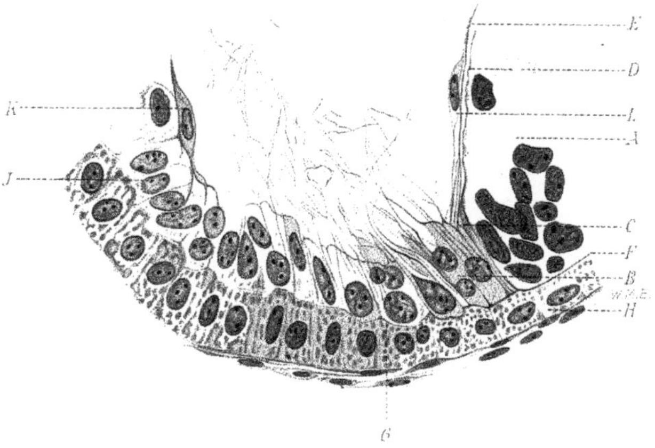
E
D
L
A
K
J
C
F
B
H
G

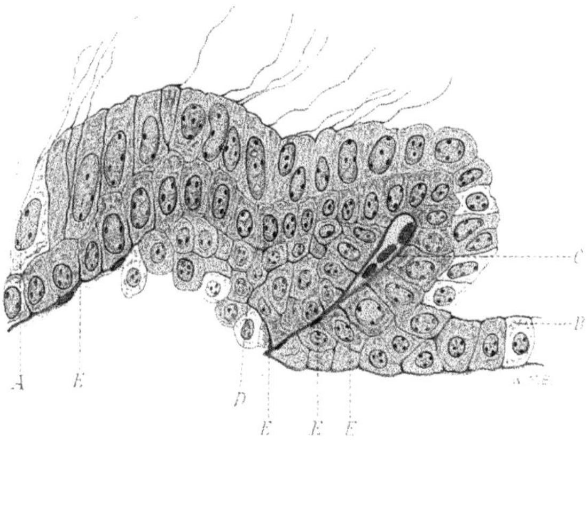
C
B
A
E
D
E
E
E

eprinted from SCIENCE, *N. S.*, *Vol. XXXVI.*, *No. 916*, *Pages 90–92*, *July 19, 1912*

THMICAL ACTIVITY OF ISOLATED HEART MUSCLE CELLS VITRO

previous communications[1,2] I pointed hat the heart muscle of chick embryos beat rhythmically for many days when nded in the media of a tissue culture rom such transplanted tissue there is an growth of cells into the surrounding . Braus[3] has repeated these experi- , using the hearts of embryo frogs and and he has found that these isolated g hearts react to electrical and chem- stimuli similar to the intact heart. s also noted that the cells which grew the hearts of cold-blooded animals were at the end of three months. Very tly, Carrel[4] by the use of the method of ted transplantation of the tissue from a re to a fresh medium (Carrel and Bur- has attempted to prolong the life and ion of heart muscle *in vitro*. His ex- ents show that the rhythm which I in fragments of embryonic chick hearts be prolonged, although intermittently, period of 85 days. The results of these iments substantiate, therefore, the former known fact, namely, that strips of heart le, both of cold and warm blooded ani- (Erlanger), will beat for some time when d in the proper media. In none of these could one rule out, however, the possi- of the existence of nerve ganglia or possible precursor in the young embry-onic hearts, which might initiate rhythn contractions.

During the present year experiments been made to determine the conditions w would prolong the life and allow the dev ment of functional activity in the cells w had grown and differentiated in the cul These experiments have shown that the n grown, cellular syncytia and the isolated si heart muscle cell can become functionall tive, beating with a rhythm similar to th the intact heart.

Pieces of the hearts of chick-embryos o ages and of young hatched chickens used. A growth of tissue, composed al entirely of muscle cells, occurred fro pieces when suspended in the media of types of cultures, (1) the ordinary han drop culture (the plasma modification the method of Harrison[5] and (2) a large fied type of culture. This apparatus is s ranged as to supply the tissues continu with fresh media and to wash away the products without in any way disturbing growing cells. I described this method in tail before the American Associatio Anatomists, December 27, 1911.[2] Serum used as the fluid medium in the latter ty culture.

Rhythmical activity of the newly g cells was noted in 3 out of 15 of the type of cultures (No. 2), and in 2 out o of the ordinary hanging drop cultures. cells were located definitely within the and had a clear cytoplasm which cont very few fat droplets. The rhythmical

urrows, M. T., 1911, *Jour. Exp. Zool.*, Vol. 3.

urrows, M. T., 1912, *Anat. Record*, Vol. 6,

raus, H., 1912, *Weiner Med. Wochschr.*, No.

arrel, A., 1912, *Jour. Exp. Med.*, Vol. XV.,

[5] Harrison, R. G., 1907, *Proc. Soc. Exp. and Med.*, 140; 1910, *Jour. Exp. Zool.*, V 787.

ity did not occur during the active outwandering of the cells but, later, after they became permanently located in a definite portion of the clot and were undergoing slow multiplication and differentiation. In one culture rhythm occurred as early as the fifth day, while in others as late as the fourteenth day of the life of the culture. The greater number of positive results in the large type of culture (No. 2) can be associated with the active and continuous growth of the tissue over a sufficient period of time. Active growth and a regular rhythm has been observed in these cultures for 30 days, while in the hanging drop culture the active growth and the regular rhythm cease after the third or fourth day. The growth then becomes gradually less and the rhythm intermittent, ceasing entirely after 10 or 18 days unless the tissue is transferred to a new medium. The method of repeated transplantation from the culture to a new medium has not as yet been sufficiently developed to allow any increase in the life and the activity of the newly grown cells. At each transfer of the tissue the actively growing and multiplying cells are destroyed and a new growth takes place from those more latently active cells in or about the tissue mass.

The original pieces of heart muscle transplanted to a tissue culture vary as to their rhythmical activity in relation to the portion of the heart from which they are taken as well as the age of the embryo. Pieces of the auricle, especially of that part situated near the entrance of the veins, taken from embryos of all ages and from young hatched chickens, beat when suspended in plasma. The pieces of the ventricle do not beat when taken from embryos older than 10 days, methods of preparation and t used.

Rhythmically beating cells ha from the contracting pieces of young embryos and from one pie tricle of a fourteen-day chick absence of movement in the ori tissue of this culture facilitate study of the delicate contraction grown cells. The syncytial n surrounded the original tissue an cell were beating rhythmically. situated far out in the clear medi all other tissues and beat with a pendent in phase from that of t The rate of all beating cells in th the same, 50 to 120 per minute, typical for rhythmical beating tricular muscle.

The experiments show: (1) which have grown and different sue culture can later assume th istic function; (2) that rhyt tion similar to that observed in heart can occur in an isolated an muscle cell; (3) that the rhyt tracting cells can be grown not pieces of hearts of young embr the heart muscle of a fourt embryo.

These experiments, therefore evidence for the myogenic theor beat.

MONTROSE T. B

ANATOMICAL LABORATORY,
CORNELL UNIVERSITY MEDICAL
NEW YORK CITY

Aus dem anatomischen Laboratorium des Cornell University Medical College, New York City (Vorstand: Prof. Dr. Ch. R. Stockard).

Rhythmische Kontraktionen der isolierten Herzmuskelzelle ausserhalb des Organismus.

Von Dr. Montrose T. Burrows.

Durch frühere Untersuchungen [1]) ist bewiesen, dass der Herzmuskel des Hühnerembryos, auf geeignete Nährböden implantiert, sich während 8 Tagen rhythmisch bewegt. Dieser Befund wurde von Braus [2]) (Unken- und Froschembryonen) und von Carrel [3]) (Hühnerembryonen) bestätigt. Aus diesen Untersuchungen geht deutlich hervor, dass die funktionelle Tätigkeit des Gewebes ausserhalb des Organismus lange Zeit erhalten bleiben kann. Eigentümlich bei den Gewebekulturen [4]) ist die Auswanderung der Zellen des ursprünglichen Gewebestückes in den umgebenden Nährboden hinein, welcher Vorgang dem eigentlichen Wachstum vorausgeht. Weitere Aufgabe wäre nun zu erforschen, ob die Funktionen solcher ausgewanderten, isoliert liegenden Zellen nach event. Teilung und Differenzierung denen des Muttergewebes gleichwertig sind. Bei der vorliegenden Untersuchung hat es sich tatsächlich herausgestellt, dass die isolierten Herzmuskelzellen rhythmische Bewegungen aus-

[1]) M. T. Burrows: Compt. rend. soc. de biol. 1910, LXIX, 291; Jour. Exper. Zool. 1911, X, 63.

[2]) H. Braus: diese Wochenschrift 1911, S. 2421 u. 2237; Wiener med. Wochenschr. 1911, No. 44.

[3]) A. Carrel, Jour. Exper. Med. 1912, XV, 516.

[4]) Der Ausdruck „Gewebekultur" wird hier in Analogie mit Bakterienkultur gebraucht. Das Prinzip der Gewebekultur besteht darin, dass man ein steril entnommenes Gewebestück auf ebenfalls sterilem Nährboden derart implantiert bezw. suspendiert, dass die Wachstumserscheinungen von Zeit zu Zeit bei Brutofentemperatur verfolgt werden können.

führen. Bei der allgemeinen Bedeutung welche diesem Resultate zukommen dürfte, soll die Methodik, sowie ihre geschichtliche Entwicklung eingehend beschrieben werden. Denn es ist im höchsten Grade wahrscheinlich, dass es auch gelingen wird, bei Anwendung geeigneter Methodik die mehr spezifischen Funktionen anderer hoch differenzierten Gewebe in vitro näher zu verfolgen.

Die erste praktische Anwendung der Gewebekulturmethode, wobei Gewebe längere Zeit ausserhalb des Organismus gehalten und ihre weitere Entwicklung gleichzeitig kontinuierlich beobachtet werden konnte, wurde von Harrison bei seinen Untersuchungen über Nervenfaserentwicklung gemacht. An früheren Untersuchungen über das Ueberleben der Gewebe und Organe hat es freilich nicht gefehlt, aber solche sind meistens von ganz anderen Gesichtspunkten aus unternommen worden. Die Versuche von Wentcher[5]), Ljunggren[6]), Carrel[7]) u. a. haben ergeben, dass es möglich ist, tierische Gewebe lange Zeit ausserhalb des Organismus am Leben zu erhalten. Bei den Untersuchungen von Ranvier[8]), Jolly[9]), Beebe und Ewing[10]) u. a. wurde hauptsächlich auf bestimmte Aeusserungen funktioneller Tätigkeit des in vitro lebenden Gewebe geachtet. Im Jahre 1897 gab Leo Loeb[11]) an, dass es ihm gelungen sei, das Wachstum des Gewebes im geronnenen Blutserum oder Agar ausserhalb des Organismus zu verfolgen. Näheres über die Technik und Resultate gibt er nicht an. Derselbe Autor (1902) beschrieb Versuche über das Wachstum der Haut. Diesmal nahm er Blöcke geronnenen Blutserums resp. Agar, spaltete dieselben, führte das Gewebestückchen dort ein und brachte das Ganze in das Unterhautzellgewebe des Tieres. Weiter untersuchte er bei der Wundheilung, wie das Epithel in den Schorf hineinwuchs.

Die erste Arbeit Harrisons[12]) auf diesem Gebiet stammt aus dem Jahre 1907. Die Methode bestand darin, dass Gewebestücke junger Froschembryonen in einem hängenden Tropfen Froschlymphe suspendiert wurden. Kurze Zeit nach dem Einbringen des Gewebestückes gerann natürlich die Lymphe und bot auf diese Weise ein festes Substrat für das weitere Wachstum des Gewebes. Mit dieser

---

[5]) J. Wentcher: Berl. klin. Wochenschr. 1894, 979; Zieglers Beitr., XXIV.

[6]) Ljunggren: D. Zeitschr. f. Chir. 1898, XLVII, 609.

[7]) A. Carrel: Jour. Exper. Med. 1910, XII, 460.

[8]) Ranvier: Traité Technique d'Histologie, Paris 1889.

[9]) Jolly: Compt. rend. soc. de biol. 1903, LV, 1266.

[10]) Beebe und Ewing: British Med. Jour. 1906, II, 1559.

[11]) Leo Loeb: Ueber die Entstehung von Bindegewebe, Leukozyten und roten Blutkörperchen aus Epithel und über eine Methode, isolierte Gewebsteile zu züchten: Chicago 1897, 41. Archiv f. Entwicklungsmechanik d. Organ. 1902, XIII, 487.

[12]) R. G. Harrison: Proc. Soc. Exper. Biol. and Med. 1907, IV, 140; Anat. Record 1908, II, 385; Harvey Lectures, Philadelphia, 1907—1908; Jour. Exper. Zool. 1910, IX, 787.

Methode hat Harrison das Wachstum des embryonalen Zentralnervensystems, des Muskels und der Haut in vitro auf schlagende Weise nachgewiesen. Bei diesen Versuchen wanderten die Zellen längs der Fibrinfäden aus dem ursprünglichen Gewebestück heraus, um sich dort im Fibringerüst weiter zu differenzieren. Aus dem Neuroblastenprotoplasma entwickelte sich selbsttätig der Achsenzylinder. Dadurch wurde die Richtigkeit der Hisschen Annahme von Harrison eindeutig bewiesen. Gleichzeitig hat Harrison die Grundlage für weitere Versuche über das Wachstum des Gewebes in vitro geschaffen.

Unter Anwendung der von Harrison angegebenen Prinzipien habe ich[13]) im Laboratorium Harrisons die Methode dadurch modifiziert und vereinfacht, dass Blutplasma statt Lymphe zur Verwendung kam. Die Anwendung von Plasma ist deswegen ein wesentlicher Fortschritt, weil man auf diese Weise geeignete Nährböden für das Gewebe verschiedener Tiere leicht gewinnen kann. Dies ist die zurzeit am meisten gebrauchte Methode der Gewebekultur. Neuerdings ist sie von Carrel und mir im Handbuch der biochemischen Arbeitsmethoden Bd. 5, Teil 2, S. 838) ausführlich wiedergegeben worden. Ich untersuchte damals das Wachstum embryonalen Gewebes vom Frosch und Huhn im Frosch- resp. Hühnerplasma und konnte die Resultate Harrisons bestätigen. Ich konnte ferner Kernteilungsfiguren häufig nachweisen. Noch wichtiger erschien mir die Tatsache, dass die funktionelle Tätigkeit des Gewebes in solchen Präparaten lange Zeit erhalten blieb. Beispielsweise führte das Herz von 60 Stunden alten Hühnerembryonen rhythmische Kontraktionen bis zum achten Tage nach der Herstellung des Präparates aus. Wegen der prinzipiellen Bedeutung dieser Tatsachen haben Carrel und ich[14]) mit Hilfe der Plasmamethode den naheliegenden Gedanken verfolgt und das Wachstum der Organ- und Gewebestücke verschiedener Tiere sowie deren Embryonen untersucht. Bei diesen Versuchen gelang zum ersten Male die Kultivierung von Organen und Gewebestücken erwachsener Tiere. Das Wachstum zahlreicher bösartiger Geschwülste (auch von Menschen) in vitro wurde gleichfalls studiert. Kurz darnach haben Lambert und Hanes[15]) ähnliche Versuche mit den Mäuse- und Rattentumoren angestellt. Dabei haben sie die wichtige Beobachtung gemacht, dass diese Geschwulst- und Gewebestücke auch in artfremdem Plasma wachsen. M. R. und W. H. Lewis[16]) haben das Wachstum embryonalen Hühnergewebes in verschiedenen flüssigen und festen Nährböden (Agar) untersucht. Sie konnten als erste das Wachstum dieses Gewebes in einfachen Salzlösungen konstatieren. In seiner ersten Mitteilung hat Harrison auf die Notwendigkeit eines festen

---

[13]) M. T. Burrows: loc. cit.; Journ. Am. Med. Ass. 1910, LV, 2057.

[14]) Carrel und Burrows: Journ. Am. Med. Assn. 1910, LV, 1379, 1554, 1732; Compt. rend. soc. de biol. 1910, LXIX, 293, 298, 299, 332; Jour. Exper. Med. 1911, XIII, 387, 571.

[15]) Lambert und Hanes: Journ. Am. Med. Assn. 1911, LVI, 33, 791; Jour. Exper. Med. 1911, XIII, 495; XIV, 129, 453.

[16]) M. R. und W. H. Lewis: Johns Hopkins Hospital Bull. 1911, XXII, 241; Anat. Record 1911, V, 277; VI, 207.

Substrats, z. B. Fibringerüst, geronnener Lymphe für die Entwicklung von Nervenfasern und Zellen schon hingewiesen. Als Bestätigung dieser Anschauung konnte ich[17]) Form und Gestaltsstörungen, wie z. B. Abrundungen und Verkleinerungen von Spindelzellen, die von Fibrinfäden losgelockert wurden, beobachten. Wenn solche Zellen wieder in Berührung mit einem festen Körper kommen, strecken sie sich alsdann zu einer länglichen oder unregelmässigen Gestalt aus. Dieser Befund wurde später von Carrel und mir an Tumorzellen bestätigt. Welche Gestalt schliesslich angenommen wird, ist von der Art des festen Substrates abhängig. Harrison[18]) hat nachträglich gezeigt, dass die Zellen, welche sich bei Anwendung eines flüssigen Nährbodens, wie bei den Versuchen von M. R. und W. H. Lewis, an dem Wachstum beteiligen, sich dem Deckglas stets anfügen. Es konnte ferner gezeigt werden, dass Spinngewebe (Harrison), baumwollene resp. seidene Fäden[19]) genügende Stütze für ein gerichtetes Wachstum in flüssigen Nährböden darbieten.

Carrel und ich[20]) haben auf die Möglichkeit sukzessiver Transplantationen von Gewebezellen aus Kulturen auf frischen Nährboden hingewiesen; sekundäre und tertiäre Kulturen erhielten wir leicht. Gelegentlich ihrer Versuche über Krebsimmunität haben Lambert und Hanes[21]) von diesem Prinzip Anwendung gemacht und konnten dabei die Zellen der Subkulturen genügend lang am Leben erhalten, um die Wachstumserscheinungen der Zellen des ursprünglichen Gewebes in einer aufeinanderfolgenden Reihe von Subkulturen auf Plasma verschiedener Herkunft zu untersuchen. Mittelst sukzessiver Transplantation hat Carrel[22]) nachher Gewebe lange Zeit hindurch am Leben erhalten können.

Gerade in der letzten Zeit ist die Plasmamethode auf die verschiedenen biologischen Probleme vielfach angewendet worden. Hier sei besonders auf die Arbeiten von Ruth[23]), Lambert[24]), Carrel und Ingebrigtsen[25]), Loeb

---

[17]) M. T. Burrows: loc. cit.

[18]) R. G. Harrison: Science 1911, XXXIV; Anat. Record 1912, VI, 181.

[19]) Carrel und Burrows: Jour. Exper. Med. 1911, XIV, 244.

[20]) Carrel und Burrows: Compt. rend. soc. de biol. 1910, LXIX, 329, 365; Jour. Exper. Med. 1911, XIII, 416.

[21]) Lambert und Hanes: Jour. Am. Med. Assn. 1911, LVI, 587; Jour. Exper. Med. XIII, 505.

[22]) Carrel: Jour. Am. Med. Assn. 1911, LVII, 1611; und 1912, loc. cit.

[23]) E. S. Ruth: Cicatrization of Wounds in vitro. Jour. Exper. Med. 1911, XIII, 422.

[24]) R. A. Lambert: The Production of Foreign Body Giant Cells in vitro. Jour. Exper. Med. 1912, XV, 510.

[25]) Carrel und Ingebrigtsen: The Production of Antibodies by Tissues living outside the Organismus. Jour. Exper. Med. 1912, XV, 287. — Carrel: Berl. klin. Wochenschr. 1912, No. 12; Jour. Exper. Med. 1912.

und Fleisher[26]), Hada[27]), Oppel[28]), Braus[29]) und Weil[30]) hingewiesen. Es sei bei dieser Gelegenheit darauf aufmerksam gemacht, dass in den Arbeiten von Hada und von Oppel die Plasmamethode zur Kultivierung der Gewebe wiederholt als die „Carrelsche Methode" bezeichnet wird. Aus dem Vorhergehenden ist aber ersichtlich, dass die wesentlichsten Fortschritte auf dem Gebiete des Gewebezellenlebens ausserhalb des Organismus fast ausschliesslich mit der Plasmamethode gewonnen worden sind. Auch ist die Plasmamethode nur eine Modifikation der Harrisonschen Methode, und die Modifikation, namentlich die Anwendung von Blutplasma statt Lymphe, ist von mir im Laboratorium Harrisons ausgearbeitet worden. Somit war Carrel auf keine Weise an der Entwicklung dieser Methodik beteiligt.

Methodik: Es wurden Präparate nach zwei Methoden angefertigt: 1. nach der Methode des hängenden Plasmatropfens (Harrison, Burrows), 2. nach einer von mir kürzlich ausgearbeiteten Methode, bei der das Gewebe fortwährend mit frischem Serum umspült wird. Beim zweiten Verfahren wird eine bessere Annäherung an die Verhältnisse im lebenden Organismus erzielt, indem dafür gesorgt wird, dass die Nahrungsstoffe kontinuierlich erneuert werden, während die Bestandteile des zellulären Stoffwechsels gleichzeitig entfernt werden. Denn es ist ohne weiteres klar, dass selbst bei Anwendung des geeignetsten Nährbodens eine Aenderung der Wachstumsgeschwindigkeit auftritt, wenn eine Anhäufung der Bestandteile des zellulären Stoffwechsels stattgefunden hat. Mit Rücksicht auf diese Faktoren habe ich diese Methode ersonnen, in der Hoffnung, dass man dadurch die Dauer des Wachstums verlängern kann, um Aufschluss über den Stoffwechsel sowie andere Aeusserungen funktioneller Tätigkeit zu gewinnen.

---

26) Loeb und Fleisher: Ueber die Bedeutung des Sauerstoffs für das Wachstum der Gewebe von Säugetieren. Bioch. Zeitschrift 1911, XXXVI, 98.

27) S. Hada: Die Kultur lebender Körperzellen. Berl. klin. Wochenschr. 1912, No. 1, 11.

28) A. Oppel: Ueber die Kultur von Säugetiergewebe ausserhalb des Organismus. Anatomischer Anzeiger 1912, XL, 464. Archiv f. Entwicklungsmechanik d. Organ. 1912, XXXIV, 132.

29) H. Braus: loc. cit.

30) Weil: Some Observations on the Cultivation of Tissues in vitro. The Jour. Med. Research 1912, XXVI, 159.

Es sei hier an das wesentlichste der Methodik erinnert[31]) (vergl. hierzu Fig. 1). Mittelst eines Dochtes wird das Medium von einem Behälter (a) durch die Kulturkammer (b) geleitet, um dann in einen anderen Behälter (c) aufgenommen zu werden. In der Kulturkammer wird der Docht in seine einzelnen Fasern, welche sich an der Oberfläche des Deckgläschens festsetzen und ein Kapillarnetz bilden, zerzupft. Das Gewebe wird in kleine Stücke zerschnitten und in das offene Netz von Baumwollfasern gelegt und hier durch das Gerinnen der hinzugefügten Plasmatropfen festgehalten. Der flüssige Nährboden bewegt sich langsam am Docht entlang durch die Kultur und sammelt sich in der Aufnahmekammer (c). Die Nährflüssigkeit an den Geweben wird auf diese Weise fortwährend verändert, ohne dass die wachsenden Zellen in irgend einer Weise gestört werden.

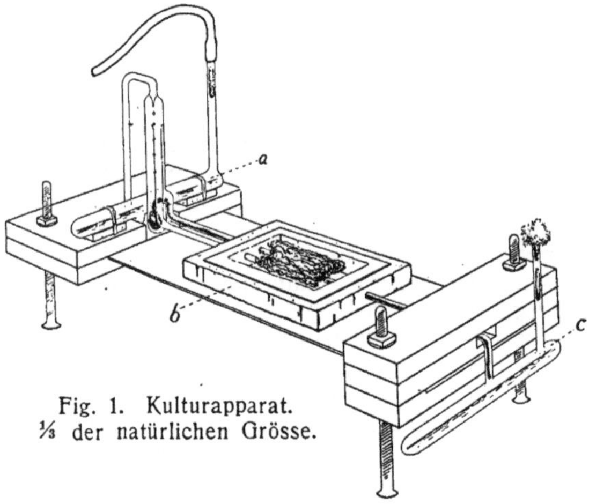

Fig. 1. Kulturapparat.
⅓ der natürlichen Grösse.

Die Zufuhrkammer (a) ist aus Glas gemacht und besitzt zwei Abteilungen. Die Lösung wird von dem horizontalen Behälter nach oben in die vertikalen Abteilungen am oberen Ende des Dochtes geführt. Hierdurch wird eine genaue Regulierung der pro Zeiteinheit benutzten Menge ermöglicht. Serum wurde als flüssiger Nährboden in dieser Reihe von Experimenten benutzt.

Herzen von Hühnerembryonen zu allen Zeiten des embryonalen Lebens, sowie auch die von ausgebrüteten Hühnchen wurden zu den Kulturversuchen verwendet. Von den jungen Embryonen wurde meist das ganze Herz, von den älteren sowie öfter auch von den jungen nur ein heraus-

[31]) Vorgetragen in der Sitzung der American Association of Anatomists am 27. Dezember 1911. Anatomical Record. VI, 141.

geschnittenes Stückchen Herzmuskel auf den Nährboden gebracht.

Das ganze Herz der 60—96 stündigen Embryonen schlägt rhythmisch in Kultur mit einer Frequenz von 50 bis 120 Schlägen in der Minute.

Ob die herausgeschnittenen Stücke pulsieren oder nicht, ist sowohl vom betreffenden Teile des Herzens, welchem sie entnommen werden, als auch vom Alter des Embryo abhängig. Ventrikelstücke von 60 stündigen bis 10 tägigen Embryonen, Stücke des Vorhofs, insbesondere aus der Nähe der Venen der Embryonen beliebigen Alters und des jungen Hühnchens zeigten rhythmische Kontraktionen. Die Ventrikelstücke von älteren Embryonen haben nicht rhythmisch pulsiert [32]). Die Frequenz des Rhythmus ist beim Vorhof grösser als beim Ventrikel; im ersten Falle beträgt sie 150 bis 220, während im zweiten Falle nur 50 bis 150 Schläge in der Minute. Der Rhythmus bei den gewöhnlichen Hängetropfenkulturen bleibt bis zum 3. oder 4. Tage regelmässig und wird später unregelmässig. Dieser intermittierende Rhythmus kann in solchen Präparaten bis zum 17. Tage dauern. In den grösseren Kulturen mit beständiger Zu- und Abfuhr frischen Serums bleibt der Rhythmus regelmässig, sogar bis zum 30. Tage.

Wachstum. Die Wachstumserscheinungen lassen sich in zwei Perioden zerlegen, erstens die der lebhaften Auswanderung der Zellen des ursprünglichen Gewebestückes in den es umgebenden Nährboden hinein; zweitens die der Teilung und Differenzierung. Die erste Periode fängt gegen Ende des ersten Tages an und dauert vom 5. Tage bis in die 2. Woche bei beiden Arten der Kulturen. Sie ist sowohl durch die Bildung eines synzytiumähnlichen Filzwerkes um das Gewebe herum als auch durch die Auswanderung einer grossen Anzahl Herzmuskelzellen (Fig. 2) gekennzeichnet. In allen Präparaten wurde die Auswanderung von Zellen, welche nach genauer Untersuchung als Herzmuskelzellen identifiziert werden konnten, beobachtet. Der Umfang und die Dauer des Wachstums ist von der Festigkeit der Schichtdicke des Nährbodens abhängig. Während der zweiten Periode erfolgt eine langsame Vermehrung und Differenzierung der an ihrem neuen Sitz ansässig gewordenen Zellen. Man erkennt das Wachs-

---

[32]) Allerdings können durch eine besondere Vorbereitung der Gewebe und der Kulturen auch Ventrikelstücke von älteren Embryonen zur Pulsation gebracht werden. Die Ursachen, die einen Rhythmus hervorrufen oder unterdrücken, werden in einer späteren Arbeit besprochen werden.

Fig. 2. Wachstum der Zellen aus einem Stück des Ventrikelmuskels eines 12 Tage alten Hühnerembryos. 8 Tage nach Herstellung des Präparates. In Hämatoxylin gefärbt.

tum an der Kernteilung und Protoplasmavermehrung der Zellen bzw. des Synzytiums.

Rhythmische Kontraktionen. In dieser Periode treten die rhythmischen Kontraktionen bei den ausgewanderten und differenzierten Zellen auf. Sie wurden in einem Fall schon am 5. Tag, in anderen Fällen erst am 14. Tage des Kulturlebens konstatiert. Die rhythmisch pulsierenden Zellen liegen stets im Fibringerüst und weisen ein durchsichtiges Protoplasma auf, welches vereinzelte Fetttropfen enthält. Unter 15 der nach der neuen Methode angefertigten Präparate, bei denen das Gewebe fortwährend mit frischem Serum umspült wurde, fanden sich rhythmische Kontraktionen neugebildeter Herzmuskelzellen bei 3 Präparaten, während nur bei 2 unter 150 der gewöhnlichen Hängetropfenpräparate dies der Fall war. Das häufigere Vorkommen funktionierender Zellen in den grösseren Kulturen hängt offenbar mit der bei diesen Kulturen länger andauernden Periode des lebhaften Wachstums zusammen. Es wurden nämlich Wachstumserscheinungen bei den grösseren Kulturen bis auf den 30. Tag nach Herstellung des Präparates konstatiert, während sie bei den gewöhnlichen Hängetropfenkulturen zwischen dem 10. bis 18. Tage schon aufhörten. Wiederholte sukzessive Transplantation der Gewebekulturen verlängern zwar im ganzen die Dauer der funktionellen Tätigkeit, aber obwohl solche Transplantation jedesmal wieder mit einem neuen Auswuchs der Zellen begleitet ist, so ist doch ein solches Verfahren keineswegs imstande, weder das Leben der an Ort und Stelle neugebildeten individuellen Zellen, noch die bei diesen auftretenden Perioden funktioneller Tätigkeit zu verlängern.

Um auf die Bewegungserscheinungen, welche an isolierten Zellen beobachtet wurden, etwas näher einzugehen, sei einiges aus dem Protokoll der 12. Kultur angeführt. In diesem Präparat hatte eine einzelne Zelle, welche weit vom ursprünglichen Gewebestücke entfernt war, sich pulsierend bewegt, und zwar mit einem Rhythmus, der von dem des Synzytiums unabhängig war.

Die isolierte Zelle ist spindelförmig. Das eine Ende ist abgerundet, während das andere zwei ausgezogene Fortsätze aufweist, welch letzteren gröbere Fibrinfäden fest anhaften. Die Lage des einzigen Kerns wird durch eine leichte Ausbuchtung des Protoplasmas in der Nähe des runden Endes verraten. Das feinkörnige oder maschige Protoplasma weist einige perinukleäre, stark lichtbrechende Granula auf.

Die Phase der Kontraktion dauert beträchtlich länger als die der Erschlaffung. Die Erschlaffung erfolgt plötzlich, und erinnert an das Zurückschnellen eines gespannten Gummibandes. Bei der plötzlichen Entspannung der Zelle wirkt möglicherweise die Elastizität der Fibrinfäden mit. Die Zelle wird bei der Kontraktion um ca. $1/5$ ihrer Länge verkürzt, während ihr kurzer Durchmesser an allen Stellen scheinbar vergrössert wird. Das Intervall zwischen den zwei Phasen ist sehr klein. Bei mittlerer Vergrösserung sind die Bewegungserscheinungen im Mikroskop leichter zu verfolgen und man kann sie mittels Kinematographie aufnehmen, wie Braus angegeben hat.

Am 14. Tage hat das Synzytium als Ganzes gleichmässig pulsiert, während am 15. und 16. Tage nur noch zwei von einander getrennte Teile des Synzytiums pulsierten; zwar war die Frequenz des Rhythmus gleich geblieben, doch trat jetzt eine Phasenverschiebung ein, infolgedessen pulsierten die zwei Teile nicht mehr synchron. Daher ist anzunehmen, dass die zwischenliegenden Zellen nicht nur ihr spontanes Kontraktionsvermögen, sondern auch ihr Reizleitungsvermögen eingebüsst hatten.

Zusammenfassung: Die Herzmuskelzellen embryonaler Hühner können, nachdem sie Teilung und Differenzierung ausserhalb des Organismus erfahren haben, ihre spezifische Funktionstätigkeit sowohl als isolierte Zellen wie auch als zusammenhängende Zellmassen wieder aufnehmen. Der Rhythmus solcher Zellen stimmt mit dem des Herzens des lebenden Tieres überein. Die rhythmische Bewegung wurde nicht nur bei den ausgewanderten Herzmuskelzellen junger, sondern auch bei denen der 14 tägigen Embryonen beobachtet. Die Stücke selber, die aus dem Ventrikel der älteren Embryonen gewonnen sind, schlagen aber nicht, trotzdem die aus solchen Stücken isoliert ausgewanderten Zellen Kontraktionen ausführen.

Durch diese Untersuchung ist demnach ein direkter Beweis für die myogene Theorie des Herzschlages gebracht worden.

Reprinted from the American Journal of Physiology.
Vol. XXXI. — November 1, 1912. — No. II.

THE EFFECTS OF ALKALOIDS ON THE DEVELOPMENT OF FISH (FUNDULUS) EGGS.

By J. F. McCLENDON.

[*From the Embryological Laboratory of Cornell University Medical College, New York City, and the U. S. Bureau of Fisheries, Woods Hole, Mass.*]

THE goal of experimental embryology is the control of development, notwithstanding the fact that the majority of attempts in this direction have been failures. The embryo results from the interaction between the egg and its environment, and we might expect that a specific change in the medium would produce a specific change in the embryo. However, the organism is capable, to a great degree, of maintaining constant conditions within itself. Take, for example, the remarkable constancy in body temperature and composition of the blood of mammals.

One mechanism for the maintenance of constant chemical conditions within the organism is evidenced in the remarkable semipermeability of living cells. Overton found that volatile anæsthetics and free alkaloid bases, which are rarely encountered by cells, penetrate easily, whereas salts, with which cells are constantly in contact, do not ordinarily penetrate. I observed that neither salts nor anions penetrate the Fundulus egg, but that kations outside may be exchanged for those within.[1]

Herbst[2] thought he had found a specific effect of lithium salts on sea urchins' eggs in the production of exogastrulæ, *i. e.*, gastrulæ in which the archenteron is evaginated instead of invaginated. However, Driesch[3] produced the same results by a rise in temperature to 30° C. In this case the archenteron, or gut, sometimes shrank and disappeared, producing a condition known as anenteria.

[1] McClendon: this Journal, 1912, xxix, p. 295.

[2] Herbst: Zeitschrift für wissenschaftliche Zoologie, 1892, lv, p. 442, and Mitteilungen der zoologische Station zu Neapel, 1895, xi, p. 136.

[3] Driesch: *Ibid.*, 1895, xi, p. 221.

Gurwitsch [4] supposed that lithium salts produced a radially symmetrical gastrula in the frog's egg. Morgan [5] showed this not to be the correct interpretation, but the chief characteristic of these embryos is that the endodermal cells are not invaginated, and hence we might call them exogastrulæ.

In opposition to the above statements, Bataillon [6] denies that lithium or other salts or sugar act otherwise than osmotically, and states that isotonic solutions all have the same effect on frog's eggs.

Stockard [7] observed that lithium chloride causes an enlarged segmentation cavity, and retards the down-growth of the blastoderm over the yolk, in Fundulus embryos. He demonstrated that this is independent of the osmotic pressure of the medium, and in a later paper [8] stated that these abnormalities are "specific for the lithium ion in its action on this egg."

On the other hand, I produced "lithium embryos" with sodium chloride, calcium chloride, ether, acetone, and dextrose.[9]

Stockard produced cyclopic or one-eyed Fundulus embryos, and at first thought the abnormality due to the specific action of the magnesium ion,[10] but later obtained similar results by the use of volatile anæsthetics, and supposed them due to the specific action of anæsthetics.[11]

However, I obtained the same results, not only with several indifferent anæsthetics, but with sodium chloride, lithium chloride, and sodium hydrate, which are considered stimulating rather than anæsthetic in their action.[12] I found the order of effectiveness of kations (added to sea water) in producing cyclopia to be $Mg < Li < Na$. Since Hedin [13] found the same order in the rate of diffusion of these ions through dead ox gut, we may suppose their action in producing cyclopia probably to be physico-chemical. This may be true also of indifferent anæsthetics, since I showed that their effectiveness in producing

[4] Gurwitsch: Archiv für Entwicklungsmechanik, 1896, iii, p. 219.
[5] Morgan: *Ibid.*, 1903, xvi, p. 691.
[6] Bataillon: Archiv für Entwicklungsmechanik, 1901, xi, p. 149.
[7] Stockard: Journal of experimental zoölogy, 1906, iii, p. 399.
[8] Stockard: *Ibid.*, 1907, iv, p. 165.
[9] McClendon: this Journal, 1912, xxix, p. 297.
[10] Stockard: Journal of experimental zoölogy, 1909, vi, p. 285.
[11] Stockard: American journal of anatomy, 1910, x, p. 369.
[12] McClendon: this Journal, 1912, xxix, p. 289.
[13] Hedin: Archiv für Physiologie, 1899, lxxviii, p. 205.

cyclopia is proportional to their effectiveness in lowering the surface tension of water, and also proportional to their toxicity. The concentrations of salts and anæsthetics producing cyclopia are very near the lethal doses, although lower than those producing "lithium embryos."

Experiments.

The majority of drugs that have a specific action on the function of parts of the human body are included in the old group of alkaloids. Since the chemistry of many of these substances is unknown, the group cannot be well defined, and we will use the name as originally applied to substances of vegetable origin with basic properties. Many synthetic substitutes of the alkaloids might well be included in the group.

I have tried a number of alkaloids on the eggs of Fundulus heteroclitus, and produced the same abnormalities with each of them. Some experiments with glucosides were begun, but were cut short by the unusually early close of the breeding season. Comparative studies were made on the eggs of the sea urchin, Arbacia punctulata.

Of the alkaloids tried, caffeine and theobromine are xanthine bases, the remainder being derivatives of pyridine and quinoline. Stovaine (chlorhydrate of dimethylaminobenzoylpentanol), a synthetic product, was also used.

The alkaloids were usually made up in centi-molecular solutions in sea water, and various strengths obtained by dilution. The free base was used except in three cases. The hydrochloride of quinine was chosen. The sulphates of strychnine and morphine were used, and made up of half strength, since each molecule liberates two molecules of the free base. These salts of the alkaloids have the advantage of being more easily dissolved, and since sea water is alkaline, the free base is completely liberated in the solutions. Overton showed that alkaloids enter living cells only in the form of the free base, hence they are effective only in alkaline or neutral solutions.

In the stronger solutions the whole embryo degenerates, but in weaker solutions certain parts are affected more than others. Organs which arise early may degenerate, those which appear later may fail to develop. The circulatory system is the most affected, and may be suppressed to a greater or less extent, or may develop and not function, or may function for a while and then degenerate. The heart and respiratory capillaries lie upon the surface of the yolk sac, and were more carefully observed than were the vessels which are obscured by the surrounding tissues of the embryo.

FIGURE 1.

FIGURE 3.

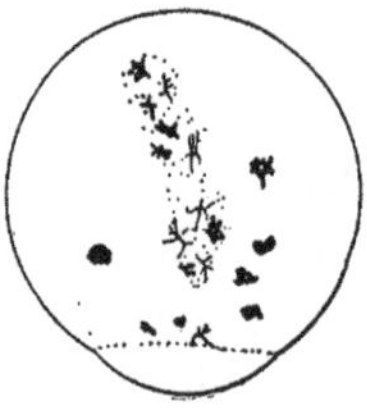

FIGURE 4.

FIGURE 1. — Normal embryo of Fundulus heteroclitus, six days old. *S*, sub-intestinal vein; *D*, duct of Cuvier.

FIGURE 3. — Fundulus egg treated with $\frac{1}{80}$ molecular caffeine, showing the degenerating embryo one week old.

FIGURE 4. — Front view of embryo from $\frac{M}{180}$ caffeine, one week old. The heart beat, but the blood did not circulate.

A glance at the normal embryo will be useful for comparison. Fig. 1 shows the yolk sac circulation in a six-day embryo, the capillaries on the ventral side being indicated by dashed lines. The caudal vein passes on to the yolk sac as the sub-intestinal vein (*S*), and breaks

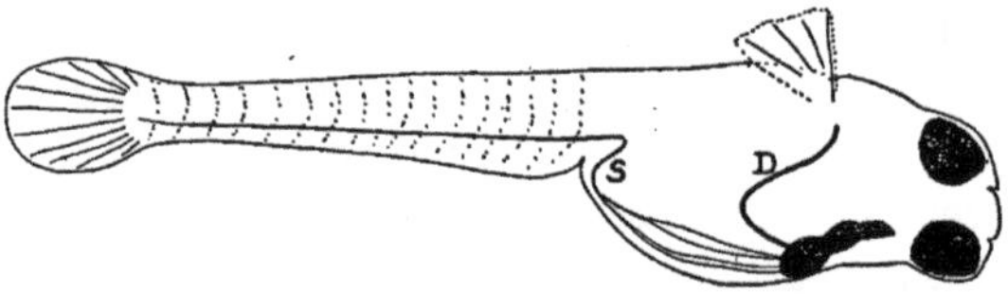

FIGURE 2. — The same embryo as Fig. 1, just hatched.

up into capillaries. The ducts of Cuvier pass out of the embryo and break up into capillaries, which anastomose with those from the sub-intestinal vein. This capillary plexus is reunited at the venous end of the heart, in front of the head of the embryo. It should be noted that in early stages the ducts of Cuvier draw their blood from the dorsal aorta through temporary connections.

As the embryo develops, the yolk is absorbed, and the capillaries on the yolk sac are gradually transformed into three large veins, the continuations of the ducts of Cuvier and the sub-intestinal vein. A transition stage is seen in an embryo just hatched (Fig. 2). The right duct of Cuvier (*D*) is shown completed, but the sub-intestinal vein (*S*) is connected with the heart by four parallel veinlets, the remains of the capillary plexus. The anastomoses between the sub-intestinal vein and ducts of Cuvier have disappeared.

In the stronger solutions of alkaloids the embryos begin to develop normally, but sooner or later the cells begin to be loosened one from

FIGURE 5.

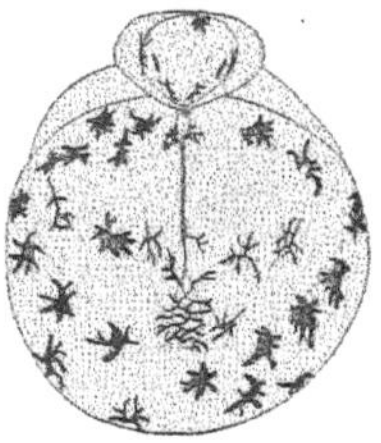

FIGURE 6.

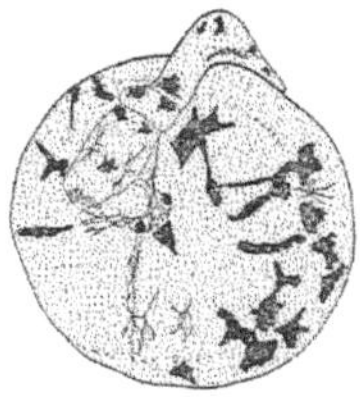

FIGURE 7.

FIGURE 5. — Same embryo as Fig. 4, ten days old.
FIGURE 6. — Embryo from $\frac{M}{80}$ caffeine solution, eleven days old.
FIGURE 7. — Cyclopic monster from $\frac{M}{100}$ nicotine solution, four days old.

another, a condition called by Roux "framboisea." Such an embryo may live a long time, but gradually undergoes de-differentiation.

In Fig. 3, which represents such an egg a week old, the stippled area represents the embryo. The spots with blunt processes represent black chromatophores, and those with slender processes, red chromatophores. The blister on the yolk is the swollen pericardial cavity, and is the only means by which the head end of the embryo may be located.

Fig. 4 represents an embryo of the same age from a weaker solution, viewed from the front. The eyes are represented by the small circles. The semicircular areas, lateral to the eyes, are the distended ear vesicles, whereas the almost circular area reaching from the eyes to the dashed line across the middle of the yolk, represents the distended pericardial cavity. The elongate body, extending from the head ventralward, is the heart. It is beating, but no blood circulates, since no hollow vessels are connected with it. Erythrocytes, represented by stipple, lie in the arterial end of the heart. The lower or venous end of the heart is covered with red chromatophores.

In this same embryo three days later the eyes are degenerating, the left eye being represented by merely a thin smear of retinal pigment

(Fig. 5). The number of chromatophores on the ventral side of the pericardium and on the heart is greater than in a normal embryo. Owing to the swelling of the pericardial cavity, the heart is greatly elongated and its middle portion is transformed into a solid cord.

Sometimes but one eye completely degenerates, resulting in a condition which I have called secondary monophthalmia asymmetrica;

FIGURE 8.

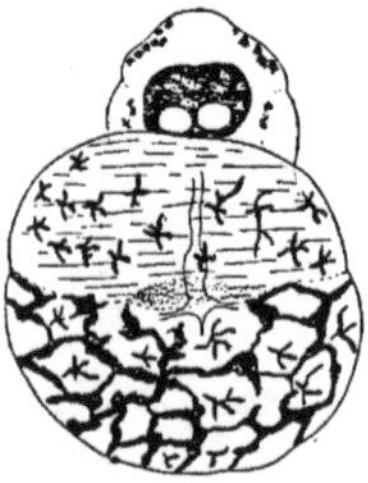

FIGURE 9.

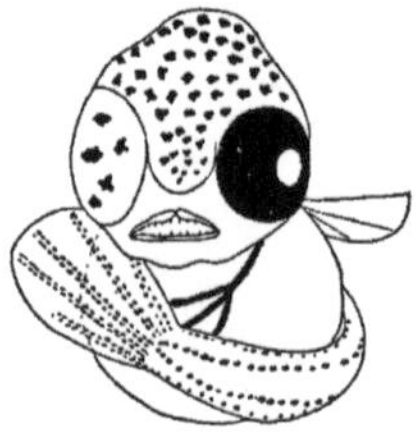

FIGURE 8. — The same embryo as Fig. 7, sixteen days old.

FIGURE 9. — Monstrum monophthalmicum asymmetricum, sixteen days old, from $\frac{M}{600}$ nicotine solution.

or the two eyes partially degenerate and then fuse together, secondary cyclopia.

Often the anterior region of the head disintegrates, and the cells wander away and become scattered over the pericardium. Such an embryo is shown in Fig. 6, and is atypical only in the fate of the eyes. The heart formed, but did not beat. The stippled area in the tail region represents a mass of erythrocytes. The pericardium was greatly distended and pressed against the ventral side of the head, to which it adhered. The anterior portion of the head disintegrated, and the eyes contracted into two spherical masses blackened by retinal pigment. By the eleventh day the eyes had fallen through the pericardial cavity and become grafted on to the heart.

Only one case of primary cyclopia occurred. This embryo, when four days old (Fig. 7), was apparently normal except for the single dumbbell-shaped eye on the ventral side of the head. The same embryo, sixteen days old, is shown from the front in Fig. 8. Although it had been removed from the nicotine solution when thirty-six hours old and had remained in pure sea water, frequently changed, for two weeks, defects began to appear in the circulatory system, and the circulation ceased entirely long before death. As shown in Fig. 8, the pericardial cavity is distended and its contents turbid, and include a mass of erythrocytes around the venous end of the heart, represented by stipple.

The only case of primary monophthalmia obtained is shown at the age of sixteen days in Fig. 9. Except for the lack of the right eye, it is apparently normal. Even the right eye socket has developed in an apparently normal manner, but is covered by skin containing chromatophores.

TABLE I

Caffeine . .	50	80–200	Quinine . .	200	400
Atropine . .	90	100	Strychnine .	...	100–1600 (Saturated)
Brucine . . .	...	400 (Saturated)	....	...	....
Nicotine . .	350	400–700	Stovaine . .	800	1600

The alkaloids, even in very weak solutions, retard development. The toxic limits, and concentrations of solutions causing the abnormalities described above, are given in Table I. A saturated solution of theobromine produced no monsters, due to its very slight solubility, but, from comparative studies on the egg of the sea urchin, we may class it with caffeine.

To give a uniform basis for comparison, the data of only those experiments in which the embryos remained in the solutions thirty-six hours, beginning with the two-cell stage, are used. The "toxic limit" is the solution in which nearly all of the eggs are dead at the end of thirty-six hours. Each number in the table is the denominator of a fraction (of a molecular solution) whose numerator is 1. The middle column gives the toxic limits, and the last column gives the concentrations which suppress the circulation in various degrees, and cause the other abnormalities described above.

Comparative experiments were made on the eggs of the sea urchin, Arbacia punctulata. These eggs are more easily obtained, develop more rapidly, and are more sensitive to changes in the medium than are Fundulus eggs. All of the alkaloids, as well as stovaine and digitalin, produced the same abnormalities.

The strengths of the solutions are shown in Table II. The eggs were not removed from the solutions and returned to pure sea water, as in the case of the Fundulus eggs. The solutions of digitalin are percentages, in case of the others the number represents the denominator of a fraction (of a molecular solution) whose numerator is 1.

In stronger solutions than those listed in the table, embryos died in segmentation stages. Even in those in the table development was enormously retarded, and the yolk granules dissolved more slowly, although ciliary activity did not appear to be reduced. Whereas

TABLE II.

Caffeine . .	500	800	Nicotine . .	3,200	6,400
Theobromine	.....	Saturated	Quinine . .	50,000	55,000
Atropine . . .	3,200	25,600	Strychnine . .	25,600	51,200
Brucine . . .	12,800	51,200	....	......	....
Cocaine . . .	6,400	25,600	Stovaine . .	12,800	102,400
Morphine . .	800	6,400	Digitalin . .	.0004%	.0001%

normal embryos, if not fed, starve to death in a few days, some embryos in the alkaloid solutions live several weeks. After the first few days a distention of the body cavity commences and may continue until the ectoderm becomes a very thin-walled vesicle.

In the solutions listed in the middle column plutei were not produced, although rudiments of the skeleton, in the form of tri-radiate stars, sometimes appeared. Some exogastrulæ were produced. In case normal invagination of the endoderm did take place, the gut was never normally differentiated, but usually degenerated into a solid mass of cells.

In the solutions listed in the last column plutei were formed, but there was an early disarrangement of the mesoderm cells, so that the resulting skeleton was abnormal. Since there was an enormous number of forms of the skeleton, they cannot be described here.

Conclusions.

It has been shown above that the very different organic compounds used, belonging to both the aliphatic and the carbocyclic series, although included in the old class of alkaloids, have the same morphological effects on the eggs of Fundulus heteroclitus. Loeb has obtained the same abnormalities in solutions of potassium cyanide and has reared similar monsters from eggs fertilized with foreign sperm.

It might be supposed that these effects follow any injury to the Fundulus egg. The distention of the pericardium follows the application of various salts. I have produced it in frog's embryos by mechanical injury. The distention of other serous cavities sometimes follows the application of certain salts and anæsthetics. Loeb prevented the heart beat with potassium salts. Some embryos in ammonium salts were observed by Stockard never to develop a heart beat, and in some in magnesium salts the circulatory system degenerated.

It may be possible that the same abnormalities can be produced by any chemical treatment. However, the quantitative data show striking differences in the effects of different substances. Using solutions of equal toxicity, almost 100 per cent of embryos in ethyl alcohol may show primary defects in the eyes, whereas such defects are seen in not more than one tenth of 1 per cent of embryos treated with alkaloids. On the other hand, the abnormalities described in the circulation may occur in nearly 100 per cent of embryos treated with alkaloids.

Thousands of eggs of Fundulus heteroclitus were placed in solutions of each of the alkaloids enumerated, and the eyes of each embryo were carefully examined, but only one case of primary cyclopia and one of primary monophthalmia asymmetrica were observed. These two occurred in the same batch of eggs treated with nicotine. I subsequently used this same concentration of nicotine on the eggs of many different females without reproducing these abnormalities.

Stockard obtained very different percentages of cyclopia in eggs treated with magnesium salts during different seasons or parts of seasons. He once thought this due to accidental variation in the concentration of the solutions. This could not have been the case at least in my experiments, as I used a finely graduated series of concentrations extending a great distance on each side of the apparent optimum for producing cyclopia. In numerous experiments covering an entire season, I failed to obtain as high a per cent of cyclopia with magnesium chloride as recorded by Stockard. By careful measurements of specific gravity, I found that the varying results were not due to differences in density of the sea water. The water I had been using was obtained directly from the sea in glass vessels, but as a control I used sea water drawn from the same pipe as that used by Stockard. This was repeated many times, and it was demonstrated that heavy metals in the water did not account for the differences.

The cause of the varying results must be that different batches of eggs have not the same tendency toward cyclopia, just as the individual eggs laid by the same female at the same time vary in this respect. Perhaps some Fundulus eggs would produce cyclopic embryos without any laboratory treatment, as I have found many cyclopic smelt embryos in the natural breeding places. It would not be safe, therefore, to draw conclusions from two individuals in many thousands. Attention should be directed rather to the quantitative data.

Reprinted from THE JOURNAL OF BIOLOGICAL CHEMISTRY, Vol. XI, No. 4, 1912

ECHINOCHROME, A RED SUBSTANCE IN SEA URCHINS.

BY J. F. McCLENDON.

(From the Embryological Laboratory of Cornell University Medical College, New York City, and the U. S. Bureau of Fisheries, Woods Hole, Mass.)

(Received for publication, March 30, 1912.)

INTRODUCTION.

My interest in echinochrome arose from studies in permeability. In the same way that haemolytic agents cause haemoglobin to leave the red blood corpuscles, so do cytolytic agents cause echinochrome to leave the cells containing it. R. Lillie is of the opinion that this is due to the action of the cytolytic agent in increasing the permeability of the cell surface.

In the elaeocytes, wandering cells of the body fluid of *Arbacia punctulata*, the cytoplasm is crowded with spherical chromatophores. Some of these may be colorless, but usually they are colored bright red with echinochrome. Similar chromatophores, though not so close together, occur in the eggs. In the unfertilized egg they are evenly distributed throughout the cytoplasm. But after fertilization, the chromatophores all migrate to the surface within half an hour. During cleavage of the egg, they are massed in the cleavage furrows. The pigment occurs also in the test of this sea urchin, and gives the animal the characteristic color, which varies from a bright red (especially in young individuals) to a dark red, and may be almost black in old specimens.

In reference to the fact that the pigment may be caused to leave the chromatophores and pass into the cytoplasm and thence into the medium, the following questions may be asked: (1) How is the pigment held in the chromatophores? (2) What is its function? (3) What is its chemical nature? The present paper is concerned with these questions.

HISTORICAL.

Echinochrome was studied spectroscopically by McMunn,[1] who found it in the elaeocytes of the sea urchins, *Strongylocentrotus lividus*, *Amphidotus cordatus*, *Echinus esculentus?* and *E. sphaera*. The spectrum showed faint absorption bands, which varied with different solvents and different reactions of the same solvent. McMunn thought that he noticed changes in the spectrum on the addition of powerful reducing agents, such as stannous chloride, and concluded that echinochrome functioned as an oxygen carrier. However, the absorption bands in its spectrum are difficult to make out except in absolute alcohol (or glycerine) and in this solvent I observed that stannous chloride caused a precipitation of the pigment, which interfered with the examination.

A. B. Griffiths[2] attempted an elementary analysis of the substance. He dried the elaeocytes and extracted them with chloroform, benzol or carbon bisulphide. On evaporation of the solvent he analyzed the substance without further purification, although evidently it contained many impurities. From four analyses, he deduced the formula $C_{102}H_{99}N_{12}FeS_2O_{12}$, which would make C = 67.8 per cent, H = 5.5 per cent, and N = 9.3 per cent. He states that on boiling with mineral acids it is transformed into haematoporphyrin, haemochromogen and sulphuric acid ($E + acid = 2C_{34}H_{34}N_4O_5 + C_{34}H_{37}NFeO_5 + H_2SO_4$). Griffiths agrees with McMunn that echinochrome is an oxygen carrier, and states that the oygen is held rather firmly, and in nature is removed only by the reducing action of the cell containing the pigment.

EXPERIMENTAL.

The pigment in the elaeocytes, eggs and tests of Arbacia, shows no absorption bands, but after extraction it shows very similar bands in its spectrum to those described for echinochrome by McMunn. He published drawings of the spectra and measured the wave lengths corresponding to the *edges* of the bands. It is well known that bands become broader as the solution is more concentrated, and for that reason I measured the wave length of a line of the spectrum corresponding as nearly as could be determined to the *center* of each band. By taking the mean between the wave lengths of the edges of the band in McMunn's data I have compared his with mine. The discrepancies may be accounted for,

[1] McMunn: *Quart. Journ. Micro. Sci.* (2), xxv, p. 469, 1885; xxx, p. 51, 1889.

[2] Griffiths: *Compt. rend. soc. biol.*, cxv, p. 419, 1892; *Proc. Roy. Soc. Edinburg*, xix, p. 117, 1892; *Physiology of the Invertebrata*, New York, 1892; *Respiratory Proteids*, London, 1897.

first by the fact that the mean is not the exact center of the band in a prism spectrum, and secondly there is a personal equation in observation. I found the pigment extracted from elaeocytes, eggs or tests to give about the same spectra, though a few isolated observations seemed to vary. These might have been due to decomposition products with different spectra.

	ETHER		ABSOLUTE ALCOHOL							H_2O			
			Neutral			+HCl		$+NH_3$		+HCl		$+NH_3$	
My data...	5296	4844	5504	5302	4844	5296	4844	5154	4844	5296	4844	5154	4844
McMunn..			5512	5128	4848	5370	4998	5205	4848				

Neither McMunn nor Griffiths succeeded in crystallizing echinochrome. Dr. A. P. Mathews had observed that on the addition of iodine in potassium iodide (KI_3) crystals form easily. In 1910 I obtained quantities of these crystals, but did not succeed in recrystallizing them without great loss by the formation of amorphous masses. The iodine compound in absolute alcohol showed an additional, but very dim band in the spectrum (wave length 5628 or 5696). It crystallized in red or orange needle-like crystals, triangular in cross section, sometimes rhombic in side view and often forming rosettes. They were but slightly soluble in water unless hot or containing acid, soluble in absolute alcohol (the rhombic crystals seeming more soluble than the needles) and slightly soluble in ether. If a solution in water is shaken with ether the latter is not colored. If an alkali is added to the KI_3 solution no crystals are formed (due to combination of the base with the echinochrome) but HCl does not prevent their formation.

Some of this iodine compound which was kept for several months in a dry state became more soluble in ether and crystallized in flat thin, red or orange rhombic plates. Perhaps the substance had decomposed with the liberation of iodine, for I succeeded in crystallizing the mother substance and obtained the same plates, in addition to red or orange needles, never triangular in cross section, but sometimes forming rosettes.

I extracted echinochrome from the tests with strong, slightly acidulated alcohol and purified it by repeated precipitation with alkali and solution in acid alcohol, and filtration.[3] Finally I dissolved the precipitate in water plus HCl, filtered and shook the solution with ether. The ether did not remove all of the echinochrome and the formation of haptogen membranes caused much loss of material. The ether was evaporated at room temperature, as heat seemed to decompose the substance. Occasionally a few crystals formed at the edges of the solution but the main mass of the residue was amorphous.

The next season (1911) I tried to purify echinochrome without the use of acids or alkalies. The body fluid of the sea urchins was allowed to clot and the elaeocytes thus obtained were placed directly into acetone, which extracted the pigment. The extract was filtered and evaporated at room temperature. The residue was washed with carbon tetrachloride (which does not easily dissolve echinochrome) to remove fats, and again dissolved in the smallest quantity of acetone and filtered to free it from traces of lecithin. This solution was evaporated, dissolved in absolute ether and filtered to remove salts, evaporated to constant weight and analyzed by Dennstedt's method. A mean of two analyses gave: C = 51 per cent, H = 7.7 per cent. The echinochrome purified by precipitating with alkali gave C = 53.3 per cent, H = 4.4 per cent, N = 1.5 per cent. The nitrogen was determined by Kjeldahl's method and therefore may not be reliable, since the constitution of the molecule is unknown. Traces of sulphur and phosphorus, possibly due to impurities were found, but no iron. The ether-soluble crystals from spontaneous decomposition of the iodine compound gave C = 57.9 per cent, H = 6.5 per cent.

It was stated above that echinochrome is precipitated by alkalies in alcohol. I precipitated echinochrome with NaOH in 95 per cent alcohol and washed in the same alcohol to remove the excess of NaOH. From the amount of NaOH that was neutralized by the pigment I concluded that it combined with from 18 to 25 per cent of Na. Analysis gave C = 31.5 per cent, H = 6 per cent, Na = 19.5 per cent. Therefore we may say that the echinochrome behaves as an acid, or else is amphoteric. The former view is

[3] Alkali does not precipitate it in water; the particular base was immaterial, ammonia was added but the presence of sea salts allowed the liberation of other bases.

supported by the fact that on passing an electric current through the aqueous (colloidal) solution, the echinochrome shows a negative charge (is anodic) and again, if histological sections are placed in such a solution the acidophile portions are stained more strongly than the remainder. In fact its behavior is very similar to that of a weak solution of eosin, except that it is very easily washed out by alcohol.

However the substance is probably amphoteric (the acid character being stronger than the basic) since its aqueous solution is precipitated by phosphomolybdic and phosphotungstic acids but not by tannin.

From the analyses given above it would seem that no one has succeeded in obtaining echinochrome in a reasonably pure state. It is very unstable and probably breaks up into a host of decomposition products all having practically the same spectrum. If it is kept in the dry state for a great length of time, or is evaporated on a bath not over 50° for a shorter time, part of it becomes insoluble in ether but not in alcohol.

When heated in the combustion tube it first melts, then boils and sublimes as crystals on the top of the tube, then very soon turns brown and chars. After being crystallized from a solution in ether the crystals often become smaller and irregular in outline. Perhaps the crystals evaporate or lose water of crystallization, but I think that both these possibilities are improbable. The crystals may decompose into an amorphous substance.

On first obtaining crystals, I feared that they were crystals of some other substance merely colored by echinochrome, but this seems impossible from later observations.

In extractions made for the purpose of studying the lipoids in Arbacia eggs, red or brown substances (echinochrome or its decomposition products) appear in every fraction, rendering analysis difficult and indicating the instability and wide solubility of the substance.

In order to test the statement that it is an oxygen carrier I separated the cells from 50 cc. of body fluid by the centrifuge, and mixed them with sea water to make 50 cc. This suspension, and 50 cc. of sea water as a control, were placed in two similar graduated tubes. The air was pumped out for six hours (until the water boiled), air was then admitted and the tubes sealed. They were shaked one-half hour and the volume of air measured at atmospheric pressure. The suspension had lost 1.25 cc., the control only 0.8 cc. In another experiment the suspension lost 0.95 cc. and the sea water 0.8 cc. It was thought that in the absence of oxygen the cells would take the oxygen from the echinochrome. However no color change could be observed with the naked eye or the spectroscope, and the greater absorption of air by the suspension may have been entirely due to oxidation in the cells. In similar experiments, with an aqueous solution of the pigment, and distilled water for a control, and using pure oxygen, the two tubes gave the same absorption, as shown by two examples:

Oxygen absorbed by H_2O	1.0 / 1.1
Oxygen absorbed by echinochrome	1.0 / 1.15

The question, how echinochrome is held in the chromatophores, cannot be fully answered. The chromatophores when free from pigment are highly refractive and stain strongly with the intravitam stain, neutral red, and when fixed they stain strongly with Delafield's haematoxylin, indicating a lipoid nature. The pigment may be in solution in the lipoid.

The fact that the spectrum is different (shows no bands) in life from the spectrum of the extract may indicate chemical combination of the pigment with the chromatophores. The fact that echinochrome stains acidophile tissue may show a possible mode of such combination, if it be found that the chromatophores contain bases. However I do not think we can rely on the spectroscopic evidence, for the absorption bands are very faint in aqueous solution unless it be alkaline, and the cells containing the pigment interfere with the passage of light and make the observation difficult. I have never seen absorption bands in echinochrome extracted from the fresh cells with distilled water. The same statement is made by McMunn. If the substance is held by chemical combination why does it come out so easily?

The same argument may be made against the possibility that the echinochrome is held in the chromatophores because it is more soluble in them than in water. When the cell is stimulated mechanically or chemically the pigment comes out of the chromatophores with explosive rapidity. The cell need not be killed to accomplish this. The mere act of normal fertilization causes some of the chromatophores in the egg to lose their pigment.

The only alternative hypothesis I know of is, that the pigment is manufactured in the chromatophore, and cannot normally get out because the surface of this body is impermeable to it. An increase in permeability of the chromatophore allows the pigment to escape. Such an increase in permeability might be due to an aggregation change in the colloids of the limiting membrane or surface film.

Echinochrome is held in the chromatophores of the sea urchin's cells probably in the same way that chlorophyll is held in the chromatophores of the green plant cell.

Reprinted from The Anatomical Record, Vol. 7, No. 2,
February 1913

PREPARATION OF MATERIAL FOR HISTOLOGY AND EMBRYOLOGY, WITH AN APPENDIX ON THE ARTERIES AND VEINS IN A THIRTY MILLIMETER PIG EMBRYO

J. F. McCLENDON

From the Anatomical Department of Cornell University Medical College, New York

THREE FIGURES

The essential of a good course in histology or embryology is good material. Fresh human material should never be allowed to go to waste, but it may be at times very inconvenient to put it up in a variety of fancy fixing fluids.

Perhaps the best general cytoplasmic fixer is formalin of 10 to 20 per cent (4 to 8 per cent formaldehyde). If material so fixed is not soaked too long in alcohol of high concentration, it may be used as fresh tissue in special technique to show fats or mitochondria. In fact the formaldehyde alone makes unsaturated fats and lipoids less soluble in clearing fluids. On the other hand, if the washing in water is omitted, the structure of resting nuclei is well enough preserved for ordinary purposes.

Commercial formalin contains formic acid, which, although developing a more beautiful nuclear structure, may begin to cytolyse the more delicate cells before they are sufficiently fixed by the formaldehyde. This is especially noticeable in erythrocytes—haemolysis, or escape of haemoglobin, occurring in parts of the tissue. Furthermore, acids swell fresh white fibrous tissue. It seems worth while, therefore, to neutralize the formol, and this may easily be done by adding slack lime ($CaCO_3$) or magnesia, and filtering.

Doctor Ferguson first called my attention to the fact that kidney swells in many fixing fluids, whereas it is commonly supposed that the majority of tissues shrink a little. Death of isolated cells as seen under the microscope may be accompanied by swelling (cytolysis) or contraction. In every case, an increase in permeability to some substances occurs, but I found that during the early stages of cytolysis of the sea urchin's egg, it remains very impermeable to salts. Dead animal or plant membranes are more permeable to water than to dissolved substances, but apparently some living cells are impermeable to water. The Fundulus egg, if transferred from sea water to distilled water, does not burst, though it is certainly not capable of resisting the enormous osmotic pressure of its internal salts. Since I found this egg to be impermeable to salts, it must also be impermeable to water (it is permeable to kations, but for every kation that comes out, the electrical equivalent must go in). If such a cell became, on death, permeable to water, the osmotic pressure of its internal dissolved substances might cause it to swell. If a Paramoecium be killed by an ordinary fixing fluid, even though it be hypertonic, the protoplasm first coagulates, then the whole animal swells a little. This may be what happens to some tissue cells, and I found that it is not always prevented by the addition of 0.9 per cent sodium chloride to the fixing fluid. Therefore I supposed the swelling due to the osmotic pressure of some contained substance of large molecule, and experimented with the addition of cane sugar to neutral formol. By this means the cytolysis of adult convoluted nephric tubule cells is prevented, and the general fixation is good except that some nuclei may be slightly shrunken. This fluid may be used for all adult tissues and embryos, and is easily prepared as follows:

Formol	100–200 cc.
Cane sugar	20–40 grams
Slack lime ($CaCO_3$) or magnesia	about 1 gram
Water to make 1 liter.	

If the shrinkage of a few nuclei is very objectionable use only 20 grams of sugar. This fluid has the advantage that tissues and embryos float in it and therefore do not become distorted.

If the whole kidney of a fetus be fixed in the above mixture or any other fixing fluid, the cells of the convoluted tubules will swell until they fill the lumen. This brings us to a well known point that is often neglected. Tissues should be cut into as thin slices or pieces as is practicable and the cells not injured in the cutting. Fetal tissues are especially delicate. They should be cut with a very sharp thin blade and lifted on the blade into the fixing fluid.

Many workers object to formalin because it "causes" a homogeneous appearance to protoplasm. The ultra microscope has shown that, aside from evident granules, living protoplasm is homogeneous, contrary to Bütschli and others. There are persons who now accept formalin for cytoplasmic fixation but say that it "does not fix nuclei well." Some structures may be seen in living nuclei. I have studied many nuclei with high powers and with the ultra microscope, yet I cannot decide what form of fixation corresponds most closely to the living structure. Both cytoplasm and nucleus of a living erythrocyte of a frog is homogeneous when examined in serum or uncoagulated plasma with the ultra microscope. Sooner or later bright points or clouds appear on or in the nucleus, but this is usually associated with change of nuclear form and is evidently due to injury.

Formaldehyde not only does not coagulate protoplasm but renders it more difficult to coagulate. It also makes lipoids less soluble in clearing fluids. However, I find an after-treatment with Müller's fluid or some other oxidising fluid necessary for the preservation of lipoids, the amount of oxidation necessary depending on whether mitochondria, myelin or fats are studied.

Ordinary staining depends on the fact that all protoplasm treated with acid, stains with acid dyes, whereas certain parts take also basic dyes. Many staining solutions contain free acid, but tissues stain more quickly if they are previously treated with acid. For this reason we put everything into the formol mixture and after a few hours transfer part of it to Bouin's fluid. This tissue is finally stained on the slide in haematin and eosin. The alum haematin lake is usually so strong that it stains in three minutes, but the eosin is so much diluted that twelve hours are required to stain and in this time smooth muscle stains less intensely than white fibrous tissue The acid in

Bouin's fluid causes the tissue to stain more brilliantly but if the fresh tissue is put into Bouin's fluid the blood in some of the vessels will be laked. Part of the material is transferred from the formol mixture to Müller's fluid and subsequently stained with iron hematoxylin to show the lipoids (mitochondria, etc.).

Ordinarily, the student is shown two dimensions of a piece of tissue or embryo, and left to imagine the third. Though whole mounts of chick embryos are handed out, cleared pig embryos, and blocks or thick sections of certain tissues are even more useful. For a solid mount, the object should be placed in a dish of balsam or damar dissolved in benzol and protected from dust until it evaporates down to sirupy consistency, then mounted in the usual way. By this means the necessity of rings or other supports to the cover glass is avoided, and drying out or great shrinkage prevented.

All of the solid mounts turn yellow with age, but a number of highly refractive fluids may be obtained that are colorless. These are listed, with their refractive indices, in Landolt-Bornstein; Behren's Tabellen; and Lee's Vade Mecum. The higher the refractive index the better, for if in any case a lower index is desired, this may be obtained by the addition of paraffin oil or xylol (or water in case of aqueous media). It may be noted here that, whereas the process of clearing in a mixture of oil of wintergreen (Gaulteria) and benzyl benzoate has been patented in Germany and is widely known under the name of the patentee, wintergreen was first used by Stieda in 1866, and the synthetic oil (methyl salicylate) recommended by Guéguen in 1898, and is noted in various books on technique.

Methyl salicylate is permanently colorless, and comparatively inexpensive, and ideal for a fluid mount. If rings are cemented on slides with shellac or liquid glue and allowed to dry, they are not loosened by the oil. Paper rings soaked in shellac or glue will do, but rings may be cut from lead pipe with an ordinary saw or a bone saw if the proper size of glass rings are not at hand. The shellac must be dry before adding the oil, which must be free from alcohol. I prefer glue.

If the tissue is hardened in alcohol, thick sections may be cut free-hand. Thick sections are often better unstained, especially if injected, and much detail may be made out by partly closing the diaphragm of the microscope. If stained with very dilute haematin containing much acid, connective tissue is colorless and cytoplasm nearly so, whereas nuclei may be readily distinguished. In this way blood vessels and glands in areolar tissue are caused to stand out sharply.

Whole mounts and thick slices are especially useful in embryology, and are a necessity unless one is contented with teaching the third dimension with models. The larger the embryo, the more attention must be paid to the clearing medium in order to distinguish internal structures. Methyl salicylate is admirable for pig embryos of all sizes and even for small fetuses. I found ethyl salicylate to be as good if not better, but it is more expensive. Canada balsam has about the same refractive index ($^{n}D = 1.535$) as methyl salicylate ($^{n}D = 1.536$), but darkens with age.

Embryos may be placed directly from absolute alcohol, benzol, xylol, toluol or chloroform into methyl salicylate, but in order to obtain the proper refractive index, the preliminary fluid must all be removed. This may be evaporated, or washed out with more wintergreen. Benzol is to be recommended because it is cheapest and evaporates out most easily. The evaporation may be hastened by an air pump, which also removes any air bubbles that may get into the specimen. These bubbles expand and are absorbed after the pump is disconnected, or by successive pumpings. An ordinary air pump will cause the benzol and air to boil out. A water-suction air pump (aspirator) will suffice but a float valve and safety bottle should be interposed between the pump and the specimen to prevent the back flow of water. An exhaustible desiccator is convenient for holding large embryos while they are being pumped out. If the cover is well ground, the oil will seal it sufficiently, and vaseline should not be used.

Most of the internal organs may be distinguished in unstained embryos by cutting down the light. The individual cells of mesenchyme, cartilage and blood may be seen; the cellular structure of the neural tube is indicated by radial striations and the larger nerves appear as bundles of fibers. Some organs in smaller embryos are made more distinct by staining with very dilute alum haematin containing a large amount of acid.

Even in quite small embryos, many of the blood vessels may be traced by the blood cells, and the large empty veins followed as cavities. However, with the smaller vessels this becomes more laborious than serial sections. On the other hand, the injection of small embryos for class use means quite an outlay of time. Therefore, it seemed necessary to find some way to fix the haemoglobin, and keep the vessels full, in order to distinguish the vessels by the color of the blood. I found that *the same method that prevents the cytolysis of nephric tubule cells prevents haemolysis.*

Living embryos are obtained, the amnion opened, the placenta squeezed to force the blood into the embryo, and the umbilicus tied or clamped. Artery clamps are too strong and pinch off the cord. (I made clamps out of wire (lower part of fig. 1) in order to avoid tying so many cords at the slaughter house. The clamp may be removed in half an hour and used again.) The embryo is dropped into the neutral-formol-sugar mixture described above, and left until thoroughly fixed. In case of a fetus, part of the skin should be torn off after the superficial blood vessels are fixed, to insure penetration of the formaldehyde. A hole may be made in the skull by slicing off a small piece tangentially or by a sagittal cut near (to the right of) the median plane. Large fetuses, unless skinned completely, will have to be scraped to remove the pigment layer.

Transfer the specimen after washing, or directly, from the fixing fluid to alcohol of about 70 per cent. After they have hardened in 80 or 95 per cent alcohol it is well to split the large specimens by a sagittal cut a little to the right of the median plane with a very thin bladed knife. The dehydration with higher alcohols should be slow enough to prevent shriveling.

By this method the blood retains its color, and although it does not take the place of injection, it is a great help to the student. I have inserted three figures to show what can be seen in such specimens.

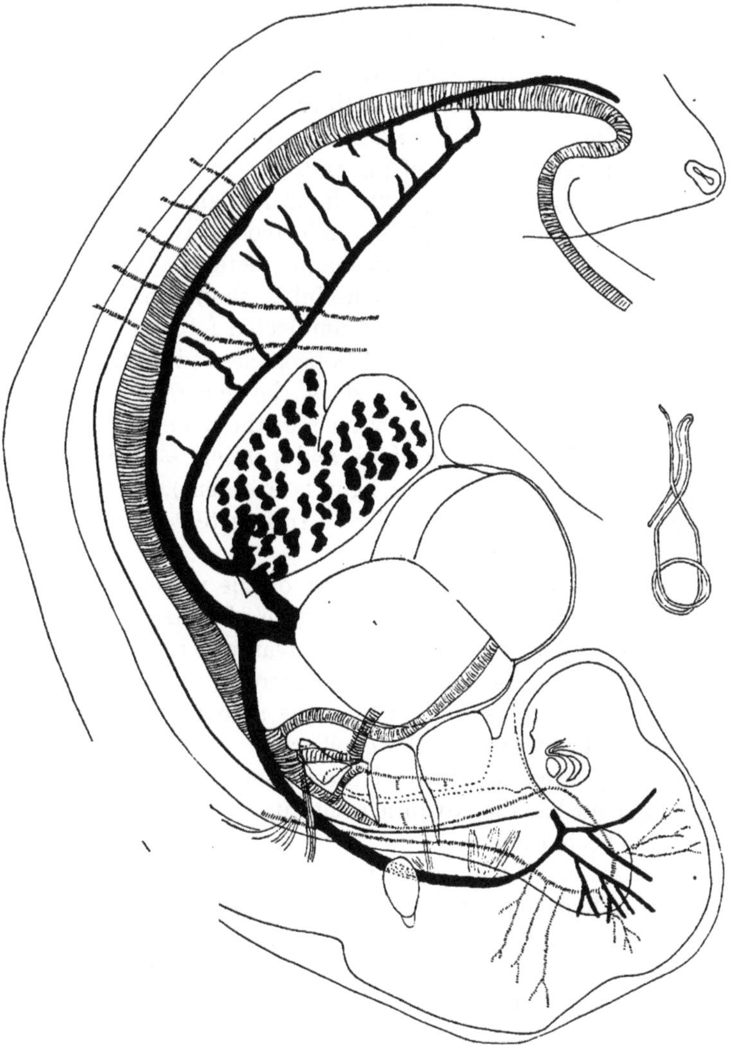

Fig. 1 Left half of a pig embryo 7 mm. long after clearing. The veins are black, the arteries cross-striated, the 5th, 7th, 8th, 10th and 11th cranial nerves are longitudinally striated, the notochord is represented by a heavy line and the fore gut by a dashed line. The sinuous line ventral to the embryo represents a wire (partly open) clamp used in clamping the umbilicus.

Figure 1 represents a pig embryo about 8 mm. long. The nerve tube, fore gut, mesonephros, liver, heart, eye and ear are clearly seen. The arterial system and part of the cardinals and subcardinals can be distinguished. The notochord is distinct, and the 5th, 7th, 8th, 10th, and 11th cranial nerve roots can be made out. Figures 2 and 3 are described in the appendix.

I have prepared hundreds of pig embryos and fetuses in this way, and also injected many with india ink and cleared them in wintergreen oil. A completely injected fetus can only be studied in comparatively thin (freehand) sections. Various degrees of partial injection are very useful to show the larger vessels, but these may be seen in the uninjected fetuses. The left side of an uninjected fetus which has been cleaved a little to the right of the median plane, will show the general circulation, except in the liver. The larger vessels in the liver may be seen by removing the lateral portions and passing a strong light through the remainder (an arc light is .excellent), or the liver may be removed and cut into slices. In injected specimens the liver is hopeless.

I washed with alcohol the blood out of the vessels of a fetus 4 inches long and cleared it in wintergeen oil, then injected it with mercury. This method has the advantage that the extent of the injection may be watched and controlled.

The injection may be limited by using a coarse granular pigment that will not go into the capillaries. A gelatine mass is not absolutely necessary to hold the pigment. A light colored opaque pigment has the advantage that it may be seen by transmitted or reflected light.

The arteries may be injected and the haemoglobin fixed in the veins, giving handsome specimens. If it is desired to show only the injection, no formalin should be used. Much of the haemoglobin may be dissolved out by putting the fresh specimen into weak alcohol or alcohol and acetic acid. All of the haemoglobin may be removed with dilute acetic acid provided an injection is used that is not affected by this acid.

APPENDIX

ON THE ARTERIES AND VEINS IN A 30 MM. PIG EMBRYO

The method of fixing the haemoglobin and clearing in wintergreen oil to show the course of the vessels has been especially successful in case of pig embryos of about 30 mm. length Figures 2 and 3 show the larger vessels of the median plane and left side of one of them. The courses of most of the vessels approach the type of the adult pig and show distinctions in topography from those in man. The common carotid artery and (right) innominate artery arise from a common trunk, the brachiocephalic artery. The posterior inferior cerebellar artery arises from the basilar instead of from the vertebral.

Notwithstanding the great development of the vena cava, the left posterior cardinal is of considerable size. The right cardinal (not figured) is smaller. The thoraco-epigastric vein is divided into two parts, one of which drains anteriorly into the internal mammary.

The vessels of the limbs could not be completely followed, but enough was seen to demonstrate that they differ very much from those in the adult.

Besides the vessels, the mouth cavity, brain, eye, endolymphatic labyrinth, lungs, mesonephros, kidney, testis and penis are outlined in the figures.

Lightning Source UK Ltd.
Milton Keynes UK
UKHW031134290119
336362UK00009B/274/P

9 781527 879119